Digital Twin Development and Deployment on the Cloud

Digital Twin Development and Deployment on the Cloud

Developing Cloud-Friendly Dynamic Models Using Simulink®/Simscape™ and Amazon AWS

Nassim Khaled

Bibin Pattel

Affan Siddiqui

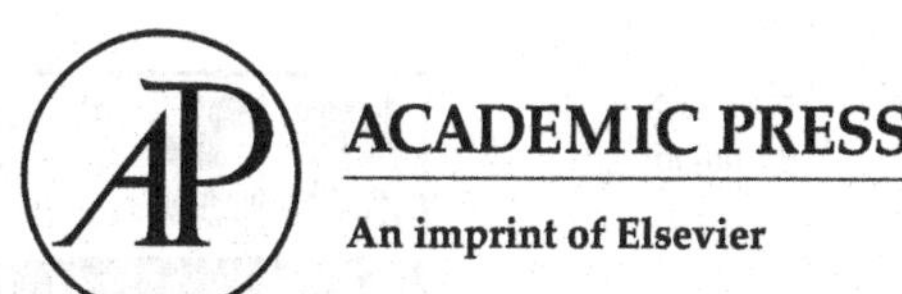

Academic Press is an imprint of Elsevier
125 London Wall, London EC2Y 5AS, United Kingdom
525 B Street, Suite 1650, San Diego, CA 92101, United States
50 Hampshire Street, 5th Floor, Cambridge, MA 02139, United States
The Boulevard, Langford Lane, Kidlington, Oxford OX5 1GB, United Kingdom

Library of Congress Cataloging-in-Publication Data
A catalog record for this book is available from the Library of Congress

British Library Cataloguing-in-Publication Data
A catalogue record for this book is available from the British Library

ISBN: 978-0-12-821631-6

For information on all Academic Press publications visit our website at https://www.elsevier.com/books-and-journals

Publisher: Mara Conner
Acquisitions Editor: Sonnini R. Yura
Editorial Project Manager: Rafael G. Trombaco
Production Project Manager: Kamesh Ramajogi
Cover Designer: Miles Hitchen

Typeset by TNQ Technologies

For MATLAB and Simulink product information, please contact:
The MathWorks, Inc.
3 Apple Hill Drive, Natick, MA, 01760-2098 USA
Tel: 508-647-7000
Fax: 508-647-7001
E-mail: info@mathworks.com

I dedicate this book to my baby daughter. Since her birth, I got motivated to work toward a brighter future for her and her generation. Tomorrow belongs to them. I hope our innovations today are motivated by a shinier tomorrow rather than a shiner dollar.
My baby, I am not a rich man. The legacy I leave to you is one of knowledge and ideas. I hope some decades from today, you can look back at this book and know that your father cared dearly about you even when he was no longer around.

Nassim

I dedicate this book to my dear Parents, who worked really hard to get quality education for myself and my brothers. I would like to also dedicate this to my dear Wife and Daughter; they always have been my inspiration and hope. I would also like to dedicate to my brothers who encourage me all the time. Thank you all for your support and being there for me whenever I needed.

Bibin

I dedicate this book to my dear parents Nadeem Parvaiz Siddiqui and Fauzia Nadeem Siddiqui for an upbringing that I am eternally grateful for. I also would like to dedicate this book to my two older brothers for their continuous guidance and support every step along the way.

Affan

Contents

Acknowledgments

For professionals like us who have full-time 8–5 jobs and a family to take care of, publishing a book is not an easy undertaking. We were fortunate over the last couple of years to have support from several parties. They were instrumental in bringing this material into production. We would like to take this opportunity to mention the main contributors and thank them for their support.

The Mathworks® sponsored this book under their Book Program. They provided us with MATLAB®, SIMULINK®, and other toolbox licenses for developing various application models and examples used in this book. We appreciate their continued support for our books.

We would like to thank Chetan Gundurao and Naresh Krishnamoorthy for contributing to Chapter 2.

We would like to thank Stephen Limbos for his contributions and experiments conducted for Chapter 6.

We would also like to thank Jisha Prakash for contributing to Chapter 7.

Elsevier provided us with a smooth process, from content submission, review to production. In particular, I would like to thank Sonnini Ruiz Yura, Kamesh Ramajogi, and Rafael Trombaco for working closely with us throughout the book project.

Added value of digital twins and IoT

1

1.1 Introduction

This is a great time to be in the modeling and simulation business. Advances in electronics and microprocessors capabilities have enabled much faster simulations on stand-alone PCs. Furthermore, cloud technologies have enabled massive amounts of simulations to be carried out on the server. It also opened the door for new types of revenue streams such as Software as a Service (as opposed to the traditional model, Software as a Product). These advances enabled MathWorks to be among the first privately owned simulation and computation software to exceed 1 billion dollars revenue.

For cloud service providers, it is even a greater time. Streaming services, data storage, data analysis, social media outlets, and IoT-related activities provided tremendous business opportunities. Microsoft Azure, Google Cloud Platform, and Amazon Web Services seem to be well positioned in the top three in terms of cloud revenue. These three are closing on 80 billion dollars revenue in 2020.

Engineering companies and design firms are reluctant to run their simulations on the cloud. The main reason for this underwhelming adaptation is rooted to the curriculum culture of engineers. Based on our experience, vast majority of engineers still believe that "cloud is an IT function." Additional reasons and excuses to the lack of adaptation by engineers are the lack of understanding the benefits of cloud servers, amount of time it takes to set up and deploy cloud simulations, security concerns, and speed of download of the massive data generated by the simulations on the servers.

Cloud simulations are a measly part of cloud revenue, and they will remain so for the near future. They are hugely underutilized. They have the ability to improve engineering practices and designs by levering the massive cloud infrastructure and existing communication, security protocols, file revision standards, and software as service model of most simulation companies. Cloud simulations also offer tremendous potential in the world of diagnostics and prognostics.

The book is aimed to promote usage of multiphysics simulation models running on the cloud to improve diagnostics and prognostics. In this book, we redefine the term digital twins as it pertains to the diagnostics and prognostics context. We believe that the existence of several definitions of digital twin is slowing down the process of its maturity. Business leaders in engineering firms need to understand "what is it?", "how can it provide value?", and "how much it costs to deploy (and maintain)?". This will help them to decide to embrace or not.

We outline the key elements needed to deploy digital twins. The hardware and software tools needed are described as well as the process of developing and deploying digital twins. Challenges and bottle necks that exist will be tackled, and we will propose current and future work-arounds.

Digital Twin Development and Deployment on the Cloud. https://doi.org/10.1016/B978-0-12-821631-6.00001-3

1.2 Motivation to write this book

The authors saw significant safety benefits that digital twins can provide to any machine/product/process. Furthermore, they saw a lack of publications related to fields of self-diagnostics and prognostics as pertaining to digital twins. This is a field that combines the understanding of the fundamental operation of electromechanical systems, automation, and controls, in addition to communication and computer science.

It is critical to provide an introduction to the background of the experts who contributed to this book. This allows the reader an opportunity to understand what forged the motivation and opinion of the authors. Seven experts collaborated closely to write this book. They took an oath to deliver an original publication in the area of simulation and cloud technologies to improve the safety of engineered products. It is a part of their mission to better the human lives around the world by spreading useful knowledge.

They have worked in the industrial, research and development, and academic world. Industries they worked in include automotive, autonomous guidance and control, battery systems, HVAC and refrigeration, electric grid control, and video and image processing. All the experts are below 40 years of age, but with a total of 60 years combined experience.

Dr. Nassim Khaled is the author of two books in the fields of controls and simulation [1,2] (both books are all-time best MATLAB e-books). He worked as an engineering manager for controls in two US engineering companies: Cummins and HillPhoenix. He is currently working as an Assistant Professor in Prince Mohammad Bin Fahd University, KSA. Dr. Khaled has more than 30 filed patent applications worldwide and 24 published US patents.

Bibin Pattel is the author of one book in the field of controls and simulation. He is currently working as a technical expert in KPIT in the field of automotive control. He is an expert in software development for diagnostics and control. Bibin has a Masters in Mechanical Engineering.

Affan Siddiqui is currently working in Cummins Emissions Solutions as a Senior Controls Engineer. He specializes in software development of control and diagnostic algorithms of diesel engines and aftertreatment systems. He has a Masters in Mechanical Engineering.

Chetan Gundurao has worked across various industries pertinent to process control and discrete automation control industries, developing solutions around software and embedded system—based closed loop and open loop control solutions over his career. Chetan is currently working as a Technical Architect for Dover Corporation. Chetan holds a Bachelor Engineering Degree in Electronics and Communication and a Masters in Computer science.

Naresh Kumar Krishnamoorthy has a Masters in Electrical Engineering. He is currently a project lead in Dover Innovation Center in India. He is specialized in diagnostics and controls.

Jisha Prakash is currently pursuing her Masters in Electrical Engineering, and her area of research is Power Electronics Optimal Control.

Stephen John Limbos is a lab technician in the College of Engineering in Prince Mohammad Bin Fahd University. He has a Bachelors in Electronics and Communications Engineering. He has extensive experience in software/hardware integration.

The authors have met and agreed that model-based diagnostics that are carried on the cloud simulations are highly underutilized in the engineering world and have the potential to bring significant safety and maintenance improvements for an array of products. This book brings a frame work to streamline the development of digital twins mainly for the diagnostics and prognostics of machines.

1.3 Digital twins

In this work, we define the digital twin that is used for the purpose of abnormality detection of a process, plant or machine. Detection of a potential abnormality that already occurred is widely referred to as diagnostics, whereas detection of potential future abnormality is referred to as prognostic. Both concepts require some form of a mathematical model that mimics the physical behavior of the system.

Companies define digital twins based on their business model. This is causing some confusion as well as limiting the benefits of digital twins. Below we mention two examples of how digital twins are defined by businesses:

MathWorks define digital twins as "an up-to-date representation, a model, of an actual physical asset in operation. It reflects the current asset condition and includes relevant historical data about the asset. Digital twins can be used to evaluate the current condition of the asset, and more importantly, predict future behavior, refine the control, or optimize operation. [3]."

Bosch defines digital twins as "connected devices— such as tools, cars, machines, sensors, and other web-enabled things—in the cloud in a reusable and abstracted way. [4] "

In this book, we will adopt the definition presented by MathWorks. The definition proposed by Bosch is one level of abstraction higher than MathWorks. We are hoping to help answer questions that might be raised by business leaders such as "what is it?", "how can it provide value?", and "how much it costs to deploy (and maintain)?".

1.4 On-board and off-board diagnostics

On-board diagnostics (OBD) is a term mostly used in the automotive world, despite the fact that it is applicable in other industries. It refers to having a processor on-board of the vehicle for the purpose of diagnosing vehicle malfunctions. Typically, there is a sensor, actuator, or component that is serving a function, and there is a virtual model that is predicting the expected outcome. The outputs of these two are compared and a diagnostic decision is issued based on the difference. Fig. 1.1 shows the traditional process to design and deploy a diagnostic in a vehicle.

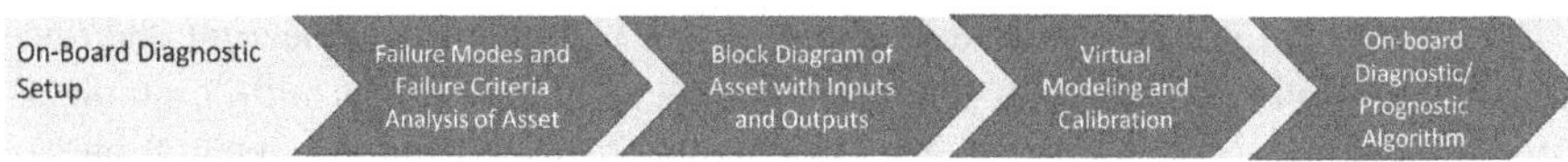

Figure 1.1 Traditional on-board diagnostic setup.

The main challenge of vehicle diagnostics is the limited processing capability of the processing unit on-board. There are usually hundreds of diagnostics running in the electronic module of the car at each 100 ms. This results in embedding limited capability digital twins on-board of the vehicle. The design and structure of these digital twins are not updated during the life cycle of the vehicle or the machine (unless there is a recall for the product). Also, these on-board digital twins do not benefit from sensory data that might not be available on-board the vehicle (such as average humidity, wind speed, or density of air in an area). Additionally, these on-board digital twins will not have access to historical data for the duration of the vehicle due to on-board memory constraints. Finally, these on-board models will not benefit from the learnings of the whole fleet because it is isolated from the rest of the fleet digital twins.

We introduce the term off-board diagnostics (Off-BD) instead of OBD to refer to the process of having the digital twins as well as the diagnostic decision taking place on the cloud or remote from the vehicle or machine. The steps to design an Off-BD are highlighted in Fig. 1.2. The failure modes of the asset are highlighted. Then a block diagram with inputs/outputs is drawn. An edge device sends the data to the cloud. A virtual model for the asset is built and calibrated (MATLAB/Simulink tools are usually used). A diagnostic algorithm is constructed. Such algorithm usually tracks the deviation of the physical asset from the virtual model. The virtual model represents the nominal behavior of the asset. Any deviations from the virtual model are deemed to be failures if they exceed a predetermined threshold.

Self-diagnostics, OBD, and Off-BD are all necessary functions for any asset. These diagnostics are meant to detect critical or primary failures in the asset. Designing such diagnostics to detect specific failure modes is an art that is not particularly taught in depth in any branch of engineering. It usually combines the knowledge of physics, observers and controllers, communication, and microprocessors.

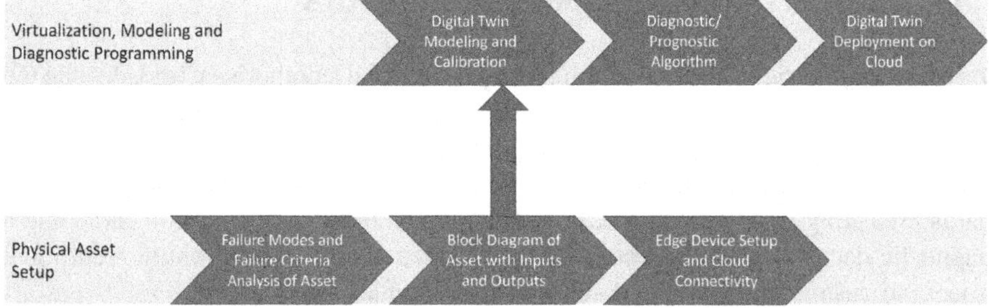

Figure 1.2 Developing and deploying digital twins based on a set of failure modes.

We believe self-diagnostics and digital twins go hand in hand. As a matter of fact, throughout this book, it is implicitly assumed that digital twins are used alongside with some form of a diagnostic. The mere display of data coming from assets remotely is not considered to be digital twinning in this work. This is why we use the term Off-BD to describe the process of having a virtual model on the cloud coupled with a usage of some failure criteria of the asset. When the output data of the virtual model and the physical data diverge per some diagnostic logic, a failure in the operation of the asset is assumed. The user has to be alerted of such failure.

In Fig. 1.3, we demonstrate how Off-BD works for one physical asset. In Fig. 1.4, we demonstrate how Off-BD works for five similar assets that belong to the same platform.

1.5 Modeling and simulation software

Modeling an oil rig, a vehicle, an aircraft, or a space shuttle in a virtual environment is a complex process. Such systems contain mechanical, electrical, structural, chemical, and electronic components. In most scenarios, separate models are built for the entire system. For example, a combustion model is built to simulate the power output, heat transfer, and emissions. A separate model is built to model the transmission dynamics. Another model is built to simulate the aerodynamics of the vehicle. Similarly, another model is constructed to simulate the robustness of the control logic to control the air-handling subsystem of the engine. These models have different fidelities and mimic partial behavior of the system. The execution time of these models can range from few seconds to few days.

Multiphysics models that represent the whole system are rare at best. Having a common solver and step size for mechanical, electrical, structural, chemical, and electronic processes make the model very difficult to handle and execute. Nevertheless, there are Multiphysics models that represent subsystems. These models are usually used for designing software and hardware components of the system.

There are many simulation softwares for Multiphysics modeling. COMSOL [5], SimScale [6], AnyLogic [7], MathWorks [8] and Ansys [9] are all powerful tools that can be used to build multiphysics models. Despite the apparent need for simulation software in automotive, aerospace, and engineering companies, revenues for modeling software companies are mediocre and seem disproportionate with the potential savings and innovation they enable. Ansys, a publicly traded company, had a revenue of 1.3 billion USD in 2018 [10]. While COMSOL, SimScale, and AnyLogic had 35, 5, and 25 million USD in revenue [9].

In this book, we focus on MATLAB® and Simulink® since it is the most-utilized software for the purpose of control and algorithm development. Industrial and research both favor the software tools provided by MathWorks. LinkedIn lists technical

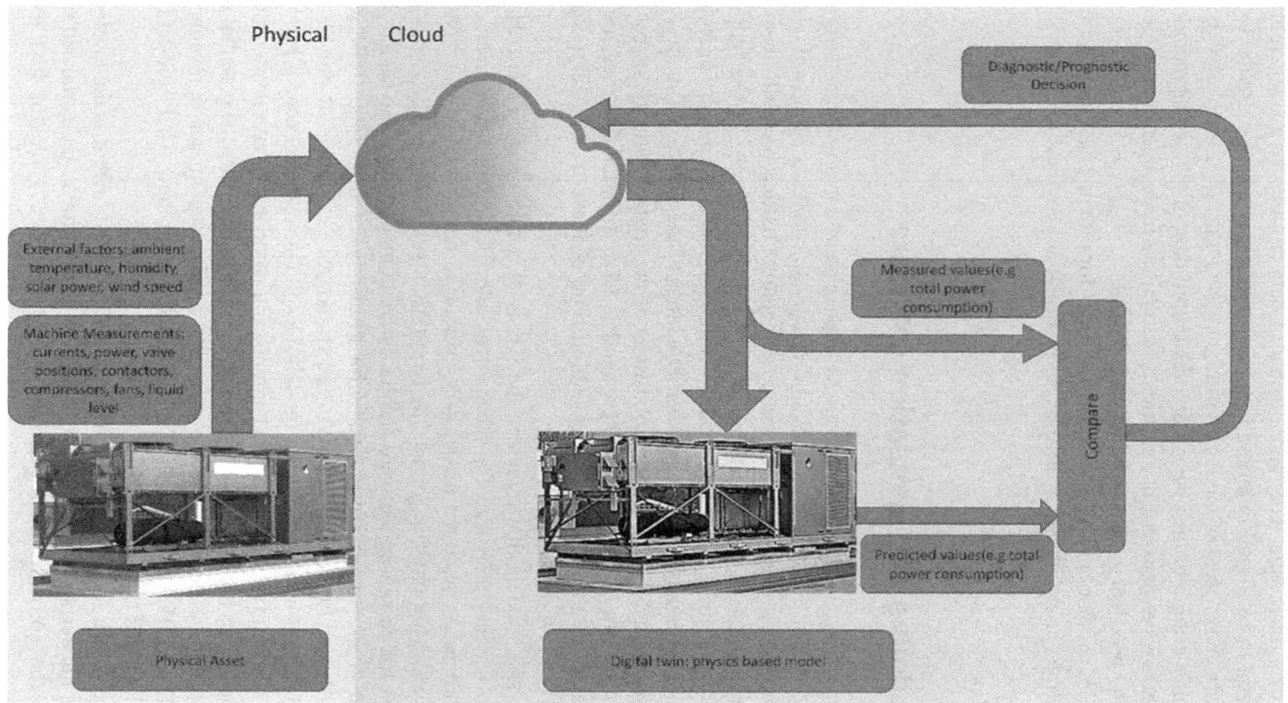

Figure 1.3 Off-BD for one asset.

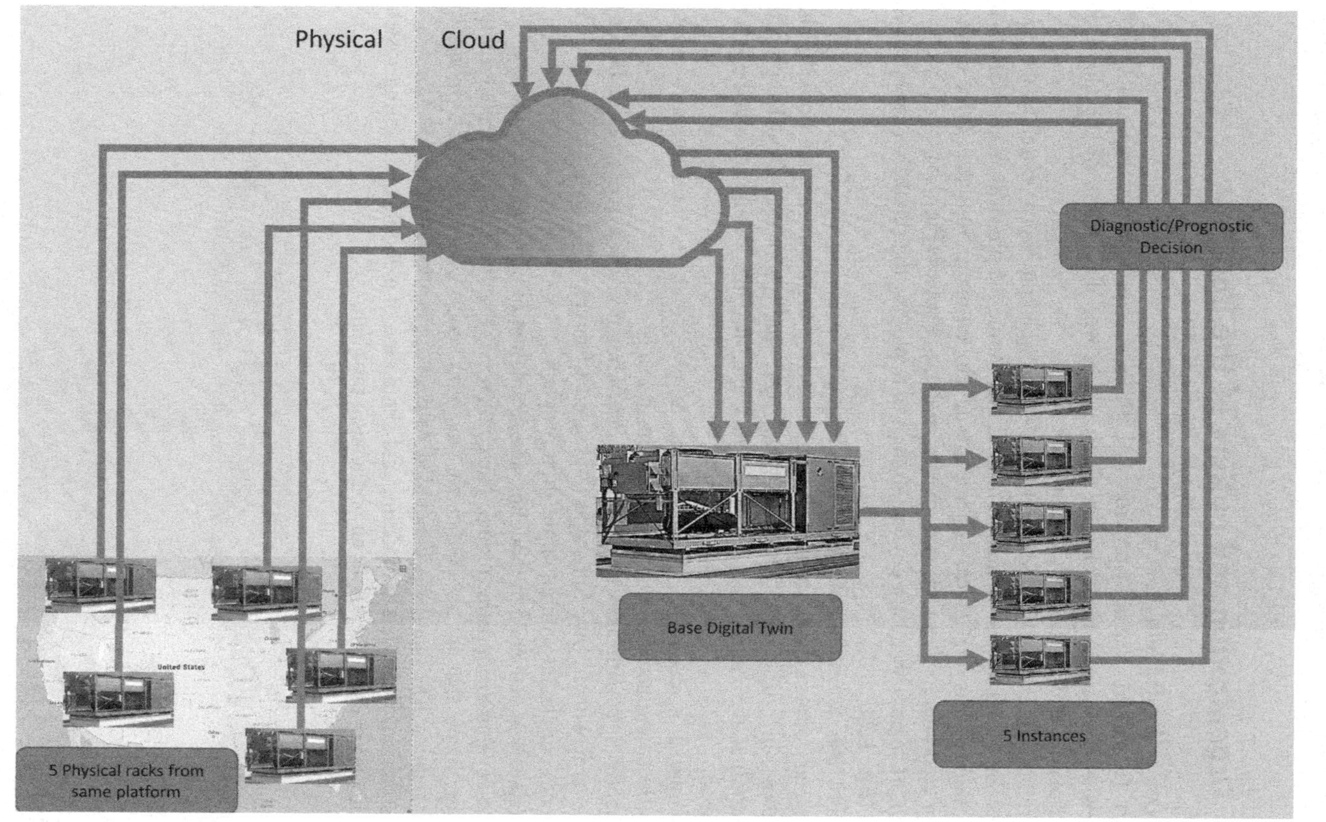

Figure 1.4 Off-BD for five assets.

computing as a top skills companies needs in 2019 [11]. In particular, they recommend learning MATLAB® to acquire the necessary skills.

1.6 Organization and outline of the book

The book is divided into three parts:

Part 1(Chapters 1 and 2) provides a high-level overview of OBD and Off-BD diagnostics and cloud technologies. They also cover the origins of the cloud, its infrastructure, and its current capabilities.

Part 2 (Chapters 3–8) is dedicated to the streamlined process proposed by the book to generate Multiphysics simulation models that mimic the physical assets. Variety of simulation models of physical systems are developed using Simscape. Edge devices and on cloud deployment are not covered in these chapters (Fig. 1.5).

Part 3(Chapters 9 and 10) covers the steps of deploying the simulation model on the cloud (Fig. 1.2).

Below is a description of the individual chapters:

Chapter 1 provides the background and motivation for writing the book. It also describes the difference between OBD and the proposed Off-BD process. Furthermore, the experience of the authors and also the outline of the book are described.

Chapter 2 is dedicated to the origins of the cloud, its infrastructure, current capabilities, and new business opportunities it brings.

Chapter 3 is dedicated to modeling and simulation of a 3 degrees-of-freedom robotic arm with a grip. Simscape™ is used to develop the dynamic model. The chapter is concluded with an application problem.

Chapter 4 is dedicated to modeling and simulation of ball on plate. Simscape™ is used to develop the dynamic model. The challenge is that the ball and the plate are two moving objects that are not connected. Custom C code was developed to model the ball dynamics. The chapter is concluded with an application problem.

Chapter 5 is dedicated to modeling and simulation of a double mass spring system. The chapter is concluded with an application problem.

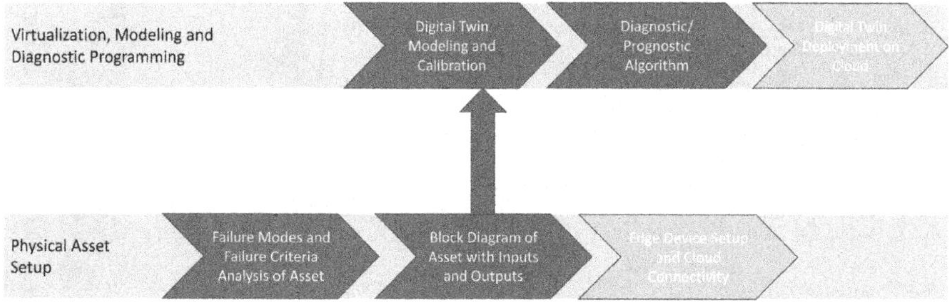

Figure 1.5 Off-BD steps covered in Chapters 3–8.

Chapter 6 is dedicated to modeling and simulation of photovoltaic cells. The model accounts for solar irradiance and temperature of the photovoltaic cells and predicts the power output of the cells. The chapter is concluded with an application problem.

Chapter 7 is dedicated to modeling and simulation of electric/hybrid vehicle. The chapter is concluded with an application problem.

Chapter 8 covers modeling, diagnostics, and fault recovery of three-phase inverter system. A model-based fault diagnostic and mitigation algorithm is implemented to diagnose open circuit, short circuit, and DC link failures. The model is used to mitigate the failure mode by adjusting the inverter drive topology without adding any additional devices.

Chapter 9 covers the entire final step in developing and deploying a digital twin of a DC motor for the purpose of diagnosing the hardware. Challenges and complexities are covered with special attention. This chapter also covers the details of the hardware platform, AWS cloud, and the use of Digital Twin to do an Off-BD for the real-time hardware. Below are the setups involved in the full Off-BD process:

1. Setting up real-time embedded controller hardware for DC motor speed control
2. Developing Simscape™ digital twin model for the DC motor speed control
3. Parameter tuning of the Simscape™ DC motor model with data from real hardware system using Simulink Design Optimization™
4. Adding cloud connectivity to real-time embedded controller hardware for DC motor speed control
5. Deploying the Simscape™ digital twin model to Amazon AWS cloud
6. Voice and email and text user interface development for the digital twin deployment

Chapter 10 covers the entire final step in developing and deploying a digital twin of a wind turbine for the purpose of diagnosing the hardware. Challenges and complexities are covered with special attention. This chapter also covers the details of the hardware platform, AWS cloud, and the use of digital twin to do an Off-BD for the real-time hardware. Below are the setups involved in the full Off-BD process:

1. Setting up real-time embedded controller hardware for the wind turbine
2. Developing Simscape™ digital twin model for the wind turbine
3. Parameter tuning of the Simscape™ wind turbine with data from real hardware system using Simulink Design Optimization™
4. Adding cloud connectivity to real-time embedded controller hardware for the wind turbine
5. Deploying the Simscape™ digital twin model to Amazon AWS cloud
6. Voice and email and text user interface development for the digital twin deployment

All the codes are available for free download by searching for the ISBN of the book in MATLAB Central. Alternatively, the reader can search for "Nassim Khaled" in MATLAB Central.

Visit the dedicated website for more visuals and latest news of the work: www.practicalmpc.com/digital-twins.

References

[1] https://www.elsevier.com/books/practical-design-and-application-of-model-predictive-control/khaled/978-0-12-813918-9.
[2] https://www.springer.com/gp/book/9781447123293.
[3] https://www.mathworks.com/discovery/digital-twin.html.
[4] https://aws.amazon.com/marketplace/pp/Bosch-Software-Innovations-Bosch-IoT-Things/B07DTJK8MV.
[5] https://www.comsol.com/.
[6] https://www.simscale.com/.
[7] https://www.anylogic.com/.
[8] https://www.mathworks.com/.
[9] https://www.ansys.com/.
[10] https://www.mathworks.com/company/aboutus.html.
[11] https://learning.linkedin.com/blog/top-skills/the-skills-companies-need-most-in-2019–and-how-to-learn-them.

Cloud and IoT technologies

2

2.1 Overview

Cloud technology spun out of software virtualization. In software virtualization, the same piece of hardware (PC) can run multiple instances of software (operating systems) efficiently. This has helped organizations to reduce on hardware costs without sacrificing performance.

The cloud is essentially a gigantic computer hardware in the form of computer server systems that are deployed by the service providers in server farms with a common virtualization software platform running on them. These servers have one goal in common: to provide simultaneous tenants/customers access to the same software using virtual replicas.

Fig. 2.1 represents a block diagram that shows a typical layout of a cloud platform with an underlying hardware on which a virtualization software layer enables dedicated virtual machines to be deployed. On top of such dedicated virtual machines, customized software and platform configurations are available, which are dedicated for more specific jobs.

The cloud computing allows infrastructure scalability and enables multiple users to utilize the same hardware and software using a "pay-as-you-use" model.

The cloud platform brought the below key advantages:

a. Significant savings in hardware infrastructure that traditionally required huge investments in form of capital expense.
b. Reduction in hardware and software maintenance cost, software license ownership cost, and IT-related services.
c. Simplification of software upgradability and scalability.
d. Digitization of software licensing.

Today's market has many cloud platform service providers like Amazon, Microsoft, and Google. Many other players offer specialized cloud services.

2.2 History of cloud

The initial years of the cloud platform offerings saw a tremendous adaption by the organizations to scale their IT infrastructure onto the cloud, which enabled them to dynamically scale the resources, add more software services, scale the hardware capacities, and enable many more IT and network security services that traditionally required huge capital investments. This is what came to be known as Infrastructure as a Service.

Digital Twin Development and Deployment on the Cloud. https://doi.org/10.1016/B978-0-12-821631-6.00002-5

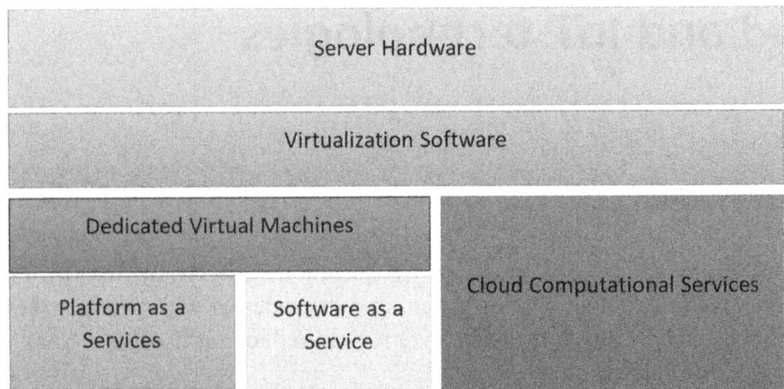

Figure 2.1 Layout of cloud platform.

The focus of cloud platform services then moved on toward what came to be known as Software as a Service (SaaS). Organizations prior to the cloud revolution invested huge capitals in the form of software licenses for which enormous overheads were incurred in maintenance of software tools. Moreover, dynamics software licensing made company reorgs much easier.

With SaaS, the companies are now able to scale the software licenses offered as cloud licenses when the need arise. Also, the SaaS offerings almost offer no limits on the data storage capabilities, which earlier required physical media upgrades. Other than solving software license issues, companies in the small- to medium-scale range of industries can afford many financial, enterprise resource planning (ERP), customer relationship management, human capital management, and other expensive software on a lease basis, which are licensed and invoiced typically on a monthly fee.

The Platform as a Service and SaaS have a very thin demarcation line between them with respect to configurations they offer.

2.3 Evolution of cloud technologies

The cloud platform services have now evolved to include

a. Infrastructure scalability at almost real time—to enable hardware scaling up/down without having to shut down the virtual PCs.
b. Redundancy systems to secure data across geological locations serving for disaster recovery.
c. Humongous storages for databases for storing customer data safe and secure (Database as a Service).
d. Developer centric tools offered as SaaS—to enable small- to large-sized teams to collaborate and develop software using cloud licensing.
e. Various collaboration tools to enable global teams to collaborate over Internet.
f. Analytics tools for data scientists—which have huge horse power computers able to churn data and provide results in almost real time.
g. Video analytics tools that are developed by experts, which are offered as services to do real-time video streaming to cloud and processing for object detection, analysis, etc.

h. Image processing and computer vision toolkits and services with standard algorithms for face recognition, text recognition, speech processing, color segmentation, and object detection, etc.
i. Security services—companies can now secure their web traffic by using services and cryptic authentication methods that use hardware keys for security signing, etc.
j. Internet of things (IoT)—to enable machines to cloud interfacing, which has opened a new dimension in predictive maintenance, remote monitoring (real time), tracking of machine performances historically, etc.

2.4 Connecting machines to the cloud

Connecting the embedded systems to the cloud platform is a challenge on its own. Embedded systems have few main challenges—limited memory availability, lower horse power, etc. Above all, the embedded systems are quite often dedicated to do a predefined critical task. To mitigate the problems with limited memory and lower horse power, many modular (layered) cloud platform communication protocols evolved, which led to enabling connectivity to cloud for embedded systems. Many cloud platform integration protocols like AMQP, MQTT, REST, etc., are lighter in memory foot prints and are more secure.

Also, in recent times, many semiconductor chip vendors now have variants of dedicated silicon chips that are meant to provide services around security layer, TCP/IP interfacing, WIFI communication interfaces, etc., and are built in to a hardware module that can be interfaced to lower power devices using traditional low-speed communication protocols.

In recent times, many organizations like Microsoft (viz. *Azure Sphere* OS) have come up with dedicated microprocessor-based solutions that allow organizations who are ambitious to integrate their embedded solutions into cloud but lack the expertise or have limited resources and time, to just use such modules and program the control methods (business logic) into them and use the preavailable infrastructure to securely connect to respective cloud services.

This cloud to device connectivity has primarily created enormous opportunities around the areas of

1. Remote monitoring of an equipment by OEMs/customers.
2. Data gathered from field have opened a broader opportunity around applications of analytics, predictive and preventive maintenance, fault detection and fail safe, remote site surveillance and supervision, easy tracking of alarms, etc.
3. Integration of analytics/data science techniques into such ecosystem has opened avenues for many predictive and automatic inventory purchase order generations (when stock is expected to go low).

2.5 Applications

2.5.1 Enterprise Resource Planning

ERP is a critical tool for production and manufacturing industries. Many organizations that had no or very minimal software investments in managing and planning the enterprise resources have now an opportunity to explore and move to SaaS model pricing for ERP software. ERP systems nowadays offer a large set of features that are licensed under monthly pricing model and come packaged with options to integrate machine data from shop floor to the ERP system. This has led the ERP systems to forecast work load, arrive at accurate pricing by monitoring the cost expenses in machine maintenance and run hours, also track inventories, monitor maintenance schedules, etc.

2.5.2 Automotive and asset management

Automotive industry was one of the early adopters of the IoT + cloud solutions. From monitoring the vehicle location, performance (ECU health, OBD data, tire pressures, oil temperatures), and other parameters on cloud (remotely) to arriving at optimal fuel efficiency (in case of fleet management companies) or monitoring the vehicle health and performance versus idle time and busy times in a day, etc., IoT and cloud technology together has enabled the fleet management/transport companies to effectively monitor resources over cloud. Logistics companies can now track shipments from dispatch to delivery using technologies like GPS integrated onto the cloud. Companies that rely on fleets in their business are moving to using cloud technologies to enable asset tracking and fleet management of their trucks/vehicles by using the cloud computing and IoT. Asset management is the new trend for industries in fuel transportation, chemical transportation, commodity delivery (especially liquids), etc. One such example includes OEMs that allowed for transportation trucks fitted with flow, level sensors on trucks integrated with the cloud to monitor thefts and efficiency of drivers, etc.

2.5.3 Equipment monitoring

Many original equipment manufacturing industries are adapting the digital techniques to monitor their equipment's performance on the cloud—real time. This has enabled the OEMs to analyze system performance, monitor health, and provide for warranty claims (equipment wrong/abuse, etc.).

Many OEMs are also moving toward use of cloud as a part of their platform for enhanced user experience.

Some good examples include a winch industry that adapted to special sensors and integrated the data to cloud to monitor system's performance and use.

A truck OEM that adapted to video processing over cloud to monitor driver behavior and alert the fleet manager if the driver was speeding above the limits on a highway, etc.

Others include pump manufacturers that integrated flow sensor data and pump parameters that are monitored on cloud for overall throughput of pump, predictive maintenance, etc.

2.5.4 Farming

IoT and cloud technologies have seen a tremendous momentum in farming, which has enabled them to monitor quality and efficiency, and as well control the operations of production from over the cloud. One example is where a chicken farm adapted to voice processing to identify chicken's clucks to identify any chicken that required medical attention.

2.5.5 Charity

A charity organization that feeds 20,000 orphans and underprivileged children had a challenge to monitor the quality of food while during transit along with vehicle parameters. They were successful in implementing a cloud platform–based solution that enabled them to track and monitor vehicle location from the destination schools to calculate the arrival times, temperature of the food throughout the transit, etc., to ensure food that reaches the destination was of good quality.

2.5.6 Biomedical

A biomedical pump–manufacturing company had a problem in hand, which required frequent update of configurations based on new fluids that were introduced for production. Each installation in the production site was similar in configurations. Every time a new fluid was to be manufactured, each pump would require the configuration parameter set and also specific sensor thresholds, etc.

Also, there was a good need to monitor the pump's performance parameters constantly for quality metrics. The company had a challenge of technicians visiting pumps every time to configure and load the new parameters, etc.

The company chose to go with IoT-based approach for each of their purposes to achieve the following:

a. Each pump was retrofit with an IoT Gateway module.
b. All configuration sets are now available over cloud.
c. Any new configuration set was also made available on cloud.
d. Each pump would then fetch any new configuration(s) from cloud via standard cloud interfaces like AMQP/MQTT.
e. So, the operator would just choose to load the configuration per job during the execution.
f. All vital parameters of pumps were tracked online over cloud for each job and hence enabling remote monitoring of each pump.
g. The same features were available across sites for the company, so it saved them time of upgrades.

The below block diagram explains the overall system functions (Fig. 2.2).

a. The pump would connect to an IoT Hub over cloud platform using an IoT Gateway hardware and standard protocols like MQTT/AMQP protocols.
b. The device would be capable to control the pump for speed and as well measure feedback for RPM.
c. The fluid's transfer functions are designed and tested using mathematical tools like MATLAB.
d. The transfer function/model obtained so is exported from MATLAB or converted into standard programming languages like C/python.

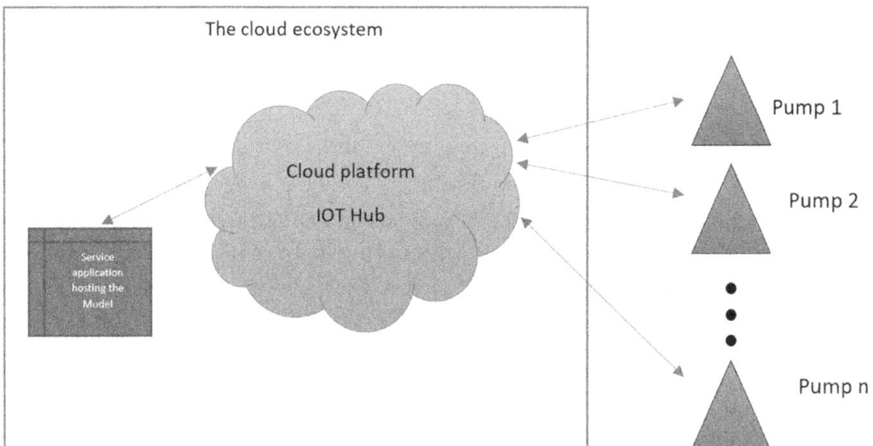

Figure 2.2 Biomedical pump application.

e. An application that would use the transfer function libraries so developed would integrate to the IoT Hub for which each device is mapped.
f. At start of the job, the custom service would get the job information, and it would look for preavailable transfer functions for that pump.
g. The custom model would then invoke on the preavailable transfer functions and apply the job-specific parameters to generate the outputs of target power and RPM.
h. The pump would receive the information via MQTT about the target power and RPM.
i. The cloud services would also monitor the health parameters like pressure, total hours of RUN, motor health, etc., and would send out alerts to stakeholders about any required replacements/servicing.

2.5.7 Miscellaneous applications

The cloud platform today offers many services around data sciences, analytics, machine learning, and other cognitive services including computer vision. These services also provide application programming interfaces (APIs) that can be used to consume such services in client applications. Such offering has enabled engineers develop analytical solutions using methods like inferential statistics, machine learning, etc., on historical data (present in cloud) to arrive at inferences and perform forecast of performance. One such application can be seen in companies like telecom service providers who use time series analysis and other inferential statistical method to arrive at customer churn prediction models, etc.

The cognitive services have evolved in the areas of artificial intelligence and machine learning technologies into providing many vision-based analytics, with primary focus on

a. Object detection.
b. Face recognition.
c. Text recognition (OCR engines).

d. Train, teach, and identify custom objects in image.
e. Identify obscene images.
f. Recognize objects in videos.
g. Speech recognition and speaker recognition.
h. Much more.

There are cognitive services over the cloud platform that are available in the form of language recognition to detect emotions expressed in scripts! One of the biggest successes is the speech recognition. One can easily develop solutions that involve speech/speaker recognition and integrate artificial intelligence into the system that would enable interactive voice-based user request processing. Furthermore, chatbots that are powered by artificial intelligence are currently a common practice of many businesses to service humans by interacting with them through interactive chat messengers.

Other offerings in data science, artificial intelligence, and machine learning include ready availability of environments with data science and machine learning toolkits that can be leased on an hourly basis for development and deployment of models into production.

There are readily available virtual machines for data science and artificial intelligence, which offer a ready-to-use environment with a storage at the backend allowing the engineers and scientists to design, develop, and eventually deploy the models into production and expose REST APIs for their services to use with other applications (or for their customers to consume the custom services).

There are widely three major players in the cloud arena: Amazon, Microsoft, and Google.

Amazon was the first (and is probably the leading) provider of various cloud platform services as offerings to its customers over cloud. Microsoft eventually picked up the race and is in line with Amazon's offerings in services they offer. One can find almost every equivalent service offered by both systems but with minor differences (be it in costs/application interface methods/ease of use, etc.). Azure leads the way in integrating microsoft-specific technologies into cloud and easy deployment of solutions, whereas Amazon has the lead in most of the services they offer, new feature introduction and support to many forms and methods of communication protocols, etc., including good support to open source software. Amazon's speech recognition has been leading the markets.

Google started its journey toward cloud platform offerings as early as the others and mostly around SaaS arena and cloud storage spaces particularly for mobile device profiles and for personal storage profiles, but Google's focus has been mostly into speech recognition, image processing, and video analytics. Google also offers similar cognitive services as the other two giants.

All the service providers also provide many services like TensorFlow services, etc., which can be integrated into the application via simple interface methods for custom image processing application requiring very high image processing capabilities.

Moreover, the cloud service providers also provide basic "monitor" services that enable users to constantly monitor field data (variables) by connecting to an IoT Hub. The monitors would allow to plot trends, arrive at histograms, etc., on variable(s) for a predefined interval of time. IoT Hub interfacing modules are the modules that interface the embedded systems to the cloud platform. These are usually called Gateways (IoT Gateways to be specific). A Gateway is a passive device that up streams all data it receives from its clients. Whereas IoT Edge Gateways usually tend to have some intelligence of filtering and forwarding only required data to the IoT Hub. This allows the customers to save on network bandwidth costs incurred due to traffic entering or stored at IoT Hubs.

2.6 Considerations for cloud services

2.6.1 Security

Cloud presents multiple Cyber-security risks. Redundant firewalls and improved communication protocols in addition to hardware requirements (machine, edge processors and servers) are all essential part of cloud services. One advantage of cloud platform is that it is constantly updated, and all services offered by the cloud platform provider use same basic security for all customers which presents an opportunity to consolidate cost of deploying security measures.

2.6.2 General Data Protection Regulation

The General Data Protection Regulation (GDPR) is a European Union regulatory body. It imposes strict restrictions on data retention of end customers. If your organization retains end customer information, it is highly likely that some of the GDPR and similar regulations apply to the data. Before choosing the right IT infrastructure for storing data and choosing location of services (where/which region the data is hosted), organizations should thoroughly go through the GDPR and other regulations to ensure end customer data are protected in terms of recommendations by such regulations.

2.6.3 Shadow IT services

Shadow IT services are basically employees who have access to the organization's cloud platform accounts and are at free will and full access to deploy resources when they want. This could lead to uncontrolled use of resources and untracked expenses. Organizations evolving over cloud should ensure proper process in place for such handoffs.

2.6.4 Cloud cost optimization

Most organizations that have moved completely to cloud have later realized that due to improper architectures and insufficient knowledge and planning have led hefty expenses could have been avoided. In an effort to reduce cost, many organizations have reduced ingress/egress traffics. They deployed computing resources only during day time. They also process as much data as possible at the client's end before streaming the data to the cloud. This reduced storage requirements significantly.

2.7 What is edge computing?

The "edge" of a network typically refers to the processor located near the machine like HVAC system, traction system in a locomotive, or a pelletizer equipment, which is being monitored or actuated. The technology has been limited to accumulating and forwarding data to the cloud. It has been underutilized thus far. What if the industrial companies could turn vast amounts of data into pragmatic intelligence, available right at the edge? This technology is essential as it is rapidly emerging as a powerful force in turning industrial machines into intelligent machines.

2.8 Edge computing versus cloud computing

Cloud and edge computing are complementary technologies. Edge computing takes the leading position in scenarios such as low latency, bandwidth, real-time/near real-time actuation, intermittent, or no connectivity, etc., and the cloud will play a more prominent role in computationally-heavy tasks, machine learning, digital twins, etc. Both options need to work in tandem to provide design choices across edge to cloud that best meet business and operational goals. We project a significant shift towards edge computing in the next decade (Fig. 2.3).

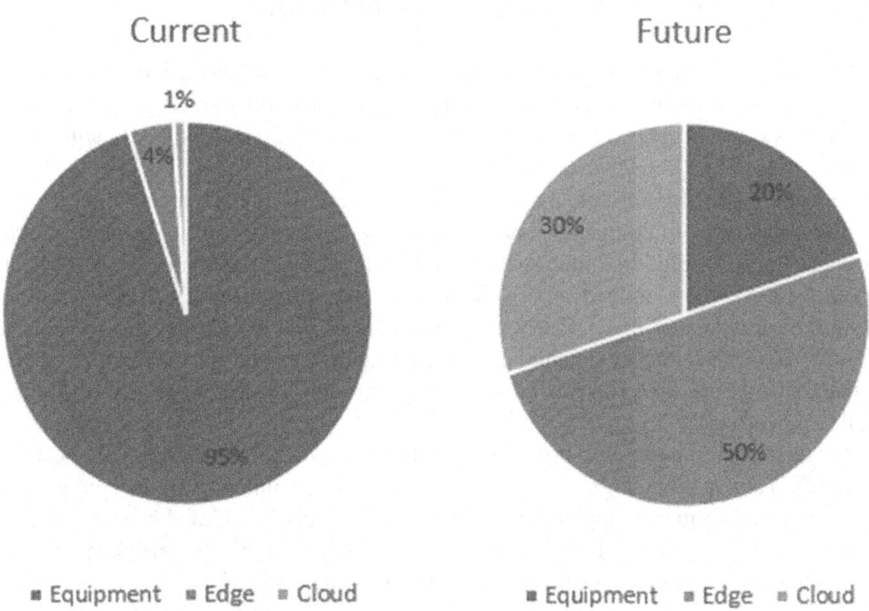

Figure 2.3 Current versus future computing overlook.

2.9 Edge and cloud computing examples

Let us look at a couple of scenarios. In an industrial context, intelligent edge machines benefit more from the growing capabilities of edge computing for real-time analytics and actuation.

For example, locomotives carry about 150 sensors that can transmit 1 billion instructions per second. Sending all that data created by the locomotive to the cloud for processing, analyzing, and actuation is not useful, practical, or cost-effective. Nowadays, applications can collect data locally, respond to changes, and perform meaningful localized analytics. Few variants of locomotive use on-board edge computing to analyze data and apply algorithms for running smarter and more efficiently, thereby improving operational costs, safety, and uptime.

Another problem generally faced in transportation is asset performance. Let us consider the following case, a company has 1000 locomotives under management. There are sensors on each locomotive tracking fuel efficiency and performance of the vehicle. Instead of real-time analytics and actuation on the machine, the data are being transferred, then stored, and forwarded to the cloud where time series data and analytics are used to track efficiency and performance of rail vehicles. The company can use these data to improve uptime of its vehicles, lower repair and fuel costs, and improve the safe operation of the vehicle. Here, edge plays an important yet minor role, while cloud is in dominant position.

2.10 Will 5G accelerate cloud computing?

In addition to big increase in speeds, 5G also presents a revolutionary decrease in latency. Latency is the time it takes for two devices on a network to react to one another. 3G networks had latency of about 100 ms; 4G is around 30 ms; while 5G will be as low as 1 ms. What does that mean for the cloud?

One of the main reasons the cloud is so beneficial in an organization is for numerous devices to connect to and transmit data with a central machine or cloud storage. For example, the cloud made transferring a large video simple—just put it on the shared drive, upload, and once it is up, the other person can download it from the same shared drive. But why go through all that if both devices can relate to only a millisecond of latency and a minimum connection speed of 20 gbps down and 10 gbps up? There is no need to go through an additional step or use an online repository.

Cloud computing needs to be forward compatible with 5G (and potentially 6G and beyond). Taking advantage of 5G for cloud will boost innovations in application varying from health care applications to autonomous vehicles. Cloud-based products and services are expected to become more reliable, faster, and efficient. These innovations will, in turn, accelerate cloud business investments.

Digital twin model creation of a robotic arm

3

3.1 Introduction

In today's industries, we find many of the heavy-lifting tasks are being worked on by large automated mechanical robotic arms. A common task associated with these robotic arms is to pick up an object and place it at a desired position. The robotic arm is a forward kinematics problem. The arm structure is made of many elements connected by linkages or joints. The application involves the use of actuators to move different elements of the arm to a specific (x, y, and z) position. The design of these arms involves the knowledge of many factors of which the important ones are the degrees of freedom, torques, and the equations of motion.

In this chapter, we will be modeling a custom-made miniature 3 degrees-of-freedom (DOF) robotic arm that was purchased from www.roboholicmaniacs.com in MATLAB Simscape™. The actuators used in this arm are stepper motors. The main inputs to this 3-DOF arm are the torque angles to these actuators, which will output the x, y, and z position of the end effectors. With regards to off-board diagnostics, we can sense the acceleration of the joints of the arm from accelerometers mounted on them for a respective torque angle input. The acceleration from the physical sensor can then be sent to the cloud and compared with the twin model simulation results for the same torque angle inputs. Fig. 3.1 shows the off-board diagnostics process where the parts highlighted in blue (dark gray in print version) are utilized in this chapter, whereas the gray ones are not. As seen in the figure, we will be focusing on the digital twin model for the arm such that a framework is provided for the reader to develop their own models. Fig. 3.2 shows the boundary diagram of the robotic arm system along with its interaction with its twin model.

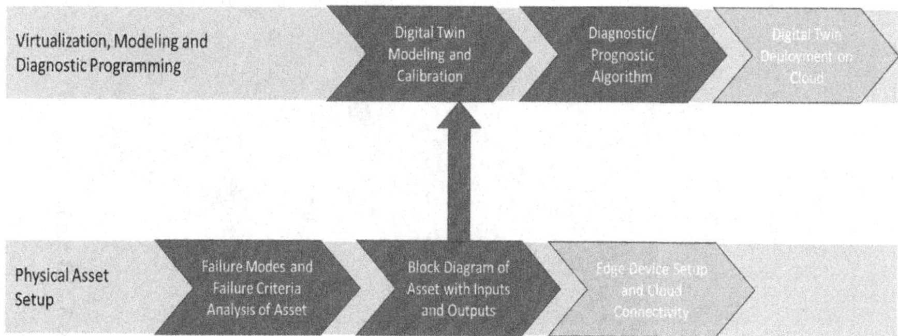

Figure 3.1 Off-board diagnostics process for 3-DOF robotic arm.

Digital Twin Development and Deployment on the Cloud. https://doi.org/10.1016/B978-0-12-821631-6.00003-7

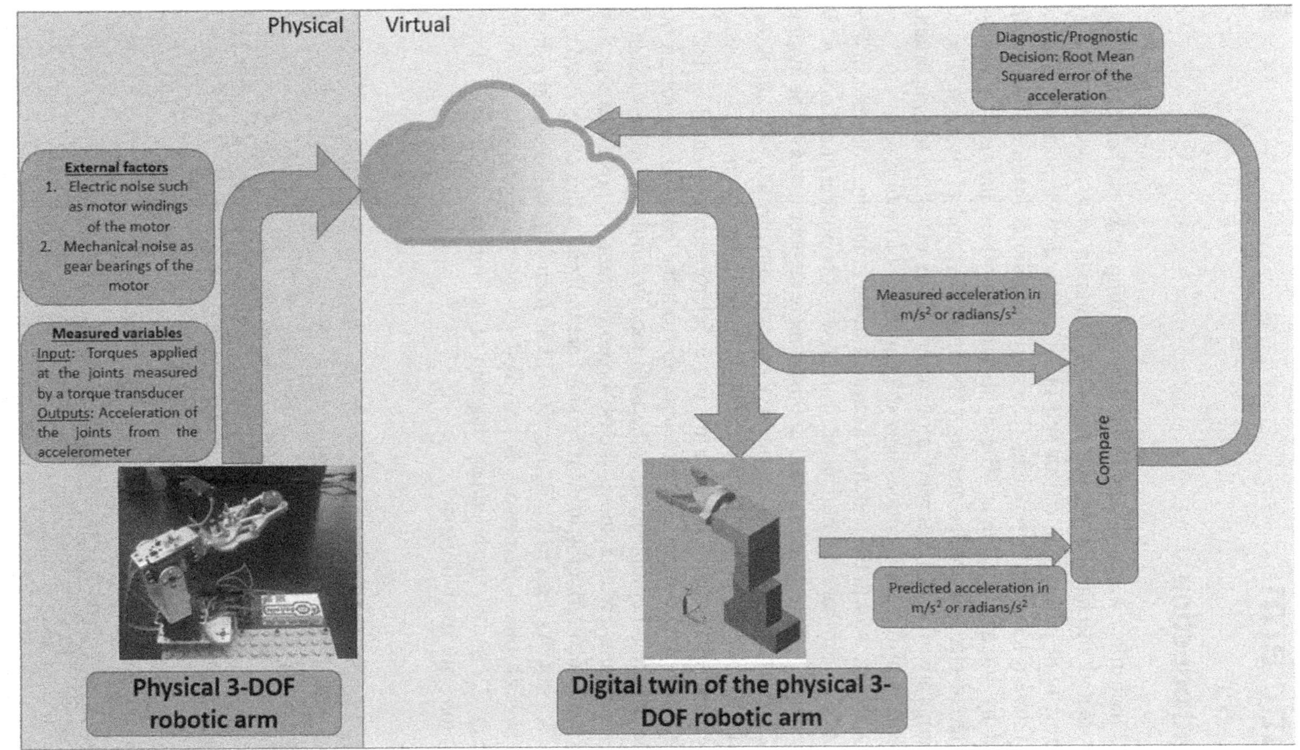

Figure 3.2 Boundary diagram of 3-DOF robotic arm.

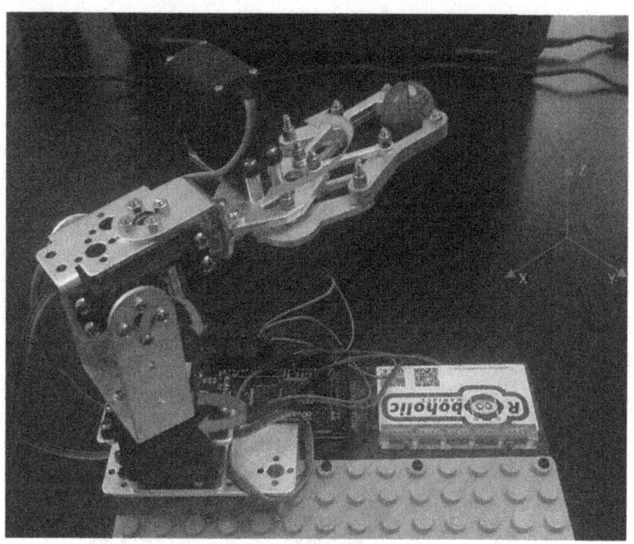

Figure 3.3 3-DOF robotic arm.

Fig. 3.3 shows the 3-DOF robotic arm we are simulating in this chapter. As shown in the figure, we have a rotary motion about the Z axis to on the base, a rotary motion about the Y axis for the upper arm, and a rotary motion about the Z axis for the gripper. Adding up these three motions constitutes a 3-DOF robotic arm. The motion is controlled by servomotors and a microcontroller. It is possible to program the motion using the Arduino software toolset.

The purpose of this chapter is to show a step-by-step guide on how to model a 3-DOF robotic arm in MATLAB Simscape™ such that you will be able to see how the robotic arm moves with the specified motion inputs.

All the codes used in the chapter can be downloaded for free from MATLAB File Exchange. Follow the link below and search for the ISBN or title of this book:

https://www.mathworks.com/matlabcentral/fileexchange/.

Alternatively, the reader can also download the material and other resources from the dedicated website or contact the authors for further help:

https://www.practicalmpc.com/.

3.2 Hardware parameters

Before we can simulate this arm in Simscape™, we must measure the dimensions of the elements such that we can simulate our arm in Simscape™.

As shown in Fig. 3.4, measure the elements of the arm by a tape measure. In this instance, we are measuring the width of the base along the Y axis. Repeat this same procedure for all the other elements of the robotic arm along the X, Y and Z axis.

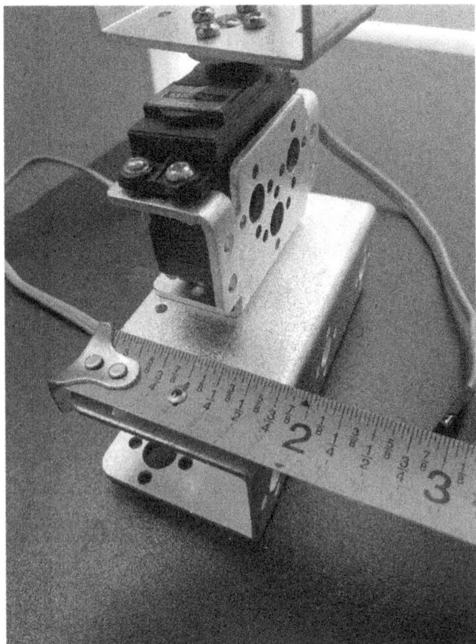

Figure 3.4 Measuring the dimensions of the robotic arm elements.

We will be approximating the shape of this robotic arm in Simscape™, therefore finer details are not required. You can approximate the gripper as a cuboid with length, width, and height.

3.3 Simulation process

The final model is presented as follows in Fig. 3.5.

Below are the detailed steps to building the Simscape™ model based on the sections shown in Fig. 3.5.

3.3.1 Initial conditions

1. Create a new **Simulink® model** and then click on the **Simulink® Library Browser** button on the toolbar as shown in Fig. 3.6
2. Navigate to **Simscape>Utilities** as shown in Fig. 3.7. Add **Mechanism Configuration** library in the model.
3. Navigate to **Simscape>Utilities** as shown in Fig. 3.8. Add **Solver Configuration** library block in the model.

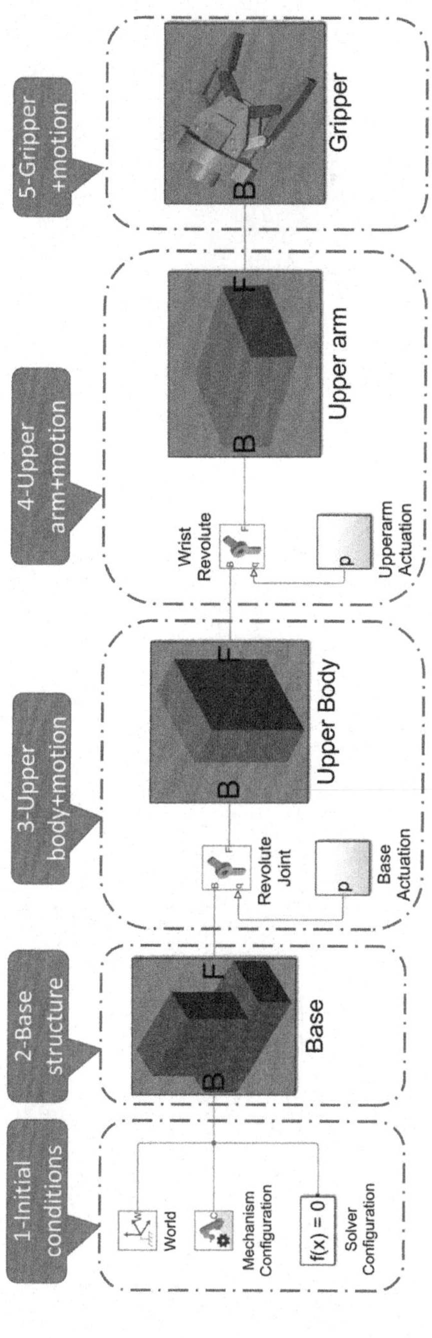

Figure 3.5 Main elements of the robotic arm model.

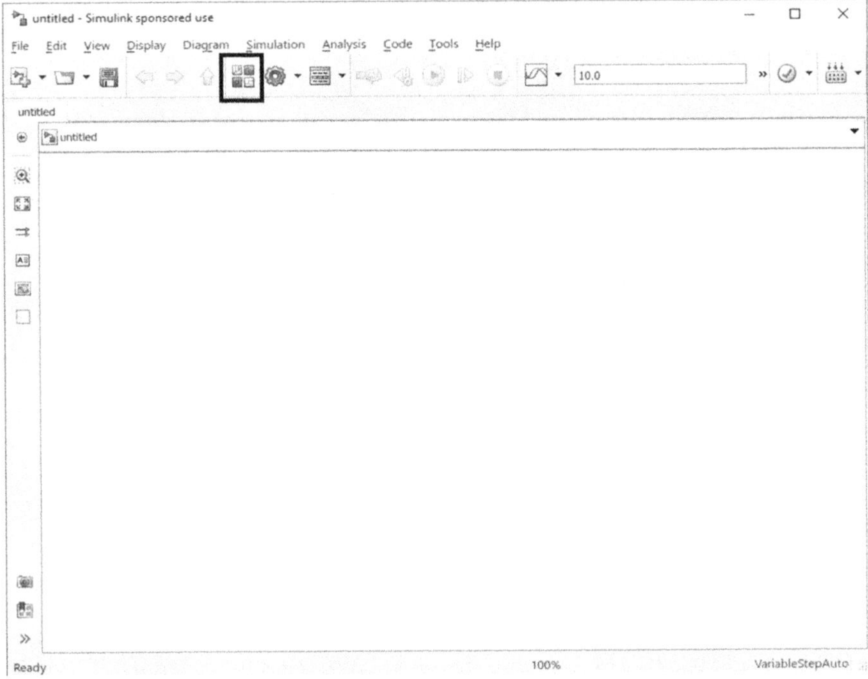

Figure 3.6 Simulink® Library Browser.

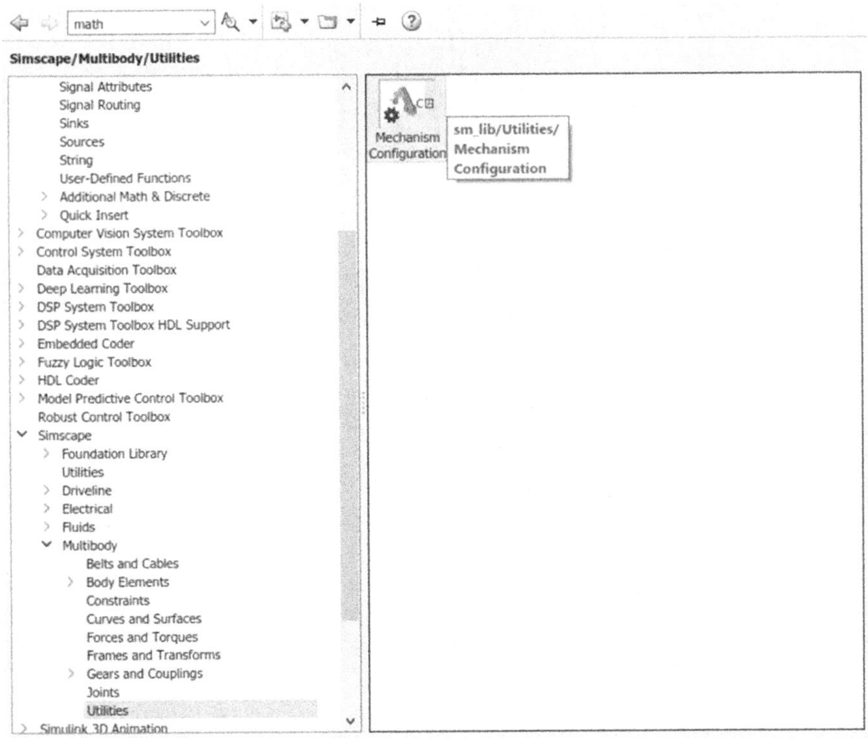

Figure 3.7 Mechanism Configuration.

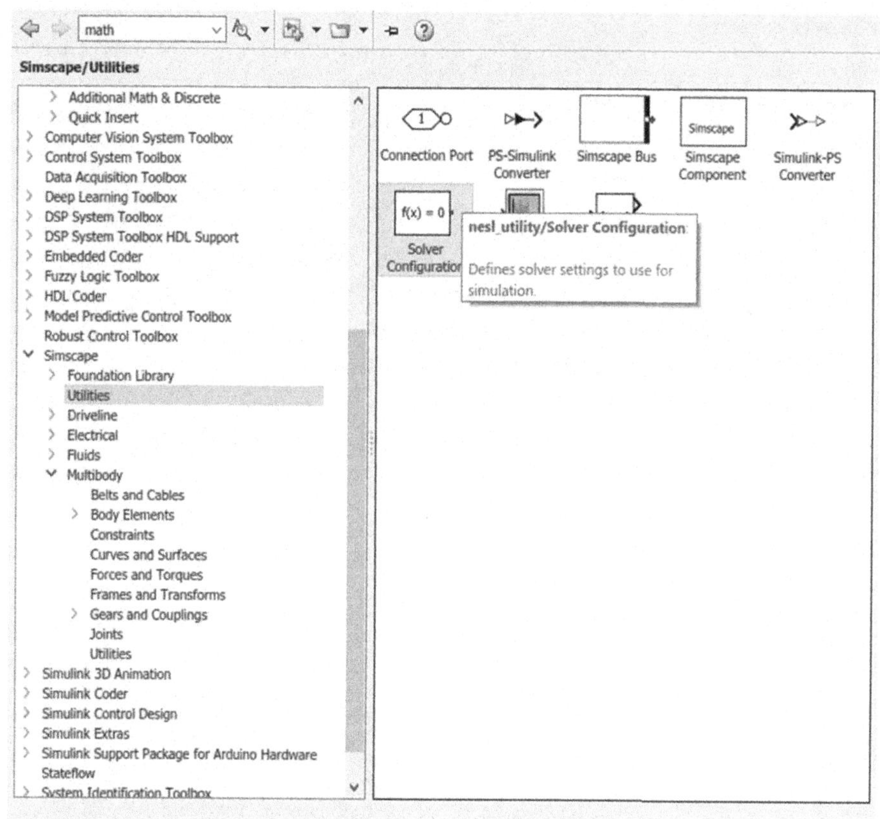

Figure 3.8 Solver Configuration block.

4. Navigate to **Simscape>Multibody>Frames and Transforms** as shown in Fig. 3.9. Add **World Frame** library block in the model.
5. Connect the ports of the **World Frame**, **Mechanism Configuration,** and **Solver Configuration** to each other to form a junction as shown in Fig. 3.10. This marks the Initial Conditions for the model.

3.3.2 Base structure

6. Go to **Simulink® library browser** and navigate to **Simscape>Multibody>Body Elements** as shown in Fig. 3.11. Add **Solid** block in the model. The **Solid** block is a part of the Base Structure of the robot. You can also give a label to this **Solid** block by clicking underneath the block and typing the name "Base." You will notice that there is an in-port "**R**" for this block. The "**R**" is the reference [X, Y, Z] coordinates for this body.

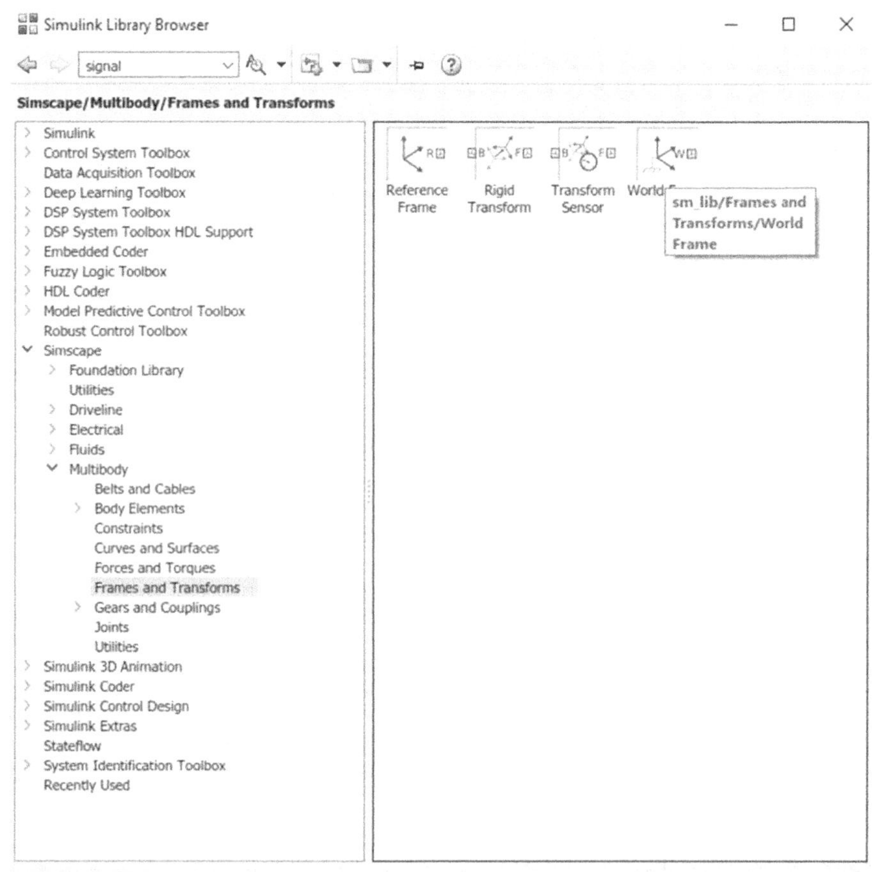

Figure 3.9 World frame.

7. Double click on the Base block until you see a dialog box appearing as shown in Fig. 3.12.
8. Take note of the dimensions shown in Fig. 3.12. As explained in Section 3.2, we can approximate the shape of the robot arms elements. In this case, a brick is used to approximate the shape of the base. The dimensions of the base are [0.09, 0.042, 0.027] m.
9. Under the **Frames** section of the dialog box in step 8, add a new **Frame** by clicking on the plus symbol on the right-hand side of the **New Frame** field. You will see a dialog box or the new frame properties as shown in Fig. 3.13. Use the default properties. It will be in the same orientation as the Frame **R.** Name this Frame **F1.** Once this **F1** has been defined, close out the dialog box for this base. You will now notice in your model that this body has now two Frames "**F1**" and "**R.**"
10. Connect Frame **F1** to the initial conditions as shown in Fig. 3.14. With **F1** connected, this base is now fixed to the world frame coordinates. Frame **R** shall be used for defining the spatial relationships between this body and the consequent body element.

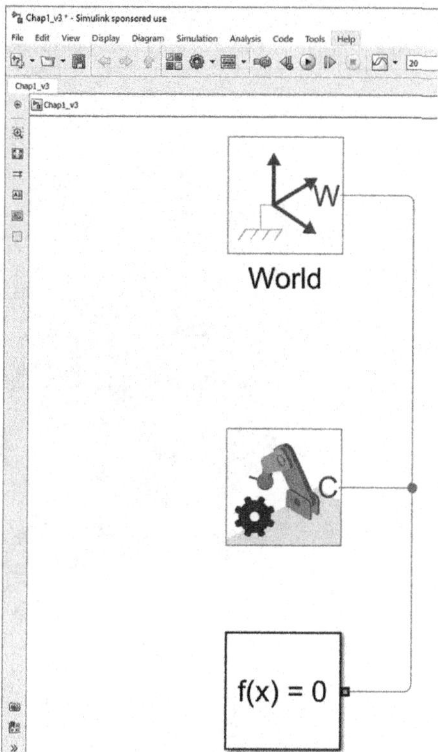

Figure 3.10 Initial conditions.

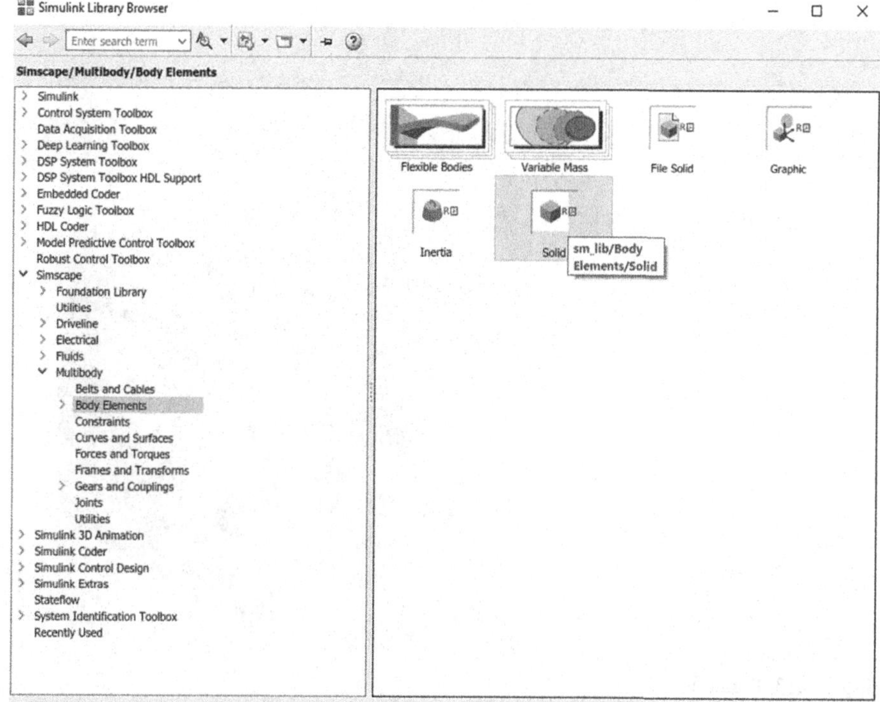

Figure 3.11 Solid library block.

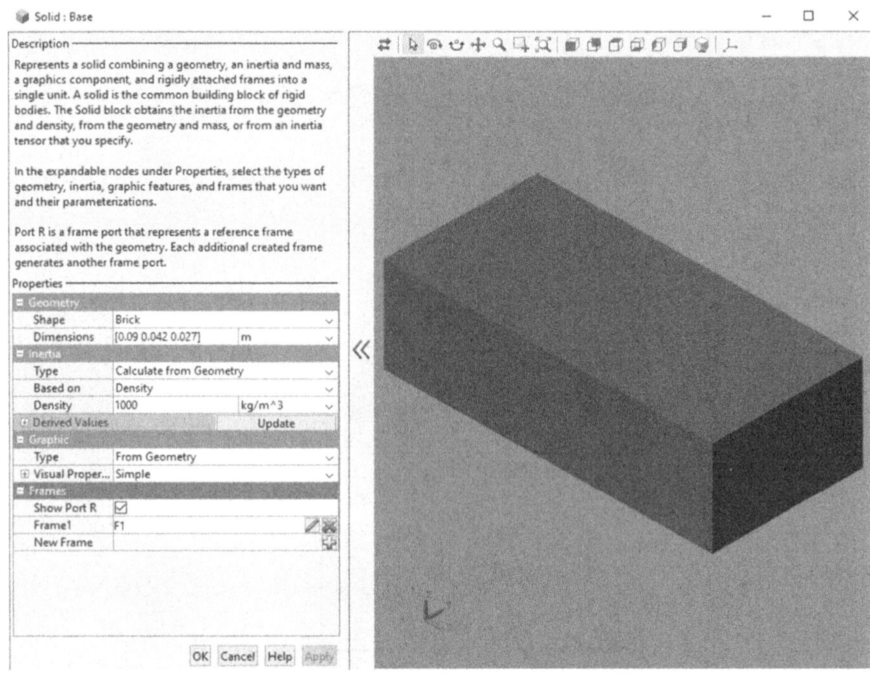

Figure 3.12 Base properties.

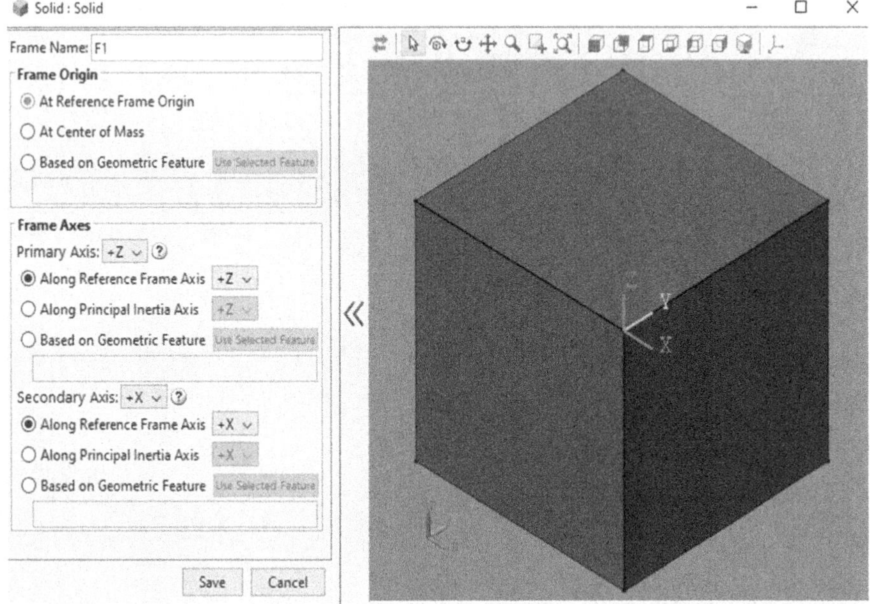

Figure 3.13 Frame F1.

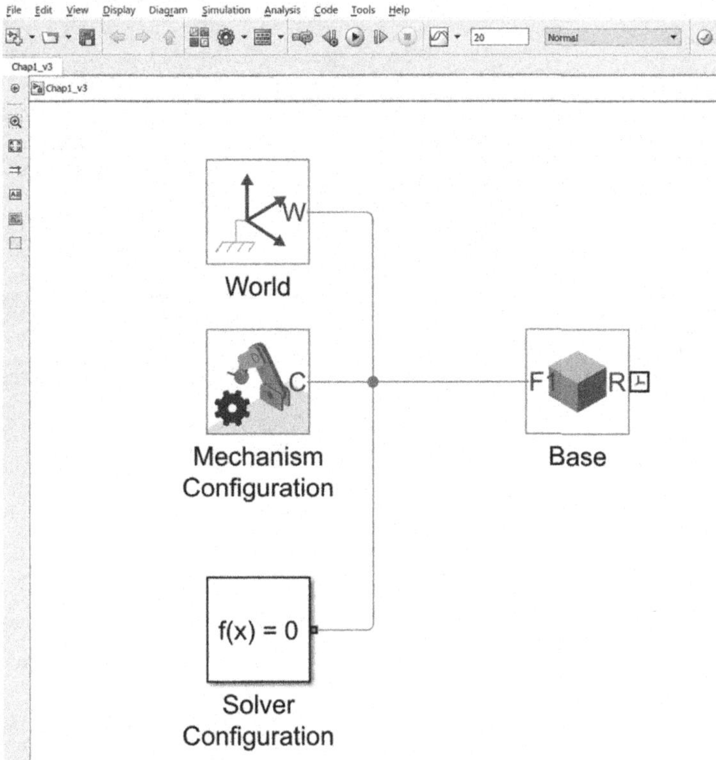

Figure 3.14 Initial conditions applied to solid block (Base).

11. Go to Simulink® library browser and navigate to **Simscape>Multibody>Frames and Transforms** as shown in Fig. 3.15. Add **Rigid Transform** library block in the model.
12. Connect the **B** (short for Base) port of the **Rigid Transform** block to the **R** port of the **Solid** block we have. Rename this **Rigid Transform** block to "Translation" as shown in Fig. 3.16.
13. Double click on this Translation block and take note of the translation offset from the bottom body base as shown in Fig. 3.17. As seen in the figure, there is no offset in the X direction but there is an offset in the −Y direction and the +Z direction. The numbers shown in the offset are taking the distance between the center of mass of the bottom base and the upper base. We can find the center of mass by halving the dimensions measured in Section 3.2 of this chapter.
14. Go to Simulink® library browser, navigate to **Simscape>Multibody>Body Elements** and add the **Solid** block in the model. Add a new frame **F1** as explained in step 9 and connect **F1** to the **F** (short for Follower) port of the Translation block as shown in Fig. 3.18.
15. Define the properties of this block by double clicking on it and inputting the values as shown in Fig. 3.19. This shall serve as our upper base. Go to Simulink® library browser, navigate to **Simscape>Multibody>Frames and Transforms,** and add **Rigid Transform** library block in the model. Connect the **B** port of the **Rigid Transform** block to the **R** port of the Body1 block we have as shown in Fig. 3.20.

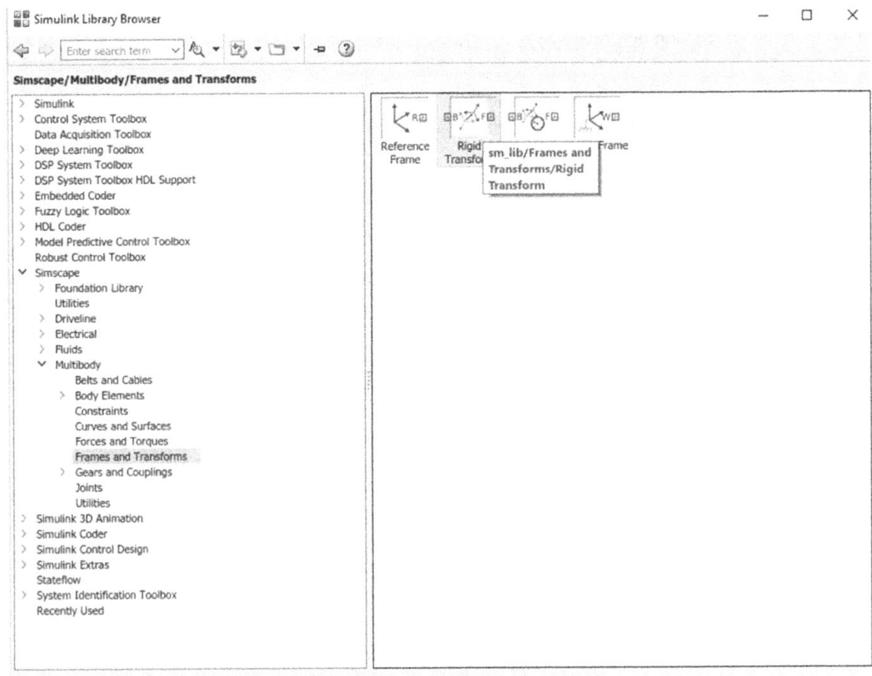

Figure 3.15 Rigid transform block.

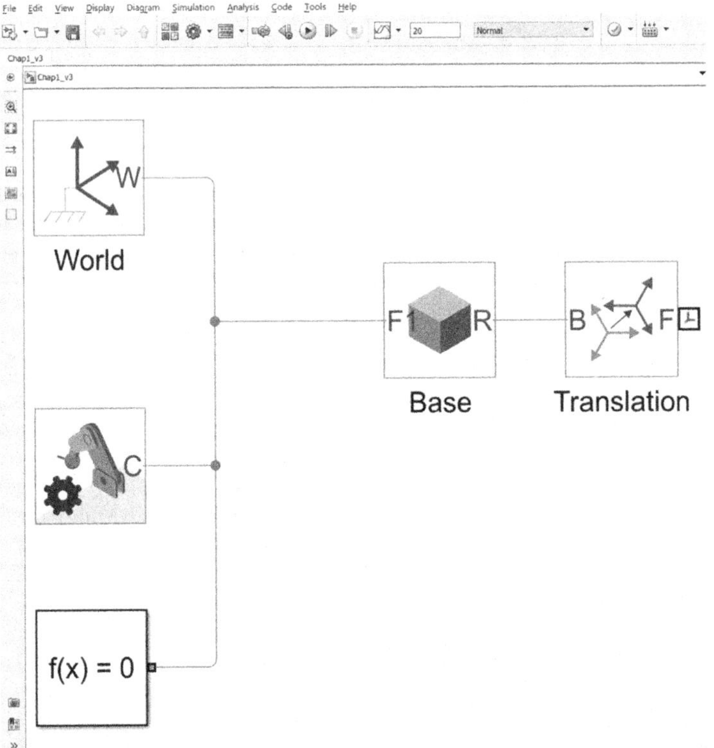

Figure 3.16 Translation from Base.

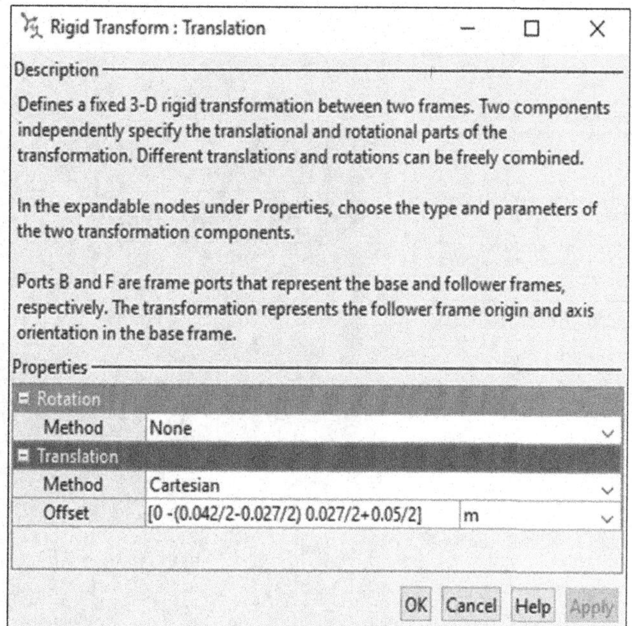

Figure 3.17 Translation from bottom body base properties.

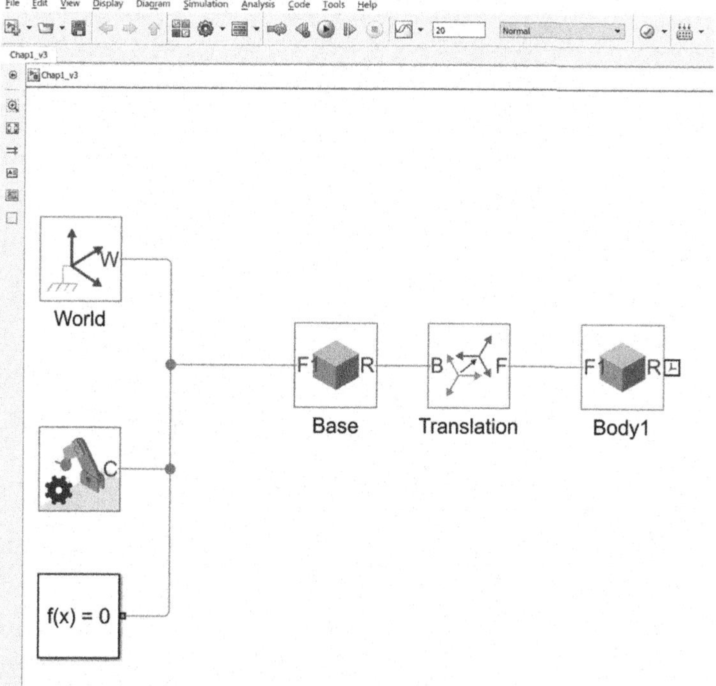

Figure 3.18 Upper body base block.

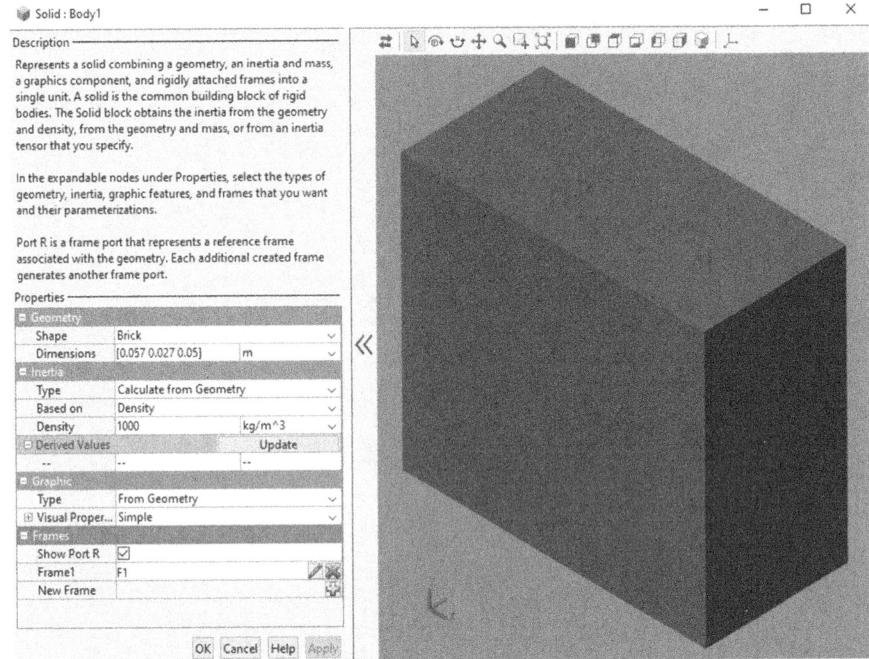

Figure 3.19 Upper base.

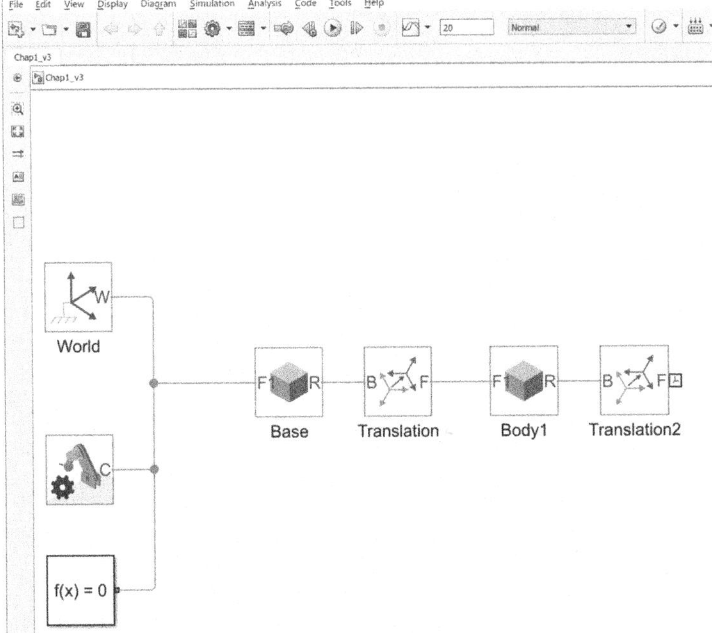

Figure 3.20 Translation from upper body base.

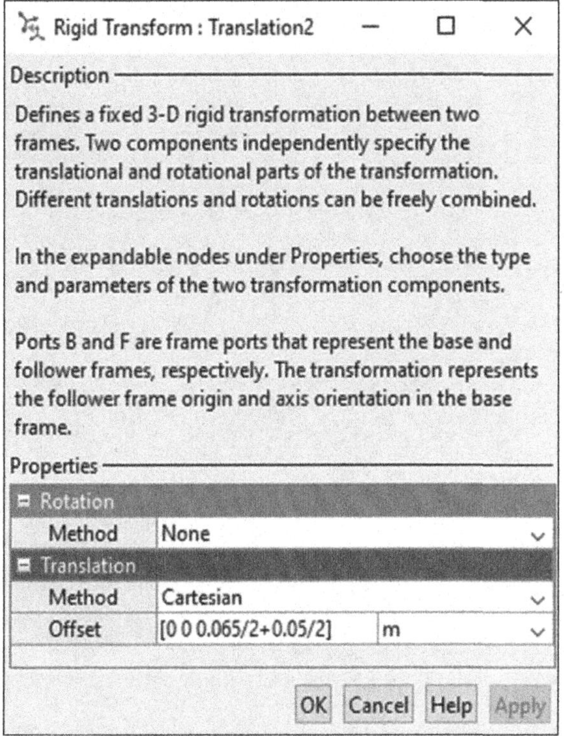

Figure 3.21 Translation from upper body base properties.

16. Double click on this Translation block and take note of the translation offset from the upper body base as shown in Fig. 3.21.

17. Highlight the Base, Translation, Body1, and Translation1 blocks, and right click and select "**Create Subsystem from Selection**" as shown in Fig. 3.22. Name this subsystem "Base." A subsystem incorporating the base of the robotic arm is shown in Fig. 3.23. With this step, we have finished making the base of the robotic arm and can now move on to the upper body.

3.3.3 *Upper body*

18. Go to **Simulink® library browser**, navigate to **Simscape>Multibody>Joints**, and add **Revolute Joint** library block in the model as shown in Fig. 3.24. This joint allows movement about the Z axis for any solid block. We will use this joint to rotate the upper body about the Z axis to simulate 1-DOF by the servomotor.

19. Connect the **B** port of the **Revolute** joint to the **F** port of the Base subsystem as shown in Fig. 3.25.

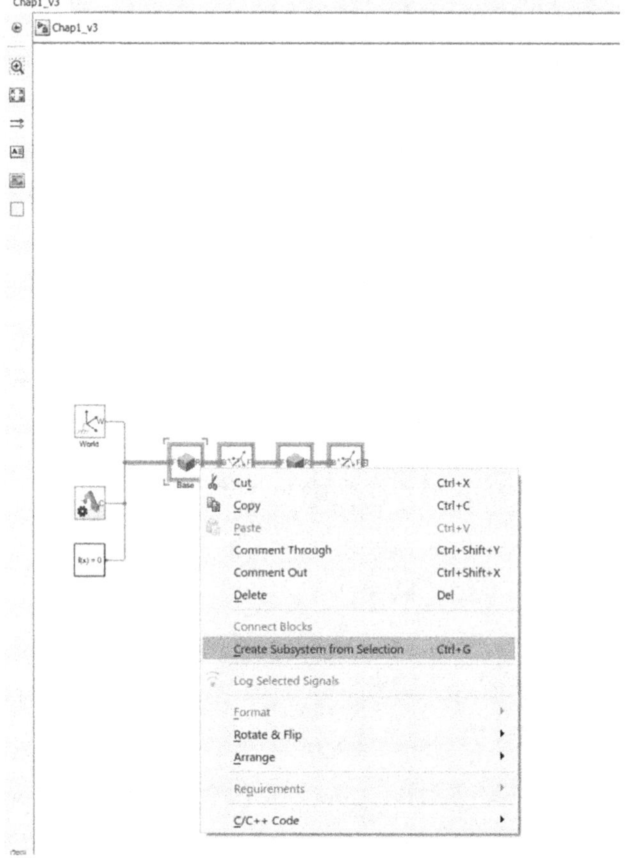

Figure 3.22 Create subsystem for base.

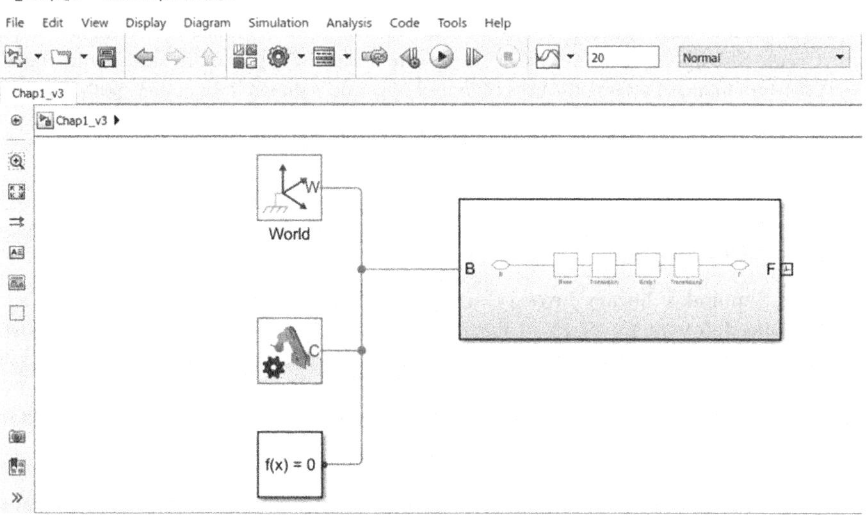

Figure 3.23 Subsystem for the base.

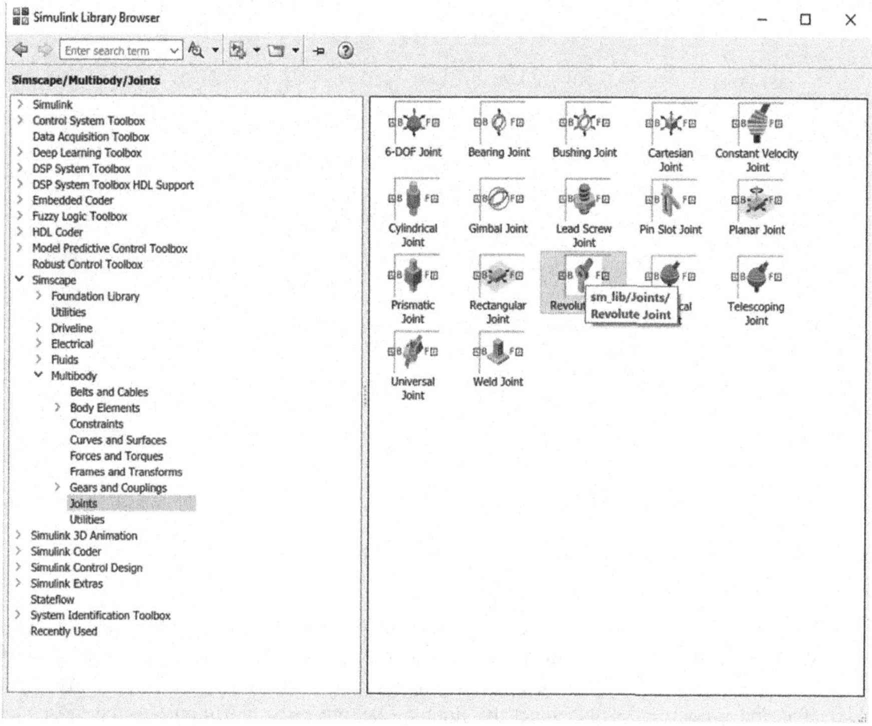

Figure 3.24 Revolute joint.

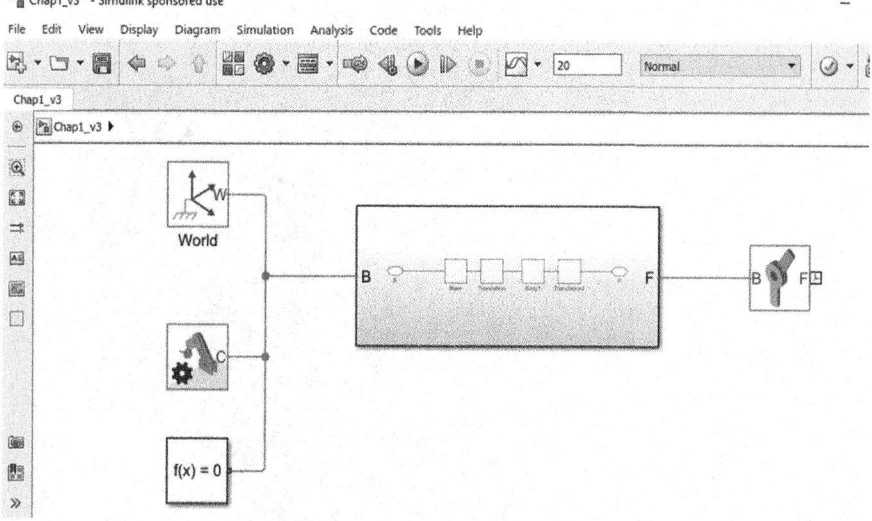

Figure 3.25 Revolute joint for the upper body.

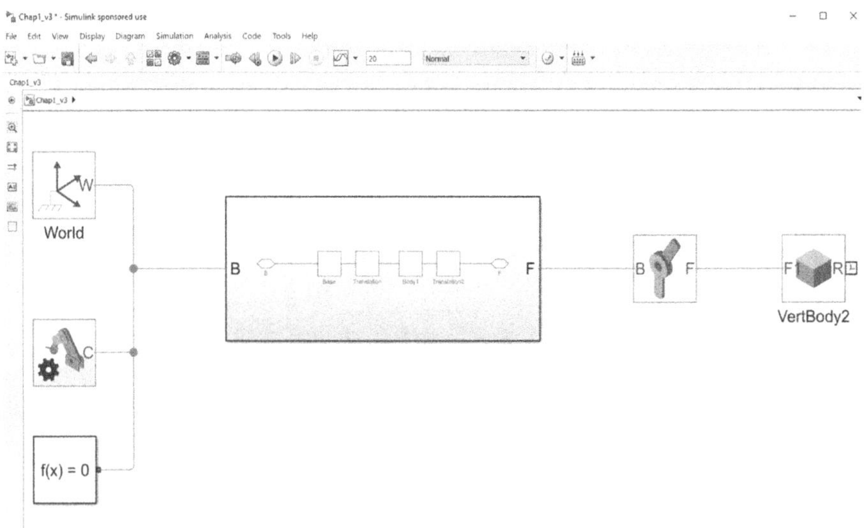

Figure 3.26 Upper Body “VertBody2” connected to Revolute Joint.

20. Go to **Simulink® library browser**, navigate to **Simscape>Multibody>Body Elements**, and add the **Solid** block in the model. Add a new frame **F1** as explained in step 9 and connect **F1** to the **F** port of the **Revolute Joint** as shown in Fig. 3.26. Name this block VertBody2.
21. Define the properties of this block by double clicking on it and inputting the values as shown in Fig. 3.27. This shall serve as our upper body.

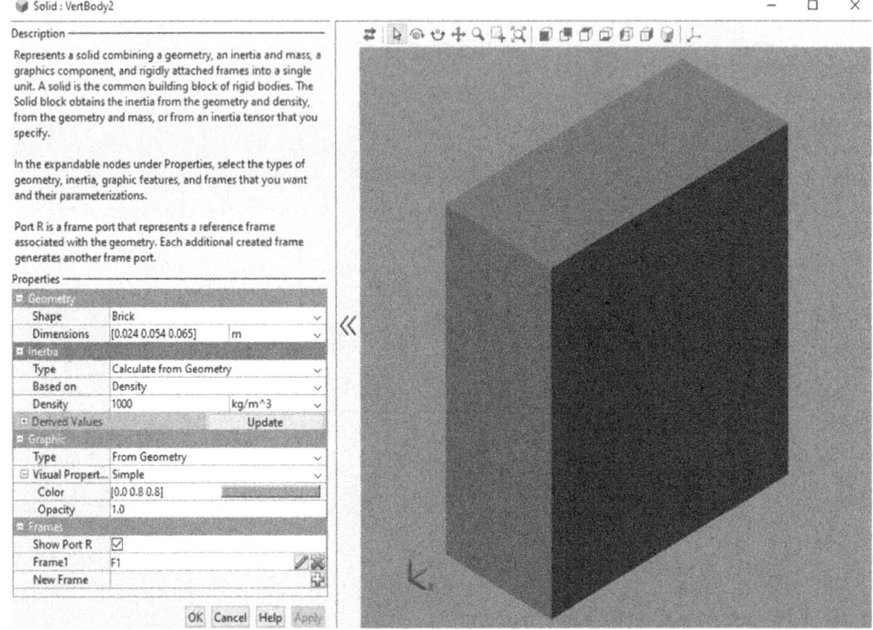

Figure 3.27 Vertbody2 properties.

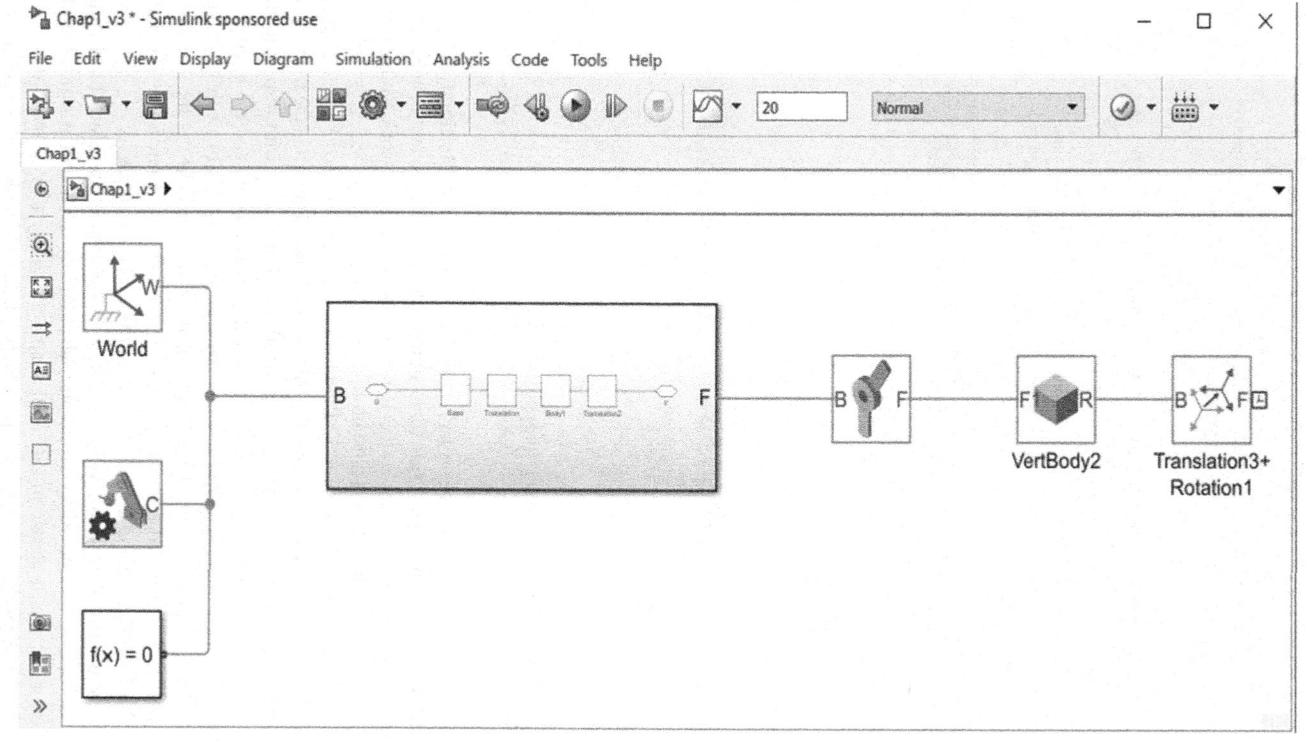

Figure 3.28 Translation and Rotation from VertBody2.

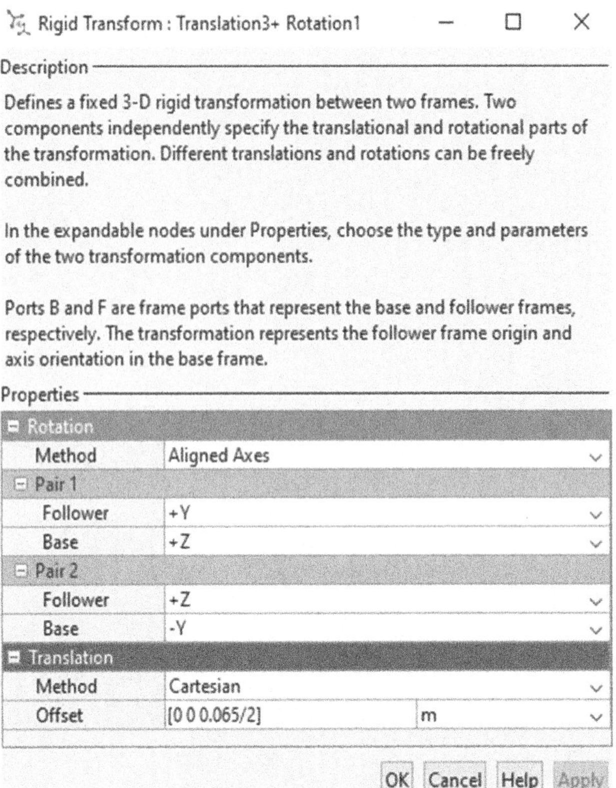

Figure 3.29 Translation3+ Rotation1 properties.

22. Go to **Simulink® library browser**, navigate to **Simscape>Multibody>Frames and Transforms**, and add **Rigid Transform** library block in the model. Connect the **B** port of the **Rigid Transform** block to the **R** port of the VertBody2 block we have as shown in Fig. 3.28. Name this block Translation3+ Rotation1.
23. Double click on this Translation 3+Rotation1 block and take note of the rotation and translation offsets as shown in Fig. 3.29. The rotation method shown in the figure is **Aligned Axes** in which we apply transforms to the follower frame or **F** with respect to the Base frame. The Base Frame is the frame used in the previous body, and the Follower Frame is the frame used in the subsequent body. Aligned axes requires two pair of frames to be transformed. In the first pair, the +Y axis of the Follower Frame is the +Z axis of the base frame. In the second pair, the +Z axis of the Follower Frame is the −Y axis of the base frame. The reasoning behind this rotation is explained in step 25.
24. Highlight the **Revolute Joint**, VertBody2, and Translation3+ Rotation1 blocks, and right click and select "**Create Subsystem from Selection**." A subsystem incorporating the base of the robotic arm is shown in Fig. 3.30. Name this subsystem "Upper Body." With this step, we have finished making the upper body of the robotic arm and can now move on to the upper arm.

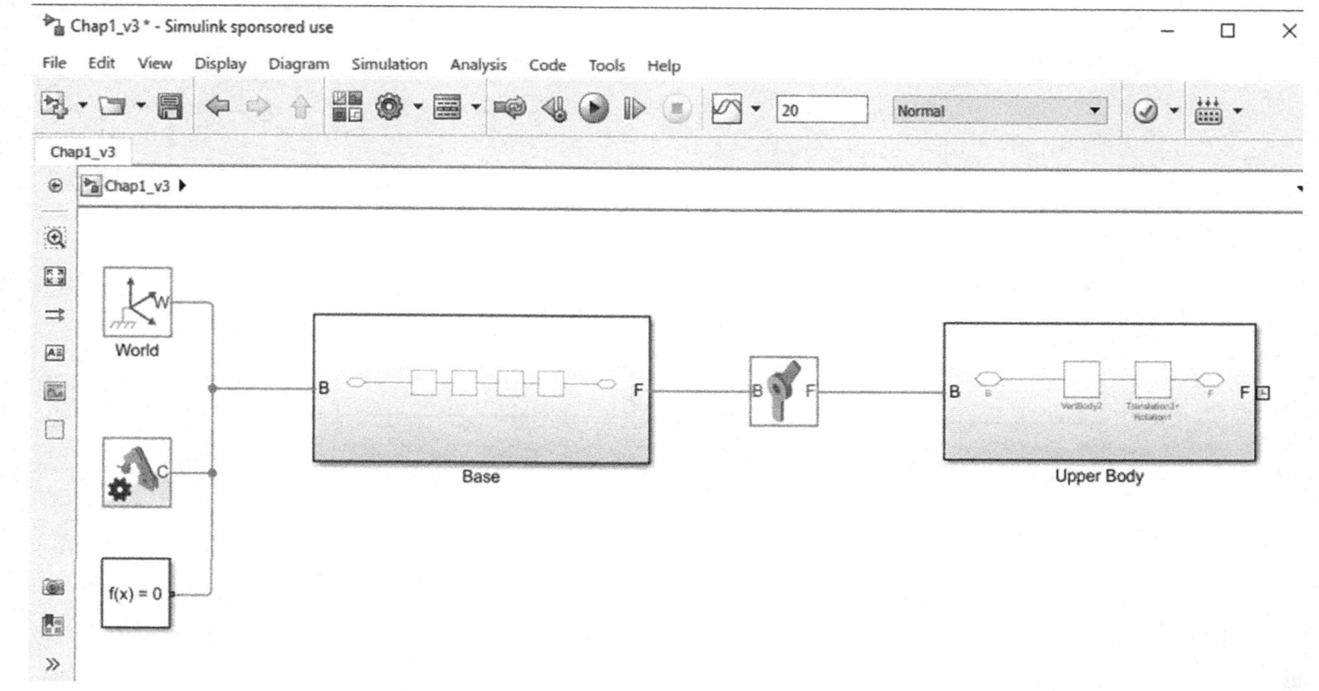

Figure 3.30 Upper body subsystem.

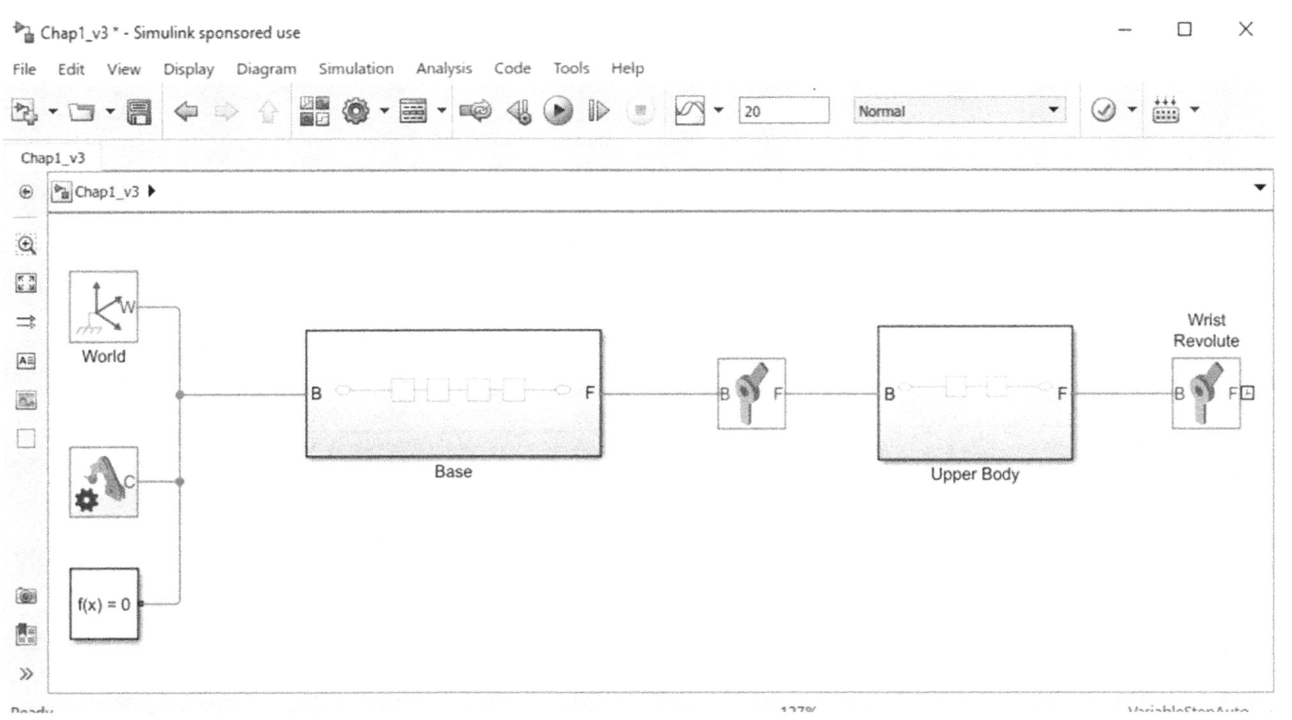

Figure 3.31 Wrist revolute for upper arm.

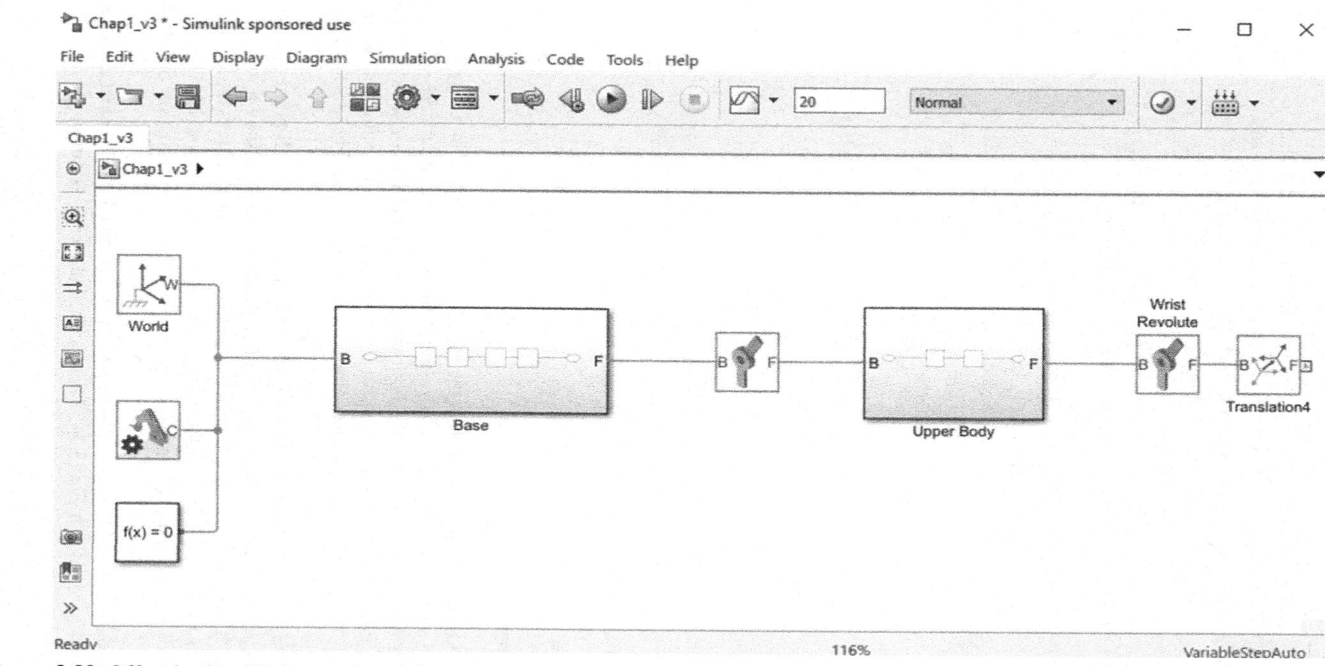

Figure 3.32 Offset in the –X for revolute joint.

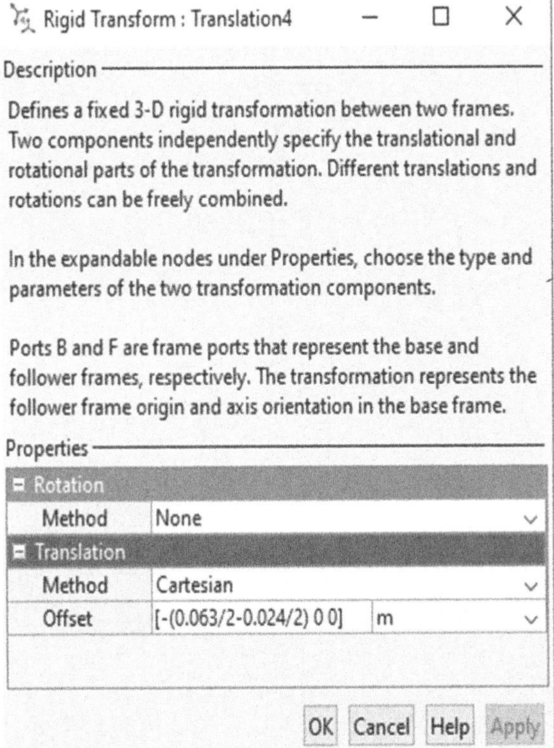

Figure 3.33 Translation4 properties.

3.3.4 Upper arm

25. Add a **Revolute Joint** library block in the model. This revolute joint is needed for the upper arm motion around the +Y axis. A **Revolute Joint** is for rotation around +Z axis but due to the transformation of +Y axis of the Follower Frame being the +Z axis of the base frame and +Z of Follower being −Y of the base, we are now able to define this **Revolute Joint** around the Y axis. Connect the **Revolute Joint** as shown in Fig. 3.31. Name it "Wrist Revolute."
26. Add a **Rigid Transform** library block in the model. Connect the **B** port of the Rigid Transform block to the **F** port of the Wrist Revolute block we have as shown in Fig. 3.32. Name this block "Translation4." This Translation is needed to move the revolute joint in the −X direction to simulate the second DOF by the servomotor about the +Y axis.
27. Double click on this Translation block and take note of the translation offset as shown in Fig. 3.33.
28. Add a **Solid** library block in the model. Add a new frame **F1** and connect **F1** to the **F** port of Translation4 as shown in Fig. 3.34. This block shall serve as our upper arm.

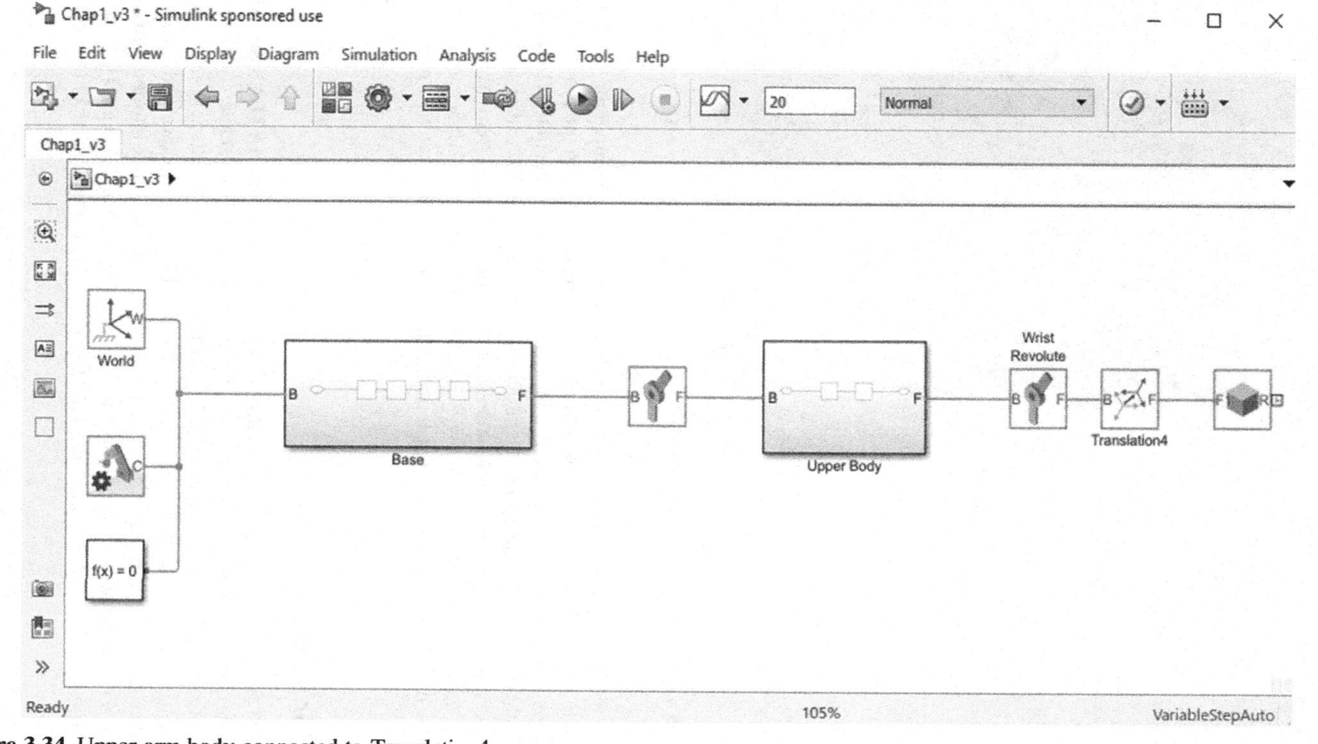

Figure 3.34 Upper arm body connected to Translation4.

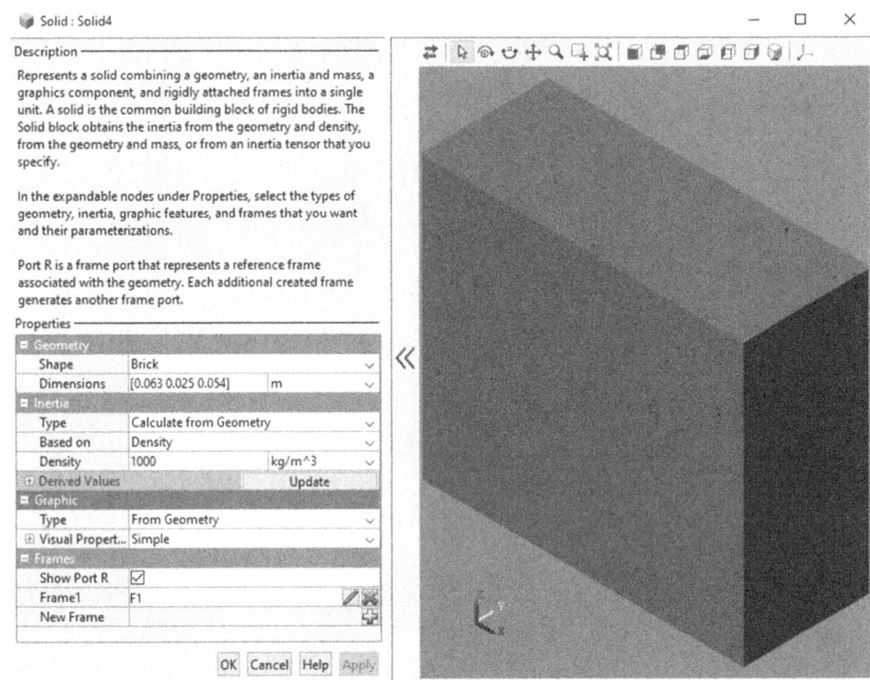

Figure 3.35 Upper arm properties.

29. Define the properties of this block by double clicking on it and inputting the values as shown in Fig. 3.35. Note that the measurement we took originally in the Hardware Parameters will have to be reflected that the Y measurement is the Z value in the **Dimension** tab.
30. Highlight the Wrist Revolute Joint, Translation4, **Solid** block, and Translation5 blocks, and right click and select "**Create Subsystem from Selection**." A subsystem incorporating the base of the robotic arm is shown in Fig. 3.36. Name this subsystem "Upper arm." With this step, we have finished making the upper arm of the robotic arm and can now move on to the gripper.

3.3.5 Gripper arm

31. We will be importing a gripper model already in .STL format in the **MATLAB Simscape™** examples. Fig. 3.37 shows the gripper model as shown in **Simscape™**. The gripper elements are the green metacarpal, four finger links (two red, 1 yellow, 1 gray), and two gray finger tips. To use this gripper, go to the **MATLAB workspace** and type "**smimport('sm_robot')**." **Simscape™** allows the importing of .STL files into the **Simscape™** world by this command. The user would need to create the solid model in a 3D CAD software and export it as an .xml file. Then the MATLAB command **smimport** shall import the xml

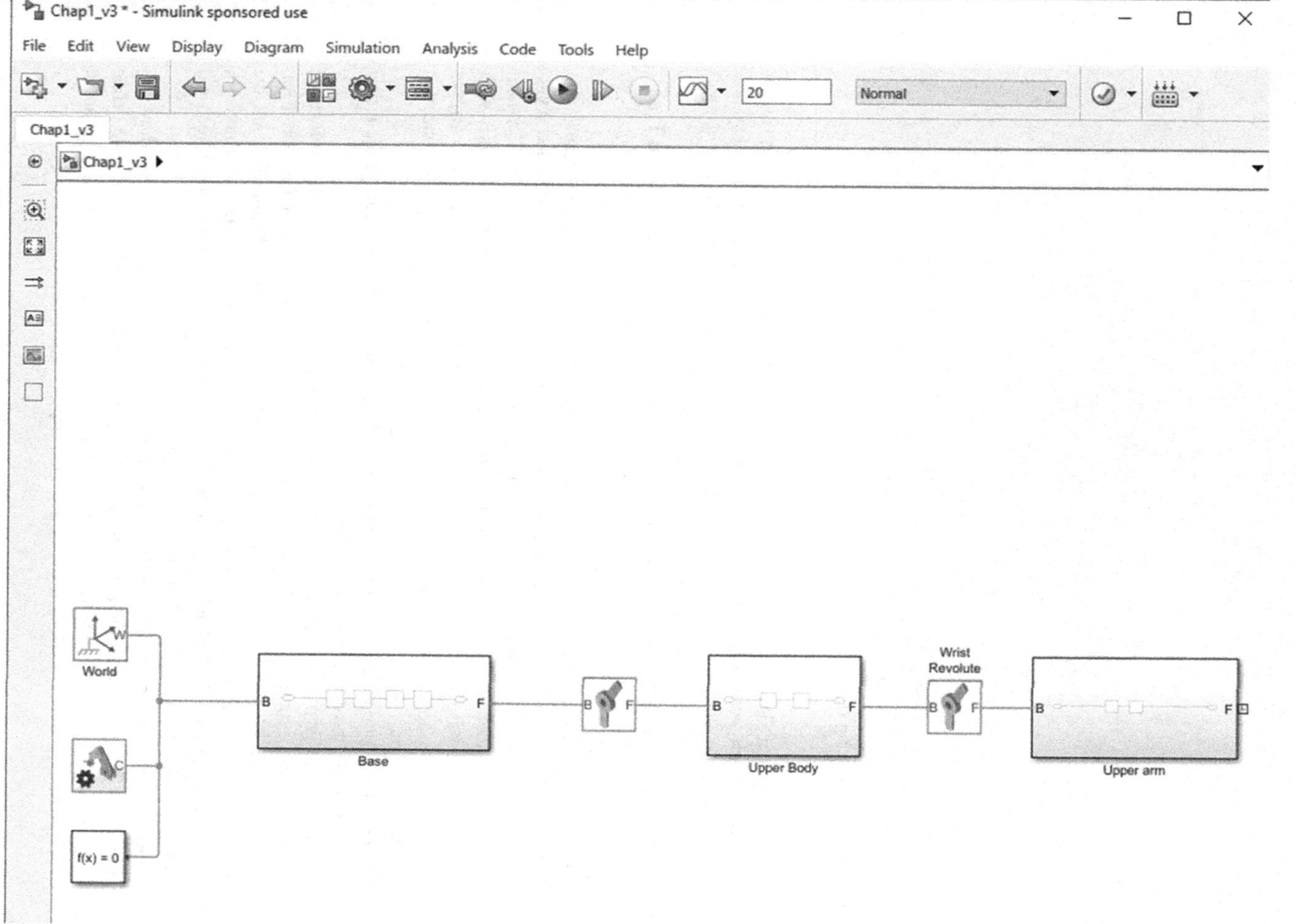

Figure 3.36 Upper arm subsystem.

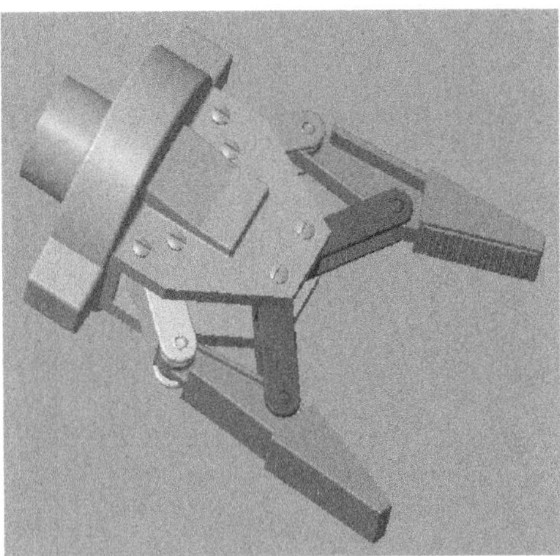

Figure 3.37 Close up of gripper.

file, turn it into a **Simscape™** model, and break it into joints and solid elements. This is what we see in the **sm_robot** model. Once the model has opened, copy the last "grip_1" subsystem block onto our model. Before we can attach this subsystem, we need to apply some rigid transforms to get the orientation of this gripper to our frame.

32. Add a **Rigid Transform** library block in the model. Connect the **B** port of the **Rigid Transform** block to the **R** port of the **F** port of the Upper arm Subsystem as shown in Fig. 3.38. Name this block "Grippertransform."
33. Double click on the Grippertransform block and take note of the rotation and translation offsets as shown in Fig. 3.39. The rotation method shown in the figure is **Aligned Axes**. The reasoning behind this is due to the fact that a .STL model has +Y axis as the Z axis in **Simscape™** such that a transform is needed to transform the follower's Y axis with respect to the base's Z axis. The translation in X axis is needed to offset the gripper from the upper arm frame which is in the center of the upper arm body and hence the "−(0.063/2)" m.
34. Connect the **F** port of the grip_1 subsystem copied from the **sm_robot** model to the **F** port of the Grippertransform block as shown in Fig. 3.40.
35. Go inside the grip_1 subsystem. Remove the Transform block close to the **Reference Frame** on the extreme left and instead add a **Rigid Transform** connected to port **F** on the left as shown in Fig. 3.41. Note: You will notice a **Reference Frame** here on the extreme left. A **Reference Frame** defines a frame to which other frames in a network can be referenced or to which blocks can be attached. **Reference Frames** are not required but serve as a modeling and design convenience.
36. Double click on this **Rigid Transform** block and take note of the rotation as shown in Fig. 3.42. The rotation method shown in the figure is **Standard Axis** in which we define a fixed angle transformation about a specified axis. In this case, it is 180 degrees about the Y axis since the .STL file was oriented 180 degrees around the Y axis.

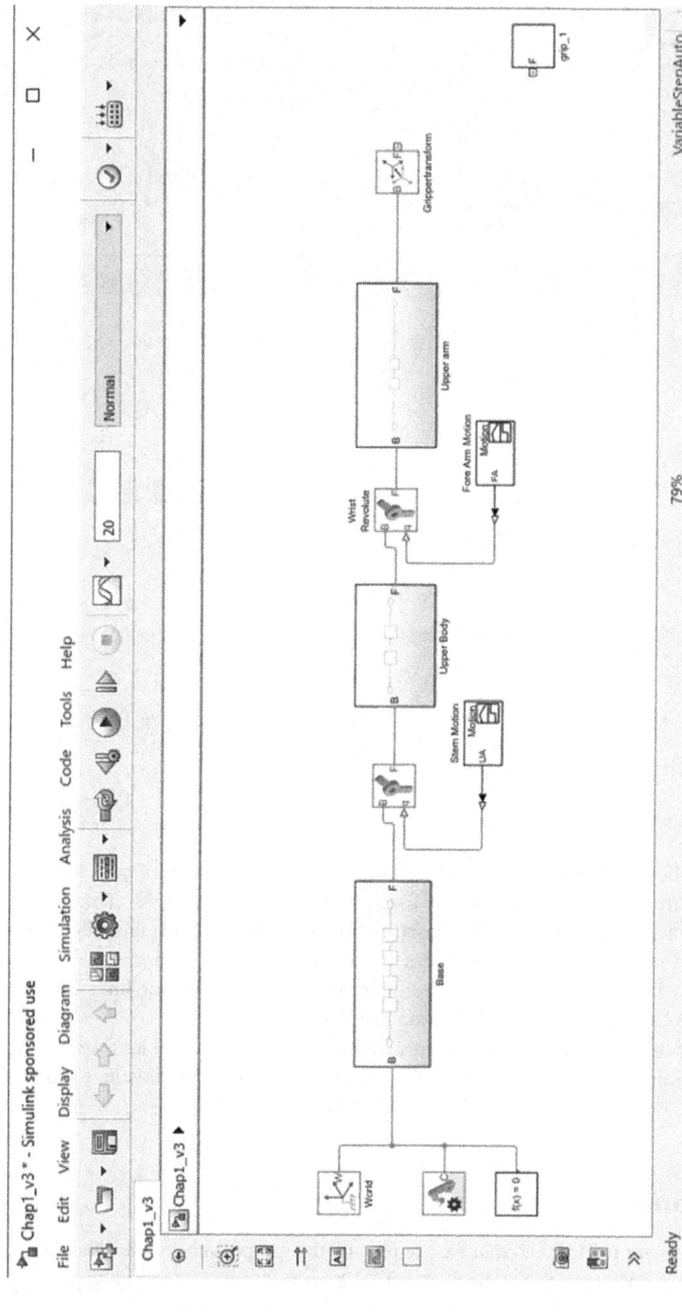

Figure 3.38 Translation + Aligned axis for gripper.

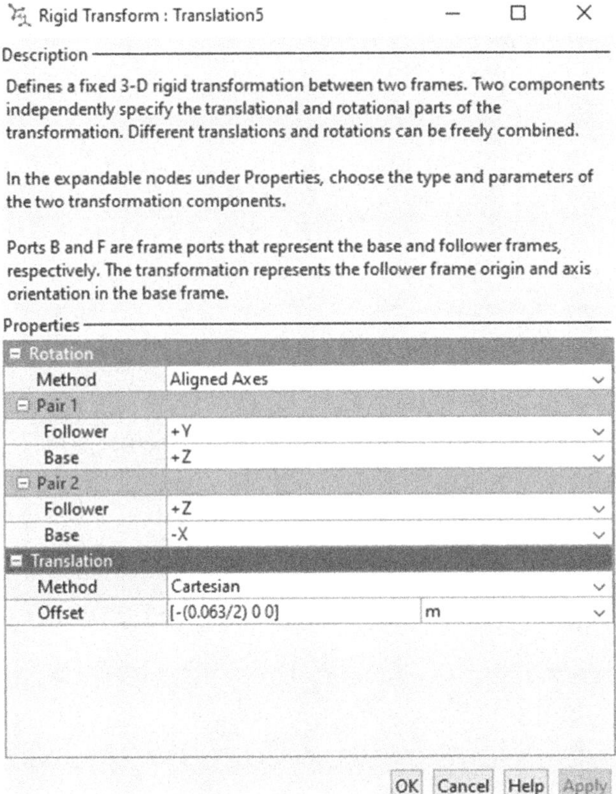

Figure 3.39 Grippertransform properties.

37. Double click on the metacarples_1_RIGID subsystem shown in Fig. 3.41. You will see now the transforms needed for the four finger links in Fig. 3 43. These already have been defined, thanks to the **smimport** command in step 32. Double click on the **Solid** block and notice the properties shown for the metacarpal shown in Fig. 3.44. All of the properties have already been defined. The other elements (finger links and finger tips) have also been defined in the same manner.
38. With this, we have oriented the constructed gripper to our model and we will now move on to defining the motions for the DOF in the model. Note we have not shown the third DOF yet; it will be explained in the next section.

3.3.6 Motion

39. Go back to our model such that we are outside the gripper subsystem. Double click on the **Revolute Joint** between the Base and the Upper body we defined in step 19. Notice the

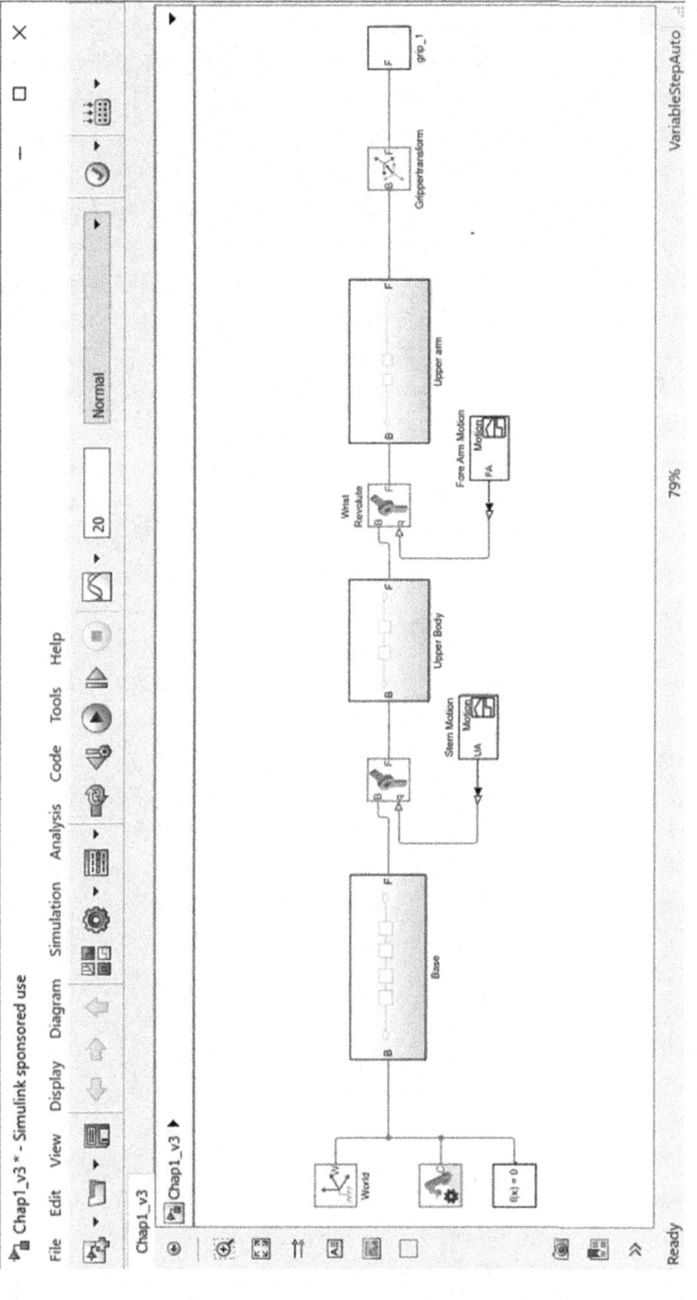

Figure 3.40 Gripper subsystem from sm_robot.xml.

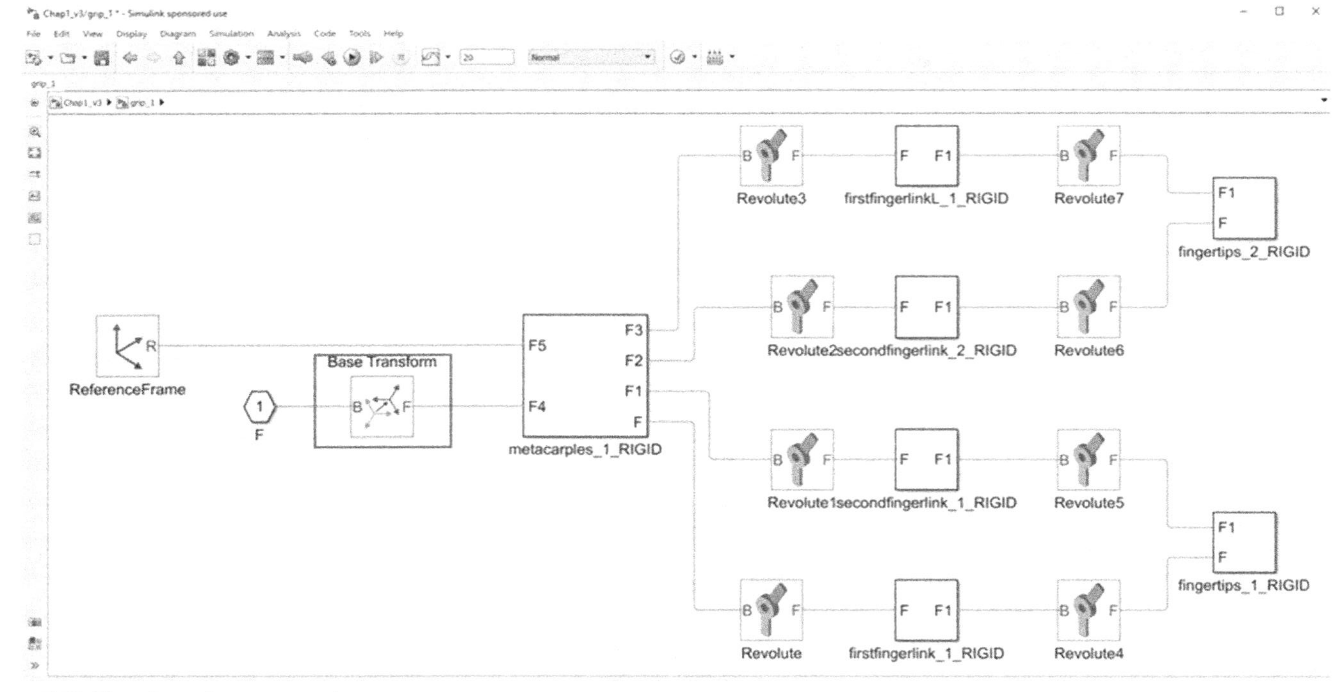

Figure 3.41 Transform for metacarpal.

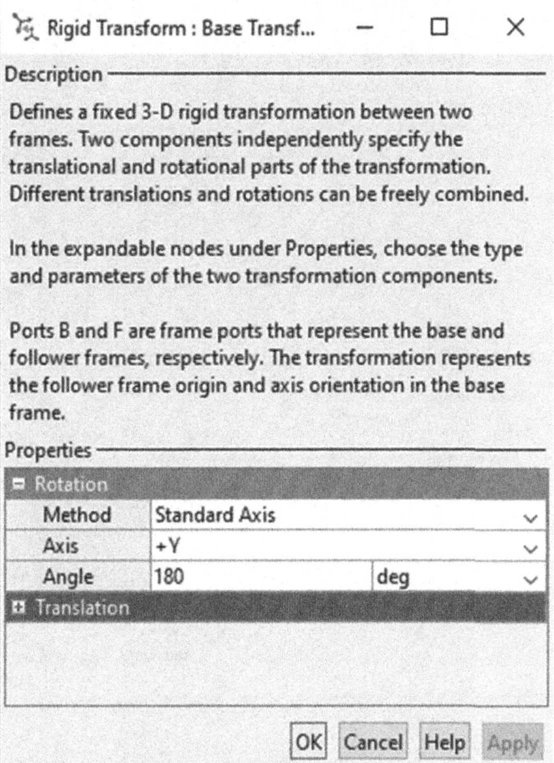

Figure 3.42 Rigid transform for metacarpal properties.

properties shown in Fig. 3.45. Note that we will be providing the input whereas the torque shall be automatically computed by **Simscape™** based on our inputs. The inputs will be in the form of degrees around the Z axis. Once you click OK, you will now see a q input to the **Revolute Joint** for the motion. Notice that there is also a tab for **Sensing** in the case when we have to monitor the outputs of the joint.

40. Go to the **Simulink® library browser** and navigate to **Simulink®>Sources** as shown in Fig. 3.46. Add **Signal Builder** library block in the model. This will be used to define the movements for the **Revolute Joint.**

41. Navigate to **Simscape>Utilities** as shown in Fig. 3.47. Add the **Simulink®-PS Converter** library model to the model. This block is used to convert the unitless **Simulink®** signal to a physical signal in **Simscape™**. As **Simscape™** uses physical signals for simulation, this is needed to convert the movement we define in the Signal builder to a unit of degrees. Notice that there is also **PS-Simulink® converter** that converts a physical unit to a unitless signal.

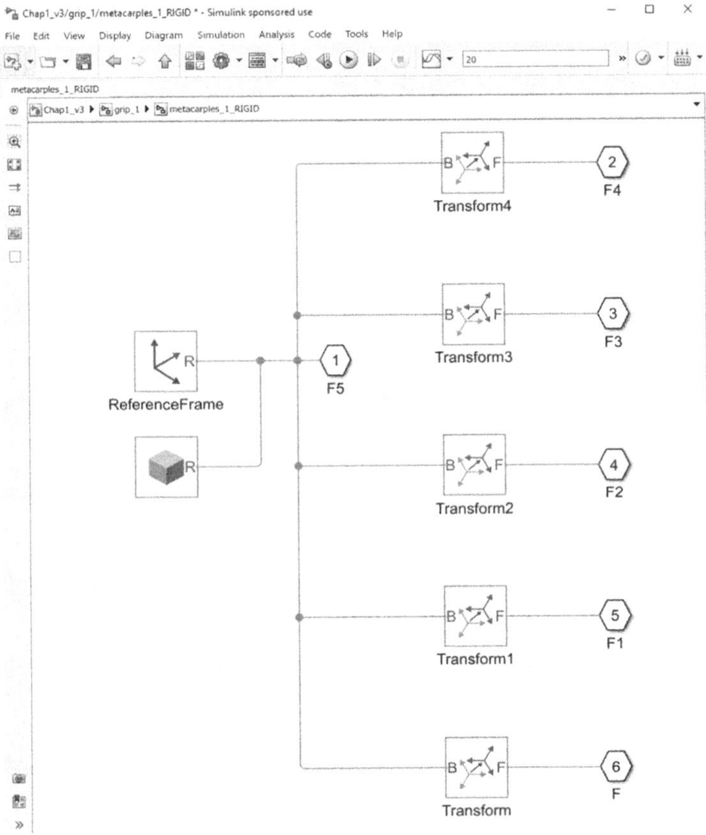

Figure 3.43 Metacarpal transform frames.

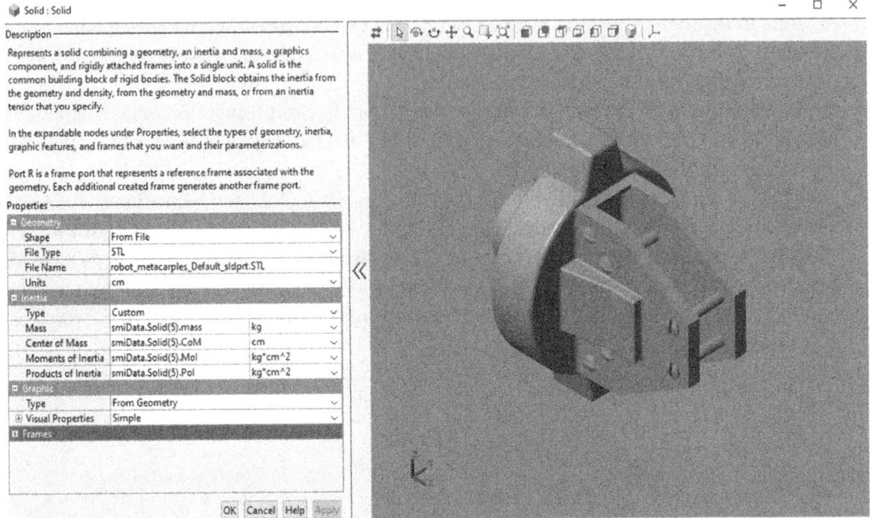

Figure 3.44 Metacarpal properties.

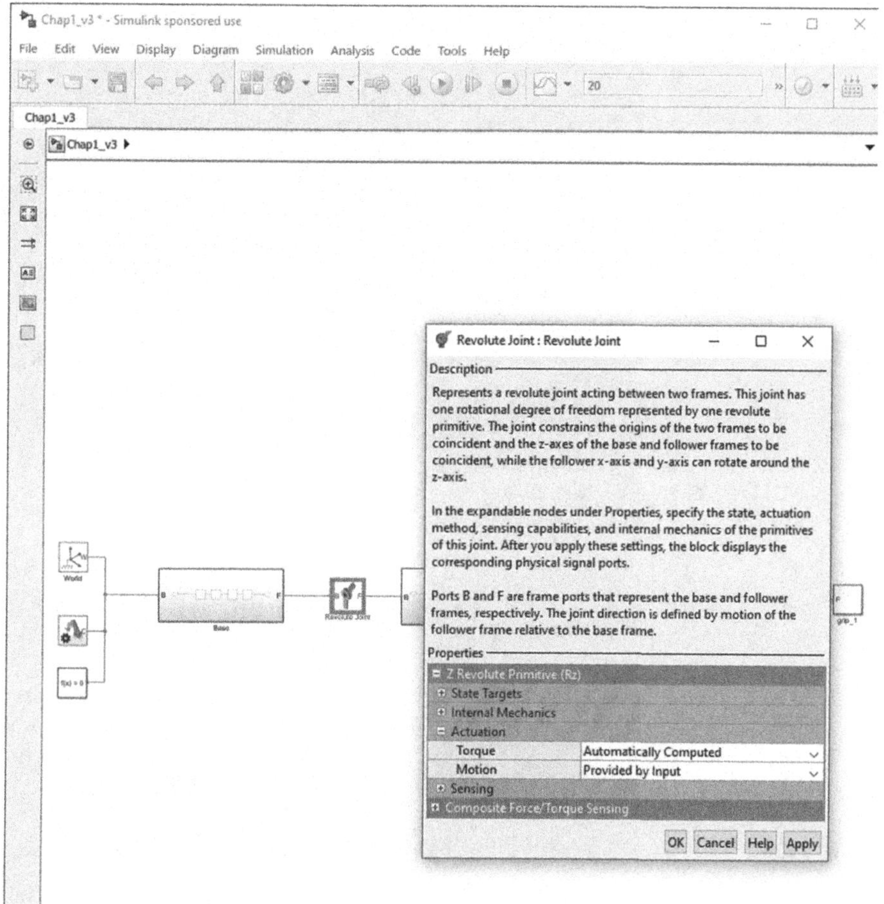

Figure 3.45 Motion for upper body revolute joint.

42. Flip the **Signal builder** and the **Simulink®-PS converter** such that their inputs are on the left side instead of the right. This is just for visual preference only. You can do so by highlighting both of the block, and then right clicking and selecting **Rotate & Flip>Flip Left-Right** as shown in Fig. 3.48. Name this **Signal builder** block as "Stem Motion."
43. Now connect the **Signal builder** to the **Simulink®-PS converter.** Then connect the **Simulink®-PS converter** to the q input of the **Revolute Joint** as shown in Fig. 3.49. We have now a means of defining the movement in degrees for the upper body motion.
44. Double click on the **Signal builder** and notice the default signal shown in Fig. 3.50. We will be editing the timestamps and the signal values of each of these six points in the blue (light gray in print version) circles.
45. Firstly, extend the time range of the inputs by right clicking and choosing "**Change Time Range**" as shown in Fig. 3.51. Set the **Max** time to 20 s. The result is shown in Fig. 3.52.

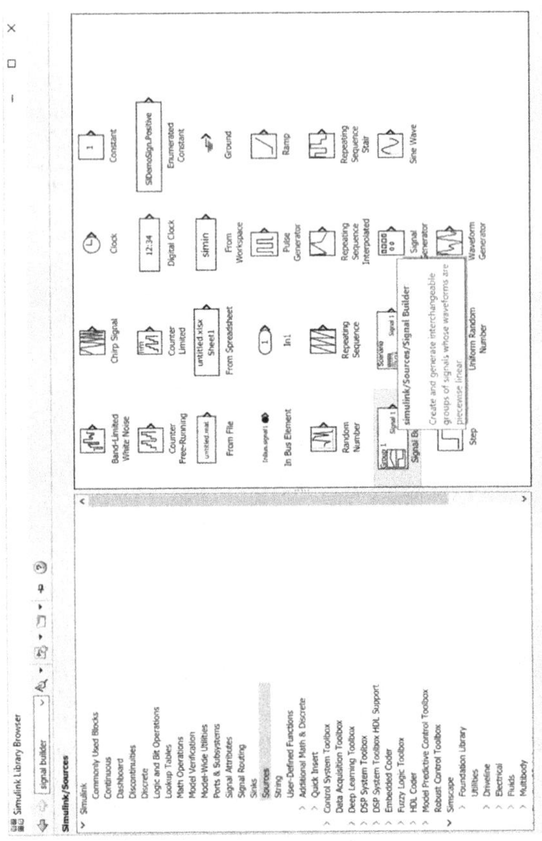

Figure 3.46 Signal builder.

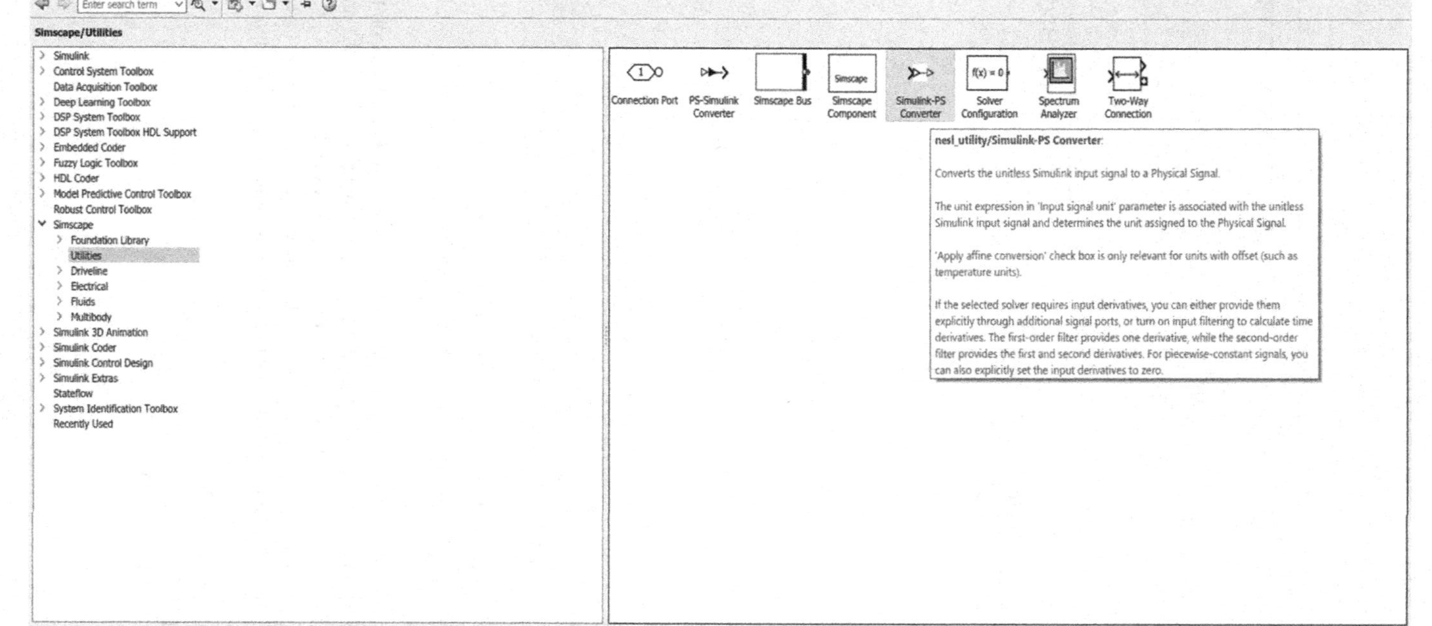

Figure 3.47 PS-Simulink® Converter.

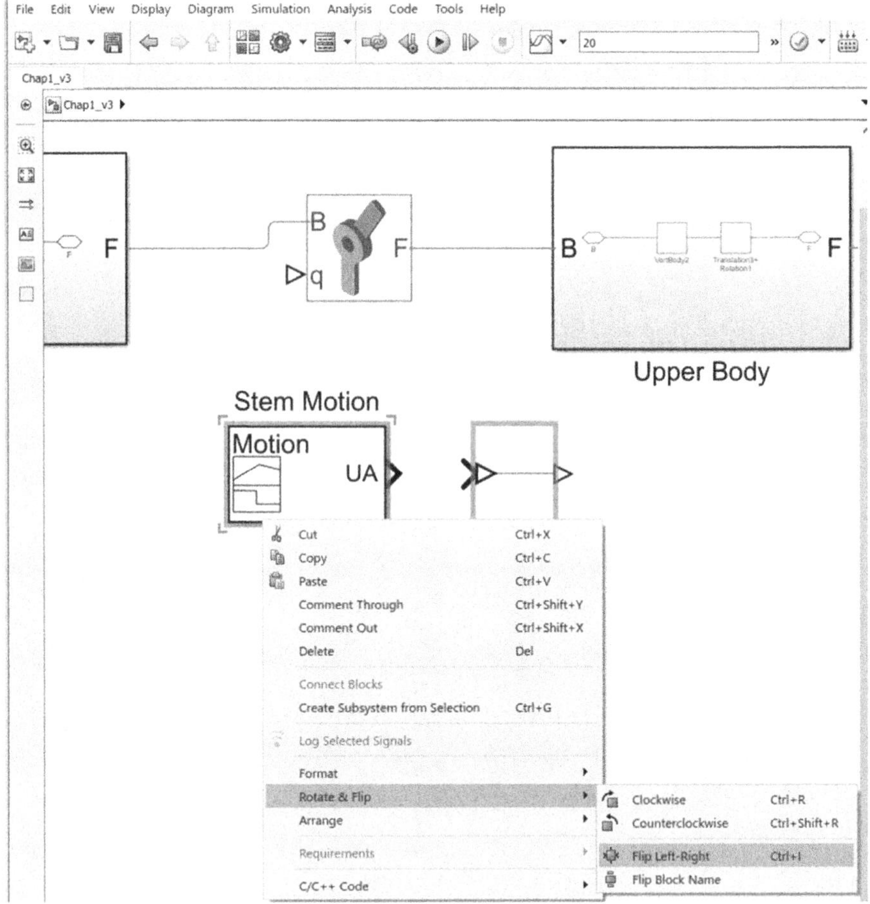

Figure 3.48 Flip the Signal builder and PS-Simulink Converter.

46. The left and right end points are already set to the value we need. Click on the fifth point (6 s, 0 degree) and change the values in the left point fields for **T** as 17.5 and **Y** as 90 as shown in Fig. 3.53. This means the signal is +90 degrees at 17.5 s.

47. Repeat step 46 for the other points. The result is shown in Fig. 3.54. With this step, we have finished defining the motion for the first degree of motion, which is the upper body around the Z axis.

   a. Fourth point: [12.5,90]

   b. Third point: [7.5,−90]

   c. Second point: [2.5,−90]

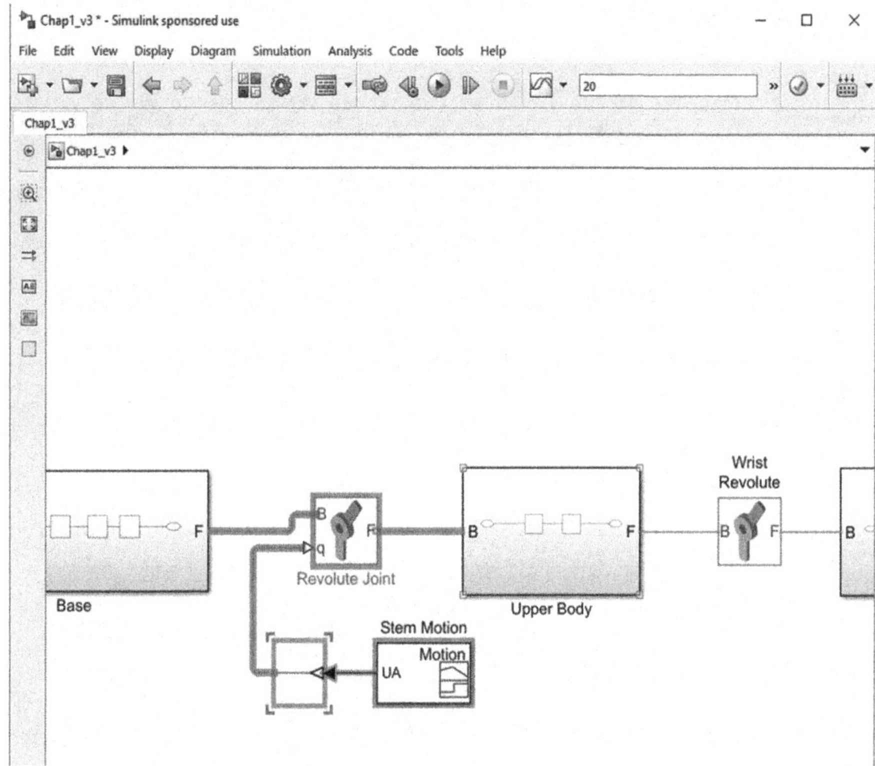

Figure 3.49 Motion for the upper body revolute joint.

48. The next step would be defining the properties for the **Simulink-PS Converter** block we added. Since we defined a unitless signal in the previous, we must convert it to physical units for the actuation of the joint. Double click on the **Simulink-PS Converter** block and type in deg to signify degrees for the unit as shown in Fig. 3.55.
49. Now we will be defining the motion for the second degree of motion about the Y axis for the upper arm. Navigate to the revolute joint that was placed between the Upper Body subsystem and the Upper arm subsystem. Double click on the **Revolute Joint** and notice the properties shown in Fig. 3.56.
50. Repeat steps 40 to 43 for definition the motion for the upper arm. Name the signal builder "Fore Arm Motion" as shown in Fig. 3.57.
51. Repeat steps 44 to 48 for defining the upper arm motion. Take note of the point coordinates below. The resulting signal is shown in Fig. 3.58. With this step, we have finished defining the motion for the second degree of motion which is the upper arm around the Y axis.
    - **a.** First point: [0,0]
    - **b.** Second point: [2.5,−45]
    - **c.** Third point: [7.5,−45]

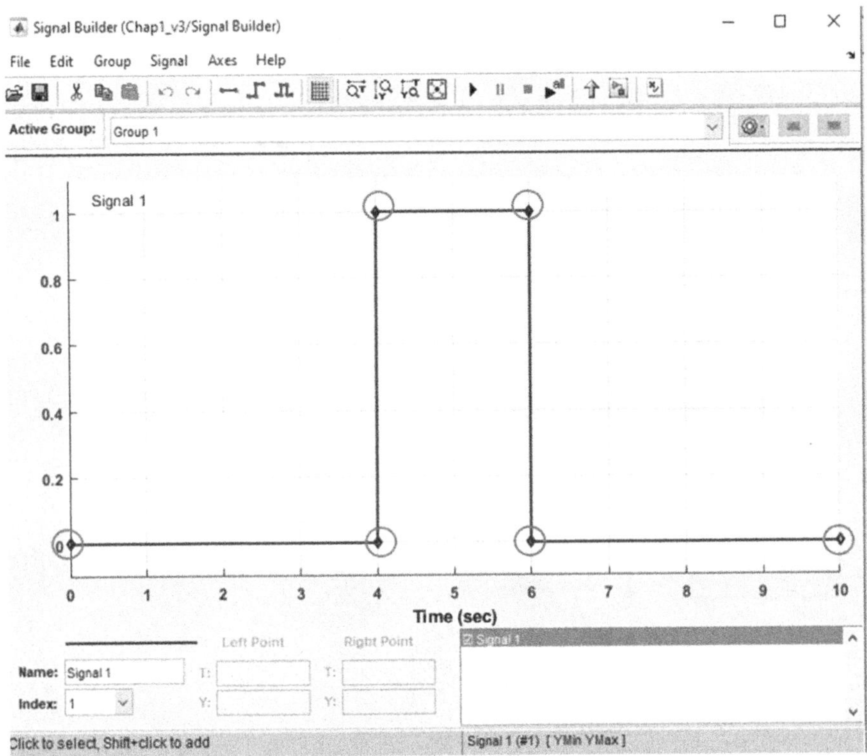

Figure 3.50 Default view of signal builder.

   d. Fourth point: [12.5,30]
   e. Fifth point: [17.5,30]
   f. Sixth point: [20,0]

52. Go inside the **"grip_1"** subsystem and double click on the Revolute1 and Revolute2 joints. These are the joints connector the finger links. These two joints will be used for the third degree of motion, which are the gripper jaws. Take note of the **Revolute Joint** properties as shown in Fig. 3.59.
53. Go to the **Simulink® library browser** and navigate to **Simulink®>Math Operations** as shown in Fig. 3.60. Add **Gain** library block in the model. We will be using this **Gain** block in the later steps.
54. Double click on this **Gain** block and set a value of −1 as shown in Fig. 3.61.
55. Add a Signal builder and two **Simulink®-PS Converter** blocks onto the model and have them connected as follows in Fig. 3.62. The **Gain** block value of −1 is used to mirror the motion for the other end of the jaw as the angles will be reversed with a gain value of −1.

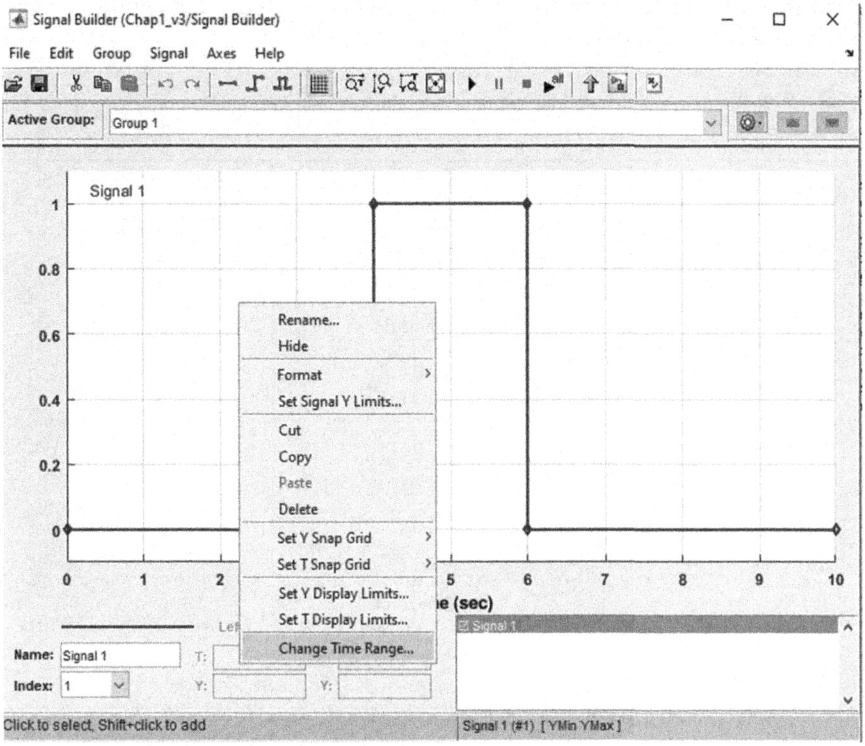

Figure 3.51 Extending the time range.

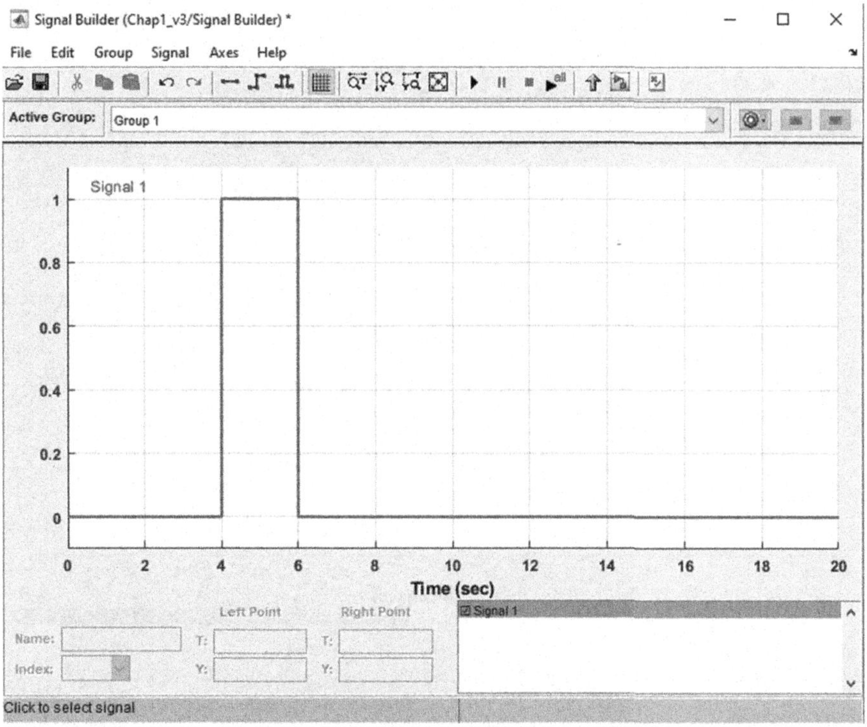

Figure 3.52 Result of extending the time range.

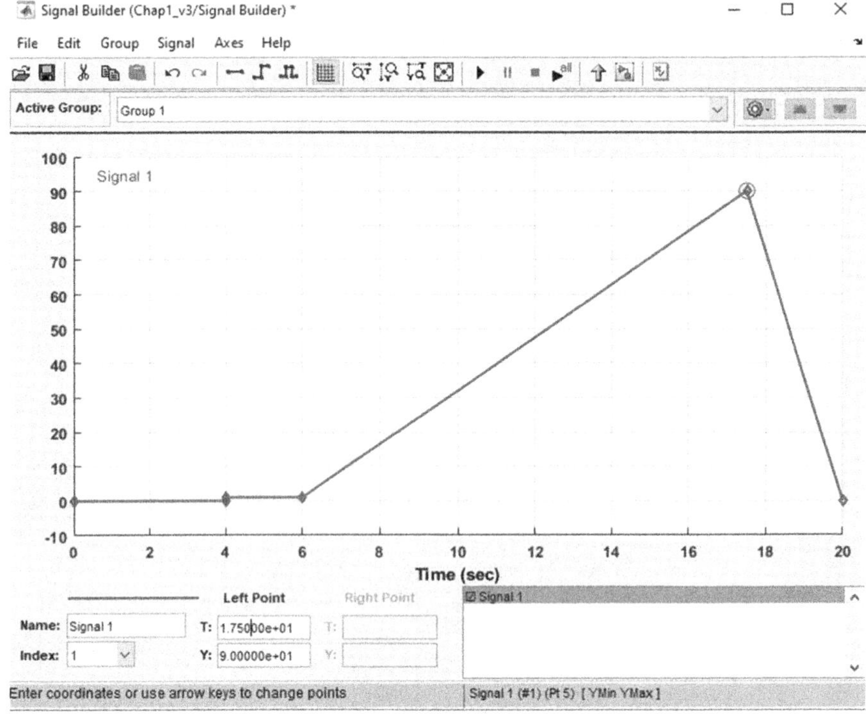

Figure 3.53 Changing the fifth point values to [17.5,90].

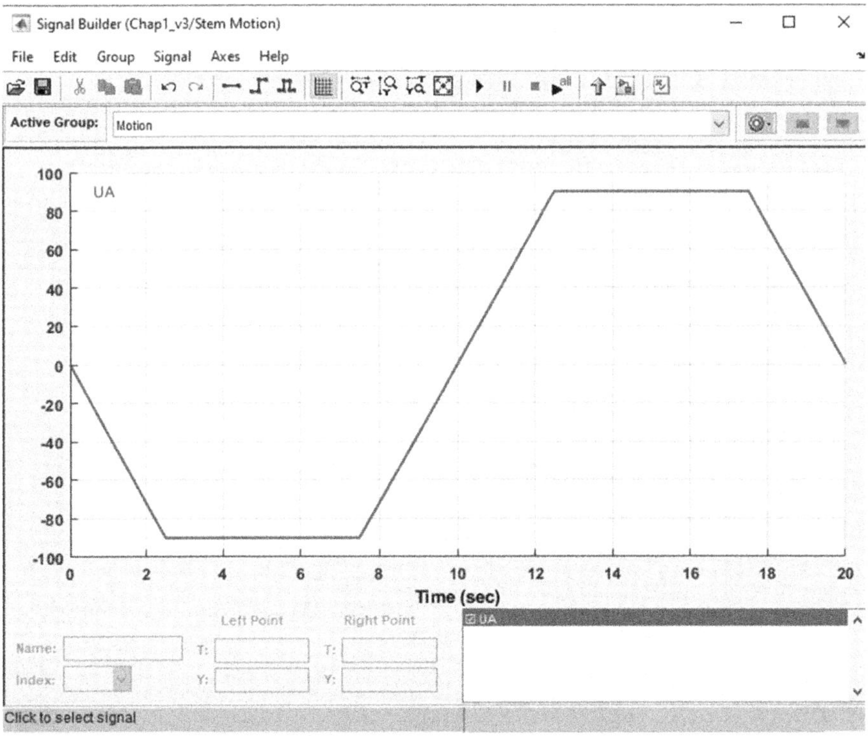

Figure 3.54 Finished signal properties.

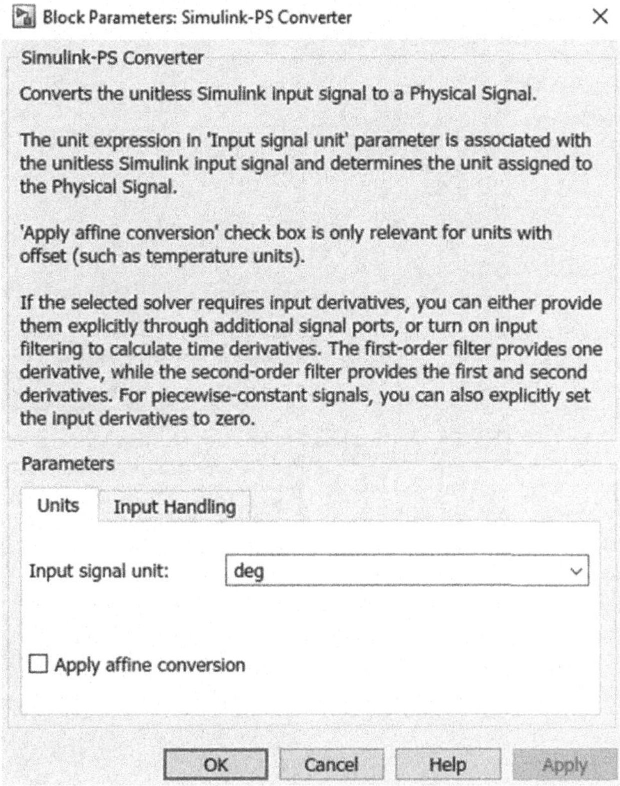

Figure 3.55 Simulink-PS Converter.

56. Repeat steps 44 to 48 for defining the gripper motion. Take note of the point coordinates below. The resulting signal is shown in Fig. 3.63. With this step, we have finished defining the motion for the third degree of motion, which is the gripper motion about the Z axis.

- **a.** First point: [0,130]
- **b.** Second point: [4.5,130]
- **c.** Third point: [5.5,150]
- **d.** Fourth point: [14.5,150]
- **e.** Fifth point: [15.5,130]
- **f.** Sixth point: [20,130]

57. With the model created for the 3-DOF arm and its motions defined, we can now simulate the model. Press the play button and see how the arm behaves with the motion provided. A final layout of the 3-DOF arm is shown in Fig. 3.64.

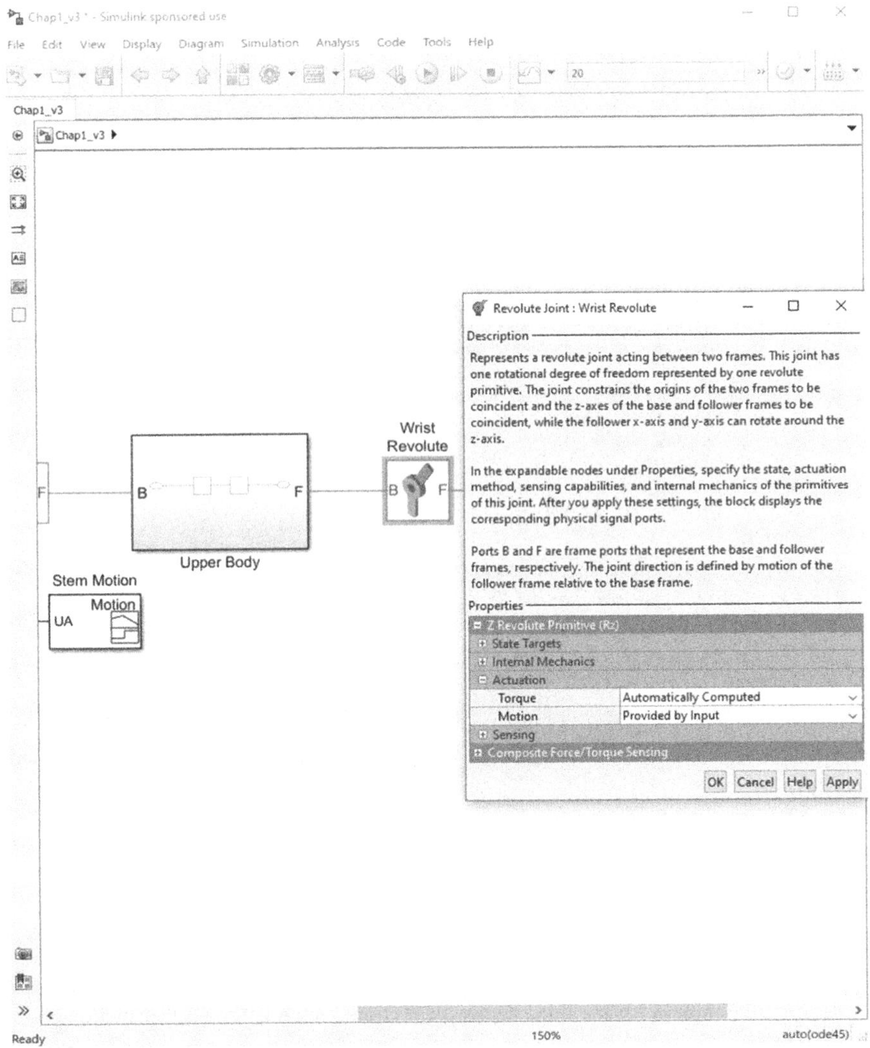

Figure 3.56 Upper arm wrist revolute motion.

3.3.7 Simulation of a failed system

58. This section shall introduce a failed system into the model and emulate the response of the failed system with the same inputs. Select the entire model except the **Solver Configuration**, **Mechanism Configuration**, and **World Frame** block and create a subsystem for it. Rename it as Digital Twin. Then copy–paste that subsystem and rename that new subsystem as Physical Asset. This new subsystem shall act as the failed subsystem. Then connect the **Solver Configuration**, **Mechanism Configuration,** and **World Frame** to the inputs of the Physical Asset subsystem. The changes can be seen in Fig. 3.65.

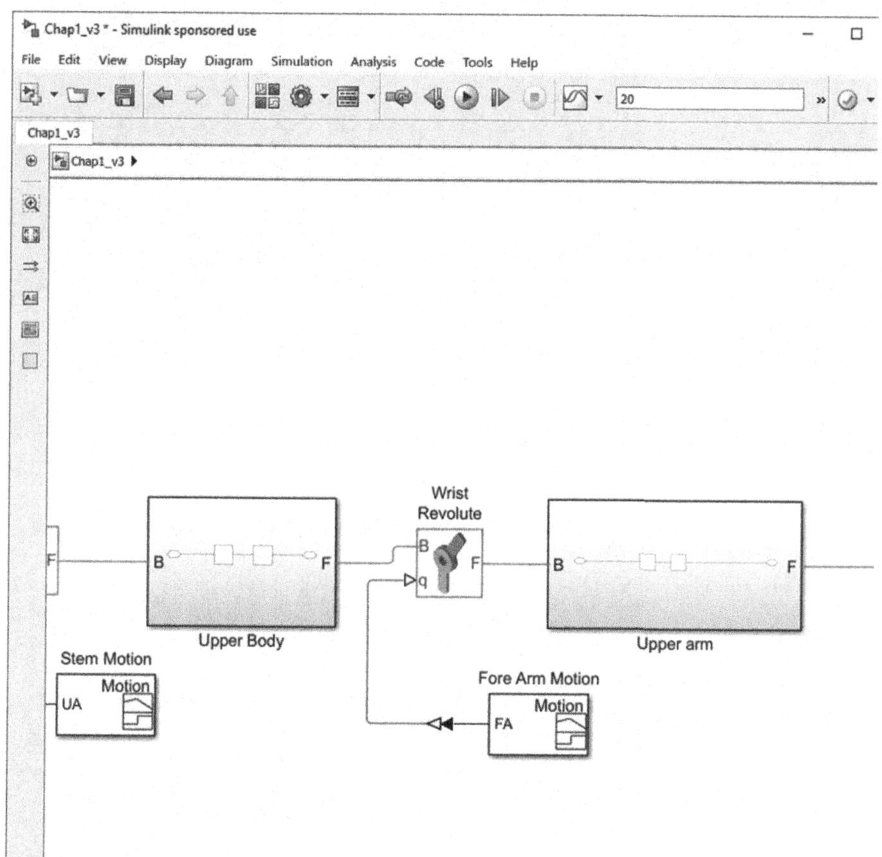

Figure 3.57 Motion for the upper arm revolute joint.

59. Go inside the **Physical Asset** subsystem and then inside the **Base Actuation** subsystem. Insert a **Gain** block before the **Simulink®-PS converter** block. Change the gain value to 0. This simulates a stuck stepper motor. Run the simulation and observe the discrepancy between both physical asset and digital twin in the Mechanics Explorer window as shown in Fig. 3.66.

3.4 Application problem

a. Add a new DOF such that the gripper is now revolving around the X axis of the world frame.

b. Add a predefined motion to this DOF such that the gripper rotates +45 degrees when the arm gripper is at the top position and −45 degrees when the arm is at the bottom position. The movement should happen before the gripper opens/closes.

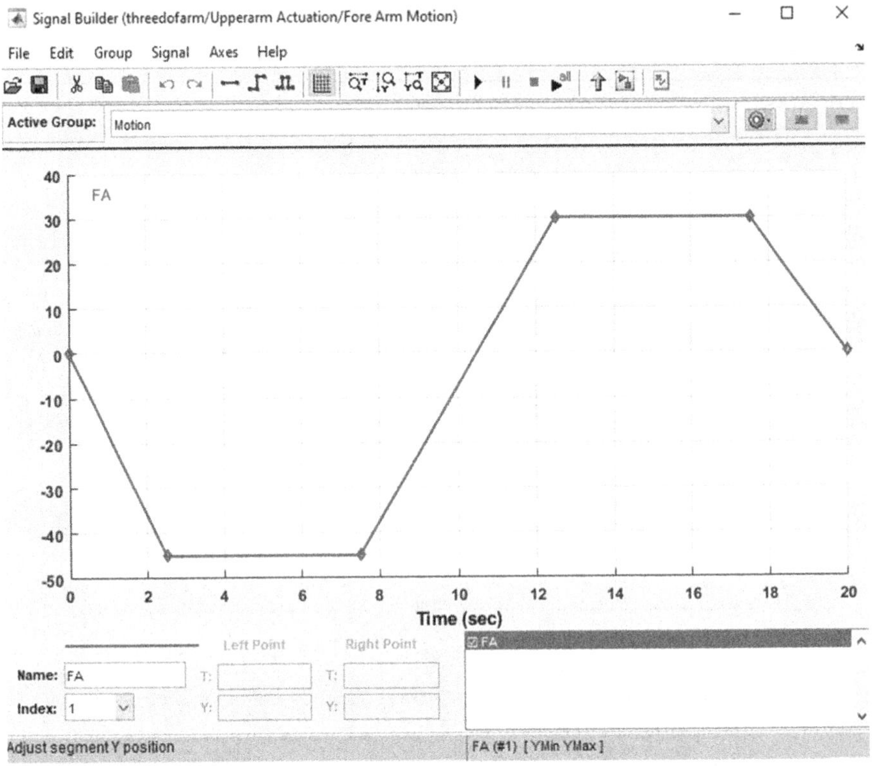

Figure 3.58 Upper arm signal motion.

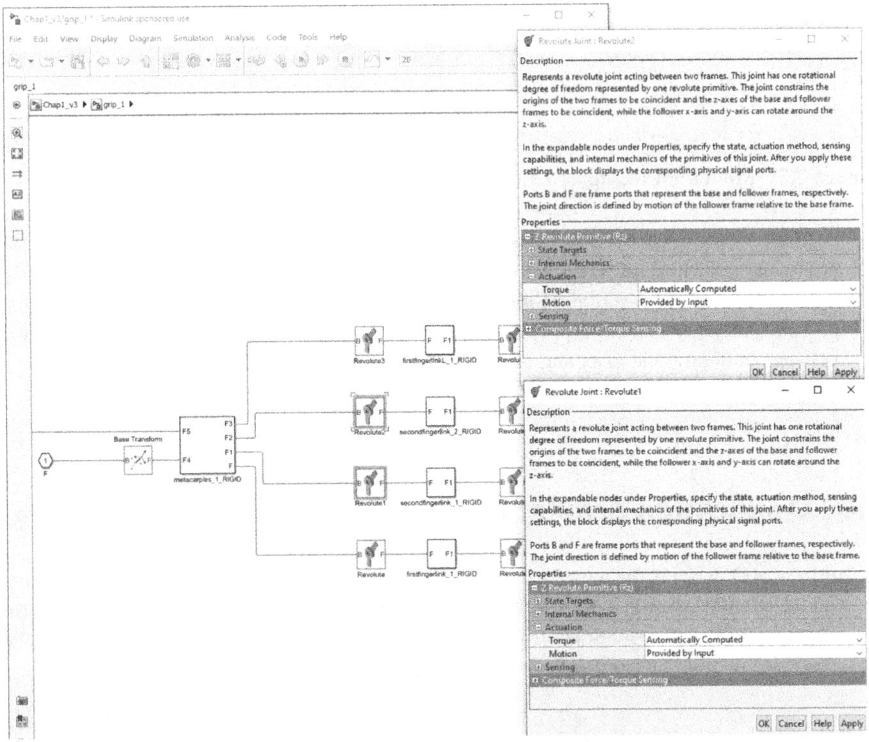

Figure 3.59 Revolute1 and Revolute2 properties.

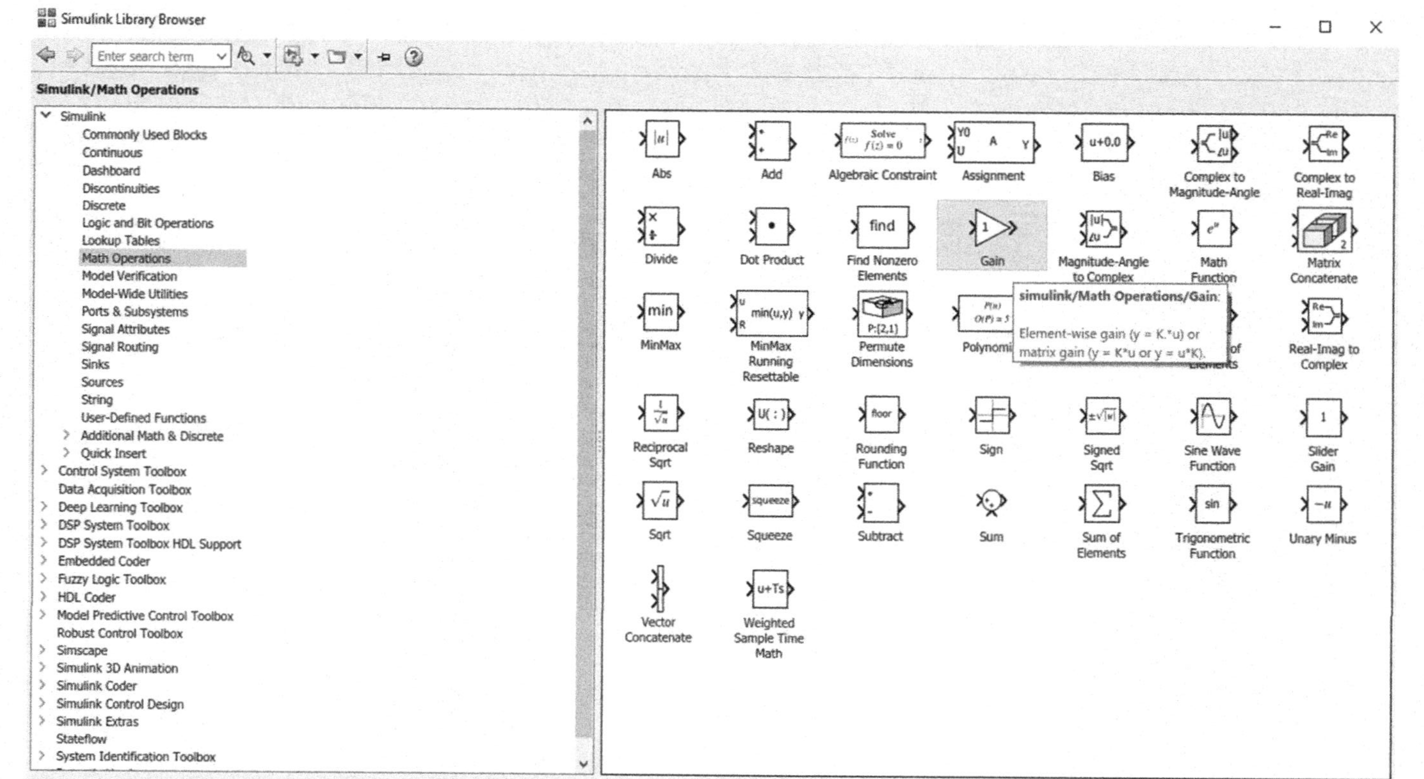

Figure 3.60 Gain library block.

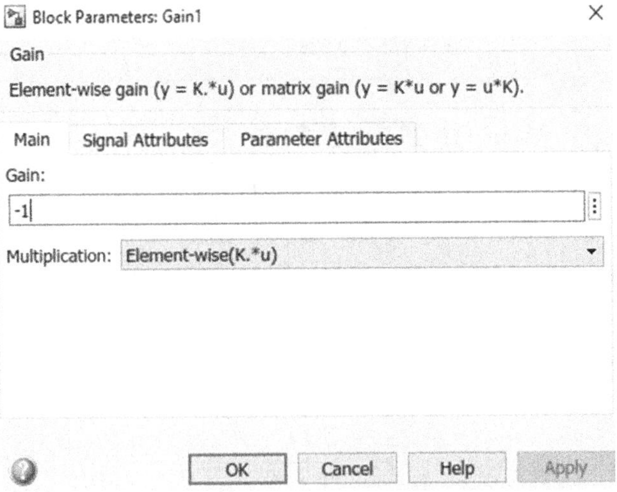

Block Parameters: Gain1

Gain

Element-wise gain (y = K.*u) or matrix gain (y = K*u or y = u*K).

Main | Signal Attributes | Parameter Attributes

Gain:

-1

Multiplication: Element-wise(K.*u)

OK Cancel Help Apply

Figure 3.61 Gain properties.

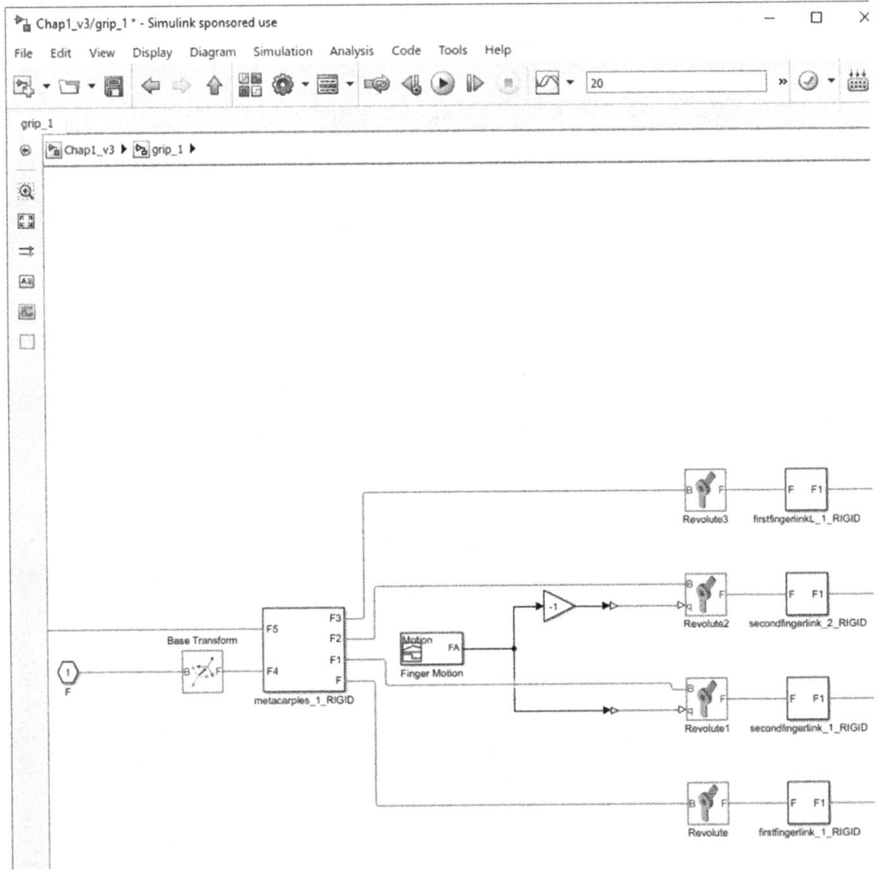

Figure 3.62 Motion for the upper arm revolute joint.

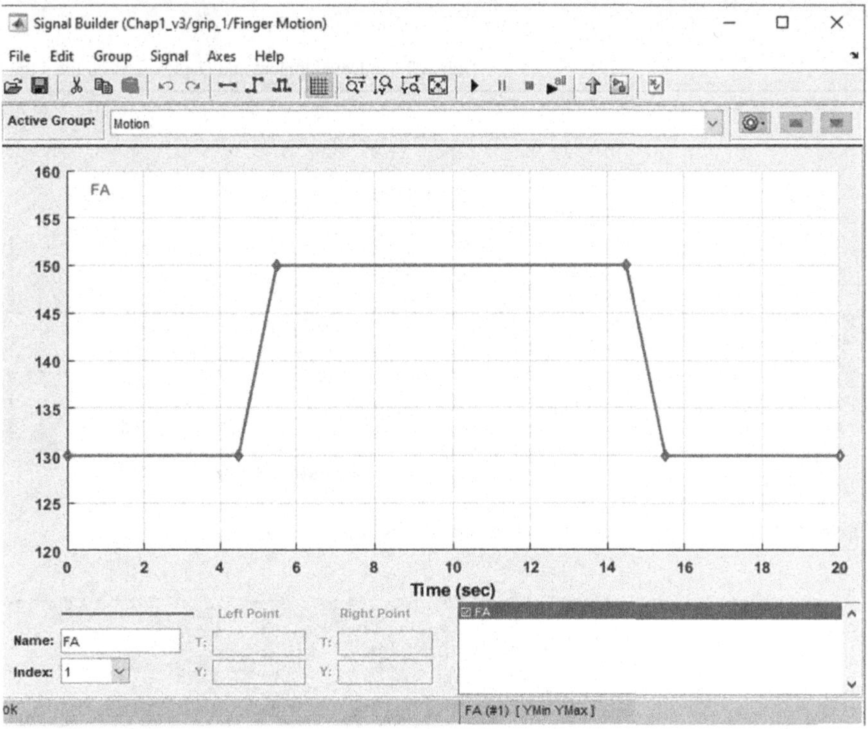

Figure 3.63 Gripper signal motion.

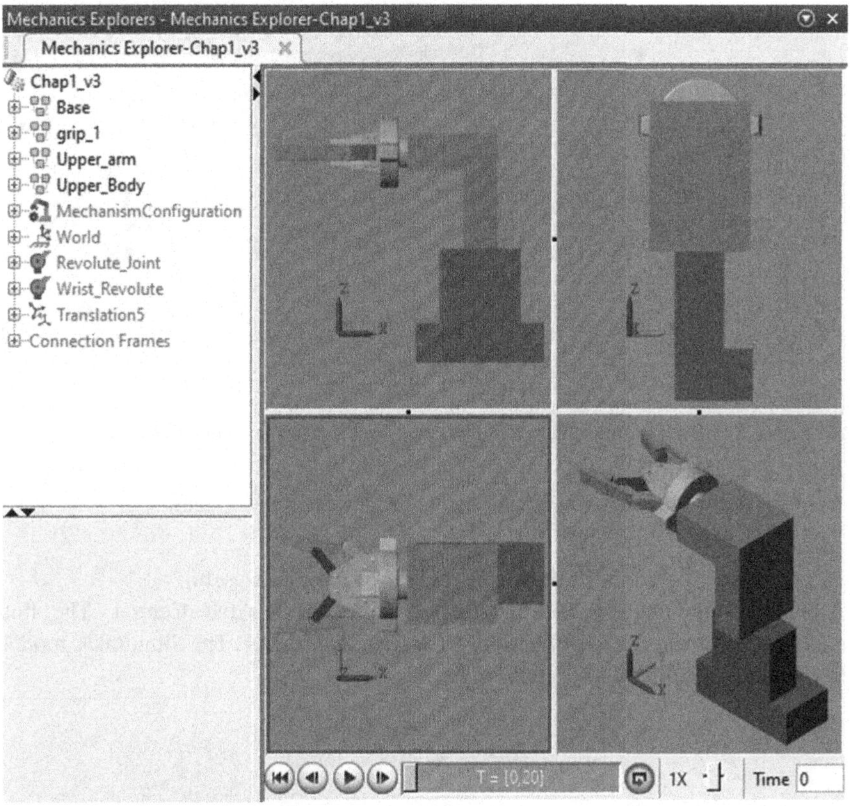

Figure 3.64 Final layout of the 3-DOF arm.

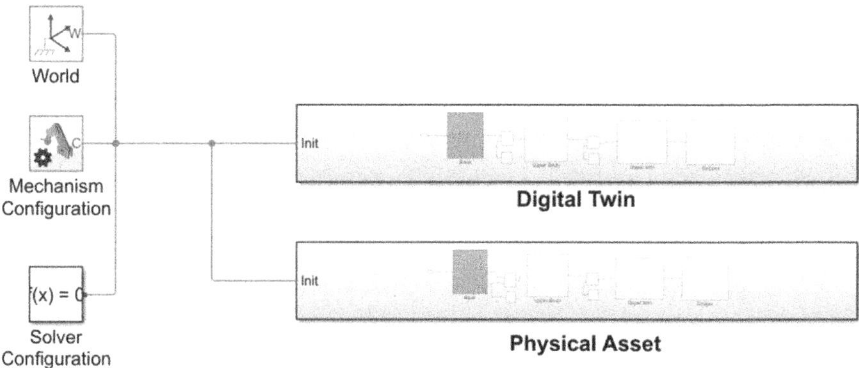

Figure 3.65 Digital Twin and Physical Asset subsystems.

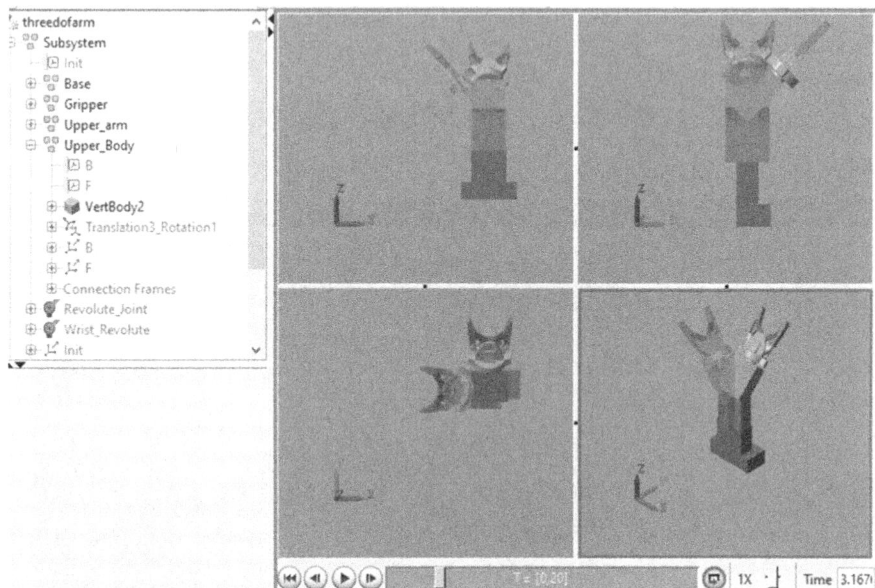

Figure 3.66 Response of physical asset (the transparent one) when compared with digital twin.

c. For each of the DOF defined, monitor the torque at each of these joints.
d. Plot the torques in a Simulink Scope.

Hint: You may find the "**Sensing**" property of the joints useful.

Download the material for the chapter from MATLAB® Central. The final Simulink® is provided in **Application_Problem_arm** folder. The Simulink® models is saved under **Application_Problem.slx**.

Ball on plate modeling

4

4.1 Introduction

Ball on a plate is a good benchmark problem that is widely used in control design and testing. It serves as a good example to build a digital twin that can be used for Off-BD of the setup. The ball on plate has two servo motors that change the angle of the plate, which in turn changes the position of the ball. The intent is to keep the ball in a prespecified position, usually the center.

In this chapter, the focus will be on creating the digital twin model for the ball on plate. Fig. 4.1 shows the Off-BD steps that will be covered in this chapter. This includes failure modes for the asset, the block diagram, modeling of the digital twin, and a preliminary diagnostic algorithm. Edge device setup, connectivity, and deployment of the digital twin on the cloud are not covered.

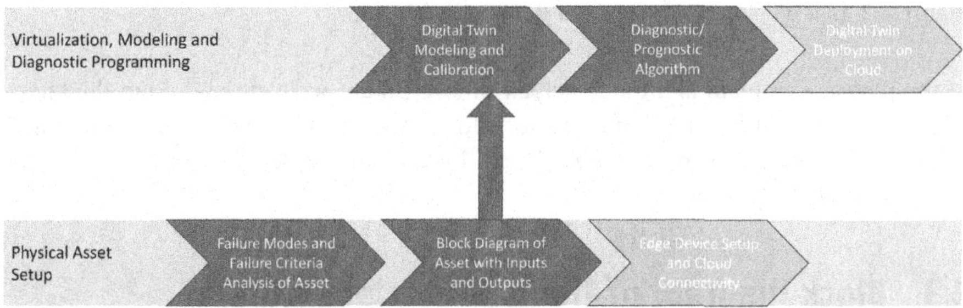

Figure 4.1 Off-BD steps covered in this chapter.

All the codes used in the chapter can be downloaded for free from MATLAB File Exchange. Follow the link below and search for the ISBN or title of this book:

https://www.mathworks.com/matlabcentral/fileexchange/.

Alternatively, the reader can also download the material and other resources from the dedicated website or contact the authors for further help:

https://www.practicalmpc.com/.

4.2 Ball on plate hardware

The ball on plate hardware was bought from RoboholicManiacs (www.roboholicmaniacs.com). The plate has two servo motors to control the x and y slope

Digital Twin Development and Deployment on the Cloud. https://doi.org/10.1016/B978-0-12-821631-6.00004-9

Figure 4.2 Ball on plate hardware from RoboholicManiacs.

of the plates. The plate is a touch screen of size 38 cm × 24 cm and 2 cm thickness (Fig. 4.2). The ball is 1 inch in diameter and its mass is 51 g. The servo motors and the crankshaft mechanism are not modeled. Instead, the angle of the plate is an input to the Simscape™ model.

4.3 Block diagram of the ball on plate system

Ball on plate is a second-order multiinput and multioutput problem. The system has two inputs (in general there are two actuators) and two outputs (x and y position of the ball) (Fig. 4.3). The three main parts are the ball, the plate, and the actuation mechanism. The later can be actuated by servo motors or magnetic induction coils. The plate is generally constrained to rotate around the x and y axis (roll and pitch angles). The actuation mechanism modifies the two angles of the plate, which causes the

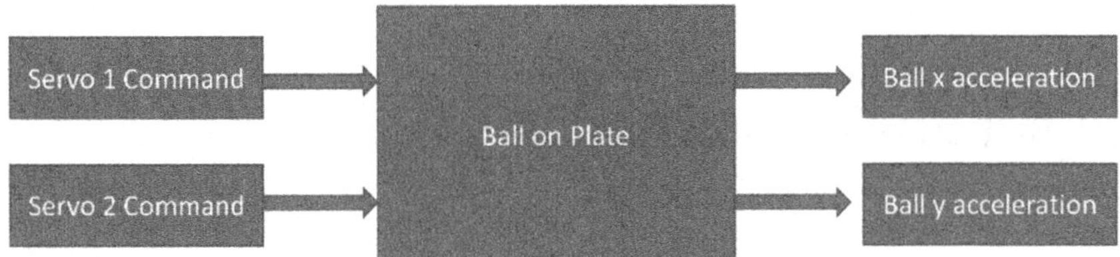

Figure 4.3 Block diagram of the ball on plate.

ball to roll due to gravity in the x and y directions. In many control problems, the objective is to maintain the ball at given coordinates despite external disturbances.

4.4 Failure modes and diagnostics concept for the ball on plate

In this chapter, we will focus on detecting failures that affect the acceleration of the ball. These failure modes include servo motor failure such as wiring issue, aging servo, or mechanical failure in the rotating mechanism. Failure could also include a stuck ball (due to dust or other obstacles). Lastly, failure in the sensing mechanism can also affect the ball acceleration.

Regardless of the servo type or sensing mechanism, the process outlined to diagnose the system should still be applicable. The basic idea relies on acceleration measurement or estimation. If the acceleration of the physical asset deviates significantly from the acceleration of the digital twin, then a fault will be set. Usually the art of designing the diagnostic lies in implementing a logic that can look for the conditions where the deviations of the physical asset from the digital twin are magnified. This ensures a robust logic that can overcome uncertainties pertaining to modeling errors, slight aging of the hardware, uneven surface where the plate is place, and unmeasured disturbances (such as the air condition in the room blowing on the ball on plate setup). When implemented properly, implementing diagnostic enable conditions produces a detectable signal-to-noise ratio.

We will enable the diagnostic when either the servo command or the acceleration of the ball is big. For few samples of time, the digital twin and physical asset average accelerations will be compared. If the difference is bigger than a threshold, then a fault will be set. Fig. 4.4 shows the diagnostic process.

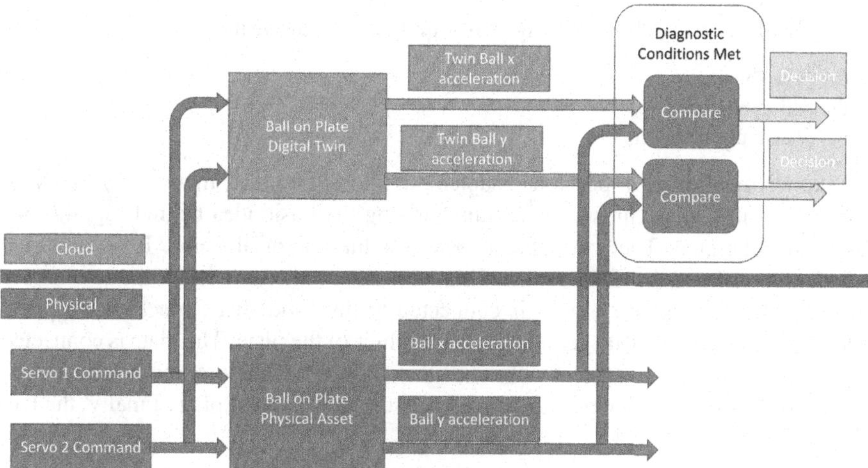

Figure 4.4 Off-BD diagnostic process for the ball on plate.

4.5 Simscape model for the ball on plate

To create the digital twin, we will use Simscape Multibody™ (formerly known as SimMechanics™). The inputs to the model will be the angles of the plate while the outputs will be x and y position of the ball. Furthermore, we will integrate the model with two PID controllers. Using Mechanics Explorer, we will watch and record the transient response of the ball. The simulation will be recorded as a video for playback and demo purposes. Fig. 4.5 shows the Cartesian coordinate as well as the rotation angles of the plate.

Figure 4.5 Cartesian coordinates fixed to the plate as well as rotation angles.

The challenge in modeling this application using Simscape™ is that unlike most of components in Simscape Multibody, ball and the plate are not connected by a joint, a belt, or are constrained to be in contact. This means the custom equations between the ball and the plate have to be developed and solved explicitly in Simulink or in user-defined subsystem and not by Simscape™ solver.

The final model is shown in Fig. 4.6. The main components are

- the plate dynamics
- the ball dynamics
- the plate/ball interaction

Like we mentioned in previous chapters, unlike Simulink®, there is no directionality of signal flow in Simscape™. Summarizing the basic idea behind Fig. 4.6, we have a world frame. The world frame serves as the universal frame. It is connected to 2 degree-of-freedom (DOF) joint. This joint allows the user to set the two angles for the plate. The universal joint is connected to the solid shape block of the plate. It contains the mass, inertia, geometry, and graphics of the plate. The plate is connected to the ball and plate interaction block. The block has the 6 DOF joint for the ball as well as the s-function that calculates the forces between the ball and plate. Finally, the ball

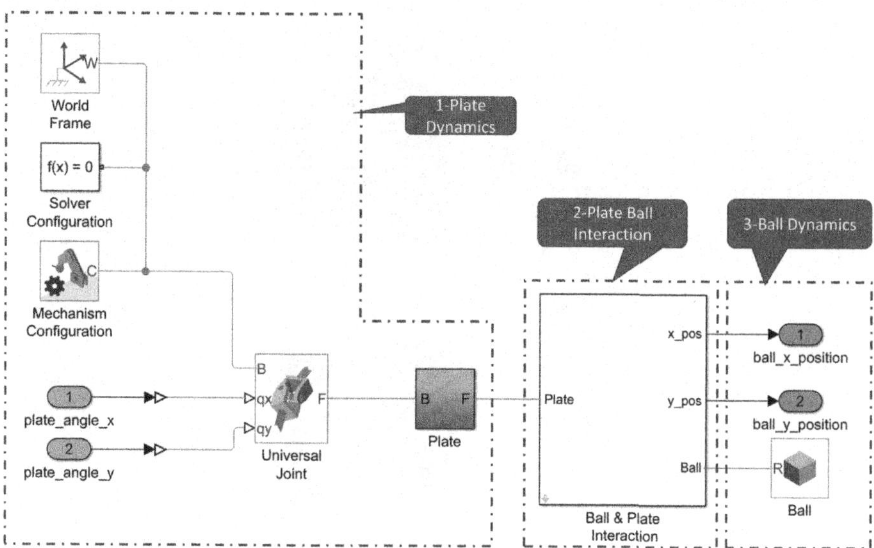

Figure 4.6 Overview of the Simscape™ model.

geometry is connected as a follower frame to the 6-DOF joint for ball. Fig. 4.7 shows the main elements of the ball on plate model.

Below are the detailed steps to building the Simscape® model:

Step 1: Create a new **Simulink® model.**
Step 2: Press the **Simulink® Library Browser** (Fig. 4.8).
Step 3: Navigate to **Simscape> Multibody Library>Frames and Transforms** and add **World Frame** block to the model (Fig. 4.9).
Step 4: Go back to **Simscape> Utilities** and add **Solver Configuration** block (Fig. 4.10).
Step 5: Go to **Multibody>Utilities** and add **Mechanism Configuration** block (Fig. 4.11).
Step 9: Go to **Multibody>Joints** and add **Universal Joint** block to the model (Fig. 4.12).
Step 10: To force the plate to have the angles as input, double click the **Universal Joint** and change the settings **Actuation>Motion>Provided** by input for both the **X Revolute Primitive (Rx)** and **Y Revolute Primitive (Ry)** (Fig. 4.13). Note that these angles need to be provided in radians.
Step 10: Connect all the previously mentioned blocks to the **Universal Joints** (Fig. 4.14). To do so, connect one block to the **Universal Joint** and connect the rest to the connecting line. Connecting **Simscape** blocks is a bit different than connecting Simulink blocks.
Step 11: To create the physical object representing the plate, add the Solid block from the **Simscape>Multibody Library> Body Elements Library** (Fig. 4.15).
Step 12: Double clock the Solid block and change the dimensions of the plate to [0.38, 0.25, 0.02] m, which represent the dimensions of the plate (Fig. 4.16).
Step 14: Change the density to 1.18 g/cm^3 since the plate is plexiglass (Fig. 4.17).
Step 15: Change the **Color** and **Opacity** to your desired values (Fig. 4.18).
Step 16: Add two connection ports to the model from the **Simscape>Utilities** (Fig. 4.19).

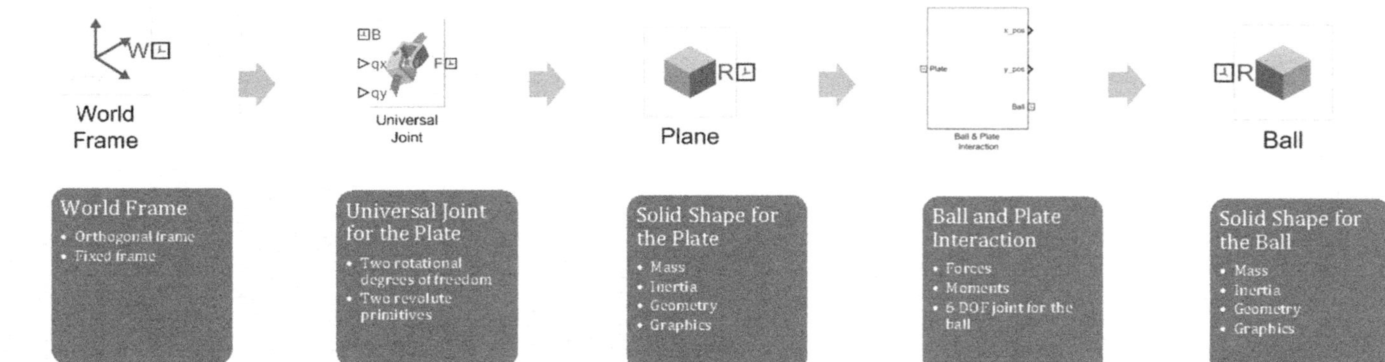

Figure 4.7 Main elements of the ball on plate model.

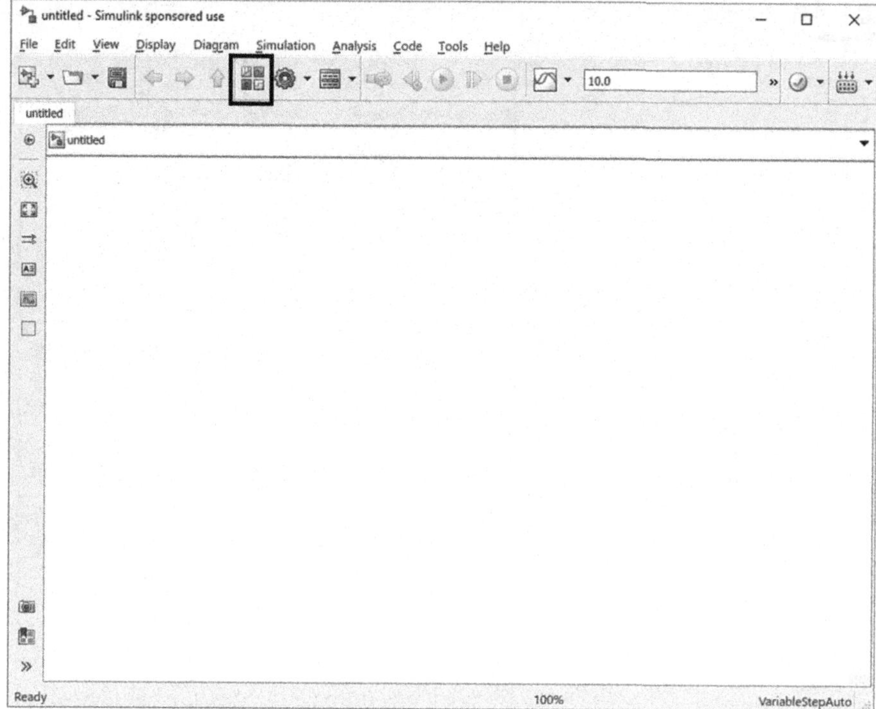

Figure 4.8 Simulink® Library Browser.

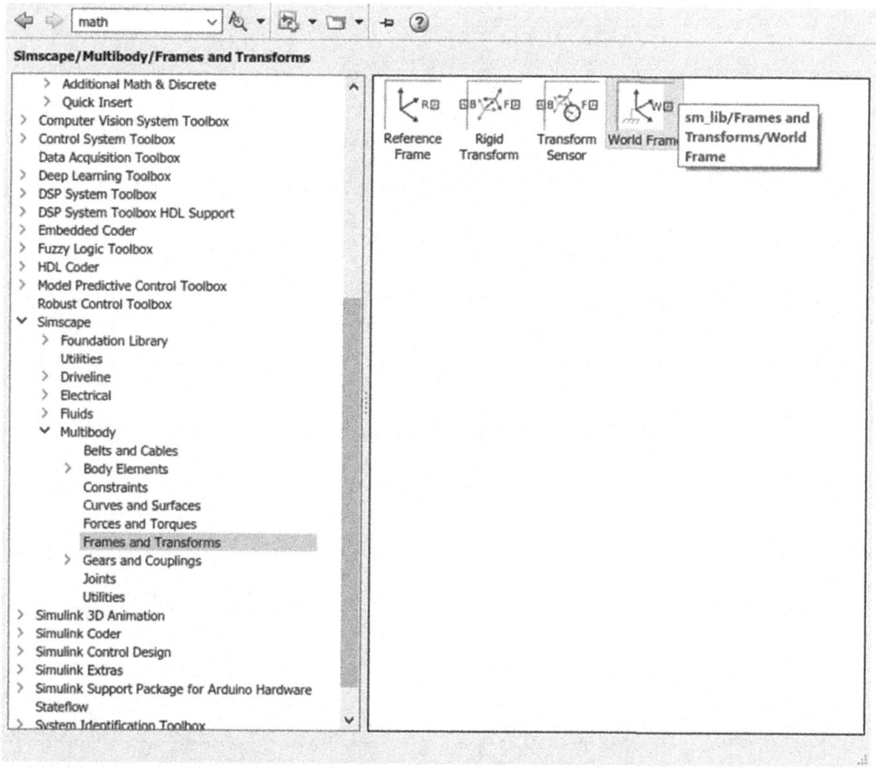

Figure 4.9 World Frame block.

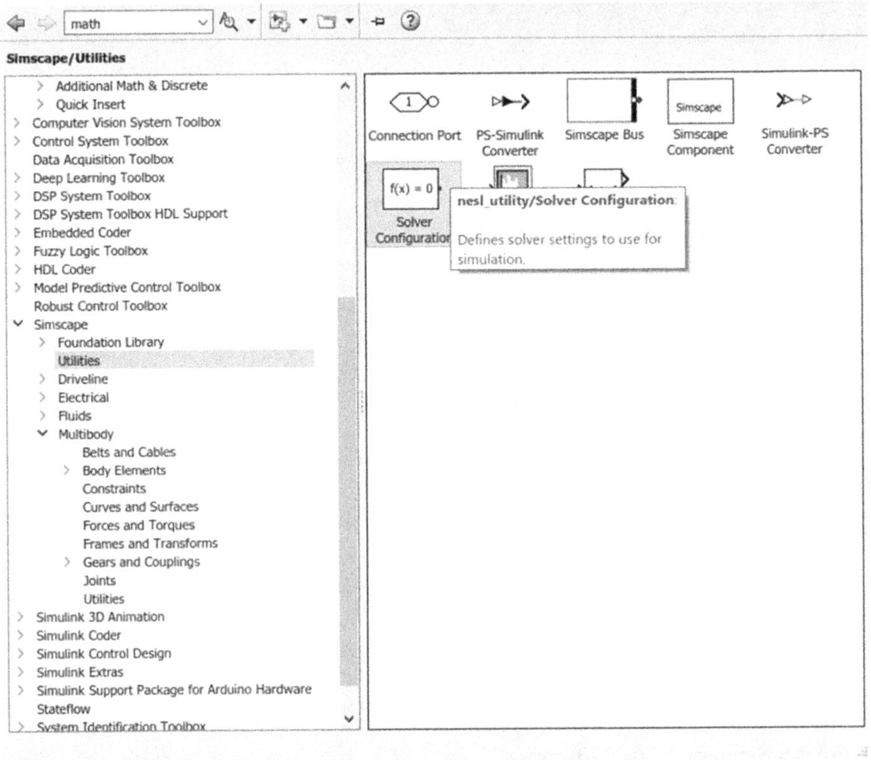

Figure 4.10 Solver Configuration block.

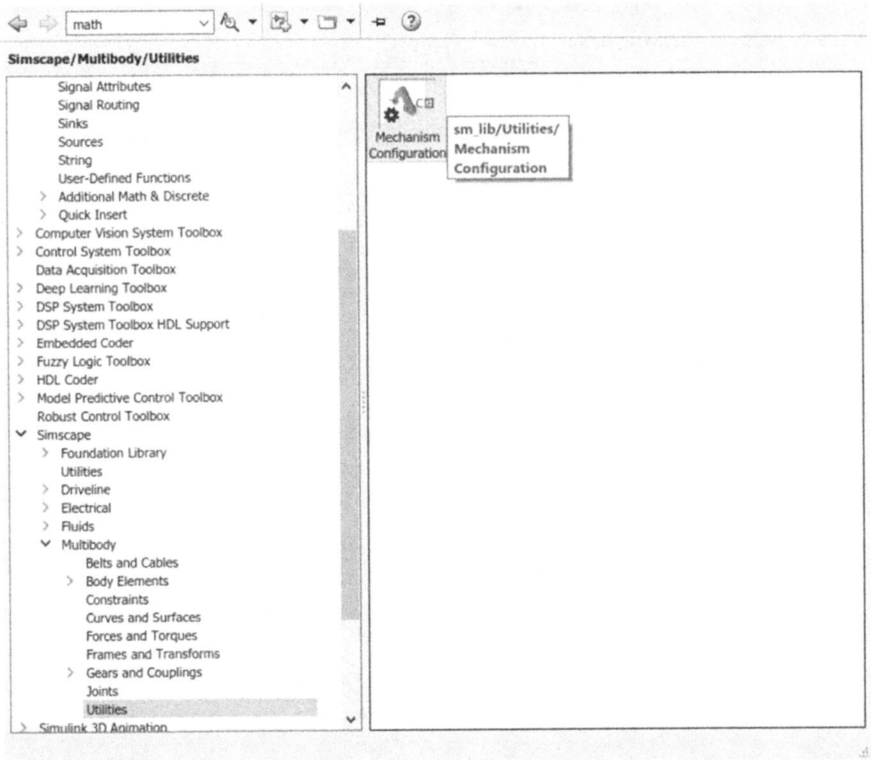

Figure 4.11 Mechanism Configuration.

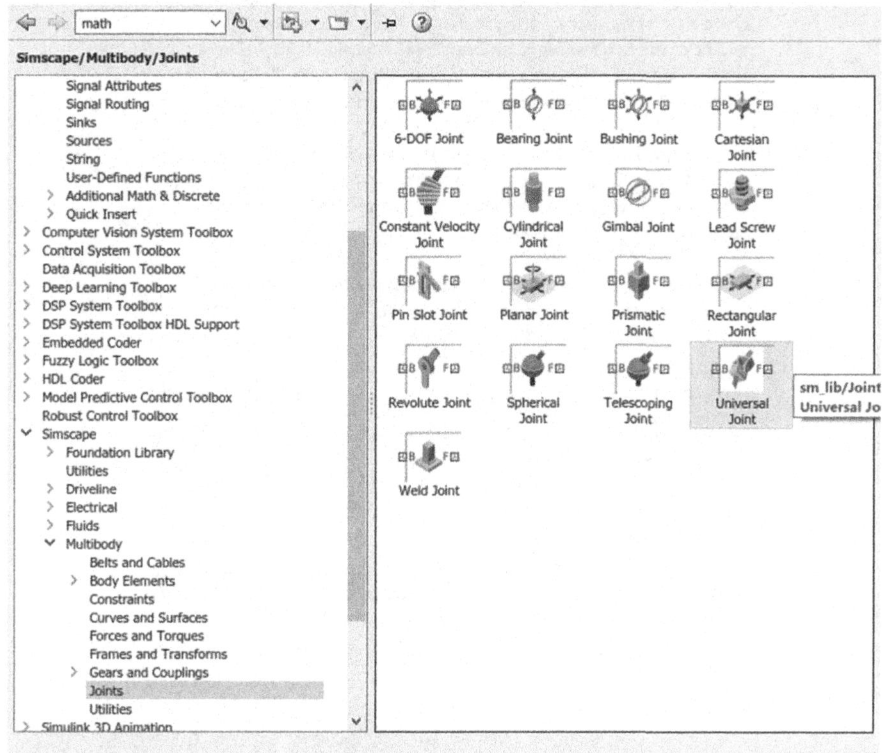

Figure 4.12 Universal Joint.

Step 17: Rename the ports **F** and **B**, respectively, by clicking once on the block and then editing the name underneath the block. Do the same for the Solid and rename it Plate (Fig. 4.20).
Step 18: For better display when we create a subsystem for the components, double click the connection port named **B**, and change the port location to the right (Fig. 4.21).
Step 19: Connect the **B** connection port to the **Solid** and then connect **F** connection port to the connection line.
Step 20: Select **F** and **B** connection ports along with the Plate and create a subsystem (Fig. 4.22).
Step 21: Rename the subsystem **Plate** and remove the input and output ports named **F** and **B.** The model is shown in Fig. 4.23.
Step 22: Connect the **Universal Joint** to the **Plate** subsystem. Note that the order of the blocks and creating the subsystem are for better presentation of the model. Both **F** and **B** connection ports are identical except for the name and location of the block.
Step 23: Go to **Simscape>Multibody>Joints** and add **6-DOF Joint** to the model (Fig. 4.24). This will be used as the reference frame for the ball. The ball exhibits both translation and rotation.
Step 24: The plate exerts friction forces (fx and fy) on the ball in the x and y directions. Additionally, the plate exerts a normal component force, fz, that constrains the ball from falling

Description

Represents a universal joint between two frames. This joint has two rotational degrees of freedom represented by two revolute primitives along a set of mutually orthogonal axes. This joint constrains the origins of the two frames to be coincident. It prohibits any relative rotation or translation in the plane of the primitives. The follower frame rotates first around the follower x-axis, then around the follower y-axis.

In the expandable nodes under Properties, specify the state, actuation method, sensing capabilities, and internal mechanics of the primitives of this joint. After you apply these settings, the block displays the corresponding physical signal ports.

Ports B and F are frame ports that represent the base and follower frames, respectively. The joint direction is defined by motion of the follower frame relative to the base frame.

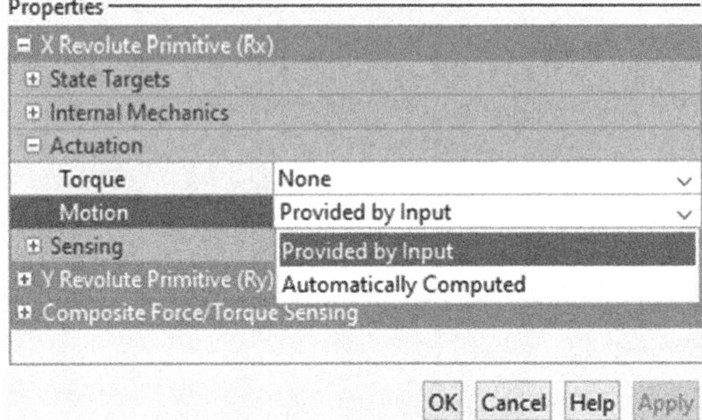

Figure 4.13 Actuation setting of the Universal Joint.

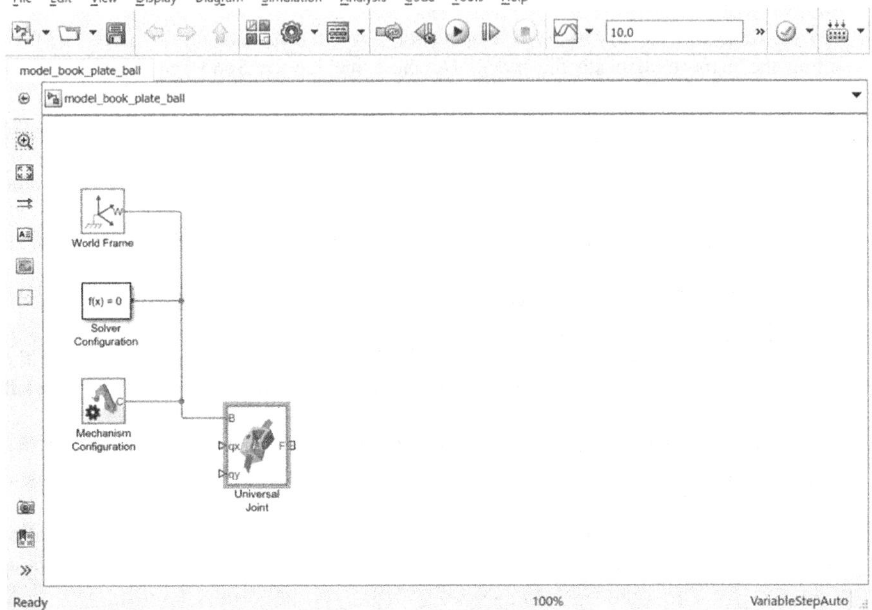

Figure 4.14 Connecting the blocks to the Universal Joint.

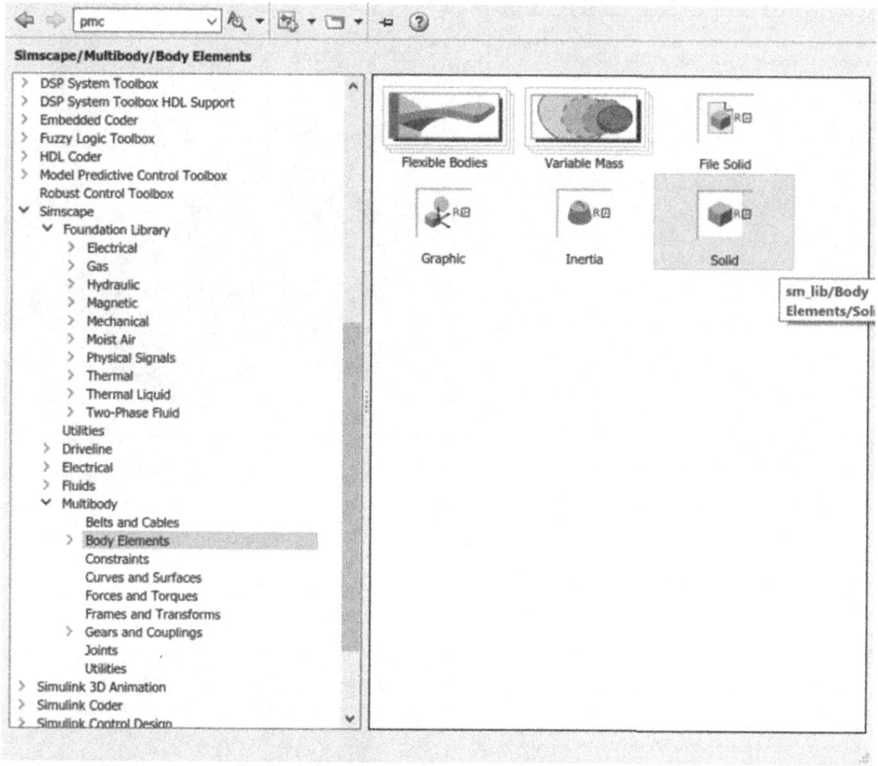

Figure 4.15 Solid bock from the Body Elements library.

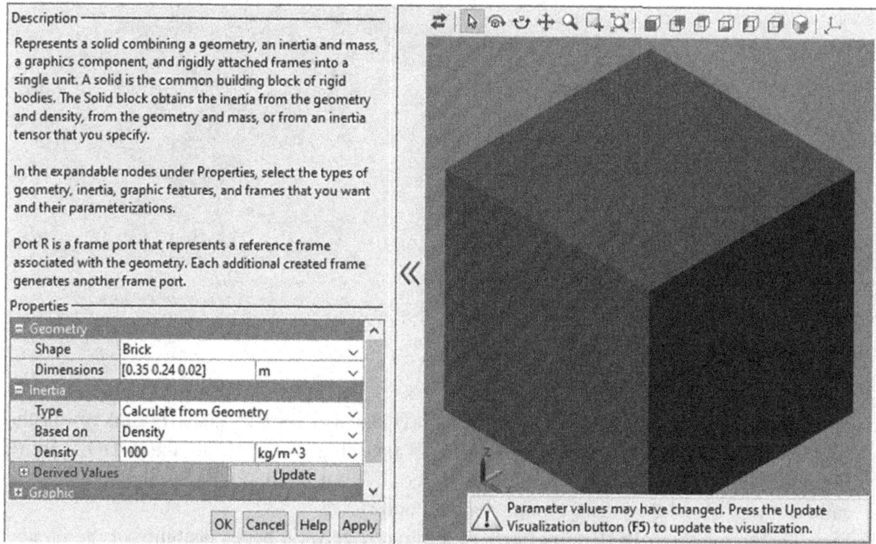

Figure 4.16 Dimensions of the plate.

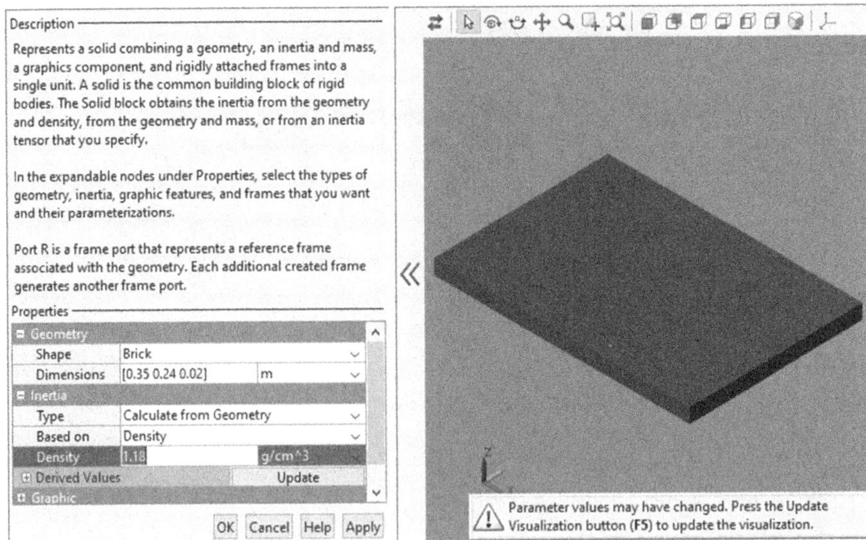

Figure 4.17 Density of the plate.

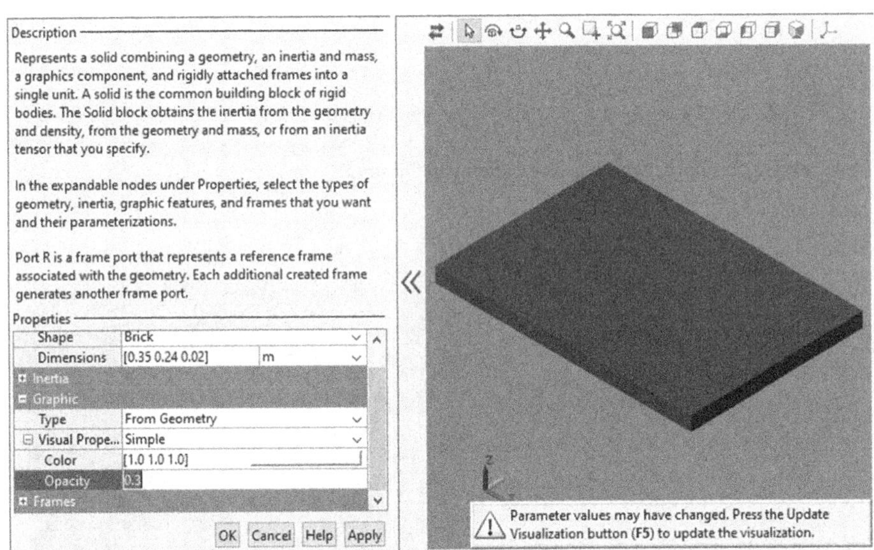

Figure 4.18 Graphical properties of the plate.

through the plate. Finally, the friction forces (fx and fy) form torques (tx and ty) on the sphere. All these forces and torques need to be setup as inputs to the **6-DOF Joint.** Fig. 4.25 shows how to setup the forces fx as an input to the joint. Similarly, fy and fz can be setup. Fig. 4.26 shows how to setup tx and ty.

Step 25: Stretch the **6-DOF** Joint block to a bigger size for a better visibility of the inputs.

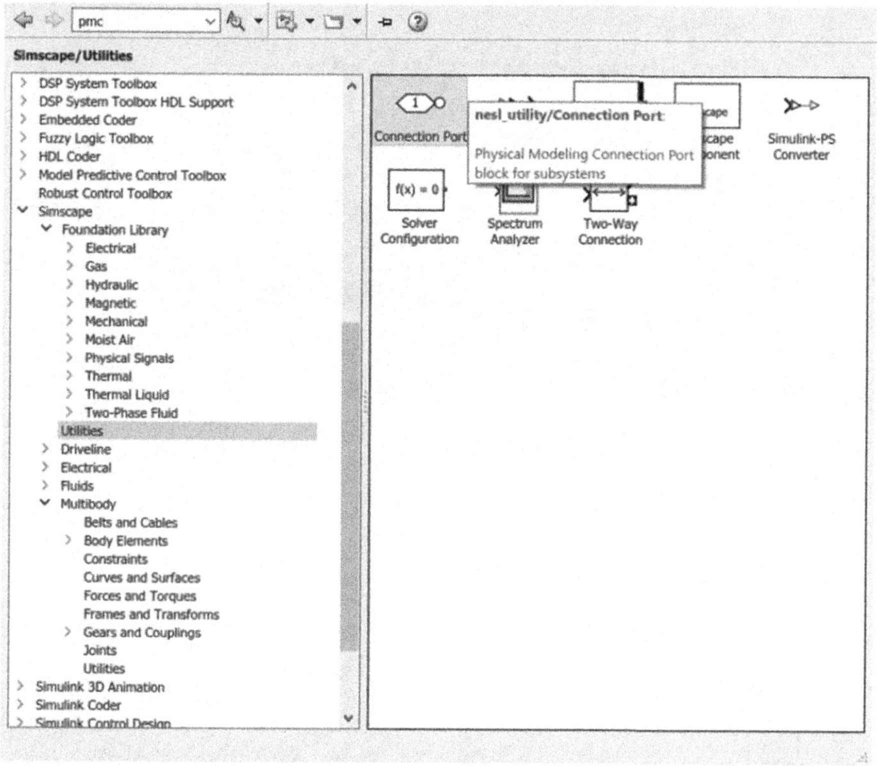

Figure 4.19 Connection ports.

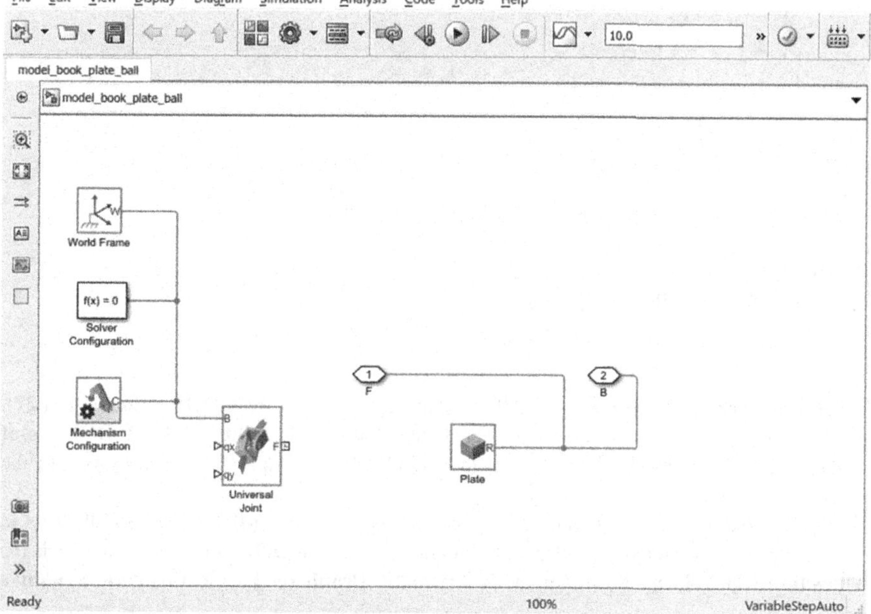

Figure 4.20 Connecting F and B connection ports to the Solid.

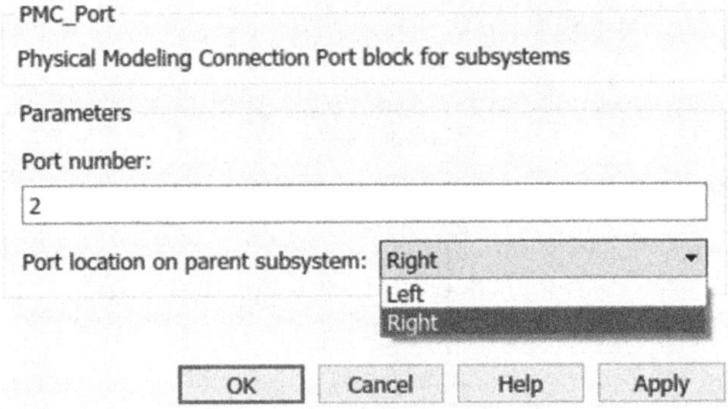

Figure 4.21 Port location.

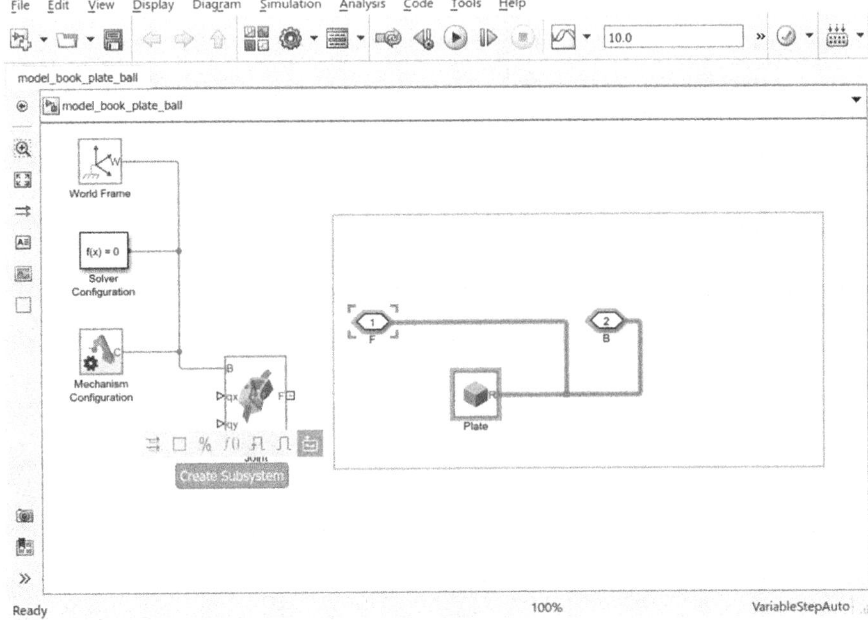

Figure 4.22 Creating Plate subsystem.

Step 26: Connect the **F** port of the Plate block to the **B** port of **6-DOF Joint** (Fig. 4.27). Alternatively, we could have connected the **Universal Joint** to the **6-DOF Joint** (Fig. 4.28). The objective is to make the **6-DOF Joint** a follower frame relative to the **Universal Joint**.

Step 27: We need a set of outputs (measurements) from the **6-DOF Joint** to allow us to compute the ball and plate interaction. Double click the **6-DOF Joint** block, and tick the **Position** and **Velocity** boxes for the **X Prismatic Primitive (Px)**, **Y Prismatic Primitive**

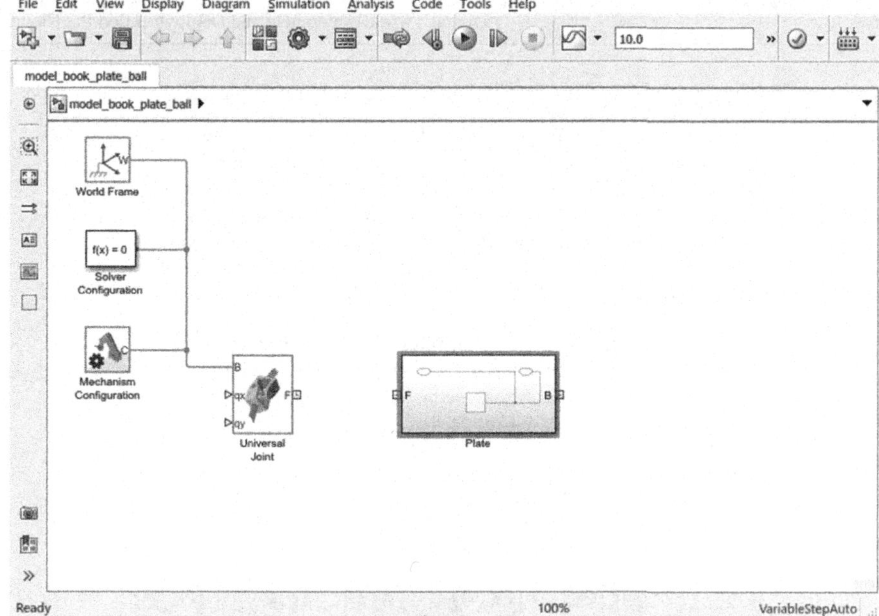

Figure 4.23 Overall model with Plate subsystem.

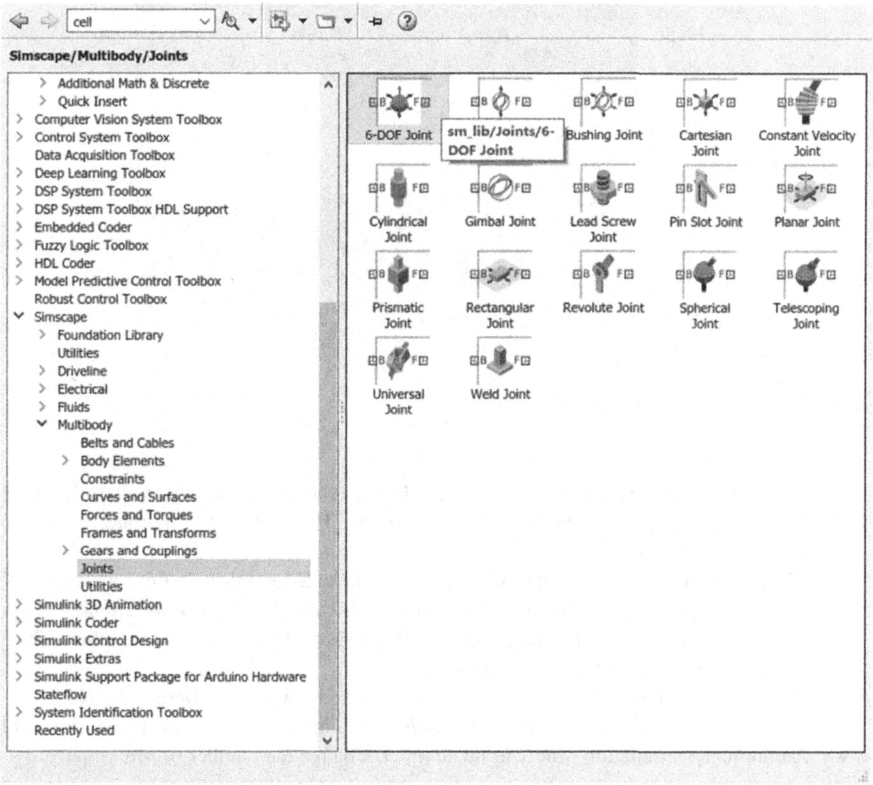

Figure 4.24 6-DOF Joint.

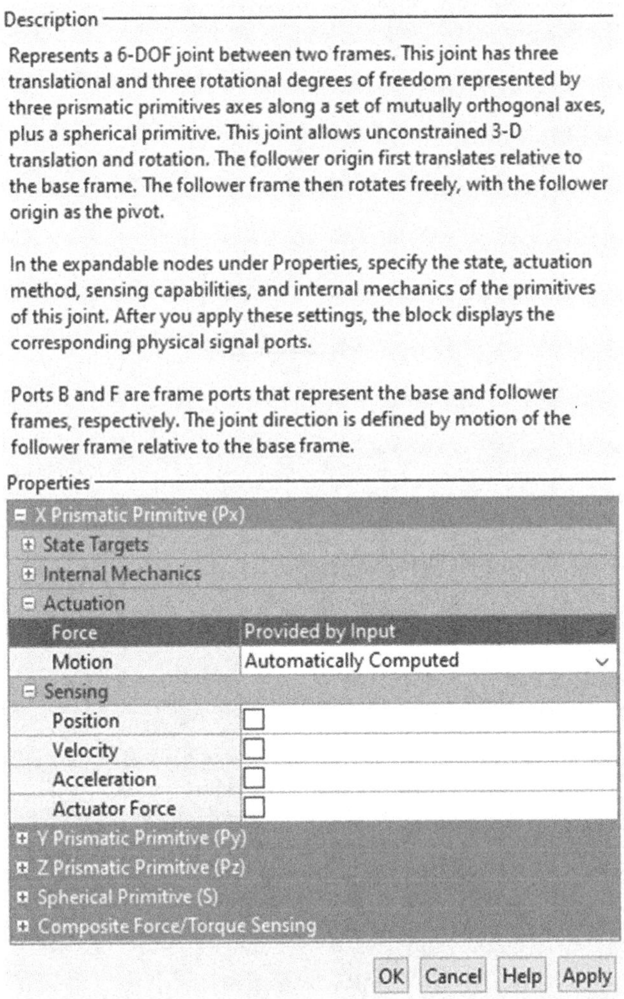

Description

Represents a 6-DOF joint between two frames. This joint has three translational and three rotational degrees of freedom represented by three prismatic primitives axes along a set of mutually orthogonal axes, plus a spherical primitive. This joint allows unconstrained 3-D translation and rotation. The follower origin first translates relative to the base frame. The follower frame then rotates freely, with the follower origin as the pivot.

In the expandable nodes under Properties, specify the state, actuation method, sensing capabilities, and internal mechanics of the primitives of this joint. After you apply these settings, the block displays the corresponding physical signal ports.

Ports B and F are frame ports that represent the base and follower frames, respectively. The joint direction is defined by motion of the follower frame relative to the base frame.

Properties

X Prismatic Primitive (Px)	
State Targets	
Internal Mechanics	
Actuation	
Force	Provided by Input
Motion	Automatically Computed
Sensing	
Position	☐
Velocity	☐
Acceleration	☐
Actuator Force	☐
Y Prismatic Primitive (Py)	
Z Prismatic Primitive (Pz)	
Spherical Primitive (S)	
Composite Force/Torque Sensing	

OK Cancel Help Apply

Figure 4.25 Setting up the force fx as an input.

(Py), and **Z Prismatic Primitive (Pz)** (Fig. 4.29). For rotational measurements, tick the **Velocity (X)** and **Velocity (Y)** for the **Spherical Primitive (S)**. These represent the rotational speed around the X and Y axis.

Step 28: From **Simulink>User-Defined Functions** add an **S-Function.** In this block, we will embed the c script, **ballplateforces.c,** that will include the calculation of the friction forces and moments of the plate on the ball in addition to the support forces in the z-direction. The c script will be discussed in the next section.

Step 29: From **Simulink> Commonly Used Blocks**, add a **Mux** and **Demux** blocks.

Step 30: The inputs for the c script, **ballplateforces.c**, are px, vx, py, vy, pz, vz, wx, and wy. Additionally, simulation time will be an input. Change the number of Mux inputs to 9.

Description

Represents a 6-DOF joint between two frames. This joint has three translational and three rotational degrees of freedom represented by three prismatic primitives axes along a set of mutually orthogonal axes, plus a spherical primitive. This joint allows unconstrained 3-D translation and rotation. The follower origin first translates relative to the base frame. The follower frame then rotates freely, with the follower origin as the pivot.

In the expandable nodes under Properties, specify the state, actuation method, sensing capabilities, and internal mechanics of the primitives of this joint. After you apply these settings, the block displays the corresponding physical signal ports.

Ports B and F are frame ports that represent the base and follower frames, respectively. The joint direction is defined by motion of the follower frame relative to the base frame.

Properties

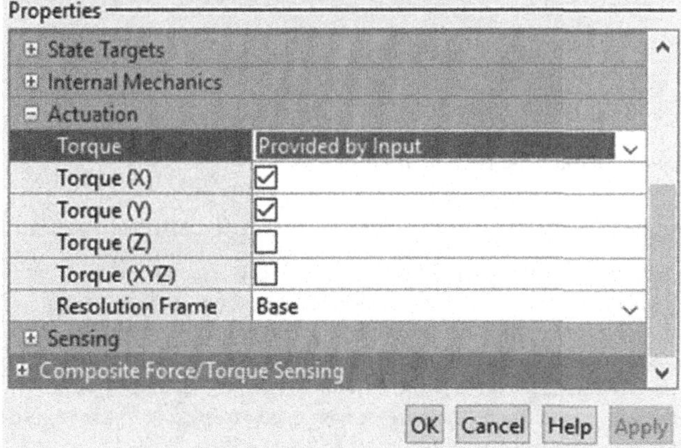

Figure 4.26 Setting up torques tx and ty.

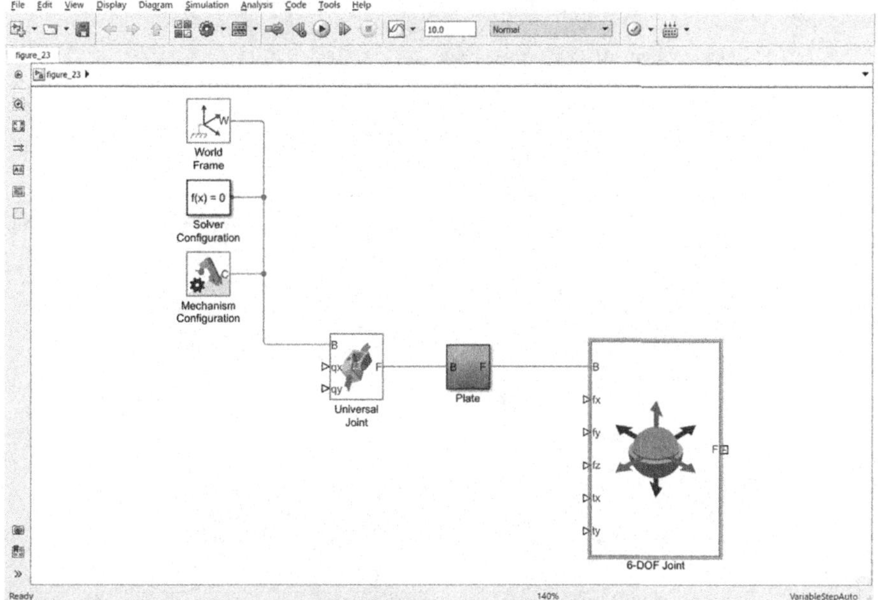

Figure 4.27 Plate and the 6-DOF Joint blocks.

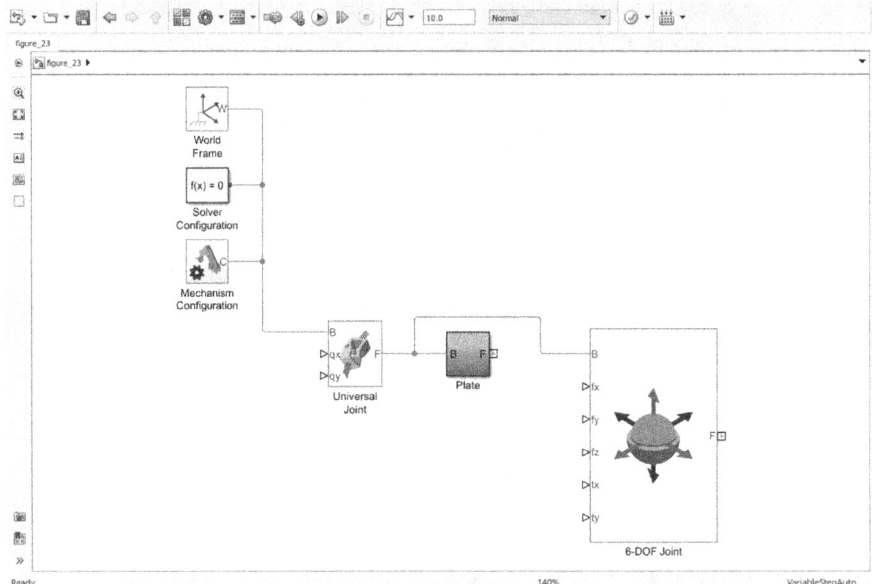

Figure 4.28 Universal Joint the 6-DOF Joint blocks.

Description

Represents a 6-DOF joint between two frames. This joint has three translational and three rotational degrees of freedom represented by three prismatic primitives axes along a set of mutually orthogonal axes, plus a spherical primitive. This joint allows unconstrained 3-D translation and rotation. The follower origin first translates relative to the base frame. The follower frame then rotates freely, with the follower origin as the pivot.

In the expandable nodes under Properties, specify the state, actuation method, sensing capabilities, and internal mechanics of the primitives of this joint. After you apply these settings, the block displays the corresponding physical signal ports.

Ports B and F are frame ports that represent the base and follower frames, respectively. The joint direction is defined by motion of the follower frame relative to the base frame.

Properties

- X Prismatic Primitive (Px)
  - State Targets
  - Internal Mechanics
  - Actuation
  - Sensing

Position	☑
Velocity	☑
Acceleration	☐
Actuator Force	☐

- Y Prismatic Primitive (Py)
- Z Prismatic Primitive (Pz)

OK Cancel Help Apply

Figure 4.29 Position and Velocity sensing.

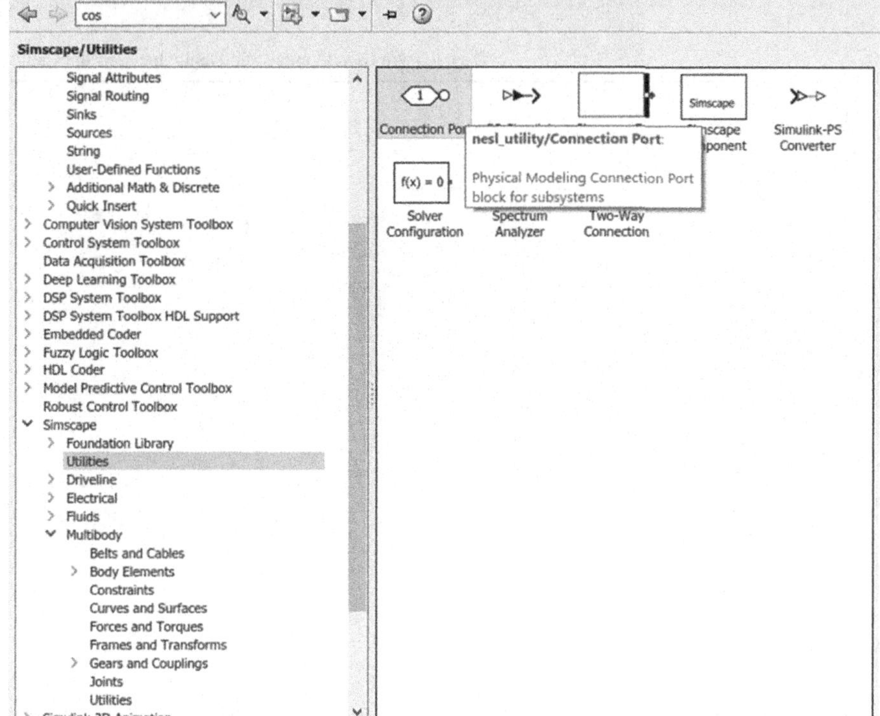

Figure 4.30 Connection Port.

Step 31: The outputs for the c script, **ballplateforces.c**, are fx, fy, fz, tx, and ty. Change the number of Demux outputs to 5.
Step 32: From **Simscape>Utilities** add a **Connection Port** (Fig. 4.30). Press **Ctrl+R** twice to rotate the port.
Step 33: From **Simscape>Utilities,** add eight **PS-Simulink Converter**s (Fig. 4.31).
Step 34: Connect the **F** port of the **6-DOF Joint** to the **Connection Port.** Similarly, connect px, vx, py, vy, pz, vz, wx, and wy to the **PS-Simulink Converters** and then to the **Mux block.** Fig. 4.32 shows the resulting connections.
Step 36: Connect the **S-Function** block to the **Demux** block.
Step 37: From **Simscape>Utilities**, add five **Simulink-PS Converters**.
Step 38: Connect the **Demux** block to the five **Simulink-PS Converter**s and connect the latter to the **6-DOF Joint** block (Fig. 4.33).
Step 39: From **Simulink>Commonly Used Blocks**, add two **Out1** blocks. Rename them **x_pos** and **y_pos** and connect them to px and py per Fig. 4.34.
Step 40: Select the components highlighted in the rectangle in Fig. 4.35 and create a subsystem. Rename the subsystem **Ball Plate Interaction**.
Step 41: Open the **Ball Plate Interaction** subsystem and rename the port **Conn2** (that is connected to **B**) as **B**. Rename the port that is connected to the **F** port as **F**.
Step 42: To create the physical object representing the ball, add the Solid block from the **Simscape>Multibody Library> Body Elements Library** (Fig. 4.36).
Step 43: Rename the **Solid** block as **Ball.** Click **Ctrl+R** twice to rotate the Solid block.
Step 44: Double click the **Solid** block. Change the shape to a sphere and the radius to 0.5 inches (Fig. 4.36).

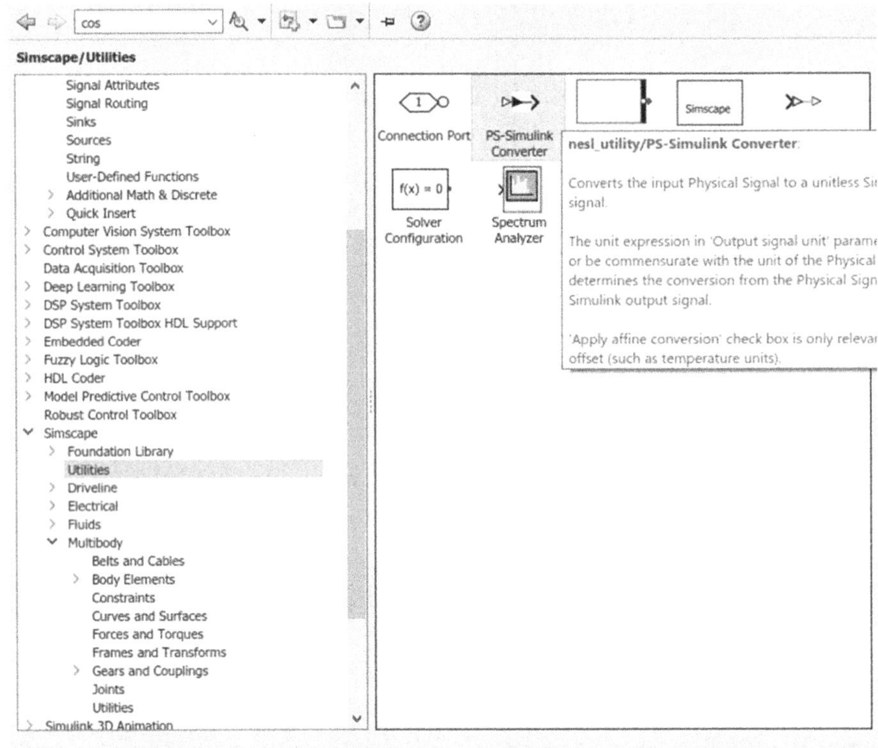

Figure 4.31 PS-Simulink Converter.

Step 45: Change the mass of the sphere to 51 g (Fig. 4.37). Click OK when done.
Step 46: Double click the block and check to see the shape changed to a sphere (Fig. 4.38).
Step 47: Delete the **F** port connected to the **Ball** port of the **Ball Plate Interaction** subsystem. Connect the **Ball** port to the **Ball** block (Fig. 4.39).
Step 48: To add the input angles to the plate, go to **Simulink>Commonly Used Blocks** and add two Constant blocks. Set the value to pi/20 for both constants.
Step 49: In order to connect the **Constant** to the **Universal Joint**, we need to add two **Simulink-PS Converters** like we did in step 37. Add two converters, connect the constants to the converters, and connect the later to the Universal Joint (Fig. 4.40).
Step 50: In **Simulation>Model Configuration Parameters**, change **max step time** to 1e-3 and the simulation **stop time** to 0.6.

In these 50 steps, the process of adding a universal frame, frame 1 and frame 2, has been outlined. Frame 1 is attached to the plate while frame 2 is attached to the ball. The inputs to the x and y rotational angles of frame 1 are constants. In the next section, the forces of the plate on the ball will be detailed. The c-script for the s-function will be discussed.

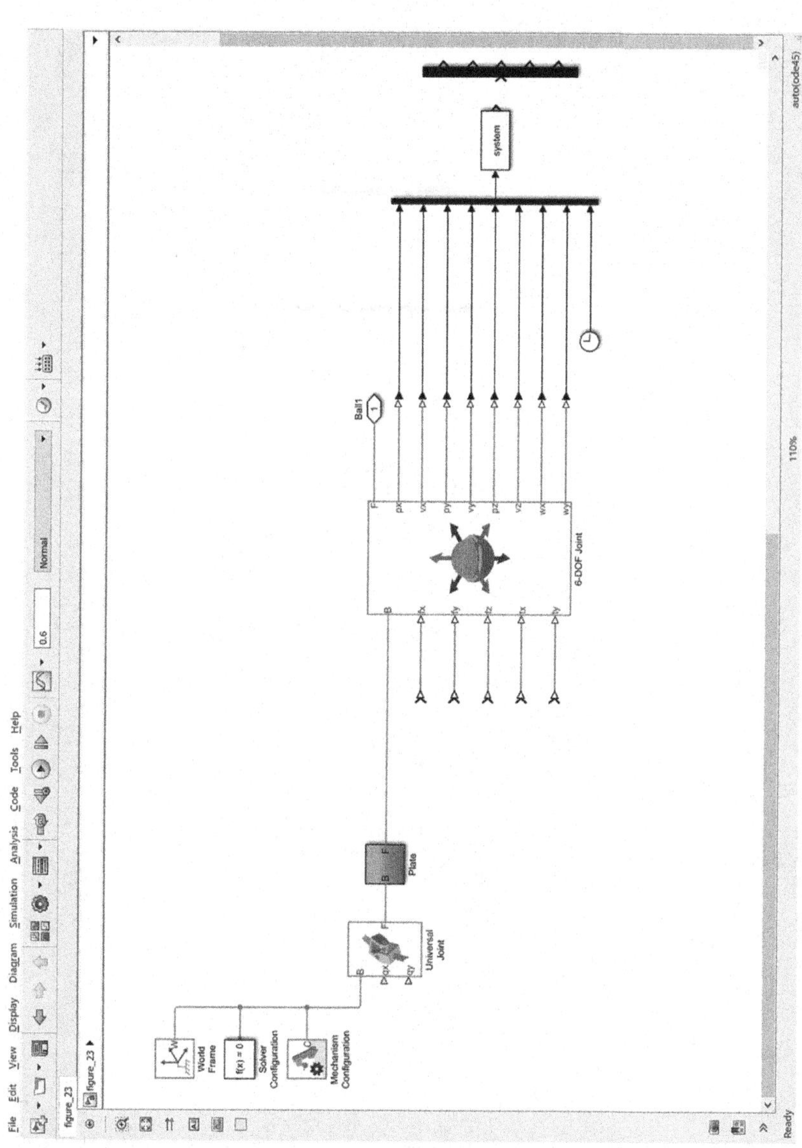

Figure 4.32 Mux block connections.

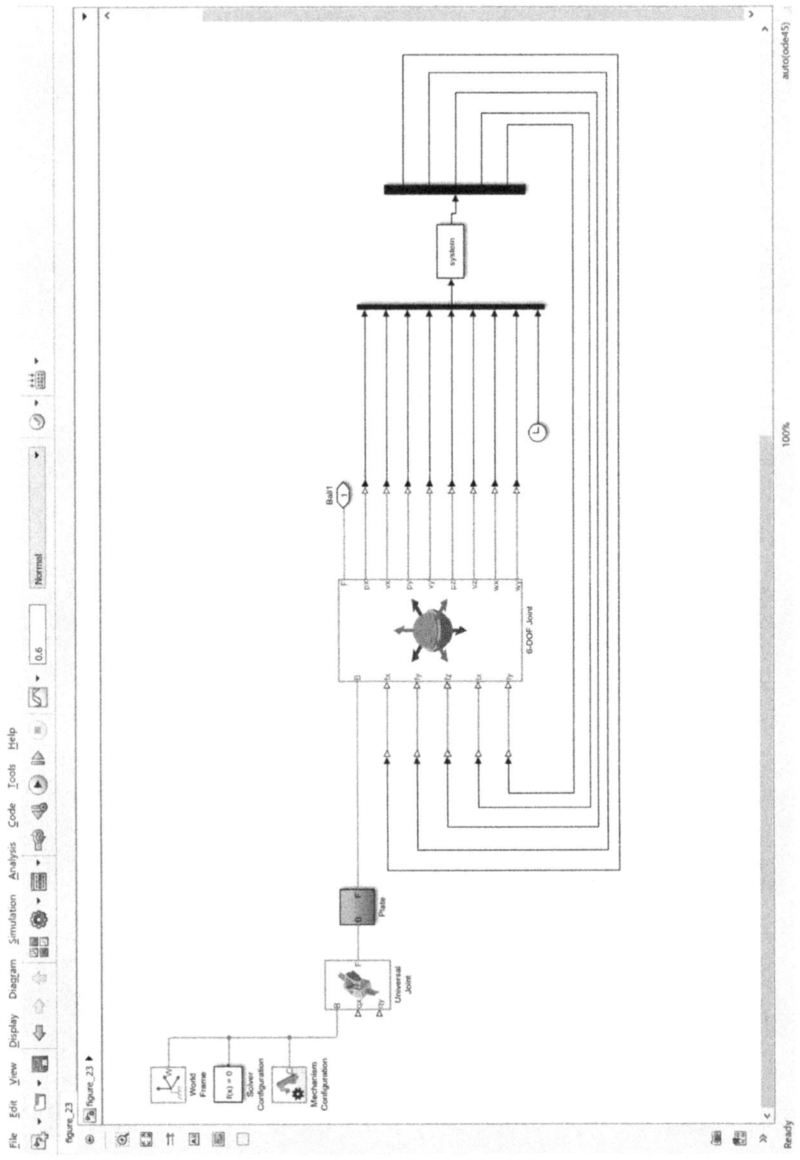

Figure 4.33 Model Connections.

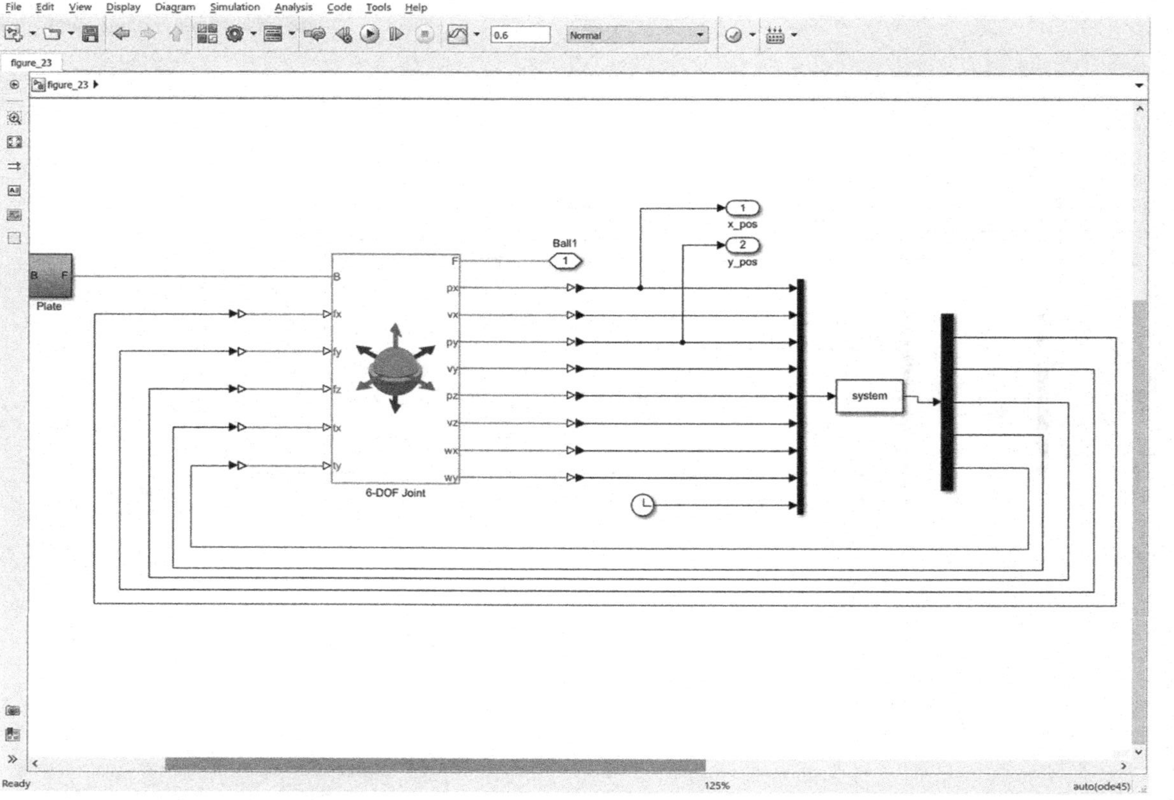

Figure 4.34 x and y positions of the ball.

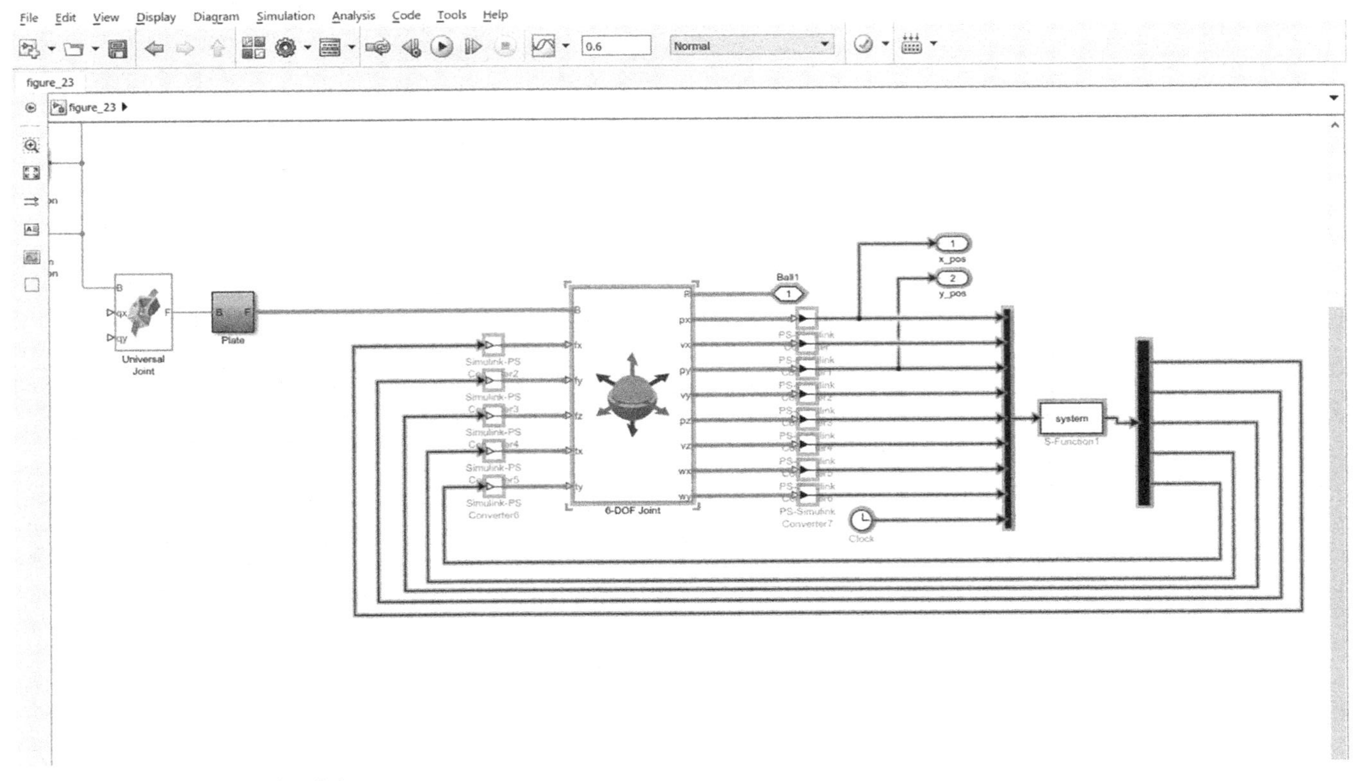

Figure 4.35 Ball-Plate Interaction Subsystem.

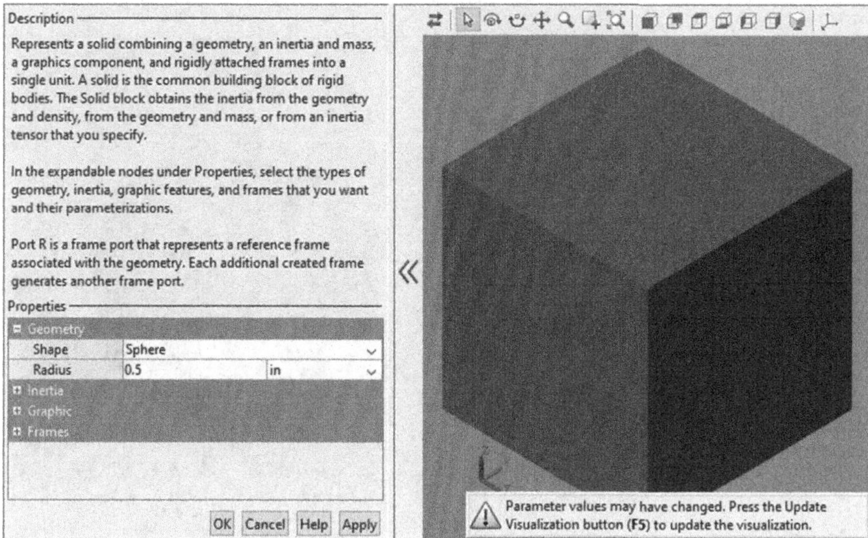

Figure 4.36 Radius of the sphere.

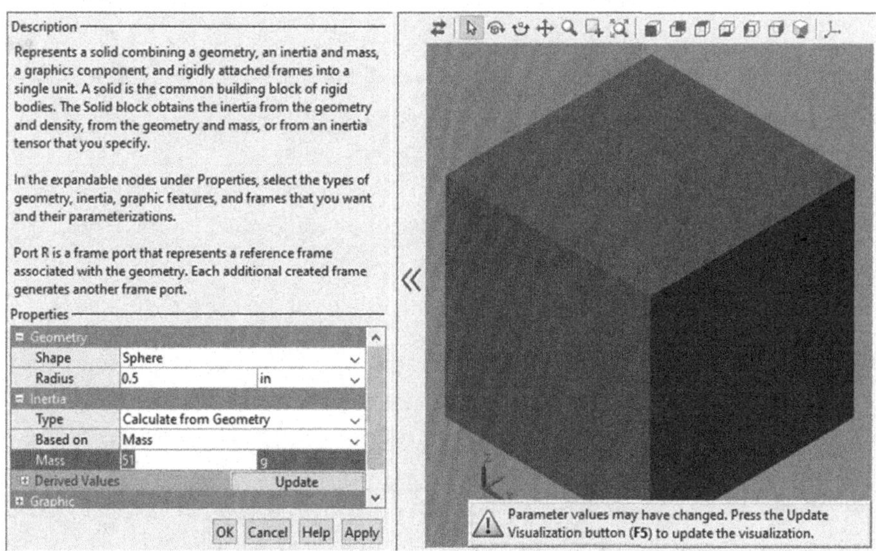

Figure 4.37 Mass of the sphere.

4.6 Ball Plate Interaction

Simscape Multibody natively handles solving the interaction between the 3D components. It can solve for the forces given motion constraints, the motion using the force constraints, or a hybrid of both. The unique element of the ball on plate is that these

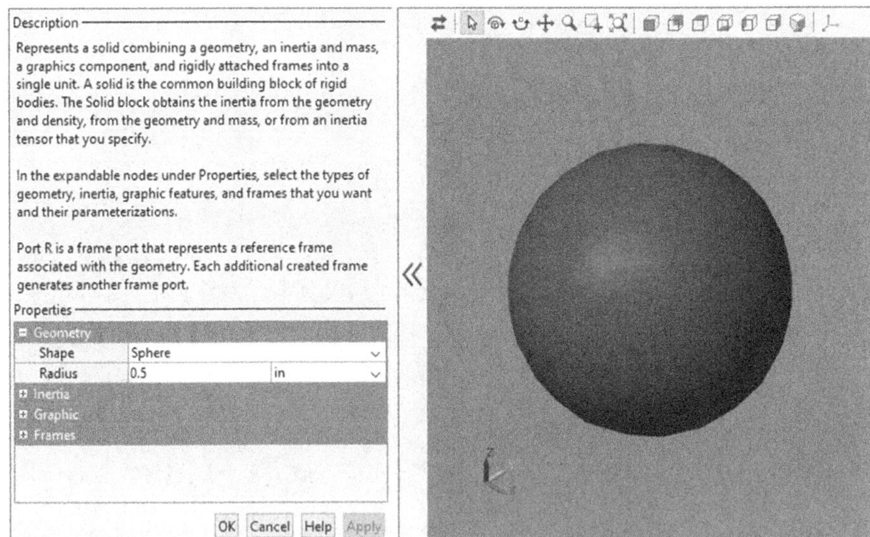

Figure 4.38 Sphere shape.

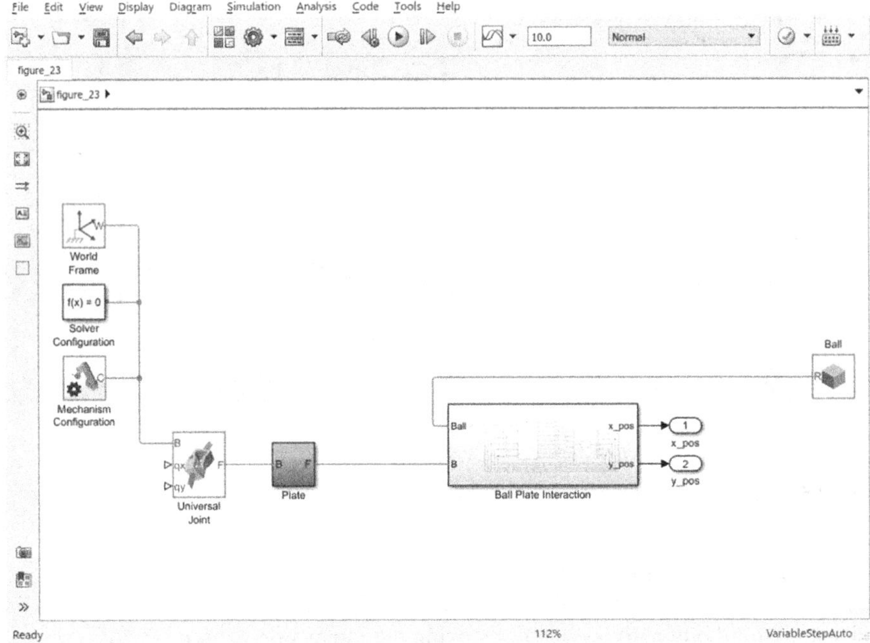

Figure 4.39 Ball Plate Interaction and Ball connection.

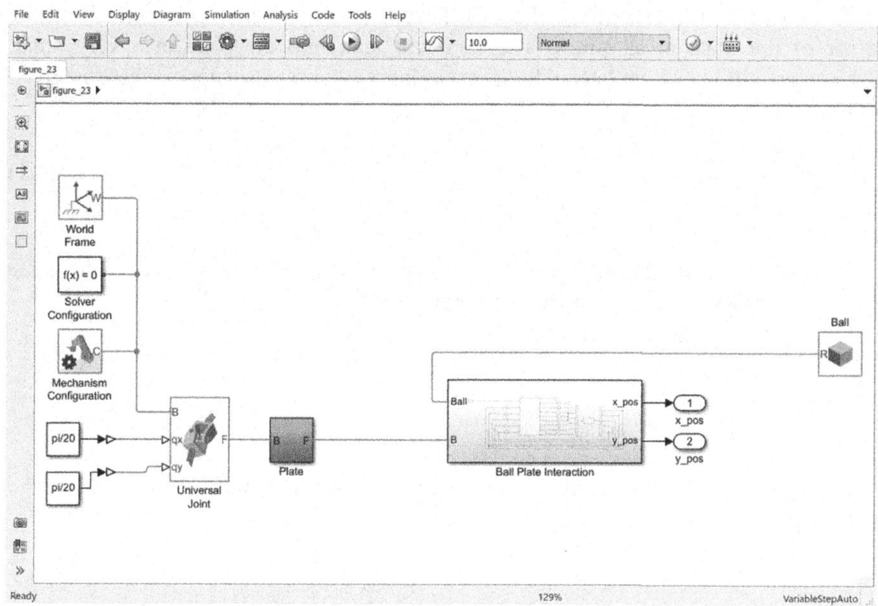

Figure 4.40 Constant angle inputs to the plate.

two 3D components are not natively connected in Simscape. This mandates calculating the interaction forces between the ball and the plate.

The first force that will be discussed is the support force that prevents the ball from sinking into the plate due to gravity (note that the force of gravity on the ball is already taken care of inherently in the solid block for the ball). This support force is the reaction force of the plate. As long as the ball is in contact with the plate, the plate will generate a force on the ball in the z direction, measured with respect to frame 1 (plate frame). This force has two components: a spring and a damper. It is an overdamped system so that the ball does not exhibit bouncing on the plate. The two components will prohibit the ball from penetrating the plate. The ball center will be distanced at $(C_{gap} + r_{ball})$ from the center of the plate (which is where frame 1 is fixed) (Fig. 4.41). In reality, it is

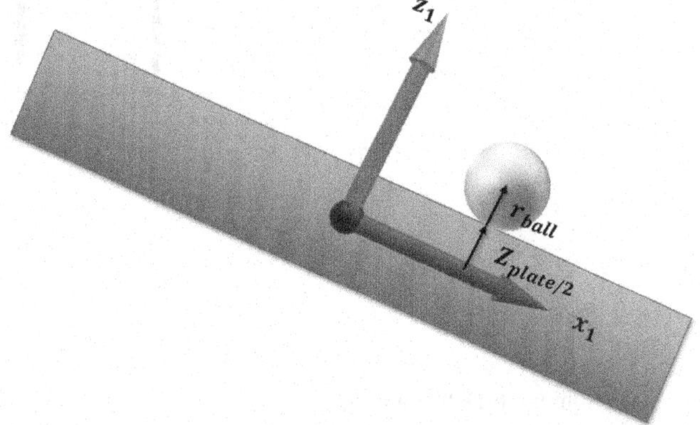

Figure 4.41 Ball distance from the center line of the plate.

hard to solve for the exact force that will keep the ball just at $(C_{gap} + r_{ball})$ from the center of the plate at all times. The ball will sink into the plate at a very small distance as will be shown later. The mass spring model for the support force is given by Eq. (4.1) as

$$F_z = -K_{pen}(P_z - C_{gap}) - D_{pen}V_z \tag{4.1}$$

where

P_z is the z coordinate of the ball center measured from the center of the plate (Fig. 4.42 shows the **Pz** output from 6-DOF joint).

C_{gap} is a constant and is defined as

$$C_{gap} = \frac{Z_{plate}}{2} + r_{ball}$$

Z_{plate} is the plate thickness and is equal to 0.02 m (2 cm from the dimensions of the plate in the first section).

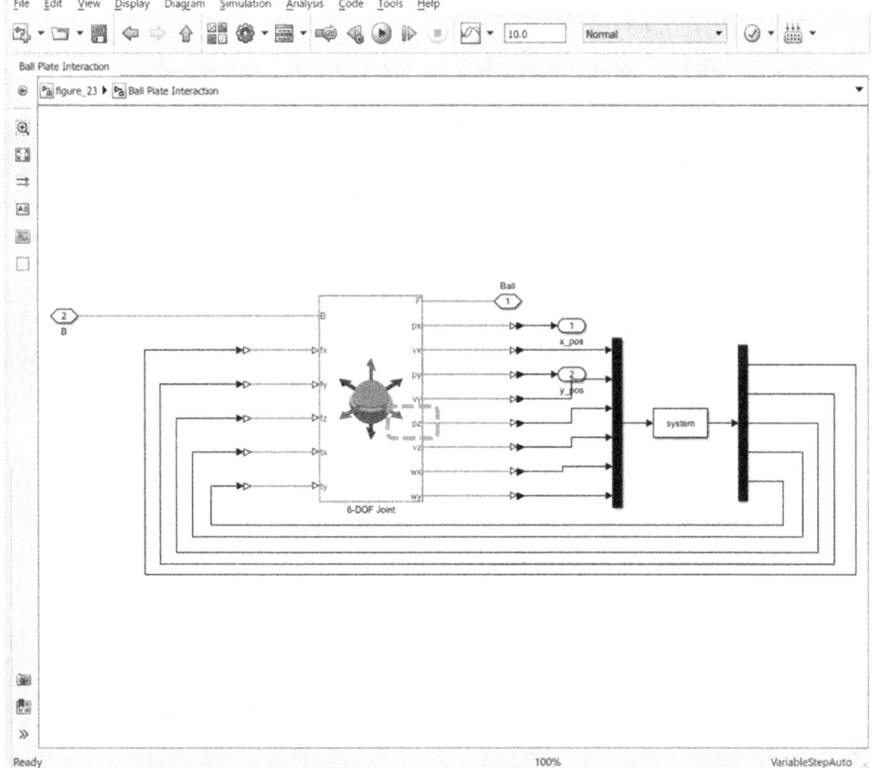

Figure 4.42 pz output from the 6-DOF Joint.

r_{ball} is the radius of the ball and is equal to 1.27 cm (½ inch from the dimensions of the ball in first section).

K_{pen} is the spring constant for the plate to guarantee an overdamped system. It is set to 800,000 N/m in the model and can be determined experimentally as will be shown later in this section.

D_{pen} is the damping coefficient. It is set to 1000 kg/s and can be determined experimentally as will be shown later in this section.

Note that F_z will be set to zero whenever the ball is not in contact with the plate.

The remaining forces and torques that the plate exerts on the ball are the two frictions in the x and y directions and their torques.

Friction force in the x and y direction is given as

$$F_{fx} = -sign(v_x) \times \mu_{static} \times F_z \tag{4.2}$$

$$F_{fy} = -sign(v_y) \times \mu_{static} \times F_z \tag{4.3}$$

where the friction coefficient in the x and y direction is constant and is set to be $\mu_{static} = 0.6$ and v_x and v_y are the velocities of the center of the ball in the x and y directions.

When the model is initialized, and speed of the ball is very small (practically zero), a special condition is implemented to prevent slipping of the ball on the plate. The friction coefficient μ_{static} in Eqs. (4.2) and (4.3) is replaced by

$$\mu_x = \mu_{static} \times \frac{v_{x-slippage}}{v_{slippage-threshold}} \tag{4.4}$$

$$\mu_y = \mu_{static} \times \frac{v_{y-slippage}}{v_{slippage-threshold}} \tag{4.5}$$

where $v_{x-slippage}$ and $v_{y-slippage}$ have the same direction of the friction forces and they are given by

$$v_{x-slippage} = -v_x + r_{ball} \times \omega_y \tag{4.6}$$

$$v_{y-slippage} = -v_y - r_{ball} \times \omega_x \tag{4.7}$$

where $-r_{ball} \times \omega_y$ and $+r_{ball} \times \omega_x$ represent the circumferential speed of the ball at point of contact with plate in the x and y directions, respectively. Using the right-hand rule of rotation, one can determine that $r_{ball} \times \omega_y$ is opposing the motion while $r_{ball} \times \omega_x$ is along the motion and thus Eqs. (4.6) and (4.7) have opposite signs.

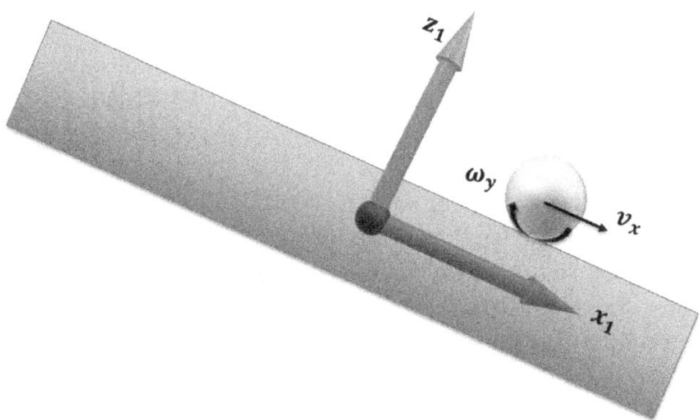

Figure 4.43 Slipping diagram for x-direction (y_1 axis is inwards).

Figs. 4.43 and 4.44 demonstrate the slippage velocities and their terms (y and x, respectively).

The torques of these forces around the center of the ball are given by multiplying the forces by the radius of the ball.

$$T_x = r_{ball} \times F_{fy} \tag{4.8}$$

$$T_y = -r_{ball} \times F_{fx} \tag{4.9}$$

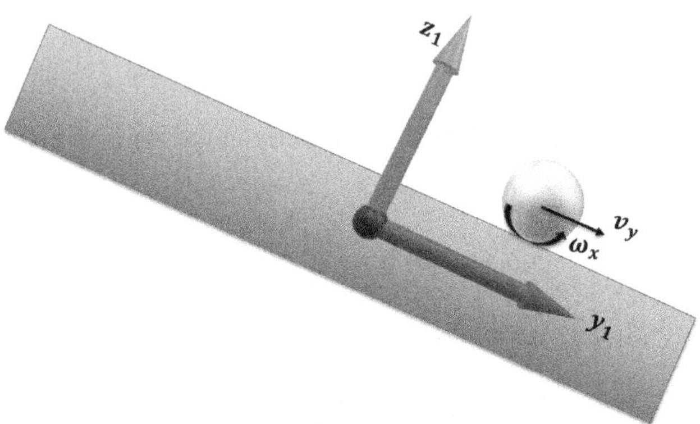

Figure 4.44 Slipping diagram for y-direction (x_1 axis is outwards).

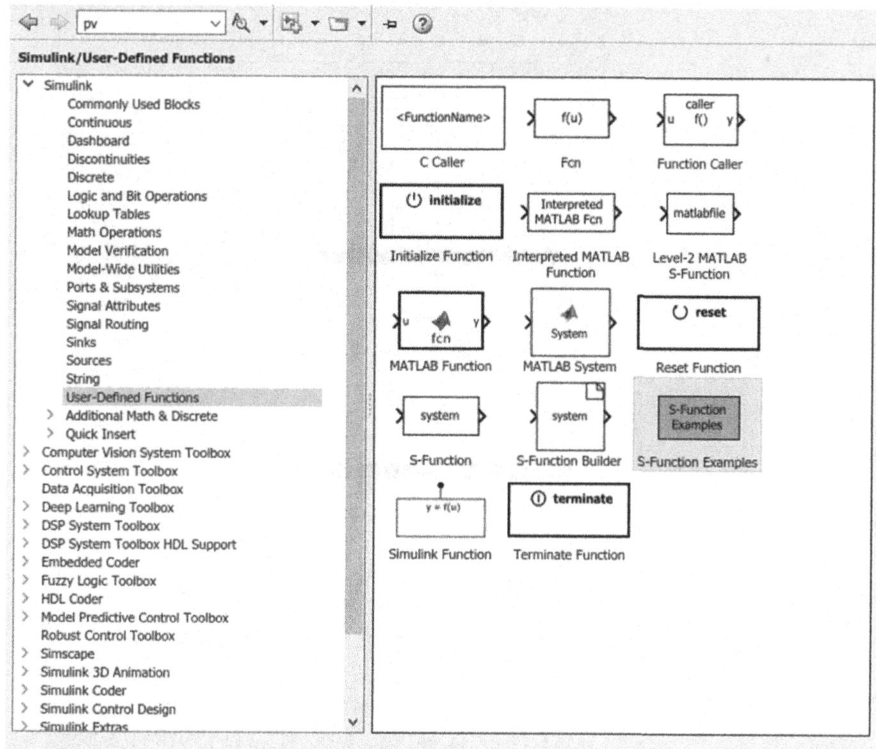

Figure 4.45 Basic C-Mex Template.

4.7 S-function for the Ball Plate Interaction

In this section, a C-script will be constructed to calculate Eqs. (4.1)–(4.9). Start by opening a C-MEX template. Go to Simulink>User-Defined Functions>S-function Examples>C-Files>Basic C-Mex Template (Fig. 4.45). Before we write the script, we will do a minor update on the model in Fig. 4.42. Instead of six inputs to the user-defined block, we will add another input, which is the simulation time (Fig. 4.46). Below is the **ballplateforces.c** script with full comments explaining the implementation of the way Eqs. (4.1)–(4.9) were implemented. Script can be downloaded from MATLAB Central by searching for the book and downloading the files for this chapter. Note that you will need to type and run **mex ballplateforces.c** in the **MATLAB®** command window if you introduce any changes in the C-script or if you face any problems running the provided Simulink model.

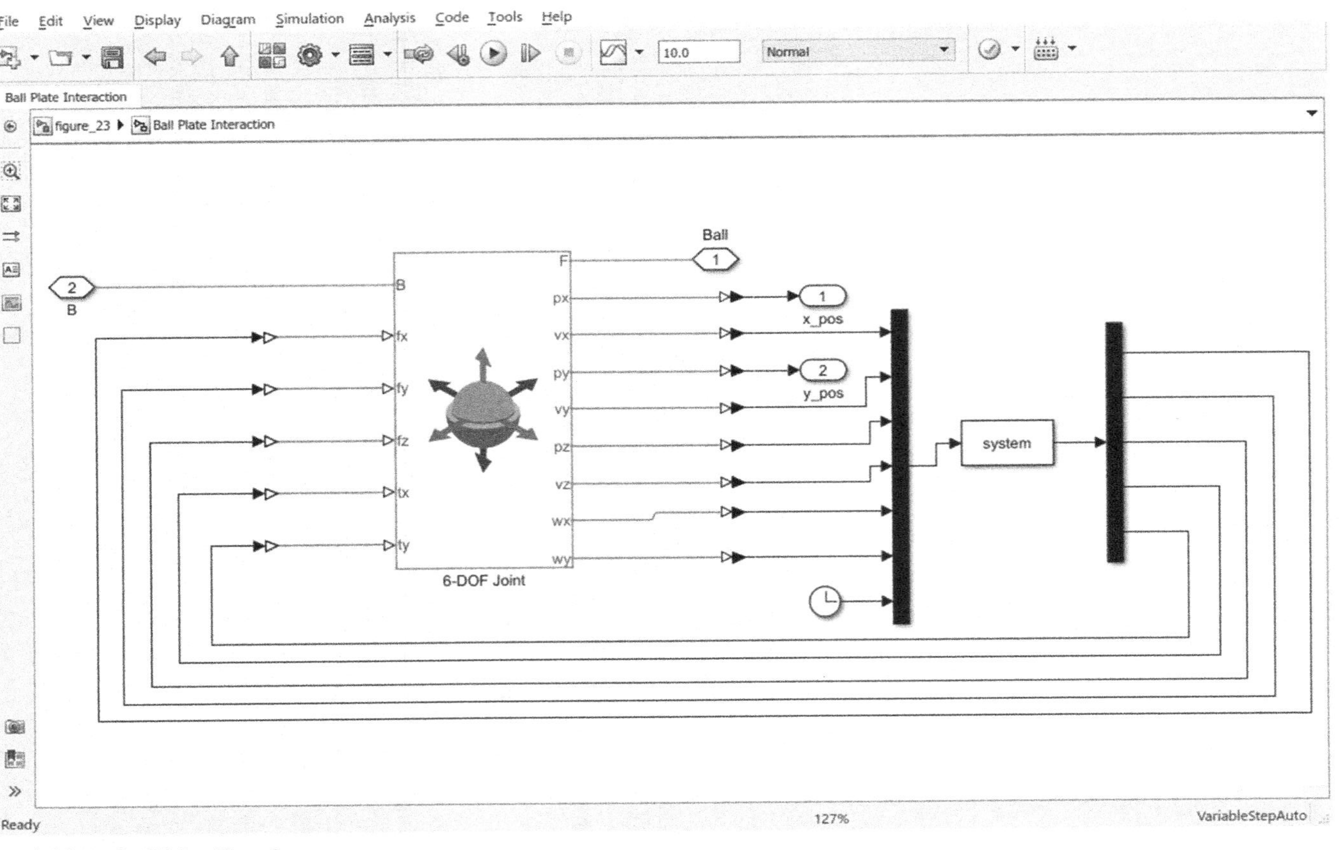

Figure 4.46 Basic C-Mex Template.

```
/* Ball Plate Forces
 *
 *
 */
#define S_FUNCTION_NAME       ballplateforces
#define S_FUNCTION_LEVEL      2
#include "simstruc.h"
#include <math.h>
#define PAR(element)         (*mxGetPr(ssGetSFcnParam(S,element)))
#define U(element)           (*uPtrs[element])
/* Parameters. */
#define xpla                  PAR(0)   /* Plane length [m]. */
#define ypla                  PAR(1)   /* Plane width [m]. */
#define zpla                  PAR(2)   /* Plane thickness [m]. */
#define rbal                  PAR(3)   /* Ball radius [m]. */
#define Kpen                  PAR(4)   /* Spring constant [N/m]. */
#define Dpen                  PAR(5)   /* Damping constant [N.s/m]. */
#define mustat                PAR(6)   /*Static friction constant [ ]. */
#define vthr                  PAR(7)   /*Friction threshold speed [m/s]. */

/* Inputs. */
#define xpos       U(0)   /* Ball position in x-direction. */
#define vx         U(1)   /* Ball speed in the x direction. [m/s] */
#define ypos       U(2)   /* Ball position in y-direction. */
#define vy         U(3)   /* Ball speed in the y direction. [m/s] */
#define pz         U(4)   /* Ball position in the z direction. [m] */
#define vz         U(5)   /* Ball speed in the z direction. [m/s] */
#define wx         U(6)   /* Ball rotation speed around the x axis.
[rad/s] */
#define wy         U(7)   /*Ball rotation speed around the y axis. [rad/s]*/
#define Tsim       U(8)   /* Simulation time. [s] */

/* Initialization. */
static void mdlInitializeSizes(SimStruct *S) {
  ssSetNumSFcnParams(S, 8);
  if (ssGetNumSFcnParams(S) != ssGetSFcnParamsCount(S)) {
    /* Return if number of expected != number of actual parameters */
    return;
  }
  ssSetNumContStates(S, 0);
  ssSetNumDiscStates(S, 0);
  if (!ssSetNumInputPorts(S, 1)) return;
  ssSetInputPortWidth(S, 0, 9);
  ssSetInputPortDirectFeedThrough(S, 0, 9);
  if (!ssSetNumOutputPorts(S, 1)) return;
```

```
  ssSetOutputPortWidth(S, 0, 5);
  ssSetNumSampleTimes(S, 1);
  ssSetOptions(S, SS_OPTION_EXCEPTION_FREE_CODE);
}
static void mdlInitializeSampleTimes(SimStruct *S) {
  ssSetSampleTime(S, 0, CONTINUOUS_SAMPLE_TIME);
  ssSetOffsetTime(S, 0, 0.0);
}
#define MDL_INITIALIZE_CONDITIONS     /* Change to #undef to remove
function */
#if defined(MDL_INITIALIZE_CONDITIONS)
  /* Function: mdlInitializeConditions
================================
  * Abstract:
  *     In this function, you should initialize the continuous and
discrete
  *     states for your S-function block. The initial states are
placed
  *     in the state vector, ssGetContStates(S) or
ssGetRealDiscStates(S).
  *     You can also perform any other initialization activities that
your
  *     S-function may require. Note, this routine will be called at
the
  *     start of simulation and if it is present in an enabled
subsystem
  *     configured to reset states, it will be call when the enabled
subsystem
  *     restarts execution to reset the states.
  */
  static void mdlInitializeConditions(SimStruct *S)
  {
  }
#endif /* MDL_INITIALIZE_CONDITIONS */
static void mdlOutputs(SimStruct *S, int_T tid) {
  real_T *y        = ssGetOutputPortRealSignal(S,0);
  real_T *x        = ssGetContStates(S);
  InputRealPtrsType      uPtrs = ssGetInputPortRealSignalPtrs(S,0);
  real_T         cgap, Fz;
  real_T         vxslippage, vyslippage;
  real_T         mux, muy, Ffx, Ffy;
  real_T         Tx, Ty;
  /* Support force parameters. */
  cgap = zpla/2+rbal;
```

```
  /* Support force Fz (eq 1) when the ball is on the plate, and
simulation time is post initialization . */
  if (pz<cgap && fabs(xpos)<xpla/2 && fabs(ypos)<ypla/2 &&
Tsim>0.001)
    Fz = -Kpen*(pz-cgap)-Dpen*vz;
  else
    Fz = 0;
  /* Slippage speed(eq 6 and 7). */
  vxslippage = -vx+rbal*wy;
  vyslippage = -vy-rbal*wx;
  if (fabs(vxslippage) <= vthr)
  /* Friction coefficient in the x direction(eq 4). */
    mux = mustat * vxslippage/vthr;
  else if(vxslippage > vthr)
  /* Friction coefficient in the x direction for positive velocity. */
    mux = mustat;
  else
  /* Friction coefficient in the x direction for positive velocity. */
    mux = -mustat;
  /* Friction force in the x direction(eq 2). */
  Ffx = mux*Fz;
  if (fabs(vyslippage) <= vthr)
  /* Friction coefficient in the y direction(eq 5). */
    muy = mustat * vyslippage/vthr;
  else if(vyslippage > vthr)
  /* Friction coefficient in the y direction for positive velocity. */
    muy = mustat;
  else
  /* Friction coefficient in the y direction for positive velocity. */
    muy = -mustat;
  /* Friction force in the y direction(eq 3). */
  Ffy = muy*Fz;
  /* Rolling torque around the x axis(eq 8). */
  Tx = rbal*Ffy;
  /* Rolling torque around the x axis(eq 9). */
  Ty = -rbal*Ffx;
  y[0] = Ffx;
  y[1] = Ffy;
  y[2] = Fz;
  y[3] = Tx;
  y[4] = Ty;
}
#define MDL_DERIVATIVES   /* Change to #undef to remove function */
#if defined(MDL_DERIVATIVES)
```

```
/* Function: mdlDerivatives
=============================
 * Abstract:
 *      In this function, you compute the S-function block's
derivatives.
 *      The derivatives are placed in the derivative vector,
ssGetdX(S).
 */
 static void mdlDerivatives(SimStruct *S)
 {
 }
#endif /* MDL_DERIVATIVES */
/* Function: mdlTerminate
=================================
* Abstract:
*       In this function, you should perform any actions that are
necessary
*       at the termination of a simulation. For example, if memory was
*       allocated in mdlStart, this is the place to free it.
*/
static void mdlTerminate(SimStruct *S)
{
}
/*==========================*
* Required S-function trailer *
*==========================*/
#ifdef   MATLAB_MEX_FILE    /* Is this file being compiled as a MEX-file?
*/
#include "simulink.c"       /* MEX-file interface mechanism */
#else
#include "cg_sfun.h"         /* Code generation registration function
*/
#endif
```

4.8 Simulation of the model

In order to simulate the ball, we need to change the name of the script in the S-function block to **ballplateforces**. Additionally, set the input parameters for the script, xpla, ypla, zpla, rbal, Kpen, Dpen, mustat, and vthr to 0.35, 0.24, 0.02, 0.0254/2, 1000, 100, 0.6, and 1e-4, respectively. Once done, run the simulation. Mechanics Explorer automatically opens up to display the two geometrical bodies, the plate and the ball, along with the frames in the system. The top left display of Fig. 4.47 shows the ball at time zero is sinking into the plate. Later in the simulation, the ball transitions into the normal position on top of the plate and rolls on the inclined plate under the impact

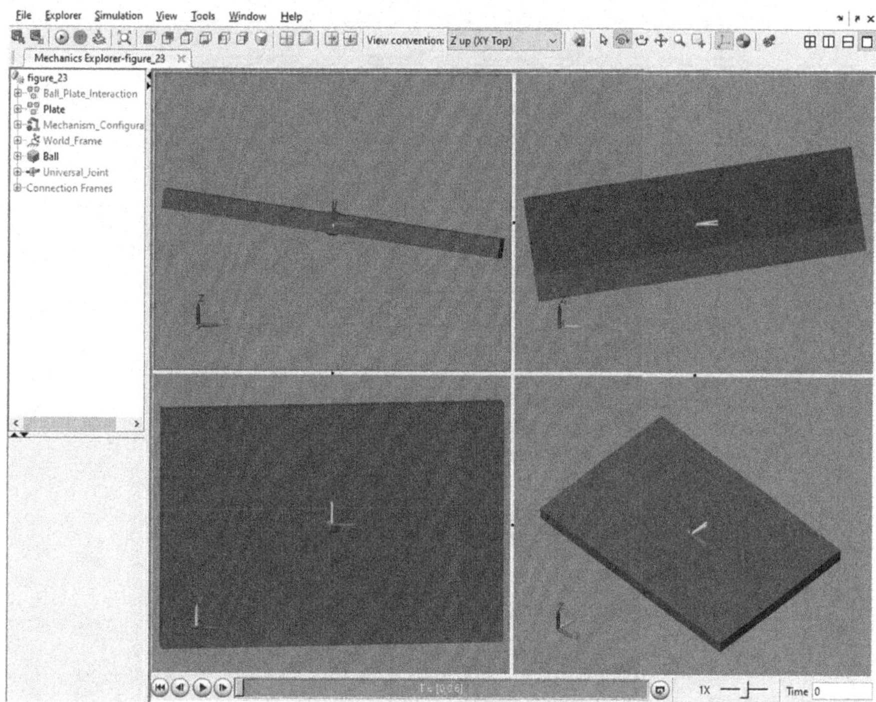

Figure 4.47 Mechanics Explorer for the ball on plate.

of gravity and friction. To set the initial position of the ball in the z direction, double click the 6-DOF Joint. Go to **Z Prismatic Primitive (Pz)>State Targets>Specify Position Target** and tick the box. Set Priority to High and Value to $C_{gap} = \frac{Z_{plate}}{2} + r_{ball}$=0.02/2 + 0.0254/2=0.0227 (m) (Fig. 4.48). Click **Apply** and **OK** when done and rerun the simulation. The ball should run on top of the plate.

It is worthwhile mentioning that the Mechanics Explorer displays the frames of the system (axis icon on the top right corner of Fig. 4.47). It is a very helpful tool in visually checking the rotational and translational motion of the bodies.

4.8.1 Application problem 1

Given the ball on plate system in Fig. 4.1:

(a) Develop two PID controllers for the x and y angles of the plate. The objective is to maintain the ball at the center of the plate. Simulation time is 30 s and execution time for the PID controllers is 0.01 s.

(b) Plot the error in x, error in y, x PID output, and y PID output.

(c) Subject the ball to an impulse force in the x-direction that is equal to 10% its weight at time=10 s. Add the force to the model and tune both PID loops. The ball needs to be within 2 cm radius of the center of the plate 7 s after the impulse (17 s from time zero).

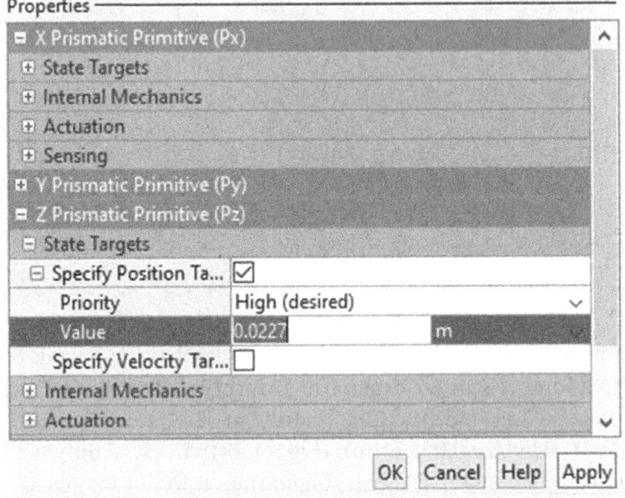

Figure 4.48 Initial ball position in the z direction.

(d) Plot the error in x, error in y, x PID output, and y PID output.

Hint: To control the x position of the ball, you will need to manipulate qy. Similarly, to control the y position of the ball, you will need to manipulate qx. Both qx and qy are in radians.

Download the material for the chapter from MATLAB® Central. The final **Simulink®** models for parts a and c are provided in **Application_Problem_2** folder. The **Simulink®** models are saved under **Application_Problem_1_a.slx** and **Application_Problem_1_c.slx.**

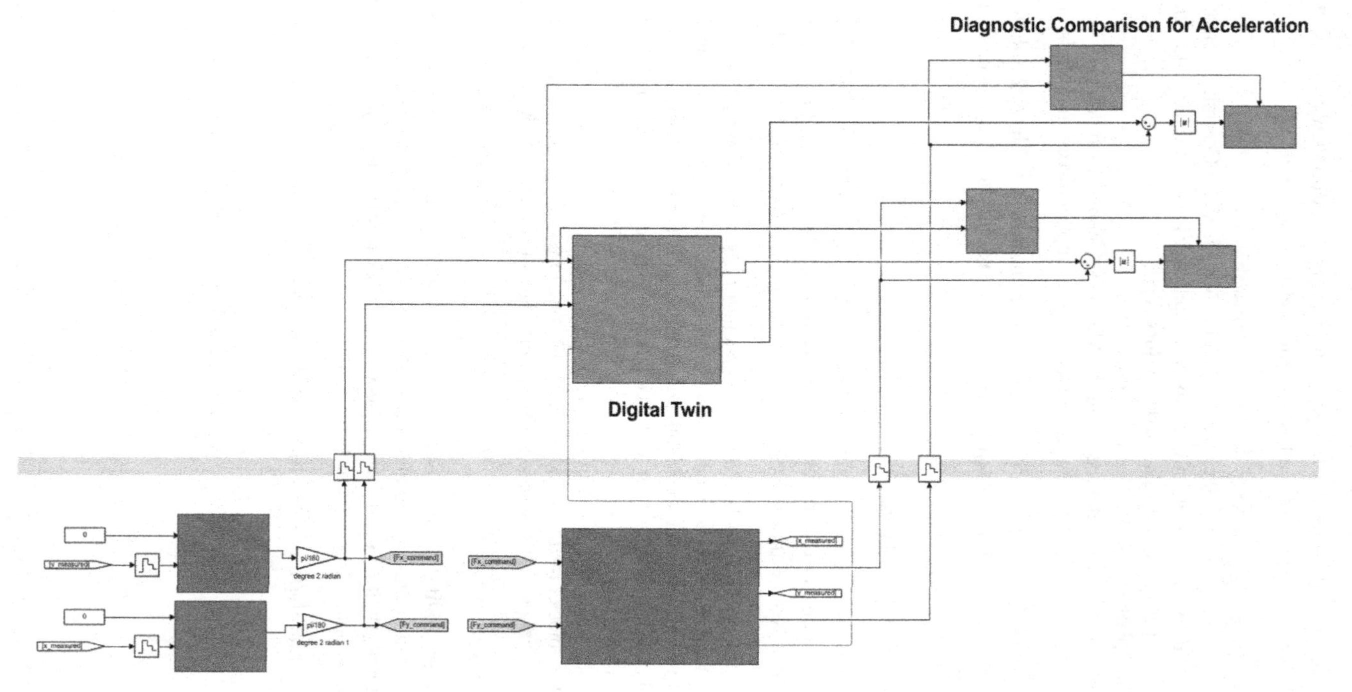

Figure 4.49 Simulated_Digital_Twin_Good_Asset.slx model.

4.8.2 Application problem 2

Replace the ball given in the chapter with square of dimensions $1.27\,\text{cm}^3 \times 1.27\,\text{cm}^3 \times 1.27\,\text{cm}^3$. The x and y angles of the plate are set to $\frac{\pi}{20}$.

(a) Revisit Eqs. (4.1)–(4.9) and see if they are still applicable
(b) Download the material for the chapter from **MATLAB®** Central. From folder **Application_Problem_2**, open **ballplateforces.c** and modify it according to the changes in introduced in part a. Save it as **squareplateforces.c**
(c) Open **ball_on_plate.slx** and change the ball to a square
(d) Type and run **mex squareplateforces.c** in the **Matlab®** command window
(e) Change the script name in the S-Function block to **squareplateforces** and run the model

The final **Simulink®** model is provided in **Application_Problem_2** folder. The **Simulink®** model is saved under **square_on_plate.slx**, and the script is saved under **squareplateforces.c.**

4.8.3 Application problem 3

Download the material for the chapter from MATLAB® Central. In **Application_Problem_3** folder, open and explore the **Simulated_Digital_Twin_Good_Asset.slx** model (Fig. 4.49). Notice that there are few changes introduced to the model:

1-the model of the plate is packaged in a subsystem for ease of readability
2-the output of the model is acceleration in the x and y directions instead of position. Position is still used to control the ball's position, but acceleration is used for diagnostic.
3-The model contains two copies of the ball on plate system: a representation of the physical hardware for the ball on plate (the lower part of the model in light blue) as well as the digital twin (the upper part of the model in green).
4-there are two blocks for the x and y diagnostic: one to enable the diagnostic and one to compute the deviation of acceleration between the digital twin and the physical asset.

Run the following exercises

(1) Run **Simulated_Digital_Twin_Good_Asset.slx** and check if there are any failures reported after the simulation finishes and plots the results
(2) Run **Simulated_Digital_Twin_Bad_x_Servo** and check if there are any failures reported after the simulation finishes and plots the results.

Hint: Notice that there is a red block in the model.

Digital twin model creation of double mass spring damper system

5.1 Introduction

Spring mass damper system is a very common scenario that is taught in mechanical engineering. Practical examples of this system are mostly seen in the suspension of a vehicle. The system consists of three elements: a spring, a damper, and a mass. The system can be used to study the response of most dynamic systems. An example of this system is shown in Fig. 5.1. When the spring is compressed by an external force, it stores potential energy, which is then released onto the mass. When the spring's energy is released, the spring will move about its equilibrium position until it reaches rest state. The damper in this system is used to dissipate some of the energy released by the spring such that the number and amplitude of the oscillations brought by the spring are reduced.

The example in Fig. 5.1 touches base on a double mass spring damper system. The objective is to understand the response of the system when an external force is introduced. In this case, we are interested to find the position and velocity of the masses. The first step is to develop a set of equations for both masses by using the conservation of energy in the x direction and then derive differential equations based on that information. As you can understand, this requires quite an extensive analytical approach for complex systems. The purpose of this chapter is to show we can model this system in MATLAB Simscape™ without the need of deriving equations on our own.

With regards to off-board diagnostics, we can sense the acceleration of the m_1 and m_2 for a disturbance input. The disturbance may be an external step input, an impulse, or due to initial conditions. The acceleration from the physical sensor can then be sent to the cloud and compared with the twin model simulation results for the same disturbance. Fig. 5.2 shows the off-board diagnostics process where the parts highlighted in blue *(dark gray in print version)* are utilized in this chapter, whereas the gray ones are not. As seen in the figure, we will be focusing the digital twin model for the double mass spring system such that a framework is provided for the reader to develop their own models. Fig. 5.3 shows the boundary diagram of the double mass spring system along with its interaction with its twin model.

All the codes used in the chapter can be downloaded for free from MATLAB File Exchange. Follow the link below and search for the ISBN or title of this book:

https://www.mathworks.com/matlabcentral/fileexchange/

Alternatively, the reader can also download the material and other resources from the dedicated website or contact the authors for further help:

https://www.practicalmpc.com/

Digital Twin Development and Deployment on the Cloud. https://doi.org/10.1016/B978-0-12-821631-6.00005-0

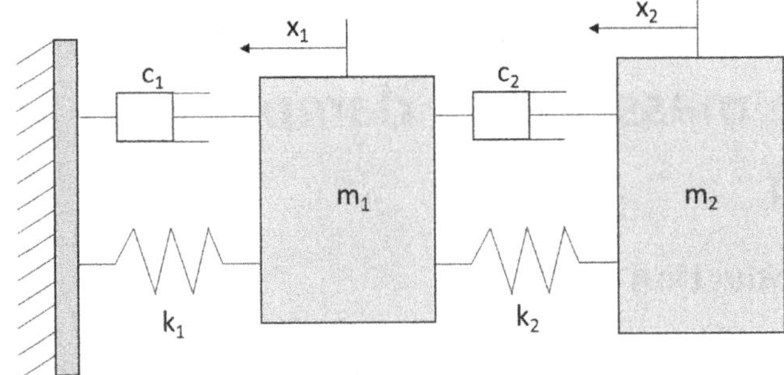

Figure 5.1 Double mass spring damper system, where k represents the spring constant of the spring, c represents the damping coefficient of the damper, and m represents the respective mass.

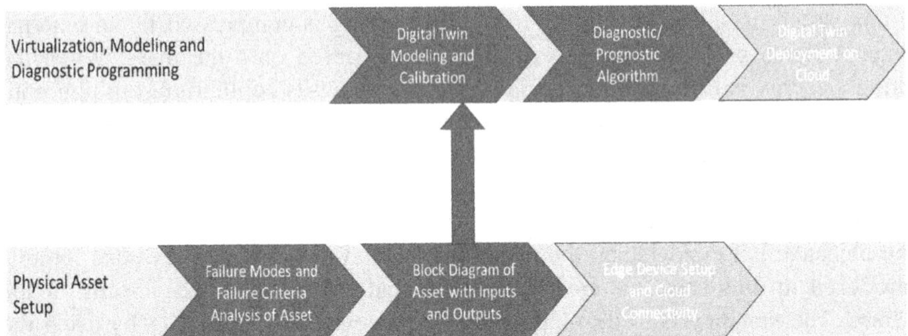

Figure 5.2 Off-board diagnostics process for double mass spring damper system.

5.2 Hardware parameters

Before we can simulate this system in Simscape™, we must determine the spring coefficients, damping coefficients, masses, starting positions, and the external forces required to introduce a dynamic behavior to this system. Take note of the value for the spring, damper, and mass parameters below:

- The spring constant shall have a value of 400 N/m.
- The damping coefficient shall have a value of 5 (N/m)/s.
- The masses shall have a value of 1 kg each.

The initial conditions for both masses should be zero. In this case, since the system is at rest, an external force is required to excite the system. In this case, a step force input of amplitude 500 N will be occurring after 1 s.

5.3 Simulation process

The final model is presented as follows in Fig. 5.4.

Below are the detailed steps to building the Simscape™ model based on the sections shown in Fig. 5.4.

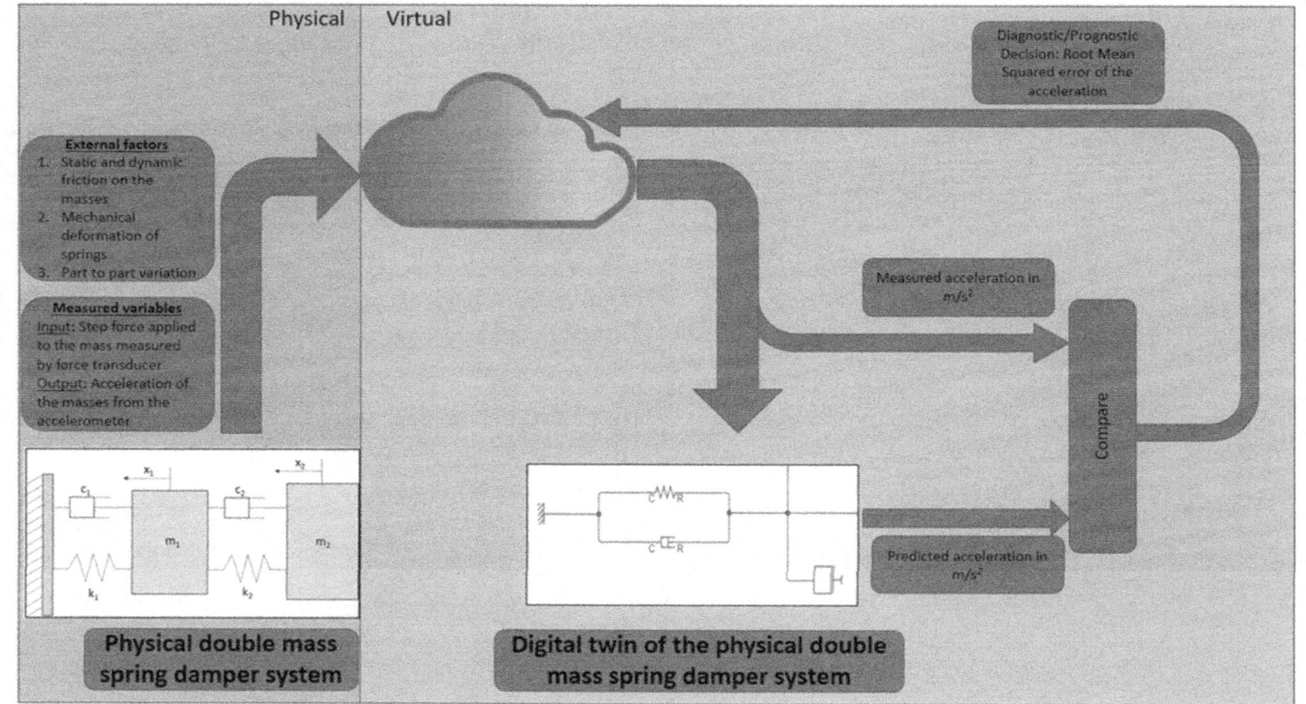

Figure 5.3 Block diagram of double mass spring damper system.

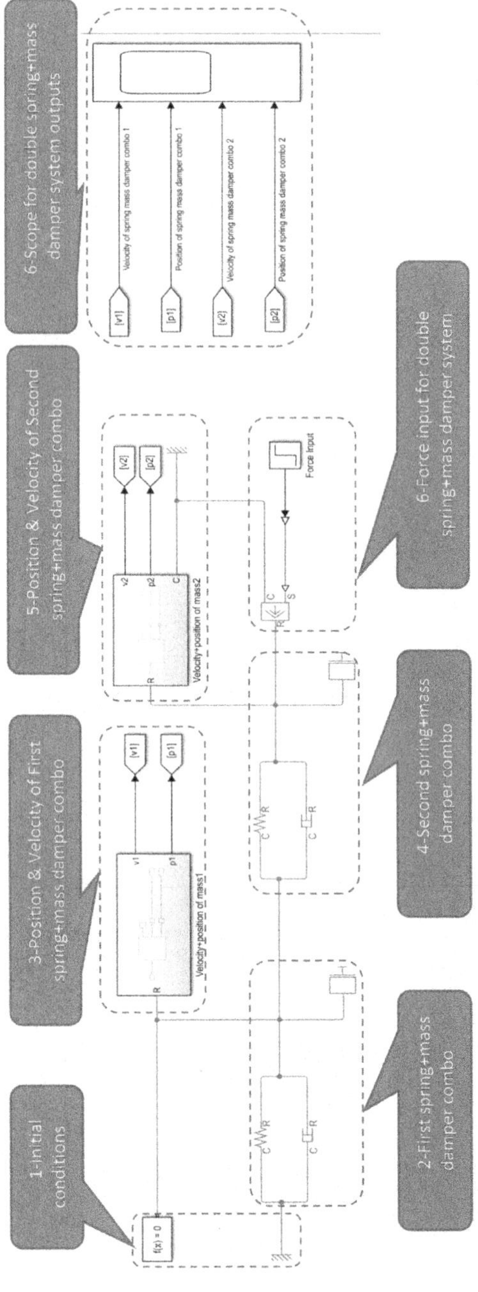

Figure 5.4 Main elements of the double mass spring damper model.

5.3.1 Initial conditions and first spring+mass damper combo

1. Create a new **Simulink® model** and then click on the **Simulink® Library Browser** button on the toolbar.
2. Navigate to **Simscape>Utilities.** Add **Solver Configuration** library block in the model.
3. Navigate to **Simscape>Foundation Library>Mechanical>Translational Elements** as shown in Fig. 5.5. Add **Mechanical Translational Reference** library block in the model.
4. Navigate to **Simscape>Foundation Library>Mechanical>Translational Elements** as shown in Fig. 5.6. Add **Translational Spring** library block in the model.
5. Navigate to **Simscape>Foundation Library>Mechanical>Translational Elements** as shown in Fig. 5.7. Add **Translational Damper** library block in the model.
6. Navigate to **Simscape>Foundation Library>Mechanical>Translational Elements** as shown in Fig. 5.8. Add **Mass** library block in the model.
7. Connect the **C** ports of the **Translational Damper** and **Translational Spring** to form a junction and then connect that junction to the **Mechanical Translational Reference** block.

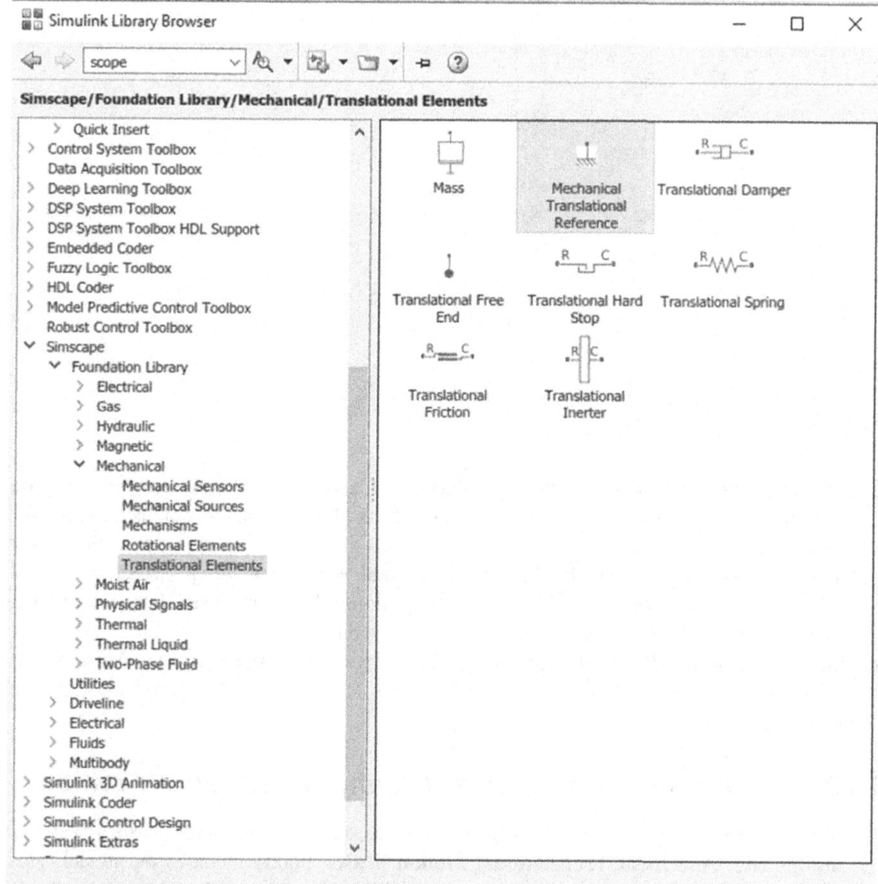

Figure 5.5 Mechanical Translational Reference.

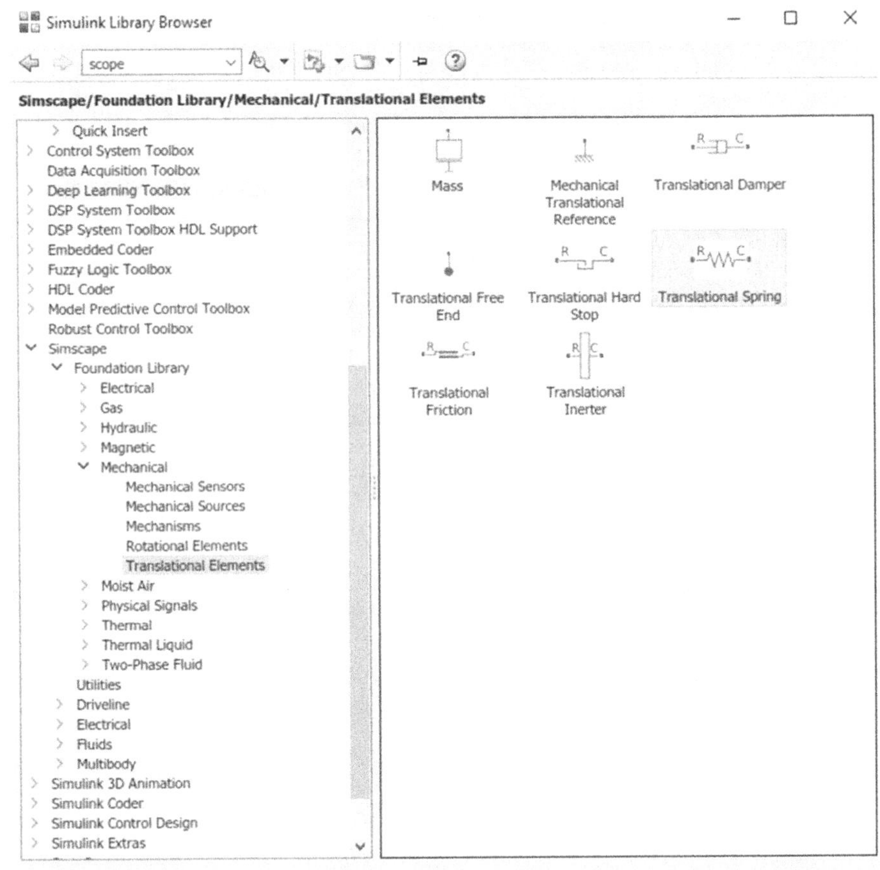

Figure 5.6 Translational Spring.

Connect the **R** ports of the **Translational Damper** and **Translational Spring** to form a junction and then connect that junction to the **Mass** block. Finally, connect the **Solver Configuration** to the junction connecting to the **Mass** block as shown in Fig. 5.9. The **Solver Configuration** and the **Mechanical Translational Reference** block marks the Initial Conditions for the model, whereas the **Translational Damper**, **Translational Spring**, and the **Mass** blocks represent the first spring+mass damper combo.

8. Double click on the **Translational Spring**, **Translational Damper**, and the **Mass** blocks and change their parameters to 400 N/m, 5 N/(m/s), and 1 kg, respectively, as shown in Figs. 5.10–5.12.

5.3.2 Position and velocity of first spring+mass damper combo

9. Navigate to **Simscape>Foundation Library>Mechanical>Mechanical Sensors** as shown in Fig. 5.13. Add **Ideal Translational Motion Sensor** library block in the model. This block will be used to monitor the position and velocity of the R ports junctions (it is sensing the position and velocity of the first spring+mass damper combo).

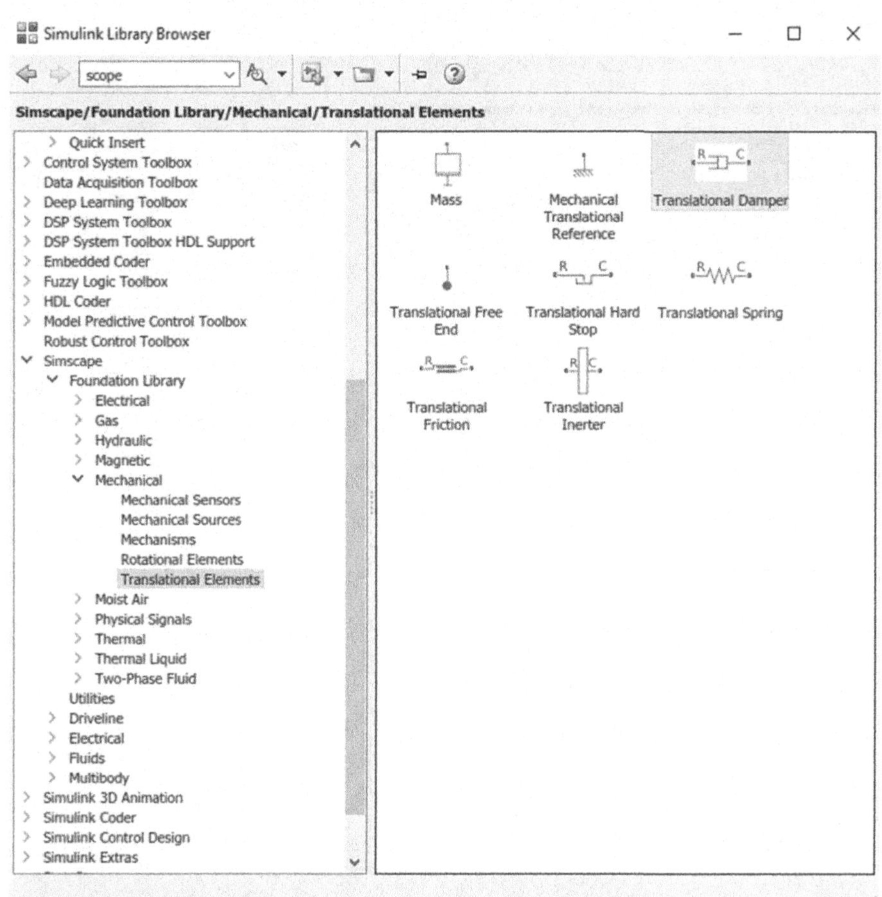

Figure 5.7 Translational Damper.

10. Repeat step 3. Flip the block left to right and place it in the model.
11. Connect the **R** port of the **Ideal Translational Motion Sensor** block to the junction formed for port **R** in step 7. Connect the **C** port of the **Ideal Translational Motion Sensor** to the **Mechanical Translational Reference** from step 10. The connections are shown in Fig. 5.14. The V and P ports represent the Velocity and Position ports of the sensor and will be used later on.
12. Double click on the **Ideal Translational Motion Sensor** block and input the initial position as 0 m as shown in Fig. 5.15. The 0 m work shall serve as our reference point.

5.3.3 Second spring+mass damper combo and position and velocity of second spring+mass damper combo

13. Repeat steps 4–12 with one exception in step 7: The **C** ports of the **Translational Damper** and **Translational Spring** should form a junction. Connect a wire from that junction to the **Mass** in the first spring+mass damper combo. Once done, follow through step 12 as shown in Fig. 5.16.

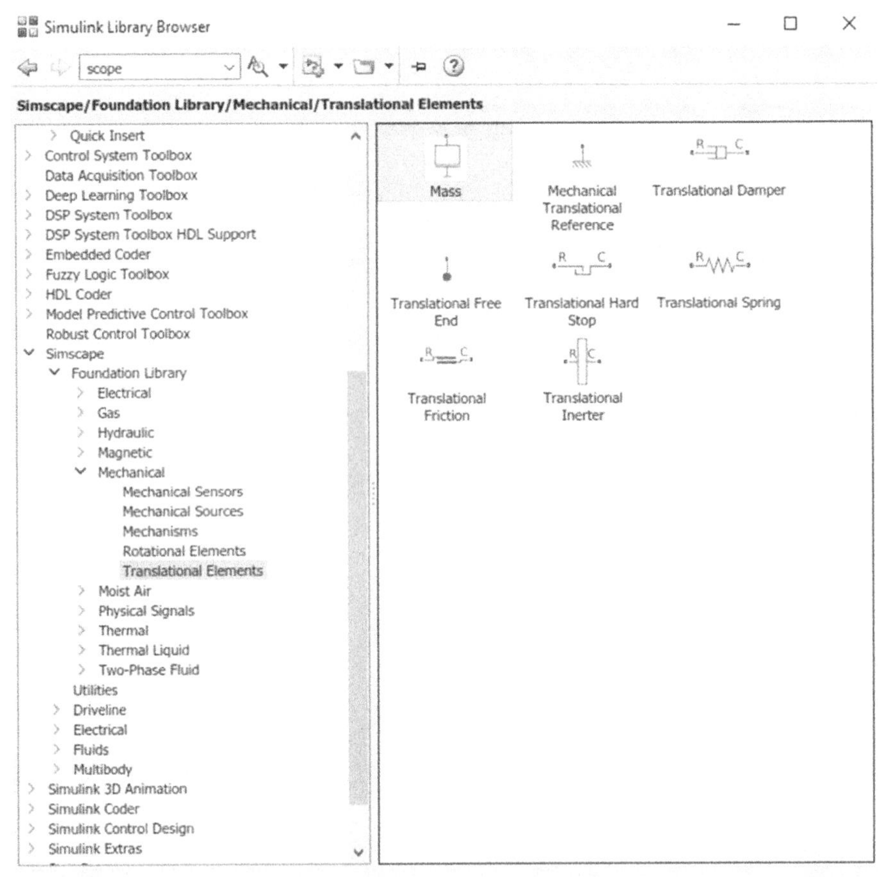

Figure 5.8 Mass block.

5.3.4 Force input for double spring+mass damper system

14. Navigate to **Simscape>Foundation Library>Mechanical>Mechanical Sources** as shown in Fig. 5.17. Add **Ideal Force Source** library block in the model. Through this block, we will be able to deliver a force input to our double mass spring system.
15. Navigate to **Simulink>Sources** as shown in Fig. 5.18. Add **Step** library block in the model. This block shall be used to deliver a unitless step signal.
16. Navigate to **Simscape>Utilities** as shown in Fig. 5.19. Add **Simulink-PS Converter** library block in the model. Through this block, we will be able convert a unitless step input to a force input for the **Ideal Force Source block.**
17. Flip the direction left-right for the **Simulink-PS Converter** and the **Step** block. Connect the **Step** block to the **Simulink-PS Converter.** Then connect the **Simulink-PS Converter** to the **S** port of the **Ideal Force Source block** as shown in Fig. 5.20.
18. Double click on the **Step** block. The step signal should be 500, which occurs after 1s as shown in Fig. 5.21.
19. Double click on the **Simulink-PS Converter** block. The unit for the step signal should be N as shown in Fig. 5.22.

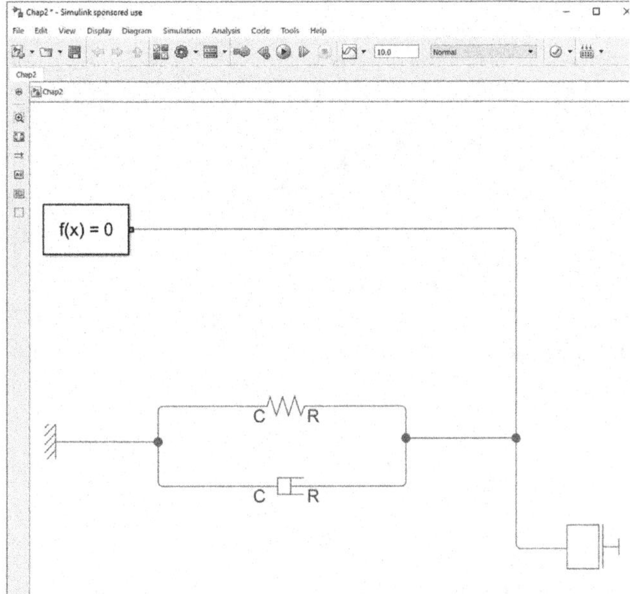

Figure 5.9 Initial Conditions and First spring mass damper combo.

Block Parameters: Translational Spring

Translational Spring

The block represents an ideal mechanical linear spring.

Connections R and C are mechanical translational conserving ports. The block positive direction is from port R to port C. This means that the force is positive if it acts in the direction from R to C.

Source code

Settings

Parameters | Variables

Spring rate: 400 N/m

OK Cancel Help Apply

Figure 5.10 Spring constant parameter.

Block Parameters: Translational Damper

Translational Damper

The block represents an ideal mechanical translational viscous damper.

Connections R and C are mechanical translational conserving ports, with R representing the damper rod, while C is associated with the damper case. The block positive direction is from port R to port C.

Source code

Settings

Parameters | Variables

Damping coefficient: 5 N/(m/s)

OK Cancel Help Apply

Figure 5.11 Damper coefficient parameter.

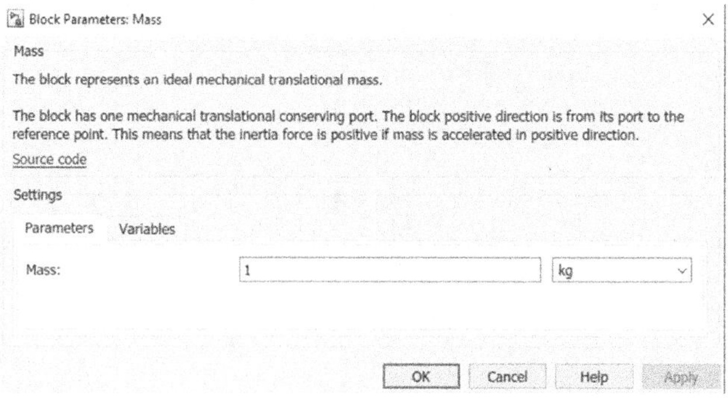

Block Parameters: Mass

Mass

The block represents an ideal mechanical translational mass.

The block has one mechanical translational conserving port. The block positive direction is from its port to the reference point. This means that the inertia force is positive if mass is accelerated in positive direction.

Source code

Settings

Parameters Variables

Mass: 1 kg

OK Cancel Help Apply

Figure 5.12 Mass parameter.

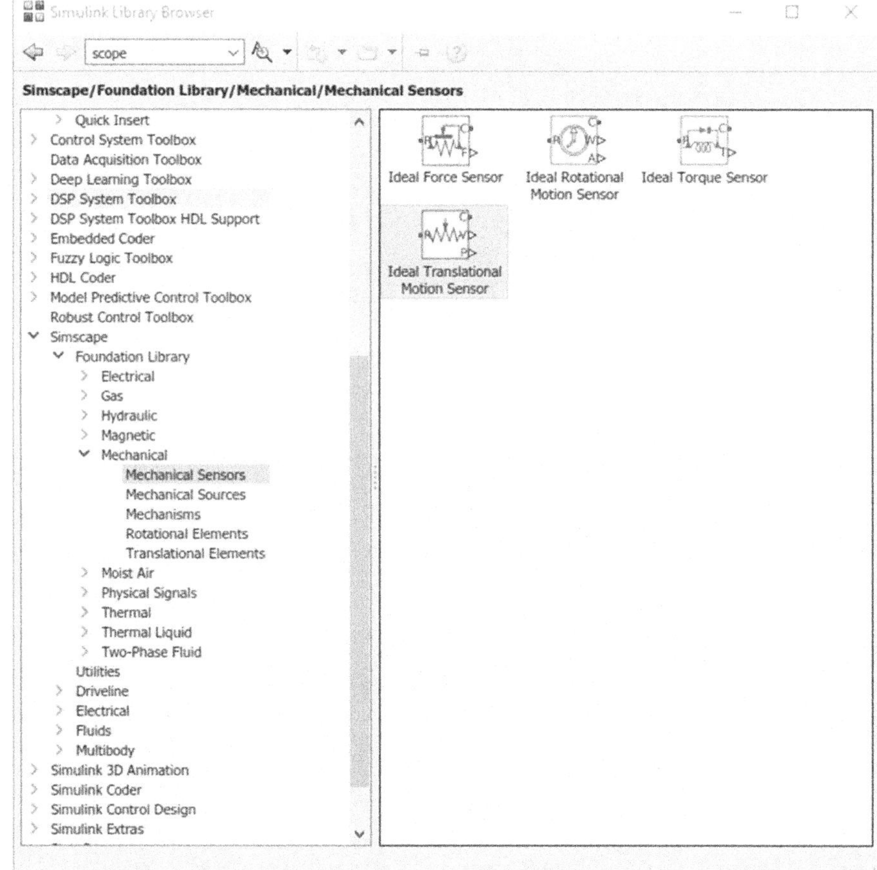

Figure 5.13 Ideal translation motion sensor.

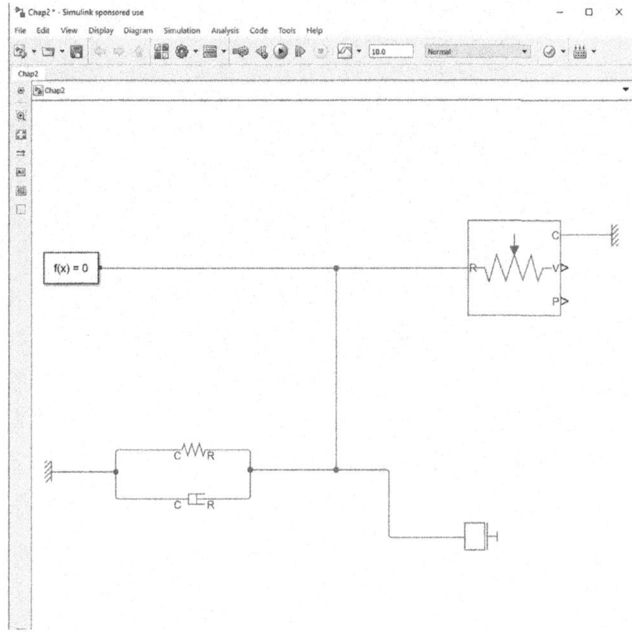

Figure 5.14 Motion sensor connected to the R junction port.

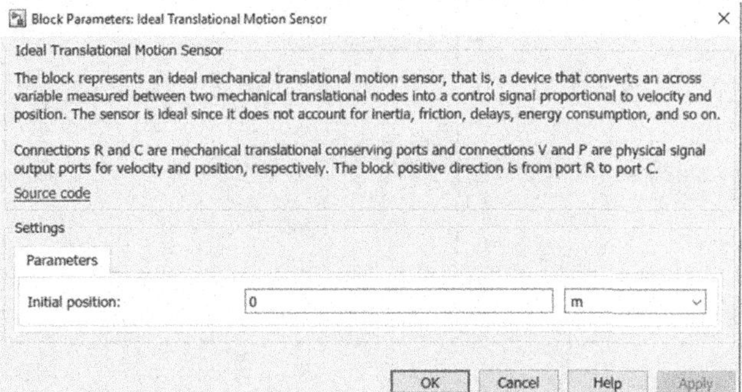

Figure 5.15 Initial position of the first spring+mass damper combo.

5.3.5 Scope for double spring+mass damper system outputs

20. Navigate to **Simscape>Utilities** as shown in Fig. 5.23. Add **PS-Simulink Converter** library block in the model. Through this block, we will be able convert the Simscape signals to Simulink form such that we display those signals on a Simulink **Scope.**
21. Navigate to **Simulink>Signal Routing** as shown in Fig. 5.24. Add four **Goto** library blocks in the model. These blocks are used for signal routing purposes.
22. Connect the **V** port of the **Ideal Translational Motion Sensor** of the first spring+mass damper combo to the **PS-Simulink Converter**, which is then connected to a **Goto** flag. Name this **Goto** flag as v1. Repeat the same for the **P** port and name that **Goto** flag as p1.

Chap2 * - Simulink sponsored use
File Edit View Display Diagram Simulation Analysis Code Tools Help
10.0
Normal
Chap2
Chap2

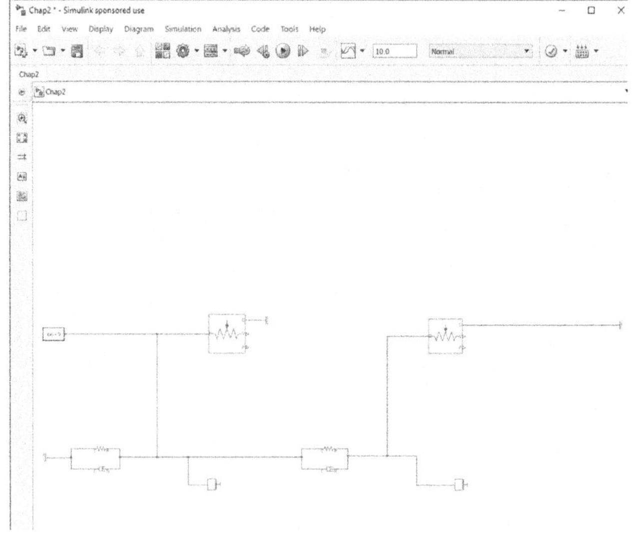

Figure 5.16 Second spring+mass damper combo.

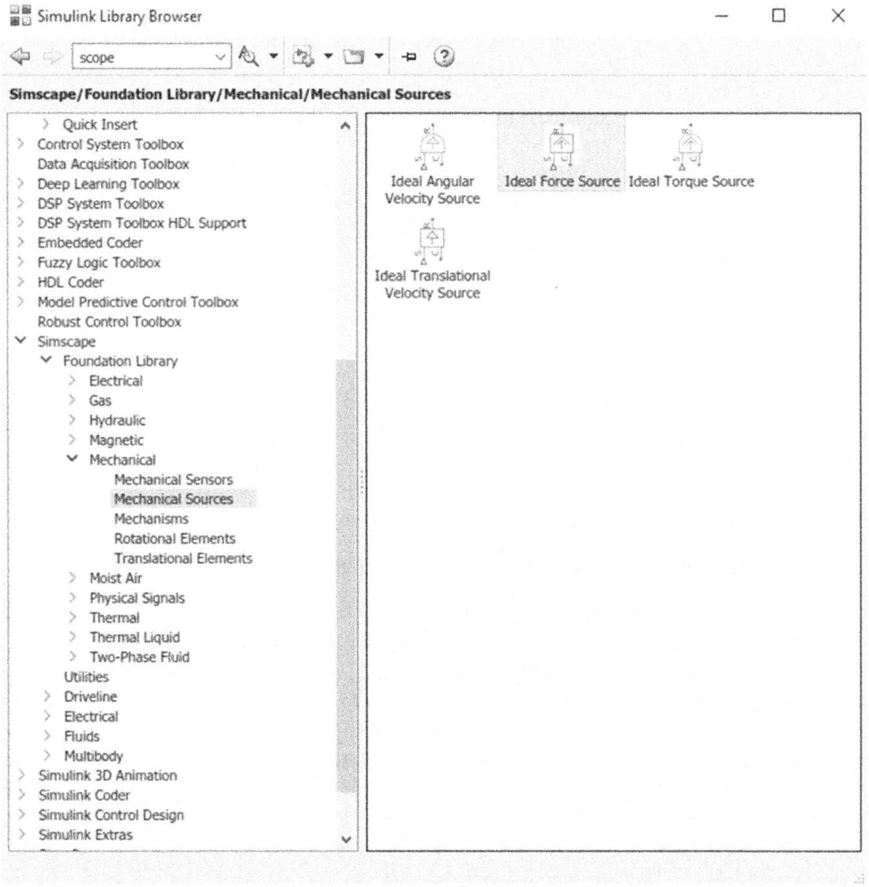

Simulink Library Browser
scope
Simscape/Foundation Library/Mechanical/Mechanical Sources
Quick Insert
Control System Toolbox
Data Acquisition Toolbox
Deep Learning Toolbox
DSP System Toolbox
DSP System Toolbox HDL Support
Embedded Coder
Fuzzy Logic Toolbox
HDL Coder
Model Predictive Control Toolbox
Robust Control Toolbox
Simscape
Foundation Library
Electrical
Gas
Hydraulic
Magnetic
Mechanical
Mechanical Sensors
Mechanical Sources
Mechanisms
Rotational Elements
Translational Elements
Moist Air
Physical Signals
Thermal
Thermal Liquid
Two-Phase Fluid
Utilities
Driveline
Electrical
Fluids
Multibody
Simulink 3D Animation
Simulink Coder
Simulink Control Design
Simulink Extras
Ideal Angular Velocity Source
Ideal Force Source
Ideal Torque Source
Ideal Translational Velocity Source

Figure 5.17 Ideal Force Source

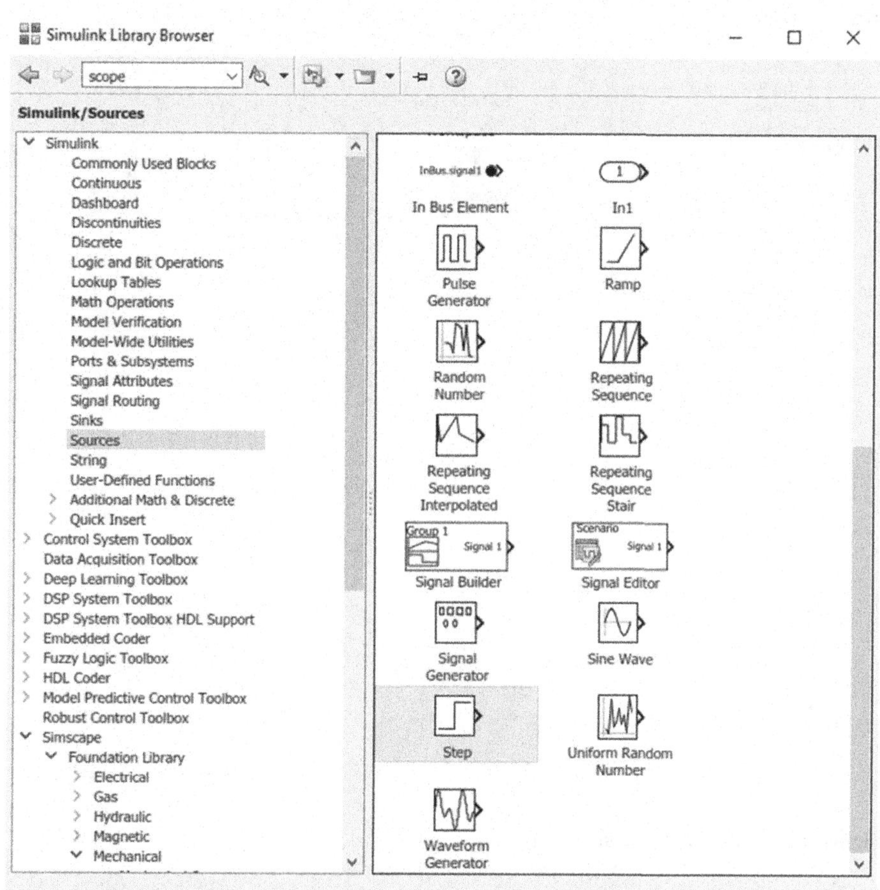

Figure 5.18 Step input.

23. Repeat the previous step for the second **Ideal Translational Motion Sensor.** The **Goto** flags for **V** and **P** ports should be renamed as v2 and p2, respectively. The final connections are shown in Fig. 5.25.
24. Double click on the **PS-Simulink Converter** coming from the **V** ports for both **Ideal Translational Motion Sensor** blocks and change the unit to m/s as shown in Fig. 5.26.
25. Double click on the **PS-Simulink Converter** coming from the **P** ports for both **Ideal Translational Motion Sensor** blocks and change the unit to m as shown in Fig. 5.27.
26. Navigate to **Simulink>Signal Routing** as shown in Fig. 5.28. Add four **From** library blocks in the model. These blocks are used for signal routing purposes.
27. Navigate to **Simulink>Sinks** as shown in Fig. 5.29. Add a **Scope** library block in the model. This block is used to display Simulink signal outputs.
28. Double click on the **Scope** block and click on the settings icon (it looks like a gear), and in the Main tab, change the number of input ports to 4 as shown in Fig. 5.30.

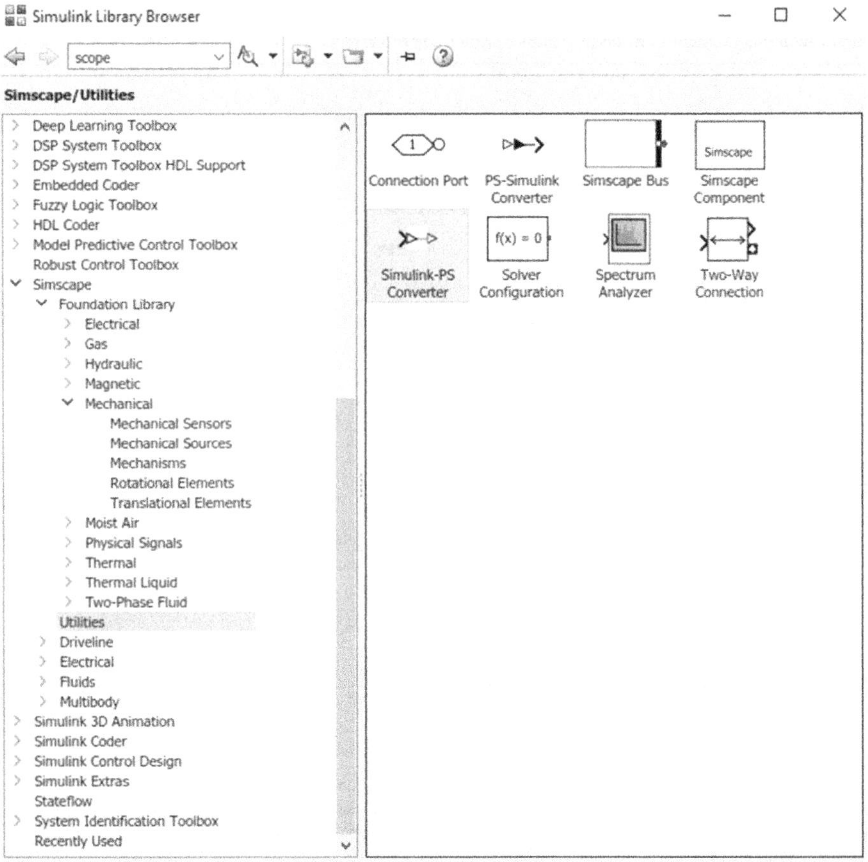

Simulink Library Browser
scope
Simscape/Utilities
Deep Learning Toolbox
DSP System Toolbox
DSP System Toolbox HDL Support
Embedded Coder
Fuzzy Logic Toolbox
HDL Coder
Model Predictive Control Toolbox
Robust Control Toolbox
Simscape
Foundation Library
Electrical
Gas
Hydraulic
Magnetic
Mechanical
Mechanical Sensors
Mechanical Sources
Mechanisms
Rotational Elements
Translational Elements
Moist Air
Physical Signals
Thermal
Thermal Liquid
Two-Phase Fluid
Utilities
Driveline
Electrical
Fluids
Multibody
Simulink 3D Animation
Simulink Coder
Simulink Control Design
Simulink Extras
Stateflow
System Identification Toolbox
Recently Used
1
Connection Port
PS-Simulink Converter
Simscape Bus
Simscape
Simscape Component
Simulink-PS Converter
f(x) = 0
Solver Configuration
Spectrum Analyzer
Two-Way Connection

Figure 5.19 Simulink-PS Converter block.

Chap2 * - Simulink sponsored use
File Edit View Display Diagram Simulation Analysis Code Tools Help
10.0
Normal
Chap2
Chap2

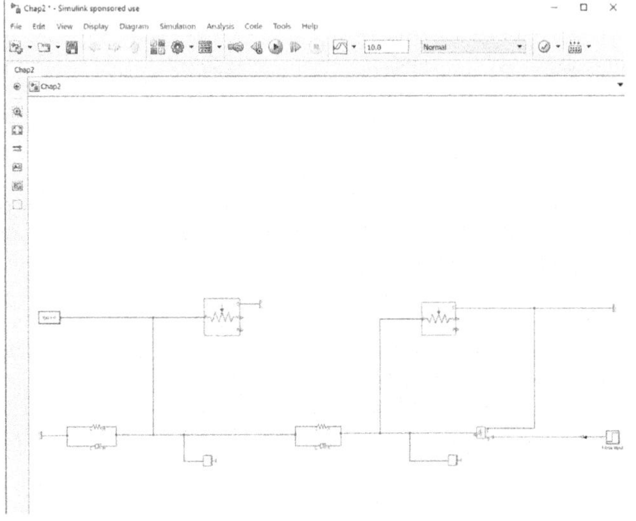

Figure 5.20 Step force input.

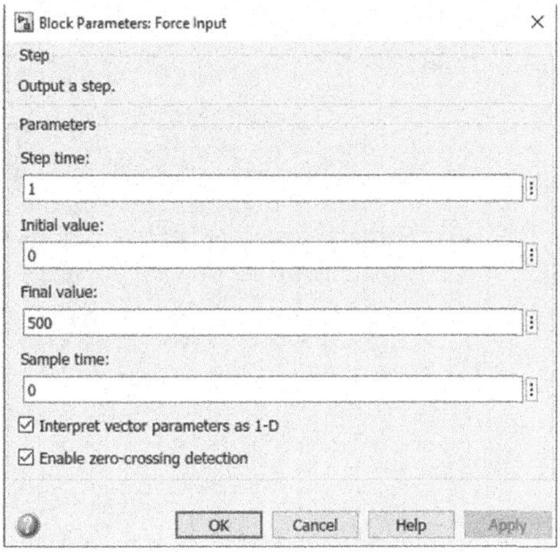

Block Parameters: Force Input
Step
Output a step.
Parameters
Step time:
1
Initial value:
0
Final value:
500
Sample time:
0
☑ Interpret vector parameters as 1-D
☑ Enable zero-crossing detection
OK Cancel Help Apply

Figure 5.21 Step signal parameters.

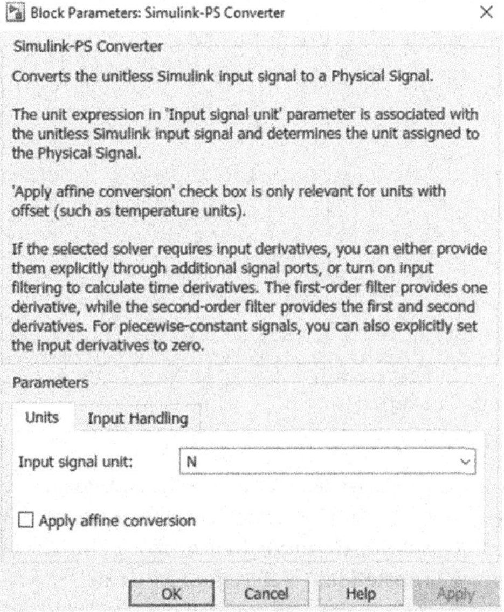

Figure 5.22 Simulink-PS converter parameters.

29. Connect the **From** blocks to the **Scope,** one to each port of the scope. Rename the **From** flags from top to bottom as v1, p1, v2, and p2 as shown in Fig. 5.31. Each **From** flag corresponds to a **Goto** flag in our model depending on the name. They are used to reduce the amount of wiring in our model.
30. Right click on the wire between the **From** flag named as **v1** to the scope and click on Properties. Change the Signal name to Velocity of spring mass damper combo 1. Repeat the same for p1, v2, and p2 and give the Signal name as Position of spring mass damper

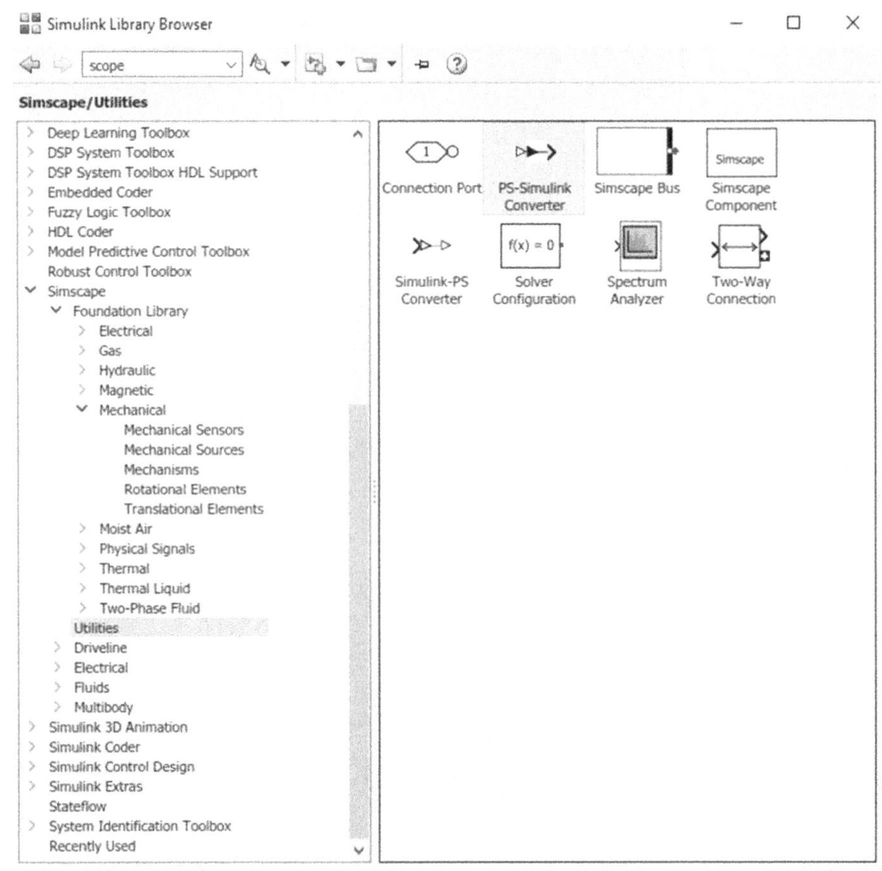

Figure 5.23 PS-Simulink Converter.

combo 1, Velocity of spring mass damper combo 2, and Position of spring mass damper combo 2, respectively. The signal names are also shown in Fig. 5.31.

31. Select the first **Ideal Translational Motion Sensor** block also including the **C** port connection to the **Mechanical Translational Reference** and create a subsystem for that selection. Rename the inport of the subsystem as **R** and the outports as v1 and p1.
32. Repeat step 31 for the second **Ideal Translational Motion Sensor** but do not include the **C** port connection in your subsystem. The outports should be v2, p2, and C. The subsystems in this step and the previous step are shown in Fig. 5.32.
33. With the model created for the double mass spring system, we can now simulate the model. Press the play button and see the position and velocity of both masses in the scope behave with a 500 N step input as shown in Figs. 5.33–5.36.

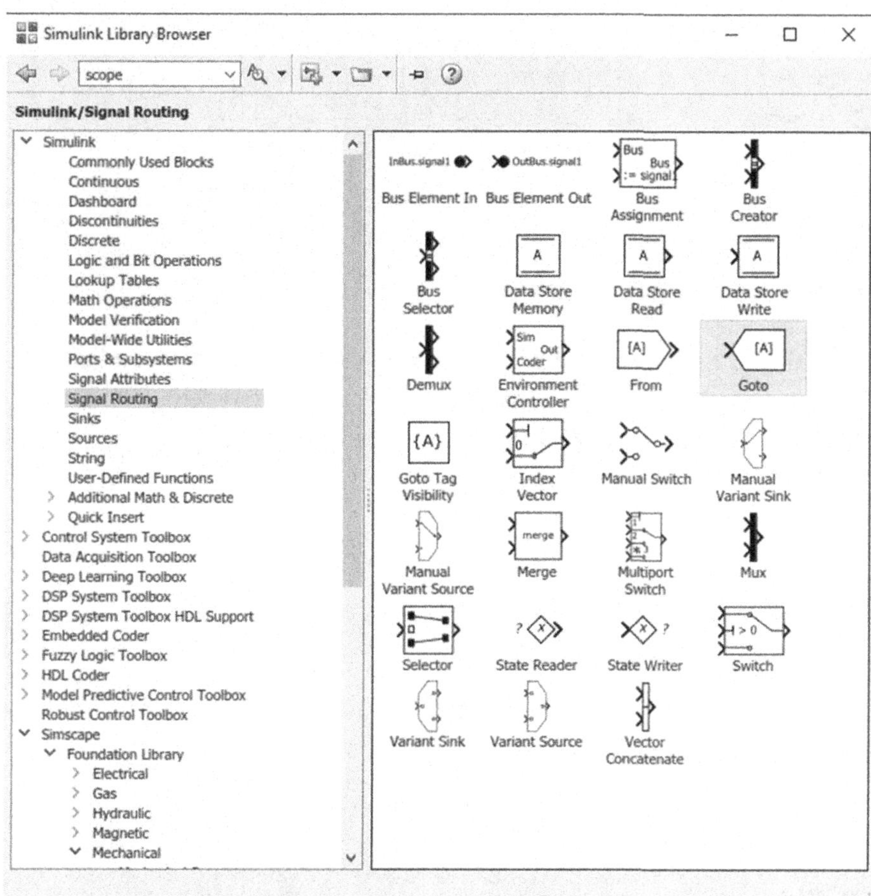

Figure 5.24 Goto library block.

5.3.6 Simulation of a failed system

34. This section shall introduce a failed system into the model and emulate the response of the failed system with the same input force. Select the entire model except the **Solver Configuration, Mechanical Translational Reference** block, and the scope and create a subsystem for it. Rename it as **Digital Twin.** Then copy—paste that subsystem and rename that new subsystem as **Physical Asset.** This new subsystem shall act as the failed subsystem. Then add a new **Solver Configuration** and **Mechanical Translational Reference block** and connect it to the inputs of the **Physical Asset** subsystem. The changes can be seen in Fig. 5.37.

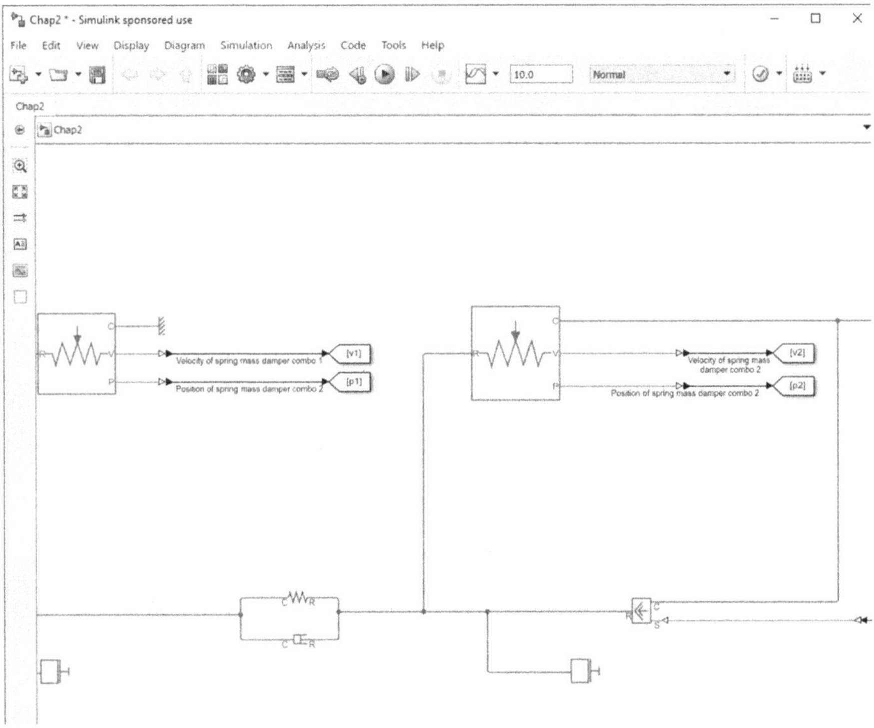

Figure 5.25 Goto flags for scope.

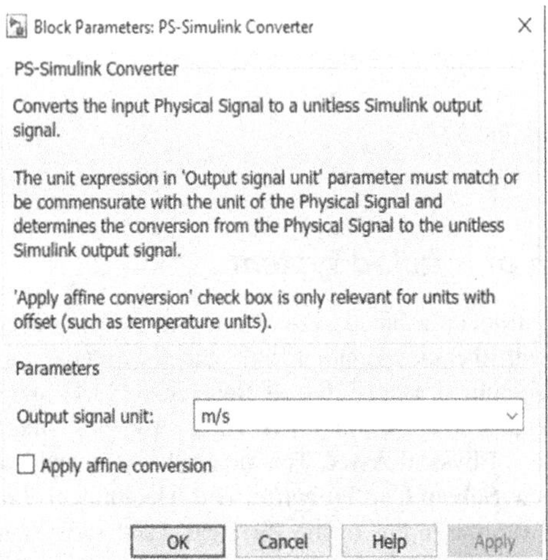

Figure 5.26 Velocity PS-Simulink converter parameters.

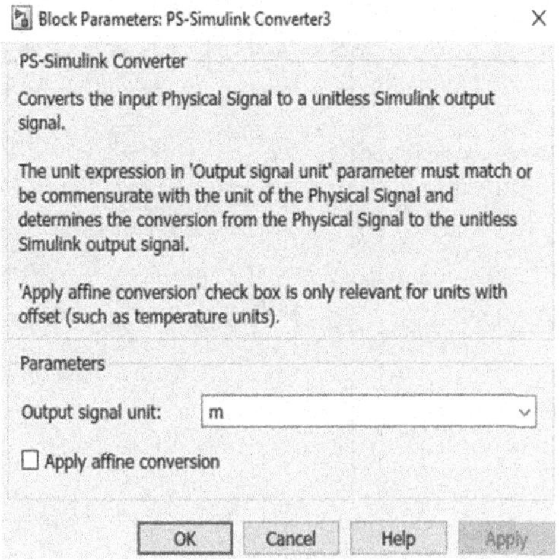

Block Parameters: PS-Simulink Converter3

PS-Simulink Converter

Converts the input Physical Signal to a unitless Simulink output signal.

The unit expression in 'Output signal unit' parameter must match or be commensurate with the unit of the Physical Signal and determines the conversion from the Physical Signal to the unitless Simulink output signal.

'Apply affine conversion' check box is only relevant for units with offset (such as temperature units).

Parameters

Output signal unit: m

Apply affine conversion

OK Cancel Help Apply

Figure 5.27 Position PS-Simulink converter parameters.

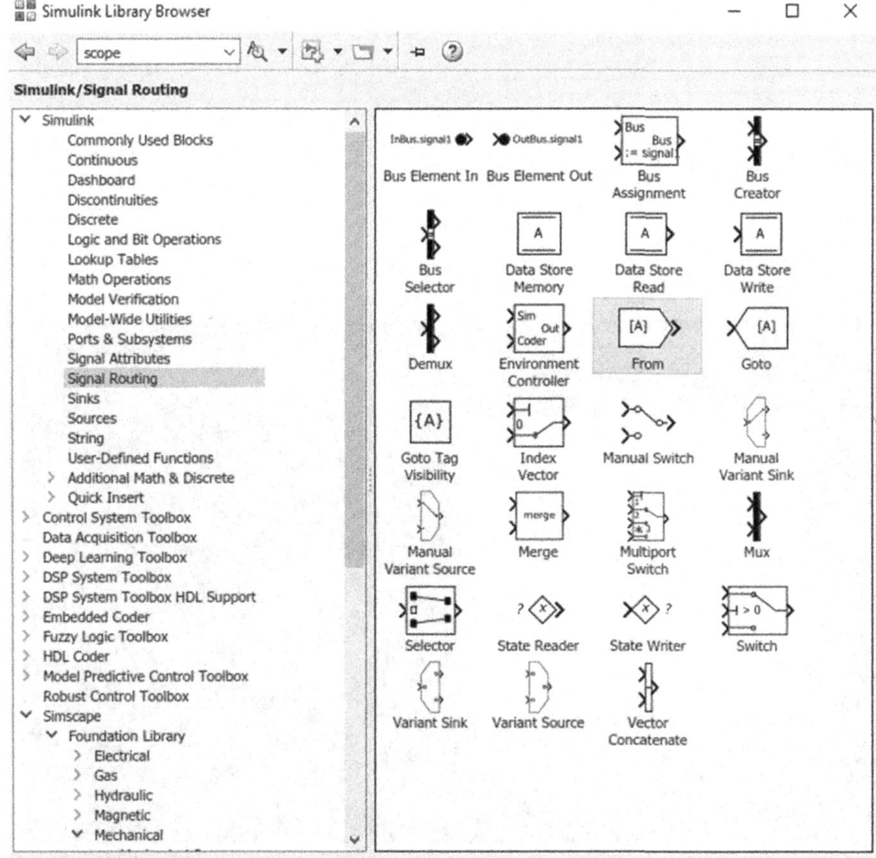

Figure 5.28 From flag.

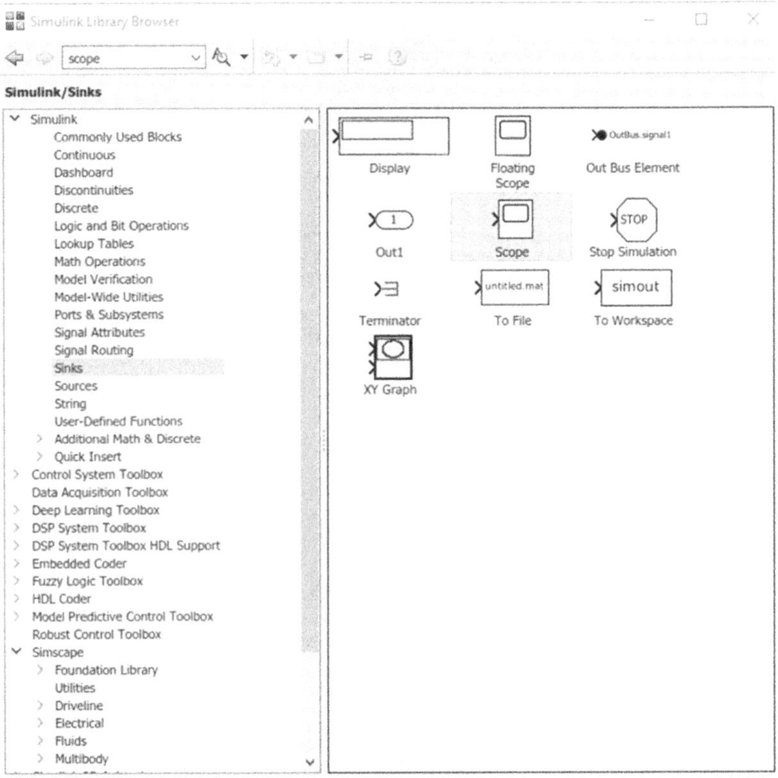

Figure 5.29 Scope block.

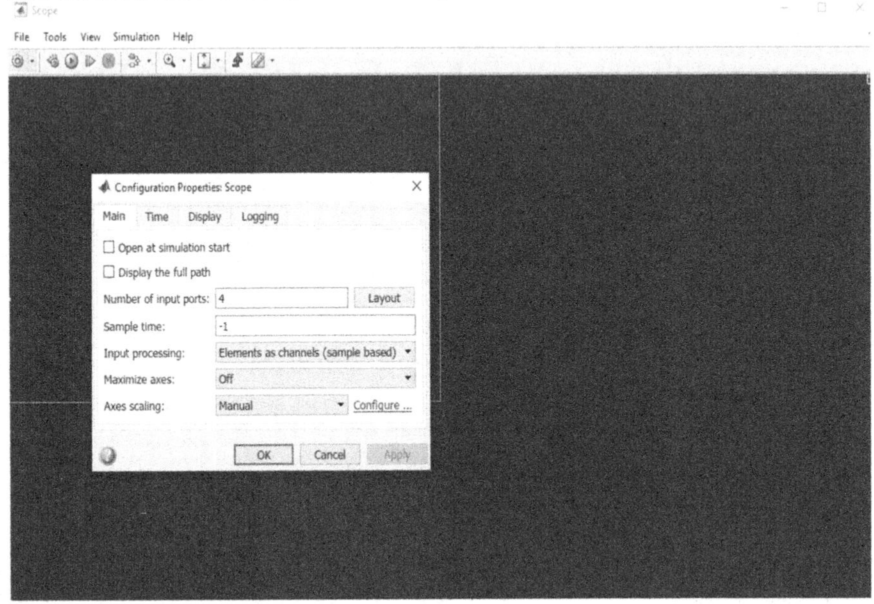

Figure 5.30 Scope number of inputs.

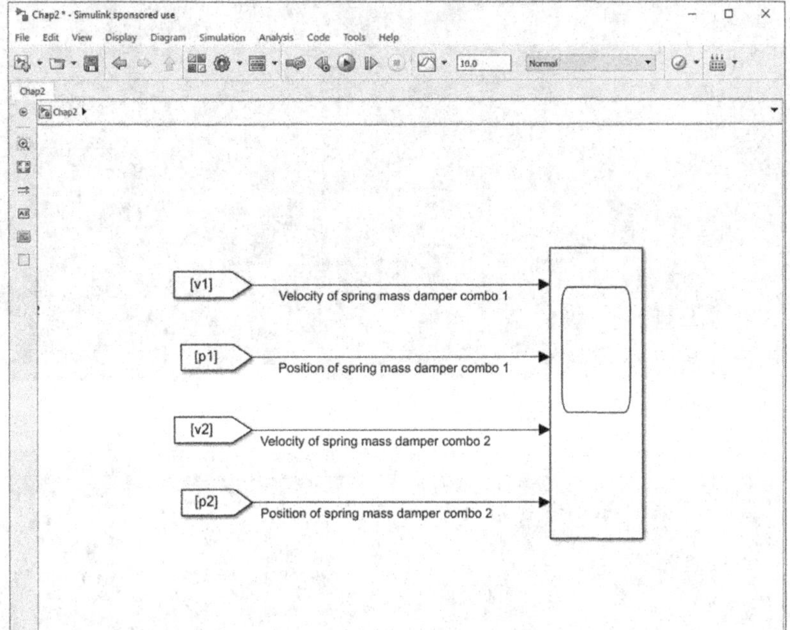

Figure 5.31 From flag to Scope.

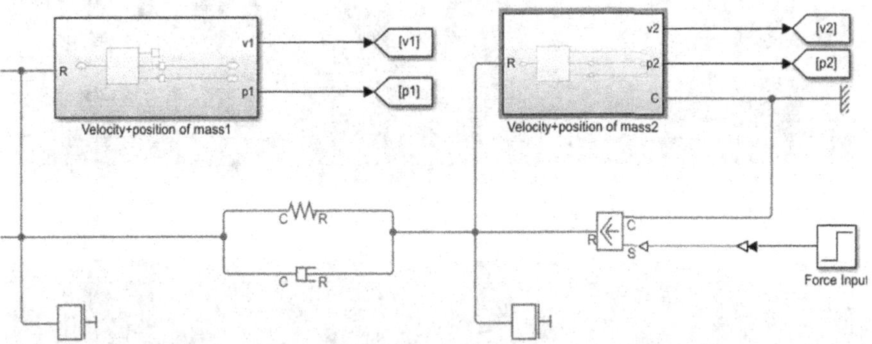

Figure 5.32 Subsystems for the first and second Ideal Translational Motion Sensor blocks.

35. Rename the **GoTo** flags inside the **Physical Asset** to add **real** before the respective signal name and then place **From** flags of the same names connected to **Mux** blocks, which are then connected to scope. The changes can be seen in Fig. 5.38.

36. Go inside the **Physical Asset** subsystem and change the spring constant of the first **Translational spring** from 400 to 100 N/m. This shall emulate a deformed spring. Run the simulation and observe the position and velocity results in the scope shown in Fig. 5.39. As seen in the scope, with a deformed spring, the impact of the oscillations are far greater.

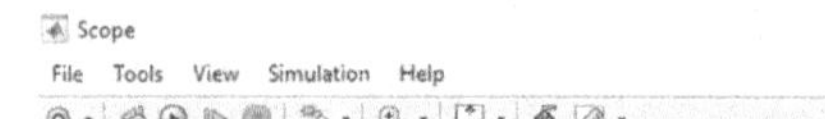

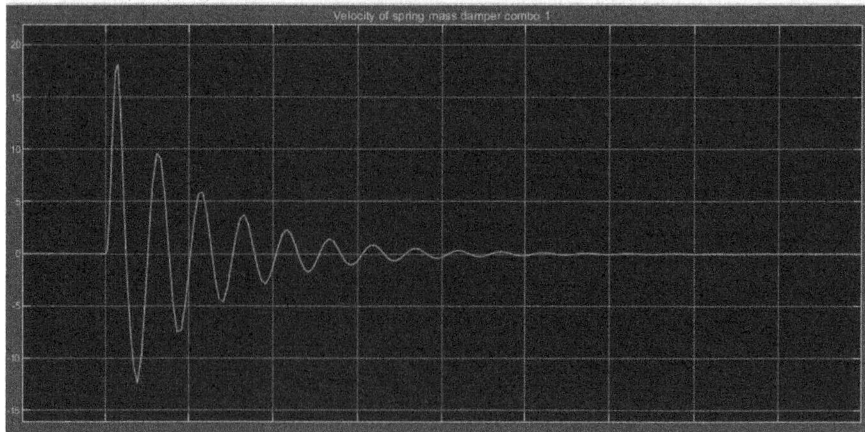

Figure 5.33 Velocity of first mass.

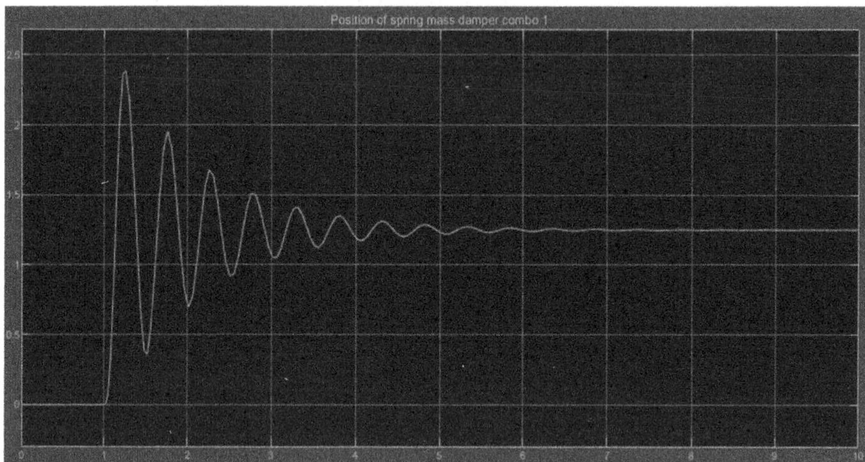

Figure 5.34 Position of first mass.

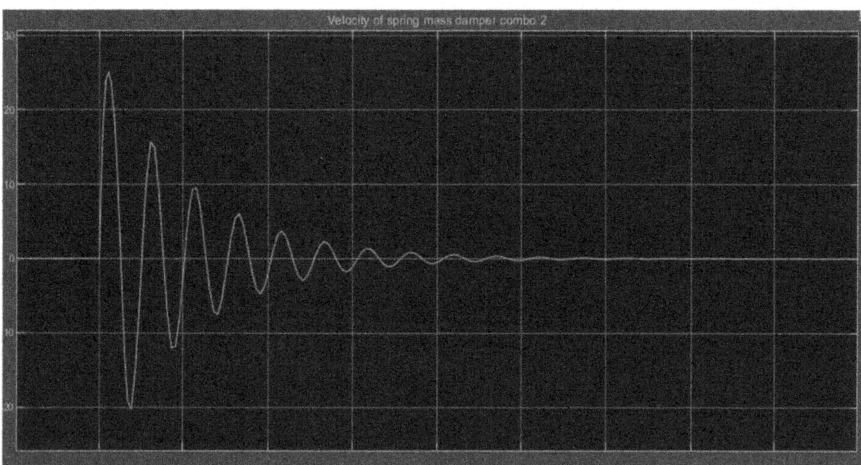

Figure 5.35 Velocity of second mass.

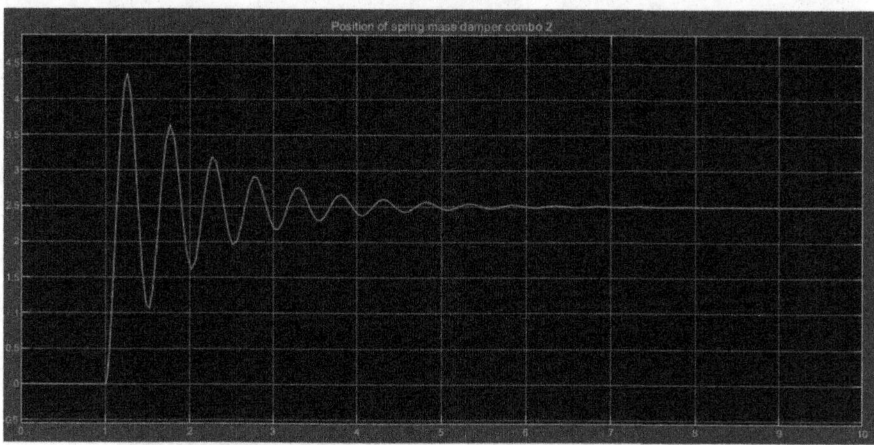

Figure 5.36 Position of second mass.

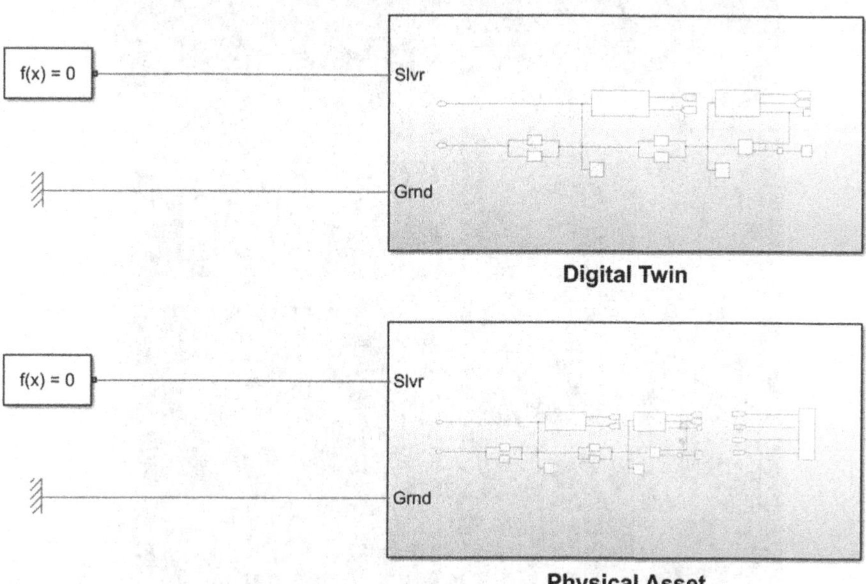

Figure 5.37 Digital Twin and Physical Asset subsystems.

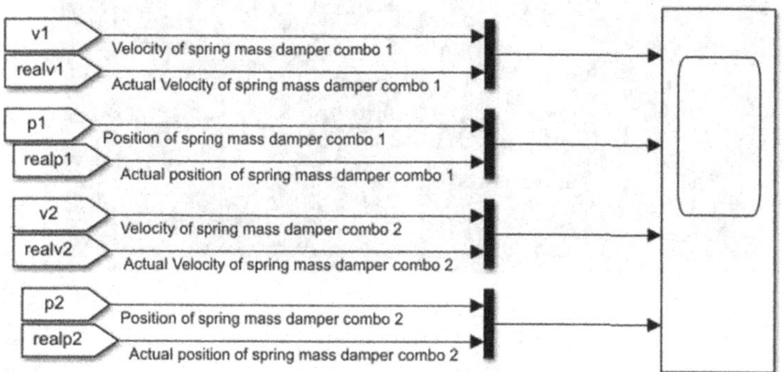

Figure 5.38 New scope connections for physical asset.

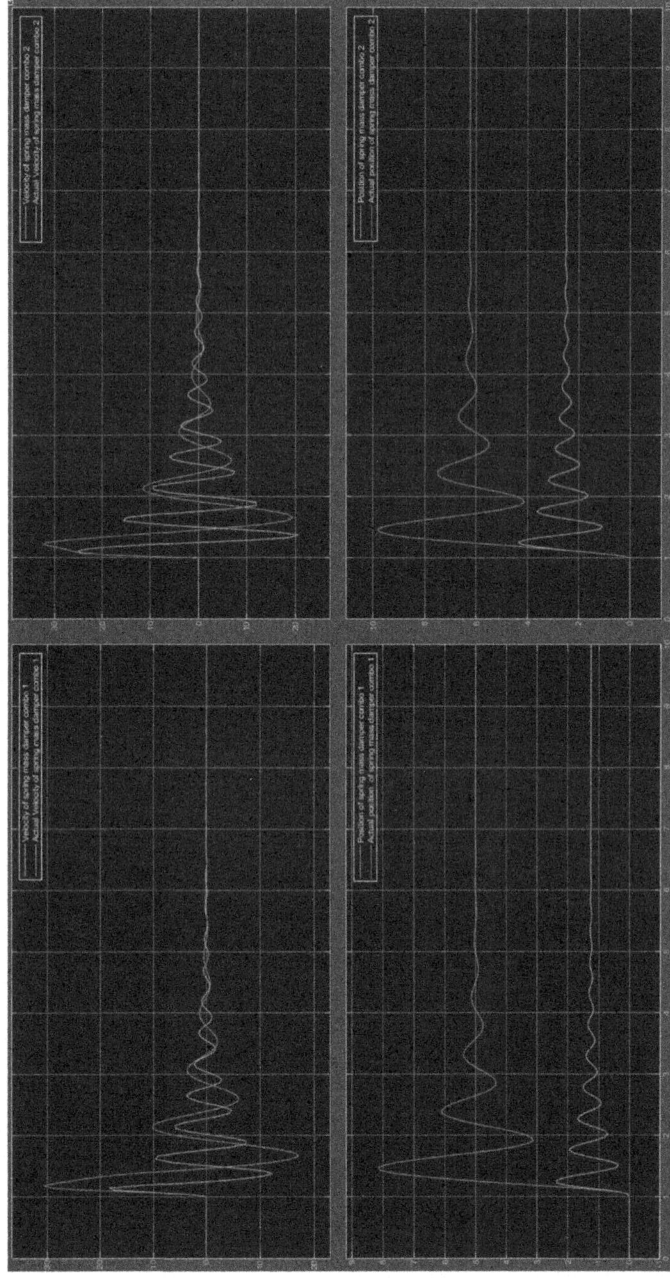

Figure 5.39 Results shown for physical asset (blue *[gray in print version]*) and digital twin (yellow *[white in print version]*).

5.4 Application problem

5.4.1 Application problem 1

Given the double spring mass damper system in Fig. 5.1, instead of using a step unit as the force input, use an impulse input. The magnitude should be 500 N of duration 10 ms.

Hint: There is no impulse input library block; you would need to get creative with step input blocks.

Download the material for the chapter from MATLAB® Central. The final Simulink® is provided in **Application_Problem_2masspringdamper** folder. The Simulink® models are saved under **Application_Problem_1.slx**.

5.4.2 Application problem 2

In this chapter we utilized Simscape library blocks from the **Simscape>Foundation Library>Mechanical** library. This time we will be visualizing the response of the system in the 3D space (**Simscape>Multibody**). Simulate only one mass spring damper combination. Once done, add a failed system with a deformed spring as previously shown in Section 5.3.6. Below are some tips to get started:

(a) The initial conditions for a 3D space is the same we used in the robotic arm chapter.
(b) Think of the joint(s) you have to use with respect to the degrees of freedom you have.

Download the material for the chapter from MATLAB® central. The final Simulink® is provided in **Application_Problem_2masspringdamper** folder. The Simulink® models are saved under **Application_Problem_2.slx.**

Digital twin model creation of solar panels

6

6.1 Introduction

This chapter focuses on creating digital twin model for photovoltaic (PV) cells. The model can be used for Off-BD of the setup. The model takes into account two external factors, mainly the solar irradiance and surface temperature of the solar panel. The system is equipped with sensors to detect the various changes of these external factors and a microcontroller to interpret and log the data to be used as input on our Simscape model.

Fig. 6.1 shows the Off-BD steps that will be covered in this chapter. This includes failure modes for the asset, the block diagram, modeling of the digital twin, and a preliminary diagnostic algorithm. Edge device setup, connectivity, and deployment of the digital twin on the cloud are not covered.

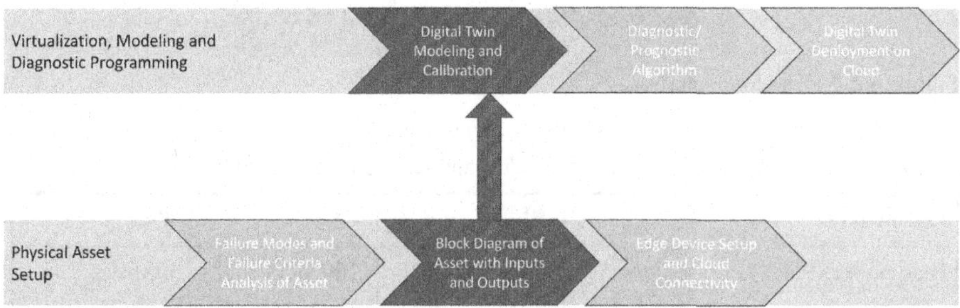

Figure 6.1 Off-BD steps covered in this chapter.

The chapter concludes with an Off-BD application to monitor 10 solar panels.

All the codes used in the chapter can be downloaded for free from MATLAB File Exchange. Follow the link below and search for the ISBN or title of this book:

https://www.mathworks.com/matlabcentral/fileexchange/.

Alternatively, the reader can also download the material and other resources from the dedicated website or contact the authors for further help:

https://www.practicalmpc.com/.

6.2 Photovoltaic hardware setup

The PV setup has three panels (PV_1, PV_2, and PV_3) (Fig. 6.2). PV_2 and PV_3 are connected in parallel. PV_1 has a power rating of 85 W and PV_2 and PV_3 are rated at 100 W of power each. Table 6.1 summarizes the electric ratings of the panels.

Digital Twin Development and Deployment on the Cloud. https://doi.org/10.1016/B978-0-12-821631-6.00006-2

Figure 6.2 PV outdoor setup.

Table 6.1 PV Electrical Ratings at 1000 W/m^2 and temperature of 25°C.

Parameters	PV_1	PV_2	PV_3
Peak power (Pmax)	85 W	100 W	100 W
Voltage (Vmp)	17 V	17 V	17 V
Current (Imp)	5 A	5.88 A	5.88 A
Open-circuit voltage (VOC)	21.5 V	21.5 V	21.5 V
Short-circuit current (ISC)	5.49 A	6.37 A	6.37 A
Minimum bypass diode	10 A	12 A	12 A
Maximum series fuse	9 A	10 A	10 A

The panels are instrumented with three identical surface temperature sensors. Furthermore, a solar radiation sensor is positioned on PV_1 to measure intensity of light (Fig. 6.3).

Fig. 6.3 shows the indoor setup of the system. The output of the solar panels on the roof is fed to the electrical panel, which is connected to the charge controller. The battery is then connected to the solar charge controller and inverter to power up AC loads such as the computers and several devices shown in the figure. The properties of the indoor setup are shown in Table 6.2. Fig. 6.4 shows the circuit schematic of the system.

Figure 6.3 PV indoor setup.

6.2.1 Block diagram of the solar panel system

Solar panel diagnostic is a multiinput and single-output problem (MIMO). The two main inputs are the solar irradiance and the surface temperature of the panel. The main output is power generation (Fig. 6.5).

6.2.2 Failure modes and diagnostics concept for the PV system

In this chapter, we will focus on failure modes that affect the power generation of the PV cells. These can include physical damage to the panel (like cracks), hardware internal components damage, dust coverage, persistent shadow, or blockage, etc.

Table 6.2 PV Electrical specifications of indoor setup.

Devices	Voltage	Current	Power
Solar charge controller	12 V DC	20 A	240 W
Battery bank	12 V DC	320 AH	3840 WH
Inverter	230 V AC	13 A	3000 W approximately

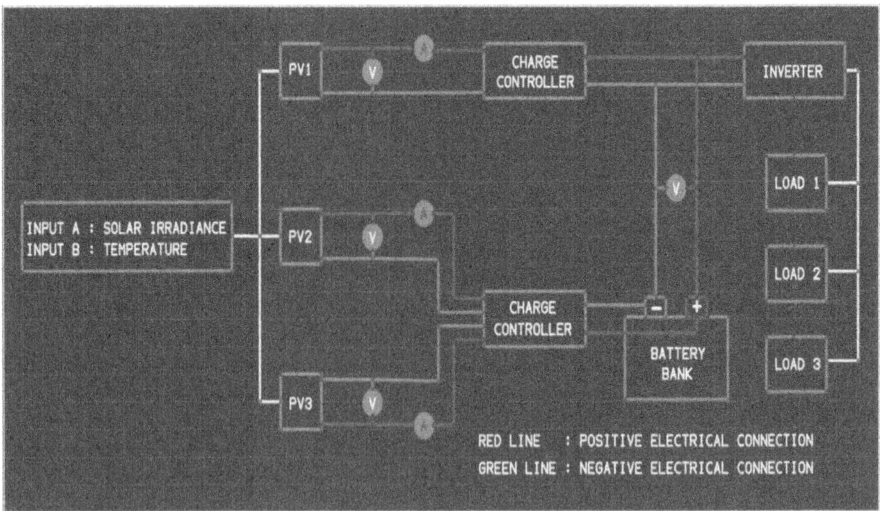

Figure 6.4 Circuit schematic of the PV experimental setup.

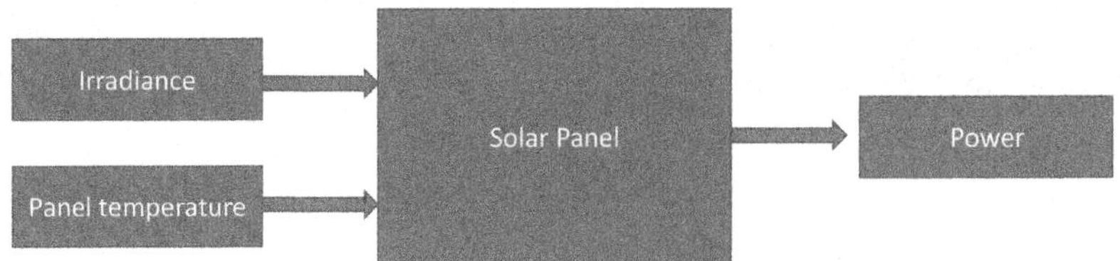

Figure 6.5 Block diagram of the solar panel system.

The measured power will be compared to the digital twin estimated power. If there is significant deviation, then an alarm is sent to the user (Fig. 6.6). Diagnostic conditions can include time of the day, level of irradiance, timer for length of failure, etc.

6.3 Experimental data collection for model creation

The data for the setup were recorded for 1 day. For the purpose of the simulation, we will use data for 6300 s. Solar irradiance, surface temperature of all three solar panels, voltage of PV_1, PV_2, and PV_3 (connected in parallel), and current of PV_1, PV_2, and PV_3 were all recorded at 4 s interval. Note that battery data (voltage, state of charge, power load) were not collected experimentally. They will be modeled in Simscape but we will not have data to validate the accuracy of the model.

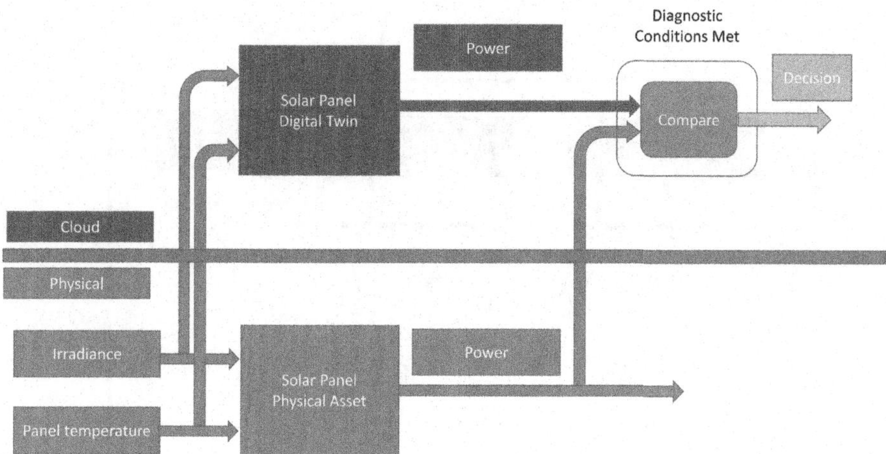

Figure 6.6 Off-BD diagnostic process for the solar panel system.

6.4 PV system simscape model

The main objective of this chapter is to use Simscape to create a complete model of the system, compare the simulated results to the actual results based on the data gathered using the sensors, and verify the accuracy of the created model. Since this chapter is mainly focused on the PVs side, the electrical loads to the system will be excluded from the model. Only the main inputs to the system, namely the solar irradiance and PV surface temperature, will be included.

Fig. 6.7 shows a high-level overview of the model. The recorded inputs, irradiance, and surface temperature are fed to the three PV panels. The outputs include the current and the voltage. These are displayed as outputs and compared to the experimental data.

In what follows, we will explain the main components used in the PV model of Fig. 6.7, how to configure them, and how to wire them.

6.5 Solar cell modeling of the PV system

The main component is the solar cell (Fig. 6.8). It can be found in elec_library/Sources. Irradiance is a direct input to all the solar cells (Fig. 6.9). The positive terminal is an input, whereas the negative terminal is an output.

Fig. 6.10 shows the cell characteristics as compared to the default Simulink parameters. Furthermore, Table 6.3 shows the values we changed from default. The column "New Value" has the name we gave the parameter. Column "Parameter Value" shows the values we set in the mask of the PV_1 panel. This was done to allow ease of changing parameters of the PV_1 model. Instead of changing individual solar cell parameters, we were able to change the parameters from the mask for all the solar cells.

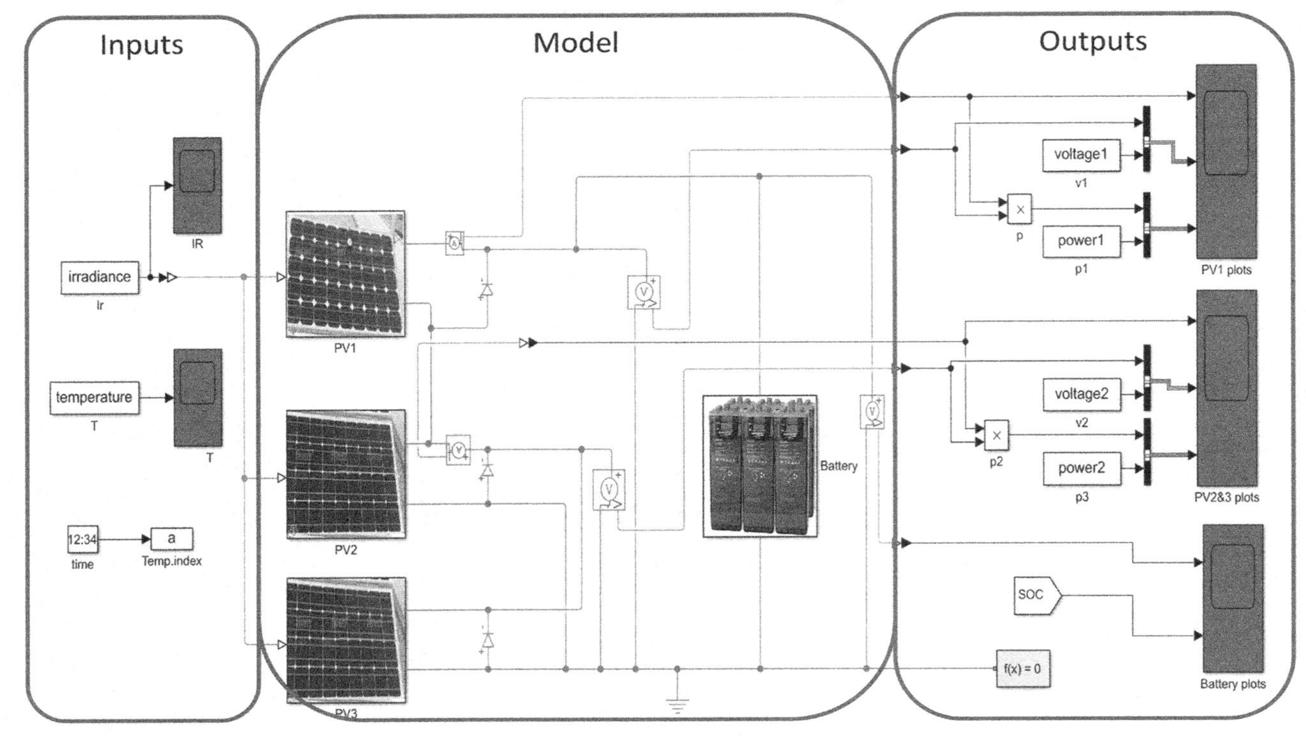

Figure 6.7 High-level overview of the model.

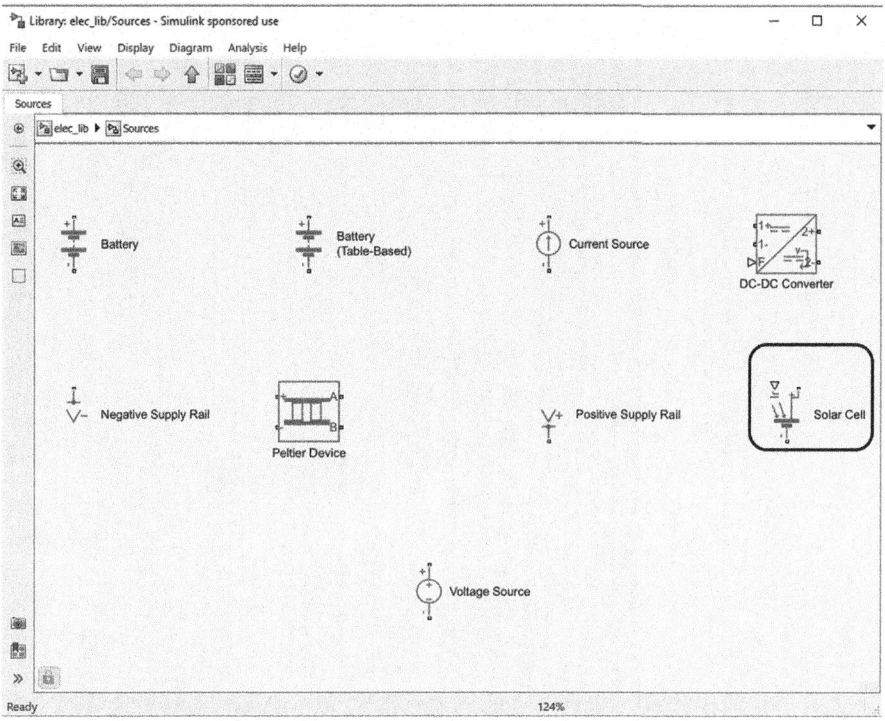

Figure 6.8 Solar cell.

Fig. 6.11 shows the mask for the PV_1 model. It will be further discussed later in this section. In particular, temperature value was set to a constant number, which is the average temperature of the solar cells recorded experimentally. For the configuration of the solar cell, we kept the number of cells to the default value of 1 (Fig. 6.12).

We set the surface temperature of the solar cell to T (Fig. 6.13). T is defined in the mask of the subsystem (Fig. 6.11). a.Data is the time index. T(a.Data) is the measured temperature at index a.Data.

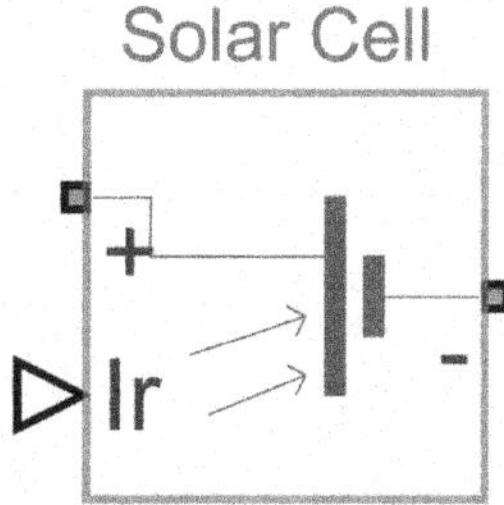

Figure 6.9 Solar cell with inputs and output.

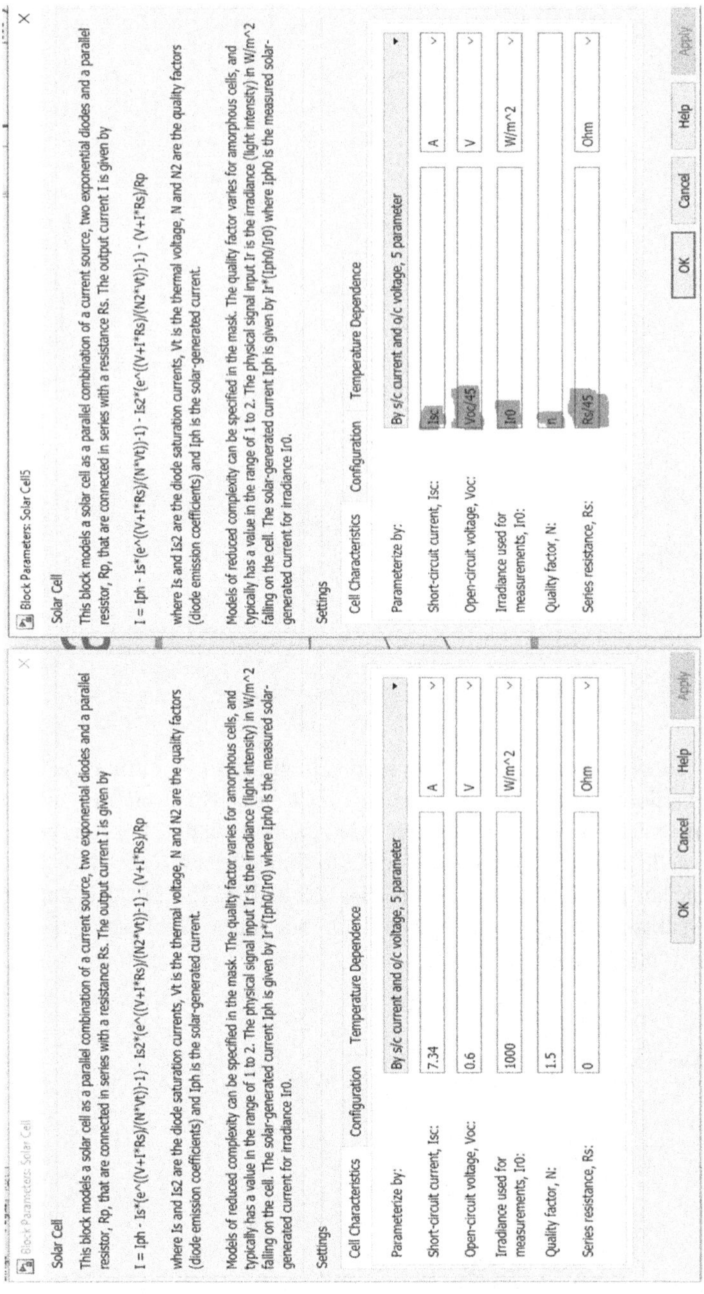

Figure 6.10 Cell characteristics of the default (left) and modified (right) solar cell.

Table 6.3 Solar cell characteristics for PV_1.

Parameter	Default	New value	Parameter value (from the mask)
Short-circuit current, Isc	7.34	Isc	5.49 for PV_1 (from Table 6.1)
Open-circuit voltage, Voc	0.6	Voc/45	21.5/45 (from Table 6.1)
Irradiance used for measurements, Ir0	1000	Ir0	Provided from experimental data as a time data
Quality factor, N	1.5	n	1.5
Series resistance, Rs	0	Rs/45	0

6.6 Solar cell modeling of the PV subsystem

We have three solar panels, PV_1, PV_2, and PV_3. For our PV_1 panel, it has $5 \times 9 = 45$ solar cells (Fig. 6.14).

We will need to have 45 solar cells configured per the previous section and connected in series (Fig. 6.15).

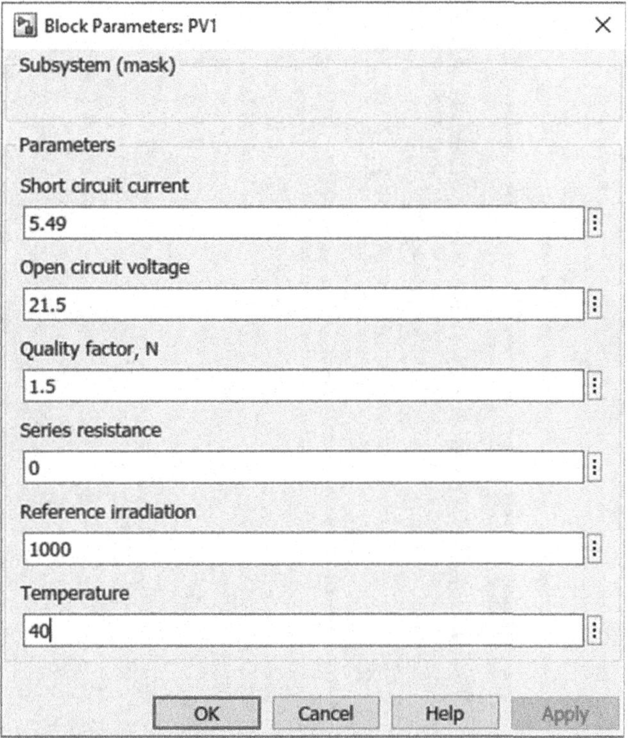

Figure 6.11 PV_1 subsystem mask.

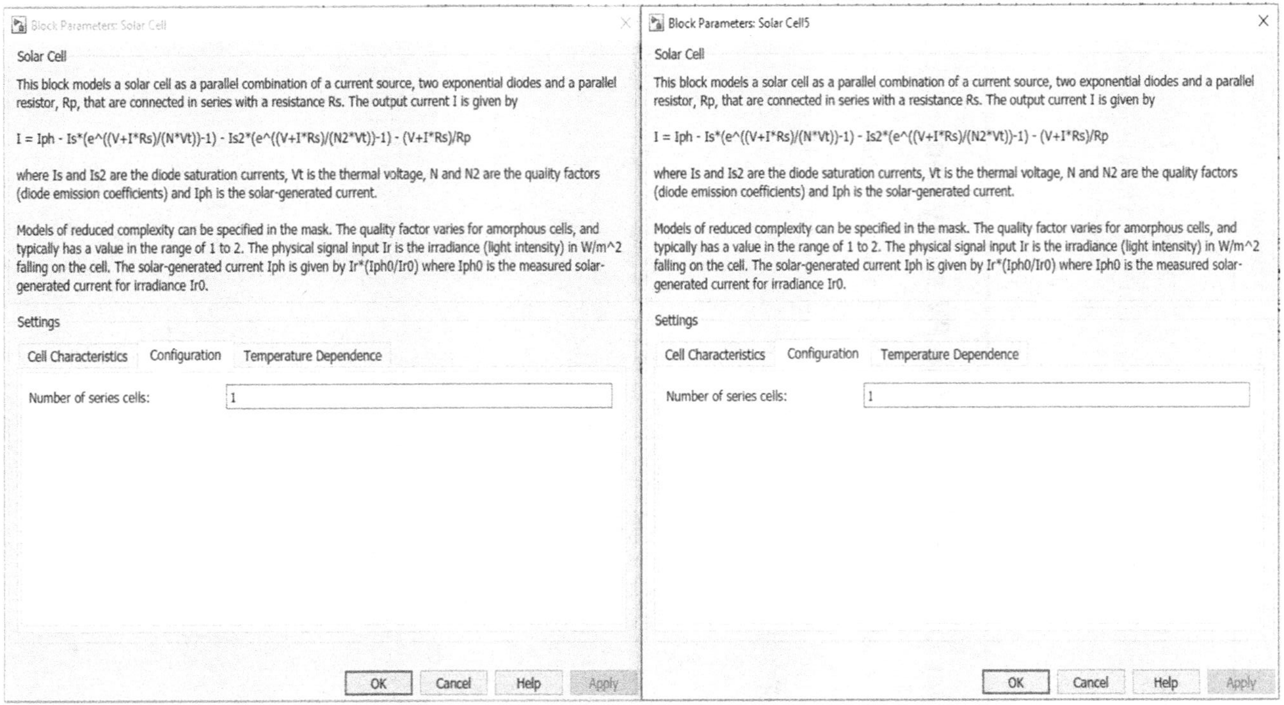

Figure 6.12 Configuration of the default (left) and modified (right) solar cell.

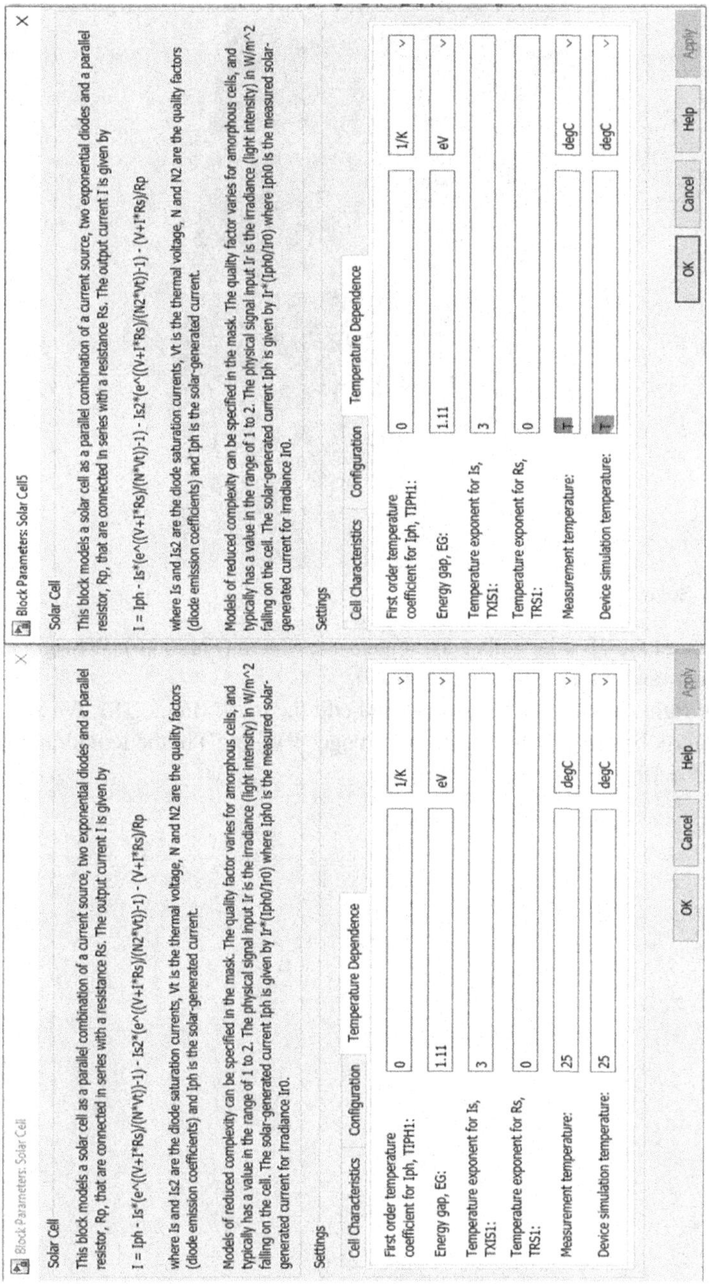

Figure 6.13 Temperature dependence of the default (left) and modified (right) solar cell.

Figure 6.14 Solar cells of the system.

We select all the 45 solar cells and create a subsystem (Fig. 6.16). We right click on the subsystem and create a mask (Fig. 6.17).

We then right click on the subsystem and edit the mask (Fig. 6.18). We add an image to the mask by writing the command: image('PV1.jpg') in the icon drawing commands (Fig. 6.19).

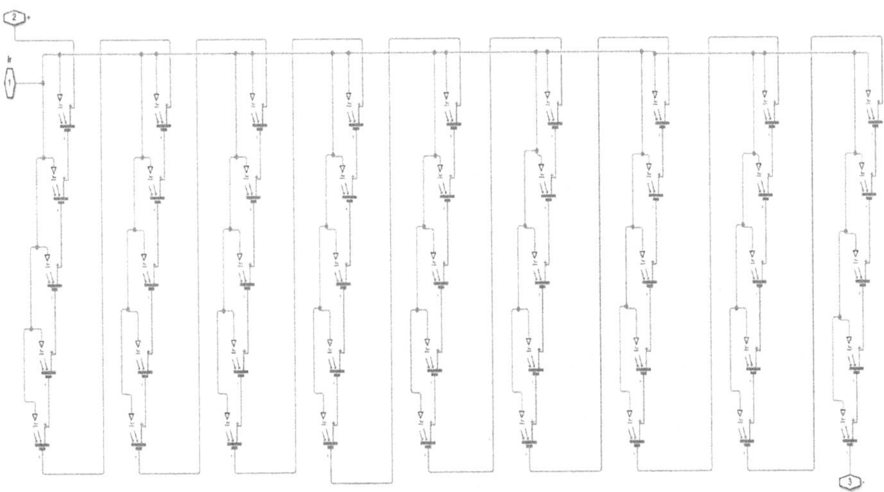

Figure 6.15 45 solar cells for PV_1 model.

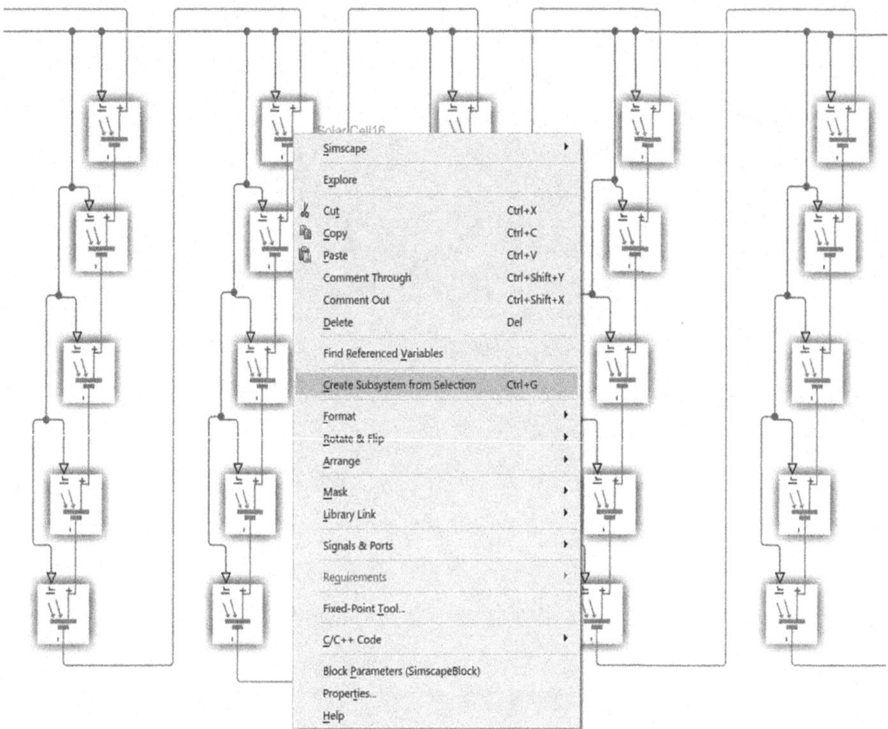

Figure 6.16 Creating a subsystem for PV_1.

We modify the mask parameters to include the parameters in Table 6.3 (Isc, Voc, Ir0, n, Rs) and the temperature (Fig. 6.20). We individually edit the values of the added parameters as per Table 6.3.

We repeat the same process for PV_2 and PV_3. They each have $4 \times 9 = 36$ solar cells. Fig. 6.21 shows the final model. Note that we added voltage and current sensors to extract the measurements from the Simscape model (Fig. 6.22). We also added diodes in the circuit per Fig. 6.23.

Similar to all Simscape models, we add a Solver Configuration block (Fig. 6.24). Note that the battery was modeled but was not tuned. This was included in case the reader wants to further diagnose the battery.

The final model can be downloaded from MATLAB Central. It can be found in **Chapter_6/ Solar_Cell_Modeling_of_the_PV_Subsystem**.

Once downloaded, run **PV_Model.slx** and the results will be plotted at the end of the simulation.

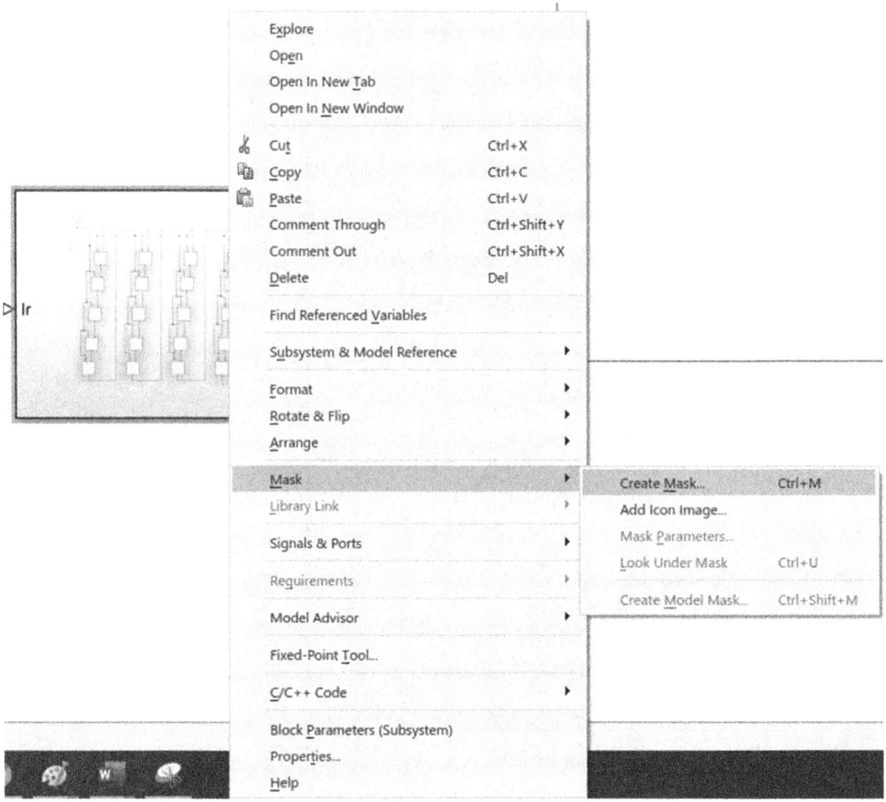

Figure 6.17 Creating a mask.

6.7 Simulation results

One of the challenges in modeling PV cells is factoring solar cell surface temperature. High temperatures can significantly deteriorate the efficiency of PV cells ([1] and [2]).Unfortunately, having temperature as a time-varying input is not possible in the current format of the solar cell block in Simscape. In [3], Khaled and Bibin proposed to use a custom block to model the impact of surface temperature based on the work of Aljoab et. al ([1] and [2]). Herein, we propose to use Simscape blocks and correct for the impact of temperature.

We set the surface temperature (last parameter of Fig. 6.11) of PV_1, PV_2, and PV_3 to 30°C and 50°C to see sensitivity of the model to temperature. As can be seen from Fig. 6.25, the results are virtually identical.

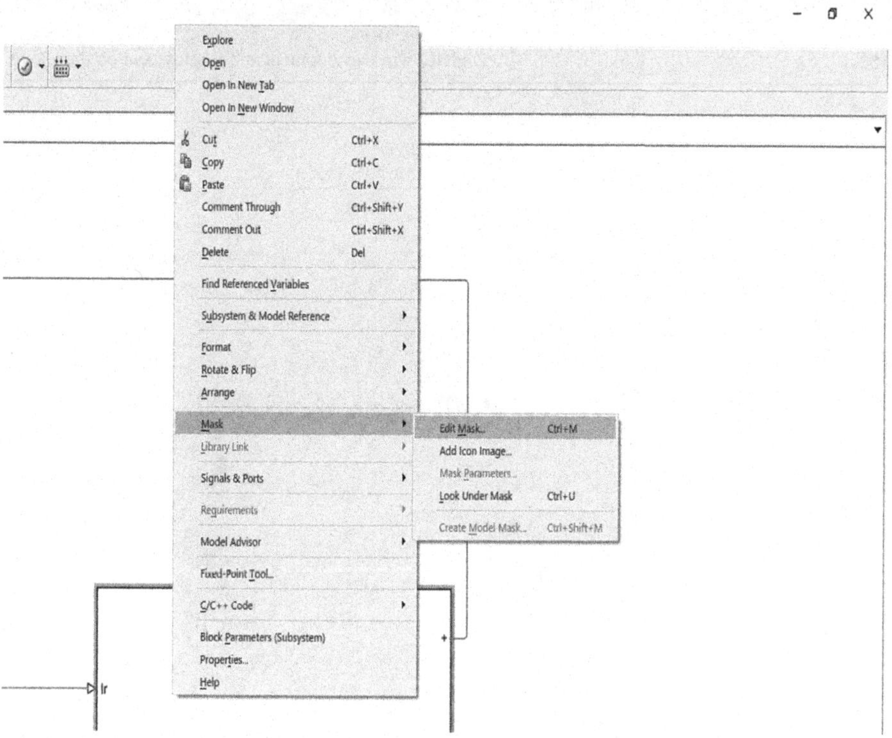

Figure 6.18 Editing the mask.

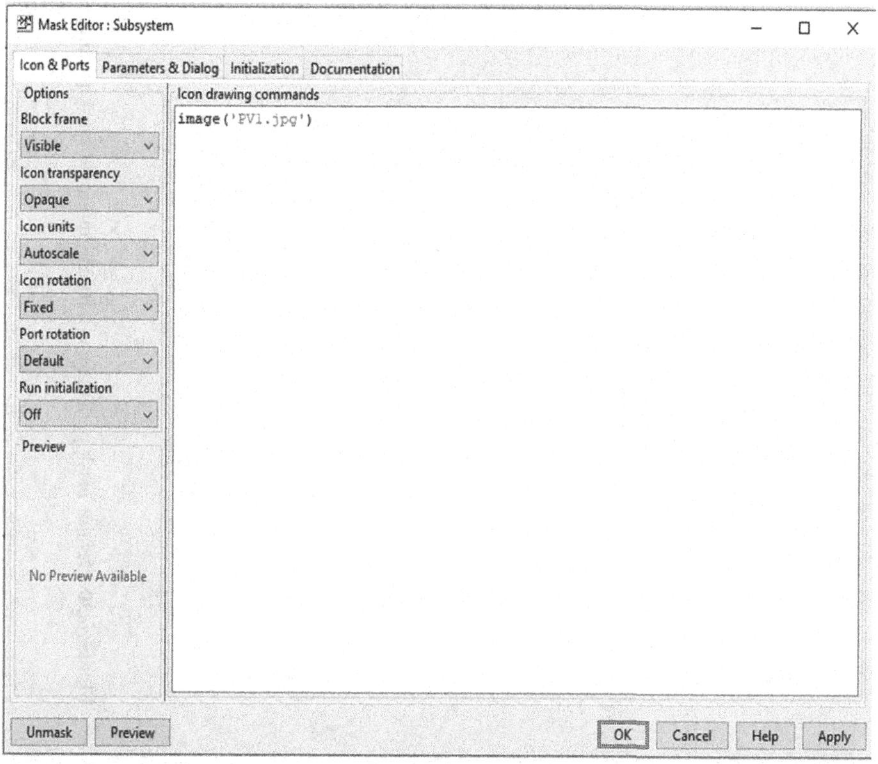

Figure 6.19 Editing the Icon and Ports of the mask.

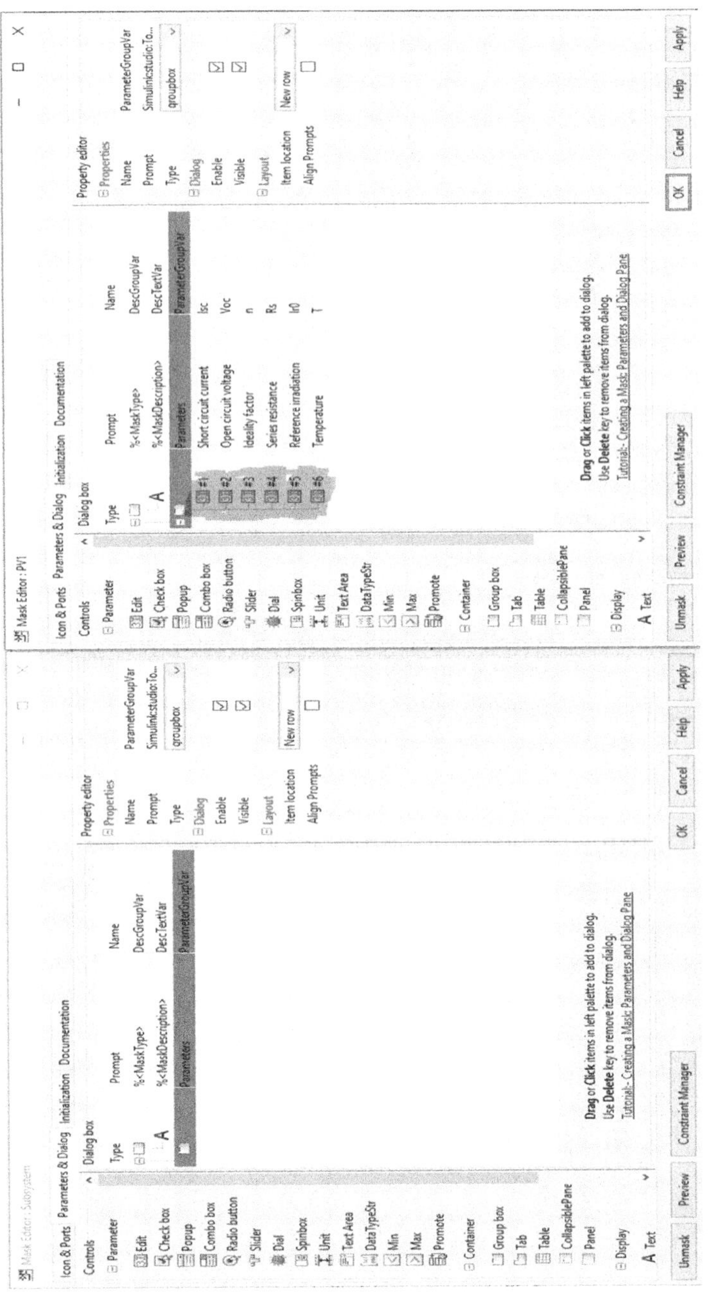

Figure 6.20 Default mask parameters (left) and modified mask parameters.

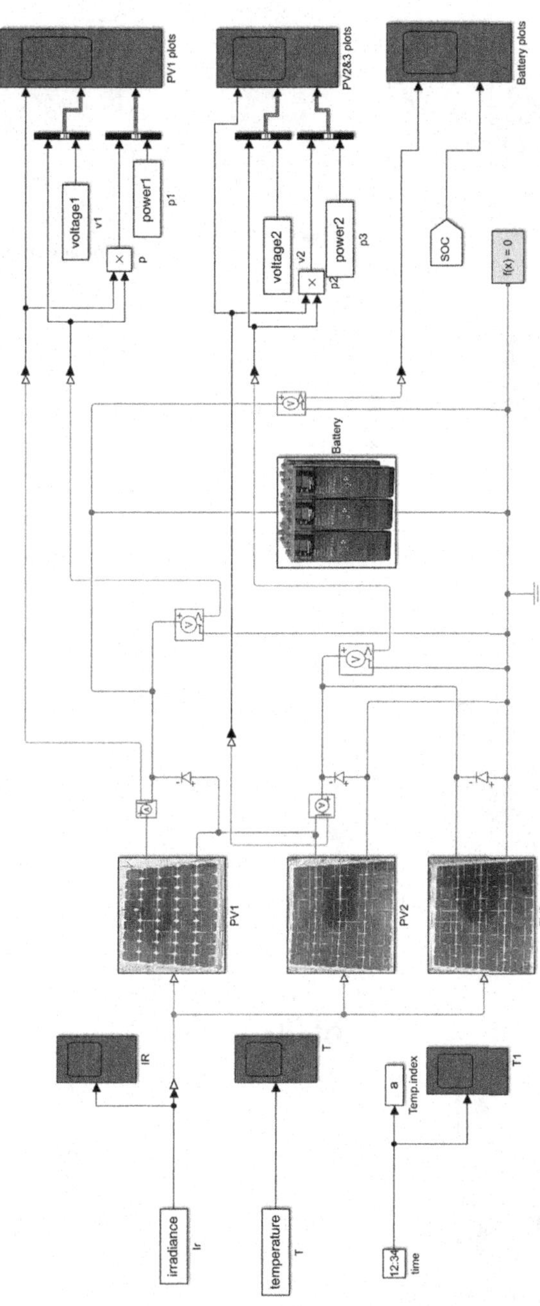

Figure 6.21 Final model.

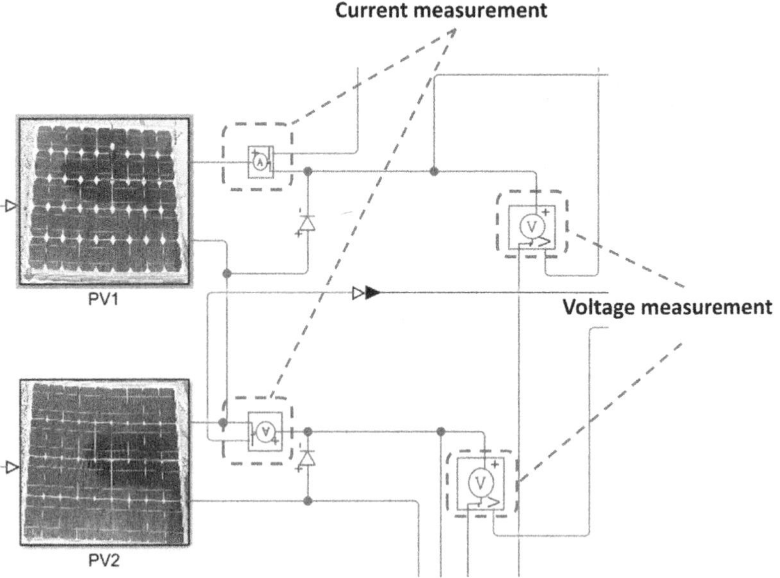

Figure 6.22 Current and voltage measurements in the system.

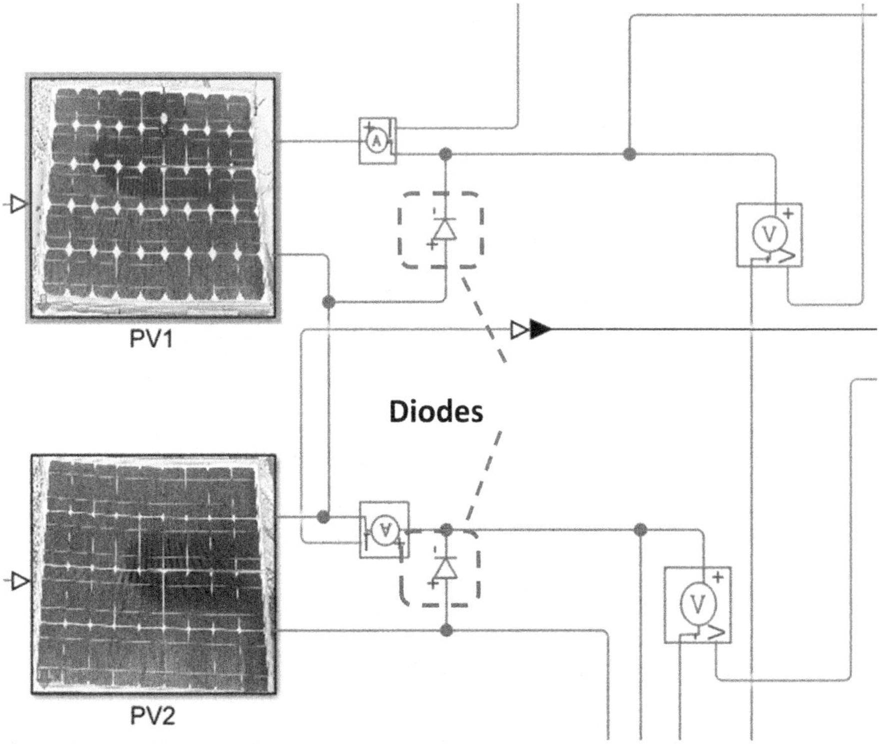

Figure 6.23 Diodes added to the system.

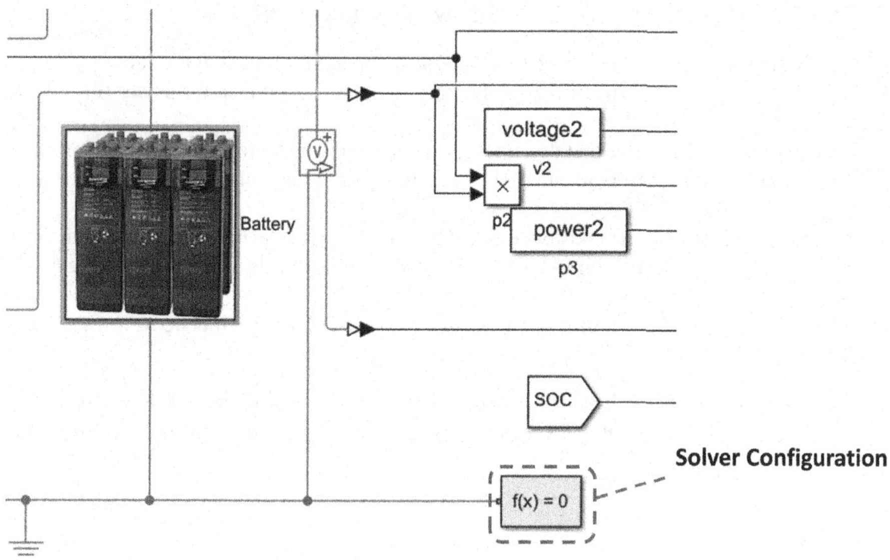

Figure 6.24 Solver Configuration block.

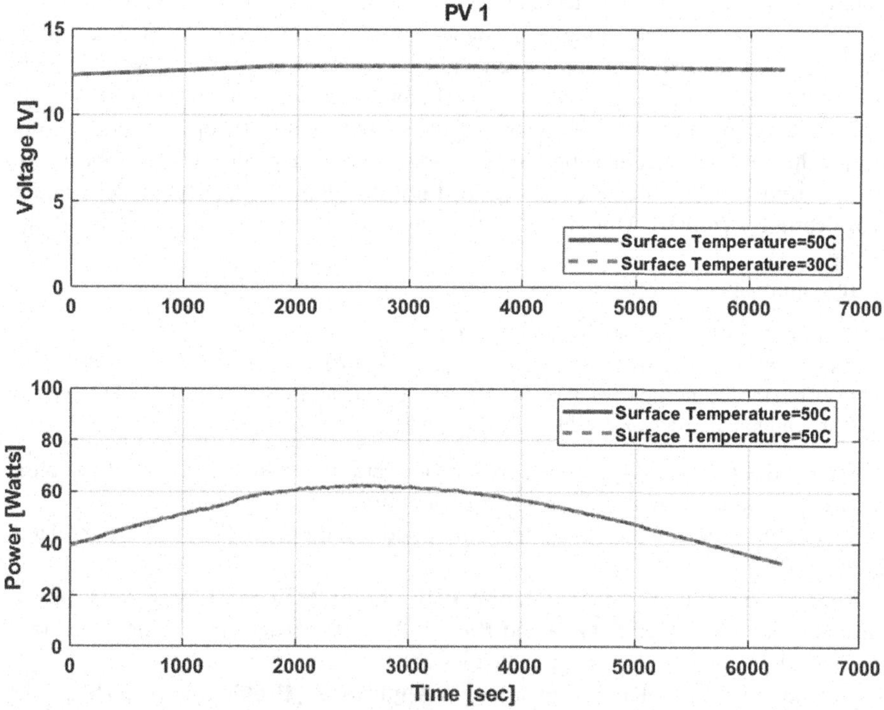

Figure 6.25 Simulation results at two surface temperatures.

To work around this problem, we follow the following steps:

1. We set the solar panel temperature to 30C. Figs. 6.26 and 6.27 show the results.
2. We compute the error between the simulation and experimental data for power (Figs. 6.28 and 6.29).
3. We fit a second-order curve that describes the power simulation error as function of solar cell surface temperature (Figs. 6.30 and 6.31) using the curve fitting tool found in the figure menu under Tools>Basic Fitting
4. We add this an offset term to power output of the simulation model by utilizing a user-defined function (Fig. 6.32) for PV_1 and PV_2 and PV_3 models. Note that the equations from the fit are used.
5. We rerun the simulation and the power results with offset look much better (Figs. 6.33 and 6.34).

The updated model with offset is saved in **Chapter_6/Simulation_Results/**

Once downloaded, run **PV_Model_Offset.slx** and the results will be plotted at the end of the simulation.

6.8 Application problem

A customer has a set of 10 identical solar panels. They are installed in a remote location. Build a diagnostic logic that can be used as along with the digital twin to detect the failure of the panels to generate adequate power. The only sensory data that will be used are irradiance. Temperature dependency is neglected due to the geographic location of the solar panels (there is not a lot of ambient temperature variations).

The Simscape model for 10 healthy panels, along with the input/output sensory data of the physical system, is provided for you in the same model. They can be downloaded from MATLAB central in **Application_Problem** folder. The model name is **Application_Problem_Model**. Figs. 6.35 and 6.36 describe the contents of the model.

Requirements:

1. Generate off-BD decision at least once a day.
2. Assume that the data are sent to the cloud once every 10 s.
3. Only run the diagnostic when there is acceptable day light (irradiance level should be greater than a minimum irradiance value for the diagnostic to run).

Hint 1: Start by quantifying the performance variation of the panels and then define failure criteria to account for part to part variability.

Hint 2: Assume that the data exist in the cloud and the diagnostic will be simulated in the cloud.

Hint 3: Compare the signals (**Power_PV1_digital_twin** and **Power_PV1_physical**), (**Power_PV2_digital_twin** and **Power_PV2_physical**), etc. Perform the comparison when conditions of the diagnostics are satisfied.

Solution of the problem can be found in **Application_Problem** folder. The model name is **Application_Problem_Solution**. Diagnostic solution is not unique.

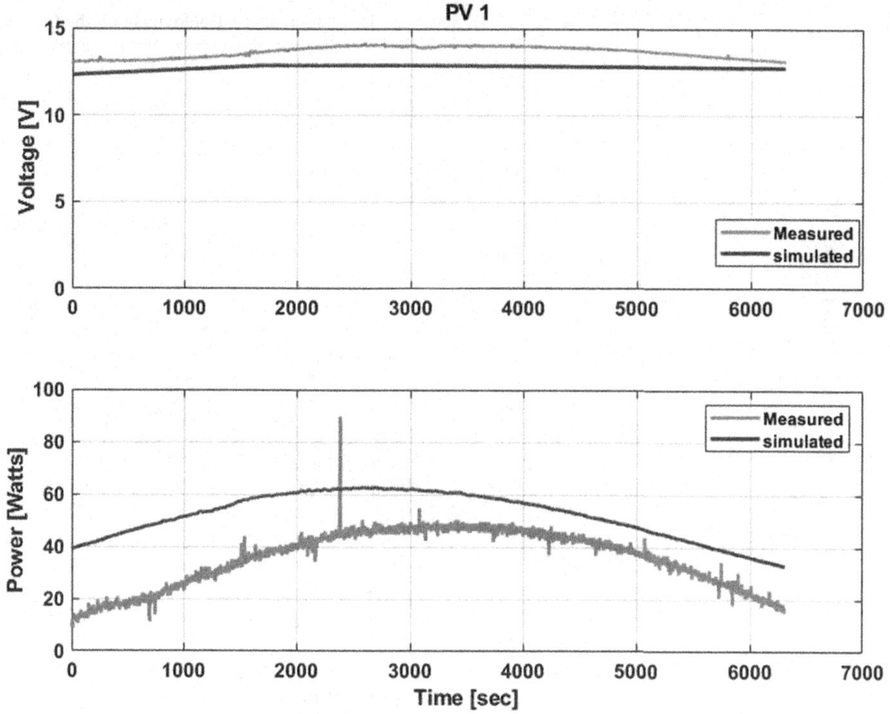

Figure 6.26 PV_1 simulation and experimental results.

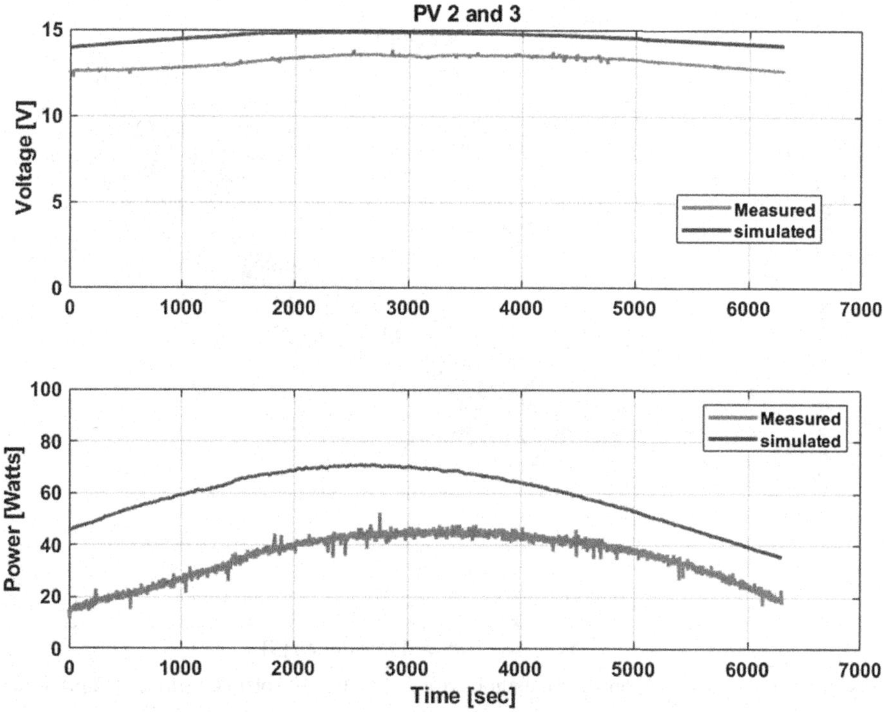

Figure 6.27 PV_2 and PV_3 simulation and experimental results.

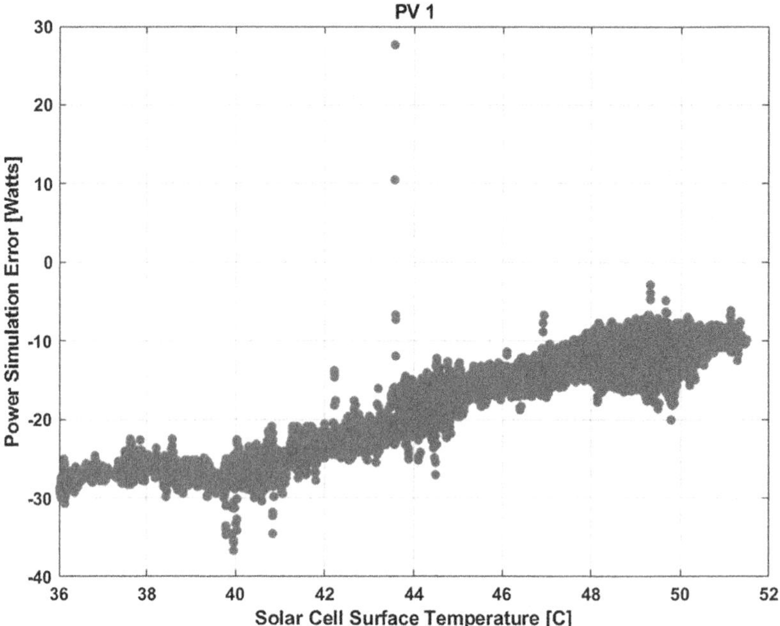

Figure 6.28 PV_1 power simulation error as function of solar cell surface temperature.

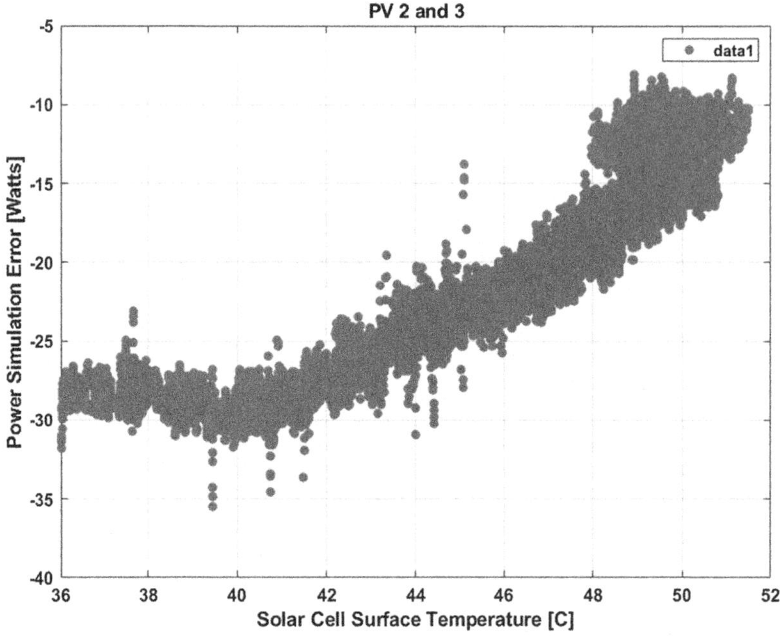

Figure 6.29 PV_2 and PV_3 power simulation error as function of solar cell surface temperature.

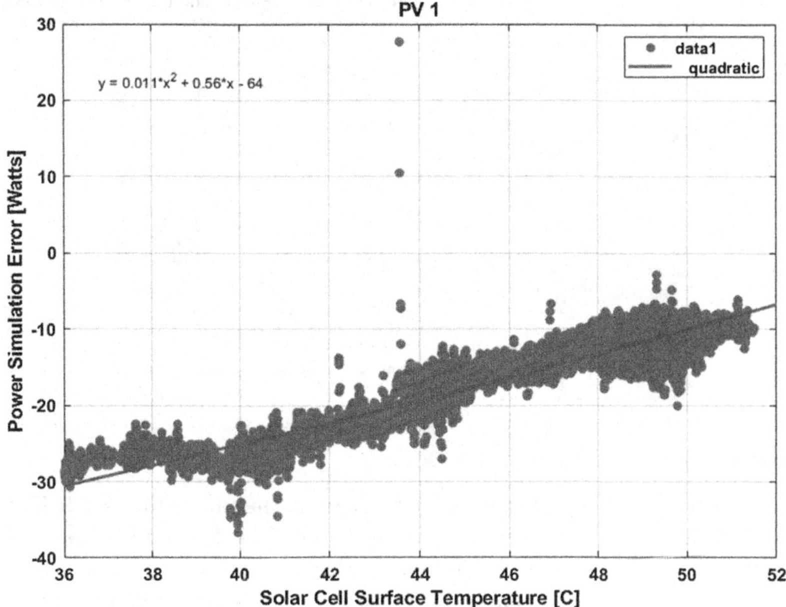

Figure 6.30 Fitting a curve for the PV_1 power simulation error as function of solar cell surface temperature.

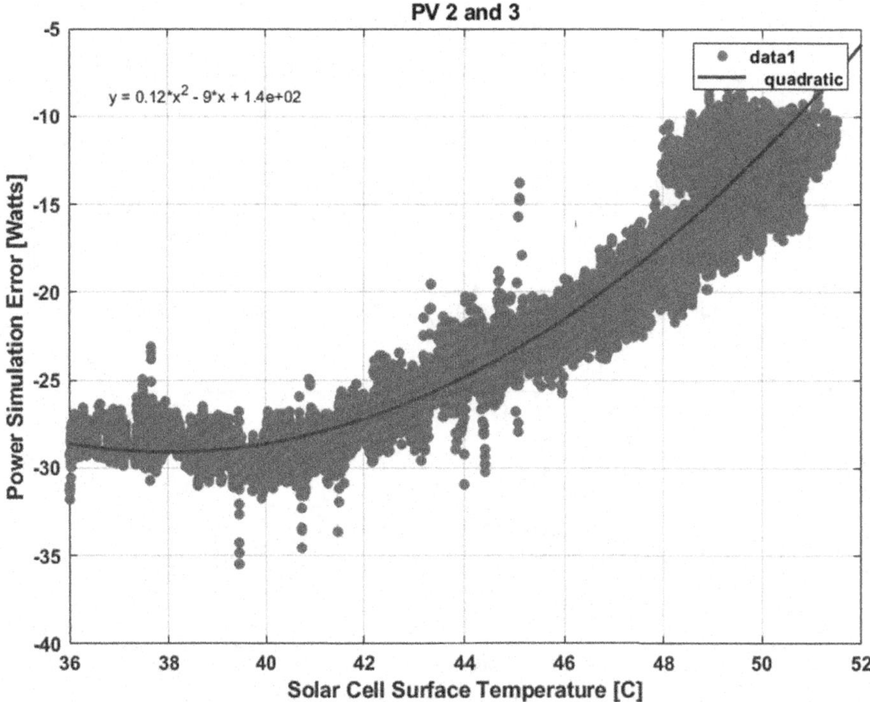

Figure 6.31 Fitting a curve for the PV_2 and PV_3 power simulation error as function of solar cell surface temperature.

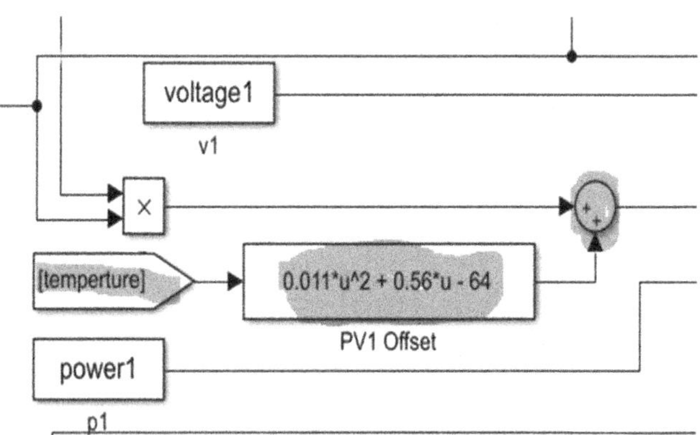

Figure 6.32 Adding offset for power for PV_1.

Codes used in the chapter can be downloaded for free from MATLAB File Exchange. Follow the link below and search for the ISBN or title of the book:

https://www.mathworks.com/matlabcentral/fileexchange/.

Alternatively, the reader can also download the material and other resources from the dedicated website or contact the authors for further help:

https://www.practicalmpc.com/.

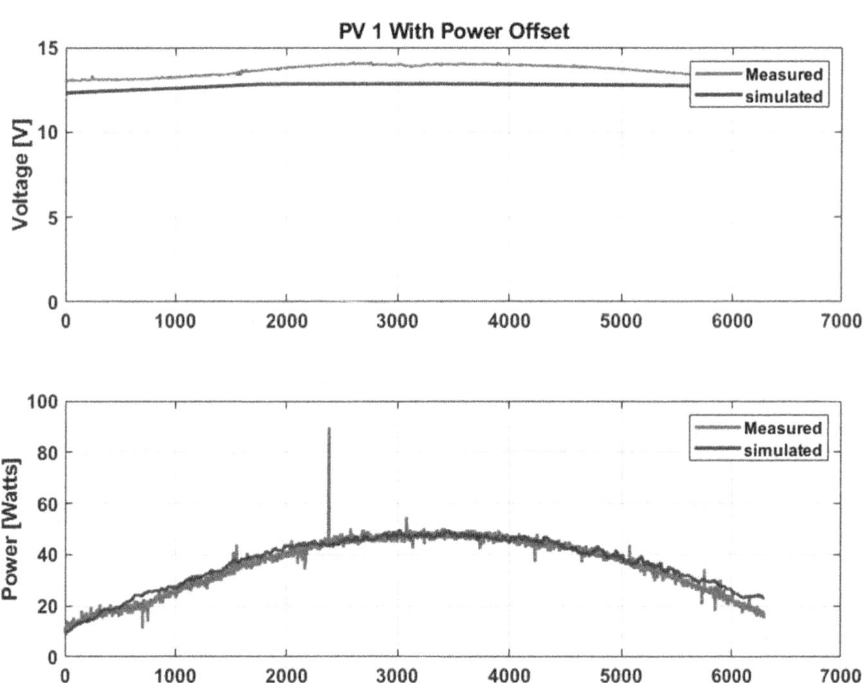

Figure 6.33 PV_1 simulation and experimental results with power offset.

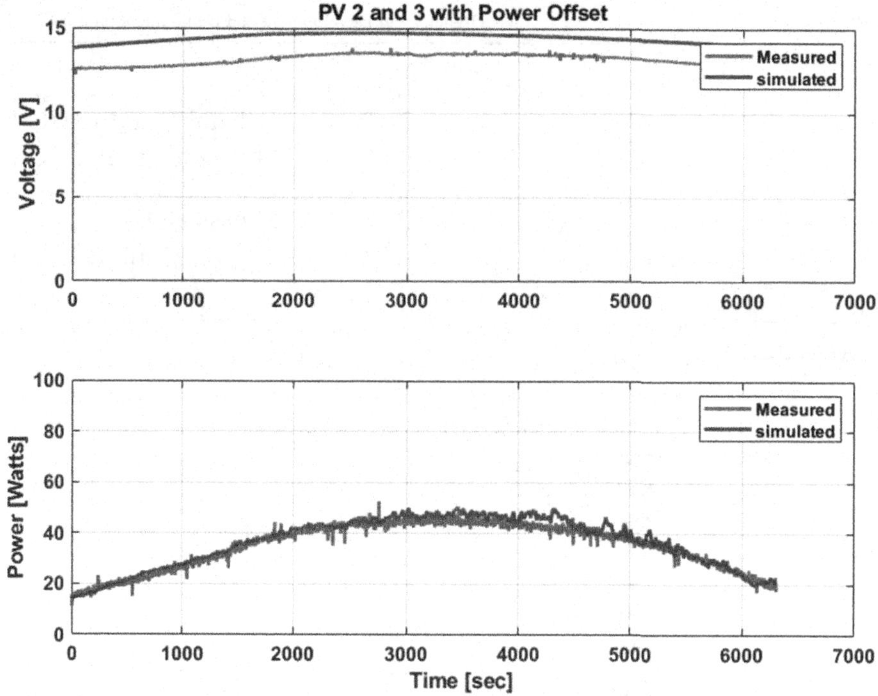

Figure 6.34 PV_2 and PV_3 simulation and experimental results with power offset.

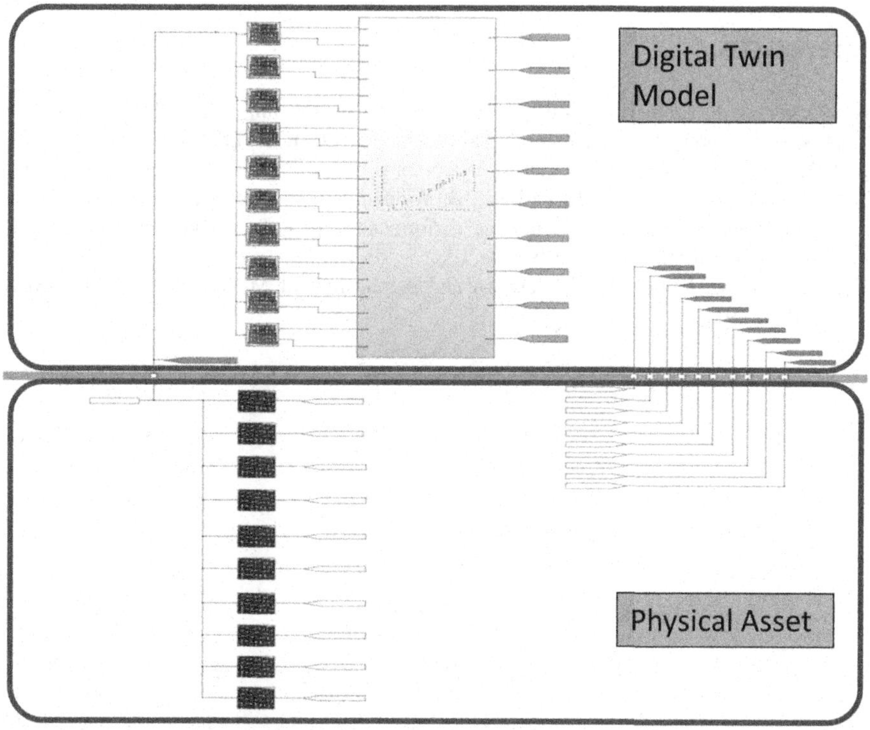

Figure 6.35 Physical and digital model.

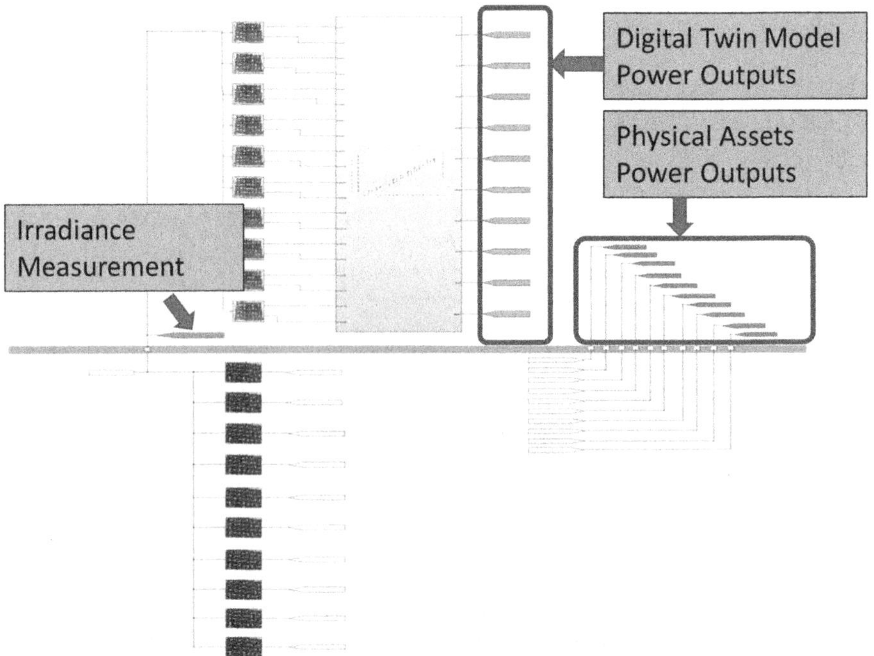

Figure 6.36 Physical and digital models input and outputs.

References

[1] S. Aljoaba, A. Cramer, B. Walcott, Thermoelectrical modeling of wavelength effects on photovoltaic module performance—part I: model, IEEE J. Photovol. 3 (July 2013) 1027–1033.

[2] S. Aljoaba, A. Cramer, S. Rawashdeh, B. Walcott, Thermoelectrical modeling of wavelength effects on photovoltaic module performance—part II: parameterization, IEEE J. Photovol. 3 (July 2013) 1034–1037.

[3] N. Khaled, B. Pattel, Practical Design and Application of MPC, Elsevier, 2018. ISBN 9780128139189.

Digital twin development for an inverter circuit for motor drive systems

7

7.1 Introduction

This chapter guides the reader through developing a digital twin model for a three-phase motor drive system using two three-leg inverters connected in series with the three-phase motor. The developed model can be used to perform off-board diagnostics of the real system's fault tolerant capability for open-circuit, short-circuit, and DC link failure modes. The model is developed using MATLAB®, Simulink, and Simscape™ Electrical and Simscape Power Systems tools. Simscape™ components like IGBT switches, diodes, voltmeter, ammeter, resistor, and inductors are used to develop two three-leg inverters. A PWM generator is created to generate the gate signals for each IGBT switches using Simulink blocks. Failure conditions are induced, and simulated behavior of the inverter configuration and results are discussed. Fig. 7.1 shows the Off-BD steps that are going to be covered in this chapter.

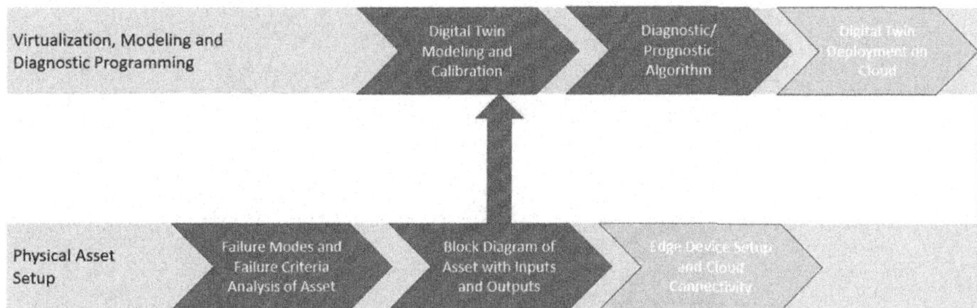

Figure 7.1 Off-BD steps covered in this chapter.

All the codes used in the chapter can be downloaded for free from MATLAB *File Exchange*. Follow the below link and search for the ISBN or title of the book:

https://www.mathworks.com/matlabcentral/fileexchange/

Alternatively, the reader can also download the material and other resources from the dedicated website or contact the authors for further help:

https://www.practicalmpc.com/

Digital Twin Development and Deployment on the Cloud. https://doi.org/10.1016/B978-0-12-821631-6.00007-4

One of the advantages of a three-phase motor drive system employing two three-leg inverters connected in series with the three-phase machine is its fault tolerance compatibility. This chapter analyzes and discusses the simulation of such three-phase motor drive system. Simulation results are obtained for different type of faults like open-circuit failure, short-circuit failure, and DC link failure. Simulation has been done on MATLAB environment with a model-based implementation of fault diagnostics and mitigation for three-phase, two three-leg inverters connected in series using MATLAB, Simulink, and Simscape. One scenario that is considered for analyzing fault tolerance capability in which the fault compensation has been achieved by reconfiguring the power converter topology with Y connection of the motor without adding any additional devices is discussed in this chapter. We will develop the Simscape model for the two three-leg inverters connected in series with the three-phase machine as shown in Fig. 7.2.

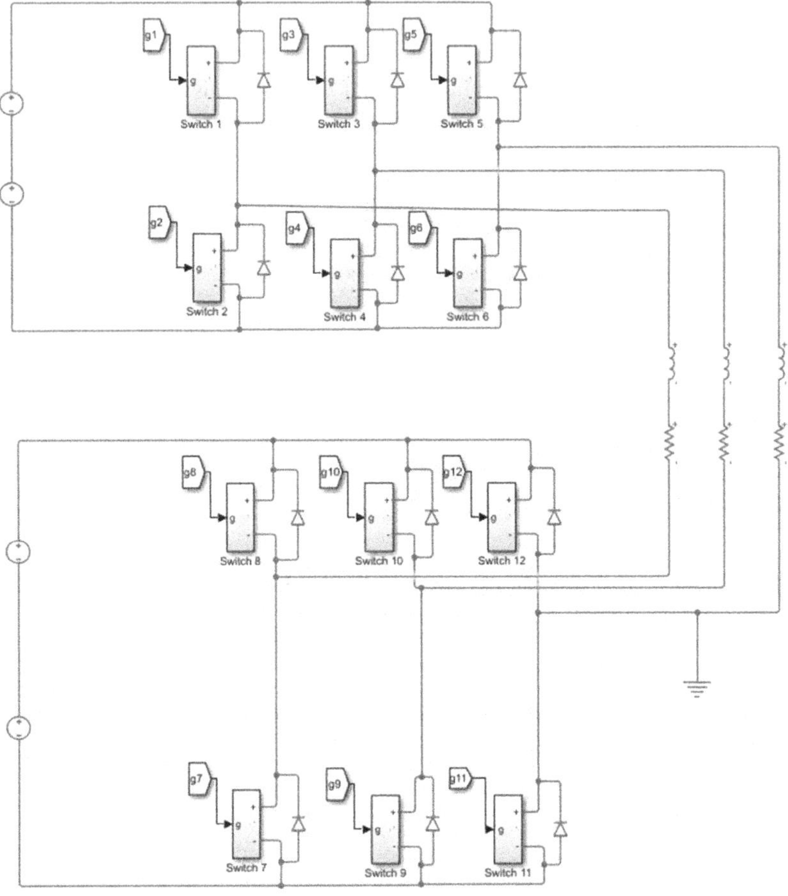

Figure 7.2 Motor drive system [1].

7.2 Block diagram of the motor drive inverter system

Fig. 7.3 shows the block diagram of the motor drive inverter system. The inputs to the system are the input DC voltage and the gate pulse signals to turn on the IGBT switches of the inverter. And the outputs to the system are the load voltages, currents for all three phases, and also the currents and voltages through IGBT switches.

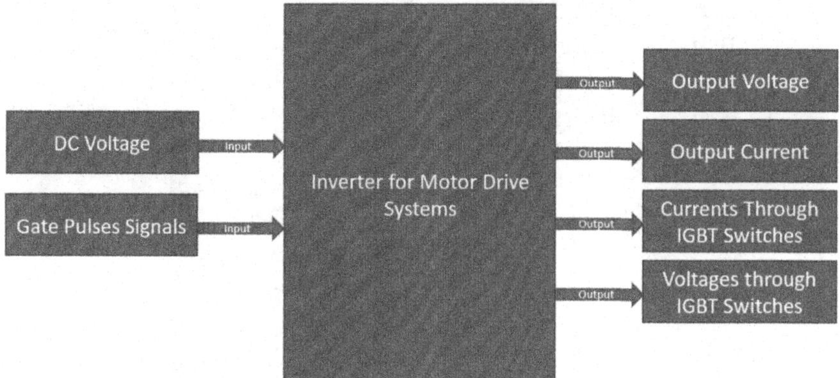

Figure 7.3 Block diagram of the motor drive inverter system.

7.3 Failure modes and diagnostics concept of the motor drive inverter system

Fig. 7.4 shows a concept block diagram for the off-board diagnostics process for the motor drive system. The same DC input voltage and the gate pulse signals that are fed to the actual inverter system can be fed to the digital twin as well. The digital twin will not be made aware about the faults introduced in the actual system such as open-circuit or short-circuit failures. So if there is a failure in the actual system with the same inputs given to both actual system and digital twin, the output voltages and currents and the currents and voltages through the IGBTs will also be different. The diagnostics algorithm will compare the difference between the signals from actual and digital twin and take a diagnostics decision about the actual system based on the comparison results.

7.4 Simscape model of the motor drive inverter system

Below are the detailed steps to building the Simscape model:

1. Create a new **Simulink model**.
2. Click on the **Simulink Library Browser** as shown in Fig. 7.5.

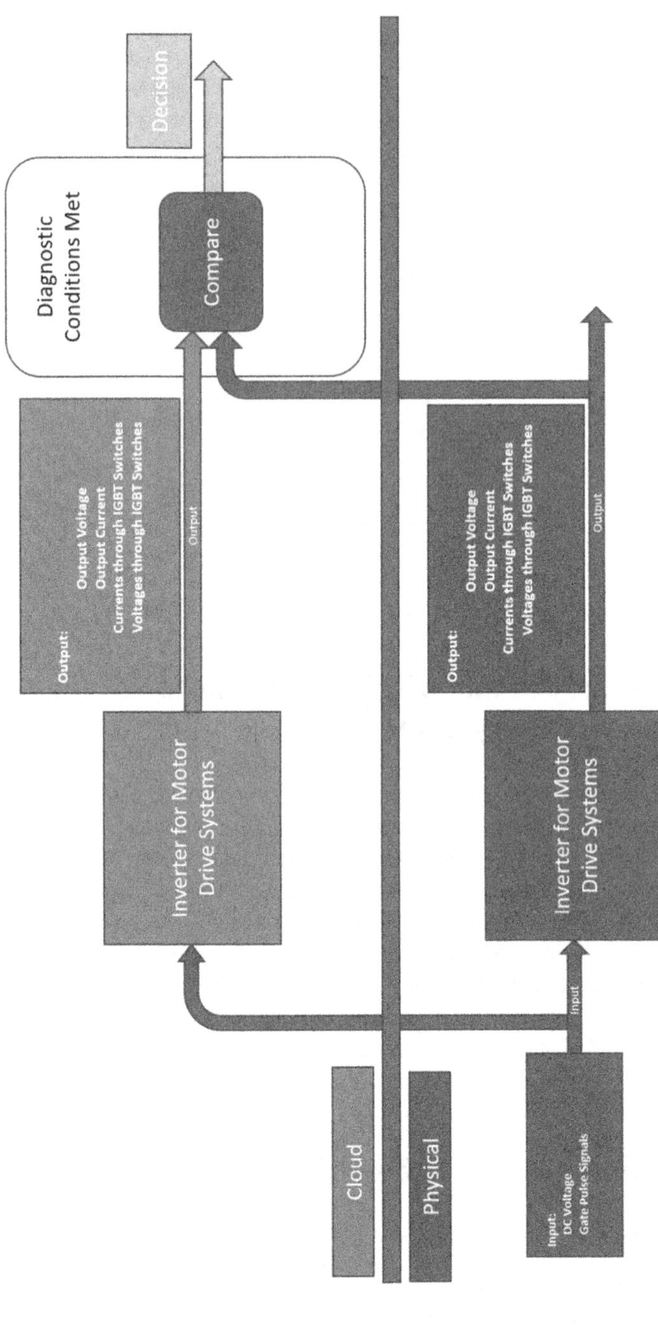

Figure 7.4 Off-BD diagnostics process for motor drive inverter system.

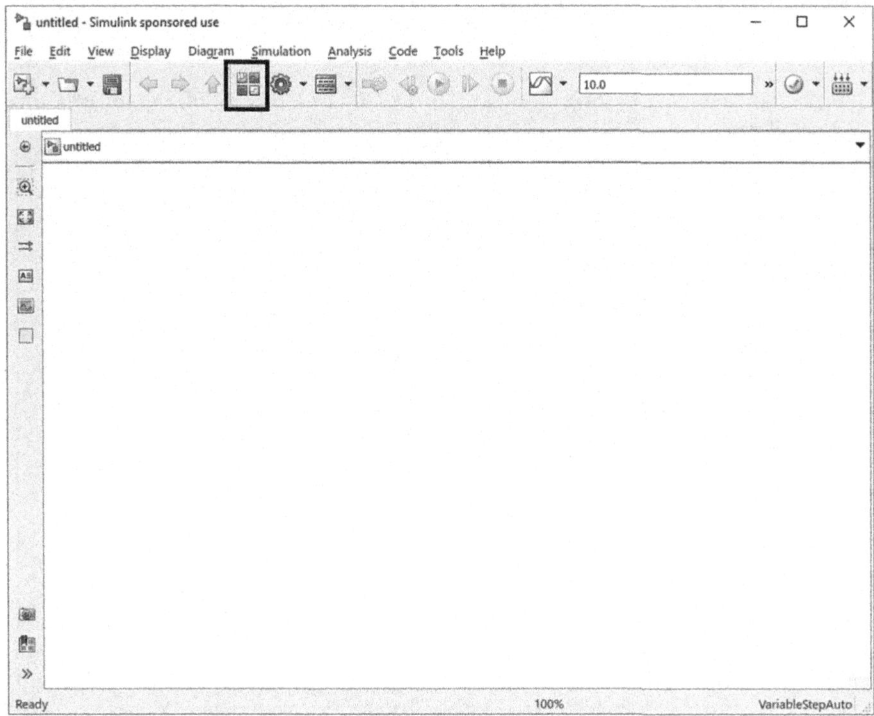

Figure 7.5 Simulink library browser.

3. Navigate to **Simscape > Electronics > Semiconductor Devices** and add **N-channel IGBT** block to the model as shown in Fig. 7.6. Update the block parameters for N-channel IGBT as shown in Fig. 7.7.
4. Go back to **Simscape > Electronics > Actuators and Drivers > Drivers** and add **Gate Driver** block (Fig. 7.8).
5. Go to Simscape > Utilities and add Simulink-PS Converter block and Connection Port block (Fig. 7.9).
6. Go to **Simulink > Signal Attributes** and add **Data Type Conversion** block to the model (Fig. 7.10).
7. Go to **Simulink > Sources** and add **In1** (input port) to the model (Fig. 7.11).
8. Add two physical modeling connection ports into the model from **Simscape > Utilities** as shown in Fig. 7.12.
9. Make all the connections with all the added blocks as shown in Fig. 7.13.
10. Select all the blocks as shown in Fig. 7.14 and right click and select "Create Subsystem from Selection." This will create a subsystem with all the blocks in it and with one input and two output ports as shown in Fig. 7.15. Name the subsystem as "Switch 1."
11. Obtain the remaining 11 switches of three-phase inverter by making 11 copies of the subsystem "Switch 1." Also replace the input port connected to the switch subsystem with "From" blocks as shown in Fig. 7.16.
12. Go to **Simscape > Foundation Library > Electrical > Electrical Sources** and add four **DC Voltage Source** blocks, two for each leg (Fig. 7.17).

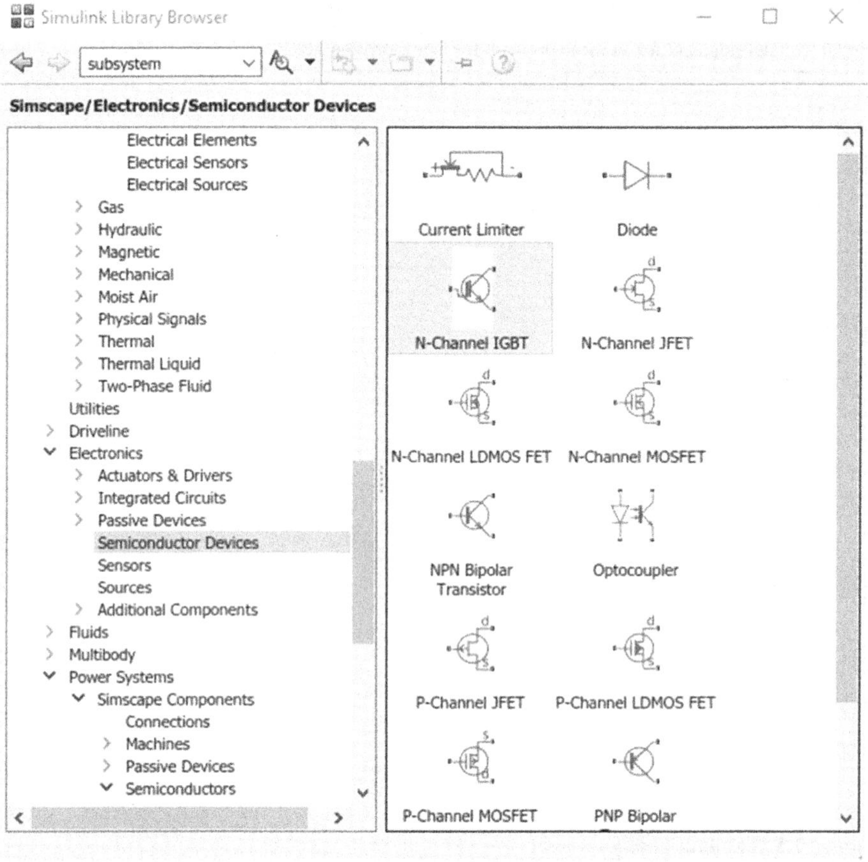

Simulink Library Browser
subsystem
Simscape/Electronics/Semiconductor Devices
Electrical Elements
Electrical Sensors
Electrical Sources
Gas
Hydraulic
Magnetic
Mechanical
Moist Air
Physical Signals
Thermal
Thermal Liquid
Two-Phase Fluid
Utilities
Driveline
Electronics
Actuators & Drivers
Integrated Circuits
Passive Devices
Semiconductor Devices
Sensors
Sources
Additional Components
Fluids
Multibody
Power Systems
Simscape Components
Connections
Machines
Passive Devices
Semiconductors
Current Limiter
Diode
N-Channel IGBT
N-Channel JFET
N-Channel LDMOS FET
N-Channel MOSFET
NPN Bipolar Transistor
Optocoupler
P-Channel JFET
P-Channel LDMOS FET
P-Channel MOSFET
PNP Bipolar

Figure 7.6 IGBT block.

Block Parameters: N-Channel IGBT
N-Channel IGBT
This block represents an N-channel IGBT. Right-click on the block and select Simscape(TM) block choices to access variant implementations.
In the default implementation the device characteristics are defined in terms of the I-V curves for different gate voltages, and dynamics are dependent on modeled junction capacitances. The I-V curves are either derived from datasheet parameters or defined directly using tabulated data.
The simplified implementation models only the on-state I-V curve for a single gate voltage, and dynamics are defined by event-based equations with turn-on, rise, turn-off and fall times specified directly. This option gives better simulation speed but with reduced modeling accuracy.
There is no integral reverse diode.
Settings
Main
Junction Capacitance
Advanced
Temperature Dependence
I-V characteristics defined by:
Lookup table (2-D, temperature independent)
Vector of gate-emitter voltages, Vge:
[-2, 6, 7, 8, 10, 12, 15, 20]
V
Vector of collector-emitter voltages, Vce:
[-1, 0, .5, 1, 1.5, 2, 2.5, 3, 3.5, 4]
V
Tabulated collector currents, Ic=fcn(Vge,Vce):
8324, 109.92, 295.67, 496.54, 702.28, 909.96, 1118.3]
A
OK
Cancel
Help
Apply

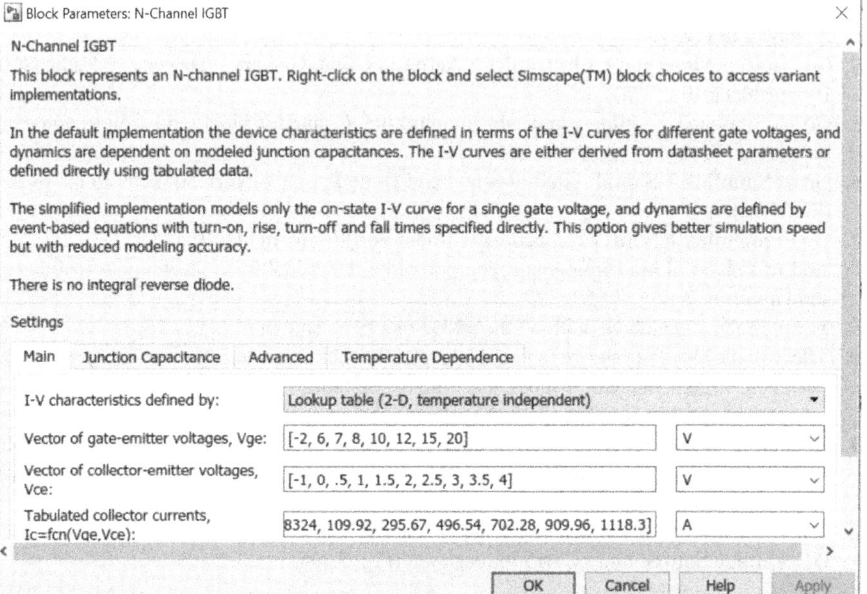

Figure 7.7 Block parameters for N-channel IGBT.

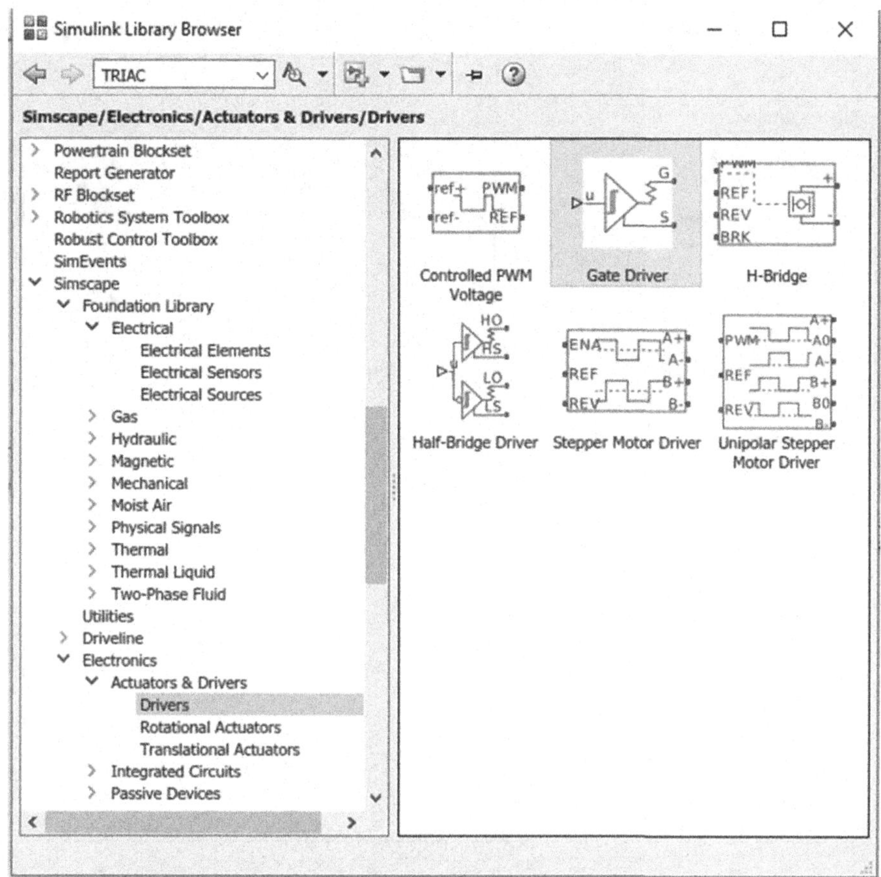

Figure 7.8 Gate driver block.

13. Setup the constant voltage parameter as 50 V for each voltage source block by double clicking on each DC Voltage Source (Fig. 7.18).
14. Connect the voltage sources to the switches as shown in Fig. 7.19.
15. Go to **Simscape > Electronics > Semiconductor Devices** and **add** 12 **Diode blocks** (Fig. 7.20) and make the connections as shown in Fig. 7.21. Update the block parameters as shown in Figs. 7.22 and 7.23.
16. Add and connect resistors and inductors to the model by selecting them from **Simscape > Foundation Library > Electrical > Electrical Elements** (Figs. 7.24 and 7.25). This RL circuit basically is considered as the load for the inverter circuit.
17. Set up the resistance and inductance at desired value for each resistor. For this model, resistance is set as 5 Ω for each resistor and inductance is set as 1e-6 H.
18. Go to **Simscape > Foundation Library > Electrical > Electrical Sensors** and add current sensors and voltage sensors to the model for reading output current and output voltage (Figs. 7.26–7.28). Also connect the sensors to Simulink® scopes by connecting the output

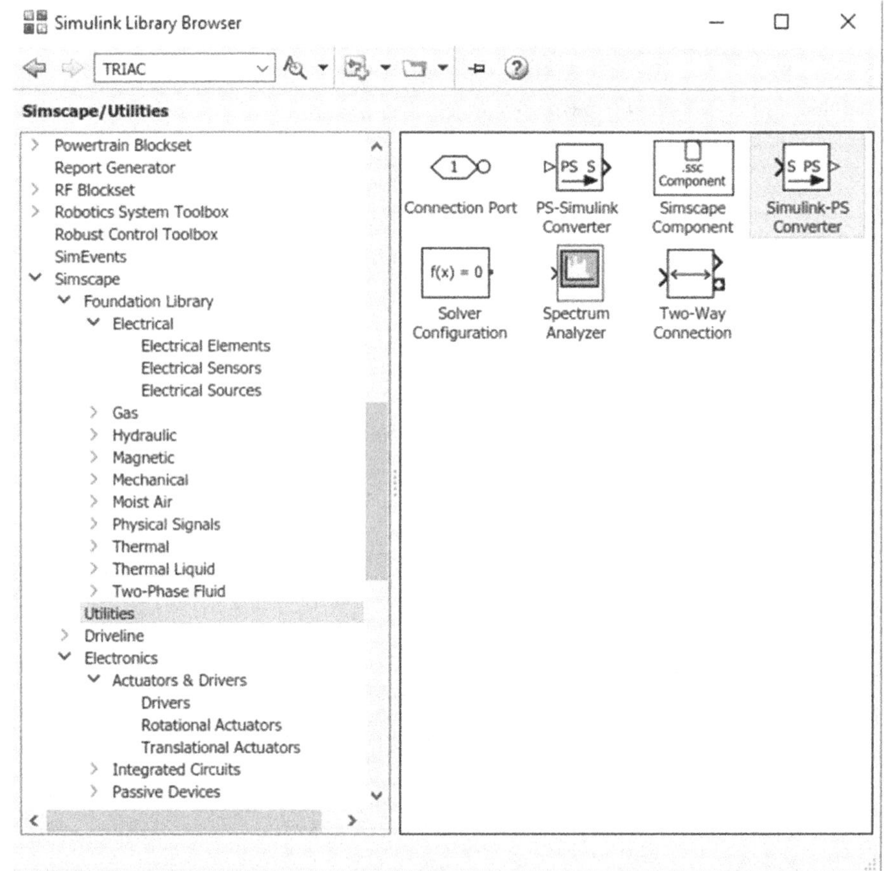

Figure 7.9 Simulink-PS converter.

from sensor to the PS-Simulink converter block which can be found at **Simscape > Utilities** (Fig. 7.27).

19. Add solver and Electrical Reference blocks to the model from **Simscape > Utilities** and **Simscape > Foundation Library > Electrical > Electrical Elements** (Figs. 7.29 and 7.30).
20. Set the sample time as 2e-6 for the solver block (Fig. 7.31).
21. Connect the solver and electrical terminator blocks as shown in Fig. 7.32.
22. Now we have to generate the gate pulse signals for firing the IGBT switches for the inverter. A PWM generator subsystem is created to generate the gate signals for each IGBT switches in the inverter model. Create a new subsystem and add digital clock to the model from **Simulink > Sources**(Fig. 7.33).
23. Add constant, gain, trigonometric function, i.e., sine, and sum blocks to the model accordingly and enter the values as shown in Fig. 7.34. This model generates the sinusoidal signals.

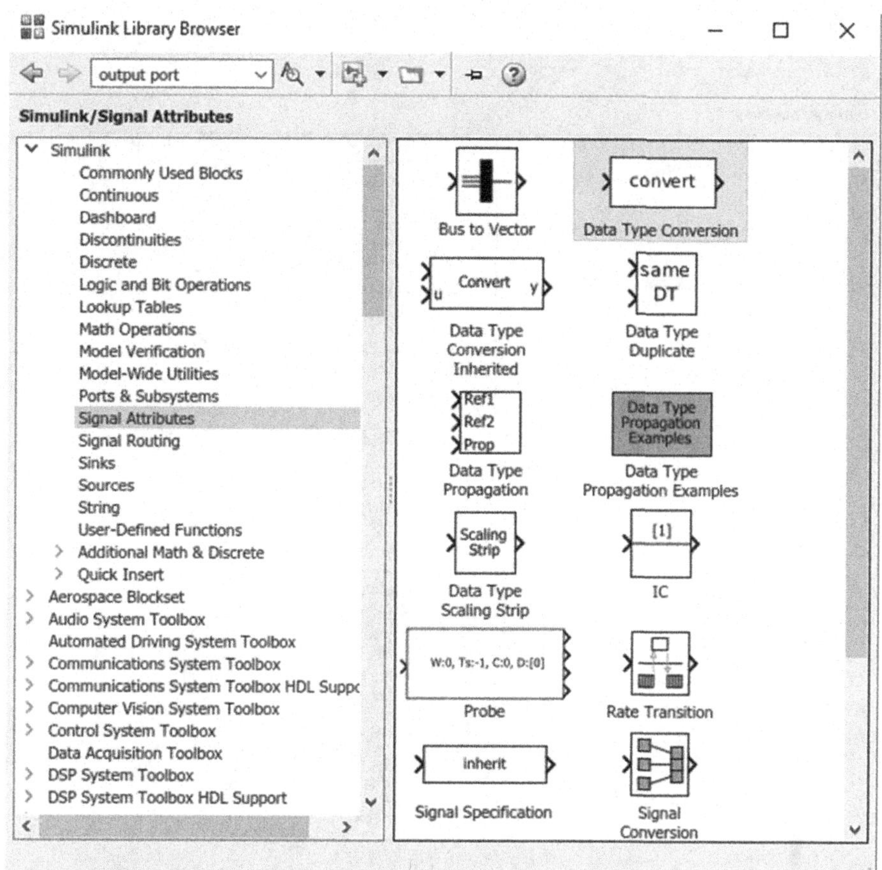

Figure 7.10 Data type conversion block.

24. Gate signals for each switches in the inverter is generated by using a PWM strategy where two high-frequency triangular waves are compared with a sinusoidal waveform. This resulted in generating output voltage with eight levels. "PWM Generator" subsystem can be created by removing the output port from the subsystem "Sin wt" and adding gain, logical operators, and triangle blocks to the model as shown in Fig. 7.35.

25. Assign the block parameter values for Triangle1 and Triangle2 blocks as shown in Figs. 7.36 and 7.37.

26. Create a subsystem and name it as PWM Generator, which outputs the gate pulses for inverter 1 and 2 as shown in Fig. 7.38.

27. The Simscape™ model design is ready. Save the model, and on the model configuration parameters window, set the simulation end time as 0.2 s, Solver Type to be Fixed, and Fixed Step Size to be 2e-6 as shown in Fig. 7.39. A fully simulatable working model is available in the attachment section under folder **Motor_Drive_System_Inverter_Model** for reference.

28. Run simulation and observe the Output Current and Output Voltage Scope blocks added in step 18. The simulated output currents and voltages for all three phases are shown in Figs. 7.40 and 7.41, respectively.

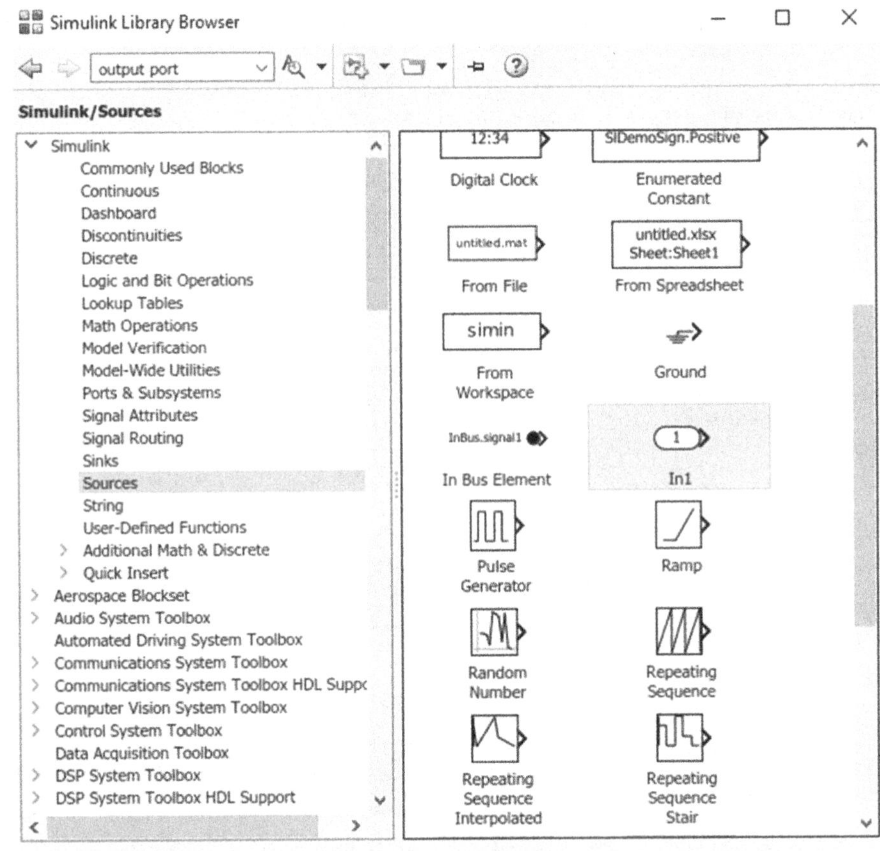

Figure 7.11 In1 (input port) block.

7.5 Fault injection and diagnostic algorithm development

In this section, we will introduce an open-circuit failure condition to one of the legs of the inverter circuit in the Simulink model and first analyze the impact of the failure on the functionality of the inverter.

1. Save the model we tested in the previous section with a different name.
2. As shown highlighted in Fig. 7.42, disconnect the line between Switch 10 and Switch 9 of the second inverter leg. Add a Step block, Simulink-PS Converter block, and a Switch block and make the connections as shown in Fig. 7.42. Configure the Step block with initial condition of 1 and a final value of 0 at time 0.06 s as shown Fig. 7.43. And also configure the Switch block with values as shown in Fig. 7.44. So initially, since the step block output is 1, it will

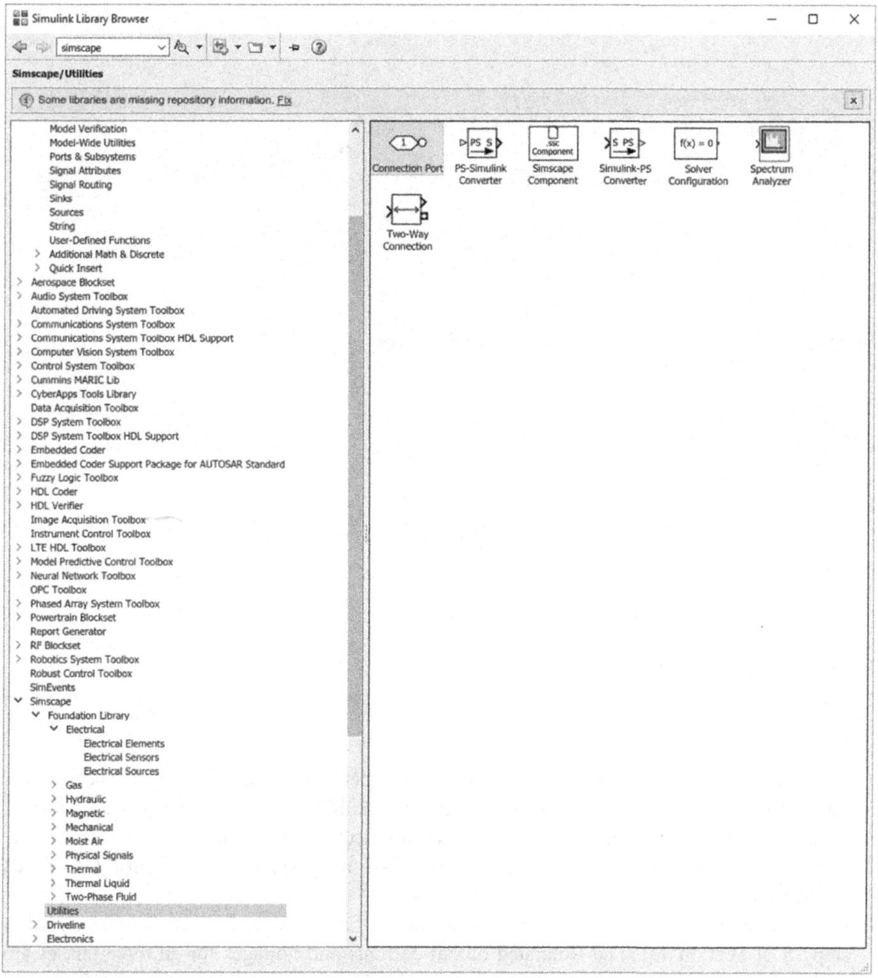

Figure 7.12 Physical modeling connection port.

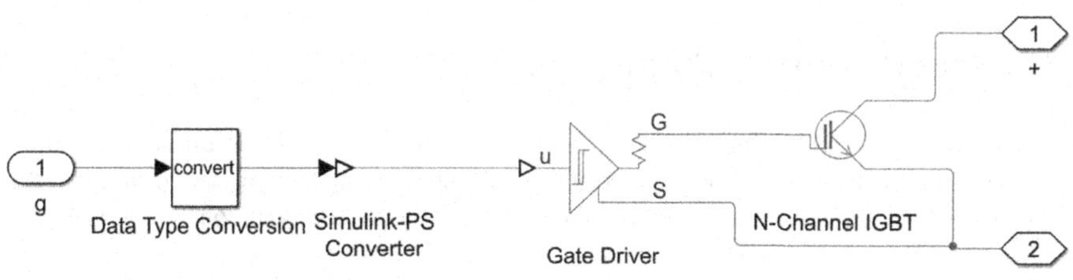

Figure 7.13 Connecting all the blocks.

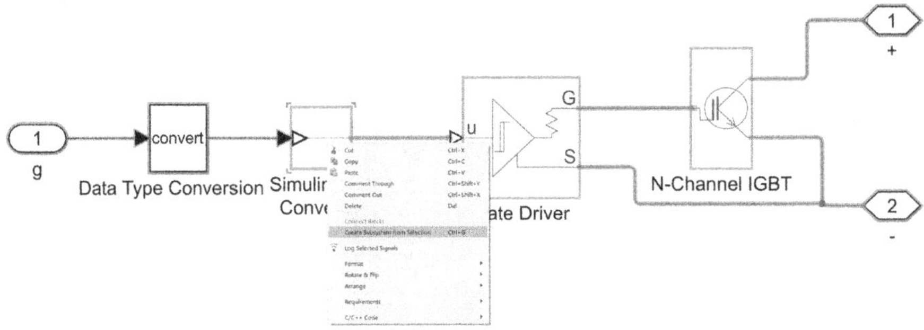

Figure 7.14 Creating subsystem with selected blocks.

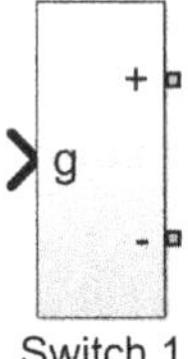

Figure 7.15 Subsystem created for switch 1.

make the switch to be closed, but at time t = 0.06 s, because the step block value becomes 0, this opens the switch and the connection between the IGBT 10 and IGBT 9 will be open.

3. A fully simulatable working model with this failure condition introduced is available in the attachment section under folder **Motor_Drive_System_Inverter_Model_with_OC_Fault** for reference.
4. Run simulation and observe the Output Current and Output Voltage Scope blocks added in step 18 of Section 7.3. The simulated output currents and voltages for all three phases are shown in Figs. 7.45 and 7.46, respectively. Note that because of the open circuit at time t = 0.06 s, the current and voltage of the second phase is significantly affected.

7.6 Application problem

1. Introduce a failure condition for a short circuit between the + and – terminals of one of the IGBT switches, using the same Step block, and switch block strategy discussed earlier in this chapter and observe the output voltage and current of the inverter and compare it to the no-fault condition.

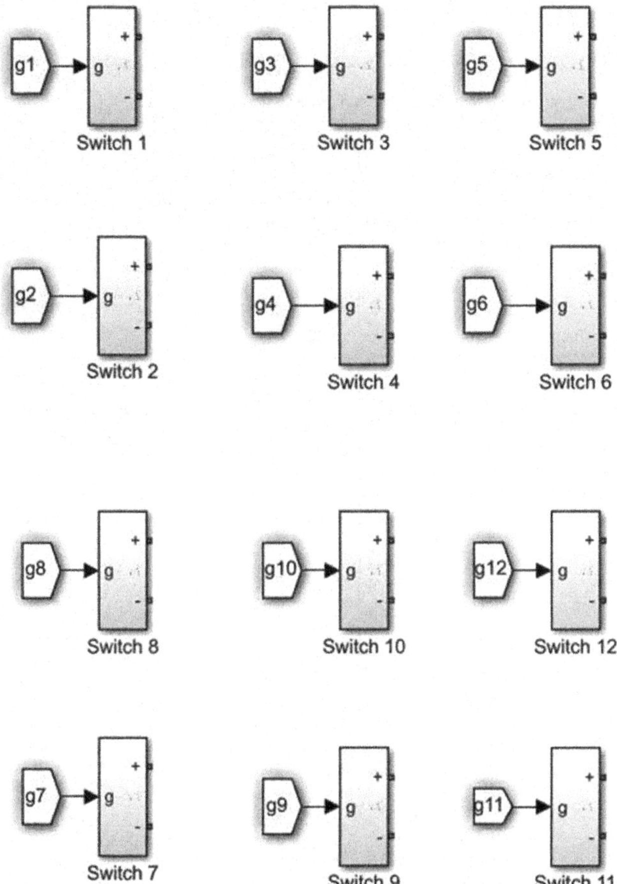

Figure 7.16 All 12 switches added.

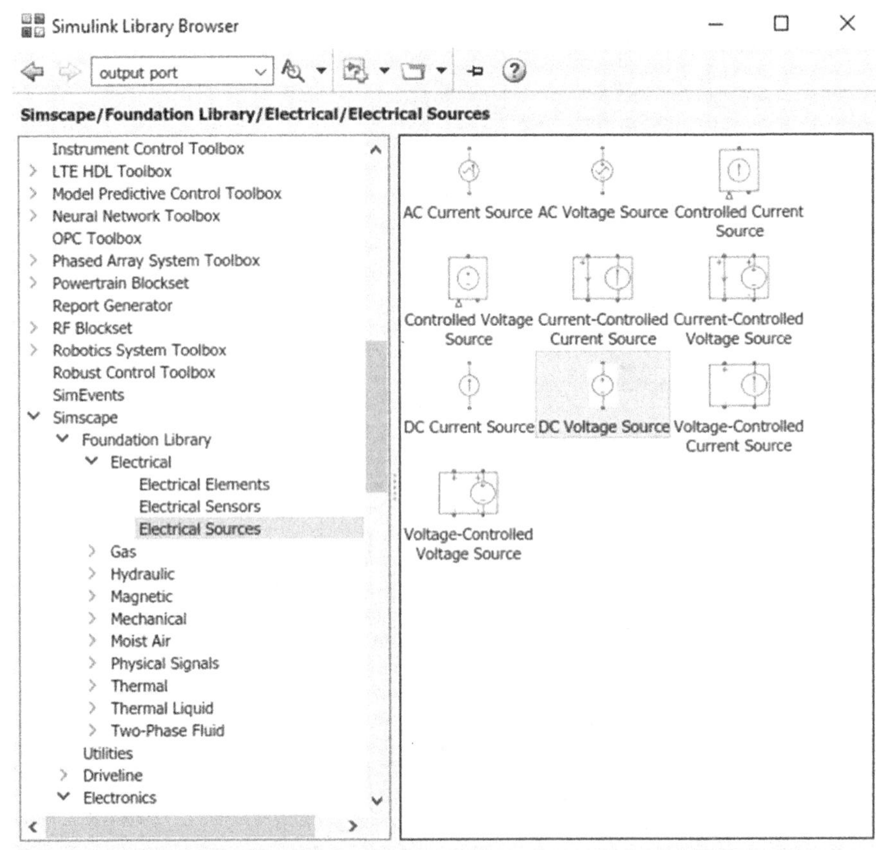

Figure 7.17 DC voltage source.

Block Parameters: DC Voltage Source2

DC Voltage Source

The ideal voltage source maintains a constant voltage across its output terminals, independent of the current flowing through the source. The output voltage is defined by the Constant voltage parameter, and can be any real value.

Source code

Settings

Parameters

Constant voltage: 50 V

OK Cancel Help Apply

Figure 7.18 Parameter setting for DC voltage source.

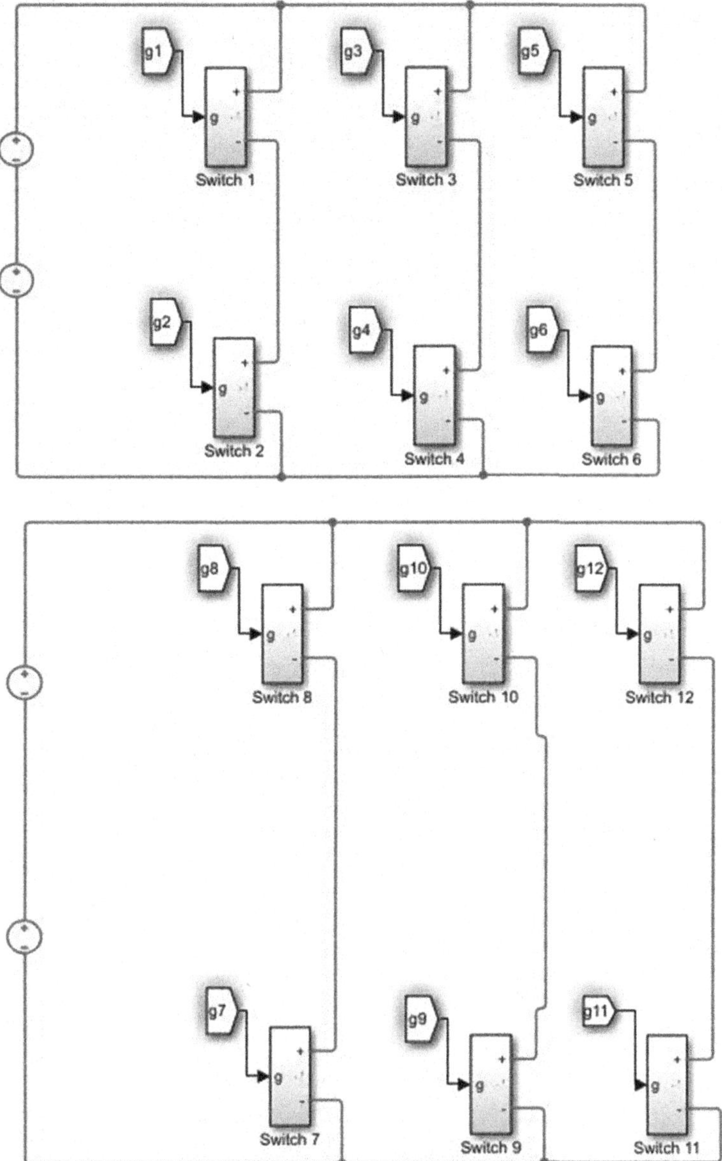

Figure 7.19 Voltage source connection to the switches.

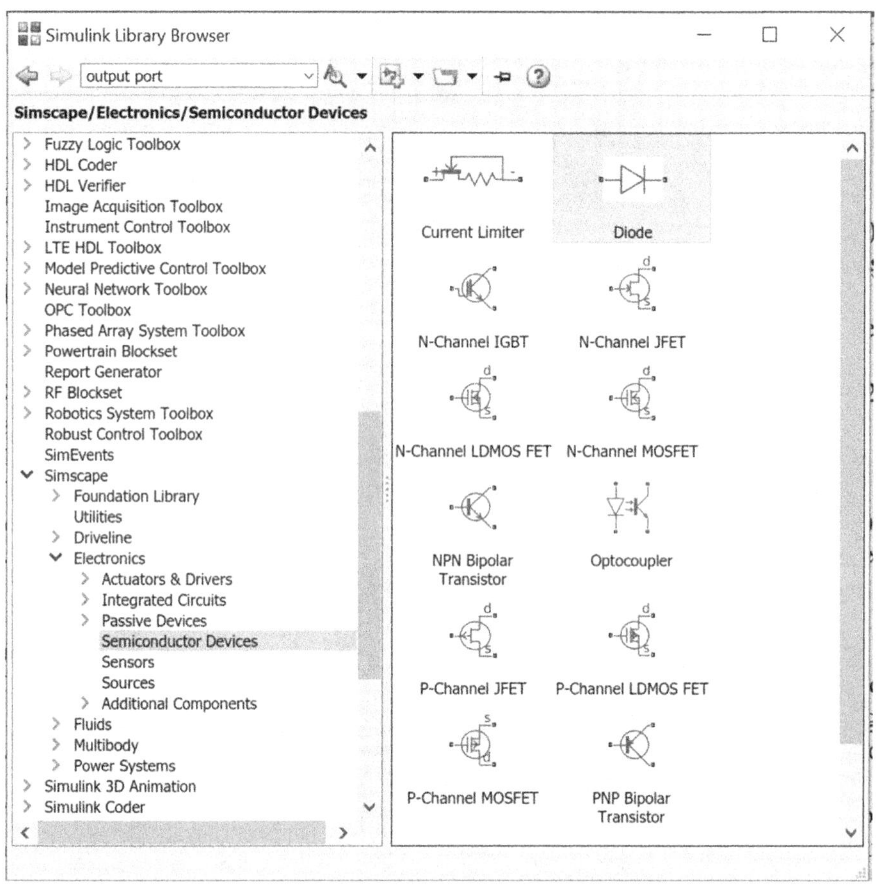

Figure 7.20 Diode block.

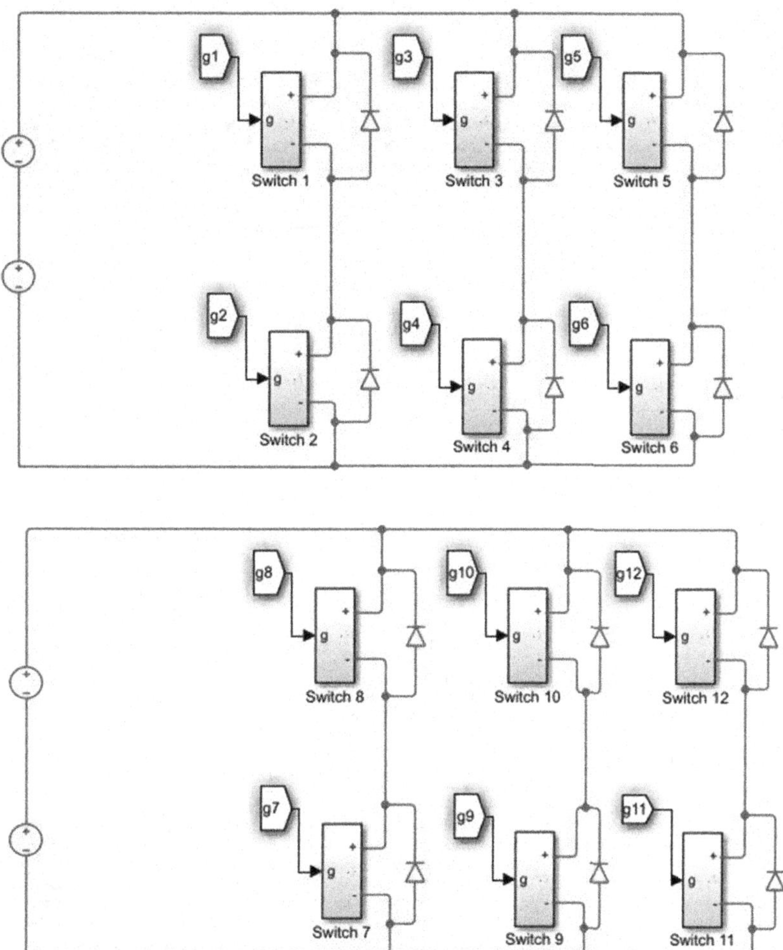

Figure 7.21 Connecting diodes antiparallel to the switches.

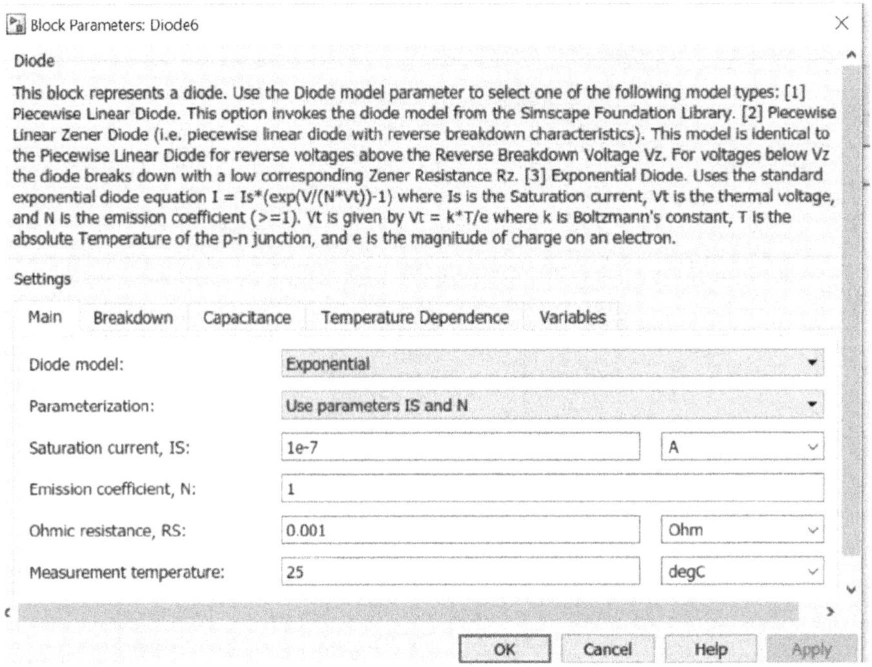

Block Parameters: Diode6

Diode

This block represents a diode. Use the Diode model parameter to select one of the following model types: [1] Piecewise Linear Diode. This option invokes the diode model from the Simscape Foundation Library. [2] Piecewise Linear Zener Diode (i.e. piecewise linear diode with reverse breakdown characteristics). This model is identical to the Piecewise Linear Diode for reverse voltages above the Reverse Breakdown Voltage Vz. For voltages below Vz the diode breaks down with a low corresponding Zener Resistance Rz. [3] Exponential Diode. Uses the standard exponential diode equation I = Is*(exp(V/(N*Vt))-1) where Is is the Saturation current, Vt is the thermal voltage, and N is the emission coefficient (>=1). Vt is given by Vt = k*T/e where k is Boltzmann's constant, T is the absolute Temperature of the p-n junction, and e is the magnitude of charge on an electron.

Settings

Main | Breakdown | Capacitance | Temperature Dependence | Variables

Diode model:	Exponential	
Parameterization:	Use parameters IS and N	
Saturation current, IS:	1e-7	A
Emission coefficient, N:	1	
Ohmic resistance, RS:	0.001	Ohm
Measurement temperature:	25	degC

OK Cancel Help Apply

Figure 7.22 Block parameters for diode 1

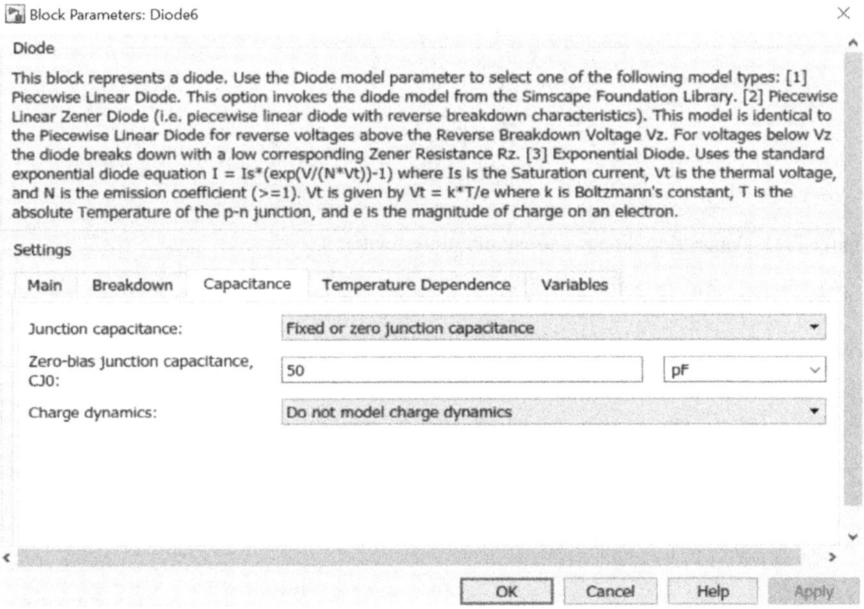

Block Parameters: Diode6

Diode

This block represents a diode. Use the Diode model parameter to select one of the following model types: [1] Piecewise Linear Diode. This option invokes the diode model from the Simscape Foundation Library. [2] Piecewise Linear Zener Diode (i.e. piecewise linear diode with reverse breakdown characteristics). This model is identical to the Piecewise Linear Diode for reverse voltages above the Reverse Breakdown Voltage Vz. For voltages below Vz the diode breaks down with a low corresponding Zener Resistance Rz. [3] Exponential Diode. Uses the standard exponential diode equation I = Is*(exp(V/(N*Vt))-1) where Is is the Saturation current, Vt is the thermal voltage, and N is the emission coefficient (>=1). Vt is given by Vt = k*T/e where k is Boltzmann's constant, T is the absolute Temperature of the p-n junction, and e is the magnitude of charge on an electron.

Settings

Main | Breakdown | Capacitance | Temperature Dependence | Variables

Junction capacitance:	Fixed or zero junction capacitance	
Zero-bias junction capacitance, CJ0:	50	pF
Charge dynamics:	Do not model charge dynamics	

OK Cancel Help Apply

Figure 7.23 Block parameters for diode 2.

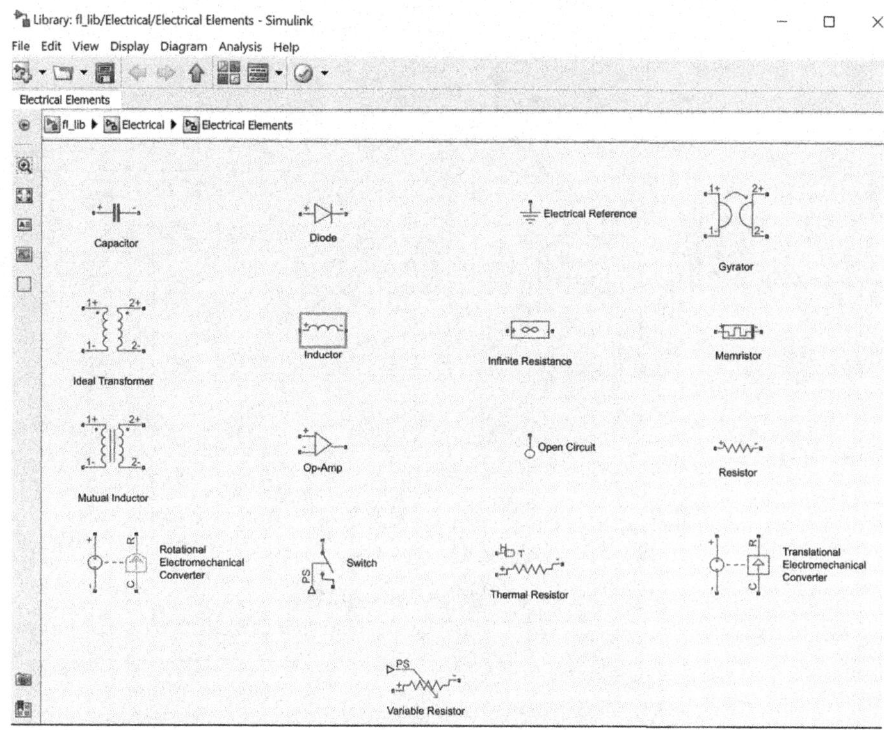

Figure 7.24 Inductor and resistor blocks.

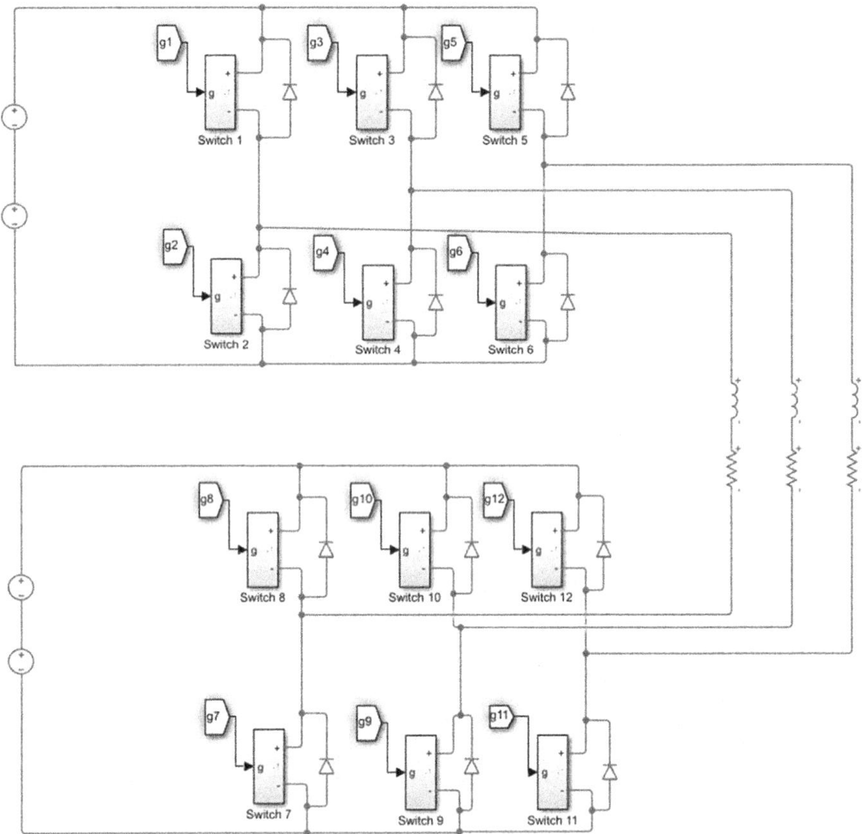

Figure 7.25 Connecting resistor and inductor blocks to the inverter model.

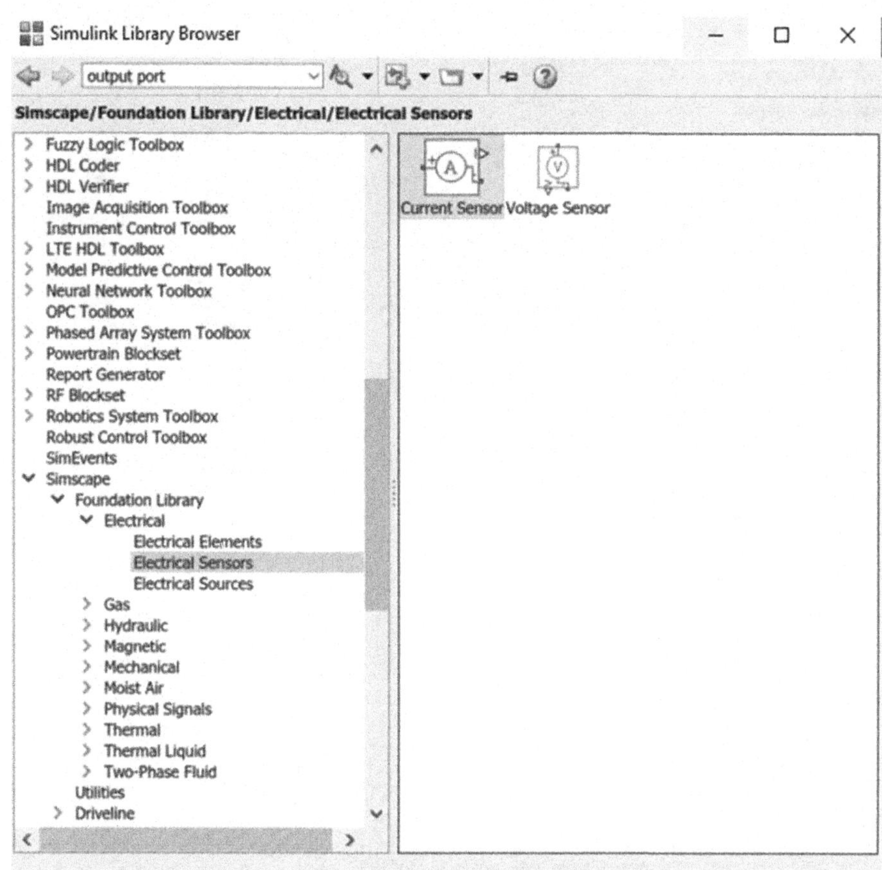

Figure 7.26 Current sensor and voltage ensor blocks.

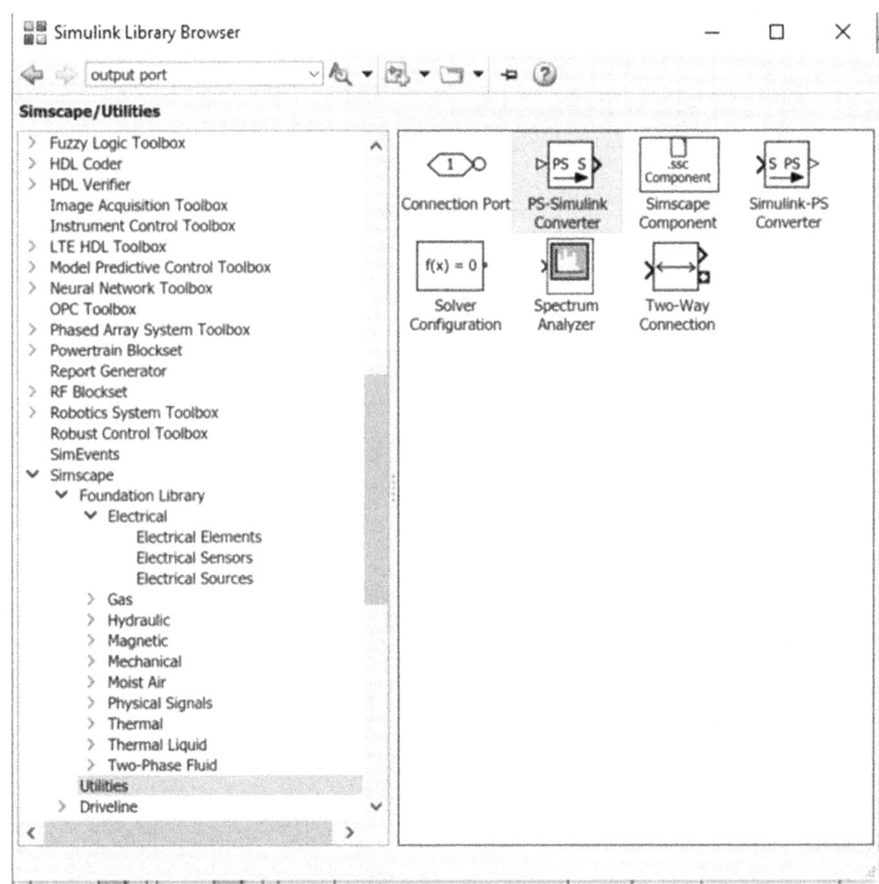

Figure 7.27 PS-simulink converter.

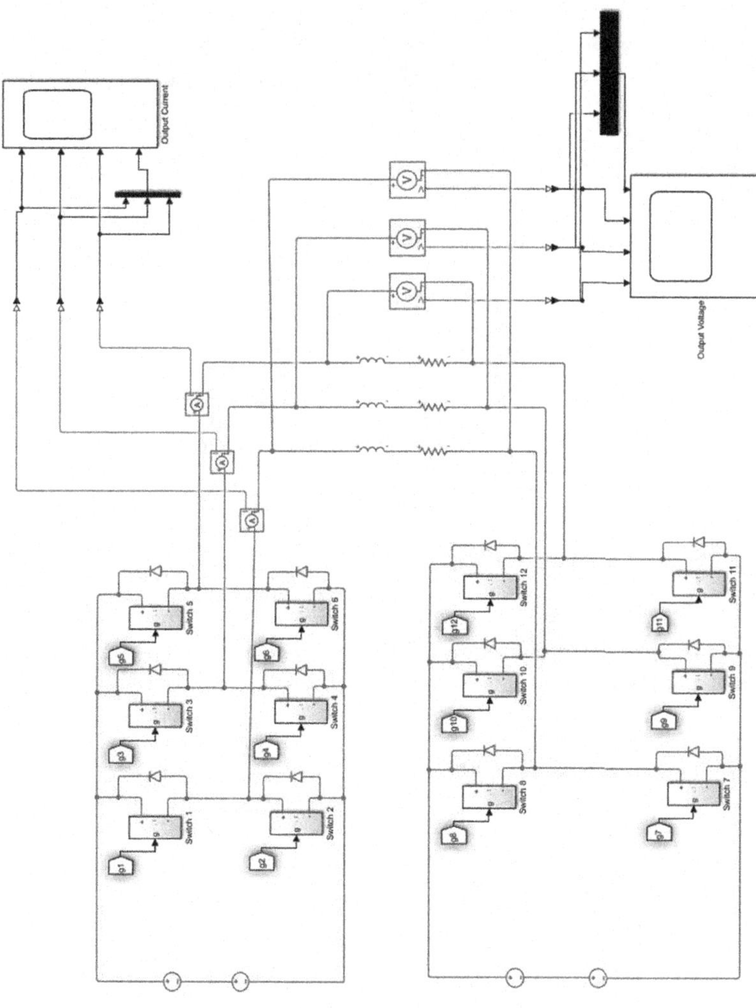

Figure 7.28 Connecting voltage and current sensors to the model.

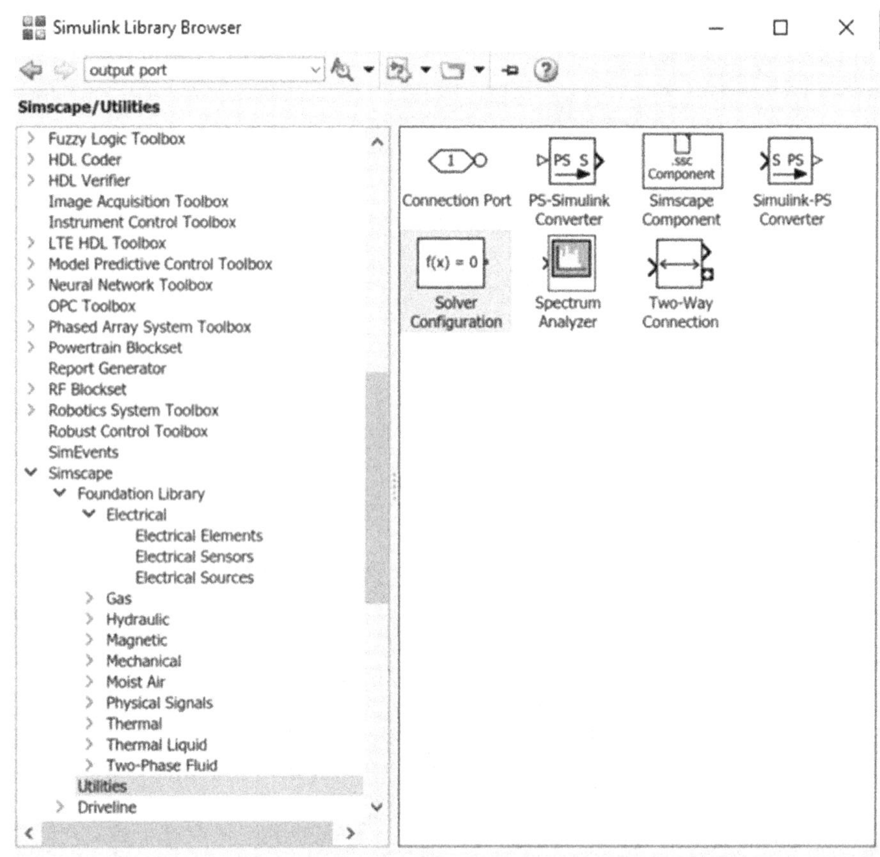

Figure 7.29 Solver block

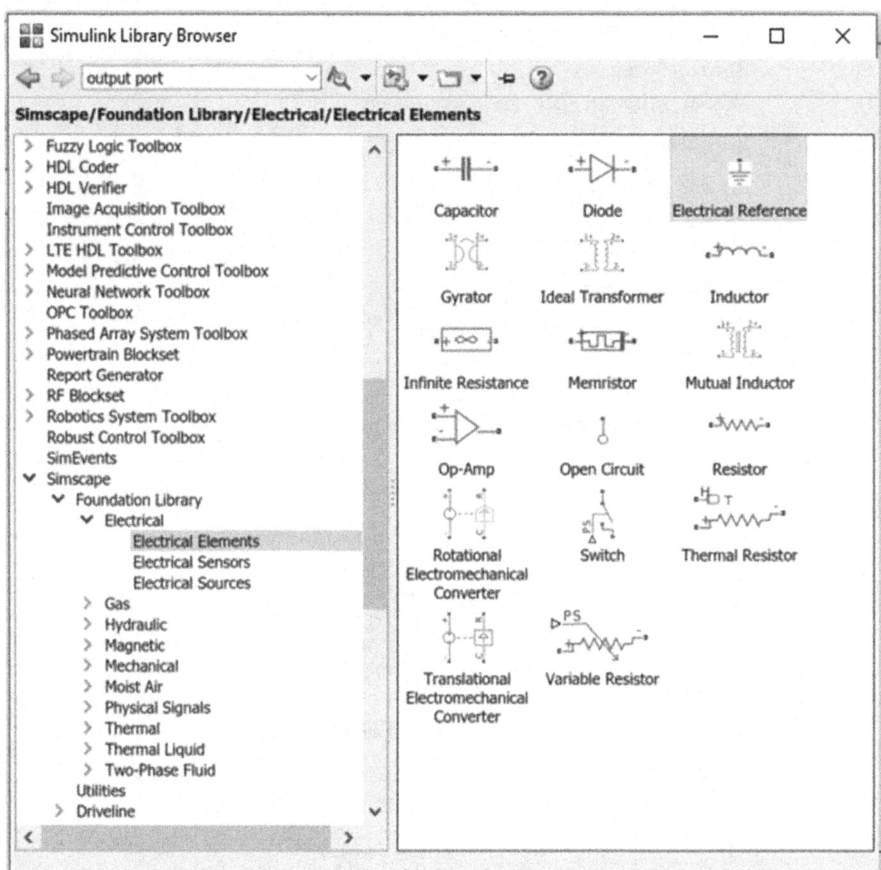

Figure 7.30 Electrical reference block.

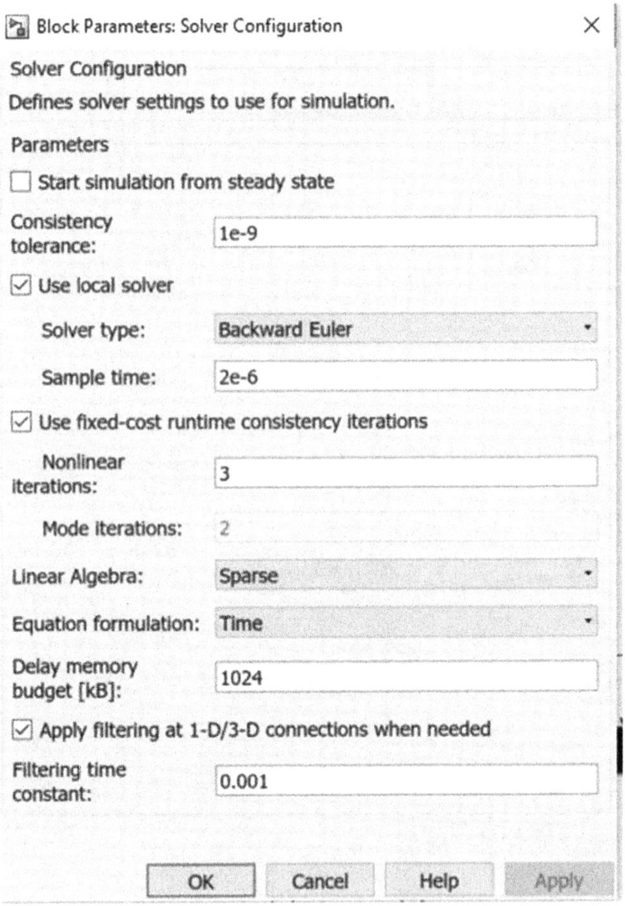

Block Parameters: Solver Configuration

Solver Configuration

Defines solver settings to use for simulation.

Parameters

☐ Start simulation from steady state

Consistency tolerance: 1e-9

☑ Use local solver

Solver type: Backward Euler

Sample time: 2e-6

☑ Use fixed-cost runtime consistency iterations

Nonlinear iterations: 3

Mode iterations: 2

Linear Algebra: Sparse

Equation formulation: Time

Delay memory budget [kB]: 1024

☑ Apply filtering at 1-D/3-D connections when needed

Filtering time constant: 0.001

OK Cancel Help Apply

Figure 7.31 Setting up the sample time parameter for the Solver block.

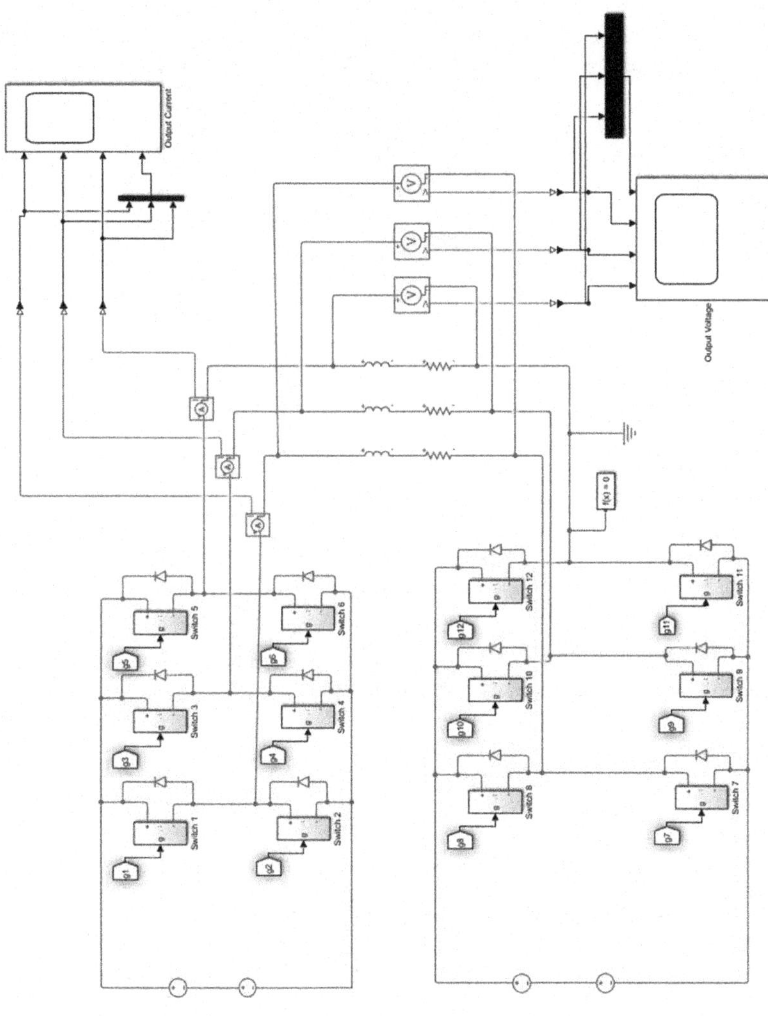

Figure 7.32 Solver and the electrical terminator block connection.

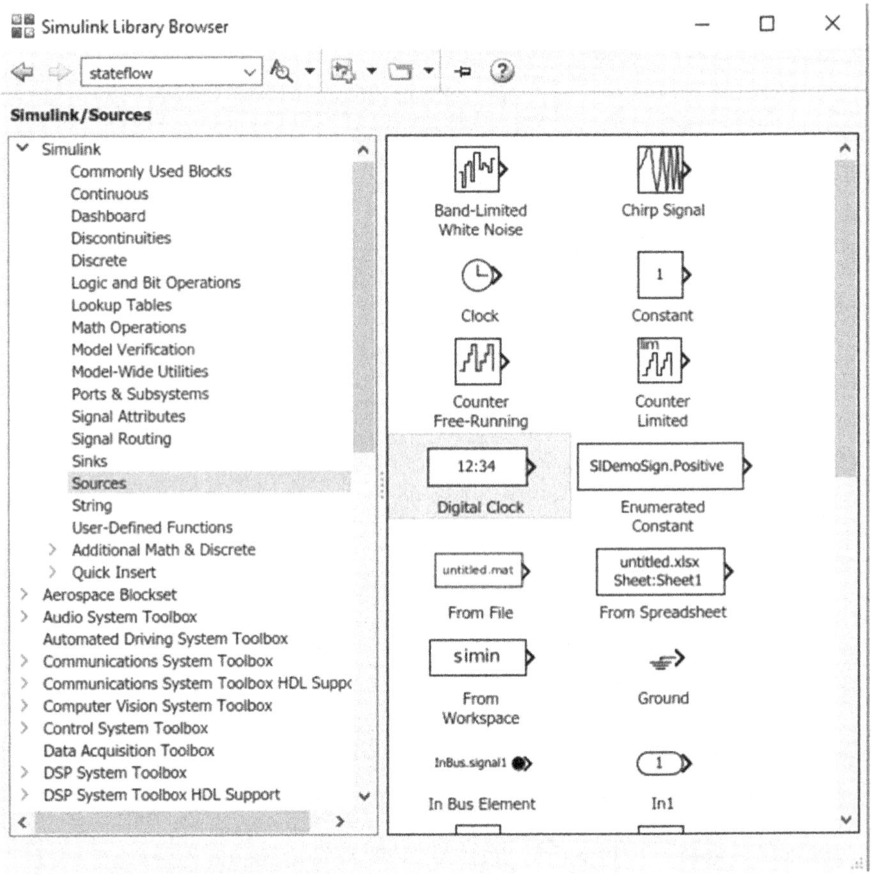

Figure 7.33 Adding digital clock block to the model.

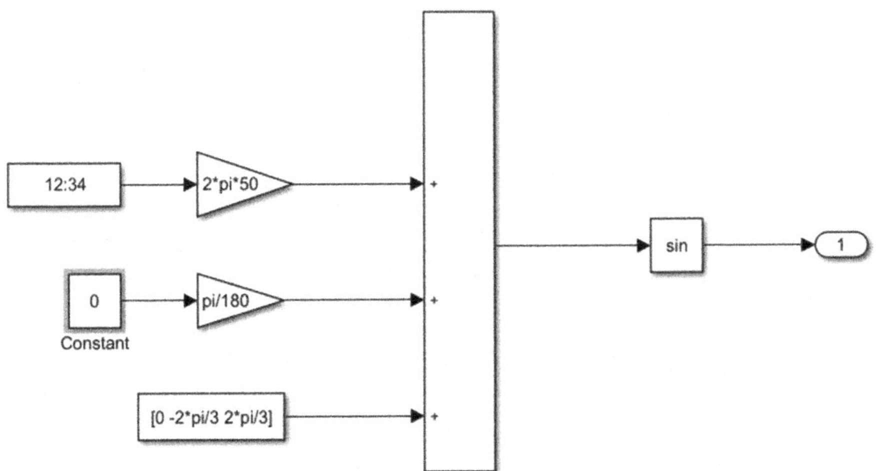

Figure 7.34 Model to generate sinusoidal signals.

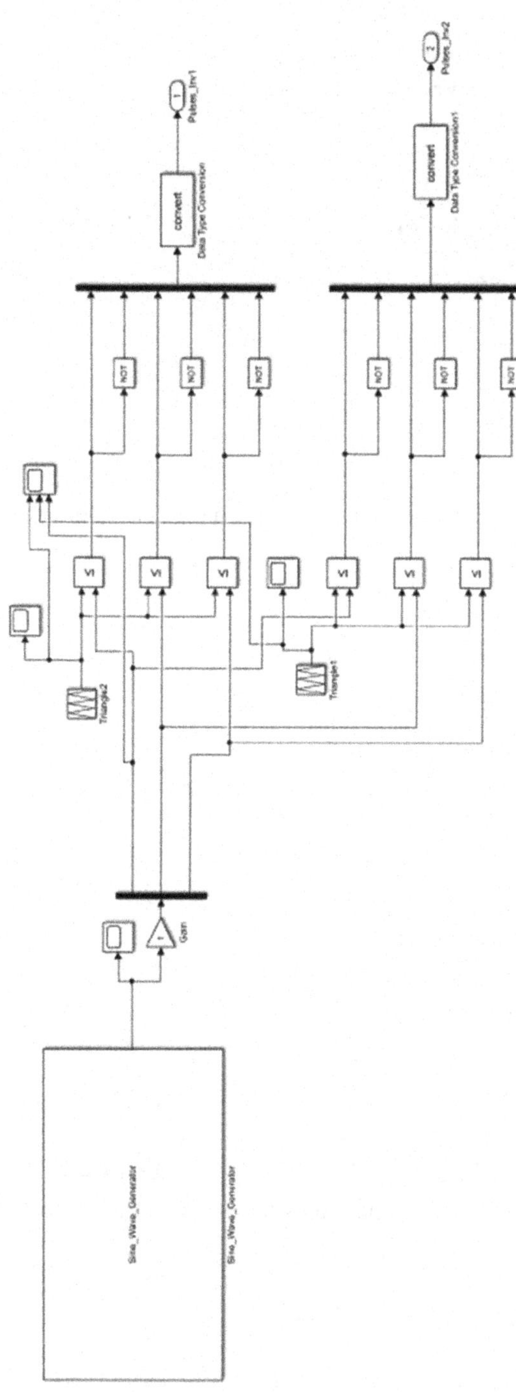

Figure 7.35 PWM generator.

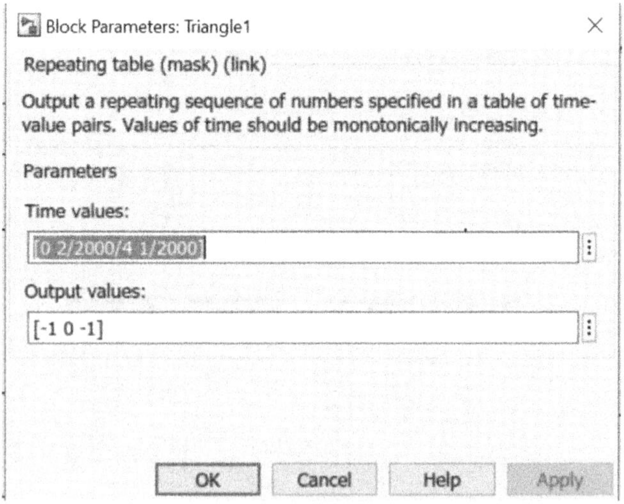

Block Parameters: Triangle1
Repeating table (mask) (link)
Output a repeating sequence of numbers specified in a table of time-value pairs. Values of time should be monotonically increasing.
Parameters
Time values:
[0 2/2000/4 1/2000]
Output values:
[-1 0 -1]
OK
Cancel
Help
Apply

Figure 7.36 Assigning block parameter values for triangle1 block repeating sequence block.

Block Parameters: Triangle2
Repeating table (mask) (link)
Output a repeating sequence of numbers specified in a table of time-value pairs. Values of time should be monotonically increasing.
Parameters
Time values:
[0 2/2000/4 1/2000]
Output values:
[0 1 0]
OK
Cancel
Help
Apply

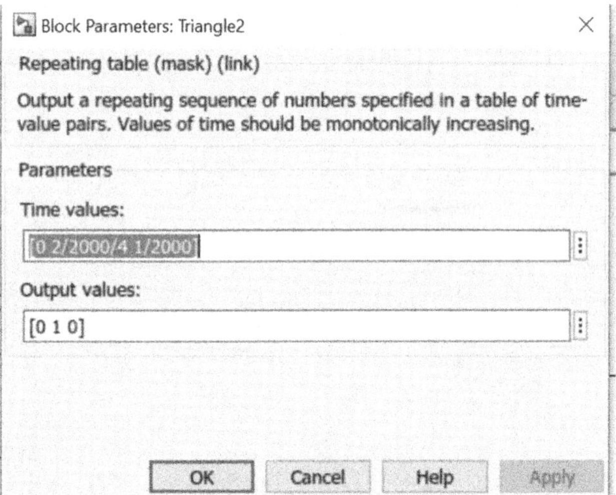

Figure 7.37 Assigning block parameter values for triangle2 block repeating sequence block.

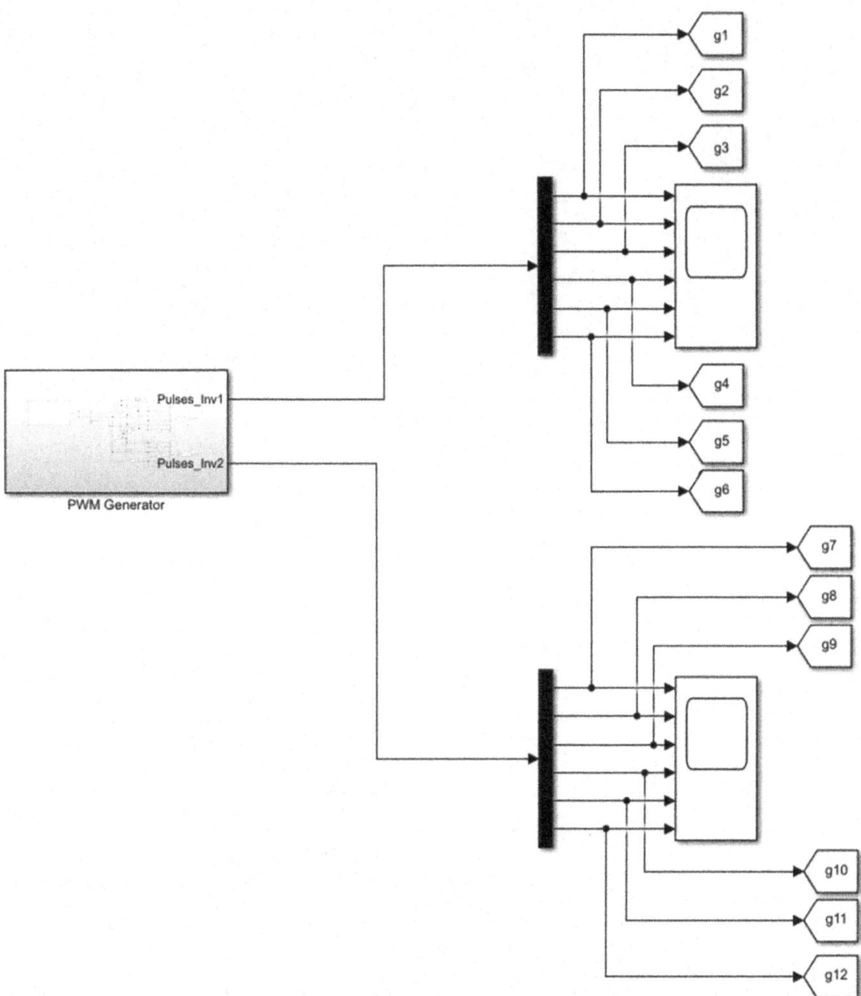

Figure 7.38 PWM generator subsystem.

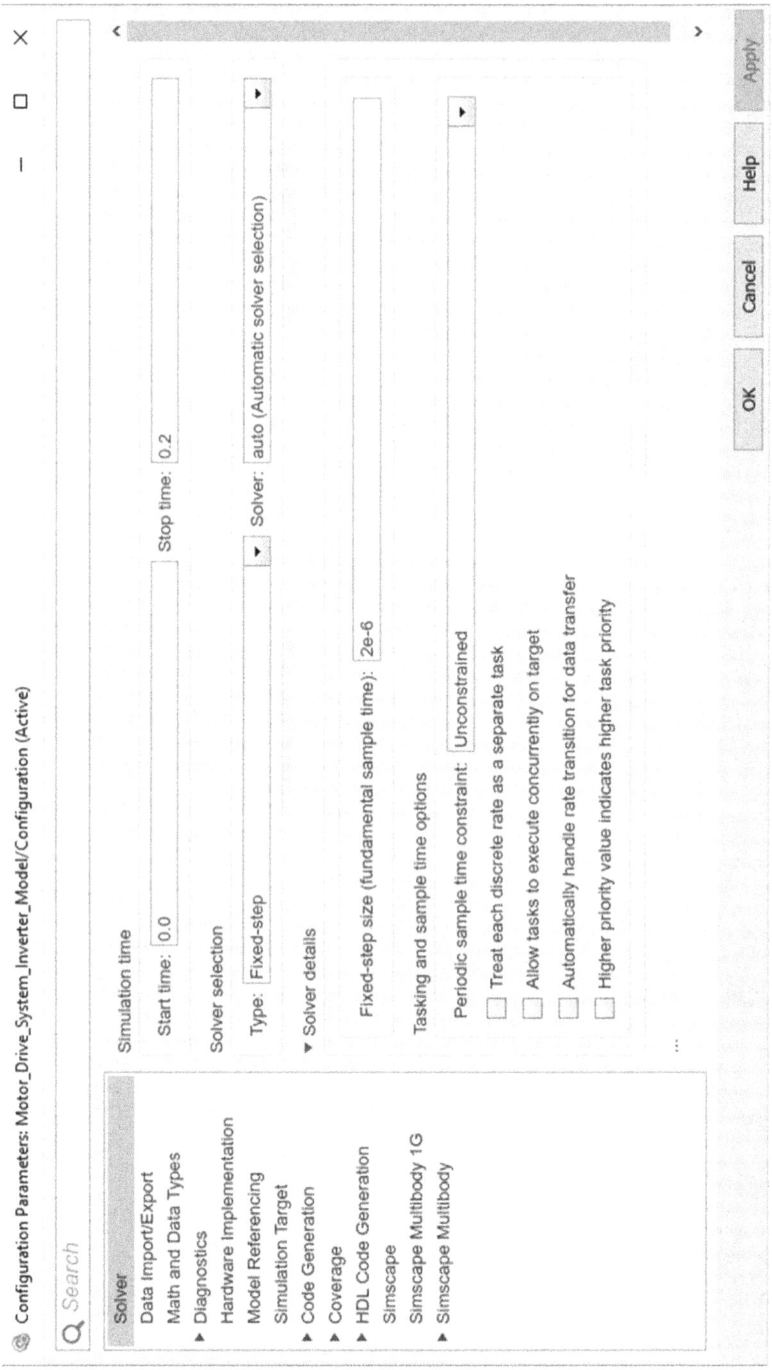

Figure 7.39 Model configuration parameters.

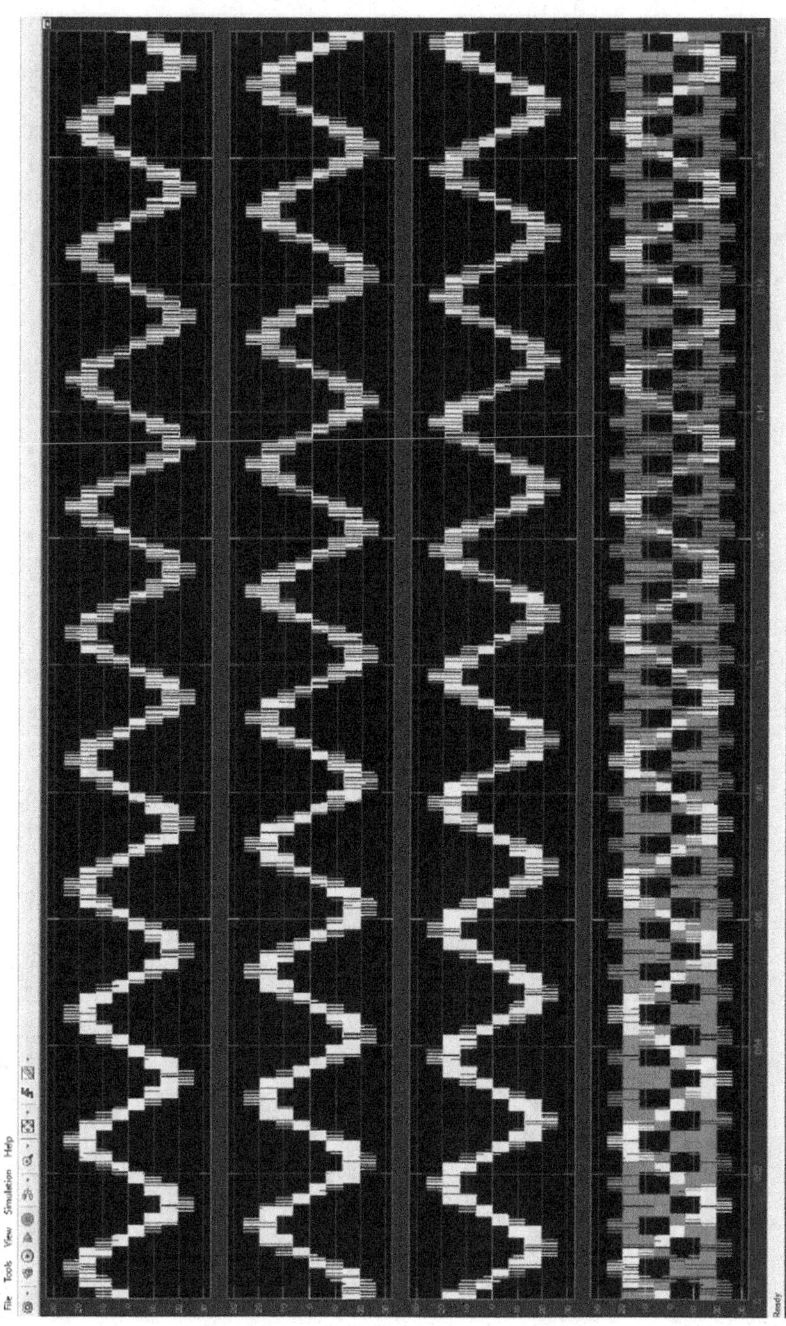

Figure 7.40 Inverter output currents from simulation.

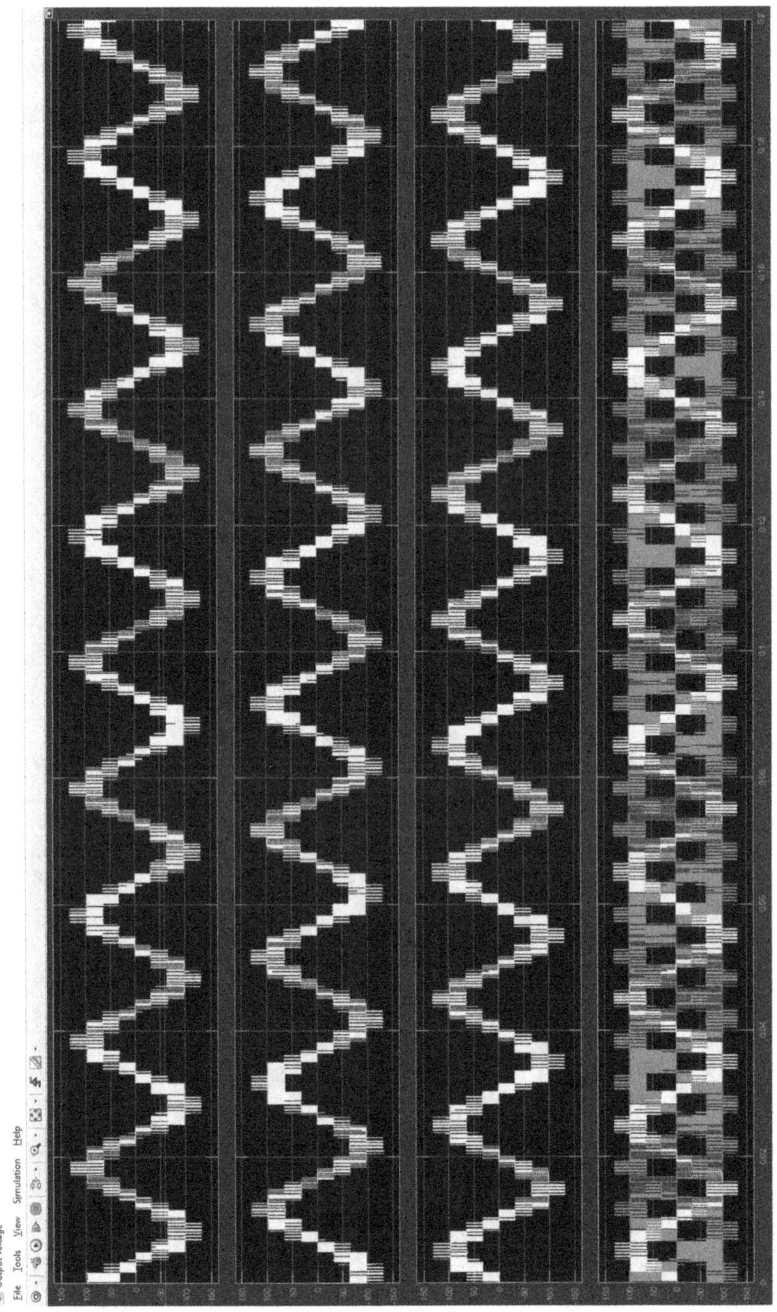

Figure 7.41 Inverter output voltages from simulation.

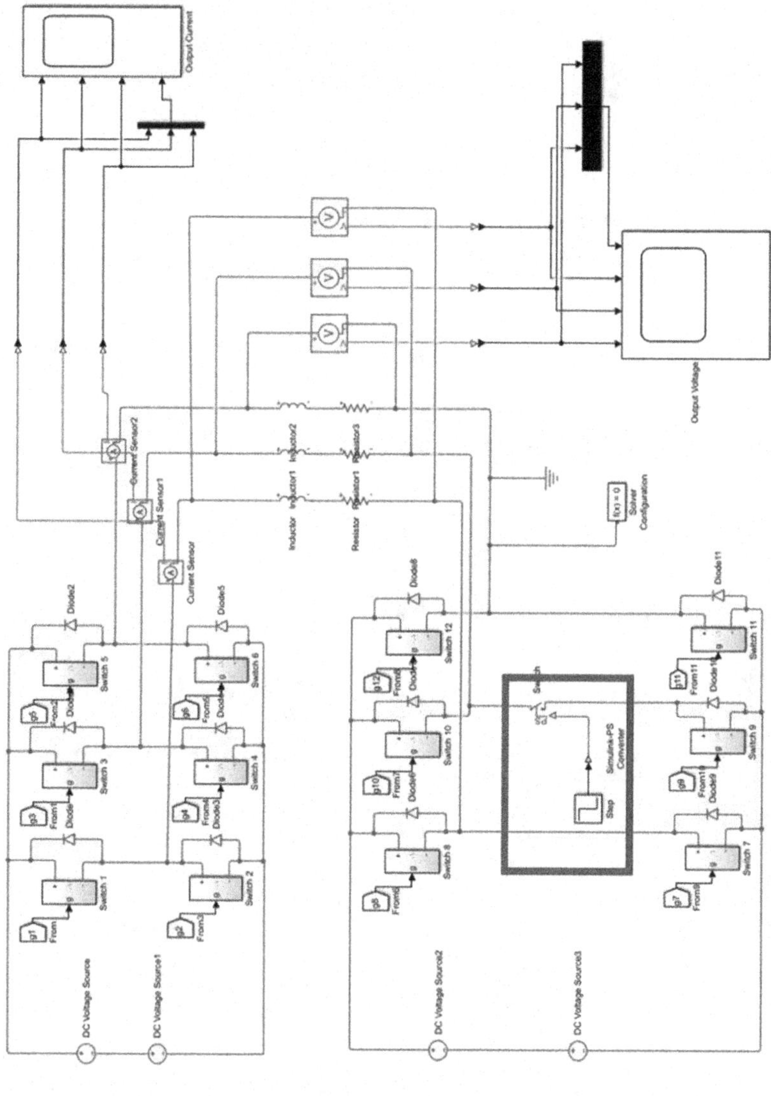

Figure 7.42 Introducing open-circuit failure to the inverter circuit.

Block Parameters: Step

Step

Output a step.

Parameters

Step time:

0.06

Initial value:

1

Final value:

0

Sample time:

0

☑ Interpret vector parameters as 1-D

☑ Enable zero-crossing detection

OK Cancel Help Apply

Figure 7.43 Configuring step input block for open-circuit failure.

Block Parameters: Switch

Switch

The block represents a switch controlled by an external physical signal. If the external physical signal PS is greater than the threshold, then the switch is closed, otherwise the switch is open.

The closed resistance is defined by parameter R_closed, and the open conductance is defined by parameter G_open. Both parameters must be greater than zero.

Source code

Settings

Parameters Variables

Closed resistance R_closed:	0.01	Ohm
Open conductance G_open:	1e-8	1/Ohm
Threshold:	0	

OK Cancel Help Apply

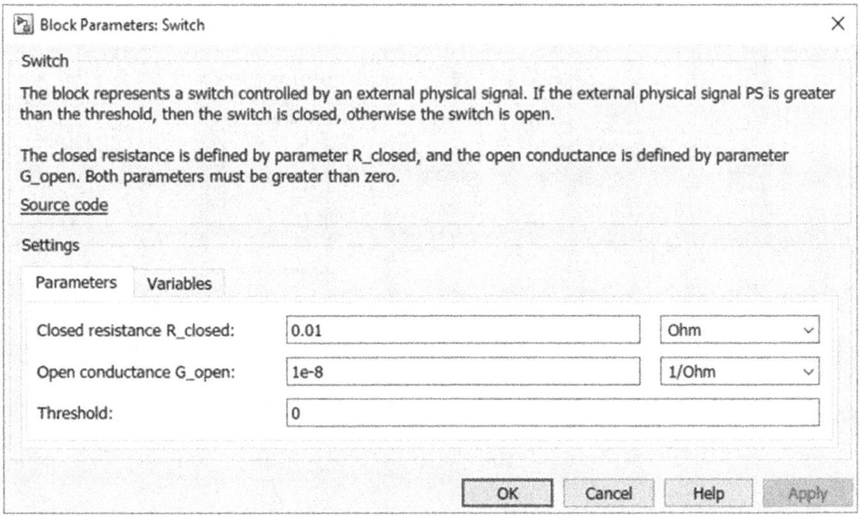

Figure 7.44 Configuring the switch block for open-circuit failure.

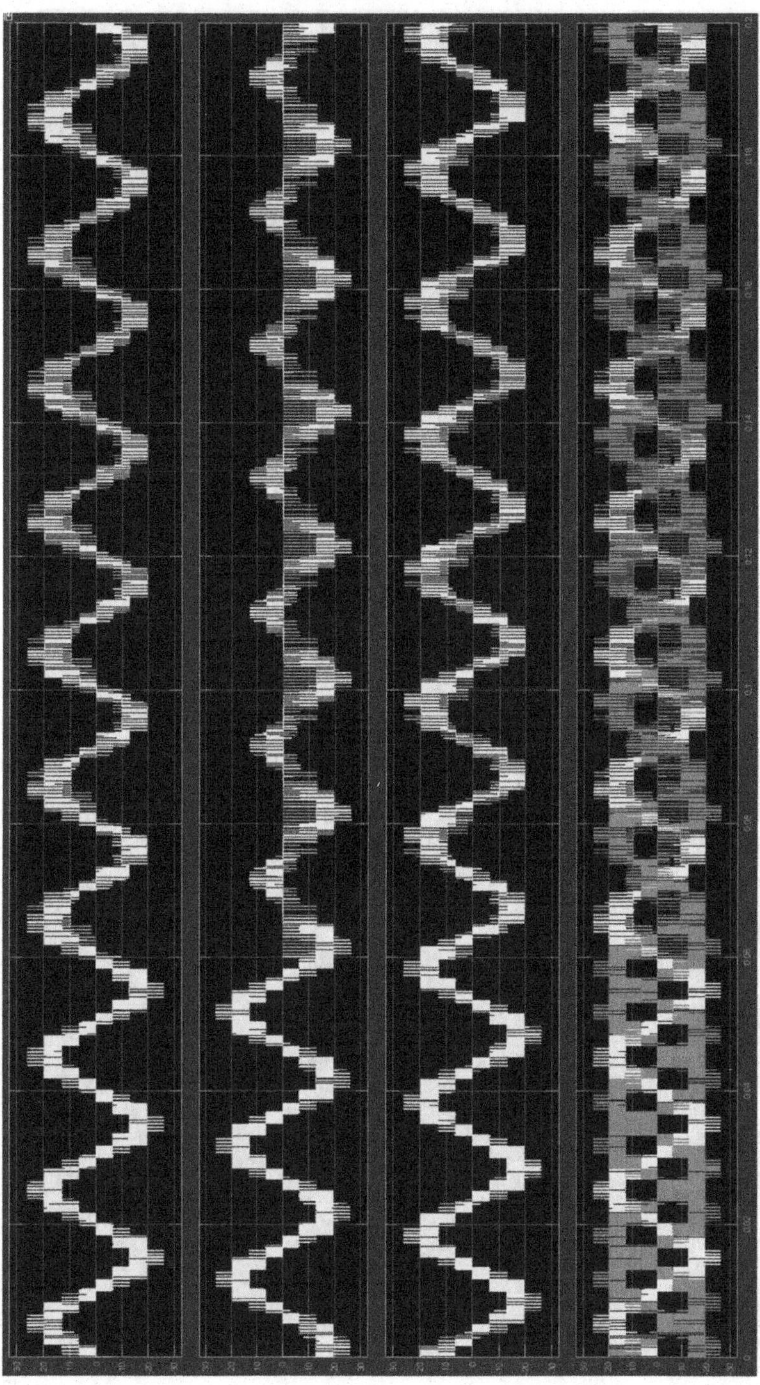

Figure 7.45 Inverter output currents from simulation with open-circuit failure.

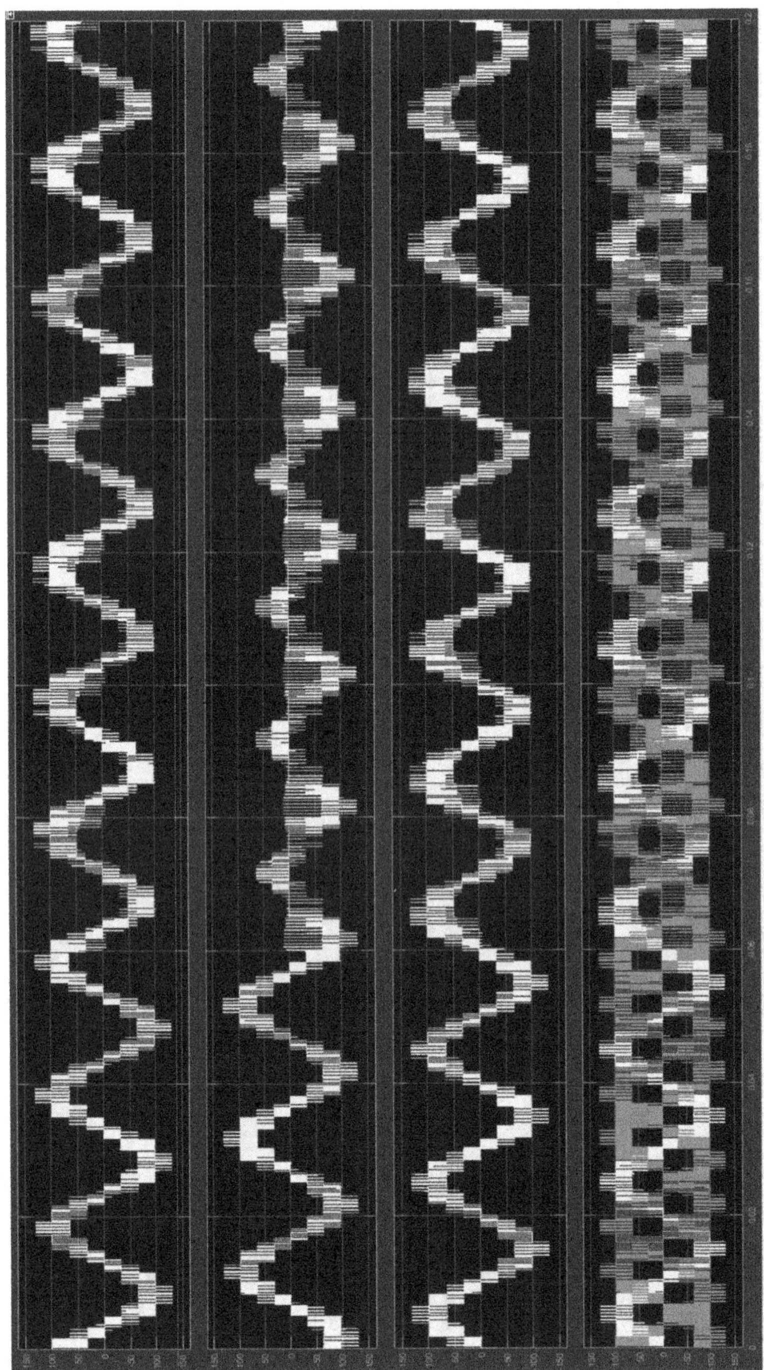

Figure 7.46 Inverter output voltages from simulation with open-circuit failure.

Reference

[1] E.C. dos Santos, E.R. da Silva, Cascade configuration, in: Advanced Power Electronics Converters: PWM Converters Processing AC Voltage, Wiley, 2014, pp. 125−171, ch. 5, sec.7.

Digital Twin development and cloud deployment for a Hybrid Electric Vehicle

8

8.1 Introduction

This chapter guides the user through the process of deploying a Hybrid Electric Vehicle (HEV) model for a Passenger Car, developed using MATLAB®, Simulink®, and Simscape™ into a Raspberry Pi hardware board in real time. Further a Digital Twin of the same HEV model is developed and deployed into the Amazon Cloud Services (AWS).The Raspberry Pi board will send the HEV system inputs, outputs, and states to the Amazon Cloud Services, and the Digital Twin model is ran concurrently on the cloud to perform off-board diagnostics (Off-BD) for detecting failures introduced in the model, which is running in Raspberry Pi. Fig. 8.1 shows the Off-BD steps that are going to be covered in this chapter. First, a couple of failure modes and failure conditions that will be detected by Off-BD is identified, then the block diagram of the Hybrid Electric Vehicle system is shown, after that the MATLAB®-, Simulink®-, and Simscape™-based model of the HEV is deployed on to the Raspberry Pi hardware, which works as the real physical asset in this case. With the inbuilt Wi-Fi capability of the Raspberry Pi, it can be configured as the Edge device to communicate to the cloud to transfer physical asset states from the model, which is running in the Raspberry Pi, to the cloud. A Digital Twin model of the same HEV system is compiled, and a Root Mean Square Error (RMSE) comparison–based diagnostic algorithm is also developed and deployed to AWS. So as the Simscape™ model of the HEV runs in the Raspberry Pi, a Digital Twin of the HEV is also running on the AWS with the same inputs collected from the Raspberry Pi, and real time diagnostic failure detection capabilities are demonstrated with Digital Twins.

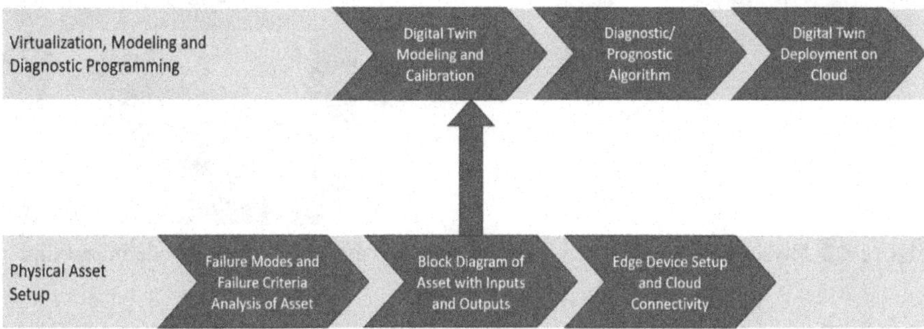

Figure 8.1 Off-BD steps covered in this chapter.

Digital Twin Development and Deployment on the Cloud. https://doi.org/10.1016/B978-0-12-821631-6.00008-6

All the codes used in the chapter can be downloaded for free from *MATLAB® File Exchange*. Follow the below link and search for the ISBN or title of the book:

https://www.mathworks.com/matlabcentral/fileexchange/

Alternatively, the reader can also download the material and other resources from the dedicated website or contact the authors for further help:

https://www.practicalmpc.com/

8.2 Hybrid Electric Vehicle physical asset/hardware setup

A Raspberry Pi computer board is used to run the MATLAB®, Simulink®, and Simscape™ model of the Hybrid Electric Vehicle in real time in a Virtual Hardware in Loop kind of mode. Authors call it Virtual Hardware in Loop because the HEV model is running on a Raspberry Pi real-time hardware (though it is not the actual physical hardware/asset like a vehicle's on-board computer because of the safety limitations to instrument the setup and collect data on a real vehicle, but it is definitely possible if we had the time and resources).We will be running the Simulink model of the HEV from a host computer in External Mode, with the target hardware selected as Raspberry Pi. The host computer and Raspberry Pi should be connected to the same Wi-Fi network as shown in Fig. 8.2.

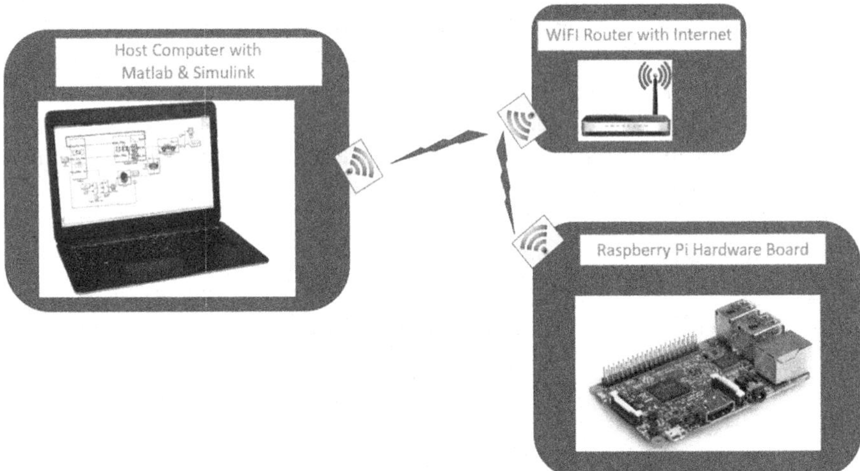

Figure 8.2 Physical asset/hardware setup running HEV model in external mode in real time.

Considering that the readers already have the host computer with MATLAB®, Simulink®, and Simscape™ already setup, and also there is a Wi-Fi router configured and available to connect, this section will focus mainly on configuring and installing necessary software for Raspberry Pi hardware needed for this chapter. The Raspberry Pi used by the Authors is Raspberry Pi 3 B+, which can be bought from various sources. The steps in this chapter can be applied on a different version of Raspberry Pi board as well as long as it supports Wi-Fi connection. One of the source links available at the time of writing is given in Ref. [1]. After getting a Raspberry Pi board follow the latest instructions from Ref. [2] to install and setup the operating system for the Raspberry Pi hardware, Authors really recommend setting up Raspbian OS, which is tested and verified in this chapter.

We will be using Python for communicating to the HEV model, which will be running on the Raspberry Pi for collecting the real-time data and also communicating to the AWS cloud. We will be first checking if Python is installed on the Raspberry Pi with the OS installation. By default, Python 2.7 should be installed with the OS. From the host computer, a remote connection can be established to the Raspberry Pi using the Putty Desktop App, which can be installed from the link [3]. After installing Putty, open up the Putty Desktop App and enter the IP address of the Raspberry Pi as shown in Fig. 8.3. The IP address of the Raspberry Pi can be obtained using the ***ifconfig*** command by directly connecting the Raspberry Pi hardware to a keyboard, and a monitor. A new window as shown in Fig. 8.4 will be popped up to enter the login credentials of the Raspberry Pi. Enter the login details used at the time of installing the OS of the Raspberry Pi.

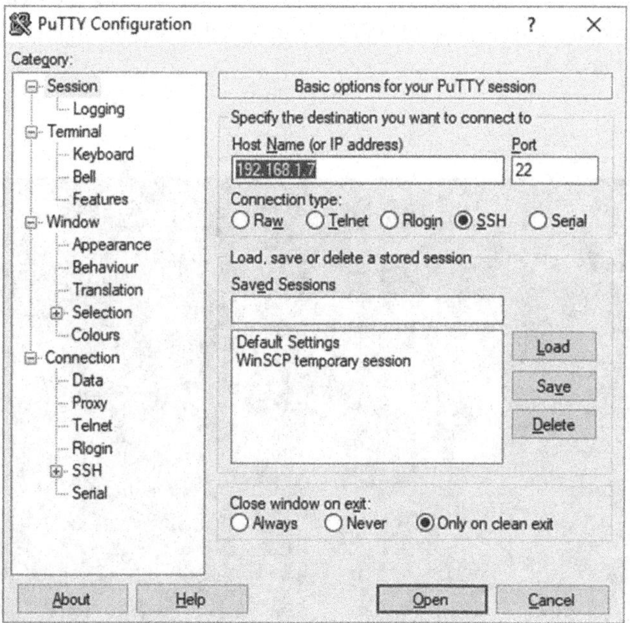

Figure 8.3 Connecting to Raspberry Pi remotely using Putty.

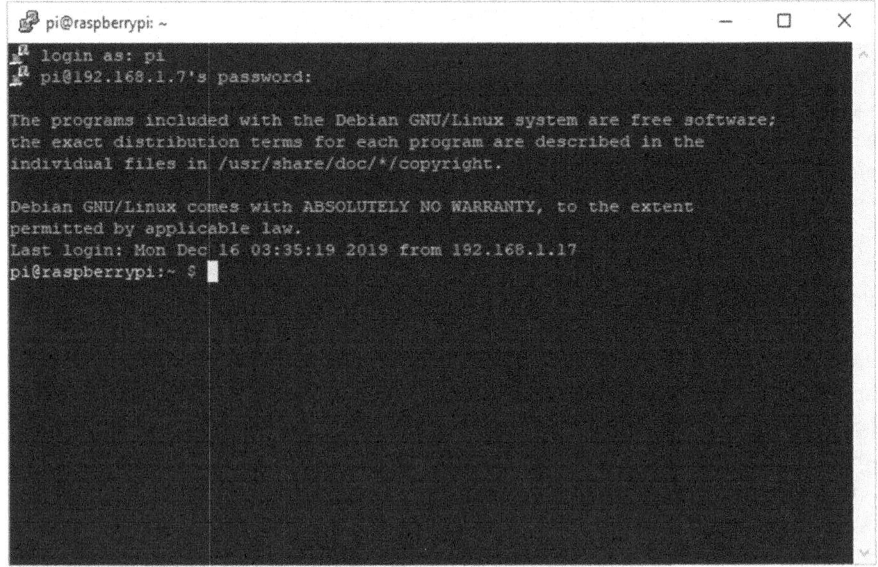

Figure 8.4 Login to Raspberry Pi remotely using credentials.

After remotely connecting to the Raspberry Pi, on the Putty command line, enter the command ***python - -version*** and ***pip - -version*** to check whether the Python software and Python package installer are installed. If installed, it should show up the versions as shown in Fig. 8.5.

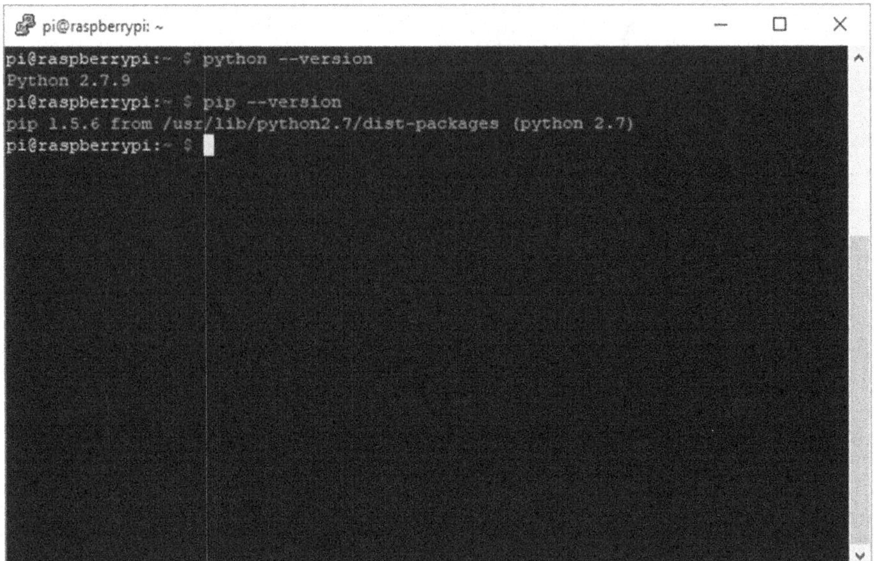

Figure 8.5 Checking Python and PIP versions.

Next we need to install the Mosquitto MQTT communication protocol broker and clients to the Raspberry Pi. MQTT is a lightweight messaging protocol designed for low-cost and low-power embedded systems. This protocol and broker are primarily responsible for receiving all messages, filtering them, deciding who is interested in it, and publishing the messages to all subscribed clients. In our example, the Simscape™ model running on the Raspberry Pi will be sending MQTT messages to the MQTT broker with the real-time inputs, states, and outputs of the HEV system. The broker will then route the message to a Python program, which has subscribed to the broker. The Python program will then buffer the messages and reformat the data and send to the AWS cloud. For installing the Mosquitto MQTT broker, enter the command ***sudo apt-get install mosquitto*** on the Putty command line as shown Fig. 8.6. After that install the Mosquitto client by entering the command ***sudo apt-get install mosquito-clients***, see Fig. 8.7. We will also need to install a Python MQTT library called Paho-MQTT. This library allows Python programs to connect to the MQTT broker and receive and publish messages. Install the Paho-MQTT using the command ***sudo pip install paho-mqtt*** as shown in Fig. 8.8. Paho-MQTT is a specific Python library, which is why in this command we used the pip.

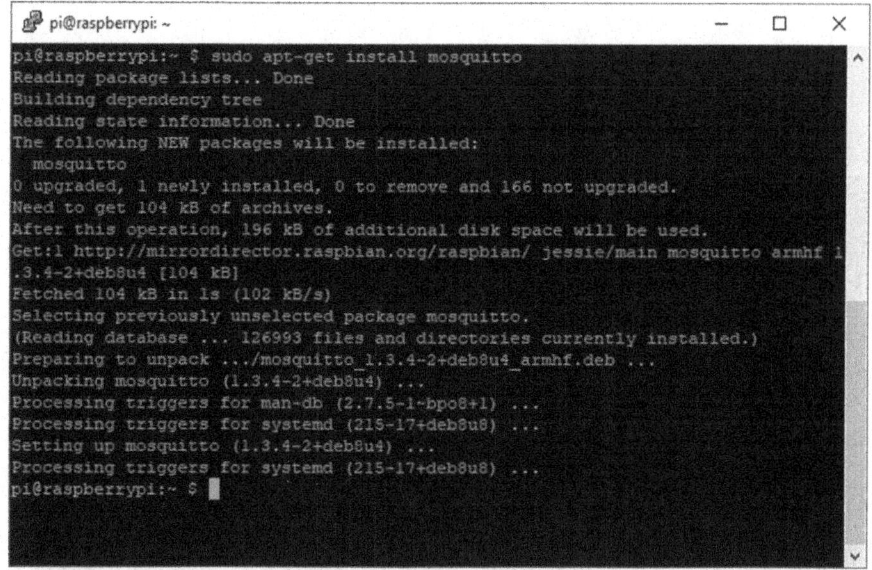

Figure 8.6 Installing Mosquitto MQTT broker.

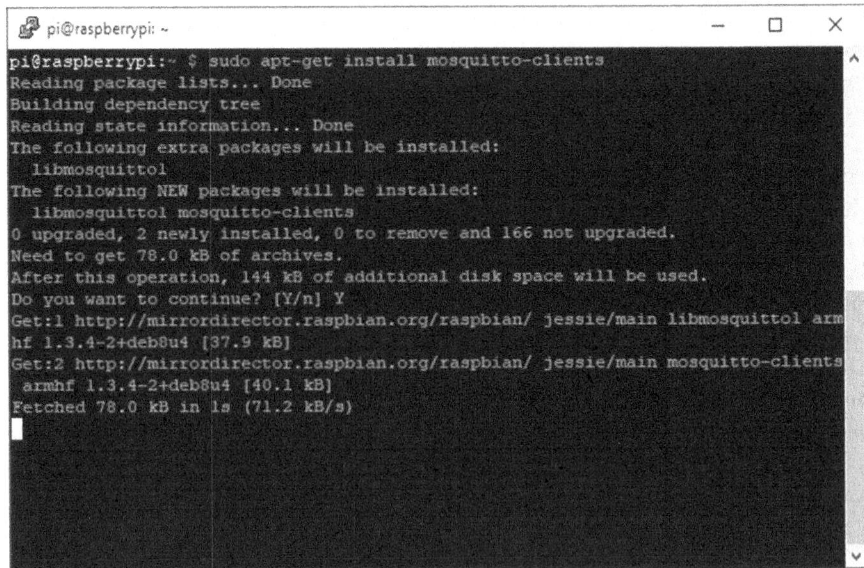

Figure 8.7 Installing Mosquitto MQTT client.

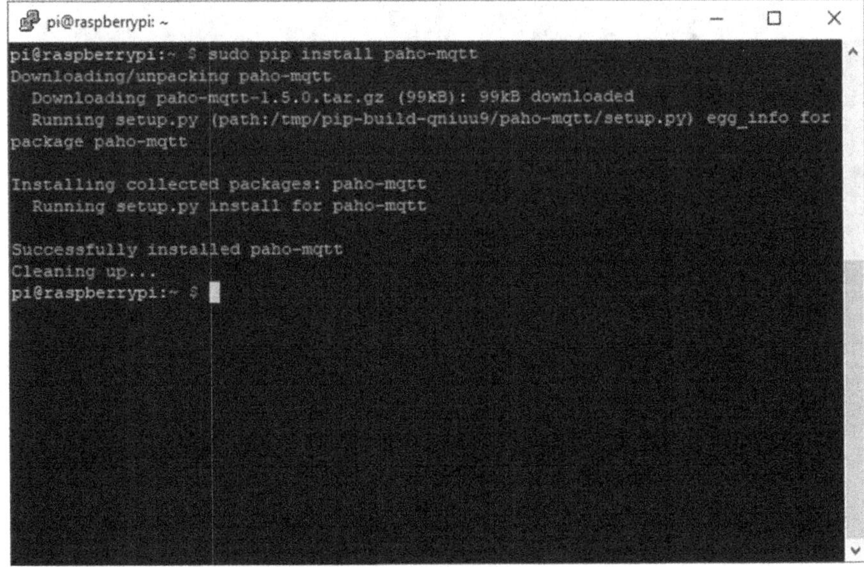

Figure 8.8 Installing Pah0-MQTT client for Python.

8.3 Block diagram of the Hybrid Electric Vehicle system

Fig. 8.9 shows the block diagram of the inputs/outputs and states of the Hybrid Electric Vehicle system. The input to the system is the target speed of the vehicle, and the vehicle controller then acts on the target speed and commands the Electric Motor, Generator, and the Engine in the HEV system accordingly to meet the target vehicle speed. So the output of the system is the actual vehicle speed. Also we are outputting the states of the system such as Motor Speed, Generator Speed, Engine Speed, and Battery SoC for Off-BD monitoring.

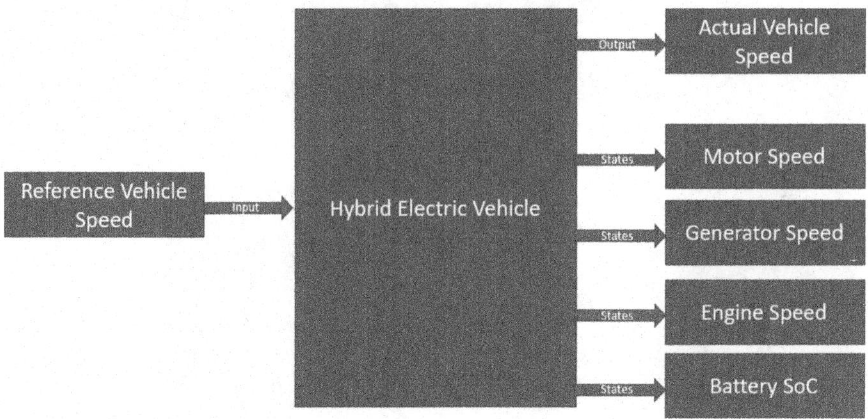

Figure 8.9 Block diagram of the Hybrid Electric Vehicle system.

8.4 Failure modes and diagnostic concept of Hybrid Electric Vehicle system

Fig. 8.10 shows the block diagram of the Off-BD Digital Twin diagnostics process followed in this chapter. The input Vehicle Speed Reference is fed to the HEV model which is running on the Raspberry Pi and also sent to the cloud along with the actual states and outputs of the system from the hardware. In the AWS cloud, a Digital Twin model of the same HEV system will be running with the same input and the states and outputs of the Digital Twin model are compared with the actual data collected from the hardware. A RMSE-based diagnostic detection and decision-making is developed to compare the actual and digital twin outputs and states. In this chapter, an example to detect the Throttle failure of the Internal Combustion Engine in the HEV system is demonstrated. A similar approach can be followed to detect failures of the other components of the system as well. A throttle failure is introduced into the model

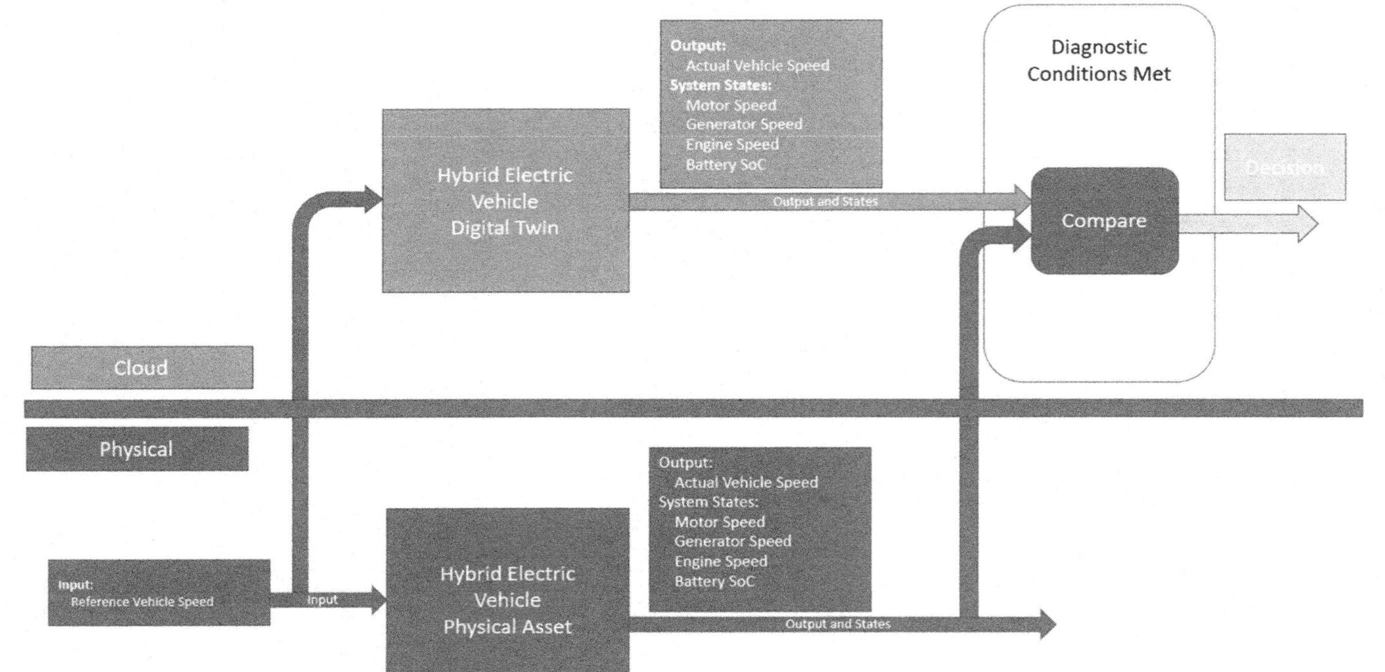

Figure 8.10 Off-BD diagnostics process for Hybrid Electric Vehicle system.

that is running in Raspberry Pi by overriding the Throttle which is commanded internally by the HEV controller for a given vehicle speed reference to zero. Because the Throttle is zero for the model which is running in Raspberry Pi, the HEV system will behave differently and its states and outputs will be different when compared to the states and outputs of the Digital Twin model for which the input is only the vehicle speed reference and no other failure conditions. This difference in the outputs and states between the Actual and Digital Twin data is identified, and if the RMSE is greater than a threshold, a failure condition is flagged and the User is notified directly from the cloud using a Text or E-mail notification.

8.5 Simscape™ model of a Hybrid Electric Vehicle system

Hybrid Electric Vehicle system is a multidisciplinary multi physicssystem including Mechanical systems such as Engine, Transmission and Powertrain, Chemical dynamics of Internal Combustion recipe of the Engine, Battery systems, and the Electrical systems including Generator, Motor, etc. Based on the Author's Automotive Industry experience, various levels of fidelity models are utilized in the industry depending on the specific application for which the models are being used for. It could be a simple map-based model for each component, a full physics-based model for all components, or a mix where some components are modeled using maps and some components with higher fidelity physics-based modeling. But the challenge always will be validating the model performance with data from the real physical system/asset and drawing a line for the usefulness of the models for the applications that are being considered. Considering the complexity of the HEV system, and all the various HEV system combinations and variations out there, the main focus in this chapter is not really to follow the step-by-step process of developing the model and validating it, instead Authors will focus more on using a validated HEV model that is readily available in MATLAB Central File Exchange portal and deploying the model to the Raspberry Pi hardware and also developing and deploying the Digital Twin on the same model to the AWS cloud and doing a real-time Off-BD.

The HEV model published in MATLAB Central link [4] is used in this chapter. This is a Series–Parallel Hybrid Electric Vehicle system model, where the vehicle drivetrain can be powered by the internal combustion engine working alone, the electric model working alone, or by both the combustion engine and motor system working together. There is a supervisory power split controller that tries to optimize the power split by trying to operate the engine at its optimal operating region. More details about the various hybrid electric configurations can be found in Ref. [5]. Fig. 8.11 shows a high-level block diagram and the flow of the power and connections in the series parallel HEV.

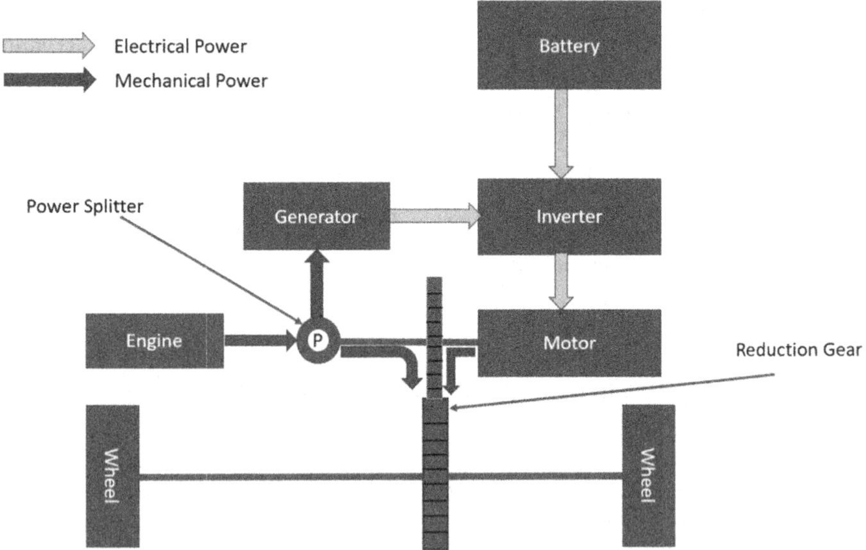

Figure 8.11 Series−Parallel Hybrid Electric Vehicle configuration [5].

The MATLAB Central HEV model by default is configured to run with a continuous time-step solver, but we have to run it with a fixed-step discrete time solver in order to run it on the Raspberry Pi hardware. So the Authors have made some necessary changes on the model downloaded from MATLAB Central to work with a discrete-time solver with a fixed step size of 0.1 s. Fig. 8.12 shows the top-level Simulink diagram of the HEV Simscape™ model. A description of various systems and components of the model is given further, as already mentioned, and the Authors did not really develop this model, so it is mainly explaining what was already in the MATLAB Central downloaded model and its purpose.

The HEV model mainly consists of the following components:

1. Vehicle Speed Controller Module

A time-based target vehicle speed is selected for the HEV system from a few set of vehicle speed profiles and fed using From Workspace blocks as shown in Fig. 8.13. Further, this target vehicle speed and the actual vehicle speed are fed to a High-Level Proportional Integral (PI) controller to derive the acceleration command or brake request to decelerate the vehicle as required as shown in Fig. 8.14. If the actual vehicle speed is less than the reference speed, the high-level controller will request more acceleration, and this acceleration request will be a low-level supervisory controller to produce the required torque either from motor, engine, or both to meet the required acceleration. But if the actual vehicle speed is higher than the reference speed, the controller will request deceleration through the brake pedal signal, which will be directly applied to the vehicle dynamics system, which will be converted to a brake torque to decelerate the vehicle.

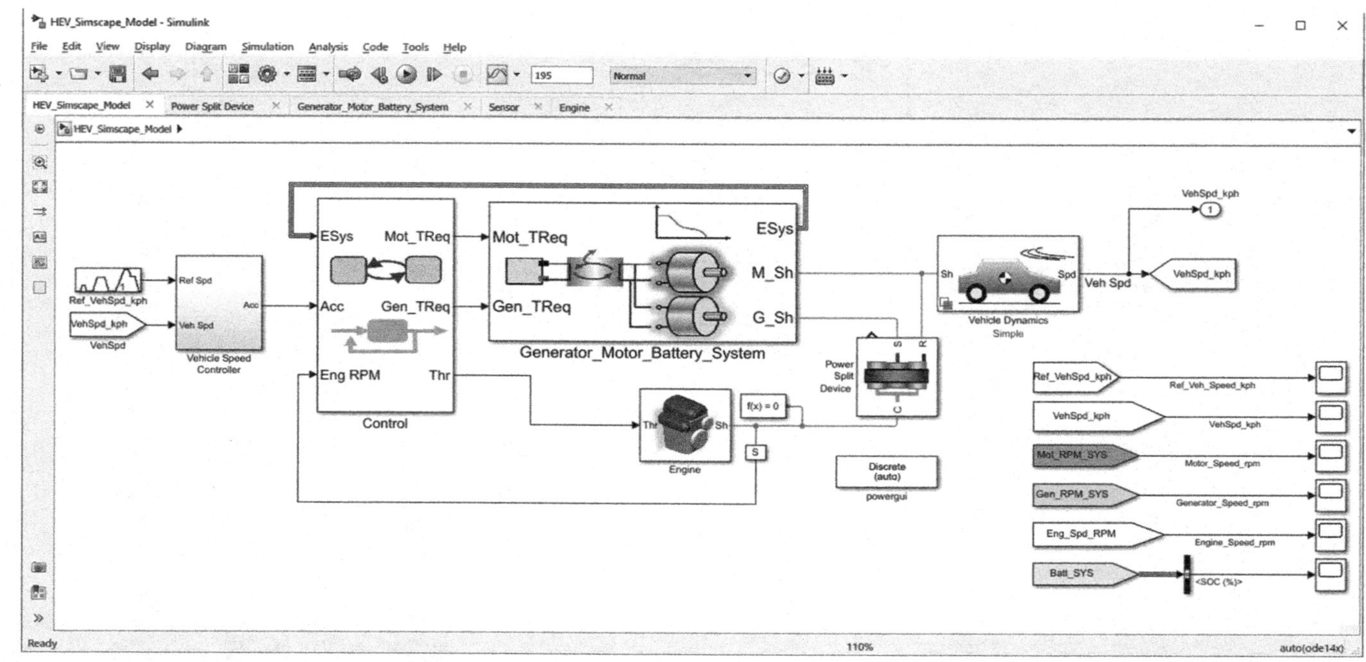

Figure 8.12 Top-Level HEV Simscape™ model.

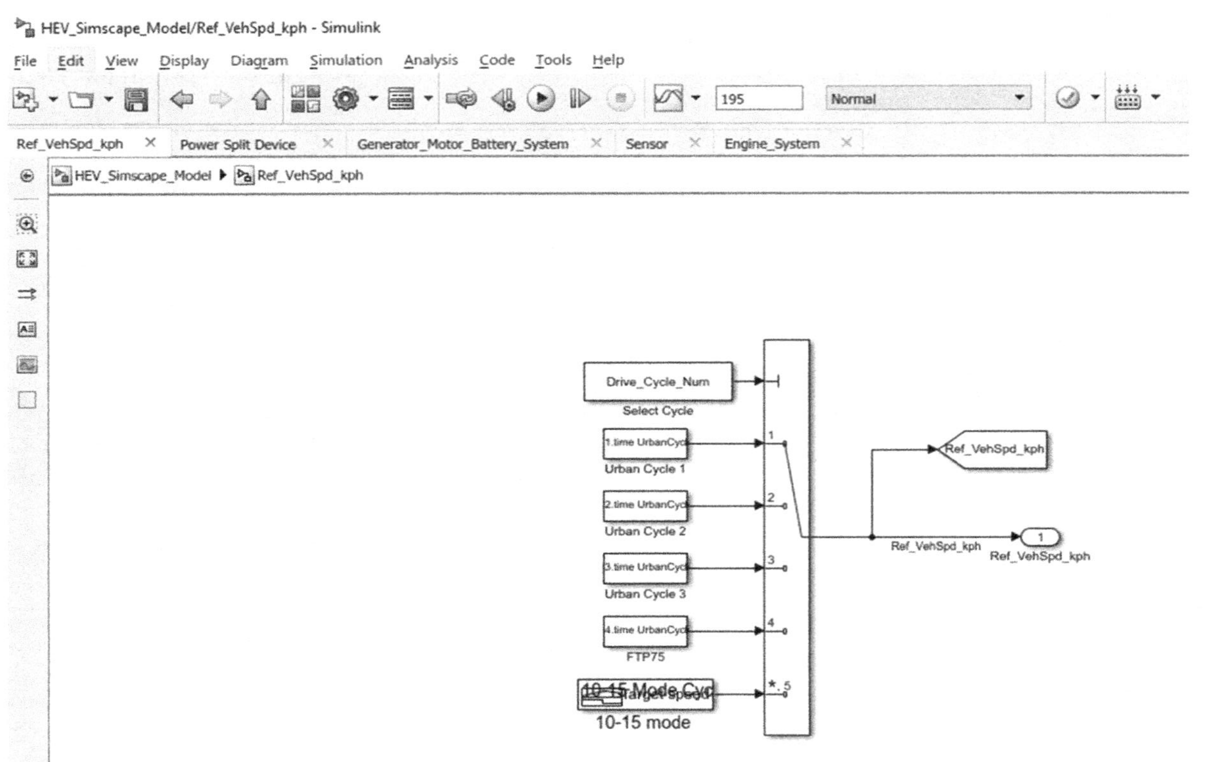

Figure 8.13 Target vehicle speed selection.

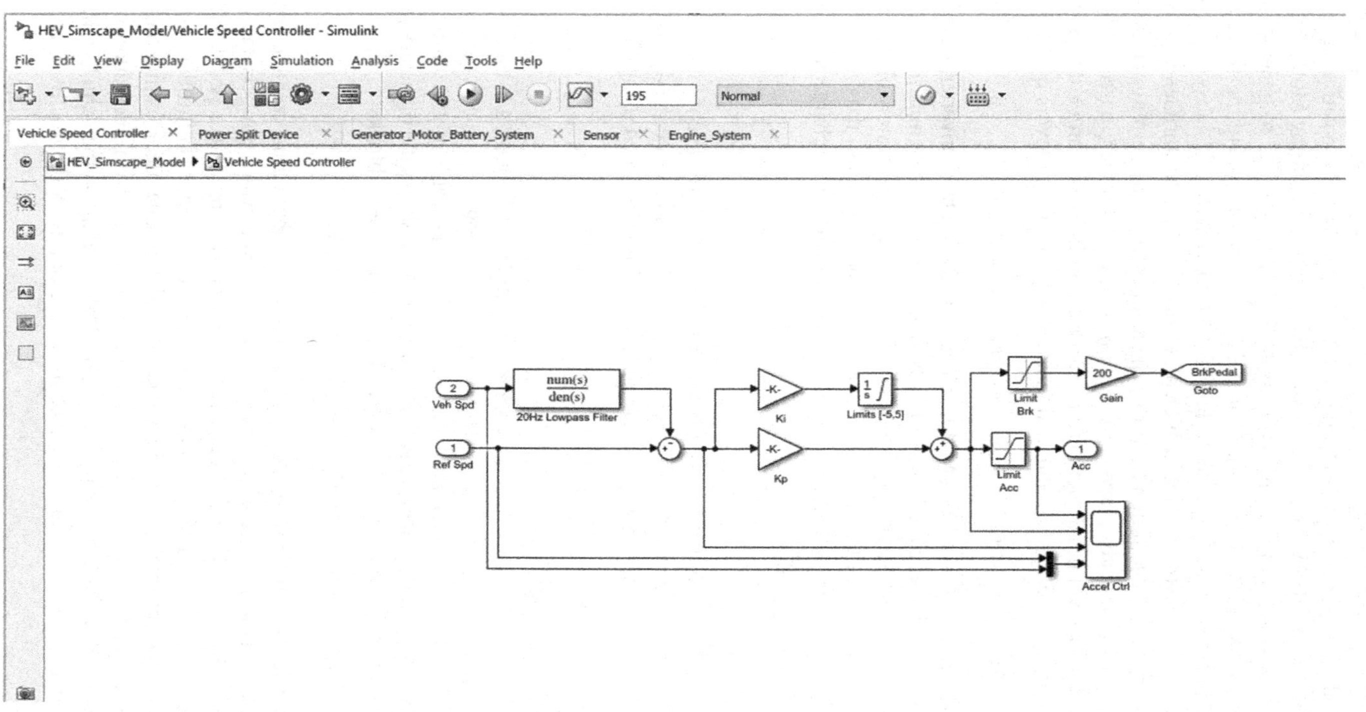

Figure 8.14 Proportional integral controller to control the acceleration and brake of HEV.

2. Supervisory Power Split Controller Module

This module or subsystem is responsible for converting the positive acceleration request of the high-level vehicle speed controller into corresponding Motor command, Generator command, and the Engine Throttle command requests to generate torques from these components to meet the required acceleration by operating these components at its optimal region. As shown in Figs. 8.15 and 8.16, this subsystem includes a mode selection logic that decides which of the devices from Motor/Generator/Engine to be operating and the required power from each based on the current operating conditions, state of the system, and acceleration demands. Further, based on selected mode, four lower-level controllers are used for controlling the torque demands from Motor, Generator, and Engine and also maintaining the charge levels of the battery.

3. Generator, Motor, DC–DC Converter, and Battery System

This subsystem includes the Motor, Generator, DC–DC Converter, and Battery system models or the so called plant models. This whole subsystem is modeled using various block sets from Simscape™ toolbox. The motor and generator subsystems take the torque requests from the respective controllers and respond to it as rotational motion to produce the required torque. The DC–DC converter will charge the battery to store the electrical energy.

4. Engine System

The engine system or the engine plant model takes the throttle command from the engine controller and convert that to a rotational motion to produce the demanded torque. See Fig. 8.17 for the engine system plant model implemented using Simscape™ blocks.

5. Power Split System

The power split device hooks the combustion engine, generator, and electric motor together and the vehicle driveline system together. This power split device is the essential part of the HEV system, and it makes the routing of power from the various power sources possible with its planetary gear system. Fig. 8.18 shows the planetary gear system logic in the HEV model implemented using Simscape™ blocks.

6. Vehicle Dynamics System

The vehicle dynamics subsystem implements the longitudinal vehicle dynamics equation given below. Here F_{wheel} is the tractive force which propels the vehicle, and F_{brake} is the braking force generated when pressing the brake pedal. Note that the BrkPedal signal is coming from the high-level vehicle speed controller subsystem, and it will be greater than zero if the vehicle speed controller decides that the vehicle needs to decelerate. $F_{drag}, F_{gravity}, F_{rolling}$ are, respectively, the forces due to aerodynamic drag, gravity, and rolling resistance, which are acting on the vehicle. The brake, drag, gravitational, and rolling resistance forces are opposing the tractive force, which

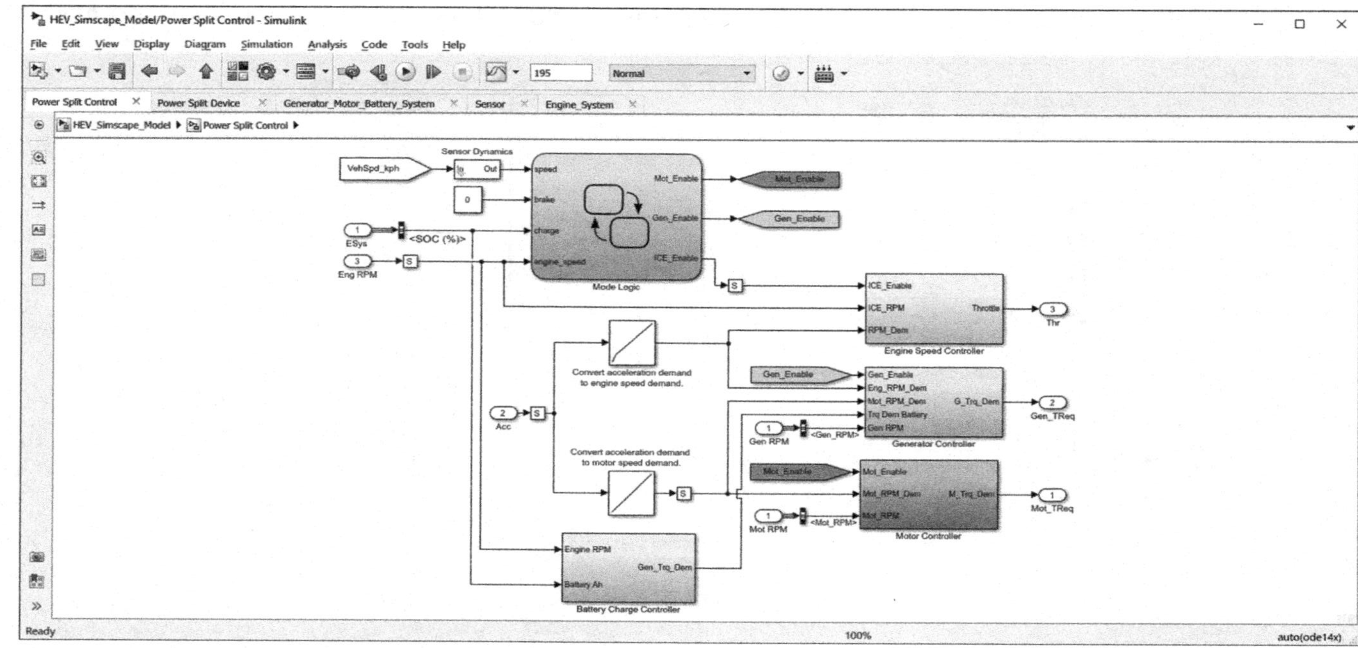

Figure 8.15 Supervisory Power Split Controller Module.

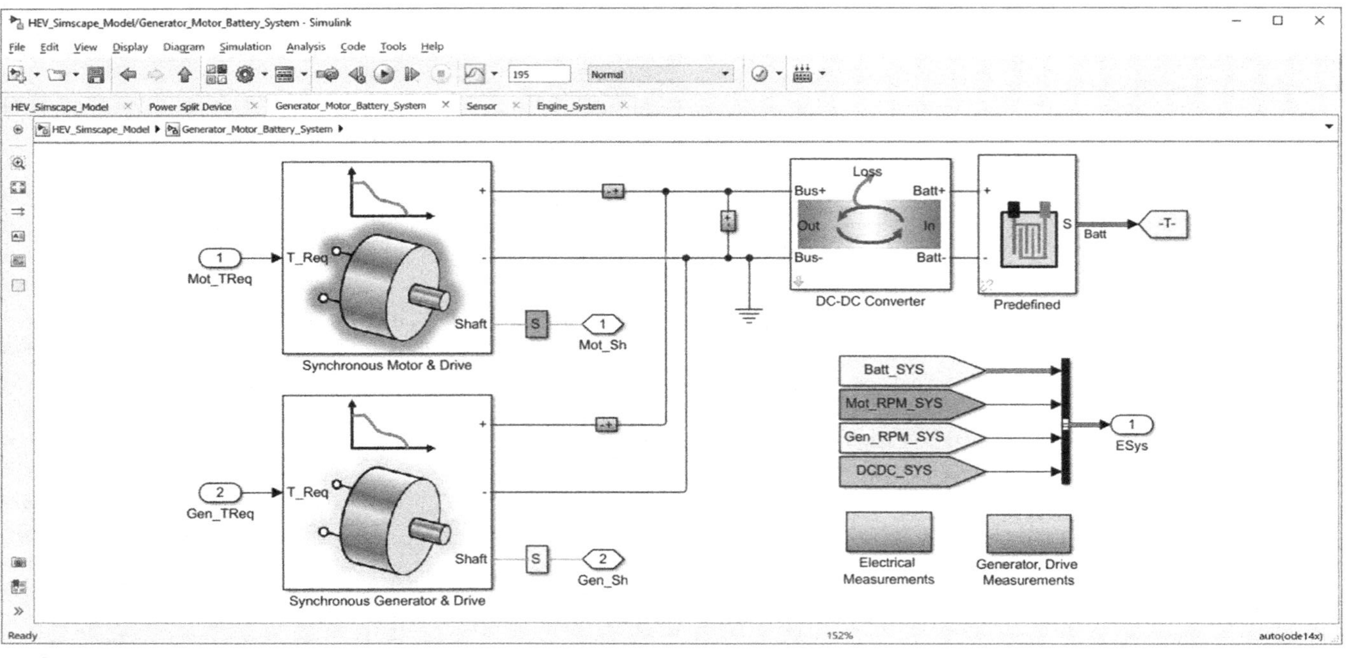

Figure 8.16 Generator, Motor, DC–DC Converter, and Battery system plant.

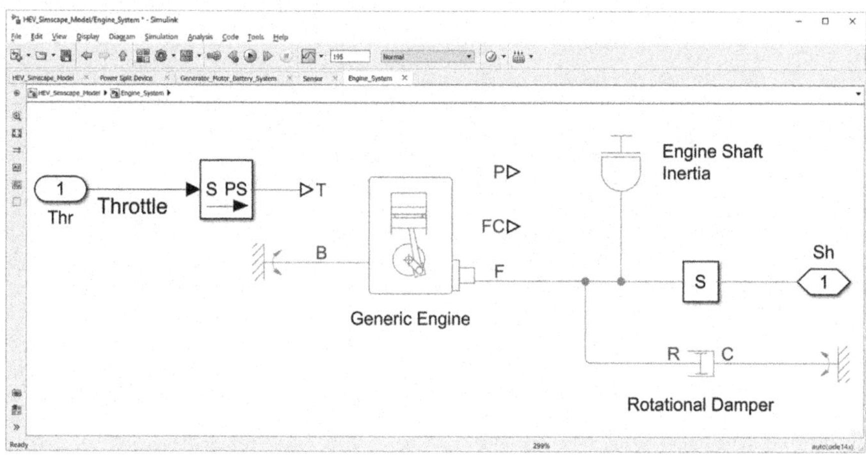

Figure 8.17 Engine system plant.

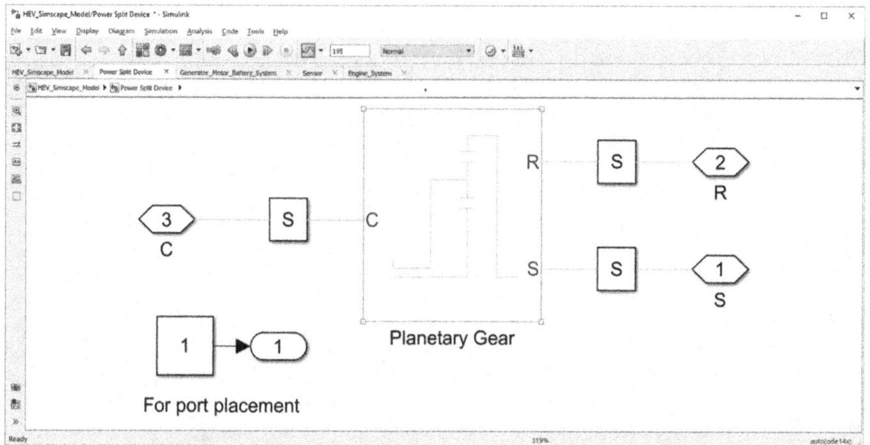

Figure 8.18 Power split planetary gear system plant.

is why all these forces are subtracted in the force balance equation. m_{eqv} is the equivalent mass of the vehicle, which is not a fixed mass, it varies based on the gear ratio mainly. $\dot{v}$ is the vehicle acceleration. Fig. 8.19 shows the vehicle dynamics subsystem implementation using Simscape™ and Simulink blocks.

$$m_{eqv} * \dot{v} = F_{wheel} - F_{brake} - F_{drag} - F_{gravity} - F_{rolling}$$

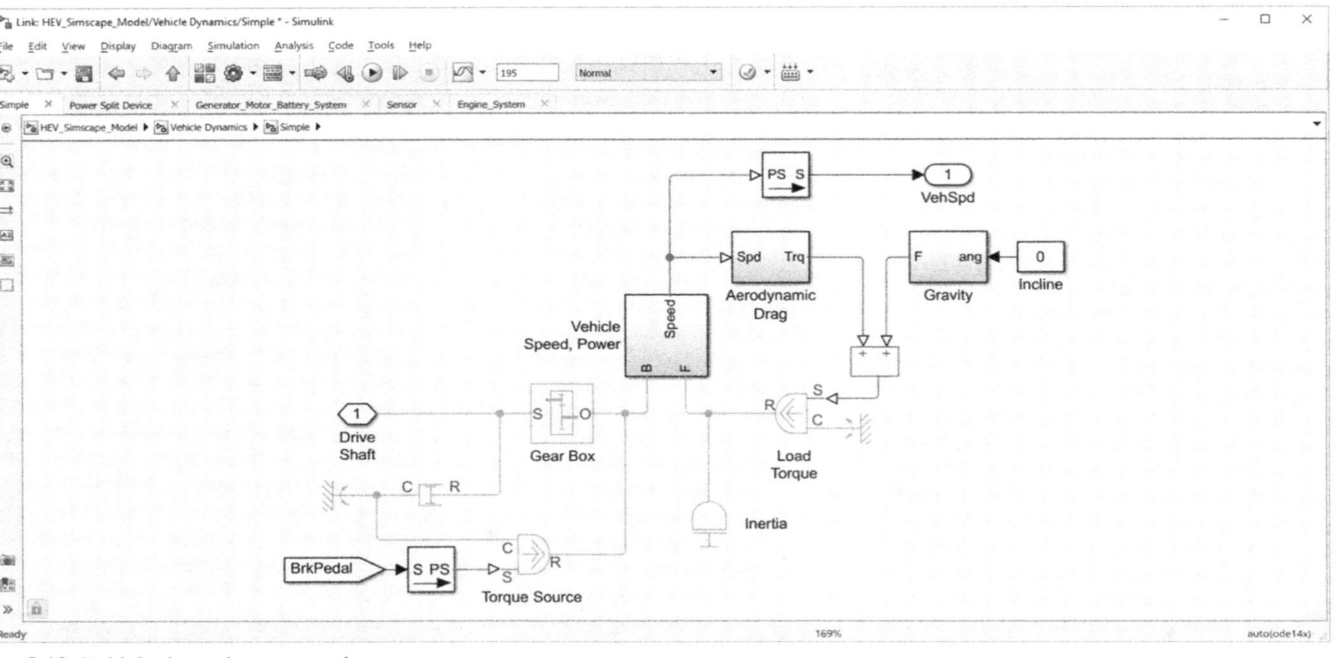

Figure 8.19 Vehicle dynamics system plant.

7. Data Monitoring and Logging

All the inputs, outputs, and states of interest of the system are logged and monitored using scope blocks from Simulink as shown in Fig. 8.20.

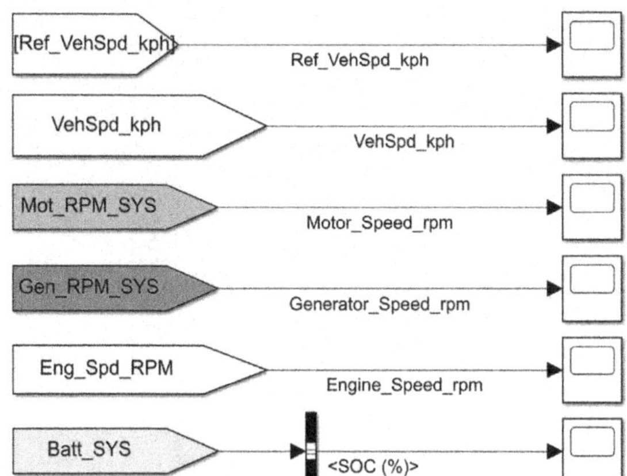

Figure 8.20 Data monitoring and logging.

Open the model HEV_Simscape_Model.slx and run the simulation from the attached folder under **Code_Files\Discrete_Time_HEV_Model** for the Chapter 8. When we open the model all the required paths will be added to the MATLAB path, and after the simulation is completed, it will plot all the key signals HEV system such as Reference and Actual Vehicle speed, Motor, Generator, and Engine Speeds and Battery SoC, etc, as shown in Fig. 8.21. It can be seen from the plot that the Engine, Motor, and Generator work together to meet the power required for vehicle speed reference. It has a pretty good vehicle speed tracking, and also during the Acceleration events, the Battery charge also depletes some, but during the steady-state and deceleration events, the controller tries to charge the battery back.

Next we will prepare to run the HEV Simscape™ on the Raspberry Pi hardware. In addition to the MATLAB® and its toolboxes already installed and used for the previous chapters, we need to install the hardware support package for Raspberry Pi for developing and deploying Simulink controls logic into the Raspberry Pi hardware. Mathworks® provides a hardware support package to develop, simulate and program algorithms, and configure and access sensors and actuators using Simulink blocks with the Simulink® Support Package for Raspberry Pi hardware. Using MATLAB® and Simulink® external mode interactive simulations, parameter tuning, signal monitoring, and logging can be performed, as the algorithms run real time on the hardware board.

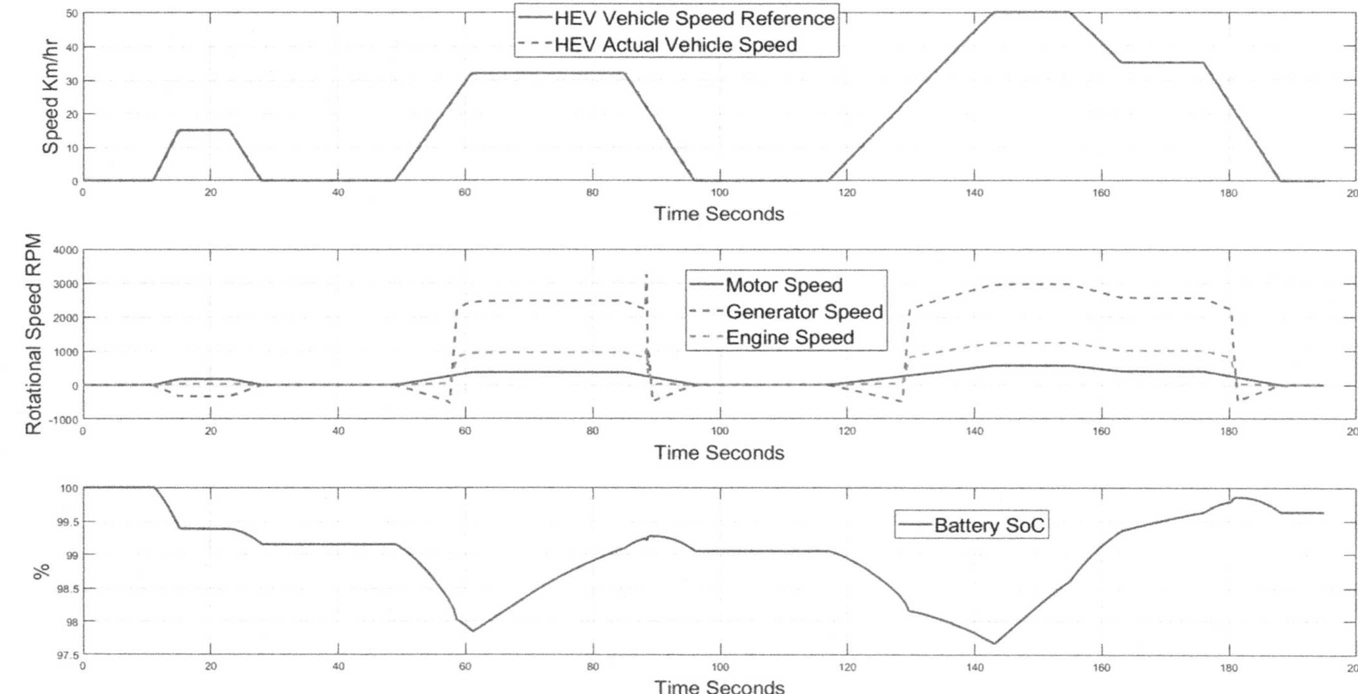

Figure 8.21 HEV vehicle simulation results for an urban cycle.

Follow the below steps to install the Simulink support package for Raspberry Pi hardware:

1. From the MATLAB® window go to Home >> Add-Ons >> Get Hardware Support Packages. Fig. 8.22 shows the Add-On Explorer GUI.
2. Click on the Simulink® Support Package for Raspberry Pi Hardware option, highlighted in Fig. 8.23, and it will guide into next window with an Install button as shown in Fig. 8.24. Click on it. For any support package installation, User needs to log on to a MathWorks account using the option shown in Fig. 8.25. Log on using the account, or create a new account if the User doesn't have it already.
3. Wait for the installation to be completed. An installation progress window is shown in Fig. 8.26. After the successful installation, it will show on the Add on Explorer that the Simulink Support package for Raspberry Pi hardware is installed as shown in Figs. 8.27 and 8.28.

Next we will make the necessary changes to the HEV model to run it on the Raspberry Pi hardware. The host computer running MATLAB and Simulink will be communicating to the Raspberry Pi hardware using the Wi-Fi protocol. We already know the IP address of Raspberry Pi when we connected to it using Putty desktop App, which will be used to connect and download the Simulink model to the Pi hardware. Open the model **HEV_Simscape_Model_Rasp_Pi.slx** from the attached folder under **Code_Files\ HEV_Model_for_Raspberry_Pi** for the Chapter 8.This is pretty much the same model we used earlier to run the desktop simulation from the host computer, but some setting changes are made to run it on the Raspberry Pi, such as the Hardware Implementation option has been set to Raspberry Pi board, simulation mode is selected to be External for running the interactive simulation on the hardware, etc. Also a simple LED blink logic is added on to the model in order to verify when this model runs on the PI hardware, and it will blink the on-board LED on the PI.

On this model based on the IP address of the Raspberry Pi hardware, user needs to update it by going to ***Simulation*** >> ***Model Configuration Parameters*** >> ***Hardware Implementation*** >> ***Target Hardware Resources*** as shown in Fig. 8.29. The login credentials also needs to be updated if they are different from the default one shown in Fig. 8.29.

After updating the IP address and login credentials, the model is ready to be deployed on to the PI hardware. Click on the simulation button from the model. It will generate code from the model, compile it for the Raspberry Pi target hardware, and generate the executable and will start running it on the PI hardware. This process can take a few minutes, but the progress can be monitored in the Simulink Diagnostic viewer as shown in Fig. 8.30. Once it is start running, we can see the green LED on the PI board will be blinking and also we can monitor the signals directly in Simulink like we did for desktop simulation. It can be noted that since this simulation is running on the hardware board, the simulation will be running in real time and it will take 195 real seconds to finish the 195 s in simulation compared to the desktop simulation.

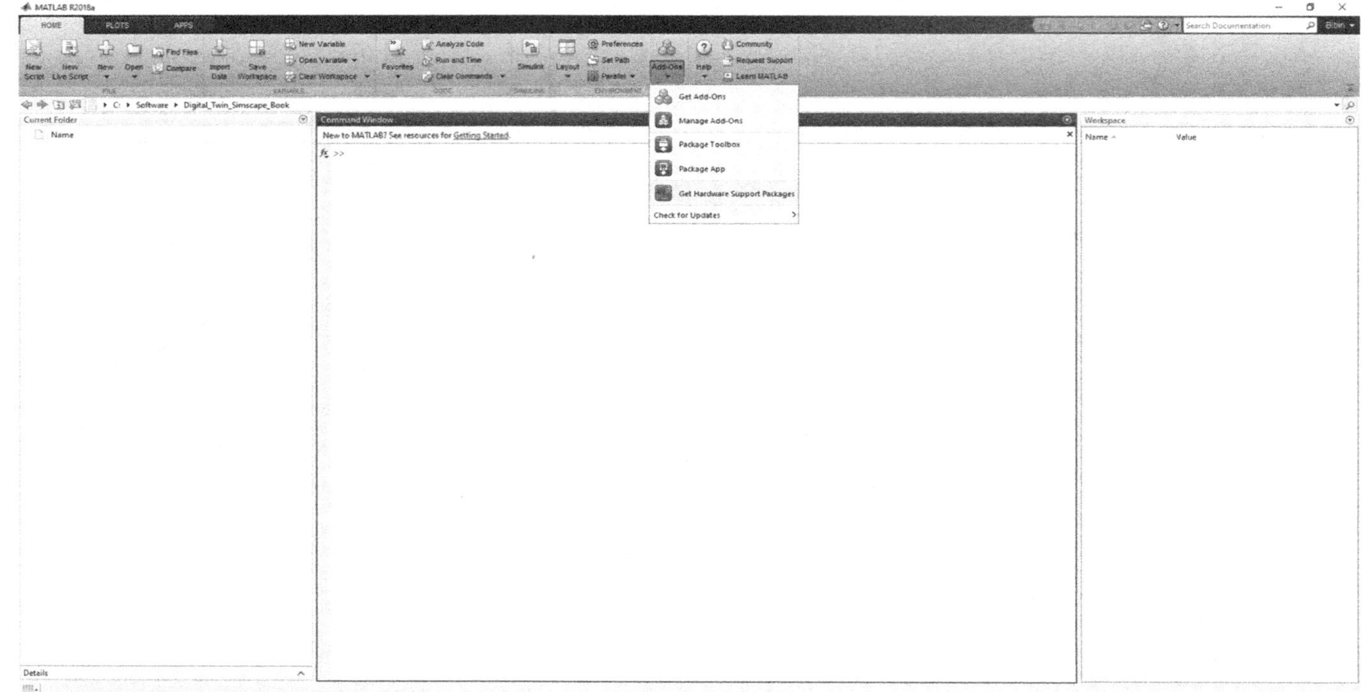

Figure 8.22 Add-on Explorer GUI.

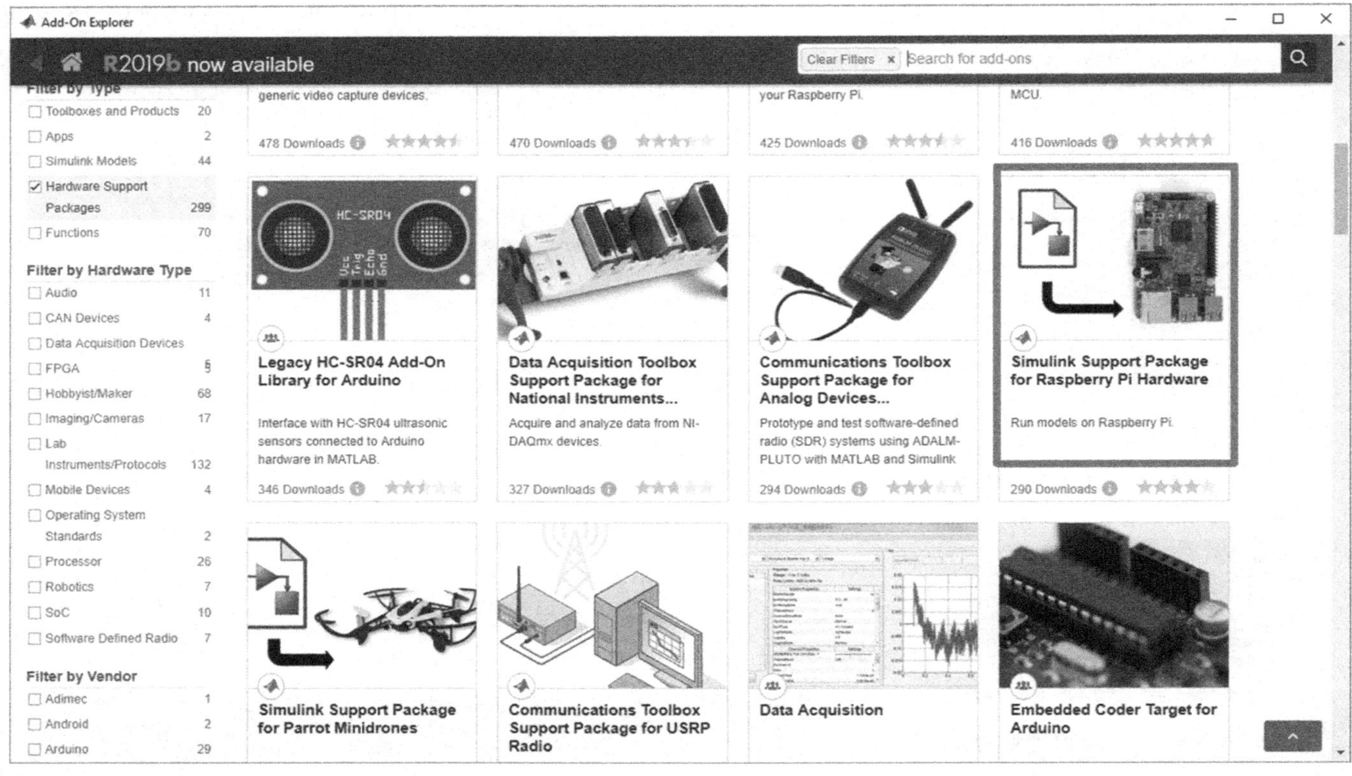

Figure 8.23 Simulink Support Package for Raspberry Pi Hardware.

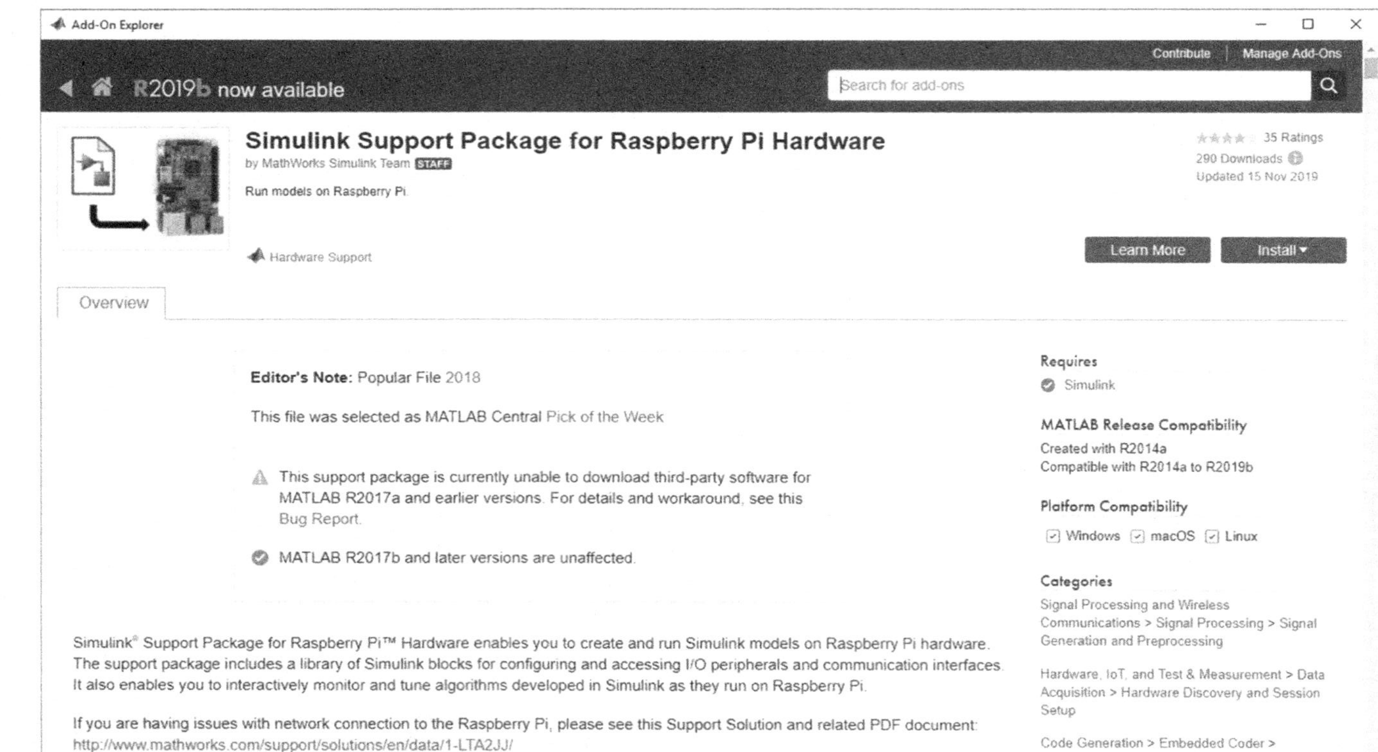

Figure 8.24 Add-On Explorer with Install Button.

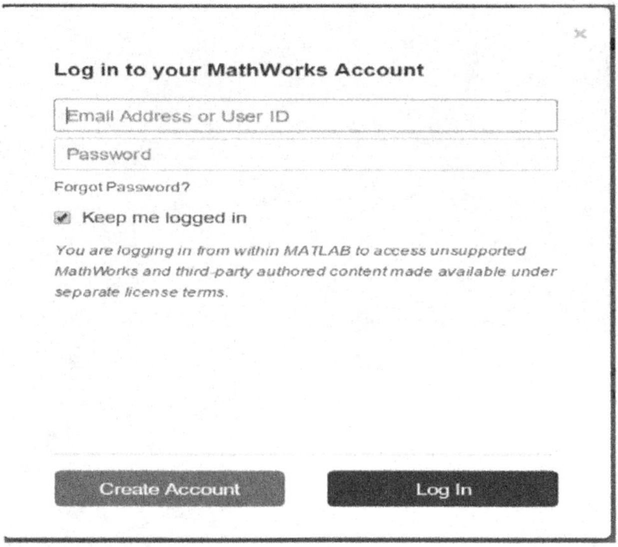

Figure 8.25 MathWorks account login.

8.6 EDGE device setup and cloud connectivity

Now that the HEV model is running in real time in the Raspberry Pi hardware, the next step is to collect the inputs, outputs, and states from the hardware and send to the AWS cloud Digital Twin for the Off-BD processing. As mentioned earlier, we will be using the MQTT communication protocol for communicating between the hardware and the AWS. So first we need to add the MQTT interface to the Simulink model. Unfortunately, in MATLAB 2018a Raspberry Pi Hardware support package, the MQTT interface is not included by default, but because of the limitation mentioned earlier for running interactive simulations on the PI hardware, we have to use MATLAB 2018a. So the Authors have included a few files that need to be copied to the Raspberry Pi Hardware package installation folder. The hardware installation folder can be easily found by typing ***which embdlinuxlib*** on the MATLAB command window.

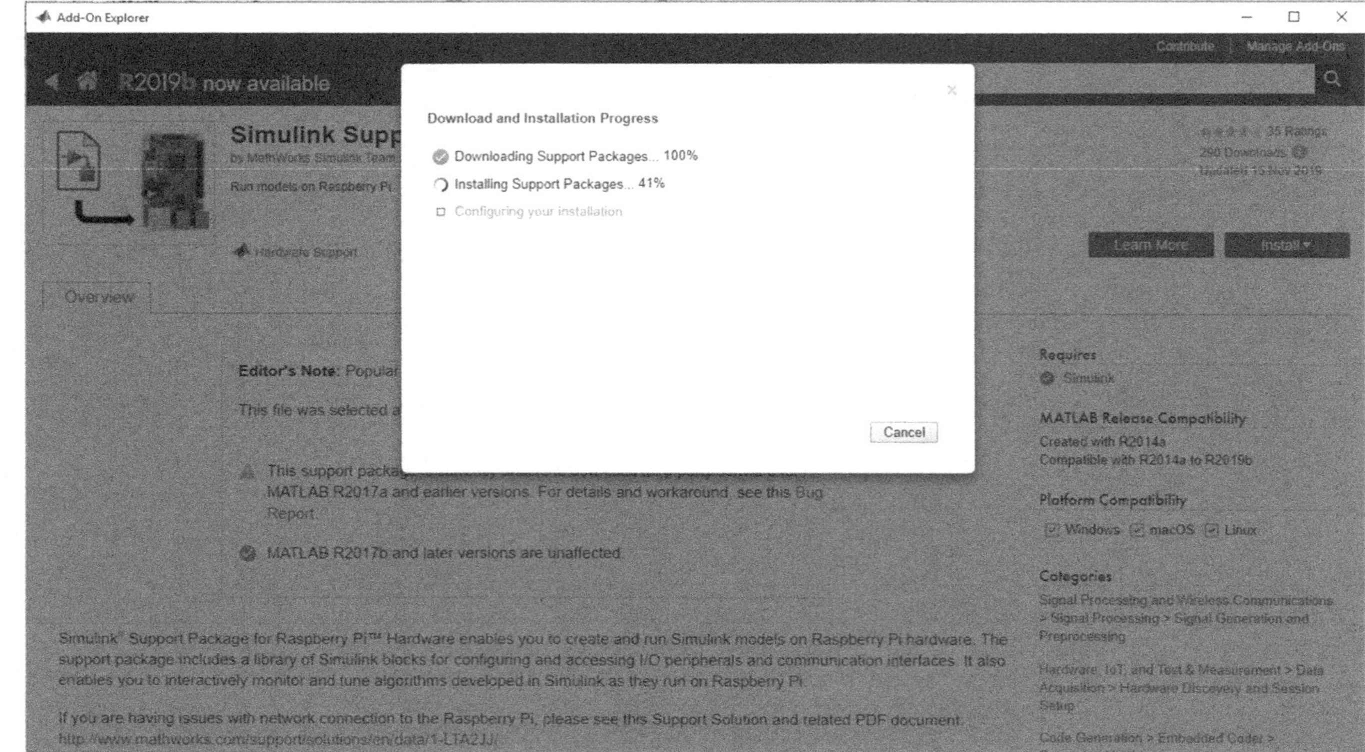

Figure 8.26 Support package installation progress.

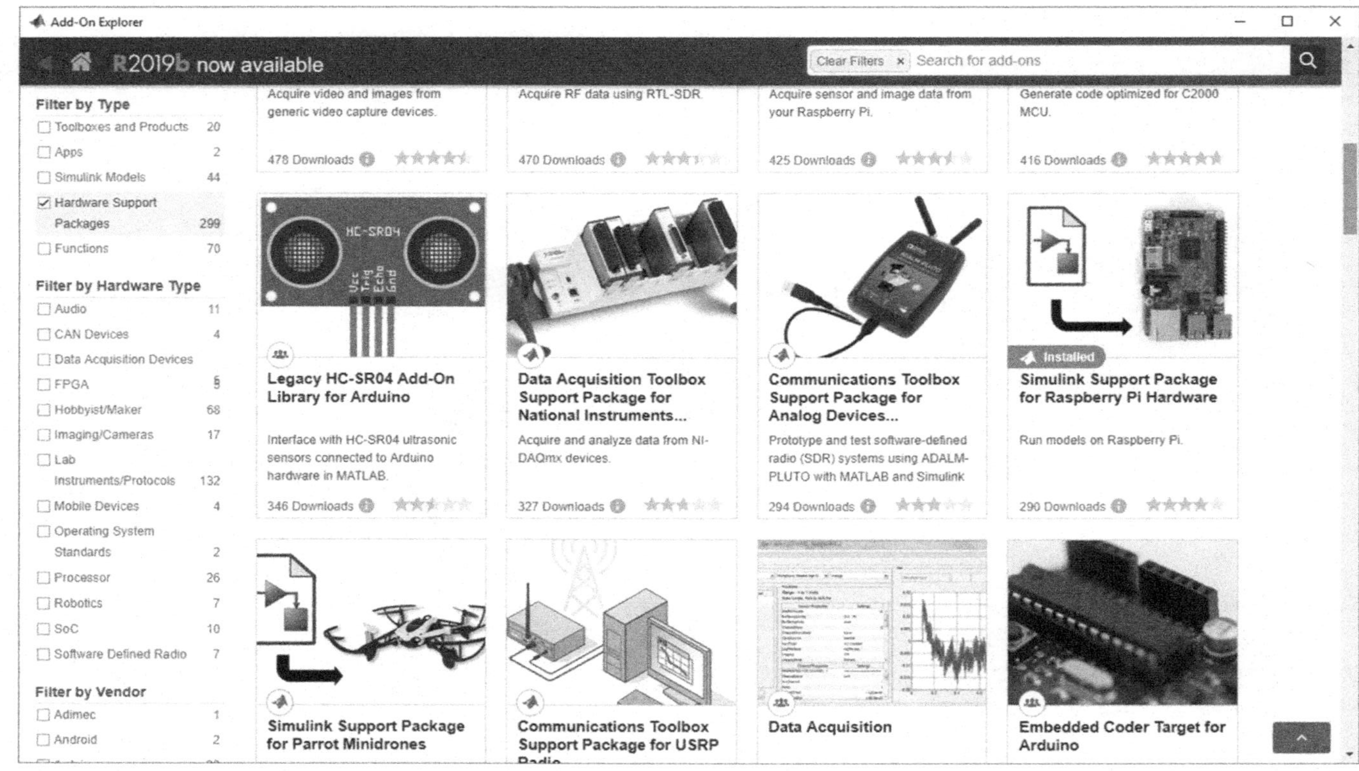

Figure 8.27 Finished installation for Raspberry Pi hardware.

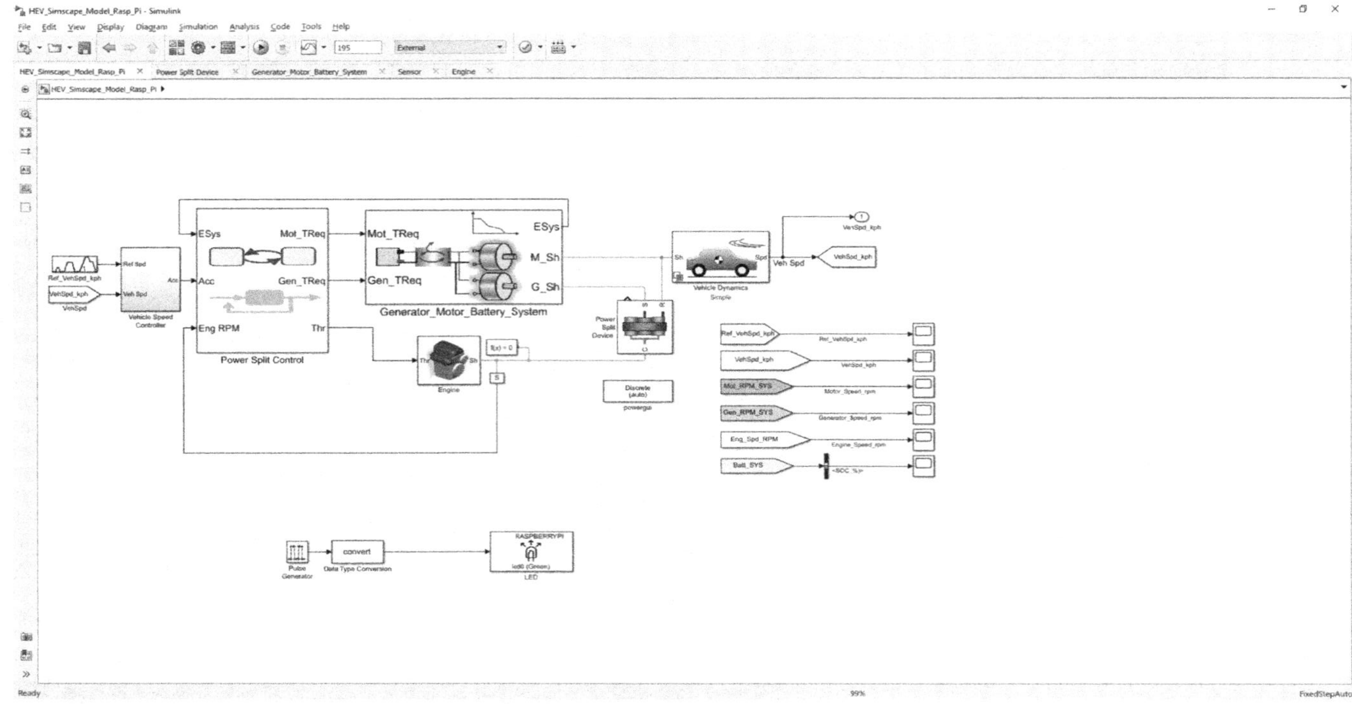

Figure 8.28 HEV Simscape™ configured to run on Raspberry Pi.

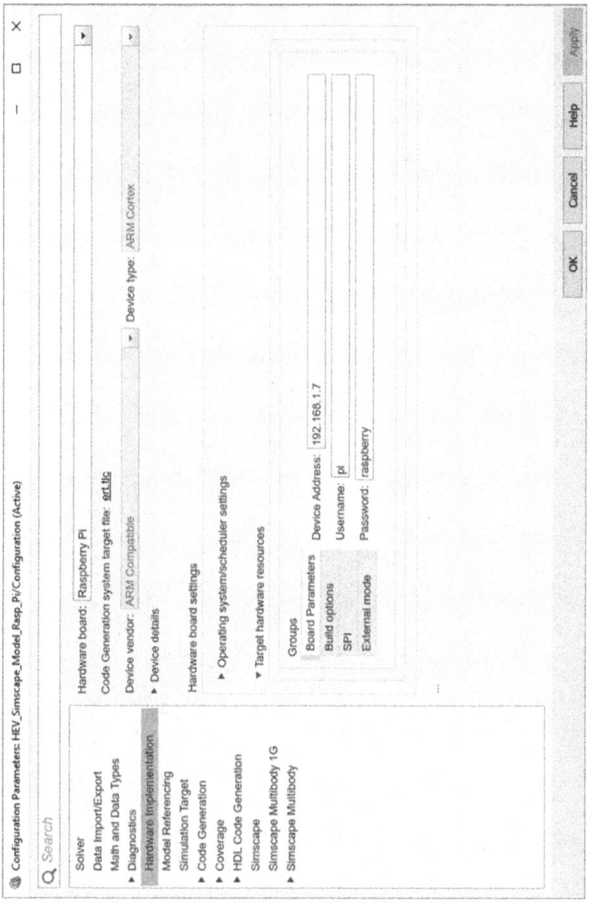

Figure 8.29 Updating the Raspberry Pi IP address and login credentials on the model.

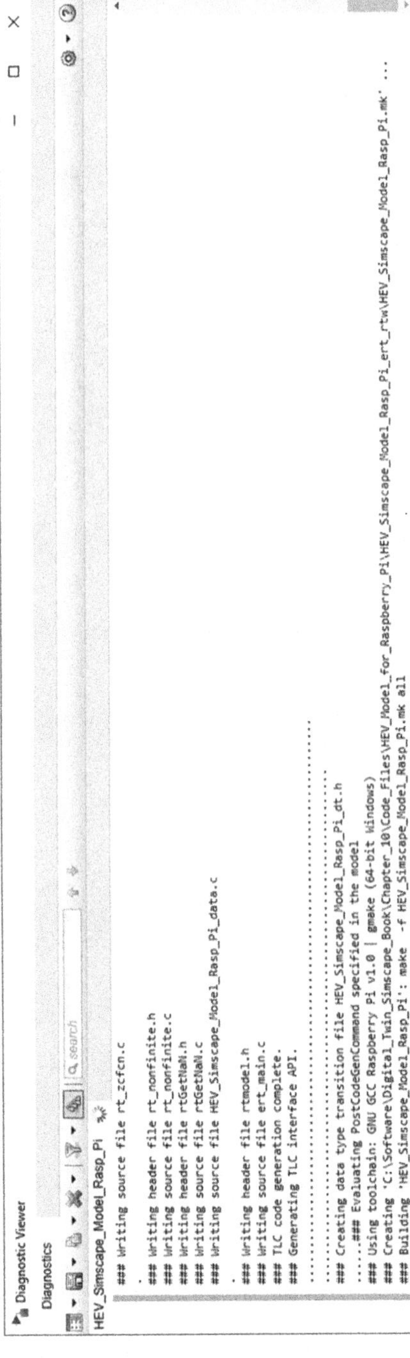

Figure 8.30 Building and downloading HEV Simulink model to Raspberry Pi hardware board.

1. Copy the **MQTTPublish.m** and **MQTTSubscribe.m** from the Chapter 8 attachment folder **Code_Files\HEV_Model_for_Raspberry_Pi_with_MQTT\MQTT_Support_Files** to the Raspberry Pi hardware installation folder **toolbox\realtime\targets\linux\+codertarget\+linux\+blocks.**
2. Copy **MW_MQTT.c** from the Chapter 8 attachment folder **Code_Files\HEV_Model_for_-Raspberry_Pi_with_MQTT\MQTT_Support_Files** to **toolbox\realtime\targets\linux\src.**
3. Copy **MW_MQTT.h** from the Chapter 8 attachment folder **Code_Files\HEV_Model_for_-Raspberry_Pi_with_MQTT\MQTT_Support_Files** to **toolbox\realtime\targets\linux\include.**

Open the model **HEV_Simscape_Model_Rasp_Pi_with_MQTT.slx** from the Chapter 8 attachment folder under **Code_Files\HEV_Model_for_Raspberry_Pi_-with_MQTT**. This model has the MQTT transmit block added to it as shown in Figs. 8.31 and 8.32. Basically all the required signals that we want to send from the hardware to the AWS cloud are connected to a Mux block in the Simulink model along with a free running counter block as shown in Fig. 8.32 and then connected to an MQTT publish block. The publish block uses the topic name **digital_twin_mqtt_-topic** for publishing the messages. This MQTT publish block is actually provided by the Raspberry Simulink library, but since MATLAB 2018a did not have the support for MQTT, this block is copied over from 2018b version of MATLAB and the PI hardware support library. The free running counter block is also connected to the MQTT block, and this counter serves as a timer indicator to buffer the data from the hardware for 5 s and then send the buffered data to AWS cloud. So the hardware sends MQTT message with the data every 0.1 s, which is the sample time of the model, and this MQTT message will be received by a Python program that is subscribed to the same MQTT topic **digital_twin_mqtt_topic**, and it buffers the data for 5 s, then formats the data to a JSON structure, and establishes a connection to AWS cloud and sends the message. Fig. 8.33 shows the high-level diagram of the cloud connection setup used in this chapter.

First we will start AWS side steps, we have to get some specific files when we setup AWS, which will then be used with the Python program for AWS−IoT connection. We will use the AWS service called AWS IoT Core in this section, which lets the connected devices securely interact with cloud or other connected devices. For more information about the AWS IoT Core services, follow the link https://aws.amazon.com/iot-core/. Fig. 9.115 shows the setup we are trying to establish in this section, which establishes a connection between ESP32 and AWS IoT Core and transmits data from Raspberry Pi hardware, which is connected to the ESP32.

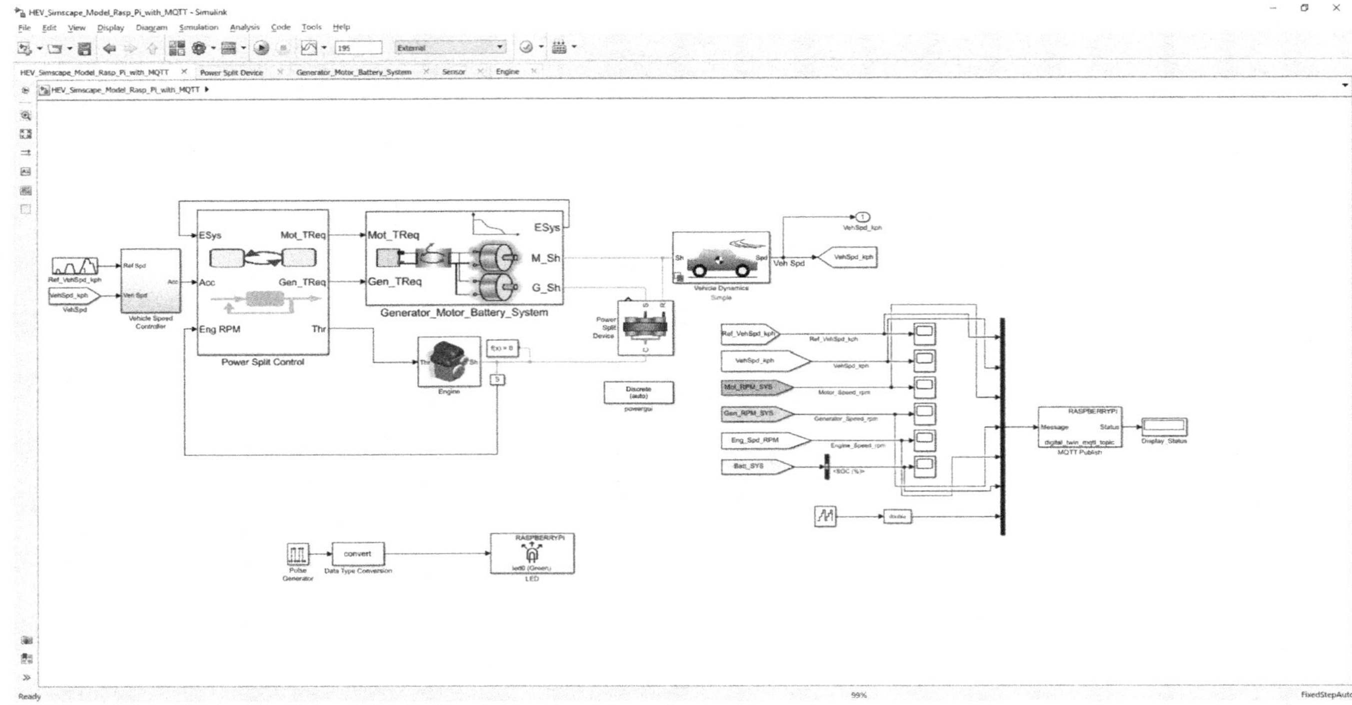

Figure 8.31 Raspberry Pi model with MQTT transmit logic.

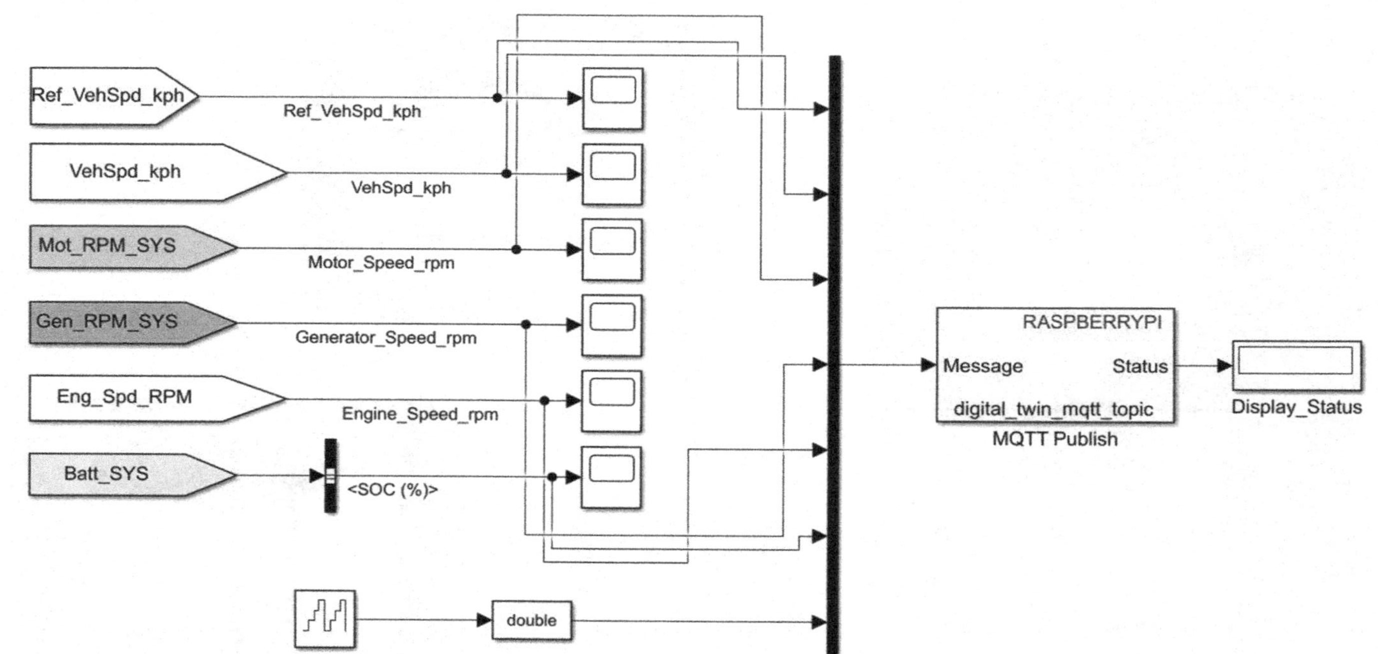

Figure 8.32 Close look at the MQTT transmit block.

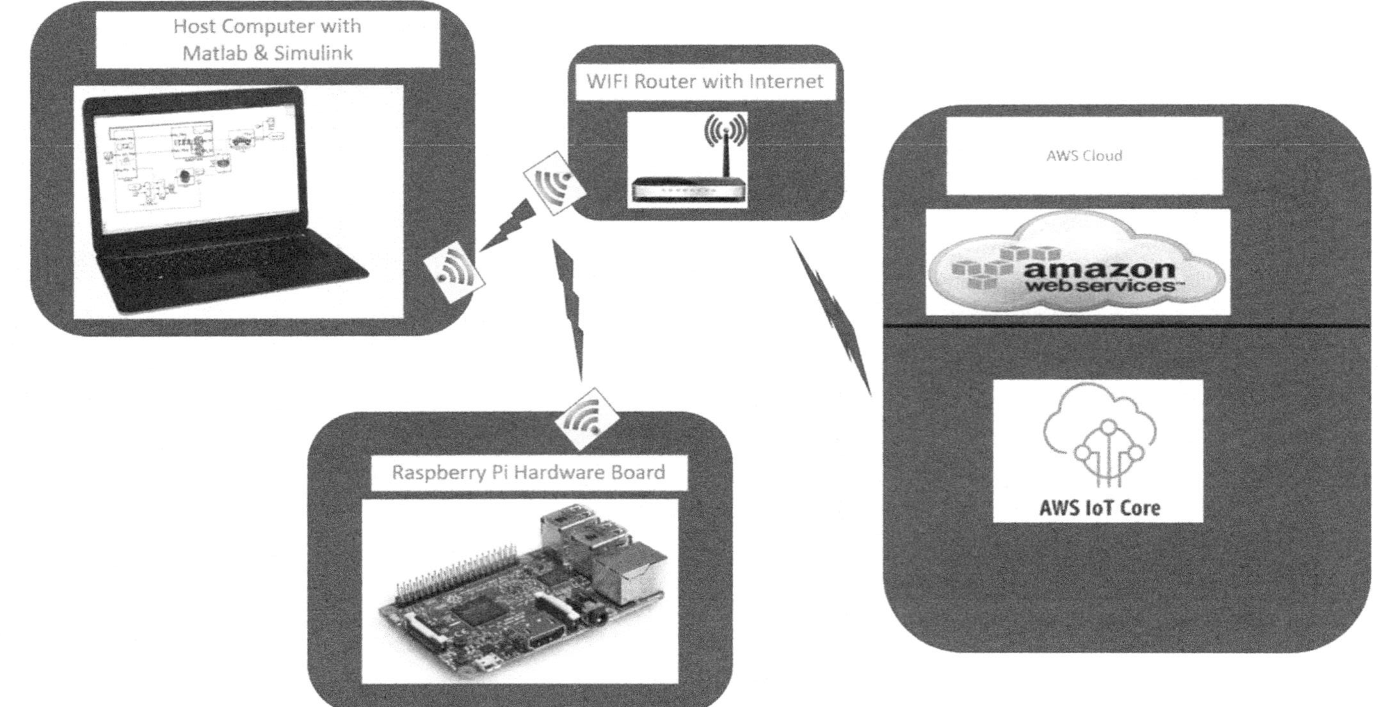

Figure 8.33 High-level diagram of the cloud connection setup.

1. As a first step, User needs to create a AWS Management Console account if you don't already have one. If you already have an Amazon account, you can use the same account information to login. If you don't have an account, please create an account. Follow this link https://aws.amazon.com/console/ on a web browser for logging in, creating a new account, etc. Use the ***Sign in to the Console*** option in the main page as shown in Fig. 8.34 and login using the credentials as shown in Fig. 8.35.
2. On the logged in Console window search for **IoT** and select the **IoT Core** as shown in Fig. 8.36. The AWS IoT window will be loaded, and it will display all the **AWS IoT Things** available in the User account. An **AWS IoT Thing** represents an **IoT** connected device, in this case, it will represent our Raspberry Pi connected to Wi-Fi and is running the HEV model in real time. The Author already has few **IoT Things** created previously on the account, which is the reason why it displays some **IoT Things** already as shown in Fig. 8.37. For security reasons, the already existing **IoT Things** are not shown in the images. We will be following similar graying out for some part of the images for protecting Author's information, but for the new additions that we are doing for this book, we will display it in the images.
3. Before creating a new **IoT Thing**, we need to create a **Policy.** A **Policy** is a JSON formatted document that allows access to the AWS services for the connected devices. We need to create a **Policy** and link it with the **IoT Thing** for the Raspberry Pi to establish a connection with the **AWS IoT Thing.** Go to **Secure >> Policies** on the left side tabbed window of the AWS Console under **AWS IoT** and click on the **Create** button as shown in Fig. 8.38.
4. On the new window enter the Policy name. This case we will name it as **hev_digital_twin_policy.** But User can choose to give a different name if needed. For the **Action** field enter **iot***, **Resource ARN** enter *, and for **Effect** check the **Allow** option and click on the Create button as shown in Fig. 8.39. The newly created Policy will show up under the Policies section as shown in Fig. 8.40.
5. The next step is to create an **AWS IoT Thing.** Goto **Manage >> Things** menu and click on **Create.** From the next window, click on **Create a Single Thing** as shown in Figs. 8.41 and 8.42.
6. Enter the name of the new **AWS IoT Thing** as shown in Fig. 8.43. We name it **hev_digital_twin_thing**, but User can choose a different name. We can leave the rest of the fields as default. Click **Next** at the bottom of the page.
7. We will use a certificate-based authentication for the Raspberry Pi device to connect to the **AWS IoT Thing** for secure connection. So we have to generate and download the certificates and use it with the Python program for the Raspberry Pi while establishing connection with **AWS IoT.** From the IoT Thing creation window, shown in Fig. 8.44, click on the **Create Certificate** option. It will generate a private key, a public key, and a certificate file as shown in Fig. 8.45. Download the certificate for the Thing, public key, private key and also the **Root CA** file to somewhere safely into the computer. We will be using these files later for programming the Raspberry PI to establish secure connection with **AWS IoT.** Next we need to **Activate** the Thing by clicking on the **Activate** button.
8. Next we need to attach the **Policy** that we have created in Step 3 and 4 to this new **IoT Thing.** Click on the Attach Policy option shown in Fig. 8.45, and in the new window select the **digital_twin_policy** and click on Register Thing as shown in Fig. 8.46.

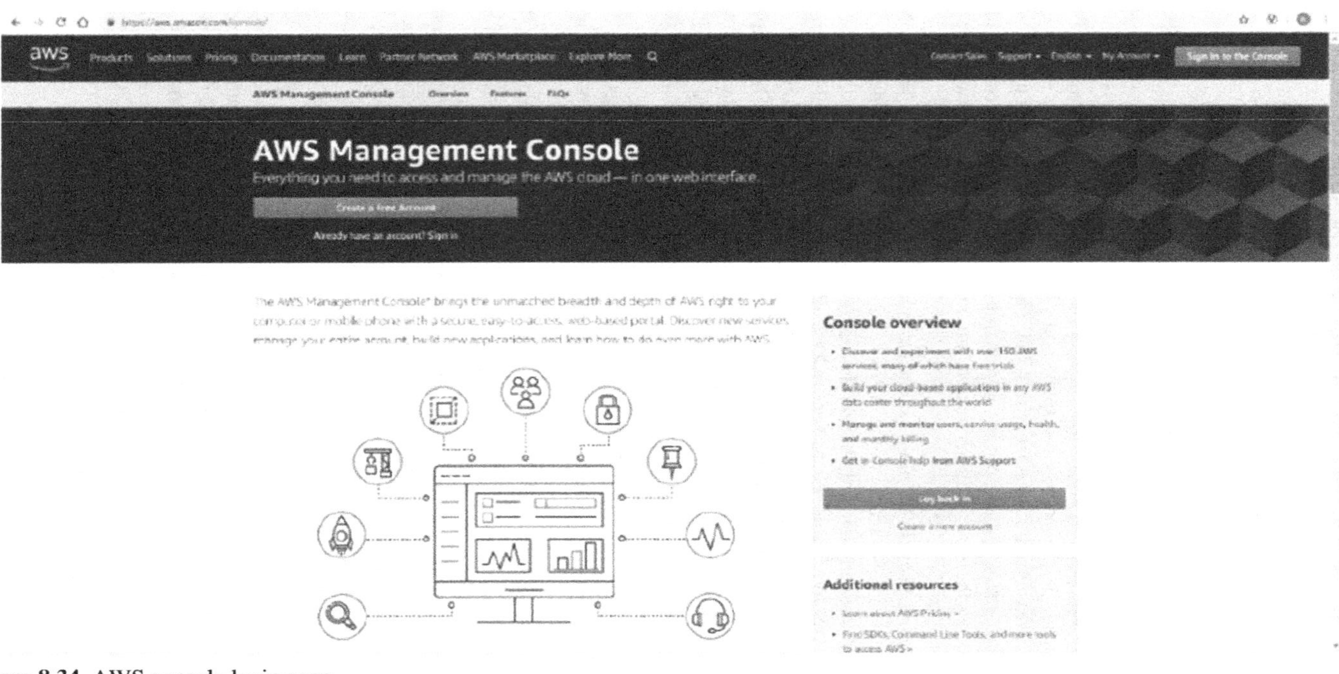

Figure 8.34 AWS console login page.

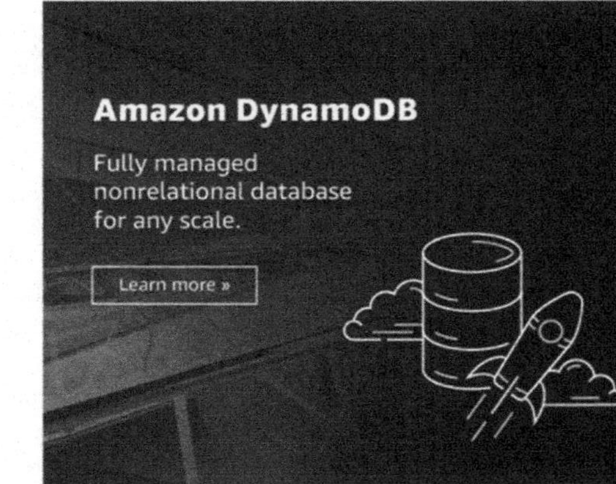

Figure 8.35 AWS console login.

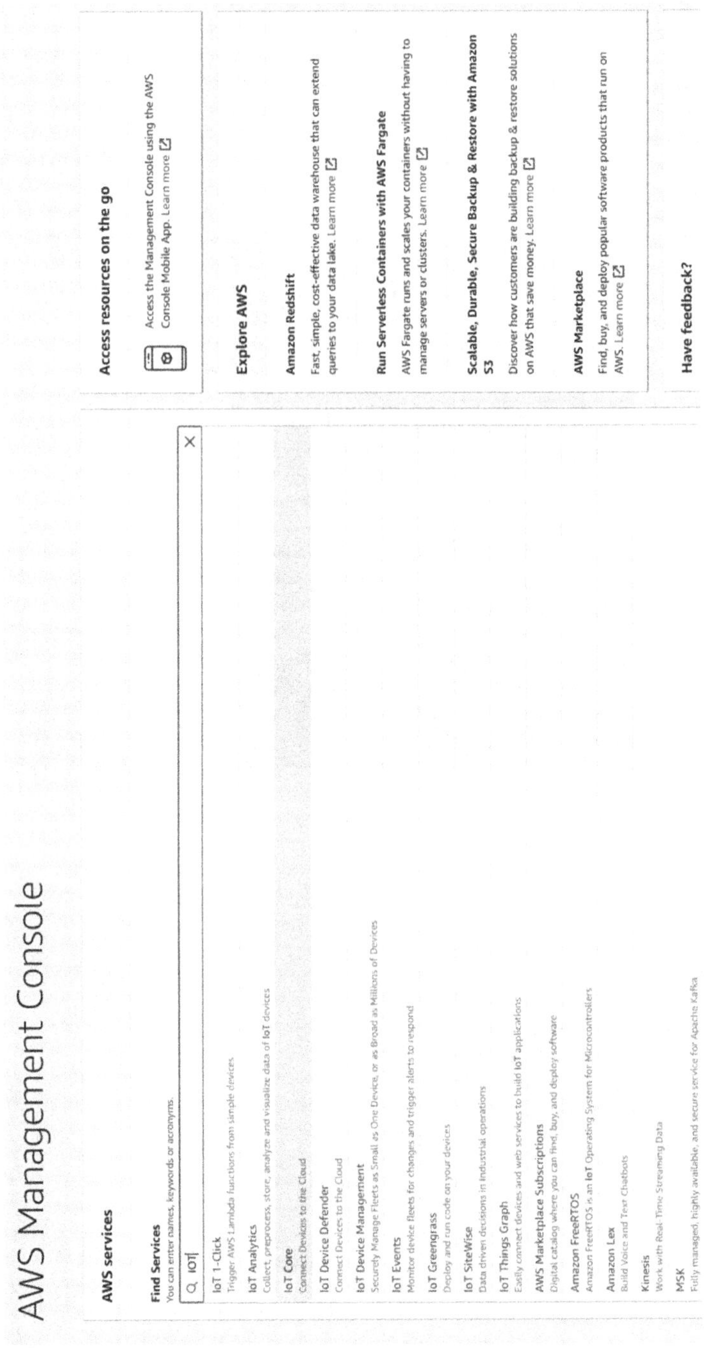

AWS Management Console

AWS services

Find Services
You can enter names, keywords or acronyms.

IOT

IoT 1-Click
Trigger AWS Lambda functions from simple devices
IoT Analytics
Collect, preprocess, store, analyze and visualize data of IoT devices
IoT Core
Connect Devices to the Cloud
IoT Device Defender
Connect Devices to the Cloud
IoT Device Management
Securely Manage Fleets as Small as One Device, or as Broad as Millions of Devices
IoT Events
Monitor device fleets for changes and trigger alerts to respond
IoT Greengrass
Deploy and run code on your devices
IoT SiteWise
Data driven decisions in industrial operations
IoT Things Graph
Easily connect devices and web services to build IoT applications
AWS Marketplace Subscriptions
Digital catalog where you can find, buy, and deploy software
Amazon FreeRTOS
Amazon FreeRTOS is an IoT Operating System for Microcontrollers
Amazon Lex
Build Voice and Text Chatbots
Kinesis
Work with Real-Time Streaming Data
MSK
Fully managed, highly available, and secure service for Apache Kafka

Access resources on the go

Access the Management Console using the AWS Console Mobile App. Learn more

Explore AWS

Amazon Redshift

Fast, simple, cost-effective data warehouse that can extend queries to your data lake. Learn more

Run Serverless Containers with AWS Fargate

AWS Fargate runs and scales your containers without having to manage servers or clusters. Learn more

Scalable, Durable, Secure Backup & Restore with Amazon S3

Discover how customers are building backup & restore solutions on AWS that save money. Learn more

AWS Marketplace

Find, buy, and deploy popular software products that run on AWS. Learn more

Have feedback?

Figure 8.36 Selecting IoT Core services from AWS console.

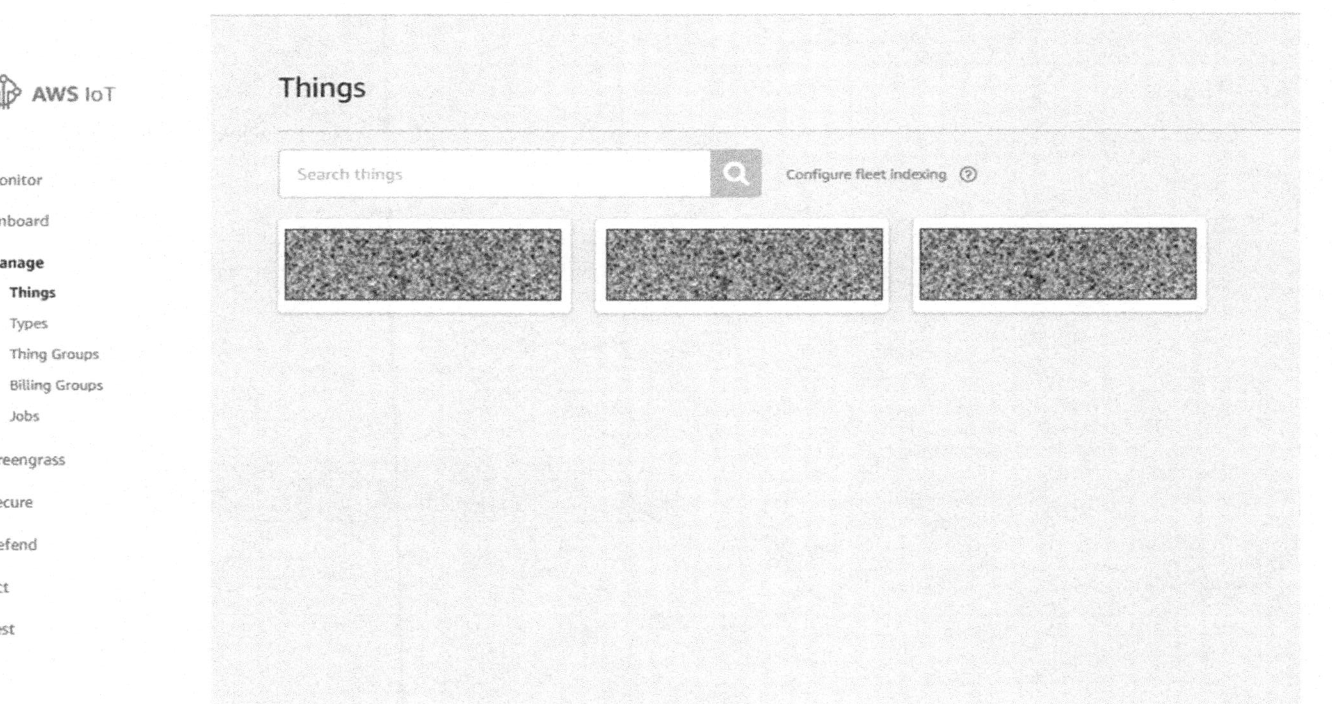

Figure 8.37 AWS IoT Things main page.

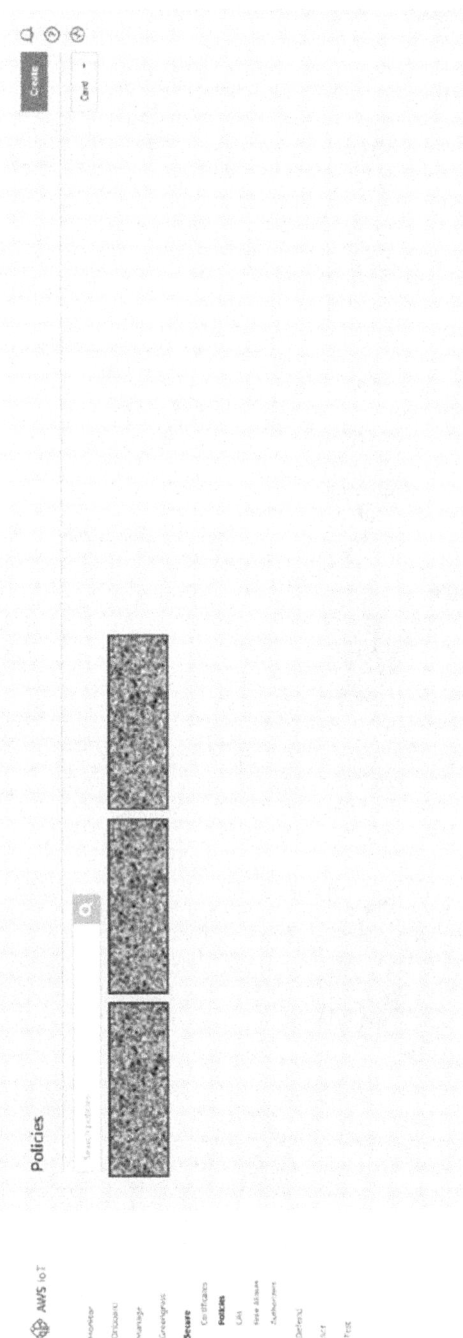

Figure 8.38 Creating a New Policy.

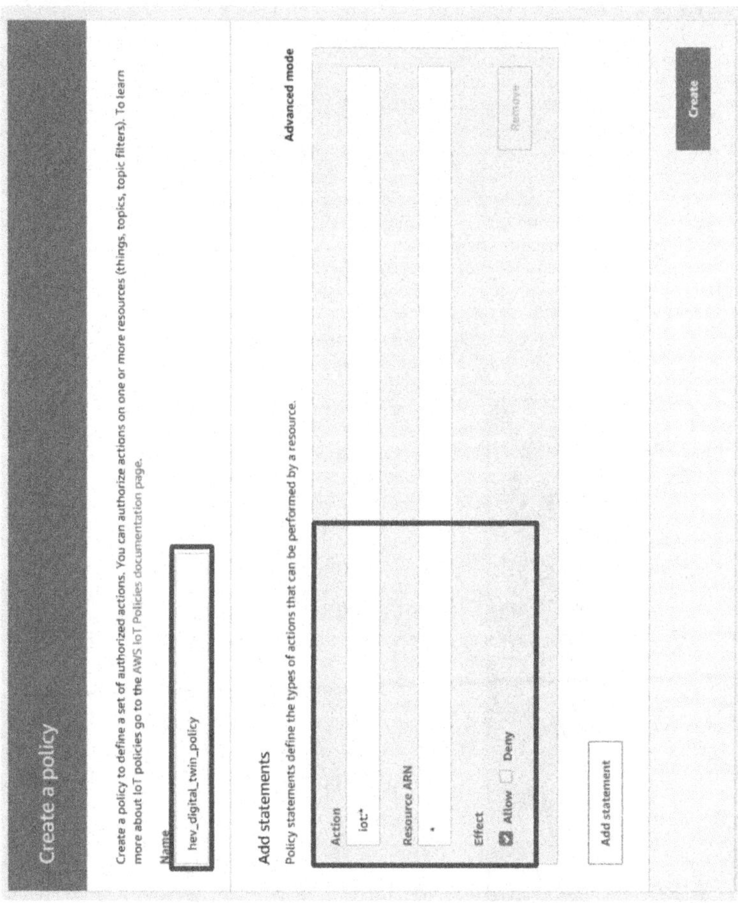

Figure 8.39 Entering New Policy Details.

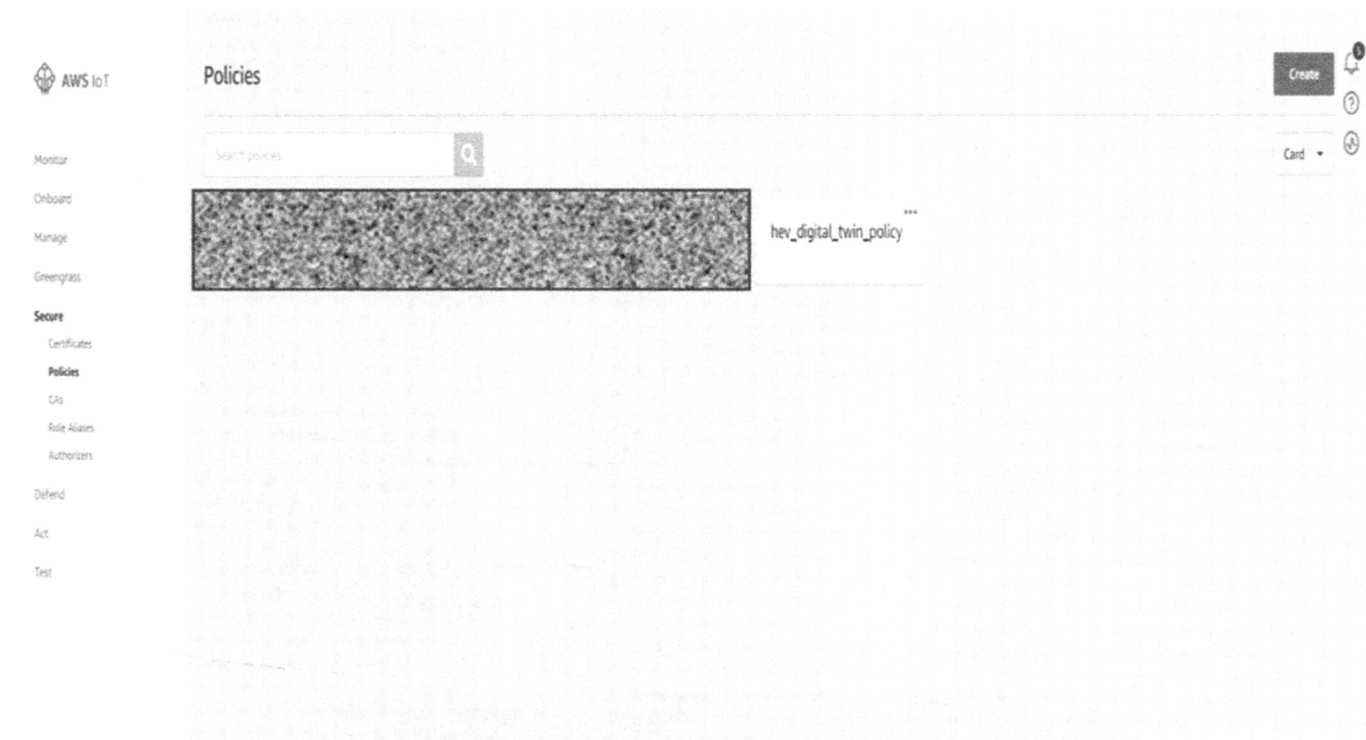

Figure 8.40 Newly created policy.

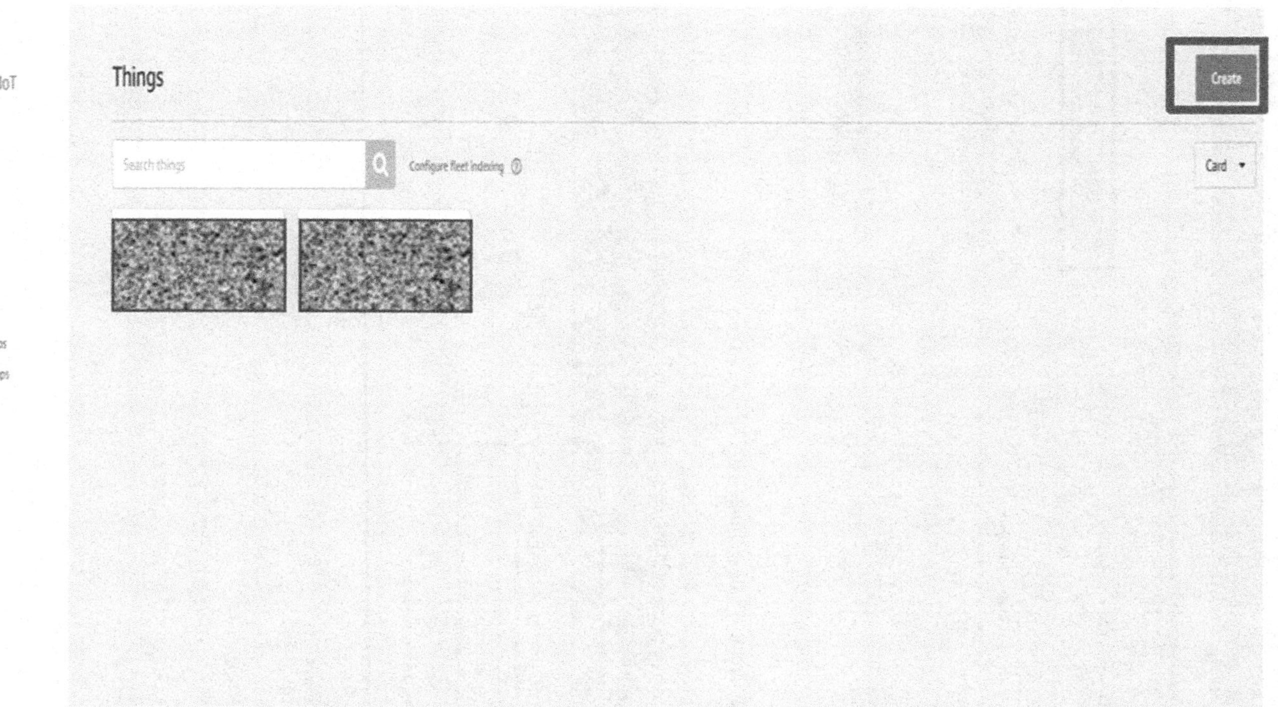

Figure 8.41 Creating an AWS IoT Thing.

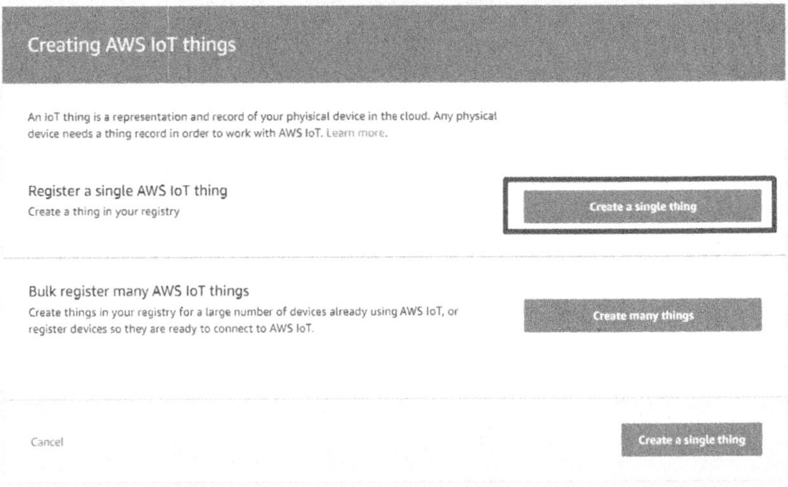

Creating AWS IoT things

An IoT thing is a representation and record of your phyisical device in the cloud. Any physical device needs a thing record in order to work with AWS IoT. Learn more.

Register a single AWS IoT thing
Create a thing in your registry

Create a single thing

Bulk register many AWS IoT things
Create things in your registry for a large number of devices already using AWS IoT, or register devices so they are ready to connect to AWS IoT.

Create many things

Cancel

Create a single thing

Figure 8.42 Create a single thing.

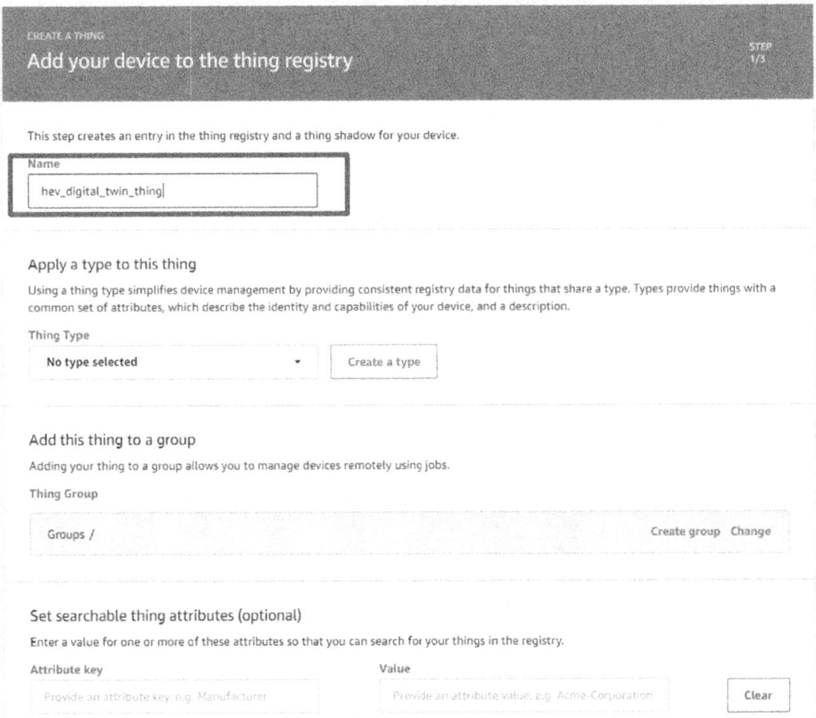

CREATE A THING
Add your device to the thing registry

STEP 1/3

This step creates an entry in the thing registry and a thing shadow for your device.

Name
hev_digital_twin_thing

Apply a type to this thing
Using a thing type simplifies device management by providing consistent registry data for things that share a type. Types provide things with a common set of attributes, which describe the identity and capabilities of your device, and a description.

Thing Type
No type selected
Create a type

Add this thing to a group
Adding your thing to a group allows you to manage devices remotely using jobs.

Thing Group
Groups /
Create group Change

Set searchable thing attributes (optional)
Enter a value for one or more of these attributes so that you can search for your things in the registry.

Attribute key
Provide an attribute key, e.g. Manufacturer

Value
Provide an attribute value, e.g. Acme-Corporation

Clear

Figure 8.43 Naming the newly created AWS IoT thing.

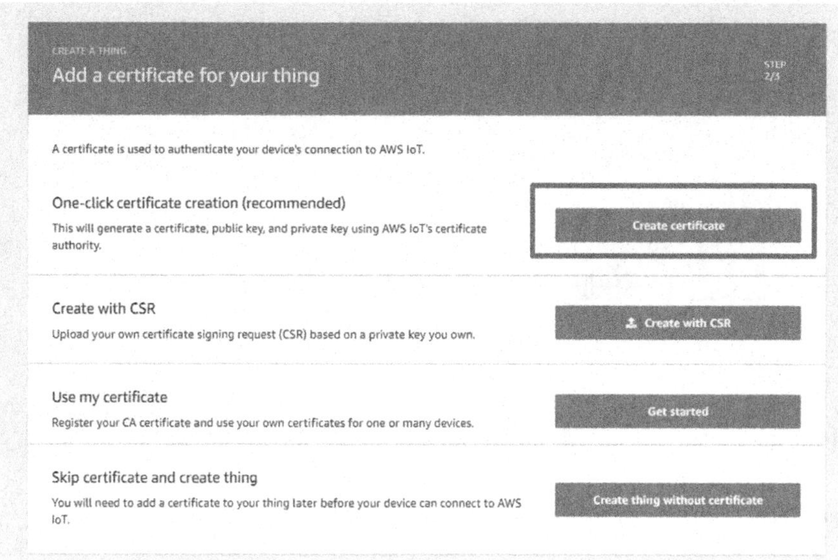

Figure 8.44 IoT Thing certificate creation window.

9. The newly created **IoT Thing** will show up in the **Manage >> Things** section. See Fig. 8.47.
10. Click on the **hev_digital_twin_thing** and go to the Interact menu as shown in Fig. 8.48 and note down the **Rest API** for the **IoT Thing**. For security reasons, the Author's API is not shown in the screenshot, but you need to copy the entire string, which will be used later for

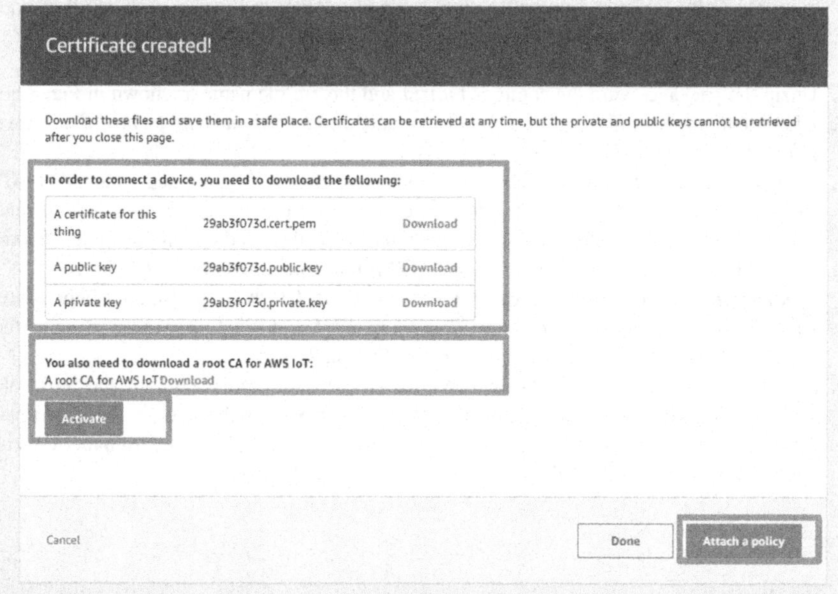

Figure 8.45 Certificates generated for IoT Thing.

Figure 8.46 Attaching a policy and registering the IoT Thing.

the Raspberry Pi Python programming. Also note down the **IoT Thing** topic name under the MQTT section. MQTT is a fast, secure, and efficient protocol that we use to communicate data between Raspberry Pi and **IoT Thing.** MQTT nodes or end points publish and subscribe to the topics and exchange information between them using the topics.

11. Next step is to install AWS IoT SDK for Python in Raspberry Pi. To connect with AWS IoT services, we will use AWS IoT Python SDK, which is built on top of Paho MQTT client library, which we installed on Raspberry PI earlier in this chapter. For installing Python SDK, download the SDK package from the link below: https://s3.amazonaws.com/aws-iot-device-sdk-python/aws-iot-device-sdk-python-latest.zip.We can download this by logging into Putty Desktop App and go into a specific folder in Raspberry Pi and then type the command ***wget*** https://s3.amazonaws.com/aws-iot-device-sdk-python/aws-iot-device-sdk-python-latest.zip as shown in Fig. 8.49.
12. Unzip the package using the command ***unzip*** and the zip file name as shown in Fig. 8.50.
13. After unzipping, install the SDK using the command ***sudo python setup.py install*** as shown in Fig. 8.51.
14. After the SDK installation is finished, we will need to copy the highlighted folder AWSIoT-PythonSDK to the folder where we will be creating and running the Python code to send data to AWS IoT. In this example, we will run the Python code from the folder ***/home/pi/Digital_Twin_Simscape_Book/Chapter_8/Python_Code_Raspberry_PI_for_AWS_-Connection***. Use the command ***cp -r AWSIoTPythonSDK//home/pi/Digital_Twin_Simscape_Book/Chapter_8/Python_Code_Raspberry_PI_for_AWS_Connection*** to copy the AWSIoTPythonSDK to the above folder. See Fig. 8.52.
15. Next on the host computer, let us create a folder aws_certificates and copy the certificate files downloaded when we created the IoT Thing in Fig. 8.45. In this case, we have renamed the files as shown in Fig. 8.53. We can keep the file names as downloaded from the AWS as well, but just need to enter the correct file names in the Python code.

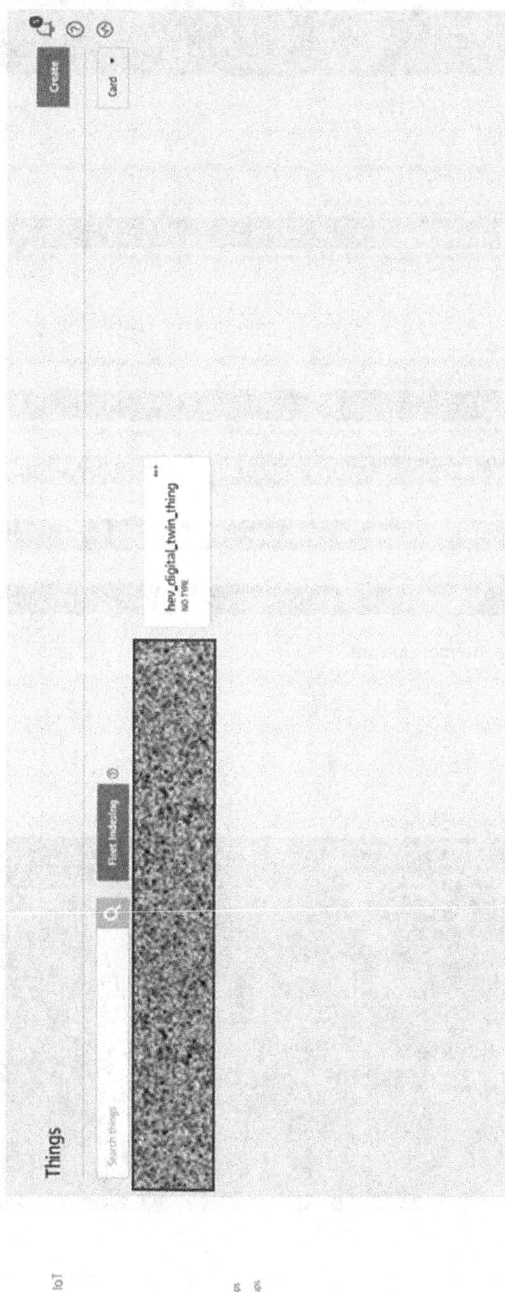

Figure 8.47 Newly created thing.

Figure 8.48 IoT Thing interact options.

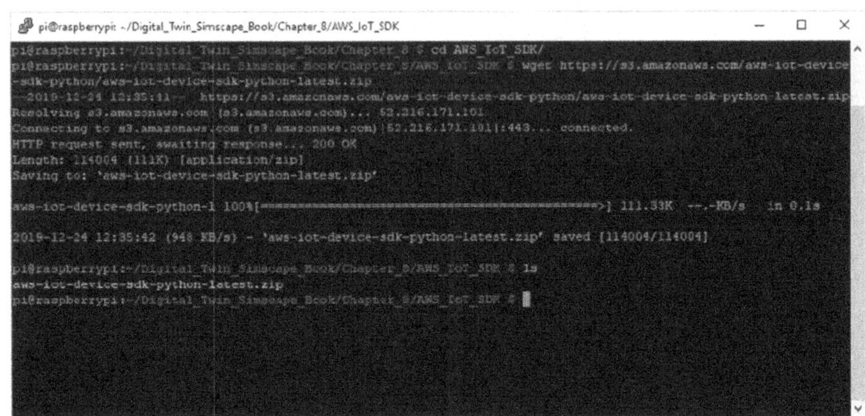

Figure 8.49 Downloading AWS IoT SDK.

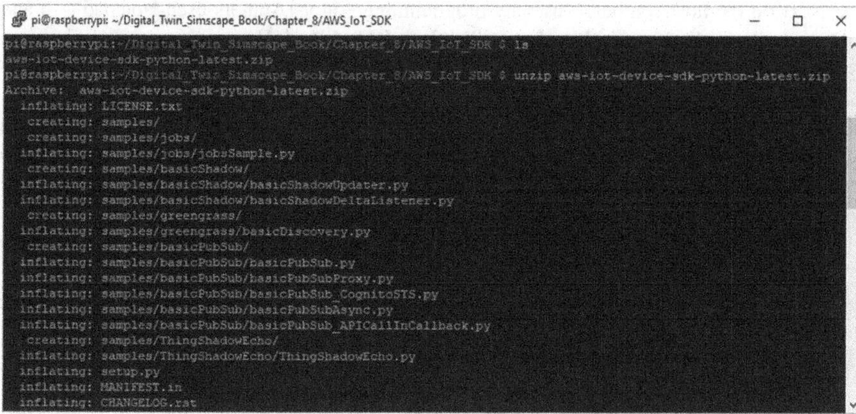

Figure 8.50 Unzip AWS IoT SDK.

```
pi@raspberrypi: ~/Digital_Twin_Simscape_Book/Chapter_8/AWS_IoT_SDK
pi@raspberrypi:~/Digital_Twin_Simscape_Book/Chapter_8/AWS_IoT_SDK $ ls
aws-iot-device-sdk-python-latest.zip  CHANGELOG.rst  MANIFEST.in  README.rst  setup.cfg
AWSIoTPythonSDK                       LICENSE.txt    NOTICE.txt   samples     setup.py
pi@raspberrypi:~/Digital_Twin_Simscape_Book/Chapter_8/AWS_IoT_SDK $ sudo python setup.py install
running install
running build
running build_py
creating build
creating build/lib.linux-armv7l-2.7
creating build/lib.linux-armv7l-2.7/AWSIoTPythonSDK
copying AWSIoTPythonSDK/MQTTLib.py -> build/lib.linux-armv7l-2.7/AWSIoTPythonSDK
copying AWSIoTPythonSDK/__init__.py -> build/lib.linux-armv7l-2.7/AWSIoTPythonSDK
creating build/lib.linux-armv7l-2.7/AWSIoTPythonSDK/core
copying AWSIoTPythonSDK/core/__init__.py -> build/lib.linux-armv7l-2.7/AWSIoTPythonSDK/core
creating build/lib.linux-armv7l-2.7/AWSIoTPythonSDK/core/util
copying AWSIoTPythonSDK/core/util/__init__.py -> build/lib.linux-armv7l-2.7/AWSIoTPythonSDK/core/util
copying AWSIoTPythonSDK/core/util/providers.py -> build/lib.linux-armv7l-2.7/AWSIoTPythonSDK/core/util
copying AWSIoTPythonSDK/core/util/enums.py -> build/lib.linux-armv7l-2.7/AWSIoTPythonSDK/core/util
creating build/lib.linux-armv7l-2.7/AWSIoTPythonSDK/core/shadow
copying AWSIoTPythonSDK/core/shadow/__init__.py -> build/lib.linux-armv7l-2.7/AWSIoTPythonSDK/core/shadow
copying AWSIoTPythonSDK/core/shadow/deviceShadow.py -> build/lib.linux-armv7l-2.7/AWSIoTPythonSDK/core/shadow
copying AWSIoTPythonSDK/core/shadow/shadowManager.py -> build/lib.linux-armv7l-2.7/AWSIoTPythonSDK/core/shadow
creating build/lib.linux-armv7l-2.7/AWSIoTPythonSDK/core/protocol
copying AWSIoTPythonSDK/core/protocol/__init__.py -> build/lib.linux-armv7l-2.7/AWSIoTPythonSDK/core/protocol
```

Figure 8.51 Installing AWS IoT Python SDK.

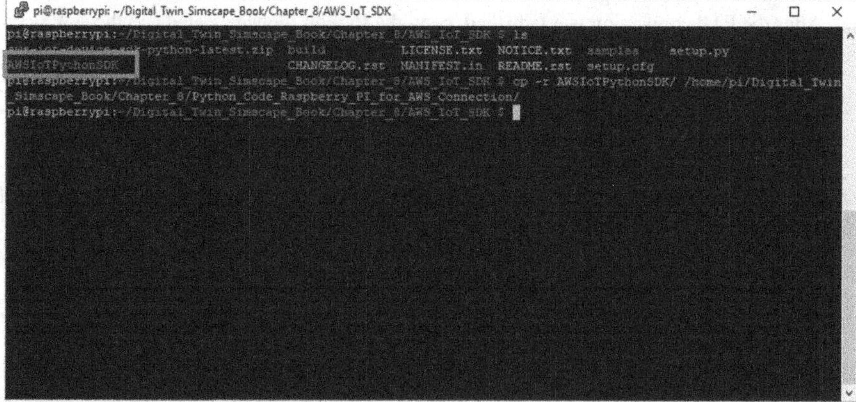

Figure 8.52 Copying the AWSIoTPythonSDK folder to the Python code folder.

16. Next on the host computer, use any programming editor (Authors used Notepad++) and create a new Python file ***Raspberry_Pi_AWS_IOT_Cloud_Connection.py***. The file name doesn't matter, User can chose any name. On the new Python file, import the required libraries first as shown in Fig. 8.54.

17. In the next section in the Python code, setup the AWS IoT Thing details. Variable ***host*** should contain the string of the REST API for the IoT Thing, which we noted in Step 10. Variable ***certPath*** should hold the path in the Raspberry Pi hardware where we will copy the AWS certificate files. Variable ***clientId*** should hold the Thing name we created, and variable ***topic*** should hold the IoT Thing topic name noted from Step 10. Then initialize the object AWSIoTMQTTClient with the AWS Thing information and point it to the certificate files for the credentials for secure connection. See Figs. 8.55 and 8.56 for more details.

18. The next section of the Python code has the main() function, where it will connect to the local MQTT broker which is running on the Raspberry Pi with IP address 192.168.1.7 and subscribe to the MQTT topic ***digital_twin_mqtt_topic***. Note that this is the same topic to which Simulink HEV model running on the PI is publishing the message with the HEV current state information. So whenever the Simulink model runs and publishes messages to the broker, the Python code which is connected to the same broker will receive the messages. Also this section of the code configures a function to be called ***on_message*** whenever the code receives an MQTT message from the broker. And the main code just waits in a while loop for the messages to be received.

19. In the next section in the Python code, we have the implementation of the callback function on_message whenever the Python code receives an MQTT message from the MQTT broker with the topic ***digital_twin_mqtt_topic***. In a nutshell, this function receives the message with the HEV model state information such as Target and Actual Vehicle speed, Motor, Generator, and Engine speed, and Battery SoC, etc., buffers the data for 5 s, and creates a JSON structure with the buffered data and sends the data to AWS IoT Thing. After that, it clears the buffer and gathers next 5 s data and continues. Detailed comments are added to the code for better understanding. See Fig. 8.57 for more details. The entire code is available under the Chapter_8 attachment ***Code_Files\Python_Code_Raspberry_PI_for_AWS_Connection***.

20. Next step is to transfer the Python code and the AWS certificate files to Raspberry Pi hardware. Users can use any FTP client software to do that. Authors used a software named WinSCP. Login to the Raspberry Pi using WinSCP using the IP address and PI login credentials as shown in Fig. 8.58.

21. After the FTP connection is established, copy the Python file we created and the AWS certificates to the Raspberry Pi folder where we have copied the AWSIoTPythonSDK. So in this example, we are copying files from ***C:\Software\Digital_Twin_Simscape_Book\Chapter_8\Code_Files\Python_Code_Raspberry_PI_for_AWS_Connection*** to ***/home/pi/Digital_Twin_Simscape_Book/Chapter_8/Python_Code_Raspberry_PI_for_AWS_Connection***. See Fig. 8.59 for the copy process details.

22. Now we can run the Python code and the HEV Simulink model together in Raspberry PI and verify that the program we wrote is making the AWS connection and sending the HEV state information to the cloud. Open Putty Desktop App, go to the folder ***/home/pi/Digital_Twin_Simscape_Book/Chapter_8/Python_Code_Raspberry_PI_for_AWS_Connection*** and run the Python program using the command ***ipython Raspberry_Pi_AWS_IOT_Cloud_Connection.py*** as shown in Fig. 8.60

23. Next step is to run the Simulink model ***HEV_Simscape_Model_Rasp_Pi_with_MQTT.slx*** from ***Code_Files\HEV_Model_for_Raspberry_Pi_with_MQTT*** back again to run the

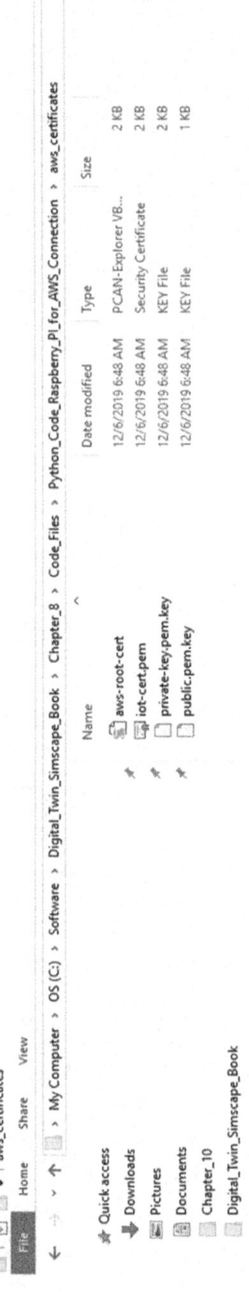

Figure 8.53 AWS certificates.

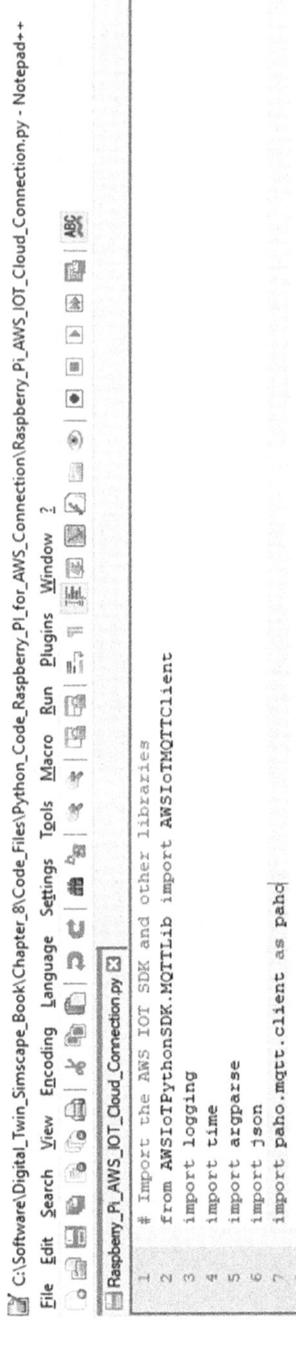

Figure 8.54 Importing libraries.

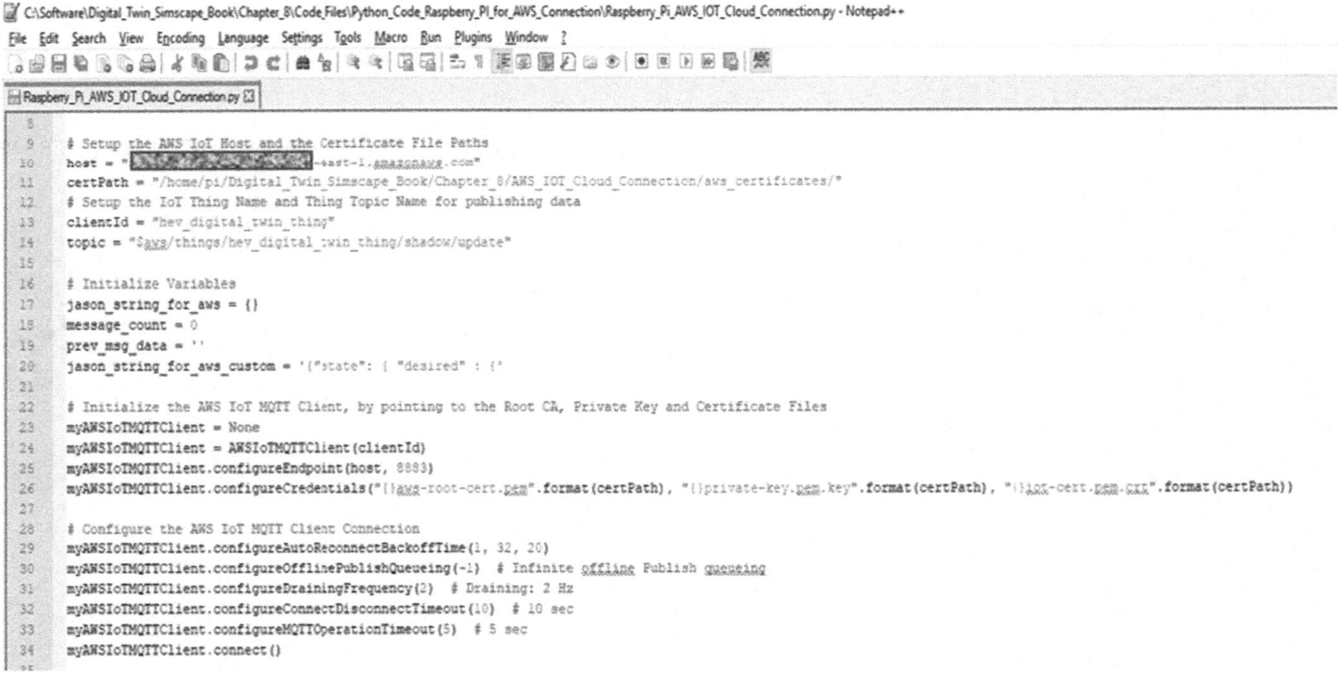

C:\Software\Digital_Twin_Simscape_Book\Chapter_8\Code_Files\Python_Code_Raspberry_PI_for_AWS_Connection\Raspberry_Pi_AWS_IOT_Cloud_Connection.py - Notepad++

File Edit Search View Encoding Language Settings Tools Macro Run Plugins Window ?

Raspberry_Pi_AWS_IOT_Cloud_Connection.py

```
# Setup the AWS IoT Host and the Certificate File Paths
host = "[illegible]-east-1.amazonaws.com"
certPath = "/home/pi/Digital_Twin_Simscape_Book/Chapter_8/AWS_IOT_Cloud_Connection/aws_certificates/"
# Setup the IoT Thing Name and Thing Topic Name for publishing data
clientId = "hev_digital_twin_thing"
topic = "$aws/things/hev_digital_twin_thing/shadow/update"

# Initialize Variables
jason_string_for_aws = {}
message_count = 0
prev_msg_data = ''
jason_string_for_aws_custom = '{"state": { "desired" : {'

# Initialize the AWS IoT MQTT Client, by pointing to the Root CA, Private Key and Certificate Files
myAWSIoTMQTTClient = None
myAWSIoTMQTTClient = AWSIoTMQTTClient(clientId)
myAWSIoTMQTTClient.configureEndpoint(host, 8883)
myAWSIoTMQTTClient.configureCredentials("{}aws-root-cert.pem".format(certPath), "{}private-key.pem.key".format(certPath), "{}iot-cert.pem.crt".format(certPath))

# Configure the AWS IoT MQTT Client Connection
myAWSIoTMQTTClient.configureAutoReconnectBackoffTime(1, 32, 20)
myAWSIoTMQTTClient.configureOfflinePublishQueueing(-1)  # Infinite offline Publish queueing
myAWSIoTMQTTClient.configureDrainingFrequency(2)  # Draining: 2 Hz
myAWSIoTMQTTClient.configureConnectDisconnectTimeout(10)  # 10 sec
myAWSIoTMQTTClient.configureMQTTOperationTimeout(5)  # 5 sec
myAWSIoTMQTTClient.connect()
```

Figure 8.55 AWS IoT—specific configuration for secure connection.

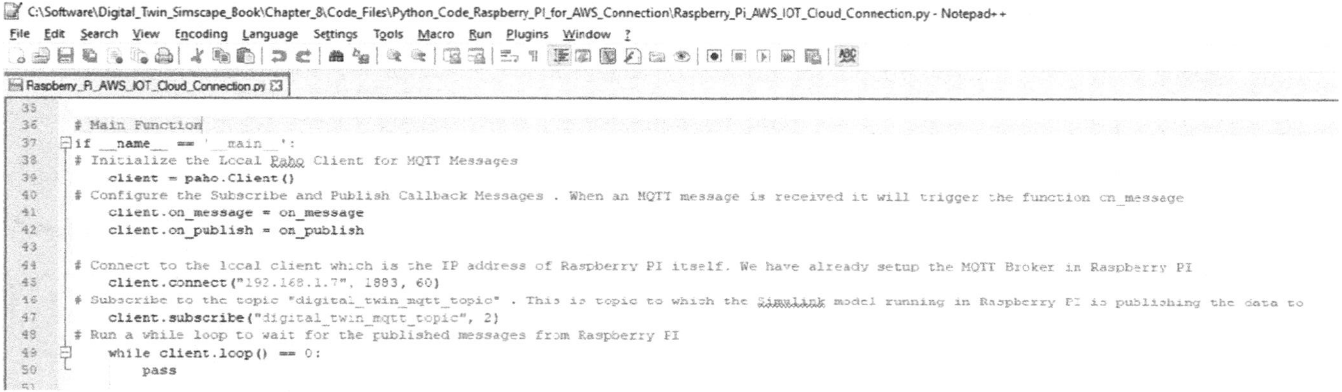

C:\Software\Digital_Twin_Simscape_Book\Chapter_8\Code_Files\Python_Code_Raspberry_PI_for_AWS_Connection\Raspberry_Pi_AWS_IOT_Cloud_Connection.py - Notepad++

File Edit Search View Encoding Language Settings Tools Macro Run Plugins Window ?

Raspberry_Pi_AWS_IOT_Cloud_Connection.py

```
# Main Function
if __name__ == '__main__':
# Initialize the Local Paho Client for MQTT Messages
    client = paho.Client()
# Configure the Subscribe and Publish Callback Messages . When an MQTT message is received it will trigger the function on_message
    client.on_message = on_message
    client.on_publish = on_publish

# Connect to the local client which is the IP address of Raspberry PI itself. We have already setup the MQTT Broker in Raspberry PI
    client.connect("192.168.1.7", 1883, 60)
# Subscribe to the topic "digital_twin_mqtt_topic" . This is topic to which the Simulink model running in Raspberry PI is publishing the data to
    client.subscribe("digital_twin_mqtt_topic", 2)
# Run a while loop to wait for the published messages from Raspberry PI
    while client.loop() == 0:
        pass
```

Figure 8.56 Main function to connect to local MQTT broker.

C:\Software\Digital_Twin_Simscape_Book\Chapter_8\Code_Files\Python_Code_Raspberry_PI_for_AWS_Connection\Raspberry_Pi_AWS_IOT_Cloud_Connection.py - Notepad++

File Edit Search View Encoding Language Settings Tools Macro Run Plugins Window ?

Raspberry_Pi_AWS_IOT_Cloud_Connection.py

```
# MQTT Message from Raspberry PI Received Callback Function
def on_message(mosq, obj, msg):
    global jason_string_for_aws
    global message_count
    global prev_msg_data
    global jason_string_for_aws_custom
# Split the received message, it normally is a comma separated string . Split and make it into a JSON array
# The Data format will be like
# data[0] => Reference Vehicle Speed    [kph]
# data[1] => Actual Vehicle Speed       [kph]
# data[2] => Motor Speed                [RPM]
# data[3] => Generator Speed            [RPM]
# data[4] => Engine Speed               [RPM]
# data[5] => Battery SoC                [%]
# data[6] => Message Count              [Number]

    data = msg.payload.split() #split string into a list
    json_array = json.dumps(data)
# There is a chance that the MQTT Broker will send the same message again and again. Try to only handle the new messages by looking at the data[6]
# which indicates the count generated from the Free Running Counter block in the simulink model. The If condition is satisfied only if the count is
# different when compared to the previous message.
    if data[6] != prev_msg_data:
# Increment the count every time a new message with a new count value is received
        message_count = message_count+1;
        print("Data is Different")
        print(message_count)
        prev_msg_data = data[6]
# When the message_count reaches 50, that means we received 5 Seconds data from the Simulink model running in Raspberry PI
        if (message_count % 50) ==0:
# If the message_count has reached 50 finalize the JSON string to be sent to AWS
            jason_string_for_aws["Data_" + str(message_count)] = json_array;
            jason_string_for_aws_custom = jason_string_for_aws_custom + '"' + 'Data_' + str(message_count) + '"' + ':' + json_array + '}}}';
# Send the JSON String to the AWS using the myAWSIoTMQTTClient.publish function
            print(jason_string_for_aws_custom)
            myAWSIoTMQTTClient.publish(topic, jason_string_for_aws_custom, 1)
            print('Published topic %s: %s\n' % (topic, jason_string_for_aws_custom))
# Initialize the new JSON string header for the next 5 seconds data
            jason_string_for_aws = {}
            jason_string_for_aws_custom = '{"state": { "desired" : {'
            message_count = 0
# If the message_count has not reached 50 keep adding the message data to the JSON string to be sent to AWS
        else :
            # print("Not Reached 50")
            print(json_array)
            # jason_string_for_aws["Data_" + str(message_count)] = json_array;
            jason_string_for_aws_custom = jason_string_for_aws_custom + '"' + 'Data_' + str(message_count) + '"' + ':' + json_array + ',';

def on_publish(mosq, obj, mid):
    pass
```

Python file length : 5,194 lines : 103

Figure 8.57 Callback function for MQTT message receive events.

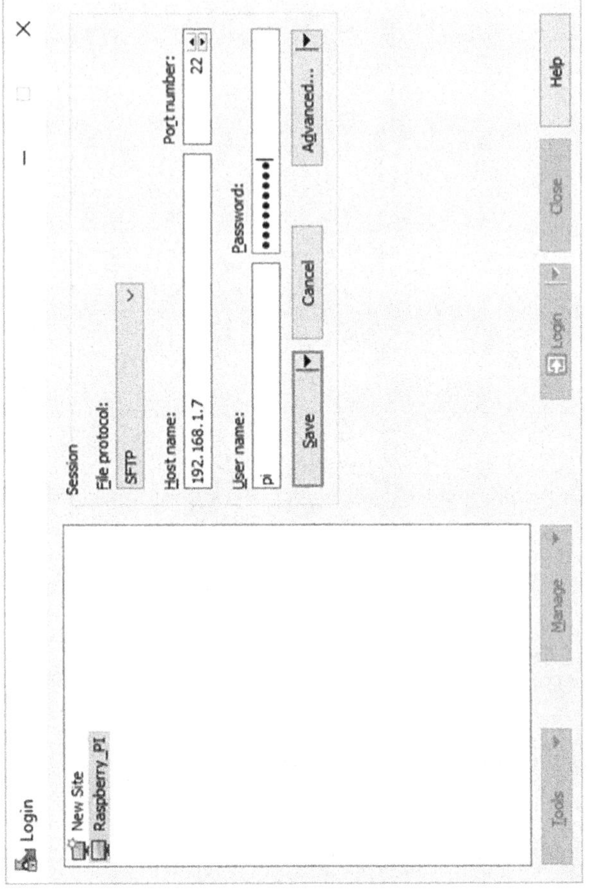

Figure 8.58 WinSCP FTP client to connect to Raspberry Pi.

Figure 8.59 Copying Python code and AWS certificates from host computer to Raspberry Pi.

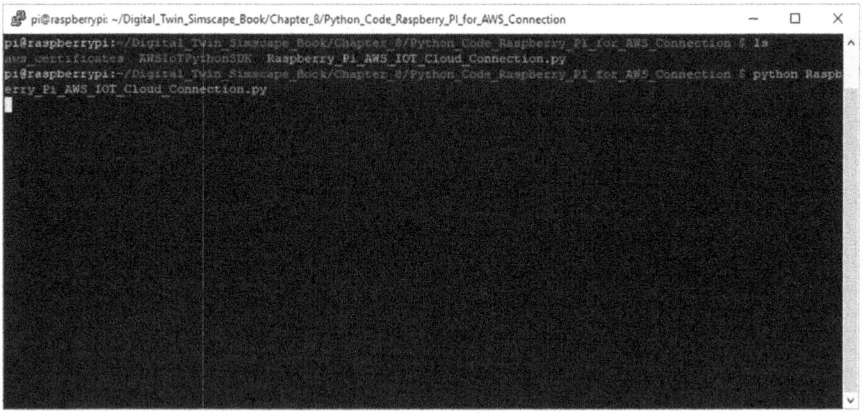

```
pi@raspberrypi: ~/Digital_Twin_Simscape_Book/Chapter_8/Python_Code_Raspberry_PI_for_AWS_Connection
pi@raspberrypi:~/Digital_Twin_Simscape_Book/Chapter_8/Python_Code_Raspberry_PI_for_AWS_Connection $ ls
aws_certificates  AWSIoTPythonSDK  Raspberry_Pi_AWS_IOT_Cloud_Connection.py
pi@raspberrypi:~/Digital_Twin_Simscape_Book/Chapter_8/Python_Code_Raspberry_PI_for_AWS_Connection $ python Raspberry_Pi_AWS_IOT_Cloud_Connection.py
```

Figure 8.60 Running the Python code.

HEV model on the Raspberry Pi. We can see that as the HEV model starts to run in the PI, on the Putty window where we ran the Python code, it starts to receive the HEV state data as shown in Figs. 8.61 and 8.62.

24. In order to verify that the published message from the Raspberry Pi is reaching AWS IoT, go to the AWS console goto ***IoT Core*** >> ***Things*** >> ***hev_digital_twin_thing*** >> ***Activity*** and it will show the same JSON structure the Python code is publishing in real time. See Fig. 8.63. It can be noted that every 5 s we receive a new JSON structure with the buffered data for the previous 5 s.

8.7 Digital Twin Modeling and calibration

So far we have completed the hardware asset setup with Raspberry Pi and its cloud connectivity to send real time data to AWS cloud. Further we will be proceeding with the cloud side work to deploy the Digital Twin model and perform Off-BD. For the HEV example, we will be using the same Simulink model deployed into the Raspberry Pi hardware to the AWS cloud as well. So no further modeling or calibration efforts are needed other than the model we already have.

8.8 Off-board diagnostics algorithm development for Hybrid Electric Vehicle system

In this section, we will develop an algorithm to diagnose and detect the Throttle failure of the Hybrid Electric Vehicle system and test the diagnostic algorithm with the Digital Twin Model. We will first test the diagnostic algorithm locally on a computer before deploying it on the cloud.

The flow chart shown in Fig. 8.64 will be used to diagnose the Throttle failure conditions where we collect the input, output, and the states data from the HEV model running in Raspberry Pi hardware, run the Digital Twin model of the HEV on the cloud with the same input data to get the expected output and states, calculate the RMSE

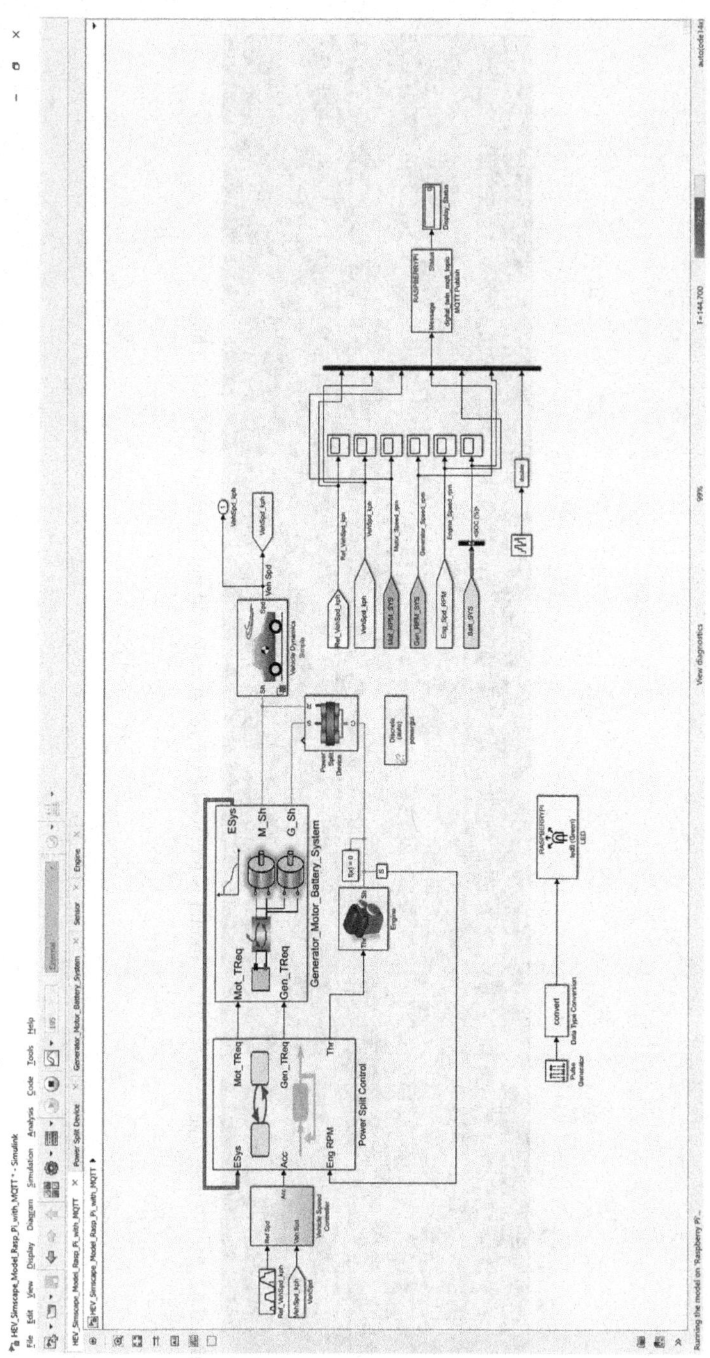

Figure 8.61 HEV Simulink model runs in Raspberry Pi external mode in real time.

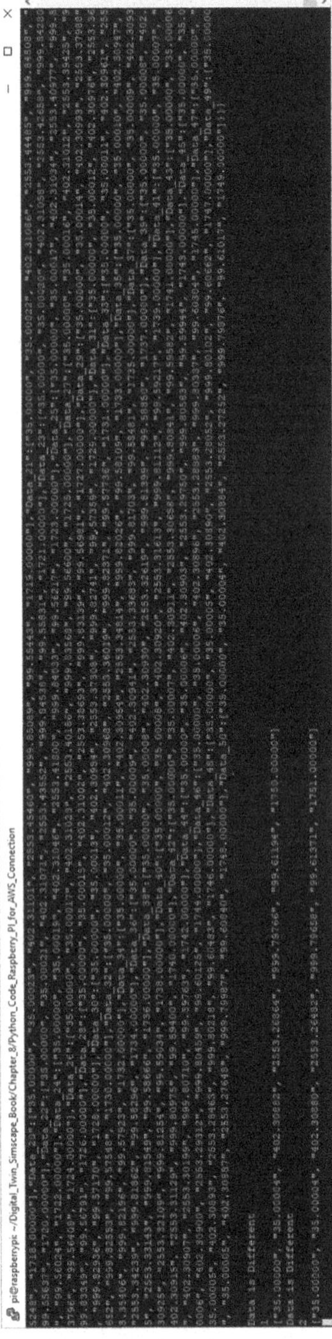

Figure 8.62 MQTT message from the HEV model is received by Python code and published to AWS IoT Thing.

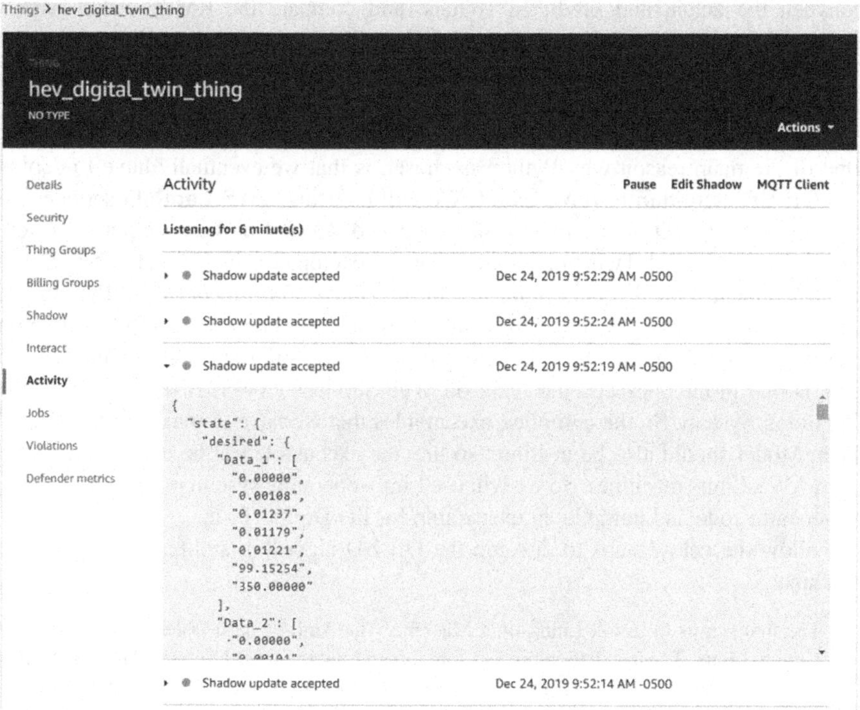

Things > hev_digital_twin_thing
hev_digital_twin_thing
NO TYPE
Actions
Details
Security
Thing Groups
Billing Groups
Shadow
Interact
Activity
Jobs
Violations
Defender metrics
Activity
Pause
Edit Shadow
MQTT Client
Listening for 6 minute(s)
Shadow update accepted
Dec 24, 2019 9:52:29 AM -0500
Shadow update accepted
Dec 24, 2019 9:52:24 AM -0500
Shadow update accepted
Dec 24, 2019 9:52:19 AM -0500
{
"state": {
"desired": {
"Data_1": [
"0.00000",
"0.00108",
"0.01237",
"0.01179",
"0.01221",
"99.15254",
"350.00000"
],
"Data_2": [
"0.00000",
Shadow update accepted
Dec 24, 2019 9:52:14 AM -0500

Figure 8.63 AWS IoT Thing receiving the HEV state information from Raspberry Pi hardware.

Off-BD Start
Get Input / Output Data from Hardware
Run the Digital Twin Model with the Hardware Input Data and get the Predicted System Output
Calculate the Root Mean Square Error (RMSE) between Predicted and Actual System Output
RMSE > Threshold
NO
YES
No Fault Detected by Off-BD
Fault Detected by Off-BD. Send Text/Email Alert
Off-BD End

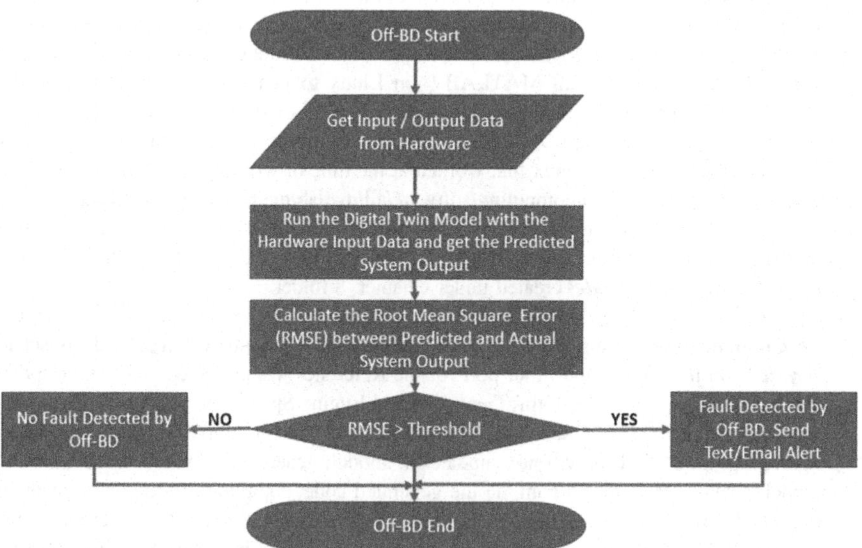

Figure 8.64 Off-BD algorithm flow chart.

between the actual and predicted signals, and compare the RMSE value with a threshold to detect if there is a failure. The E-mail/Text notification will not be tested in this section but will be covered later when we deploy the Off-BD and Digital Twin to the AWS cloud.

We will be running the Off-BD algorithm using Python programming language. One of the main reason why Python is chosen is that we eventually have to deploy this Off-BD algorithm to AWS cloud. We will be using **AWS Lambda** services for running the Off-BD algorithm on the cloud and **AWS Lambda** supports Python. Note that the Digital Twin model we have is developed using MATLAB®, Simulink®, and Simscape™, so in order to call the Digital Twin model from Python, we will have to generate "C" code from the Simscape™ model and compile it into an executable and then call the executable from Python program. One other thing to note is that in the back end the Amazon Web Services (AWS) is running on Linux Operating System. So the compiled executables that we have to make for the Digital Twin Model should also be in Linux so that the executable can be deployed and ran from AWS Linux machines. So we will use Linux operating system just for this section to generate code and compile an executable for the Digital Twin.

Follow the below steps to develop the Off-BD algorithm and test it locally on a machine:

1. The first step is to install Linux on a Machine. The Authors have tested the process with Ubuntu Linux. The installation process for Ubuntu Linux will not be covered in this chapter; there are plenty of online resources available for Linux installation. Authors recommend to install Ubuntu as a Dual-Boot setup if the User already has Windows installed on the machine. Dual-Boot installation allows Users to switch between Windows and Linux when required by just restarting one operating system and selecting others from the boot up menu. One of the working link to install Ubuntu at the time of writing of this chapter is here https://itsfoss.com/install-ubuntu-1404-dual-boot-mode-windows-8-81-uefi/.
2. The rest of the steps in this section is using Linux. So make sure Linux is installed and setup. We will also need to install MATLAB® on Linux to generate code and compile the Simscape™ Digital Twin Model. Once again, since it is outside the scope of the book, we will not go through the MATLAB® installation process in Linux. An instruction link to install MATLAB® on Linux that worked at the time of writing of this chapter is given here https://www.cmu.edu/computing/software/all/matlab/matlabinstall-linux.html.
3. Open MATLAB® in Linux and create a new folder and copy the HEV Model Simscape™ model used earlier and its initialization MAT file as shown in Fig. 8.65. The new folder is ***Digital_Twin_HEV_Model*** created under Chapter_8 folder.
4. Open the Simscape™ HEV model and make sure the System Target File under the **Model >> Configuration Parameters >> Code Generation >> System Target File** is set to "***ert.tlc***." On the model, add input port for the Reference Vehicle Speed and output ports for Actual Vehicle Speed, Motor, Generator and Engine Speed, and Battery SoC, etc., as shown in Fig. 8.66. Then click on the "***Build***" button on the Simulink Model Menu as highlighted in Fig. 8.66. This will now update the model, generate "***C***" source code from the model, create a Makefile to compile the generated code, compile the code, and create an executable application from the Model logic. MATLAB® will show the progress of this process as shown in Fig. 8.67. Since we are trying this on a Linux Machine, it will create a Linux executable application. The "***C***" code will be generated into a folder named **Model_Name_ert_rtw** under the working folder and the name of the executable will be same as

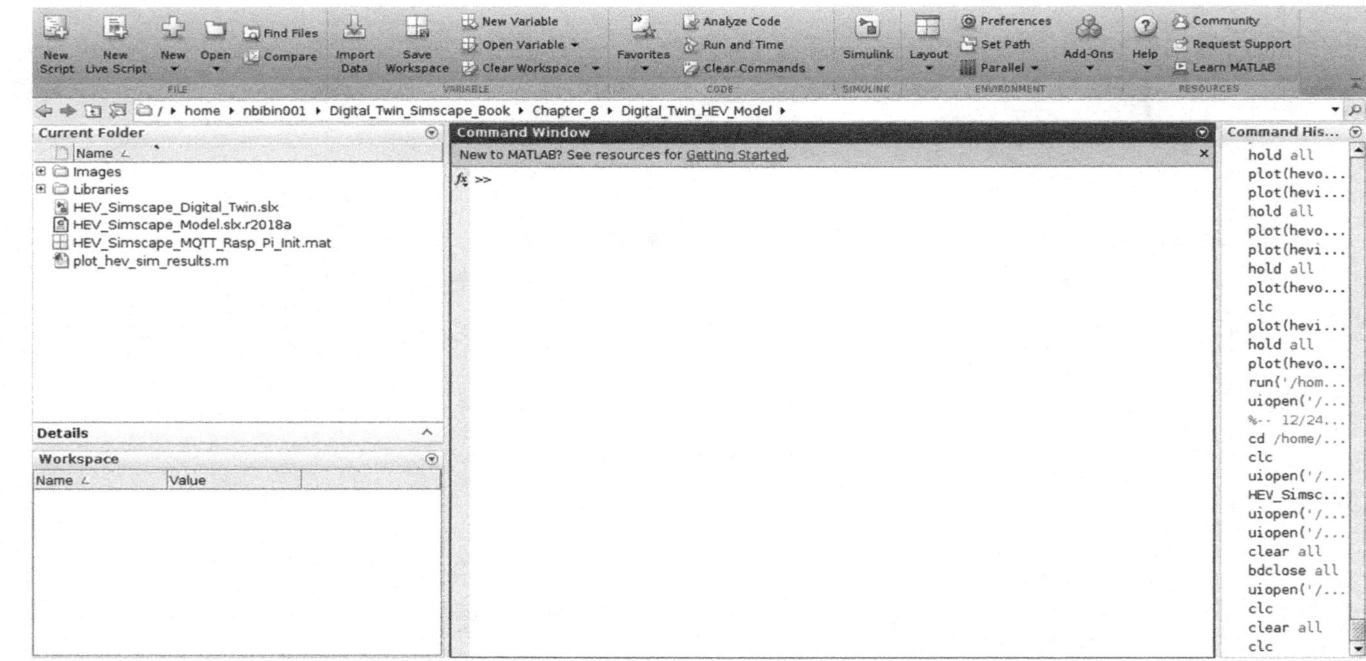

Figure 8.65 Copy Simscape™ HEV model and initialization M file to Linux MATLAB folder.

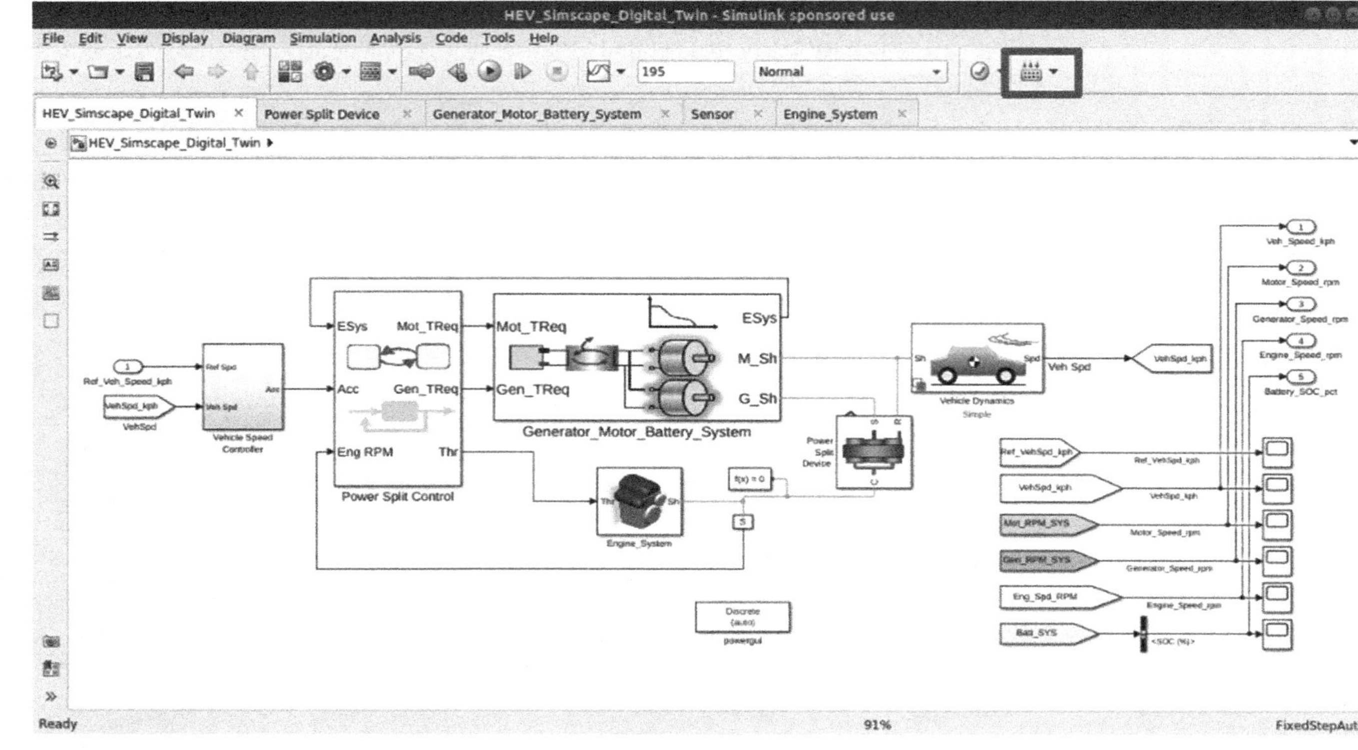

Figure 8.66 HEV model configured for Digital Twin codegen and compiling.

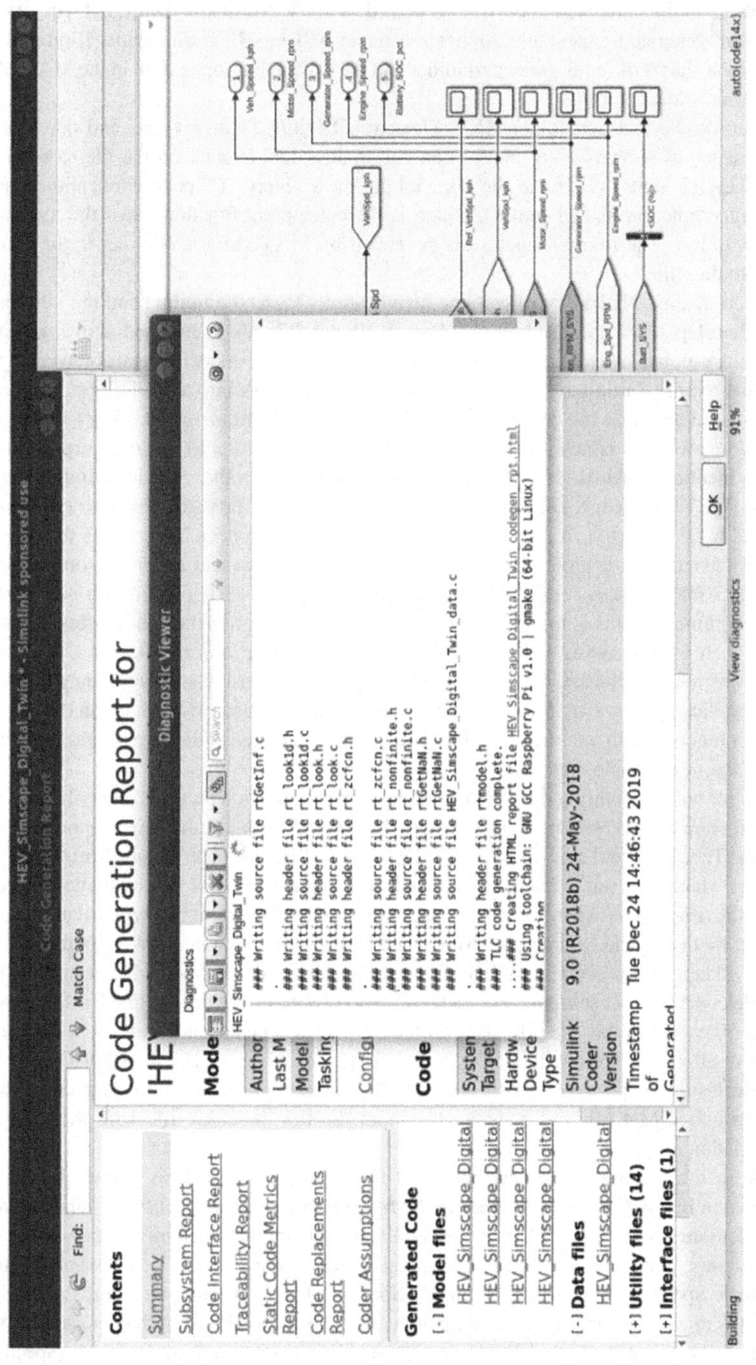

Figure 8.67 Model codegen and build progress.

that of the model name. In this case, our Model Name is **HEV_Simscape_Digital_T-win.slx**, so the generated code will be placed in **HEV_Simscape_Digital_Twin_ert_rtw** and the generated executable application name will be **HEV_Simscape_Digital_Twin**. Fig. 8.68 shows the code generated folder and the executable application in the MATLAB® working folder.

5. Go inside the codegen folder **HEV_Simscape_Digital_Twin_ert_rtw** and open the file ert_main.c as shown in Fig. 8.69 in an editor. Just double click on the file to open it in MATLAB® editor itself. As the User might know, every "***C***" code application requires an entry function named **main()**, which is the entry point function when the application starts to run. The **main()** function is generated by MATLAB® Code Generator into this **ert_main.c** file.
6. The ert_main.c file has two functions mainly in it. One is the **main()** and the other is the **rt_OneStep()** function as shown in Figs. 8.70 and 8.71. As mentioned above, **main()** is the entry point function, and this function initializes the model states if any using the **Model_Name_initialize** function and just enters and stays into a **while loop.** So we can see that even when the application runs, it is not really running our model logic yet in this framework as is now. So we will edit the **main()** function to suit our purpose in later steps mentioned below. The **rt_OneStep()** function calls another function **Model_Name_-step** (**HEV_Simscape_Digital_Twin_step()** in this case; this step function contains the actual logic that we have implemented in the Simulink® model. We can see the **Model_-step()** function uses global data structures to receive input and return the output. When MATLAB® generates code, the **rt_OneStep()** function call is commented out from the **main()** function, this is mainly because the Users can integrate the **Model_step()** function with their own operating system, embedded target software, or scheduler, etc., and doesn't really have to use the **ert_main.c** in their application software. The way we are going to run this application is we get the input data, we run the application, which will run the **Model_-step()** function with one sample of input data, get its output, and then we run the application with the next sample data again.
7. Now we will start editing the ert_main.c to suit our application purpose. We will change the main() function to read an input file "**hev_input.csv**" from a Linux user temporary folder **/tmp**. The **/tmp** folder is selected for the input and output files because**/tmp** is the only folder which is writable from AWS. The **hev_input.csv** file will be automatically created into the **/tmp** folder when we receive data from Hardware system, we will discuss that later when we deploy the Digital Twin Model and Off-BD algorithm into AWS cloud. This input file will contain the actual data collected from Raspberry Pi hardware when it runs the HEV model, one line for each sample of data at every 0.1 s, for 5 s. After the first 5 s, there will be total 50 entries in this file. The data for the next 5 s will be appended to this hev_input.csv file, so after 10 s, there will be 100 entries in the file and so on. For the HEV Digital Twin model testing purpose, we ran the HEV desktop simulation model and save the output work-space into a Mat file and generate the input file hev_input.csv file with the data from simulation output.
8. First in the **ert_main.c**, a new handwritten C function **Parse_CSV_Line()** is added as shown in Fig. 8.72 to read the lines in the **hev_input.csv** file and splits the values between the delimiter comma in each lines. Add this function above the **main()** function. So essentially, this splits and returns the individual signals for Reference Vehicle Speed, Actual Vehicle Speed, Motor, generator Engine Speed, Battery SoC, and the count variable.
9. Next is to add a portion of code to open the **hev_input.csv** file from the path**/tmp/hev_in-put.csv** and read the actual Reference Vehicle Speed into an array Ref_Veh_Speed_kph [].This code needs to be added inside the main() function as shown in Fig. 8.73.

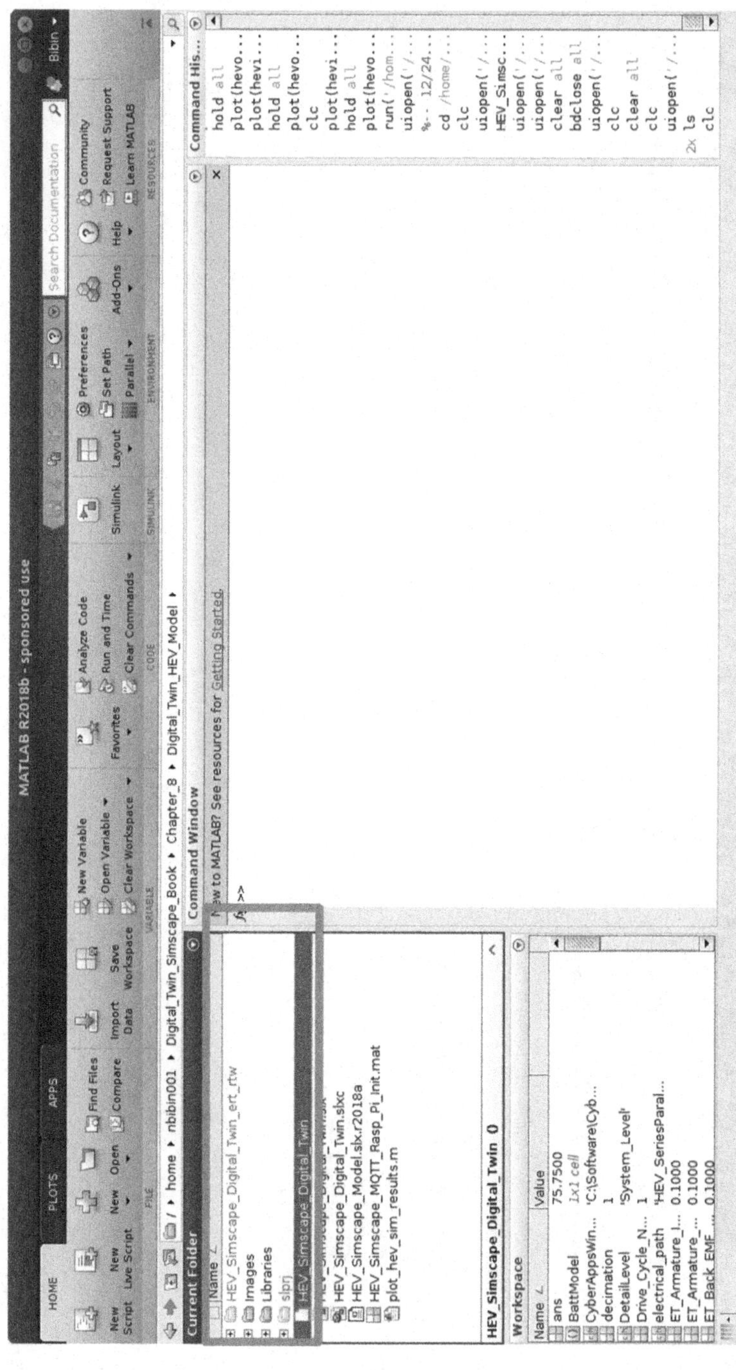

Figure 8.68 Generated code folder and executable application.

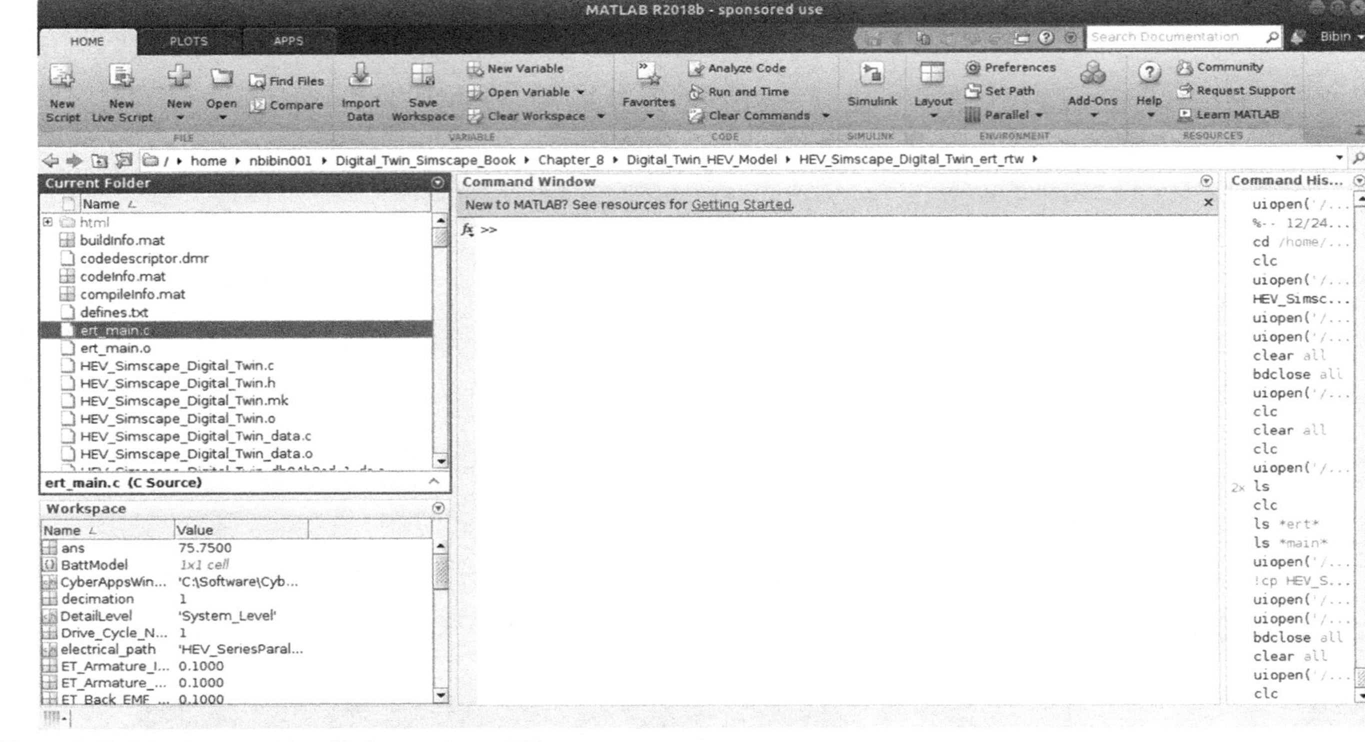

Figure 8.69 Selecting ert_main.c file from codegen folder.

```
 */
int_T main(int_T argc, const char *argv[])
{
  /* Unused arguments */
  (void)(argc);
  (void)(argv);

  /* Initialize model */
  HEV_Simscape_Digital_Twin_initialize();

  /* Simulating the model step behavior (in non real-time) to
   *  simulate model behavior at stop time.
   */
  while ((rtmGetErrorStatus(HEV_Simscape_Digital_Twin_M) == (NULL)) &&
         !rtmGetStopRequested(HEV_Simscape_Digital_Twin_M)) {
    rt_OneStep();
  }

  /* Disable rt_OneStep() here */

  /* Terminate model */
  HEV_Simscape_Digital_Twin_terminate();
  return 0;
}

/*
 * File trailer for generated code.
 *
 * [EOF]
```

Figure 8.70 main() function in ert_main.c.

```
void rt_OneStep(void)
{
  static boolean_T OverrunFlag = false;

  /* Disable interrupts here */

  /* Check for overrun */
  if (OverrunFlag) {
    rtmSetErrorStatus(HEV_Simscape_Digital_Twin_M, "Overrun");
    return;
  }

  OverrunFlag = true;

  /* Save FPU context here (if necessary) */
  /* Re-enable timer or interrupt here */
  /* Set model inputs here */

  /* Step the model for base rate */
  HEV_Simscape_Digital_Twin_step();

  /* Get model outputs here */

  /* Indicate task complete */
  OverrunFlag = false;

  /* Disable interrupts here */
  /* Restore FPU context here (if necessary) */
  /* Enable interrupts here */
}
```

Figure 8.71 rt_OneStep() function in ert_main.c.

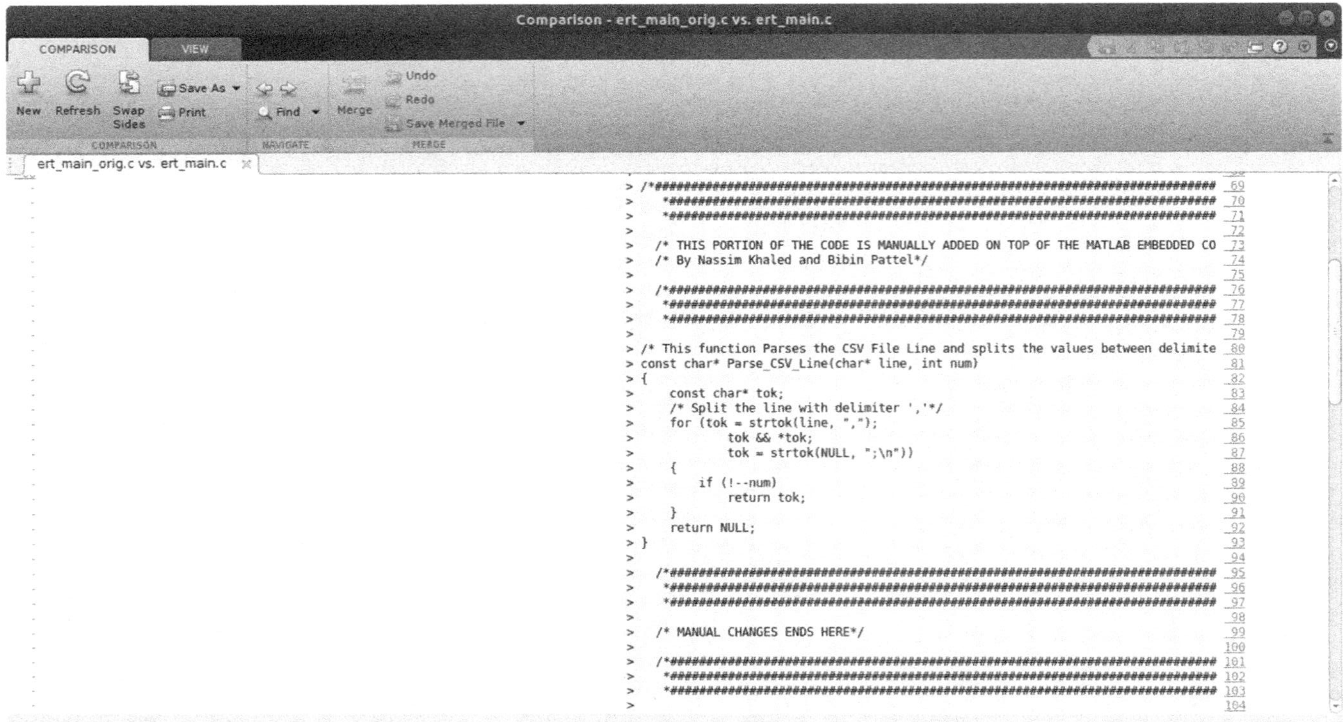

Figure 8.72 Function to parse CSV file lines and get the input file hev_input.csv

10. Next is to take out the infinite **while()** loop function from **main()** function of the **ert_main.c**. The **Model_step()** function will be called explicitly, so we don't need this **while()** loop. Also a loop is ran for the number of lines in the input file to call the step function **HEV_-Simscape_Digital_Twin_step()**. The Input array values are assigned to the global input data structure **HEV_Simscape_Digital_Twin_U.Ref_Veh_Speed_kph**. The outputs will be stored into the global data structure **HEV_Simscape_Digital_Twin_Y**. These structure definitions can be found in the **HEV_Simscape_Digital_Twin.c** file. An output file hev_output.csv will be created in the /tmp folder with the input and outputs collected when running the Digital Twin model. See Fig. 8.74.
11. The changes to the ert_main.c file is completed now. Save and close the file. From the **HEV_Simscape_Digital_Twin_ert_rtw** folder, look for the Makefile with the extension "**HEV_Simscape_Digital_Twin.mk**." Copy this file to a new file named Makefile using the Linux command ***!cp HEV_Simscape_Digital_Twin.mk Makefile*** from MATLAB command prompt. We can try the same from a Linux command window as well without the "**!**" at the beginning. "**!**" tells the MATLAB command window that what follows is an operating system command. Since we have changed the ert_main.c file, let us just delete the previously created object file ert_mian.o using the rm command ***!rm ert_main.o***. Now rebuild the executable application using the command **!make -f Makefile**. Since only the ert_-main.c file is changed and all the other files are same, the **make** command only recompiled **ert_mian.c**. All of the commands tried in this step and their output are shown in Fig. 8.75. The recompiled executable application **HEV_Simscape_Digital_Twin** is shown in Fig. 8.76.
12. Let us just make a quick MATLAB®-based program to test the executable application. The MATLAB® program will take the HEV desktop simulation data collected by running the simulation on the host computer, as if it is coming from the actual hardware in real time, crate the "**hev_input.csv**" file copy it to **/tmp** folder, call the executable application which will run with the input Reference Vehicle Speed data from the **hev_input.csv**, and create the "**hev_output.csv**" file under **/tmp** folder. The MATLAB® program will then look at the "**hev_output.csv**" file and get the predicted signal values, and to calculate RMSE between the actual and predicted data, make plots for the each of the signals, plot the error between them, and also calculate and report the RMSE between the actual and predicted values. See the MATLAB® program below. The Mat file HEV_Simscape_Model_Sim_Results.mat is created by saving the MATLAB workspace after running the model simulation in the host computer. These data are used to create the **"hev_input.csv"** file.

```
% Book Title: Digital Twin Develepent and Deployment On Cloud Using Matlab
% Simscape
% Chapter 8
% Authors: Nassim Khaled and Bibin Pattel
% Last Modified 12/14/2019
%%
% Create input file from the simulation data
% Load the Simulation Data
clc
clear all
load HEV_Simscape_Model_Sim_Results.mat
```

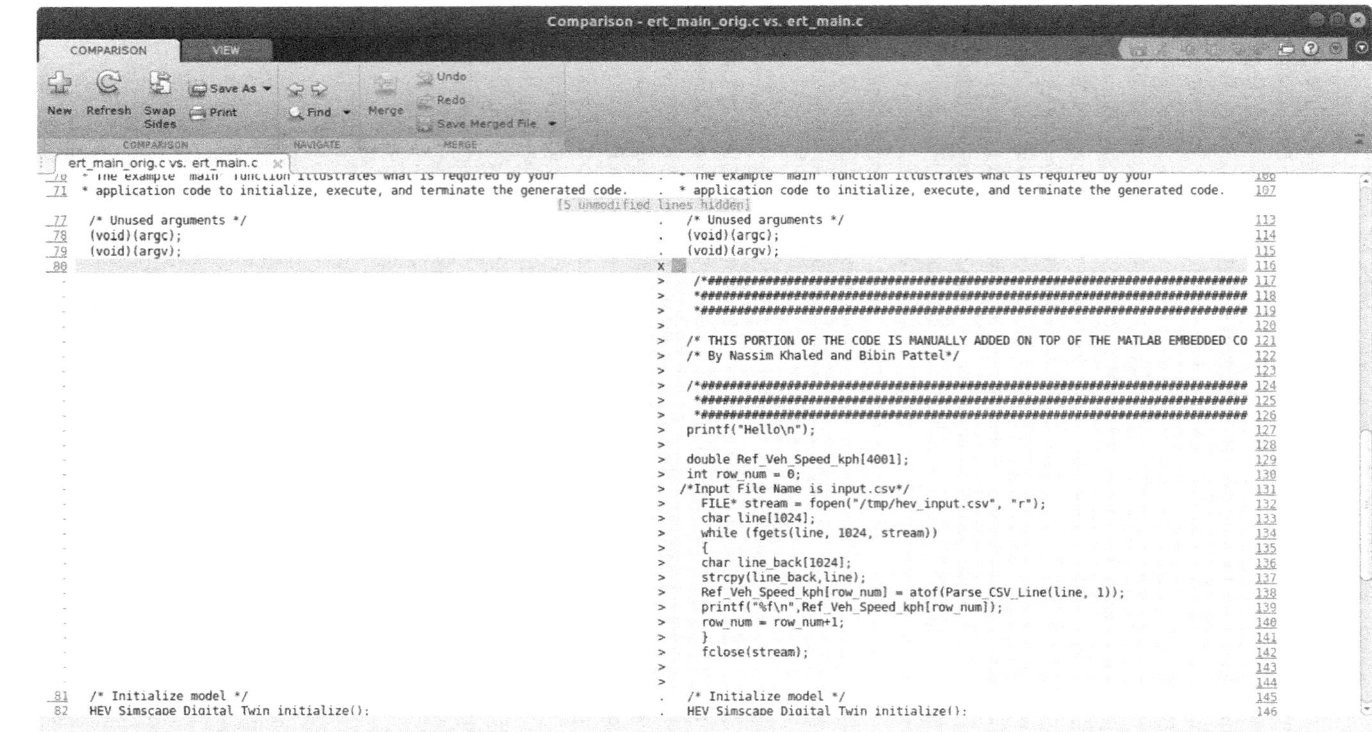

Figure 8.73 Reading input.csv and storing into Input array.

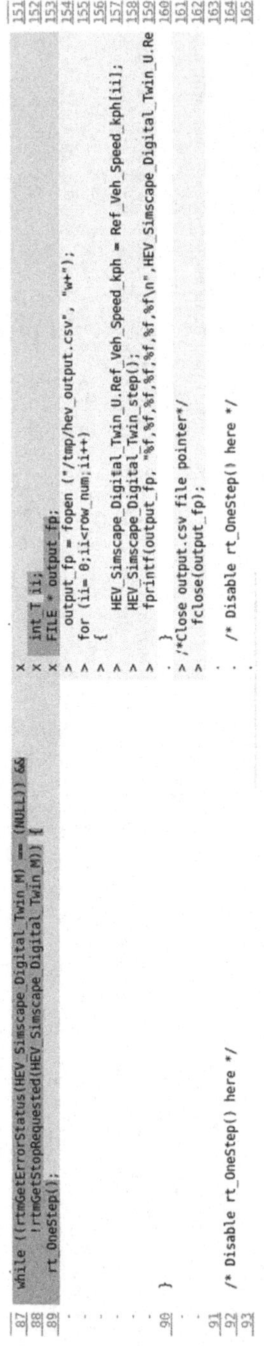

Figure 8.74 Removing while() loop from ert_mian.c and calling the model step function.

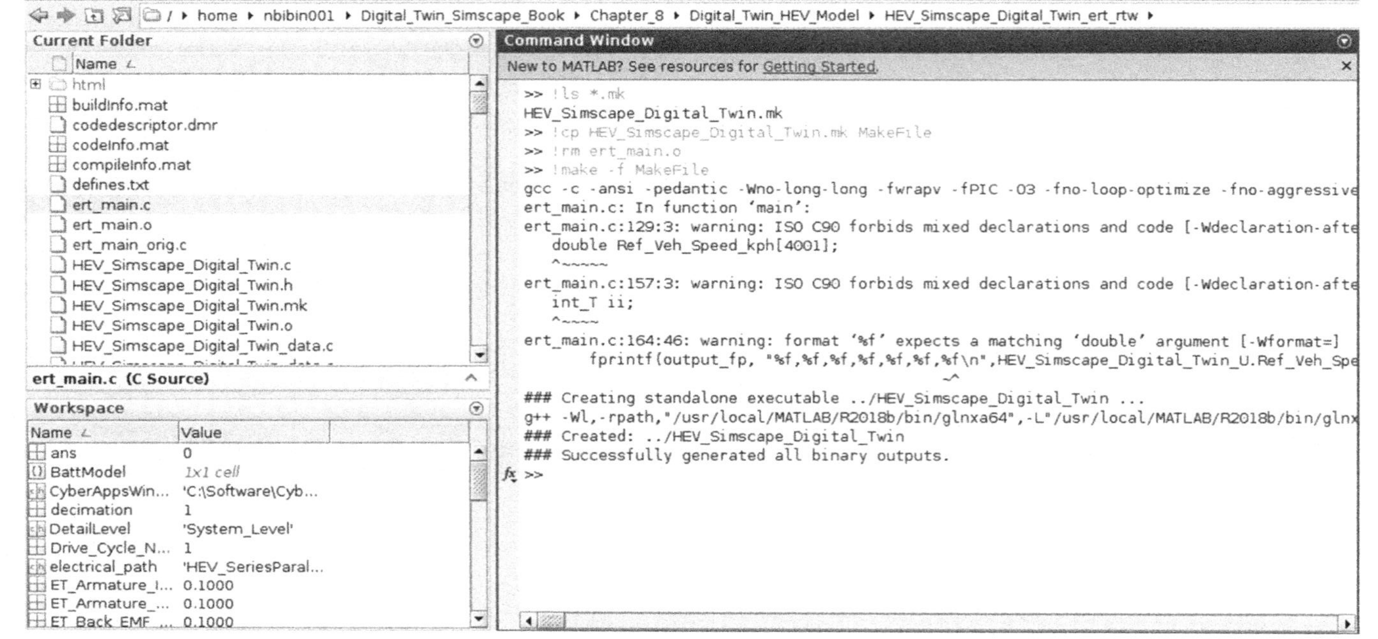

Figure 8.75 Recompiling the application with the updated ert_mian.c.

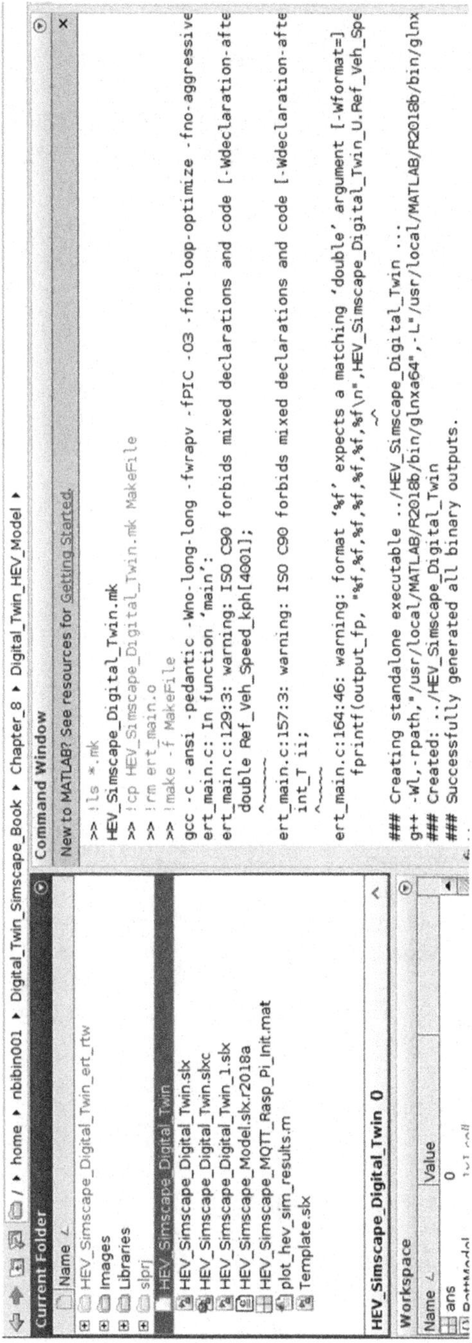

```
/ ▸ home ▸ nbibin001 ▸ Digital_Twin_Simscape_Book ▸ Chapter_8 ▸ Digital_Twin_HEV_Model ▸

Current Folder
Name
HEV_Simscape_Digital_Twin_ert_rtw
Images
Libraries
slprj
HEV_Simscape_Digital_Twin
HEV_Simscape_Digital_Twin.slx
HEV_Simscape_Digital_Twin.slxc
HEV_Simscape_Digital_Twin_1.slx
HEV_Simscape_Model.slx.r2018a
HEV_Simscape_MQTT_Rasp_Pi_Init.mat
plot_hev_sim_results.m
Template.slx

HEV_Simscape_Digital_Twin ()

Workspace
Name        Value
ans         0

Command Window
New to MATLAB? See resources for Getting Started.
>> !ls *.mk
HEV_Simscape_Digital_Twin.mk
>> !cp HEV_Simscape_Digital_Twin.mk MakeFile
>> !rm ert_main.o
>> !make -f MakeFile
gcc -c -ansi -pedantic -Wno-long-long -fwrapv -fPIC -O3 -fno-loop-optimize -fno-aggressive
ert_main.c: In function 'main':
ert_main.c:129:3: warning: ISO C90 forbids mixed declarations and code [-Wdeclaration-afte
   double Ref_Veh_Speed_kph[4001];
   ^~~~~~
ert_main.c:157:3: warning: ISO C90 forbids mixed declarations and code [-Wdeclaration-afte
   int_T ii;
   ^~~~~
ert_main.c:164:46: warning: format '%f' expects a matching 'double' argument [-Wformat=]
       fprintf(output_fp, "%f,%f,%f,%f,%f,%f,%f\n",HEV_Simscape_Digital_Twin_U.Ref_Veh_Spe
                                             ~^
### Creating standalone executable ../HEV_Simscape_Digital_Twin ...
g++ -Wl,-rpath,"/usr/local/MATLAB/R2018b/bin/glnxa64",-L"/usr/local/MATLAB/R2018b/bin/glnx
### Created: ../HEV_Simscape_Digital_Twin
### Successfully generated all binary outputs.
```

Figure 8.76 Recompiled executable Application.

```
fid = fopen('hev_input.csv','w+');
for i =1:1951
fprintf(fid,'%s,%s,%s,%s,%s,%s,%s\n',
num2str(Ref_VehSpd_kph(i,2)),num2str(VehSpd_kph(i,2)),
        num2str(Motor_Speed_rpm(i,2)),
num2str(Generator_Speed_rpm(i,2)),num2str(Engine_Speed_rpm(i,2)),
        num2str(Battery_SOC_pct(i,2)),num2str(i-1));
end
fclose(fid);
% Copy the hev_input.csv file to the /tmp folder
!cp hev_input.csv /tmp
pause(1)
% Run the Digital Twin HEV Compiled Model
!./HEV_Simscape_Digital_Twin
pause(1)
% Digital Twin Model creates the output file hev_output.csv.Copy that
file
% to the current folder
!cp /tmp/hev_output.csv .
pause(1);
%%
actual_data = csvread('hev_input.csv');
model_predicted_data = csvread('hev_output.csv');
% Loop through to plot and calculate the RMSE
% i = 1 for Reference Vehicle Speed
% i = 2 for Actual Vehicle Speed
% i = 3 for Motorator Speed
% i = 5 for Engir Speed
% i = 4 for Genene Speed
% i = 6 for Battery SoC
title1 = {'Reference Vehicle Speed','Actual Vehicle Speed','Motor
Speed','Generator Speed','Engine Speed','Battery SoC'};
title2 = {'Reference Vehicle Speed Error','Actual Vehicle Speed
Error','Motor Speed Error','Generator Speed Error','Engine Speed
Error','Battery SoC Error'};
for i =1:6
  % Plot the individual signals Actual Vs Predicted
    figure(1);
    subplot(320 +i)
    plot(actual_data(:,i),'linewidth',2);
    hold all
    plot(model_predicted_data(:,i),'--','linewidth',2);
    title(title1{i},'Fontsize',18);
```

```
    % Plot the Corresponding Signal Errors Actual - Predicted
      figure(2);
      Error = actual_data(:,i) - model_predicted_data(:,i);
      subplot(320+i)
      plot(Error,'linewidth',2);
      title(title2{i},'Fontsize',18);
    % Calculate the Root Mean Squared for Error for Each Signal
      Squared_Error = Error.^2;
      Mean_Squared_Error = mean(Squared_Error);
      Root_Mean_Squared_Error(i) = sqrt(Mean_Squared_Error);
end
figure(1)
legend('Actual Data','Predicted Data');
figure(2)
legend('Error Between Actual Data and Predicted Data');
Root_Mean_Squared_Error
```

13. Fig. 8.77 shows the "hev_input.csv" file created from the MATLAB® test script above. First column is the Reference Vehicle Speed, second column is the Actual Vehicle Speed, third, fourth, and fifth columns are the Motor, Generator, and Engine Speeds, respectively, sixth column is the Battery SoC, and last column contains a count variable.
14. Figs. 8.78 and 8.79 show the plots generated from the MATLAB® test script comparing the actual versus Digital Twin predicted signals of HEV, and also the calculated error. Fig. 8.80 shows the RMSE values between actual and Digital Twin predicted values. It can be seen that though the actual and Digital Twin predicted signals are almost matching everywhere, but for the Generator Speed and Engine Speed some deviations are noted, which caused the RMSE for those to signals to be higher than expected. Ideally, we expect the RMSE value to be close to zero because we are running the same model with same data. Authors could not really explain why the mismatch occurs in few points.
15. Now let us repeat the Step 12 with a desktop simulation model in which the Throttle going to the Engine subsystem is connected through a Gain block of value 0, which essentially makes the Engine system see a throttle value of 0 always. This is the way we introduced throttle failure conditions in this example. There could be many different ways though. See Fig. 8.81 for the introduced Throttle Failure. After running the desktop host computer simulation, save the Mat file and load it in the above MATLAB program and run it.
16. Now when the MATLAB code runs, the input Reference Speed is same for both the desktop simulation model and the Digital Twin model, but for the desktop simulation model, the throttle to the Engine system is made 0, whereas Digital Twin model is not aware of this, so it will be make a prediction as if there is no throttle fault, and we will compare the actual and predicted signals and calculate the RMSE of the signals. Check, Figs. 8.82–8.84 for the results. It can be noted that even with the Throttle failure, the vehicle speed reference is still met with the power from the Battery and Motor. The Engine, Generator, and the Battery SoC signals are now a lot different than the expected values, which is reflected in the RMSE value calculations as well. So by comparing the RMSE values to a certain threshold, we can detect the Throttle failure condition.

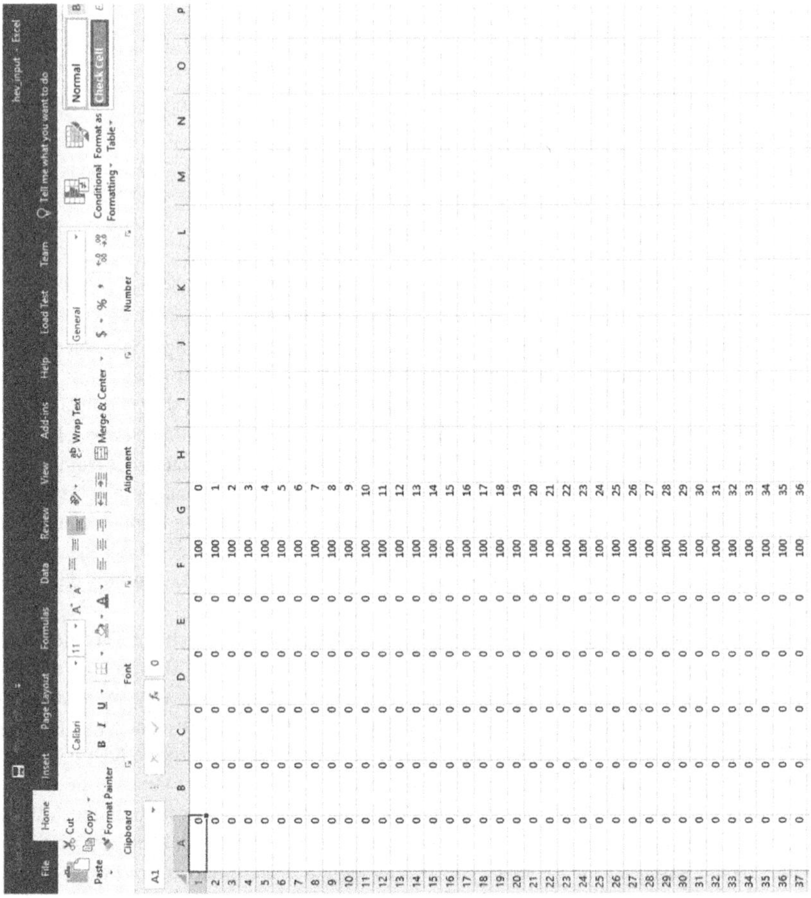

	A	B	C	D	E	F	G
1	0	0	0	0	0	100	0
2	0	0	0	0	0	100	1
3	0	0	0	0	0	100	2
4	0	0	0	0	0	100	3
5	0	0	0	0	0	100	4
6	0	0	0	0	0	100	5
7	0	0	0	0	0	100	6
8	0	0	0	0	0	100	7
9	0	0	0	0	0	100	8
10	0	0	0	0	0	100	9
11	0	0	0	0	0	100	10
12	0	0	0	0	0	100	11
13	0	0	0	0	0	100	12
14	0	0	0	0	0	100	13
15	0	0	0	0	0	100	14
16	0	0	0	0	0	100	15
17	0	0	0	0	0	100	16
18	0	0	0	0	0	100	17
19	0	0	0	0	0	100	18
20	0	0	0	0	0	100	19
21	0	0	0	0	0	100	20
22	0	0	0	0	0	100	21
23	0	0	0	0	0	100	22
24	0	0	0	0	0	100	23
25	0	0	0	0	0	100	24
26	0	0	0	0	0	100	25
27	0	0	0	0	0	100	26
28	0	0	0	0	0	100	27
29	0	0	0	0	0	100	28
30	0	0	0	0	0	100	29
31	0	0	0	0	0	100	30
32	0	0	0	0	0	100	31
33	0	0	0	0	0	100	32
34	0	0	0	0	0	100	33
35	0	0	0	0	0	100	34
36	0	0	0	0	0	100	35
37	0	0	0	0	0	100	36

Figure 8.77 hev_input.csv file created from the MATLAB test script.

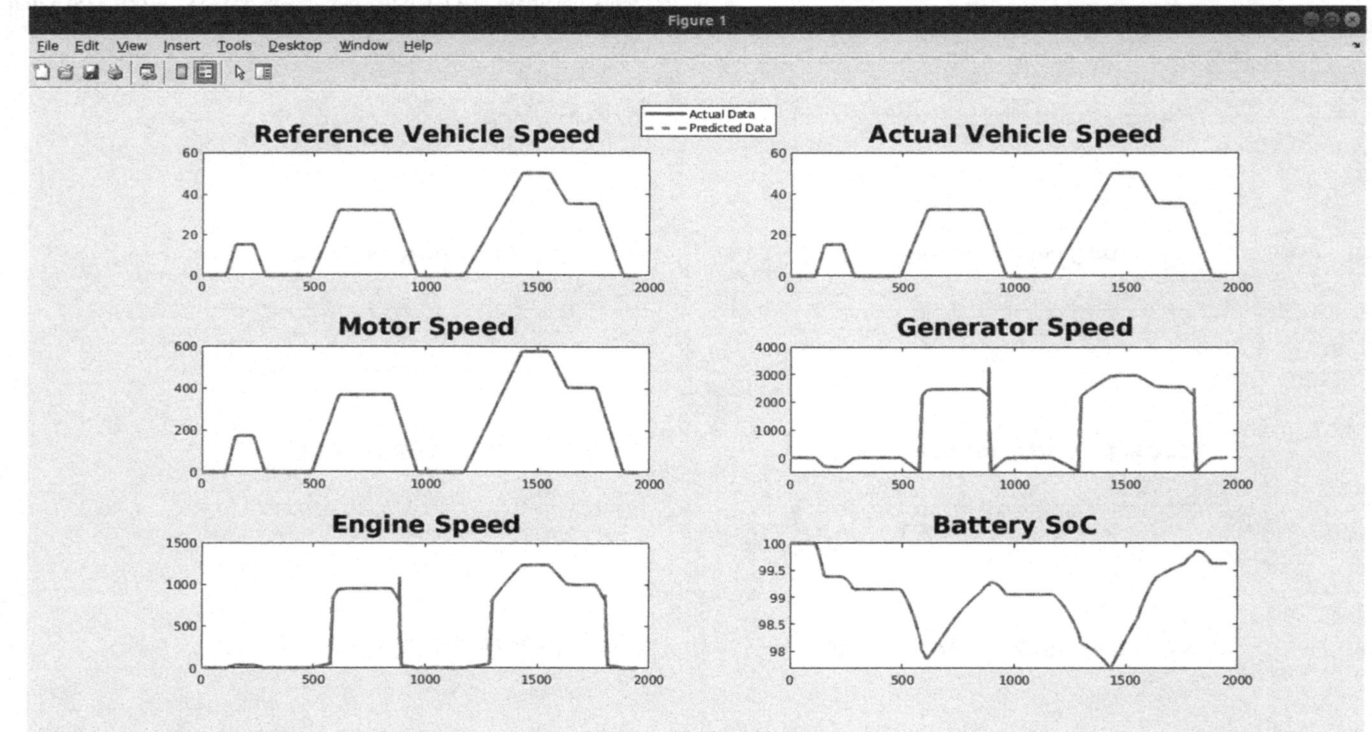

Figure 8.78 Comparing actual versus Digital Twin predicted values for HEV.

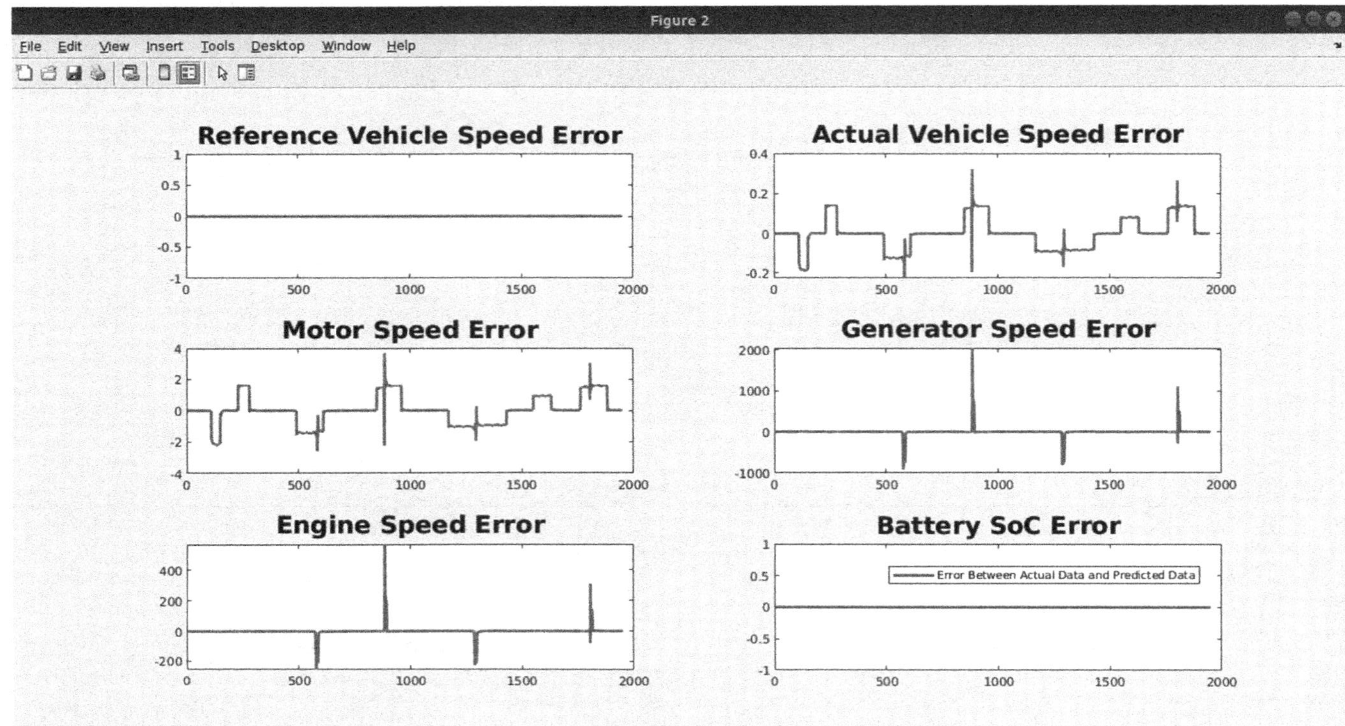

Figure 8.79 Error between actual and Digital Twin predicted values of HEV.

```
>> Root_Mean_Squared_Error

Root_Mean_Squared_Error =

          0    0.0749    0.8612  110.0032   30.6989    0.9114
```

Figure 8.80 RMSE values of the corresponding six signals.

8.9 Deploying digital twin Hybrid Electric Vehicle system model to cloud

In this section, we will deploy the compiled executable of the HEV Digital Twin model and the diagnostic algorithm to AWS cloud. We will be configuring and using three AWS in this section, Simple Notification Service (SNS), and AWS Lambda Functions and Amazon S3 Bucket. SNS is used to send Text/E-mail messages to the subscribers, and the subscriber details can be configured. We can create topics for SNS and add subscriber information such as cell number, E-mail ID. etc., so that the subscribers will be notified when the topic event is triggered. AWS Lambda is an event-driven serverless computing platform that runs a specific code, which we can develop and deploy. This code will run in response to a specific event, and in our case, the event is triggered by the AWS IoT Core when it receives data from the Raspberry Pi. The Amazon S3 Bucket is used as a means to store the data received from the Raspberry Pi Hardware. Because the HEV system simulation/running depends highly on the initial conditions or states, we need to reinitialize the model with the exact same states after every 5 s when we get the data from the hardware. This can be done, but adds a lot of complexity, so instead what we do is we just keep a history of all the data received from the hardware, and every time the Digital Twin model is ran with the entire input right from the beginning. So this approach takes away the state initialization problem. The very first initial condition is same for the hardware and the Digital Twin model, so for the same inputs the hardware and the Digital Twin model should behave the same if there are no fault conditions present. So we use Amazon S3 bucket to store the previous data, as the Lambda function doesn't have memory or there is no way to store data. Please check the AWS documentation for more details about these services. Fig. 8.85 shows the high level diagram, where the AWS IoT Core will be triggering a Lambda function written in Python, which will run the Digital Twin model, run the Off-BD algorithm, and make a diagnostic decision and trigger the SNS service to notify the subscribers about the status of Off-BD algorithm decision. The Lambda function interacts with Amazon S3 Bucket for storing and retrieving data for every run. We will first get started with the SNS configuration; please follow the below steps:

1. On the AWS Management Console, search for SNS and select the SNS as shown in Fig. 8.86.
2. Click on the ***Create Topic*** button. See Fig. 8.87.

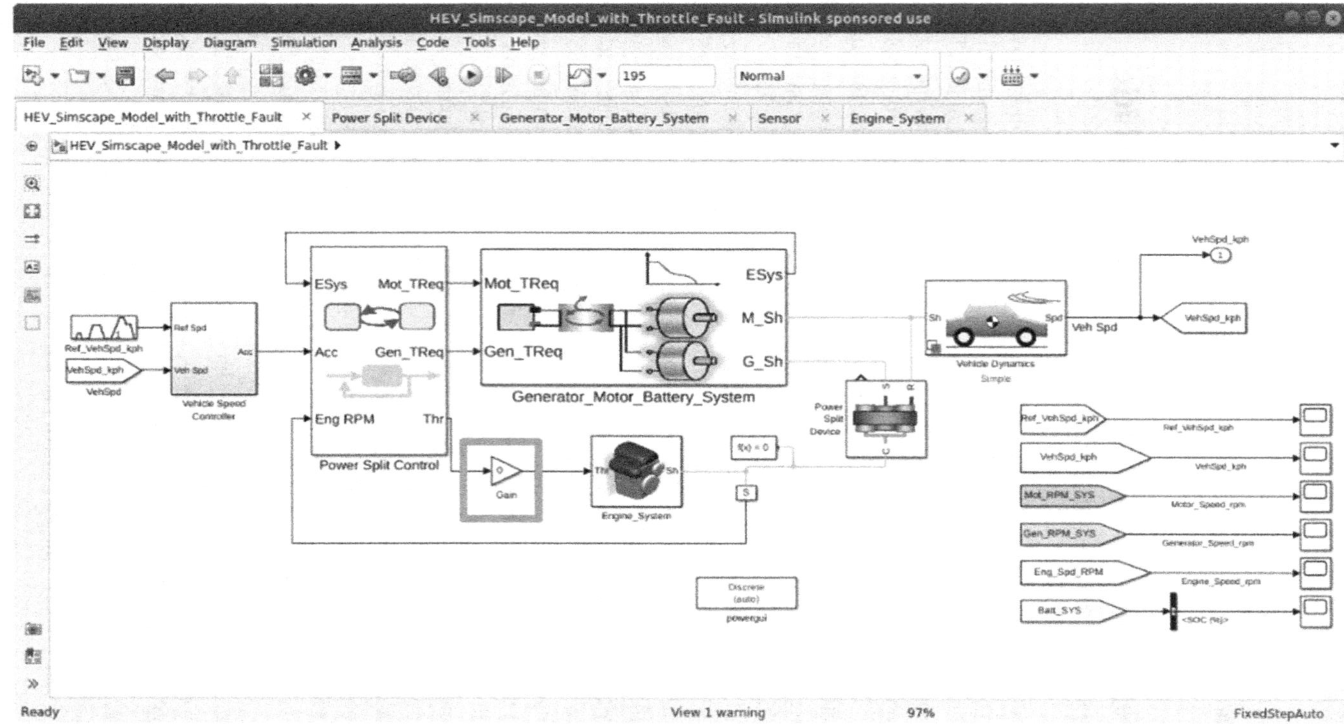

Figure 8.81 Host computer simulation model with throttle failure introduced.

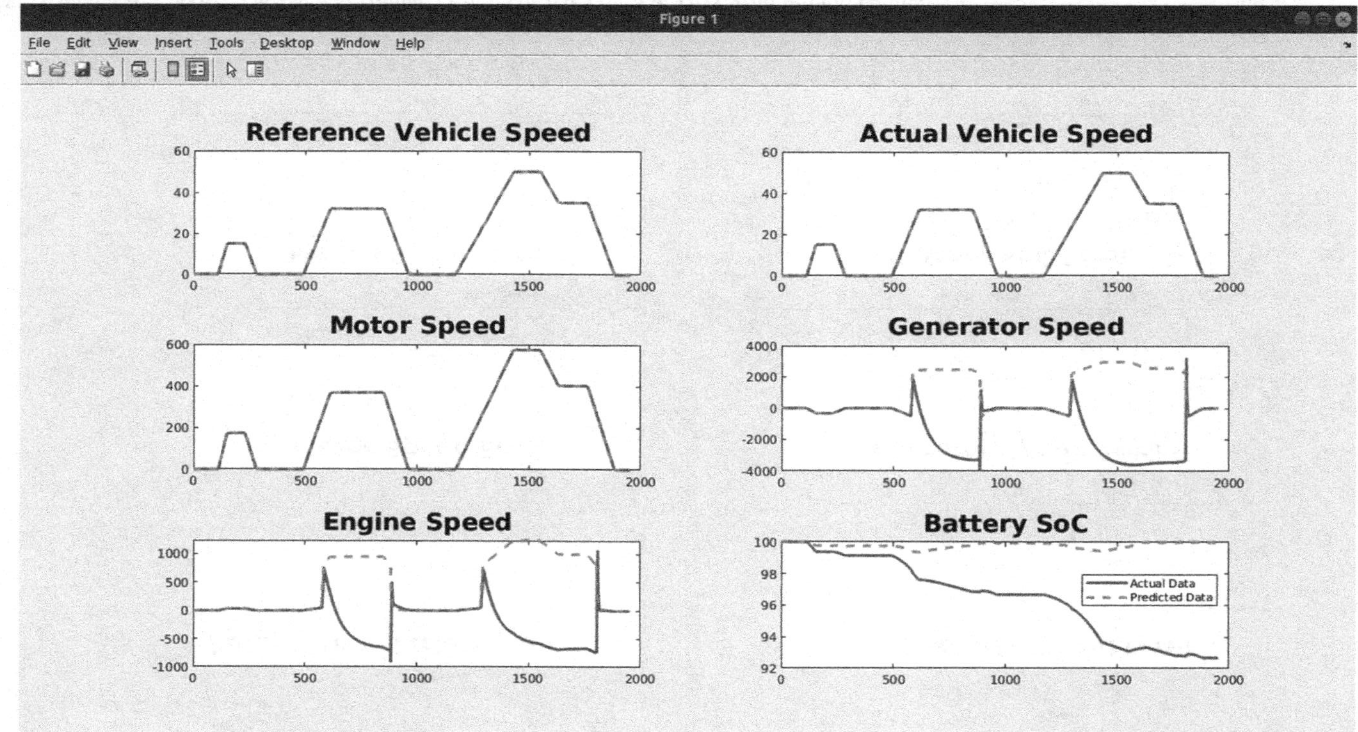

Figure 8.82 Comparing actual versus Digital Twin predicted values for HEV with throttle failure.

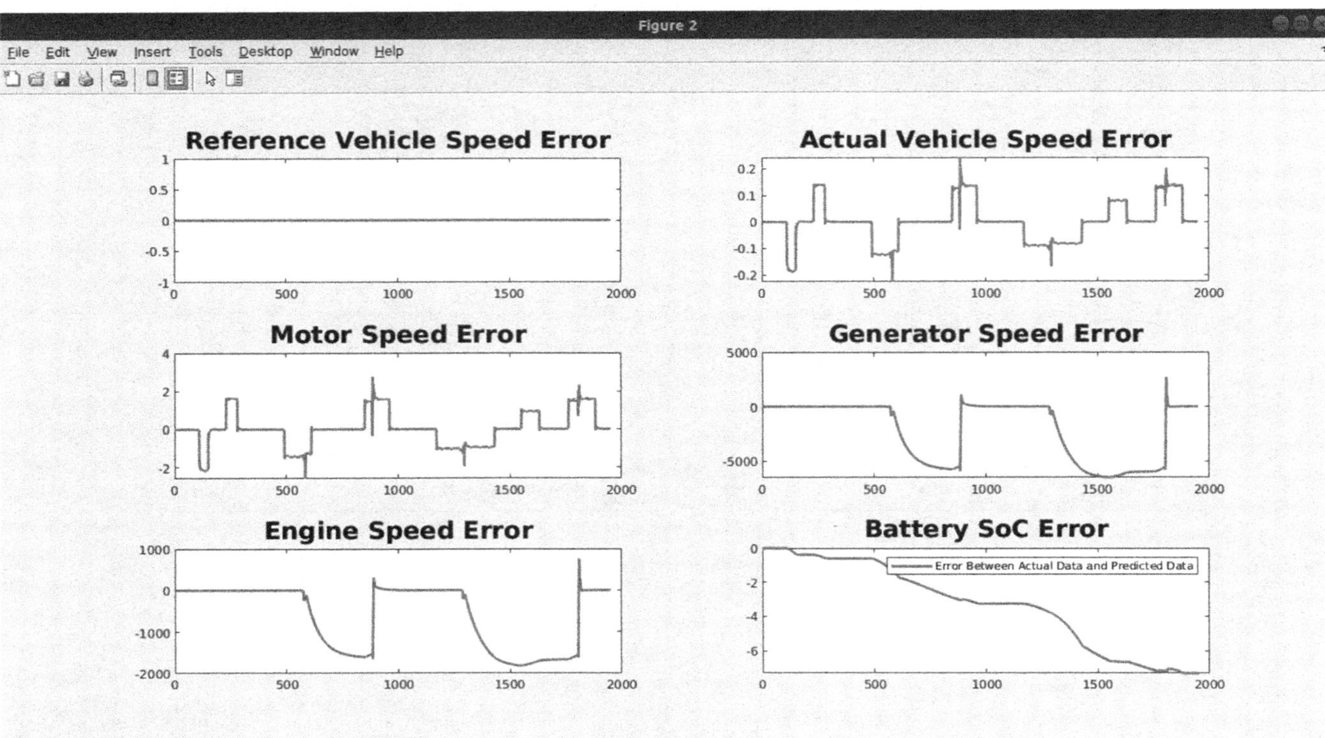

Figure 8.83 Error between actual and Digital Twin predicted values of HEV with throttle failure.

```
>> Root_Mean_Squared_Error

Root_Mean_Squared_Error =

   1.0e+03 *

         0    0.0001    0.0009    3.4804    0.9667    0.0041
```

Figure 8.84 RMSE values of the corresponding six signals with throttle failure.

3. Give a Name to the topic on the Create topic form. We used **digital_twin_topic** in this example. Also give a Display name, which will be displayed in the SMS and E-mail. Click on the **Create topic** button to finish creating the new SNS topic as shown in Fig. 8.88.
4. The newly created topic is shown in Fig. 8.89. Now we need to create a Text and E-mail subscriptions for this topic. Click on the Create subscription button as shown in Fig. 8.89. Note down the topic ARN here, we will be using it later in the Lambda function to interact with the SNS.
5. On the new ***Create subscription*** form, enter first the SMS protocol for Text subscription and enter the Cell number starting with + and Country Code and click the ***Create subscription*** button. For example **+1<10 Digit Cell Number>** for the United States. After the SMS subscription is created, use the same Create subscription button for creating the E-mail subscription from the Drop down menu and enter the E-mail ID to which we want to get the notification. Check Figs. 8.90–8.92 for details for creating Text and E-Mail subscriptions.
6. Though we created the subscriptions, it can be seen that under the **Subscriptions** menu the status shows Pending confirmation as shown in Fig. 8.93. When we created the subscriptions, AWS will automatically send an email to the email address given in the subscription with a link to confirm the subscription. Check the email and confirm the subscription by clicking on the link.
7. Once you click on the link in the E-mail, it will display the Subscription Confirmation message window. And under the Subscriptions menu in AWS SNS, we can see both the SMS and E-mail subscriptions show up with the Confirmed status. See Figs. 8.94 and 8.95. We can test these newly created topics are working by selecting the checkbox against the Topic ID and clicking the ***Publish message*** button highlighted in Fig. 8.95. Enter the Message body in the new form and publish it; based on the Topic ID we selected, it will send Text or E-mail accordingly.
8. Now let us configure the AWS Lambda Functions to run the Digital Twin Model and Off-BD algorithm and also notify the user using the SNS service configured in above steps. Search for lambda in the AWS Management Console and select the Lambda services. See Fig. 8.96.
9. On the AWS Lambda Console, click on the ***Create function*** button as shown in Fig. 8.97.
10. We will be developing and deploying the Lambda function in Python programming language. There are other options also available like Node js, etc., and depending upon the User's expertise, different language options can be selected in the ***Create function*** form. Give a name to the new Lambda function, we chose the name as hev_digital_twin and select the ***RunTime*** as **Python 3.7** and click on the ***Create function*** button as shown in Figs. 8.98 and 8.99.
11. AWS will create a new default template function as shown in Fig. 9.174.

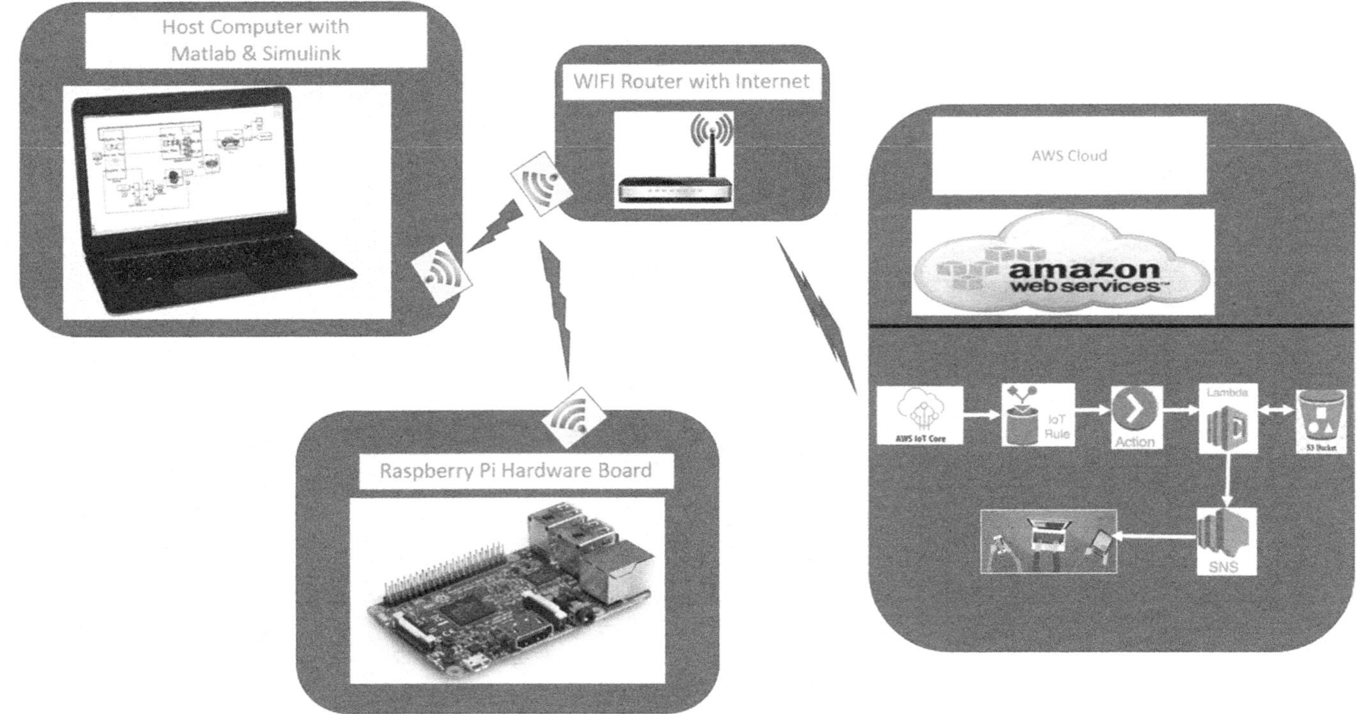

Figure 8.85 Digital Twin deployment high level diagram.

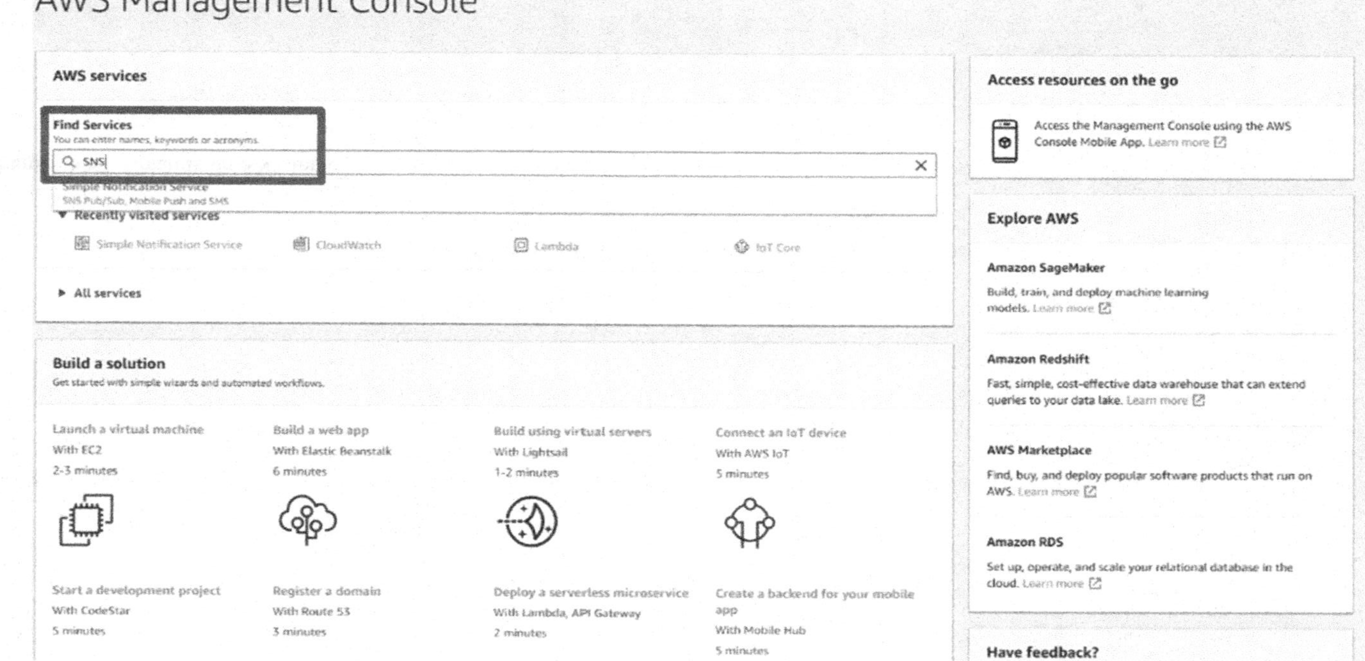

Figure 8.86 Launching Simple Notification Service (SNS).

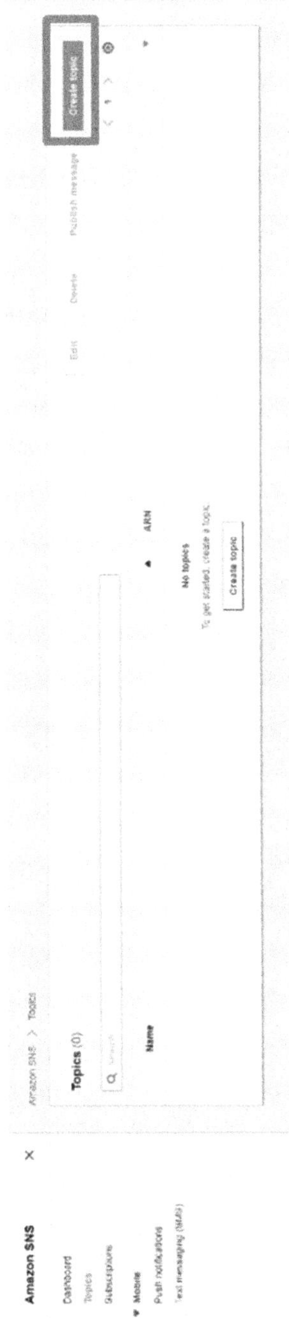

Figure 8.87 Creating an SNS topic.

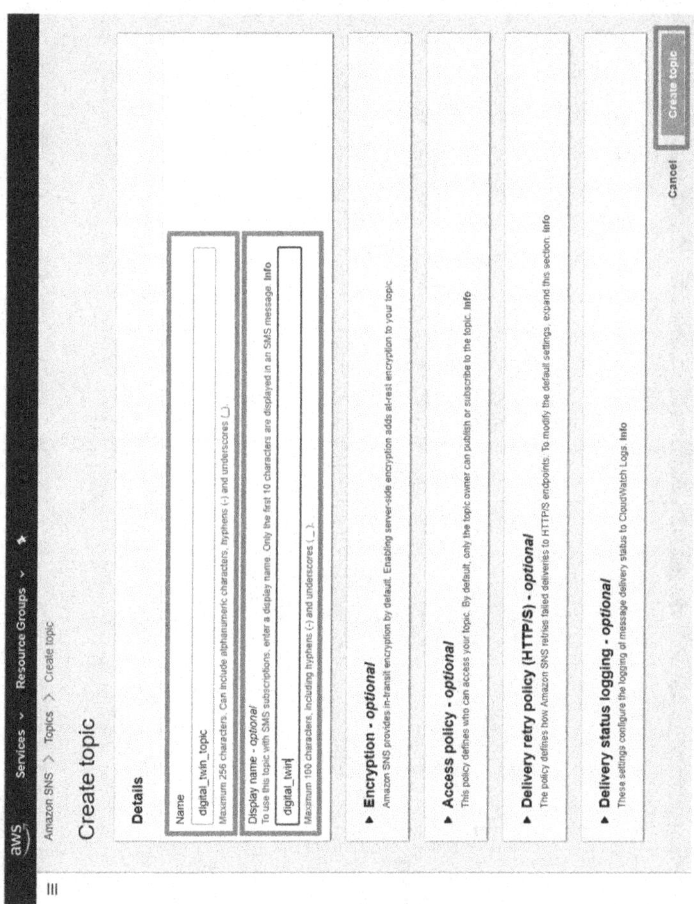

Figure 8.88 New SNS topic details.

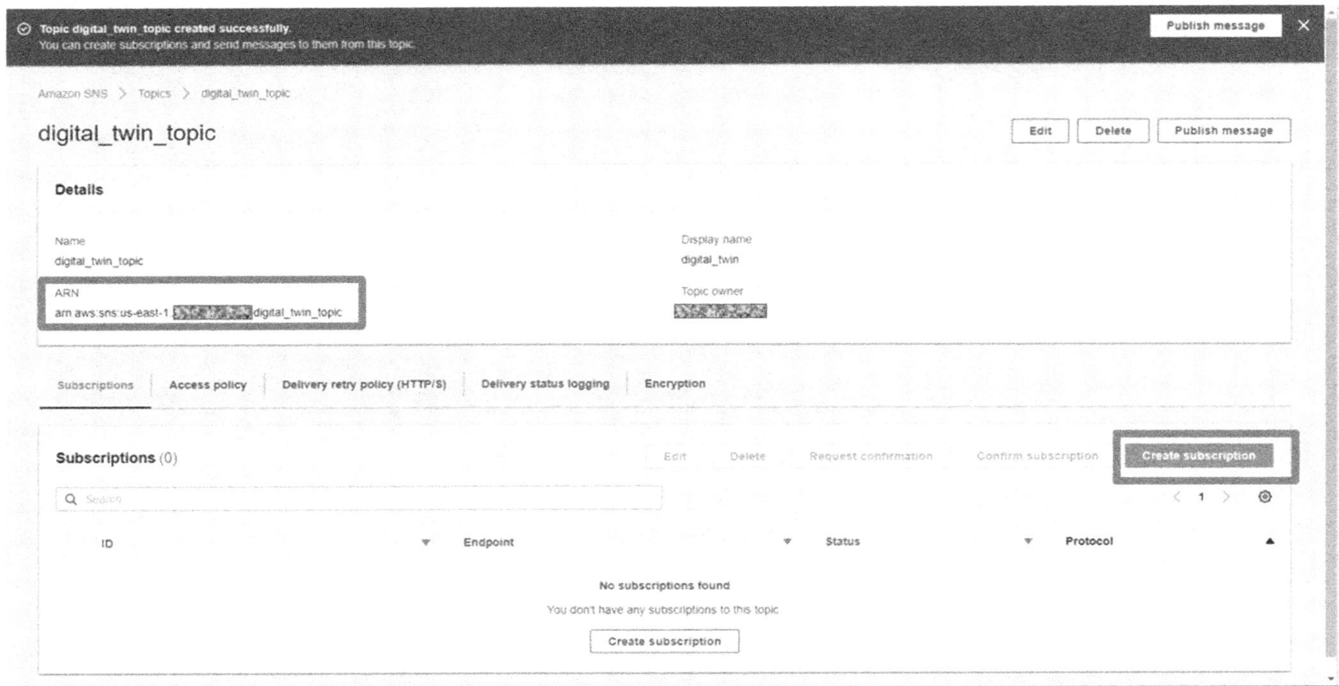

Figure 8.89 Newly created SNS topic.

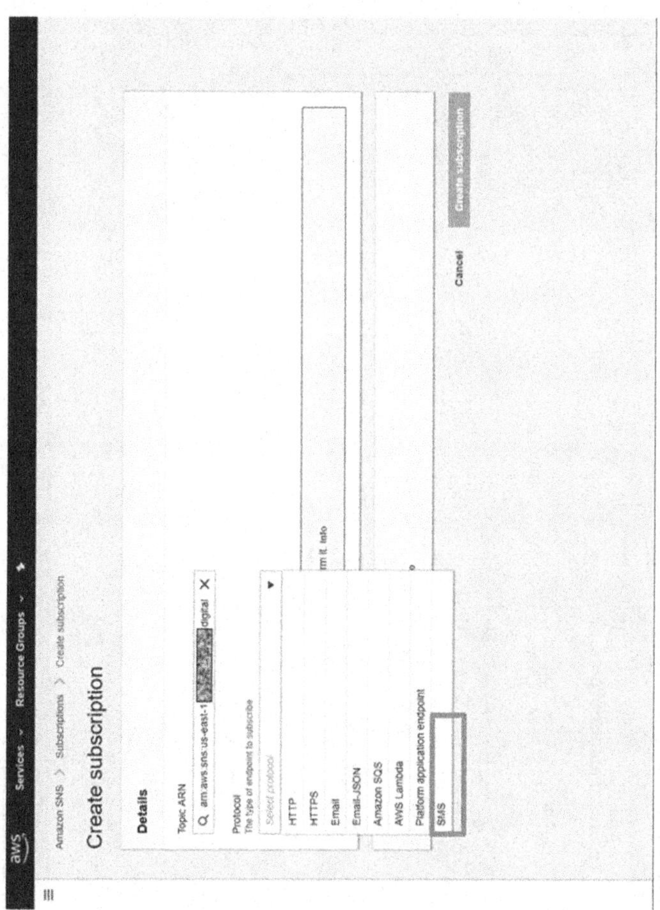

Figure 8.90 Creating subscription for SNS topic.

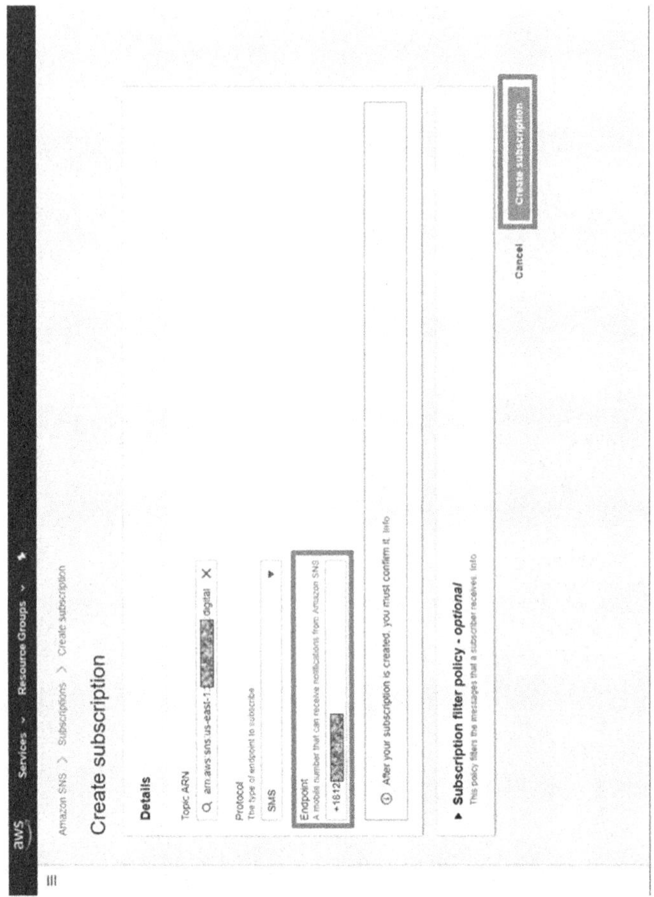

Figure 8.91 Creating text/SMS subscription.

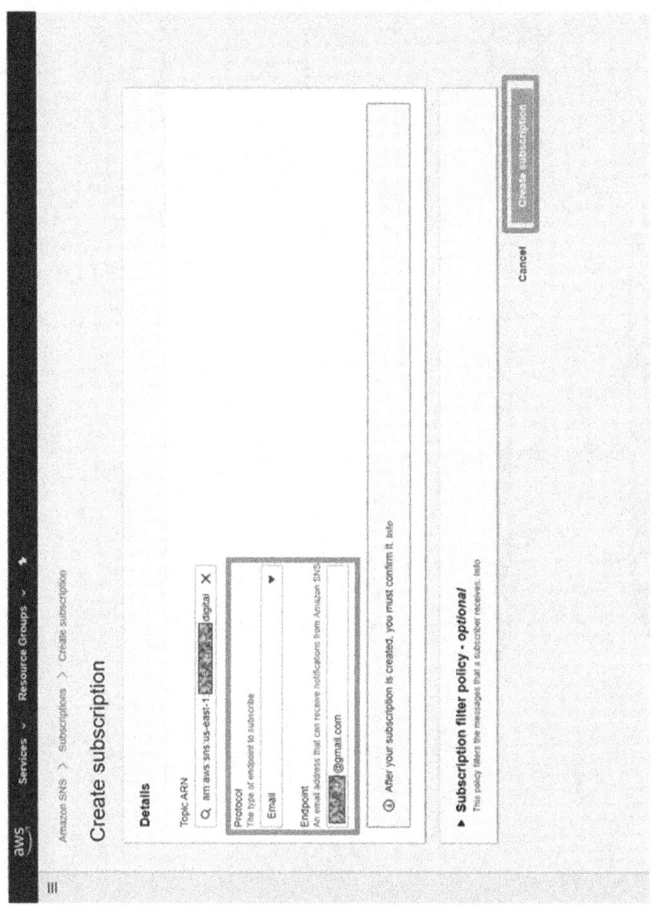

Figure 8.92 Creating E-mail subscription.

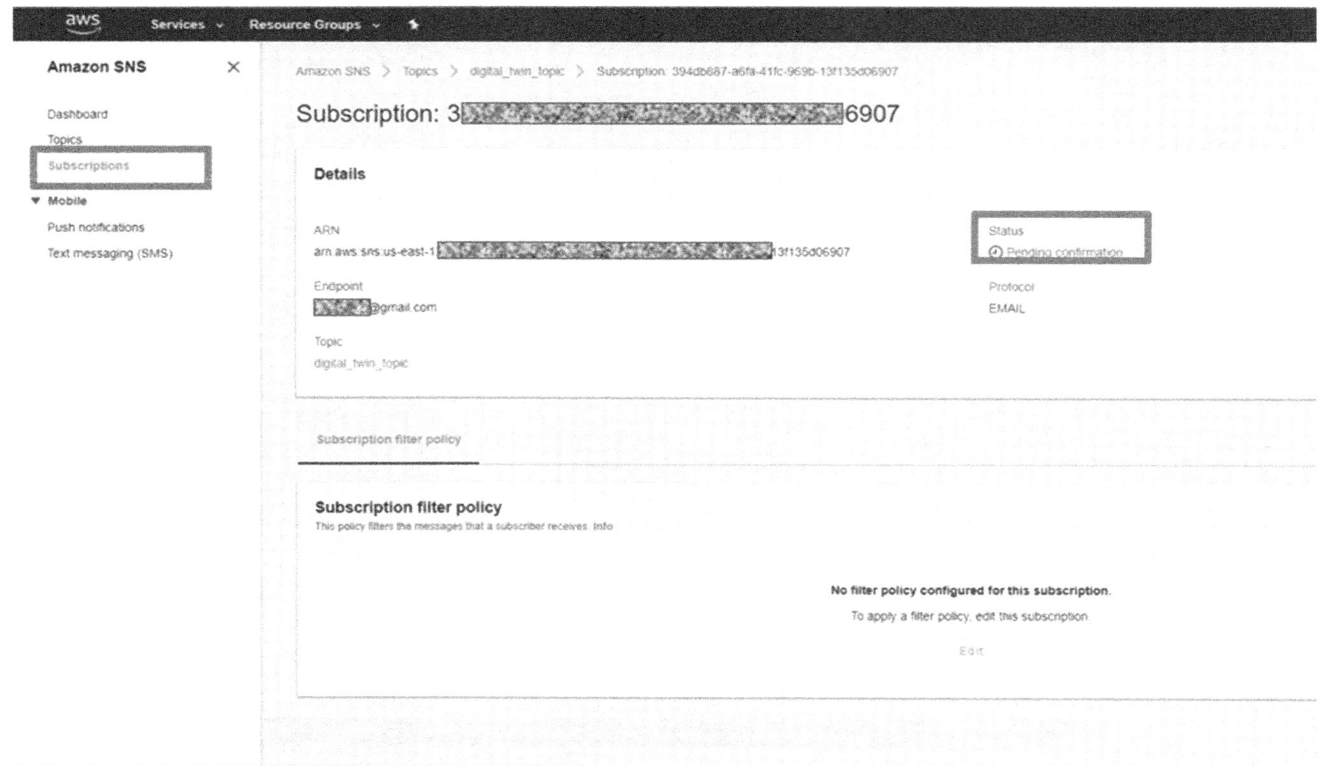

Figure 8.93 Pending subscription.

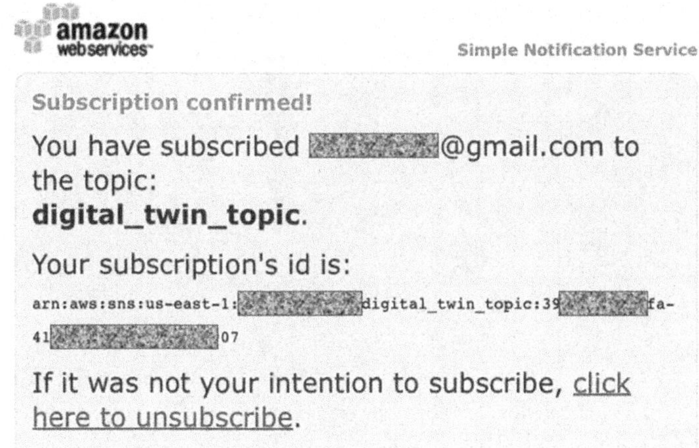

Figure 8.94 Subscription confirmation.

12. Toward the bottom of the new Lambda function window, there is an option to give the role that defines the permissions for the Lambda function. Drop down the menu and select "**Create a new role from AWS policy templates**" as shown in Fig. 8.100. What we want to do is allow the Lambda function to publish messages to the SNS topics that we created and configured in the above steps. The permission has to be explicitly given to the lambda function; otherwise, it will not be able to interact with the SNS service.
13. The new **Create role** form will show up, which involves a four-step process. On the first step, select the ***AWS service*** option and click on the ***Next: permissions*** button as shown in Fig. 8.101.
14. In the second step as shown in Figs. 8.102 and 8.103, search for SNS and S3 in the ***Filter policies*** search bar and select the ***AmazonSNSFullAccess*** and ***AmazonS3FullAccess*** policies and click on the ***Next: Tags*** button.
15. Next Adding the Tags is an optional step. We can skip it.
16. The final step is the Review, see Fig. 8.104. On the Review window, give a Role name, we chose **hev*_digital_twin_role*** in this case. We can give any name to the Role. It can be seen that the policies ***AmazonSNSFullAccess*** and ***AmazonS3FullAccess*** we selected in the previous step is attached to this role. The idea is we will attach this role to the Lambda function, which will give the Lambda function access to the SNS and S3 Bucket services and publish Text/E-mails to the topics and store and retrieve data as needed. Finish the role creation by clicking on the ***Create role*** button.
17. The new Role will be created and shown under the AWS service Identity and Access Management or called as IAM. The new Role that we created is shown under the IAM Console as shown in Figs. 8.105 and 8.106.
18. After the new Execution Role is created go back to the AWS Lambda function console, under the lambda function we created earlier, select the "***Use an existing role***" option and browse and select the new role we created named ***hev_digital_twin_role.*** If the User chose to give a different name for the role in the above step, chose the role name accordingly. See Fig. 8.107 for details.
19. Save the Lambda function by clicking on the ***Save*** button at the top as shown in Fig. 8.108.
20. Next step is to add a trigger for the Lambda function, to decide when will the Lambda function run. Click on the Add trigger button shown in Fig. 8.109. We will be triggering the Lambda function whenever the IoT Thing receives data from the Raspberry Pi Hardware.

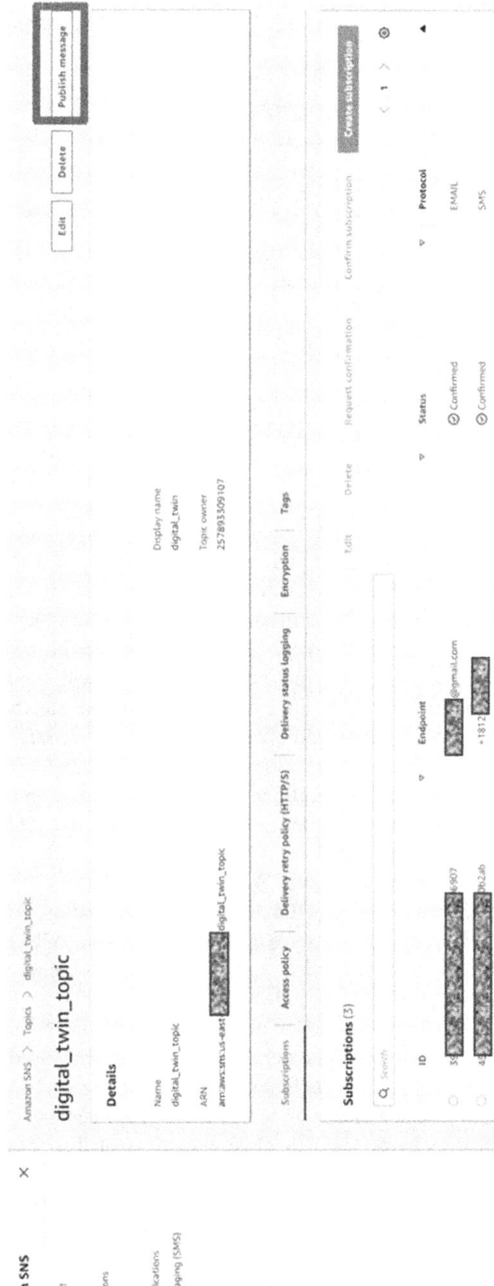

Figure 8.95 Confirmed subscriptions.

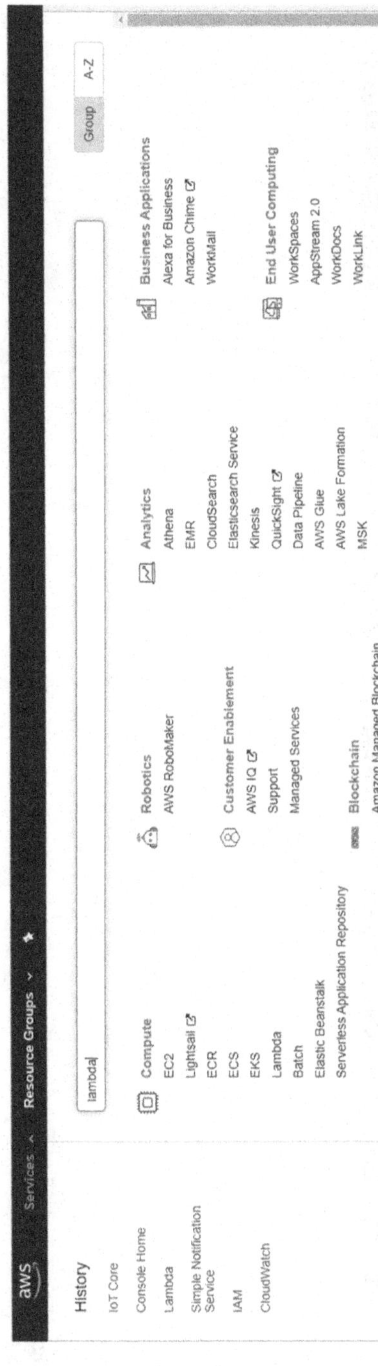

Figure 8.96 Launching AWS Lambda functions.

Figure 8.97 Creating Lambda function.

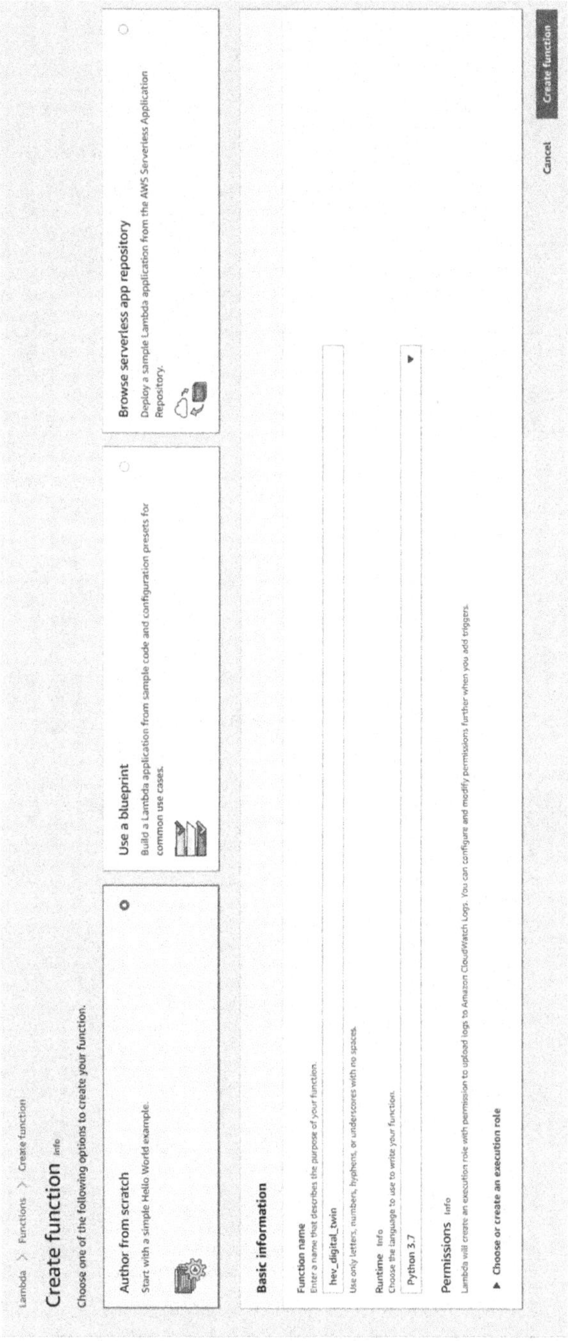

Figure 8.98 New Lambda function details.

Figure 8.99 Default Lambda function body.

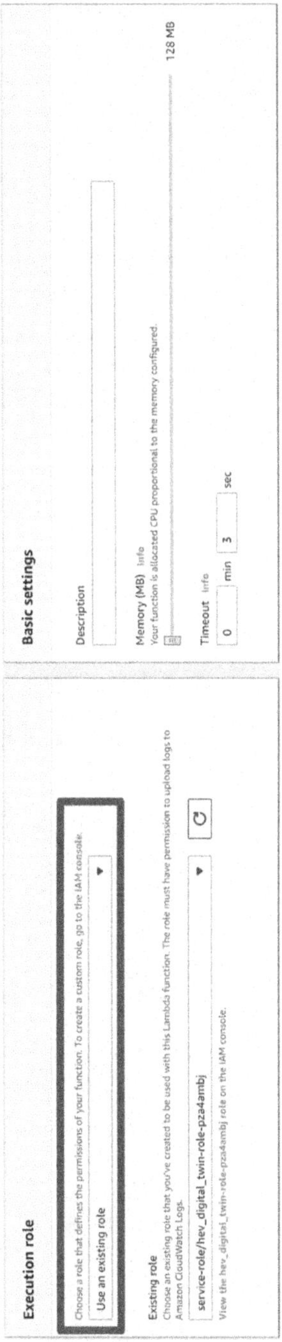

Figure 8.100 Creating an Execution Role.

Figure 8.101 Create a new role.

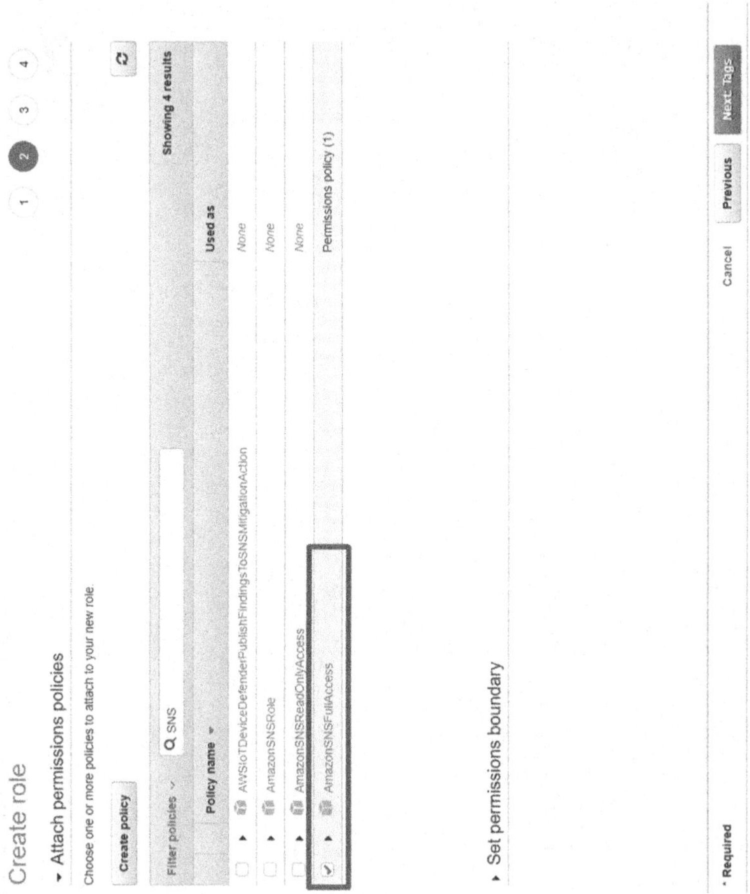

Figure 8.102 Attach SNS permission policy to the new role.

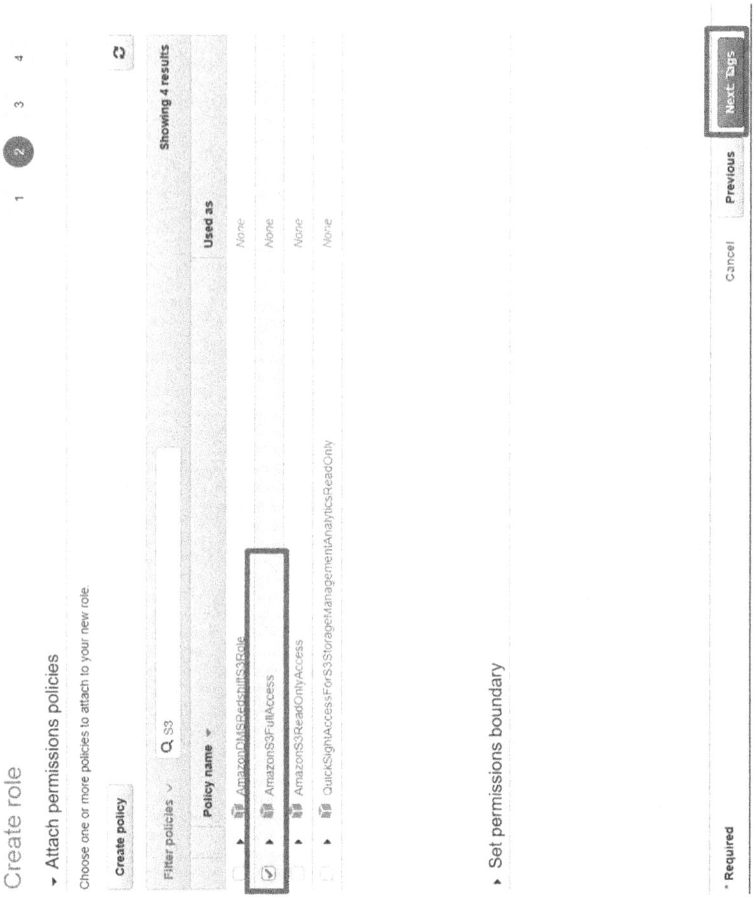

Figure 8.103 Attach S3 bucket permission policy to the new role.

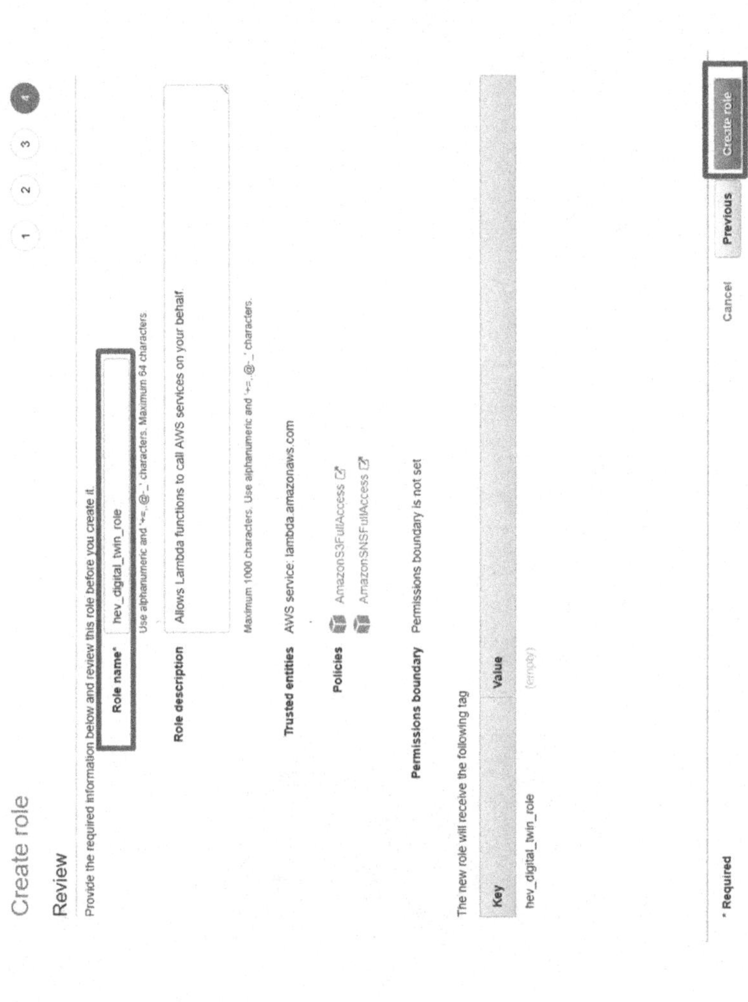

Figure 8.104 Naming the new role.

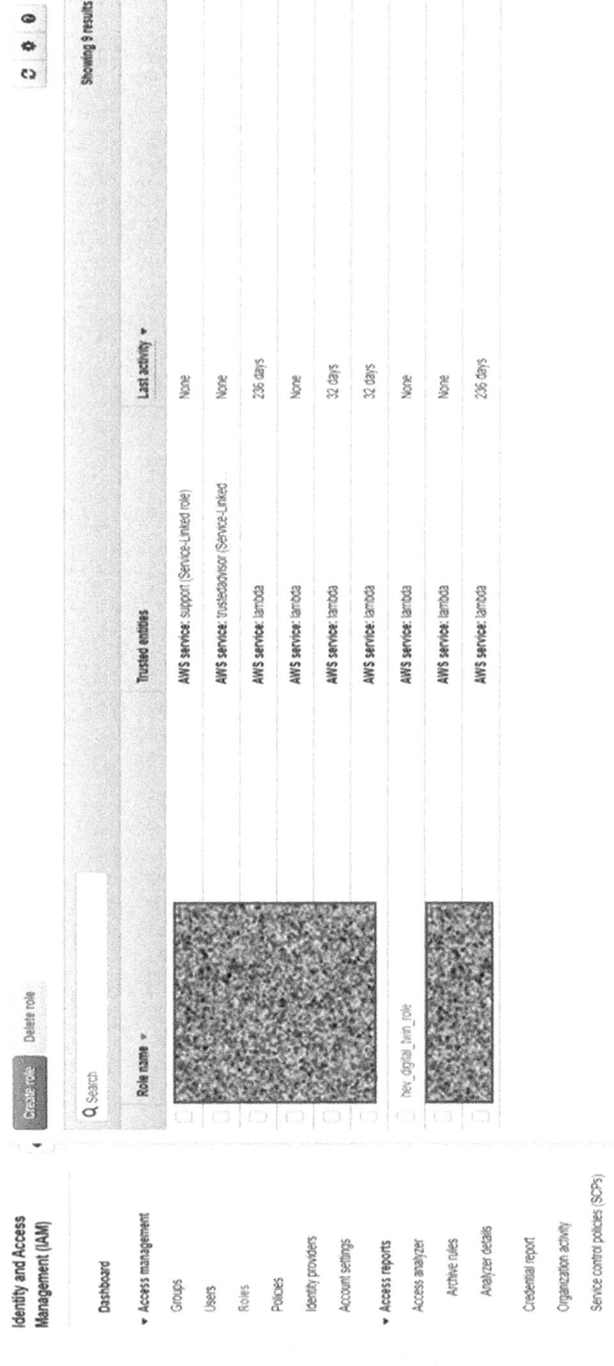

Figure 8.105 New role created.

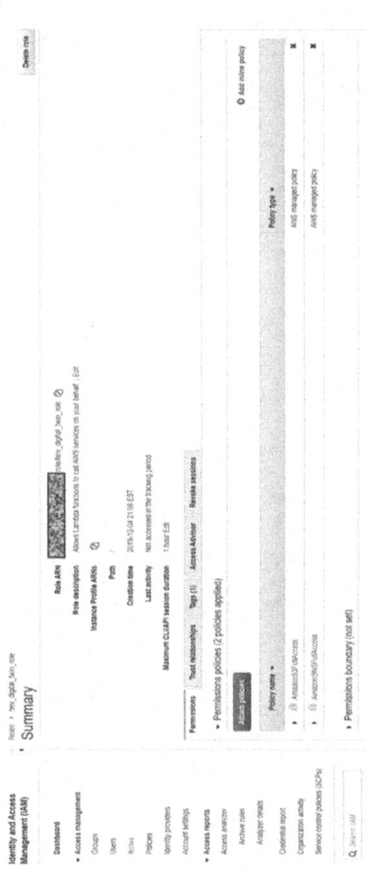

Figure 8.106 New role created showing the SNS and S3 policies attached.

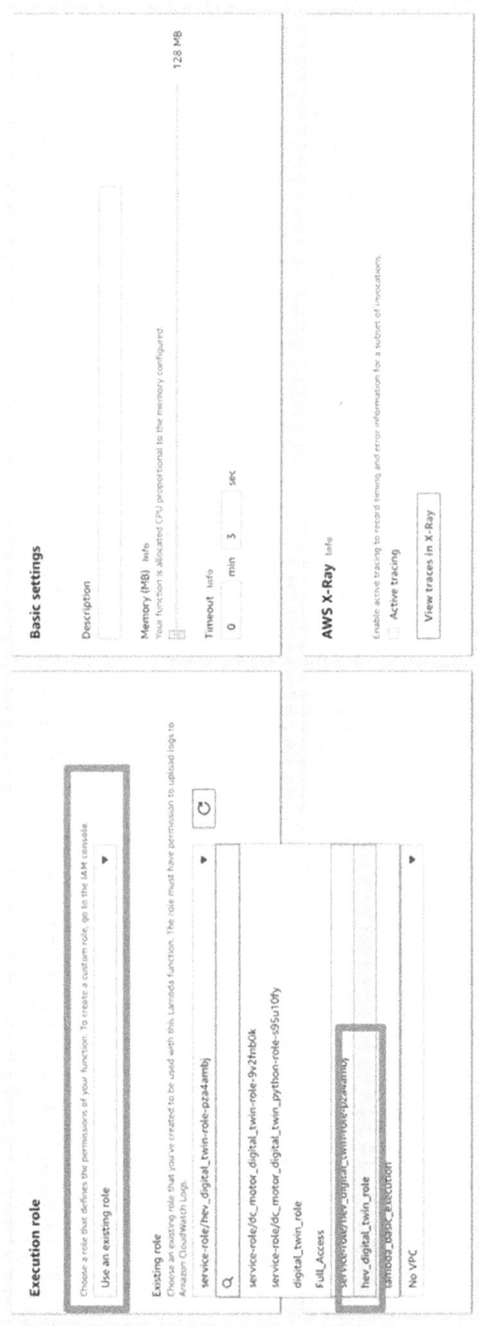

Figure 8.107 Selecting the new Execution Role in the Lambda function.

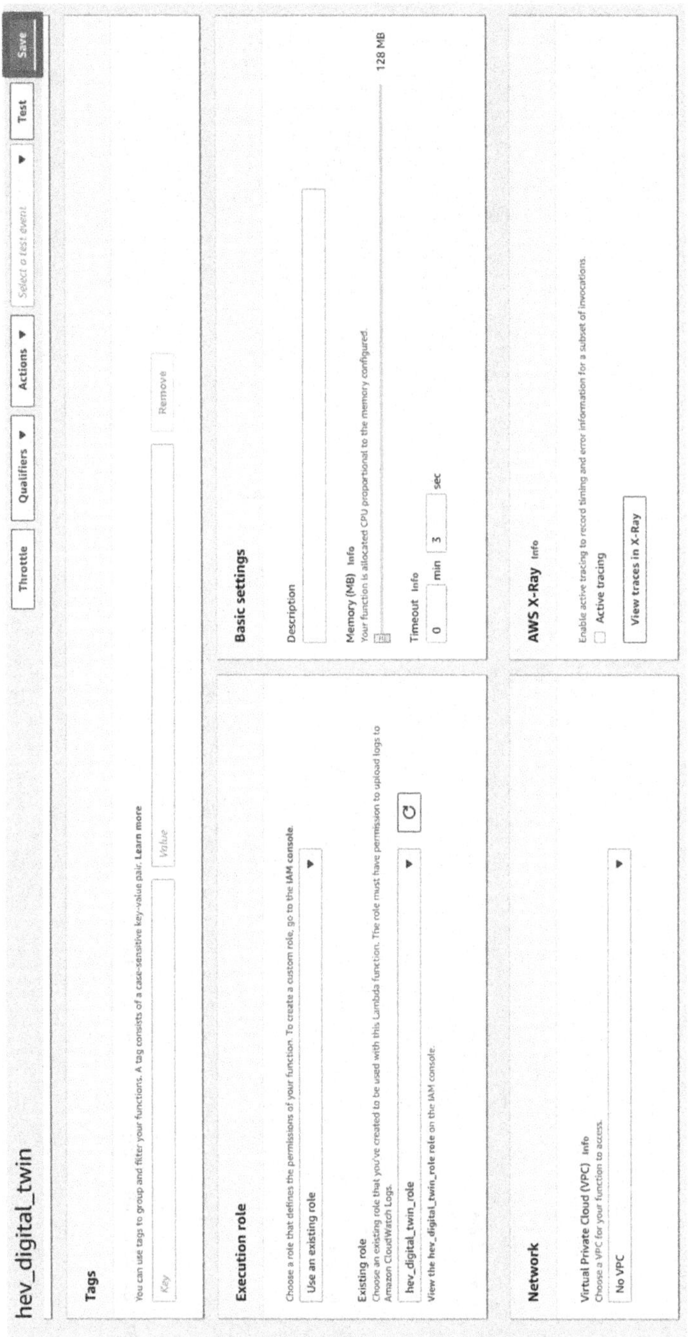

Figure 8.108 Save Lambda function with updated Execution Role.

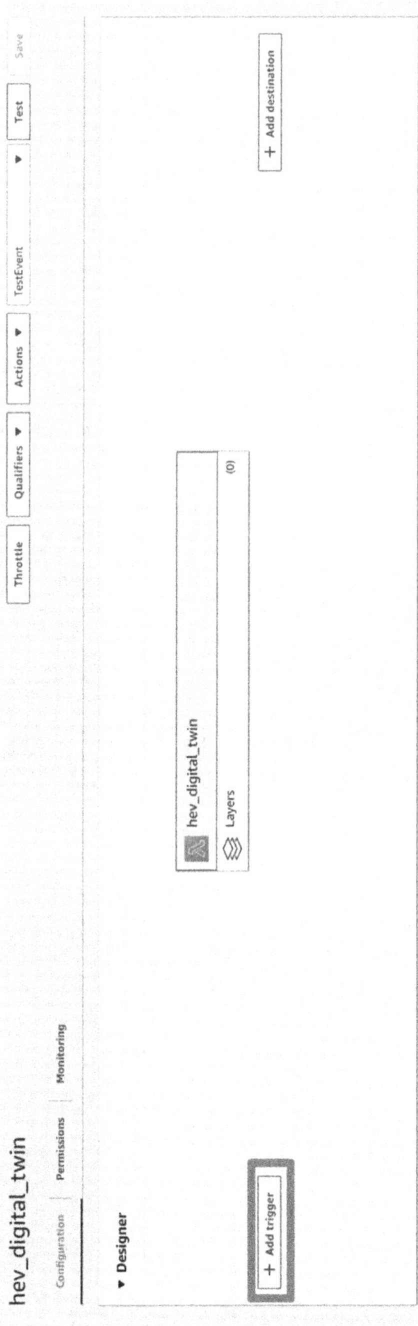

Figure 8.109 Adding trigger to Lambda function.

21. The Add trigger window will be opened as shown in Figs. 8.110 and 8.111. Select the AWS IoT from the dropdown as the trigger source. We need to create a Custom IoT Rule and use it for the trigger. An IoT routes the IoT Thing updates to a specific AWS service. In this case, we are going to create a rule to trigger the AWS Lambda service. Check the box "***Custome IoT rule***" and select the option "***Create a new rule***." Give a name to the new rule, in this case we used ***hev_digital_twin_rule***. The ***Rule description*** is optional. In the ***Rule query statement***, enter the string **SELECT * FROM '$aws/things/hev_digital_t-win_thing/shadow/update/accepted'**. This query statement will pull all values from the IoT Thing update and pass it to the AWS Lambda. Also check the "***Enable trigger***" option and click on the ***Add*** button.

22. The Lambda function will now show the AWS IoT as the Trigger source as shown in Fig. 8.110.

23. Next step is developing the Lambda function. At this point, we only have a skeleton template Lambda function, so we will have to expand on that. For convenience of testing the Lambda function along with the Digital Twin compiled executable and to package the Lambda function and Digital Twin executable to deploy to AWS, we have developed the Lambda function in Linux. From the Ubuntu Linux (which we used for compiling the Digital Twin model earlier), open an editor and name the file as **lambda_function.py**. At the top of the Python script, add all the packages we will be using in this program. So here the package **json** is used to deal with the incoming json string from AWS IoT, **os** is used to call the compiled Digital Twin executable, **boto3** is used to interact with the AWS SNS, AWS S3 Bucket, **csv** for creating csv file with incoming data, and **math** is for some math calculations for **Off-BD** algorithm. See Fig. 8.112.

24. Next we will expand the Lambda handler function. We can see from Fig. 8.99 that when we create the AWS Lambda function it creates a default handler function named **lambda_handler.** The input argument **event** contains the JSON string that we sent from the Raspberry Pi Hardware module with the HEV state data while the model runs on the hardware. As shown in Fig. 8.113, we will parse the JSON string and store it into a comma separated input.csv file.

25. A logic is added to identify whether the Lambda function is getting triggered for first time during a particular run of an HEV cycle in the hardware. If it is triggered for the first time, the data received by the Lambda function will be treated as the very first 5 s data from the hardware, and if it is not the first time, Lambda function is triggered in the HEV cycle run, then we have to combine the last 5 s data with all the previously received data for running the HEV Digital Twin. So for the very first time we run Digital Twin with first 5 s data and store all the data information from the hardware into Amazon S3 storage bucket, when the next 5 s data are received from the hardware we first append the previously stored inputs with the newly received inputs and then run the Digital Twin model. So the Digital Twin model runs in a way that it is always starting from the same initial condition and states when the Hardware starts to run. This approach takes out the hassle of reinitializing the states of the Digital Twin HEV to the state it was before 5 s when the new data arrives. It is possible, but it can be more challenging. As shown in Fig. 8.114, the input_data[6] in the received data will contain the free running counter information from the HEV

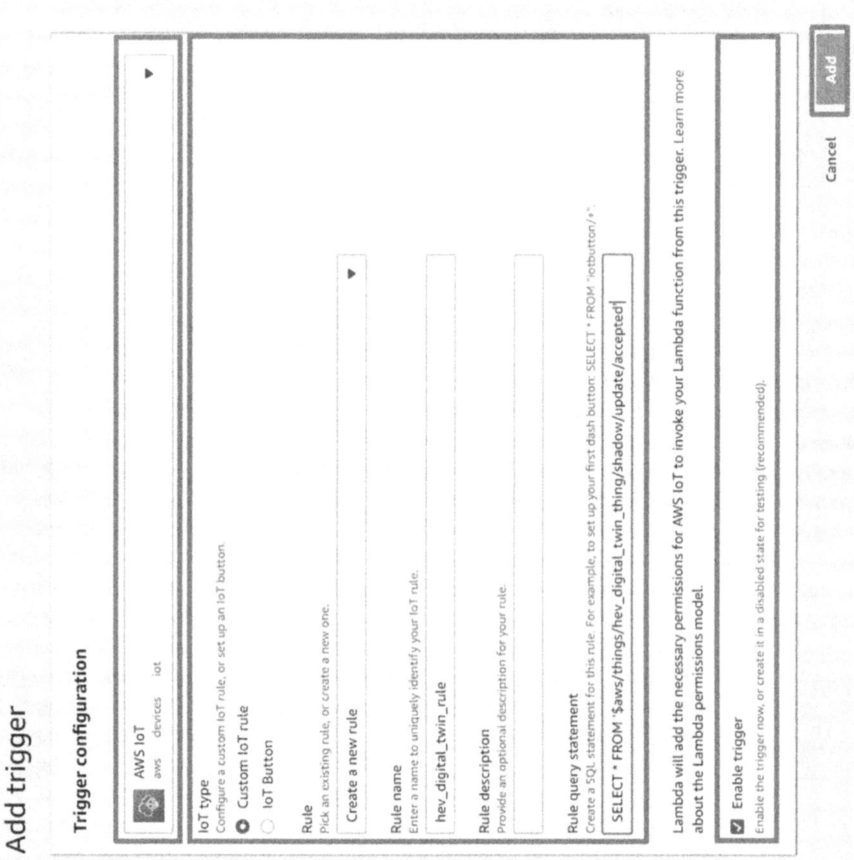

Add trigger

Trigger configuration

AWS IoT
aws devices iot

IoT type
Configure a custom IoT rule, or set up an IoT button.
Custom IoT rule
IoT Button

Rule
Pick an existing rule, or create a new one.
Create a new rule

Rule name
Enter a name to uniquely identify your IoT rule.
hev_digital_twin_rule

Rule description
Provide an optional description for your rule.

Rule query statement
Create a SQL statement for this rule. For example, to set up your first dash button: SELECT * FROM "iotbutton/+".
SELECT * FROM '$aws/things/hev_digital_twin_thing/shadow/update/accepted'

Lambda will add the necessary permissions for AWS IoT to invoke your Lambda function from this trigger. Learn more about the Lambda permissions model.

Enable trigger
Enable the trigger now, or create it in a disabled state for testing (recommended).

Cancel Add

Figure 8.110 Adding a trigger for the Lambda function from the AWS IoT.

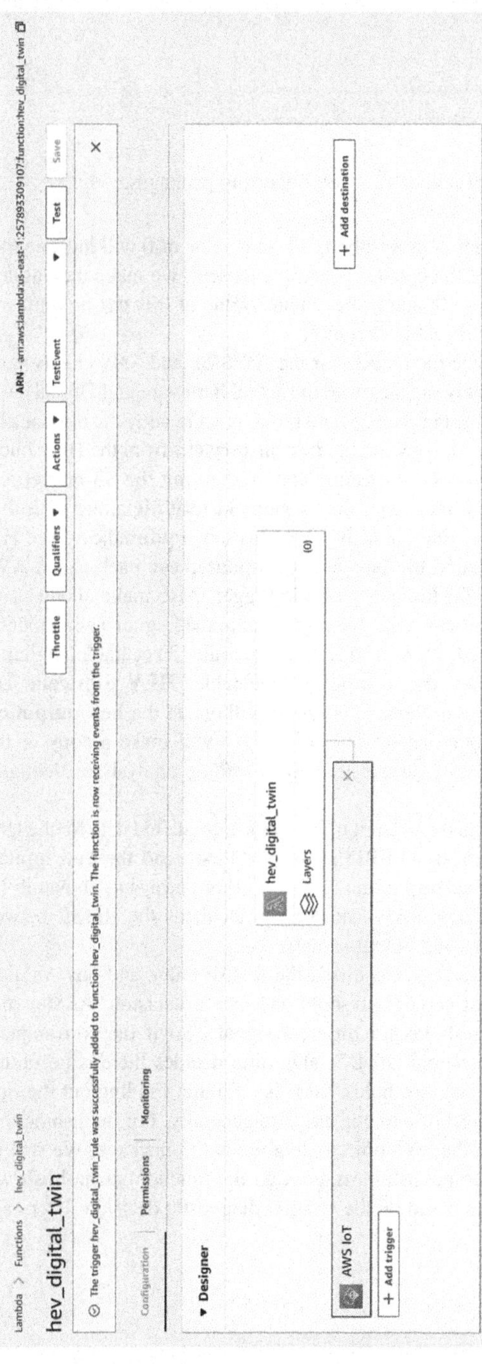

Figure 8.111 Lambda function showing the AWS IoT trigger added.

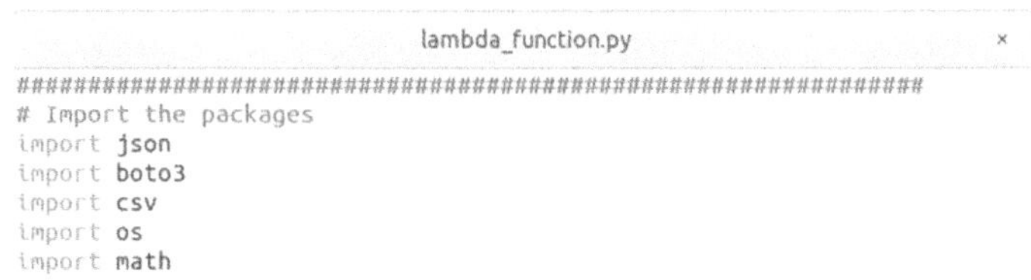

```
lambda_function.py
########################################################################
# Import the packages
import json
import boto3
import csv
import os
import math
```

Figure 8.112 Updating Lambda function importing packages.

Simulink model running in Raspberry Pi, so a value of 0 will indicate the data are from the first 5 s of run. So if the counter value 0 is present, we make the flag to get previous data from the S3 bucket to 0, and if the counter value of 0 is not present, we make the flag to get previous data from S3 bucket to 1.

26. Next we will initialize the objects for the AWS S3 and SNS clients and also initialize the arrays to store the HEV signals from the actual hardware and Digital Twin. See Fig. 8.115.

27. Next we use the flag set in Step 25, and if flag is set to copy the previous history information from the S3 bucket, we copy the file **hev_history.csv** from the HEV bucket named **digital-twinbucket**. The process of creating and configuring the S3 bucket is explained in later steps. Combine the inputs from the previous history file along with the new data for the past 5 s received into one file named **hev_history_updated.csv**. See Fig. 8.116.

28. Copy the newly created file **hev_history_updated.csv** back to the AWS S3 bucket as a history information for the next Lambda trigger. Also make another copy of this file and name it as **hen_input.csv** to be given as input to the Digital Twin model. Once the **hev_input.csv** file is created, we will trigger the compiled executable Digital Twin model. Note that we are calling the compiled executable **HEV_Simscape_Digital_Twin**. See Fig. 8.117. When the executable is run, it will create the **hev_output.csv** file in the **/tmp** folder as we have seen and tested before. We will make a copy of that file also to the AWS S3 bucket if in case we need to do offline analysis or debugging of the Digital Twin model.

29. Now we have the actual and Digital Twin model predicted data for the HEV System, and we are ready to perform the Off-BD algorithm. First, read the **hev_input.csv** and **hev_output.csv** files from the **/tmp** folder and store it into arrays as shown in Fig. 8.118.

30. Next as shown in Figs. 8.119 and 8.120, calculates the RMSE between the actual and predicted HEV states and output signals.

31. The Off-BD algorithm has calculated the RMSE value and now we need to compare the RMSE value against certain threshold and take a decision. As shown in Fig. 8.121, we picked different thresholds for different signals. So if the corresponding RMSE values are less than the threshold Off-BD, algorithm decides there is no failure detected, and if the RMSE exceeds the thresholds, there is a failure. Finding out the right threshold value may involve a little bit of a tuning exercise generally. For publishing the Off-BD decision to the SNS, we use the SNS object using the boto3 package. We will use the topic ARN noted from Step 4 to publish message. Call the function **sns.publish** with the topic ARN and custom message based on the Off-BD diagnostic decision. User can edit the Message

```
##################################################################
# Lambda Handler function which runs when IoT Trigger Happens
def lambda_handler(event, context):
    # TODO implement

##################################################################
# Create a file "input.csv" under /tmp folder and write the input and output values
# separated by commas in each row with the newly recieved data from the AWS IoT.
# There will be 50 data samples for coming in for the past 5 Seconds
    with open('/tmp/input.csv', 'w') as writeFile:
        writer = csv.writer(writeFile)
        for loop_index in range(1,51):
            temp_str = "Data_" + str(loop_index)
            input_list = event['state']['desired'][temp_str]
            writer.writerows([[(input_list[0]),(input_list[1]),(input_list[2]),(input_list[3]),(input_list[4]),(input_list[5]),
(input_list[6])]])
```

Figure 8.113 Updating Lambda function reading trigger input data.

```
# If the index number in the data entry if it is 0, that means it is the start of the
# HEV simulation in the Raspberry PI, if so we dont have to get the previous stored data
# file from AWS S3 Bucket. If the count is not zero that means this is not the first time
# the Lambda function is triggered for this HEV simulation in the Raspberry PI. So we have
# to set a flag to retrieve the previously stored data from S3 Bucket.
            if loop_index ==1:
                temp_first_val = float(input_list[6])
                if int(temp_first_val) ==0:
                    get_file_from_s3_bucket = 0;
                else :
                    get_file_from_s3_bucket =1;
            #print(input_list)
    writeFile.close()
```

Figure 8.114 Updating Lambda function decision logic to check start of run or not.

```
# Setup the AWS S3 Bucket and SNS Clients and also SNS Topic ARN for sending messages
    s3 = boto3.resource('s3')
    sns = boto3.client(service_name="sns")
    topicArn = '[illegible]:digital_twin_topic'
# Initialize the arrays for storing actual signals from hardware and digital twin predicted signals
    data_count =0;
    Actual_Veh_Speed = [];Actual_Motor_Speed = [];Actual_Generator_Speed = [];Actual_Engine_Speed = [];Actual_Battery_SOC = [];
    Digital_Twin_Veh_Speed = [];Digital_Twin_Motor_Speed = [];Digital_Twin_Generator_Speed = [];Digital_Twin_Engine_Speed =
[];Digital_Twin_Battery_SOC = [];
```

Figure 8.115 Updating Lambda function initializing SNS and S3 clients and arrays to store actual and Digital Twin data.

string however they want, this is the message we will see in the TEXT and E-mail messages when the Lambda function runs. The Lambda function development is now finished.

32. Save the above developed Lambda function into a new folder. In this case, we named the folder to be **Digital_Twin_Lambda_Function** under the folder **Digtal_Twin_Simscape_-book/Chapter_8**. Copy the Digital Twin compiled executable **HEV_Simscape_Digital_Twin** from the previous section also to this folder. Fig. 8.122 shows a set of Linux commands to give the necessary executable permissions and packaging of the Lambda function and Digital Twin executable. The commands are explained below. Open a Linux terminal and run the below commands in steps.
    a. The ***ls*** command shows the folder has the compiled executable **HEV_Simscape_Digital_Twin** and the lambda function **lambda_function.py.**
    b. We need to give full read/write/executable permissions to all files in this folder for the AWS to be able to run it. Use the command ***sudo chmod -R 777 <folder_name>***.
    c. Run the command ***ls -l.*** This will list all the files and folders in the current folder and their permissions. Everything should be showing ***rwxrwxrwx***.
    d. Now package the lambda function and compiled executable using the command ***zip bundle.zip lambda_function.py HEV_Simscape_Digital_Twin***. This will zip the files ***lambda_function.py*** and ***HEV_Simscape_Digital_Twin*** into a file named ***bundle.zip***.
    e. After the bundle.zip is created, once again repeat the Step c to allow read/write/executable permissions to the ***bundle.zip*** file as well.
    f. The ***ls*** command will show the newly created ***bundle.zip*** file as well. Now we are ready to deploy the Lambda function to AWS and do the final testing.

```
# if the Lambda function is not triggered for the first time in this HEV Raspberry PI run cycle that is
# if time is greater than first 5 seconds, copy the previously stored CSV file hev_history.csv from the Amazon S3 Bucket
# named digitaltwinhevbucket. Download the hev_history.csv to the /tmp folder
    # Get File from S3 Bucket
    if get_file_from_s3_bucket:
        s3.meta.client.download_file('digitaltwinhevbucket', 'hev_history.csv', '/tmp/hev_history.csv')
# Open a new file hev_history_updated.csv for writing in the /tmp folder
        with open('/tmp/hev_history_updated.csv', 'w') as writeFile:
            writer = csv.writer(writeFile)
# Open the recently downloaded history file from S3 bucket hev_history.csv for reading
            with open('/tmp/hev_history.csv','r') as csvfile:
                plots = csv.reader(csvfile, delimiter=',')
# Copy all entries from hev_history.csv to hev_history_updated.csv
                for row in plots:
                    writer.writerows([[(row[0]),(row[1]),(row[2]),(row[3]),(row[4]),(row[5]),(row[6])]])
                    Actual_Veh_Speed.append(float(row[1]))
                    Actual_Motor_Speed.append(float(row[2]))
                    Actual_Generator_Speed.append(float(row[3]))
                    Actual_Engine_Speed.append(float(row[4]))
                    Actual_Battery_SOC.append(float(row[5]))
                    data_count = data_count +1;
            csvfile.close()
# Now open the input.csv file which is created with fresh sample of 50 data points from the past
# 5 Seconds and append that dat points to hev_history_updated.csv file
            with open('/tmp/input.csv','r') as csvfile:
                plots = csv.reader(csvfile, delimiter=',')
                for row in plots:
                    writer.writerows([[(row[0]),(row[1]),(row[2]),(row[3]),(row[4]),(row[5]),(row[6])]])
                    Actual_Veh_Speed.append(float(row[1]))
                    Actual_Motor_Speed.append(float(row[2]))
                    Actual_Generator_Speed.append(float(row[3]))
                    Actual_Engine_Speed.append(float(row[4]))
                    Actual_Battery_SOC.append(float(row[5]))
                    data_count = data_count +1;
            csvfile.close()
```

Figure 8.116 Updating Lambda function combining previous history input data with newly received data.

```
# Copy the hev_history_updated.csv back to the AWS S3 Bucket
        s3.meta.client.upload_file('/tmp/hev_history_updated.csv', 'digitaltwinhevbucket', 'hev_history.csv')
# Copy the file hev_history_updated.csv from /tmp to /tmp/hev_input.csv. The hev_input.csv will be the input file
# for running the Digital Twin Model. And it contains all the previous data along with the past 5 seconds data
        cmd = 'cp /tmp/hev_history_updated.csv /tmp/hev_input.csv'
        so = os.popen(cmd).read()
# Run the Digital Twin Model HEV_Simscape_Digital_Twin
        cmd = './HEV_Simscape_Digital_Twin'
        so = os.popen(cmd).read()
# Collect the /tmp/hev_output.csv file and copy to the AWS S3 Bucket
        s3.meta.client.upload_file('/tmp/hev_output.csv', 'digitaltwinhevbucket', 'hev_output.csv')
```

Figure 8.117 Updating Lambda function running Digital Twin model.

```
# Open the /tmp/input.csv and /tmp/hev_output.csv files for signals comparison between Actual data
# from Raspberry PI hardware and Digital Twin predicted values
        with open('/tmp/input.csv','r') as csvfile:
            plots = csv.reader(csvfile, delimiter=',')
            for row in plots:
                Actual_Veh_Speed.append(float(row[1]))
                Actual_Motor_Speed.append(float(row[2]))
                Actual_Generator_Speed.append(float(row[3]))
                Actual_Engine_Speed.append(float(row[4]))
                Actual_Battery_SOC.append(float(row[5]))
                data_count = data_count +1;


    with open('/tmp/hev_output.csv','r') as csvfile:
        plots = csv.reader(csvfile)
        for row in plots:
            Digital_Twin_Veh_Speed.append(float(row[1]))
            Digital_Twin_Motor_Speed.append(float(row[2]))
            Digital_Twin_Generator_Speed.append(float(row[3]))
            Digital_Twin_Engine_Speed.append(float(row[4]))
            Digital_Twin_Battery_SOC.append(float(row[5]))
```

Figure 8.118 Updating Lambda function reading input and output files for actual and predicted data.

```
# Calculate the Error Between Actual and Predicted Speeds
    Veh_Speed_Error = [None]*data_count
    Veh_Speed_Squared_Error = [None]*data_count
    Veh_Speed_Mean_Squared_Error = 0
    Veh_Speed_Root_Mean_Squared_Error = 0

    Motor_Speed_Error = [None]*data_count
    Motor_Speed_Squared_Error = [None]*data_count
    Motor_Speed_Mean_Squared_Error = 0
    Motor_Speed_Root_Mean_Squared_Error = 0

    Generator_Speed_Error = [None]*data_count
    Generator_Speed_Squared_Error = [None]*data_count
    Generator_Speed_Mean_Squared_Error = 0
    Generator_Speed_Root_Mean_Squared_Error = 0

    Engine_Speed_Error = [None]*data_count
    Engine_Speed_Squared_Error = [None]*data_count
    Engine_Speed_Mean_Squared_Error = 0
    Engine_Speed_Root_Mean_Squared_Error = 0

    Battery_SOC_Error = [None]*data_count
    Battery_SOC_Squared_Error = [None]*data_count
    Battery_SOC_Mean_Squared_Error = 0
    Battery_SOC_Root_Mean_Squared_Error = 0
```

Figure 8.119 Updating Lambda function initializing arrays for Root Mean Square Error calculation.

33. From the Linux operating system itself, open the web browser, the AWS Management Console, and the Lambda function **hev_digtal_twin** we created. Select the option from the drop down "**Upload a .zip file**" as shown in Fig. 8.123. Browse and select the ***bundle.-zip*** file we packaged earlier as shown in Fig. 8.124.

```
# Calculate the Root Mean Squared Error Between Actual and Predicted Data
  for index in range(data_count-1):
      Veh_Speed_Error[index] = Actual_Veh_Speed[index] - Digital_Twin_Veh_Speed[index];
      Veh_Speed_Squared_Error[index] = Veh_Speed_Error[index]*Veh_Speed_Error[index];
      Veh_Speed_Mean_Squared_Error = Veh_Speed_Mean_Squared_Error + Veh_Speed_Squared_Error[index];
      Motor_Speed_Error[index] = Actual_Motor_Speed[index] - Digital_Twin_Motor_Speed[index];
      Motor_Speed_Squared_Error[index] = Motor_Speed_Error[index]*Motor_Speed_Error[index];
      Motor_Speed_Mean_Squared_Error = Motor_Speed_Mean_Squared_Error + Motor_Speed_Squared_Error[index];
      Generator_Speed_Error[index] = Actual_Generator_Speed[index] - Digital_Twin_Generator_Speed[index];
      Generator_Speed_Squared_Error[index] = Generator_Speed_Error[index]*Generator_Speed_Error[index];
      Generator_Speed_Mean_Squared_Error = Generator_Speed_Mean_Squared_Error + Generator_Speed_Squared_Error[index]
      Engine_Speed_Error[index] = Actual_Engine_Speed[index] - Digital_Twin_Engine_Speed[index];
      Engine_Speed_Squared_Error[index] = Engine_Speed_Error[index]*Engine_Speed_Error[index];
      Engine_Speed_Mean_Squared_Error = Engine_Speed_Mean_Squared_Error + Engine_Speed_Squared_Error[index]
      Battery_SOC_Error[index] = Actual_Battery_SOC[index] - Digital_Twin_Battery_SOC[index];
      Battery_SOC_Squared_Error[index] = Battery_SOC_Error[index]*Battery_SOC_Error[index];
      Battery_SOC_Mean_Squared_Error = Battery_SOC_Mean_Squared_Error + Battery_SOC_Squared_Error[index]

  Veh_Speed_Mean_Squared_Error = Veh_Speed_Mean_Squared_Error/data_count;
  Veh_Speed_Root_Mean_Squared_Error = math.sqrt(Veh_Speed_Mean_Squared_Error)

  Motor_Speed_Mean_Squared_Error = Motor_Speed_Mean_Squared_Error/data_count;
  Motor_Speed_Root_Mean_Squared_Error = math.sqrt(Motor_Speed_Mean_Squared_Error)

  Generator_Speed_Mean_Squared_Error = Generator_Speed_Mean_Squared_Error/data_count;
  Generator_Speed_Root_Mean_Squared_Error = math.sqrt(Generator_Speed_Mean_Squared_Error)

  Engine_Speed_Mean_Squared_Error = Engine_Speed_Mean_Squared_Error/data_count;
  Engine_Speed_Root_Mean_Squared_Error = math.sqrt(Engine_Speed_Mean_Squared_Error)

  Battery_SOC_Mean_Squared_Error = Battery_SOC_Mean_Squared_Error/data_count;
  Battery_SOC_Root_Mean_Squared_Error = math.sqrt(Battery_SOC_Mean_Squared_Error)
```

Figure 8.120 Updating Lambda function calculating Root Mean Squared Error for actual and predicted HEV states and output.

```
# Create a Message String for Email and Text
    msg_str1 = 'HEV Digital Twin Off-BD Detected a Problem... Root Mean Square Error for Vehicle Speed = ' +
str(Veh_Speed_Root_Mean_Squared_Error) + ' for Motor Speed = ' +  str(Motor_Speed_Root_Mean_Squared_Error) + ' for Generator Speed = ' +
str(Generator_Speed_Root_Mean_Squared_Error) + ' for Engine Speed = ' +  str(Engine_Speed_Root_Mean_Squared_Error) + ' for Battery SOC = ' +
str(Battery_SOC_Root_Mean_Squared_Error)

    msg_str2 = 'No Problem Detected by HEV Digital Twin Off-BD... Root Mean Square Error for Vehicle Speed = ' +
str(Veh_Speed_Root_Mean_Squared_Error) + ' for Motor Speed = ' +  str(Motor_Speed_Root_Mean_Squared_Error) + ' for Generator Speed = ' +
str(Generator_Speed_Root_Mean_Squared_Error) + ' for Engine Speed = ' +  str(Engine_Speed_Root_Mean_Squared_Error) + ' for Battery SOC = ' +
str(Battery_SOC_Root_Mean_Squared_Error)

# Compare the RMSE Values against some threshold to determine if there is Diagnostics Failure conditions
# detected . If Failure Detected send SNS notification to send Email and Text Alerts
    if (Veh_Speed_Root_Mean_Squared_Error > 1 or Motor_Speed_Root_Mean_Squared_Error > 10 or Generator_Speed_Root_Mean_Squared_Error > 150 or
Engine_Speed_Root_Mean_Squared_Error > 150 or Battery_SOC_Root_Mean_Squared_Error >1):
        sns.publish(
        TopicArn = topicArn,
        Message = msg_str1
        )
        print(msg_str1)
    else:
        sns.publish(
        TopicArn = topicArn,
        Message = msg_str2
        )
        print(msg_str2)

    return {
        'statusCode': 200,
        'body': json.dumps('Hello from Lambda!')
    }
```

Figure 8.121 Updating Lambda function Off-BD detection algorithm with SMS/text SNS alert.

```
nbibin001@nbibin001-HP-Pavilion-x360-Convertible-15-br0xx: ~/Digital_Twin_Simscape_Book/Chapter_8/Digital_Twin_Lambd...
File Edit View Search Terminal Help
nbibin001@nbibin001-HP-Pavilion-x360-Convertible-15-br0xx:~/Digital_Twin_Simscape_Book/Chapter_8/Digital_T
win_Lambda_Function$ ls
HEV_Simscape_Digital_Twin  lambda_function.py
nbibin001@nbibin001-HP-Pavilion-x360-Convertible-15-br0xx:~/Digital_Twin_Simscape_Book/Chapter_8/Digital_T
win_Lambda_Function$ sudo chmod -R 777 /home/nbibin001/Digital_Twin_Simscape_Book/Chapter_8/Digital_Twin_L
ambda_Function/
nbibin001@nbibin001-HP-Pavilion-x360-Convertible-15-br0xx:~/Digital_Twin_Simscape_Book/Chapter_8/Digital_T
win_Lambda_Function$ ls -l
total 3384
-rwxrwxrwx 1 nbibin001 nbibin001 3452072 Dec 24 18:45 HEV_Simscape_Digital_Twin
-rwxrwxrwx 1 nbibin001 nbibin001   12211 Dec 26 08:28 lambda_function.py
nbibin001@nbibin001-HP-Pavilion-x360-Convertible-15-br0xx:~/Digital_Twin_Simscape_Book/Chapter_8/Digital_T
win_Lambda_Function$ zip bundle.zip lambda_function.py HEV_Simscape_Digital_Twin
  adding: lambda_function.py (deflated 78%)
  adding: HEV_Simscape_Digital_Twin (deflated 66%)
nbibin001@nbibin001-HP-Pavilion-x360-Convertible-15-br0xx:~/Digital_Twin_Simscape_Book/Chapter_8/Digital_T
win_Lambda_Function$ sudo chmod -R 777 /home/nbibin001/Digital_Twin_Simscape_Book/Chapter_8/Digital_Twin_L
ambda_Function/
nbibin001@nbibin001-HP-Pavilion-x360-Convertible-15-br0xx:~/Digital_Twin_Simscape_Book/Chapter_8/Digital_T
win_Lambda_Function$ ls -l
total 4540
-rwxrwxrwx 1 nbibin001 nbibin001 1179940 Dec 26 08:42 bundle.zip
-rwxrwxrwx 1 nbibin001 nbibin001 3452072 Dec 24 18:45 HEV_Simscape_Digital_Twin
-rwxrwxrwx 1 nbibin001 nbibin001   12211 Dec 26 08:28 lambda_function.py
nbibin001@nbibin001-HP-Pavilion-x360-Convertible-15-br0xx:~/Digital_Twin_Simscape_Book/Chapter_8/Digital_T
win_Lambda_Function$
```

Figure 8.122 Linux command line showing packaging the Lambda function and Digital Twin executable.

34. After the bundle.zip file is successfully uploaded, click on the ***Upload*** and ***Save*** button as shown in Fig. 8.125.
35. We can see, the AWS Lambda function editor now shows our updated Lambda function, and also the Digital Twin compiled executable. See Fig. 8.126.
36. As a last step, we just need to create and configure the AWS S3 bucket which will be used to store and reuse the HEV data in the Lambda function. From the AWS service console, select the **Storage >> S3** as shown in Fig. 8.127.
37. Click on the **Create Bucket** button as shown in Fig. 8.128. On the new window as shown in Fig. 8.129, give a name to the S3 bucket. In this case, we give the name to be **digitaltwin-hevbucket**, note that this should be the same name we used earlier in the Lambda function. Click on Next.
38. On the **Configure Options** screen, we can leave all the settings to be default as shown in Fig. 8.130. Click **Next.**
39. Next on the **Set Permissions** screen, we can check the box "**Block all public access.**" Only we will be using the S3 bucket from Lambda function, so we don't have to grand any public access. See Fig. 8.131, and click **Next**, review, and finalize the S3 bucket creation as shown in Fig. 8.132.
40. Congratulations!!!We are ready to test the whole Off-BD algorithm for the HEV system now.

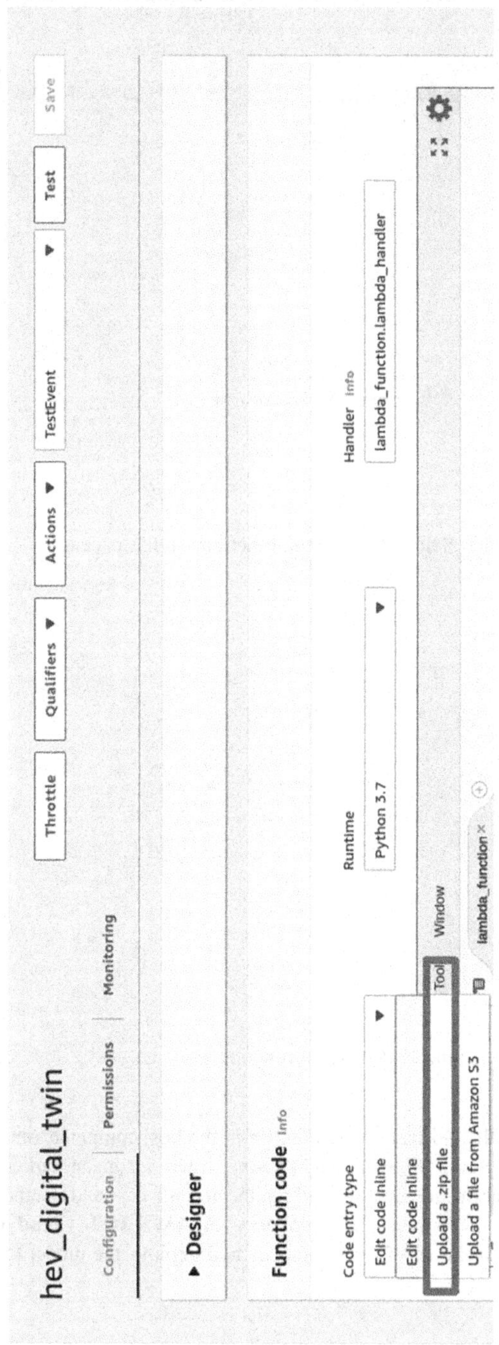

Figure 8.123 Uploading the packaged Lambda function to AWS.

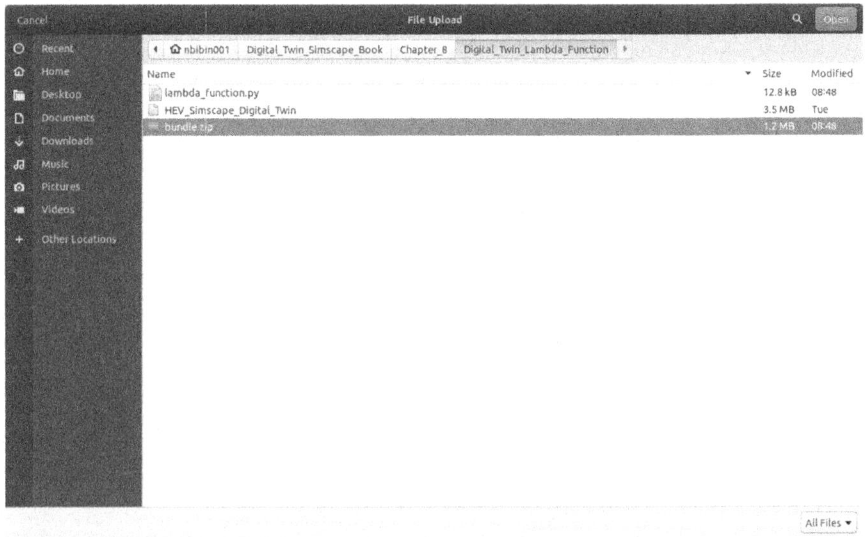

Figure 8.124 Browse and Select the Lambda function zip file package.

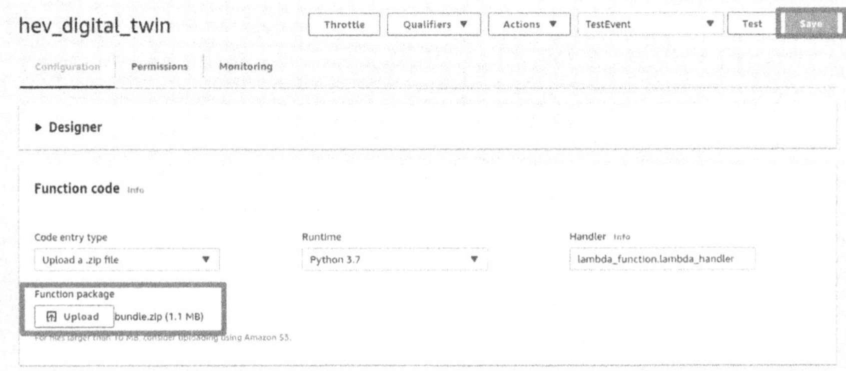

Figure 8.125 Saving the uploaded Lambda function package.

41. First, we will test the No-Fault condition. From the host computer, run the model **HEV_-Simscape_Model_Rasp_Pi_with_MQTT.slx**, which we made earlier, that will run on the Raspberry Pi hardware. First, we will run the no-fault case and test the Off-BD with Digital Twin. Also run the Python code **Raspberry_Pi_AWS_IOT_Cloud_Connection.py** on the Raspberry Pi, while MATLAB compiles and deploys the model to PI. As the model

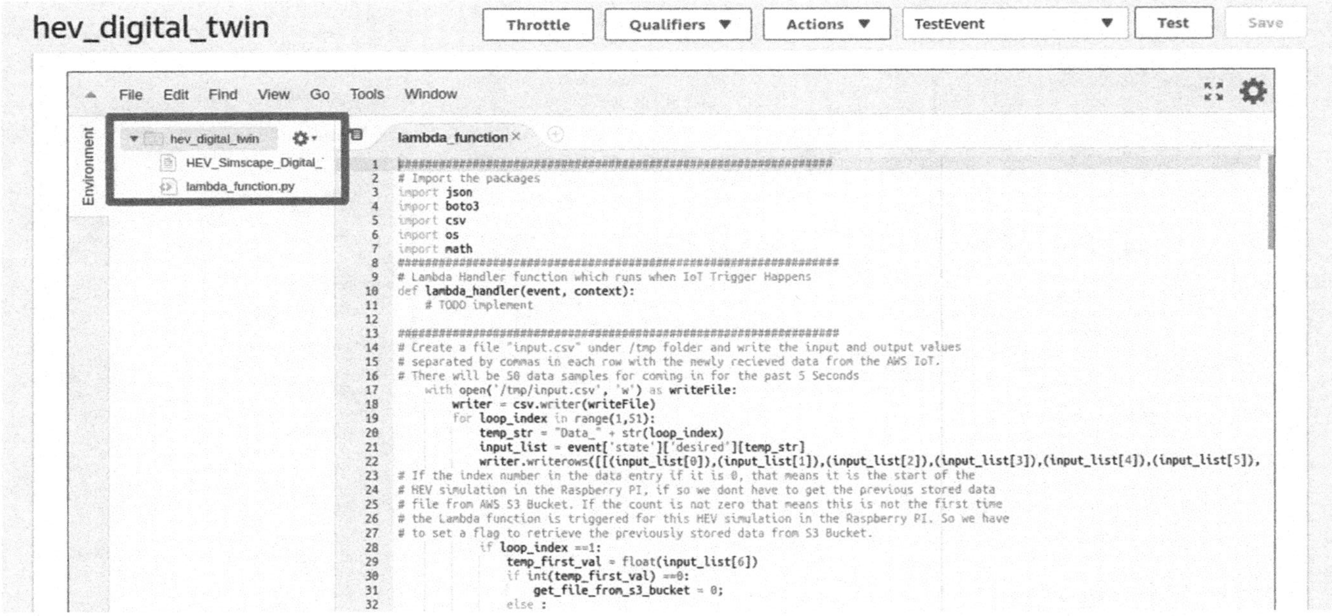

Figure 8.126 Lambda function inline editor showing the uploaded function and executable.

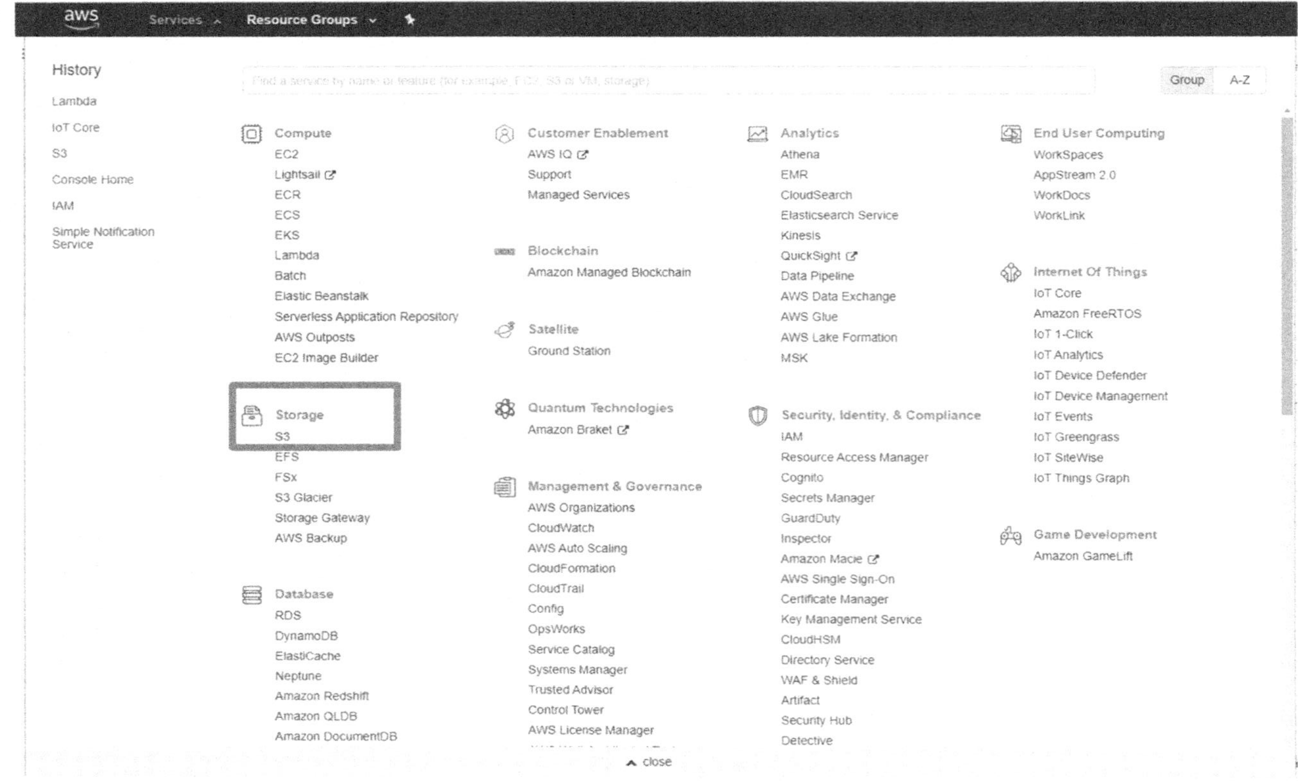

Figure 8.127 Selecting the S3 bucket service from AWS console.

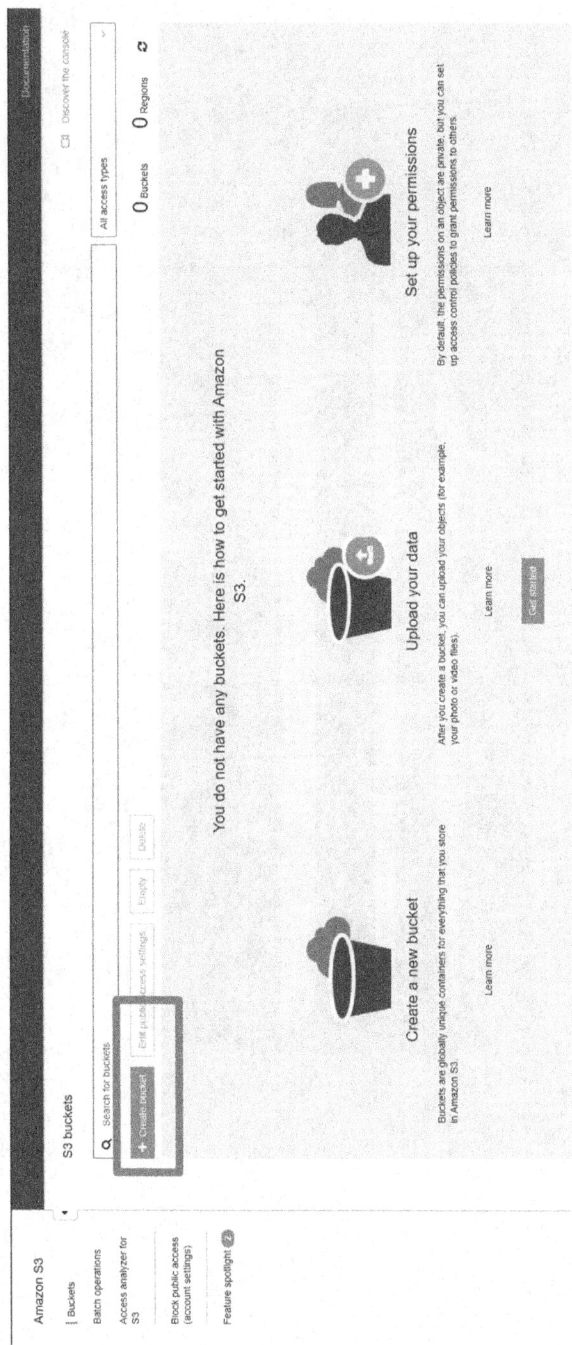

Figure 8.128 Creating a new S3 bucket.

Figure 8.129 Naming the AWS S3 bucket.

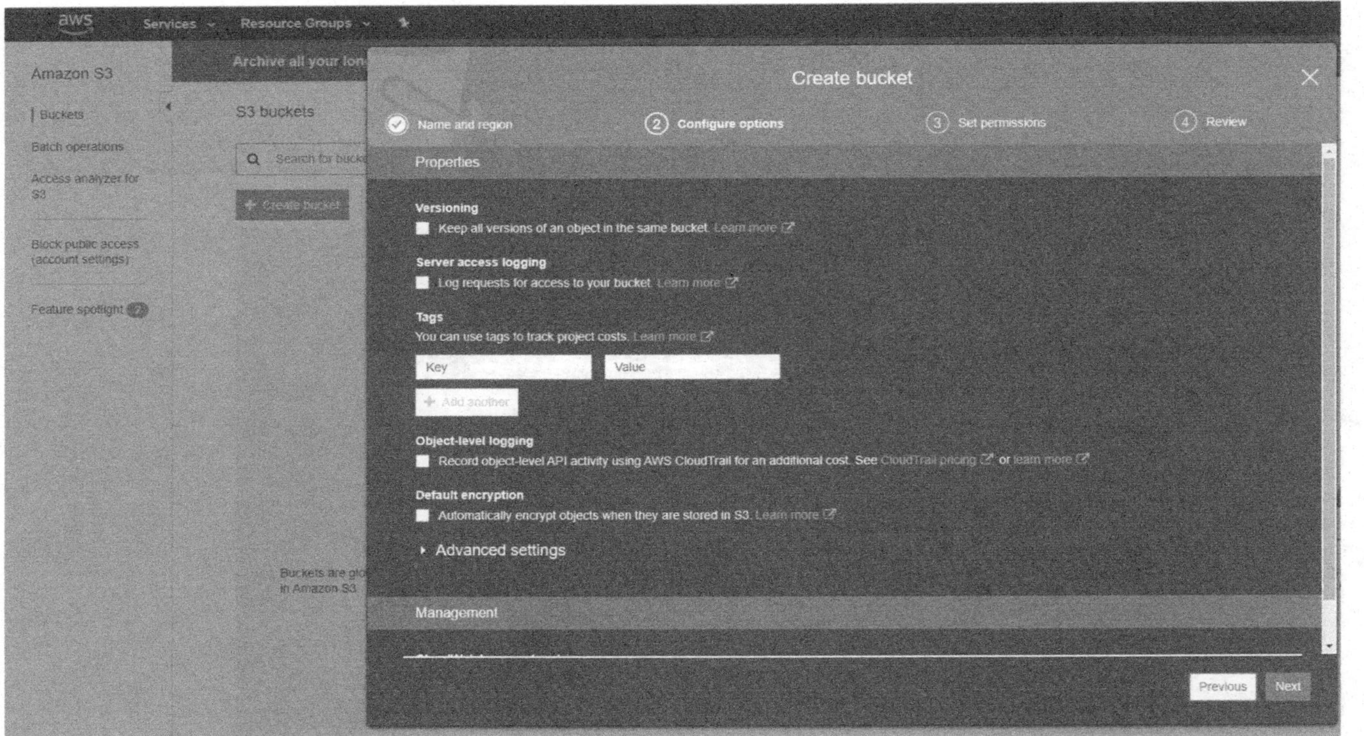

Figure 8.130 AWS S3 Configure options.

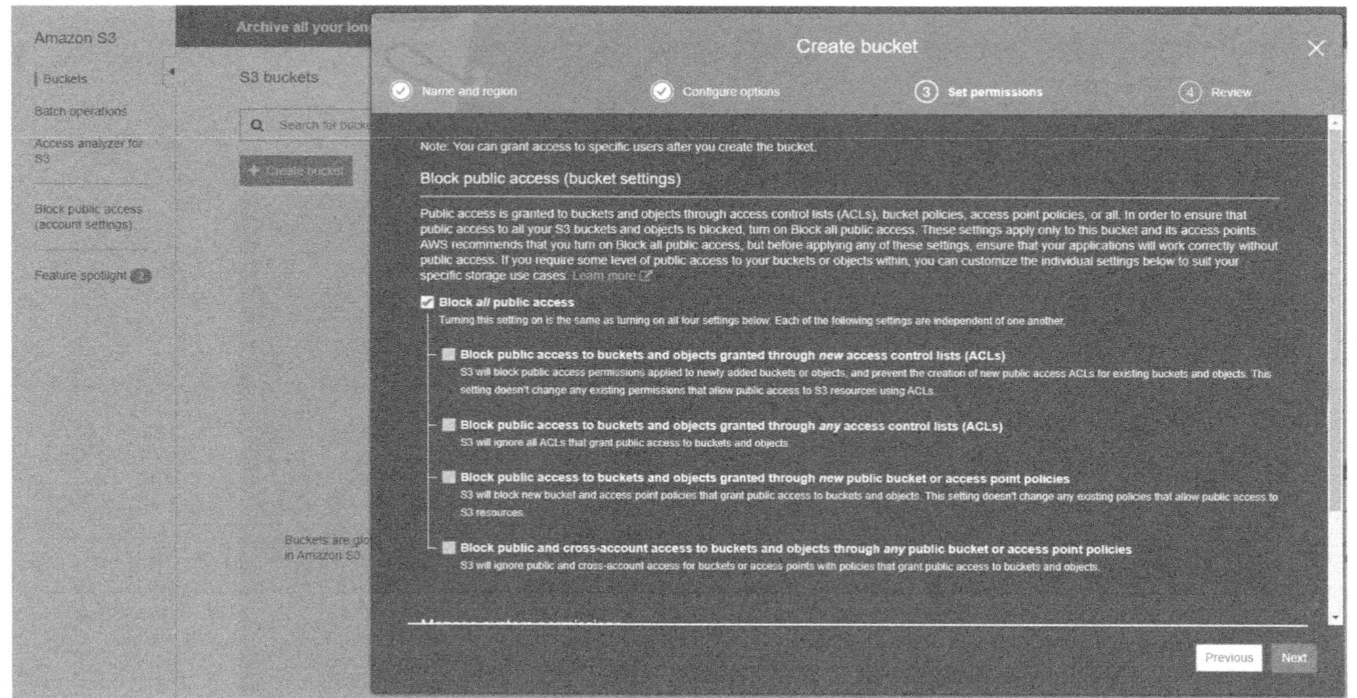

Figure 8.131 Permissions for the AWS bucket.

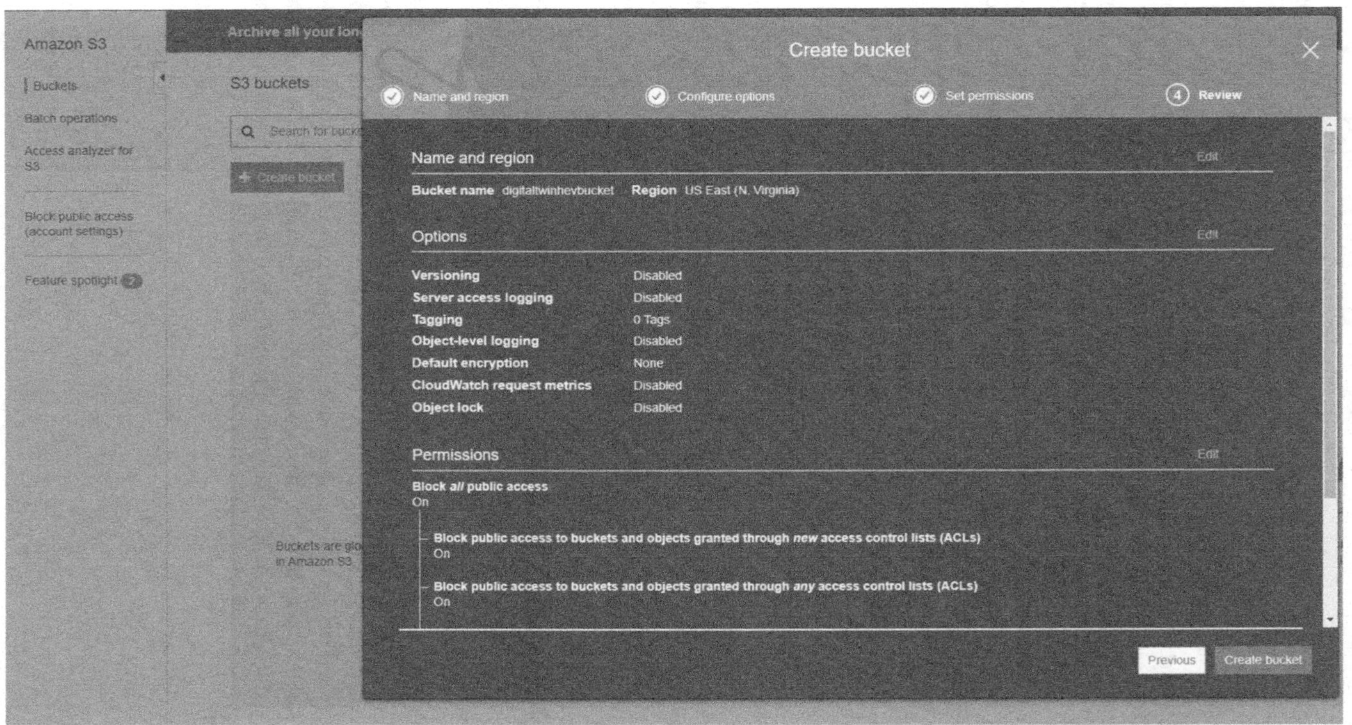

Figure 8.132 Review and finalize S3 bucket creation.

digital_twin 39 AWS Notification Message - No Problem Detected by HEV Digital Twin Off-BD... Root ...

Figure 8.133 No Fault Condition E-mail notifications from the Digital Twin about the Off-BD status.

starts running, we can see from the SNS service we configured in the Lambda function will send Text and E-mail Notifications on the Off BD status based on the Digital Twin prediction, decision logic in the Lambda function. Figs. 8.133 and 8.134 show the E-mail messages received from the Digital Twin notification service, note in this case no failure condition is detected.

42. Next we will test the Engine Throttle Fault condition. From the host computer, run the model **HEV_Simscape_Model_Rasp_Pi_with_MQTT_Throttle_Fault.slx**, which we made earlier, that will run on the Raspberry Pi hardware. As we have seen earlier, this model has the introduced fault condition for the Engine Throttle signal. Also run the Python code **Raspberry_Pi_AWS_IOT_Cloud_Connection.py** on the Raspberry Pi while MATLAB compiles and deploys the model to PI. As the model starts running, we can see from the SNS service we configured in the Lambda function will send Text and E-mail Notifications on the Off BD status based on the Digital Twin prediction, decision logic in the Lambda function. Figs. 8.135 and 8.136 show the E-mail messages received from the Digital Twin notification service, note in this case, since only the HEV system running on Raspberry Pi sees the faulty throttle but not the digital twin, the signals between two systems vary significantly, which will be detected by the Digital Twin Off-BD algorithm.

8.10 Application problem

1. Develop an Off-BD process to detect a battery capacity failure of the HEV system.

Hint: In order to simulate a failed battery, we can change the HEV model battery variable rated capacity variable **HEV_Param.Battery_Det.Rated_Capacity** on the workspace initialization Mat file from 8.1 to a low value and repeat the whole process. For the Digital Twin model that needs to be deployed on to the cloud, keep the Rated Capacity same as original. So the digital twin is not aware of the reduced battery capacity, and it should be able to detect the difference in the signals between actual system and the digital twin predicted system.

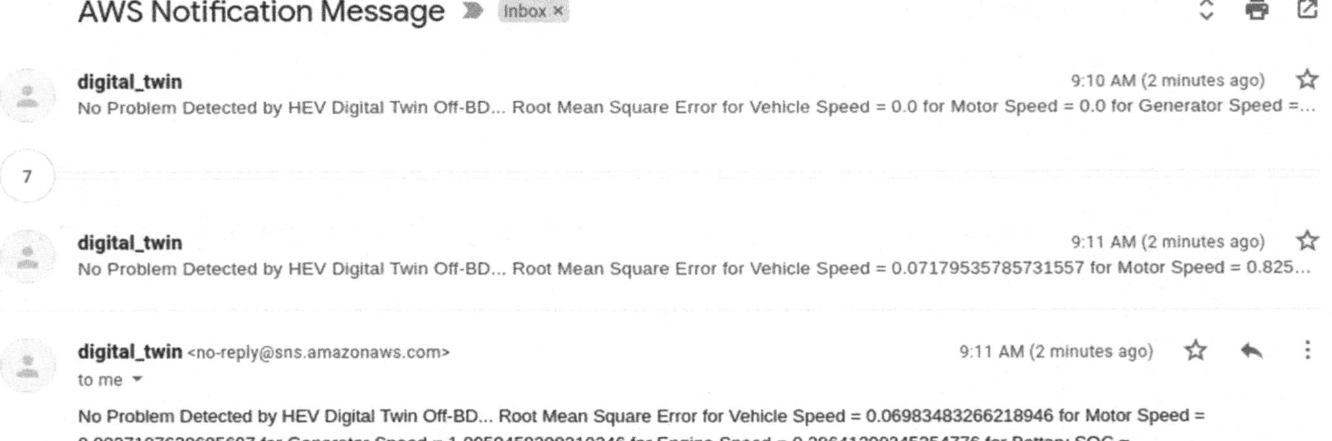

Figure 8.134 No Fault Condition detailed E-mail message information from the Digital Twin about the Off-BD status.

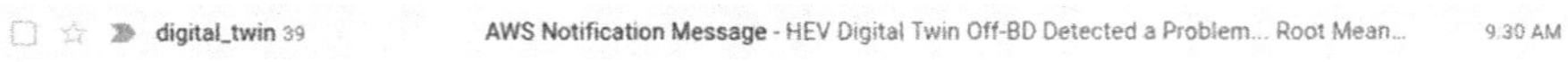

digital_twin 39 AWS Notification Message - HEV Digital Twin Off-BD Detected a Problem... Root Mean... 9:30 AM

Figure 8.135 Engine Throttle Fault Condition, E-mail Notifications from the Digital Twin about the Off-BD status.

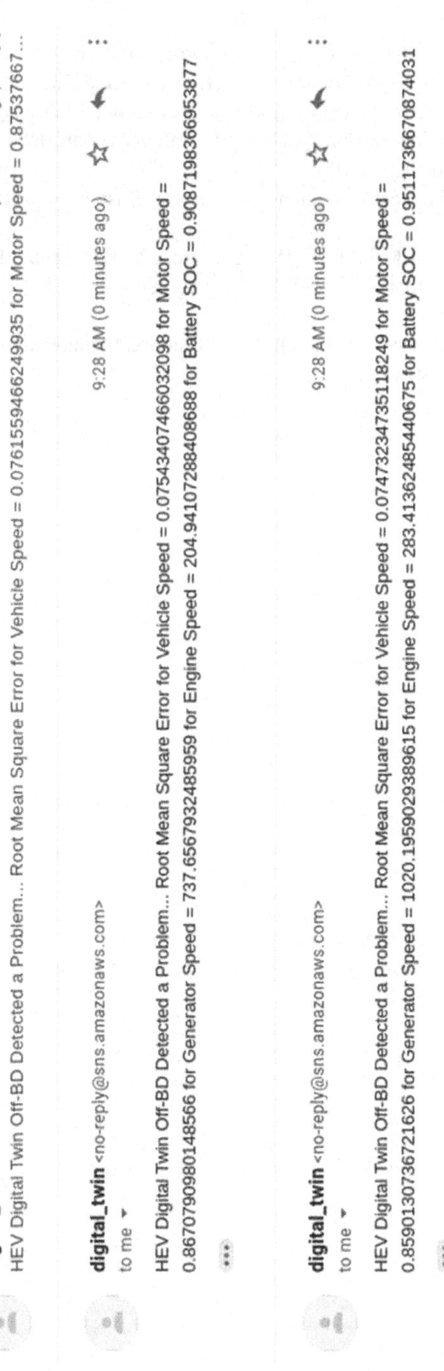

Figure 8.136 Engine Throttle Fault Condition, detailed E-mail message information from the Digital Twin about the Off-BD status.

References

[1] Raspberry Pi 3 B+ Hardware. https://www.amazon.com/CanaKit-Raspberry-Power-Supply-Listed/dp/B07BC6WH7V/ref=sr_1_3?crid=1B02XIHFO3BQK&keywords=raspberry+pi+3b+plus&qid=1576465330&sprefix=raspberry+pi+3%2Caps%2C186&sr=8-3.

[2] Setting up Operating System for Raspberry PI. http://www.mathworks.com/matlabcentral/fileexchange/39354-device-drivers.

[3] Installing Putty on the Host Computer to connect to Raspberry Pi Remotely. https://www.putty.org/.

[4] HEV Matlab® Simulink® Simscape™ Model from Matlab Central File Exchange. https://www.mathworks.com/matlabcentral/fileexchange/28441-hybrid-electric-vehicle-model-in-simulink.

[5] Hybrid Electric Vehicle Types. http://autocaat.org/Technologies/Hybrid_and_Battery_Electric_Vehicles/HEV_Types/.

Digital Twin Development and cloud deployment for a DC Motor Control embedded system

9.1 Introduction

This chapter walks the reader through performing the Off-Board Diagnostics (Off-BD) for a DC Motor Controller Real-Time hardware using its Digital Twin Deployed on the Amazon Web Services (AWS) cloud. The Off-BD requirement is to diagnose the rotational speed of the DC motor, and if there is any malfunction detected, the Off-BD algorithm is supposed to send E-mail/Text alerts to the User. An Arduino Mega microcontroller with MATLAB Simulink software is used to control the DC Motor in real time and send real-time sensor and actuator data to the AWS cloud through an ESP32 Wi-Fi controller module, which can establish cloud connectivity using Wi-Fi hotspot with Internet connectivity. A Digital Twin model of the DC Motor is developed and tuned to match closer to the real-time DC motor behavior using MATLAB®, Simulink®, and Simscape™ and is deployed into the AWS cloud. The Off-BD algorithm which is running on the cloud will receive data from the real-time hardware, trigger the Digital Twin with the input data to predict the expected DC Motor speed, compare it with the actual Motor speed, and make a decision about the actual DC Motor Hardware state. The Off-BD algorithm will also notify the User via E-mail/Text if a malfunction is detected. AWS services such as Internet of Things, Lambda, Simple Notification Service (SNSs), etc., are utilized in this chapter. Below listed are the detailed steps followed in this chapter:

1. Physical Asset Setup: Setting up Real-Time Embedded Controller Hardware and Software for DC Motor Speed Control
2. Open-Loop Data Collection and Closed-Loop PID Controller Development for DC Motor Hardware
3. Developing Simscape™ Digital Twin model for the DC Motor Speed Control
4. Parameter Tuning of the Simscape™ DC Motor Model with Data from DC Motor Hardware Using Simulink Parameter Estimation™
5. Adding Cloud Connectivity to Real-Time Embedded Controller Hardware for DC Motor Speed Control
6. Deploying the Simscape™ Digital Twin model to the AWS Cloud
7. Voice, E-mail, and Text User Interface Development for the Digital Twin Deployment

All the codes used in the chapter can be downloaded from *MATLAB® File Exchange*. Follow the below link and search for the ISBN or title of the book:

https://www.mathworks.com/matlabcentral/fileexchange/

Digital Twin Development and Deployment on the Cloud. https://doi.org/10.1016/B978-0-12-821631-6.00009-8

Alternatively, the reader can also download the material and other resources from the dedicated website or contact the authors for further help:

https://www.practicalmpc.com/

Figs. 9.1 and 9.2 shows the block diagram of the inputs/outputs of the DC Motor Speed Control System Hardware. The input to the system is the PWM command to control the voltage to the DC Motor and the output of the system is the actual rotational speed of the Motor.

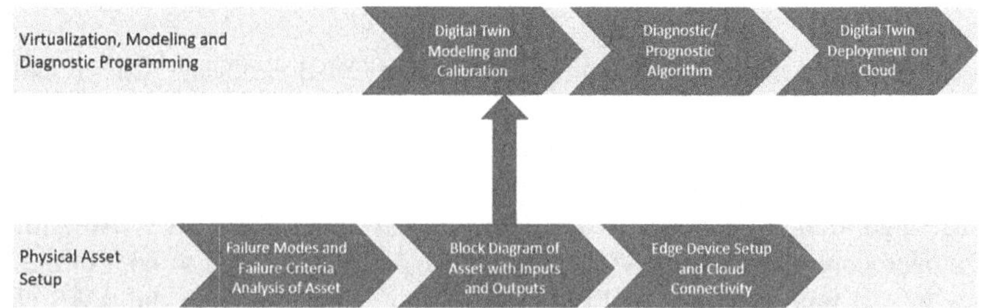

Figure 9.1 Off-BD steps covered in this chapter.

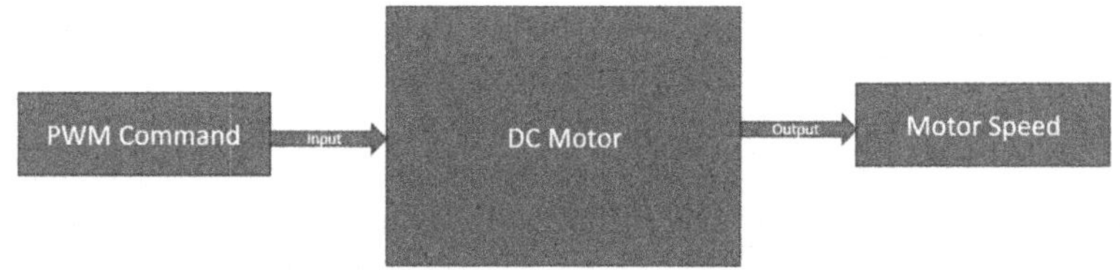

Figure 9.2 Block diagram of the DC Motor speed control system.

Fig. 9.3 shows the block diagram of the Off-BD Digital Twin diagnostics process followed in this chapter. The input PWM command fed to the Physical DC Motor

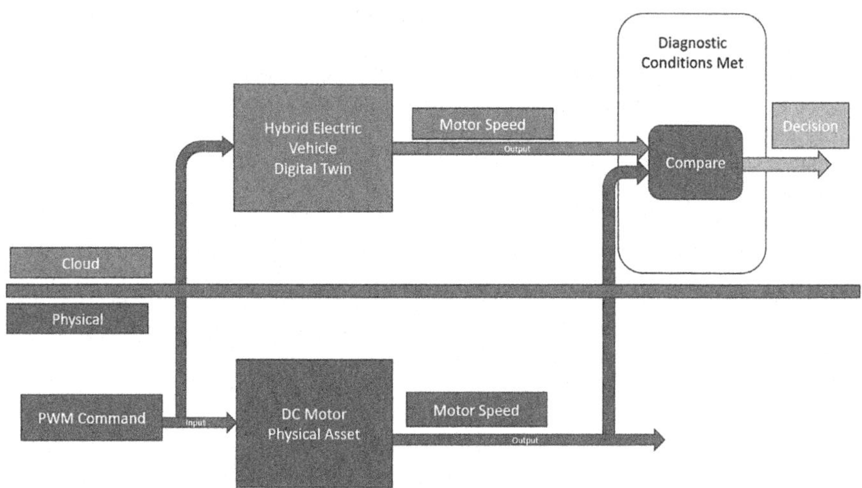

Figure 9.3 Off-BD diagnostics process for DC Motor speed control system.

Controller is also sent to the AWS cloud. On the AWS cloud, a Digital Twin model of the DC Motor system, which is parameter tuned to match the real asset behavior, will be running with the same PWM input from the physical asset. The output Motor speed of Digital Twin model is compared with the actual data collected from the hardware. A Root Mean Square Error (RMSE)–based diagnostic detection and decision-making is developed to compare the actual and digital twin outputs. In this chapter, a concept to detect the Motor Power Supply failure condition using the Off-BD is demonstrated. The power supply failure of the physical asset motor will cause the sensed motor speed to be different from the Digital Twin predicted motor speed. This difference in the outputs between the Actual and Digital Twin data is identified, and if the RMSE is greater than a threshold, a failure condition is flagged and the User is notified directly from the cloud using a Text or E-mail notification.

9.2 Setting up Real-Time Embedded Controller Hardware and Software for DC Motor Speed Control

The hardware setup for this experiment is bought from the online Educational Robotics store ***RoboholicManiacs***. Their website link is provided here [1]. Users can buy fully assembled DC Motor Controller hardware and software from this website. If the User is interested in building the hardware setup by their own, all the details is listed in Section 9.2.1. This hardware setup needs to be expanded, for cloud connectivity which will be explained later in this chapter. Fig. 9.4 shows the basic hardware setup with DC Motor, Arduino Mega Microcontroller, LCD Display, etc., packaged in a box (delivered as is from ***RoboholicManiacs***).

Figure 9.4 DC Motor Controller with Arduino Mega.

9.2.1 Hardware requirements and familiarization

Table 9.1 shows all the hardware parts required to replicate the setup described in this chapter. The fully assembled hardware setup used in this chapter can also be purchased from the dedicated website of this book: https://www.practicalmpc.com/.

Table 9.1 Hardware parts required for setting up the example in this chapter.

Item	Description
Arduino Mega	The Arduino Mega is a microcontroller board based on the ATmega2560.
Breadboard (2 numbers) and jumper wires	Breadboard and jumper wires are used to make the required electric connection between the hardware parts.
DC motor with quadrature encoder	12 V DC motor, with integrated quadrature encoder for speed sensing.
MOSFET (IRF 3205)	MOSFET is used as a switch, controlled by the digital PWM output of the Arduino to connect and disconnect motor to the battery power source.
Diode (IN4007)	Diode is used to prevent the motor's back emf causing damage to the circuit.
12 V DC adapter or 12 V DC battery	Either a DC battery or an AC–DC adapter can be used to power the DC motor.
ESP32 Wi-Fi module	ESP32 is a Wi-fi module used to establish connection with AWS cloud and DC motor hardware to exchange data.
3.3–5 V logic level converter module for Arduino	Arduino Mega and ESP32 communicate over wired serial connection. Arduino works on 5 V and ESP32 works on 3.3 V, so a voltage level shifter is required when interfacing Arduino and ESP32.

Figs. 9.5 and 9.6 show the block diagram and electrical connections of the hardware setup.

9.2.2 Software requirements

In addition to the MATLAB® and its toolboxes already installed and used for the previous chapters, we need to install the Hardware support package for Arduino for developing and deploying Simulink controls logic into the Arduino hardware. MathWorks® provides a hardware support package to develop, simulate and program algorithms, configure, and access sensors and actuators using Simulink blocks with the Simulink® Support Package for Arduino® Hardware. Using MATLAB® and Simulink® external mode interactive simulations, parameter tuning, signal monitoring, and logging can be performed, as the algorithms run real time on the Arduino board.

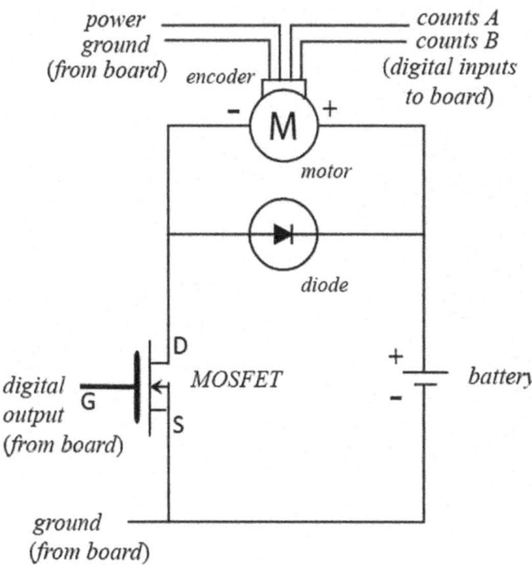

Figure 9.5 Hardware setup and connection block diagram [1].

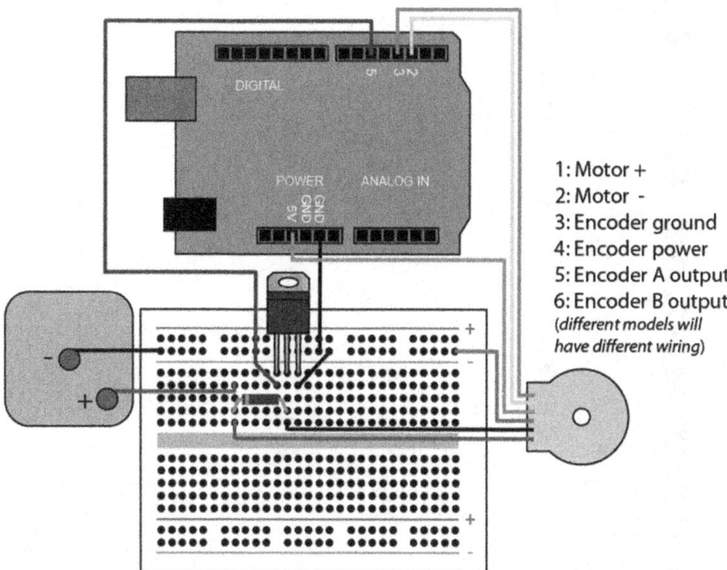

Figure 9.6 Connecting Arduino to DC Motor [1].

Follow the below steps to install the Simulink support package for Arduino hardware:

1. From the MATLAB® window go to Home >> Add-Ons >> Get Hardware Support Packages. Fig. 9.7 shows the Add-On Explorer GUI.
2. Click on the Simulink® Support Package for Arduino hardware option, highlighted in Fig. 9.8, and it will guide into next window with an ***Install*** button as shown in Fig. 9.9. Click on it. For any support package installation, User needs to log on to a MathWorks account using the option shown in Fig. 9.10. Log on using the account, or create a new account if the User doesn't have it already.

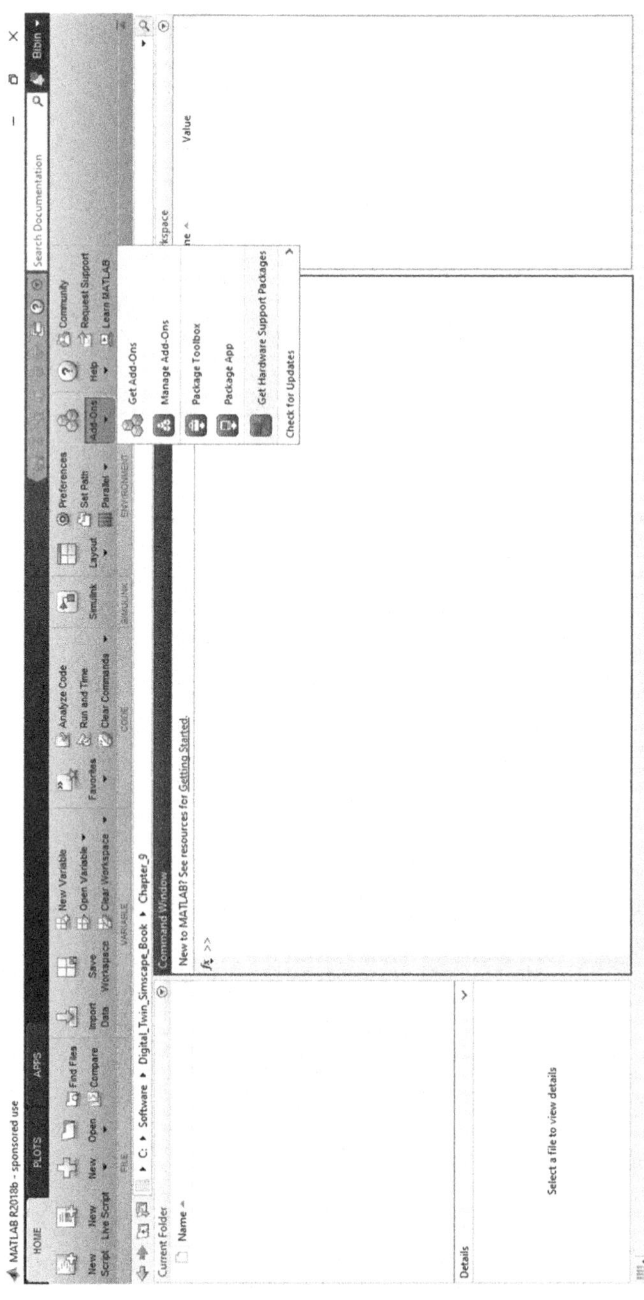

Figure 9.7 Add-On Explorer GUI.

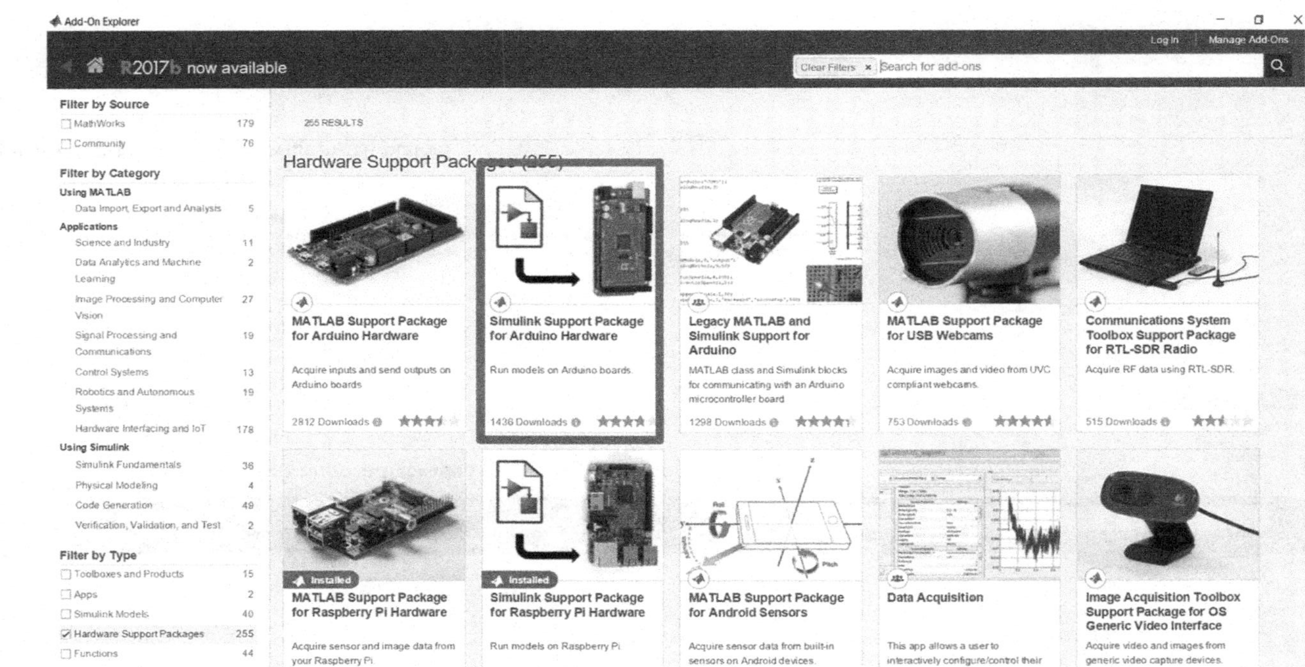

Figure 9.8 Simulink Support Package for Arduino Hardware.

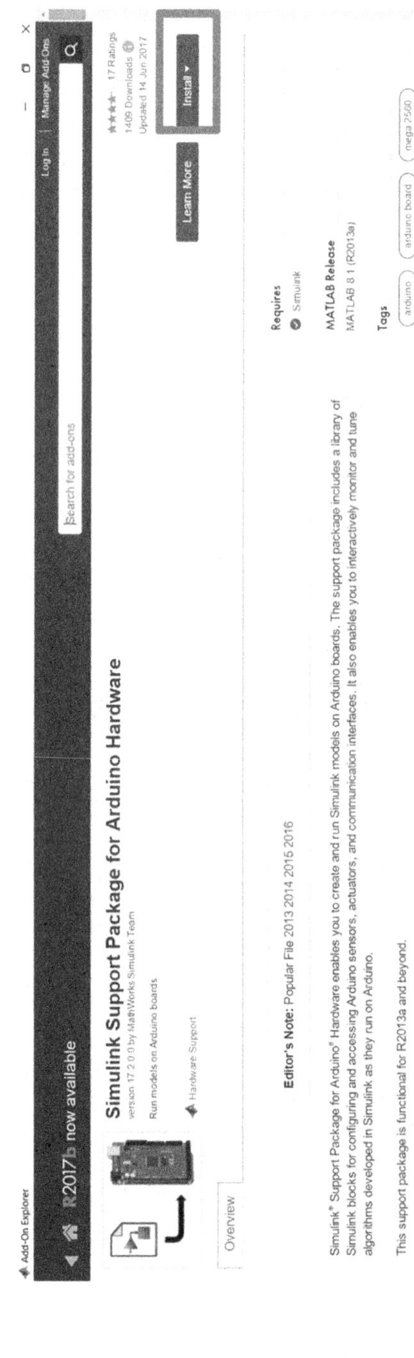

Figure 9.9 Add-On Explorer with Install button.

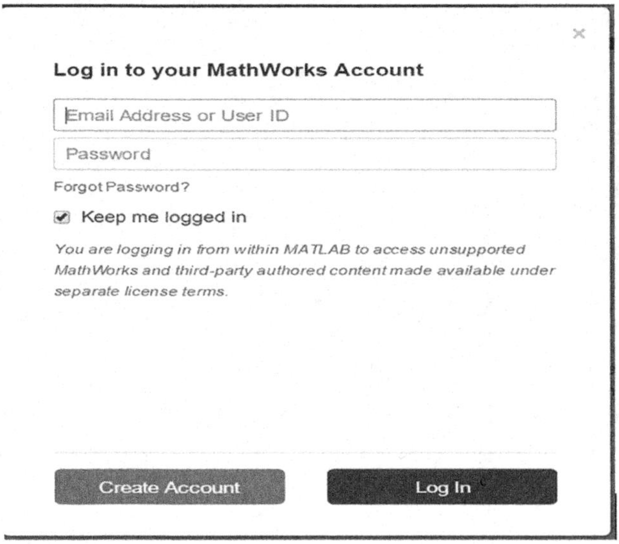

Log in to your MathWorks Account
Email Address or User ID
Password
Forgot Password?
Keep me logged in
You are logging in from within MATLAB to access unsupported MathWorks and third-party authored content made available under separate license terms.
Create Account
Log In

Figure 9.10 MathWorks Account login.

3. Wait for the installation to be completed. An installation progress window is shown in Fig. 9.11. After the successful installation, it will show an option to open the examples available in the Support package. Selecting that option will take the User to the page with various examples as shown in Fig. 9.12.

Download and Installation Progress
Downloading Support Packages... 100%
Downloading Third-Party Packages... 64%
Installing Support Packages...
Installing Third-Party Packages
Configuring your installation

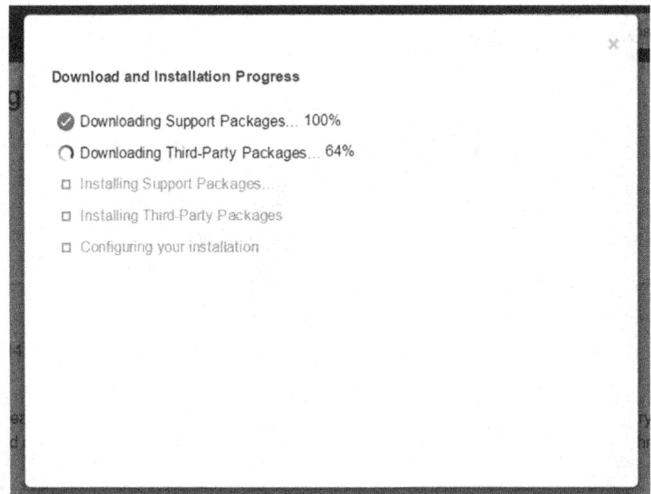

Figure 9.11 Support Package Installation Progress.

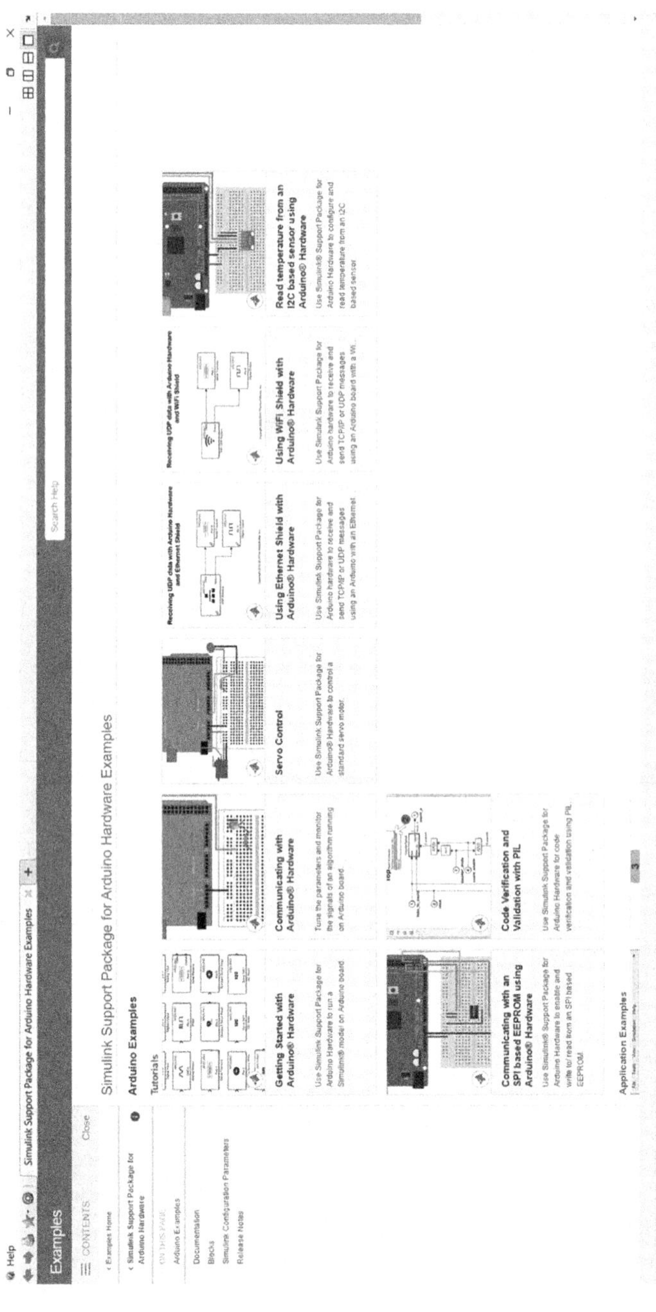

Figure 9.12 Link to example projects available in the Support Package.

4. User can make sure the support package is installed properly, by typing ***aduinolib*** in the MATLAB command window. It will open up the library model with various blocks as shown in Fig. 9.13.

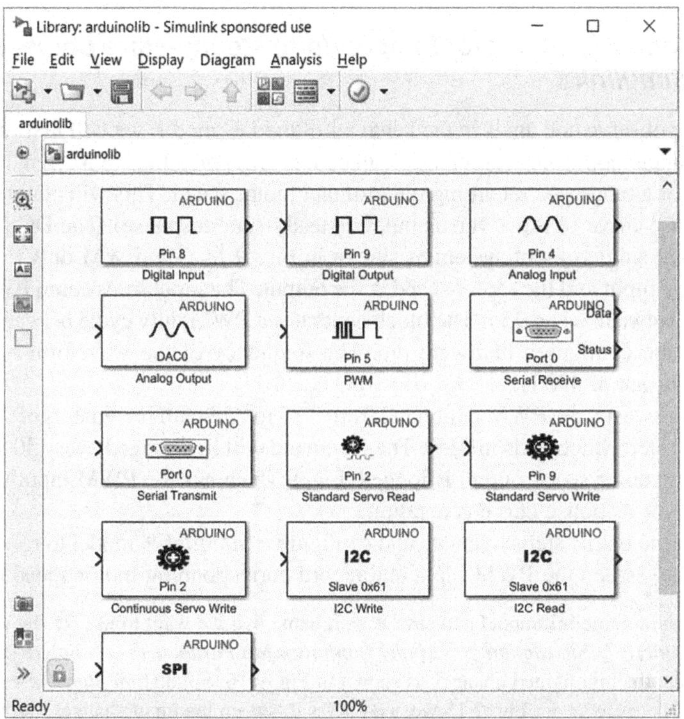

Figure 9.13 Arduino I/O Library.

9.3 Open-Loop Data Collection and Closed-Loop PID Controller Development for the DC Motor Hardware

The speed of the motor is controlled using the digital Pulse-Width Modulated (PWM) output of the Arduino Mega microcontroller. The PWM output is used to control the MOSFET, which acts as switch to connect and disconnect the 12 V DC power to the DC Motor. Though the motor is turned On and Off continuously using the PWM output from the Arduino Mega, because of its inertia and friction it doesn't go to full speed when the PWM duty cycle command is 100% or go to stop immediately when the PWM duty command goes to 0%. The motor dynamics (inertia and friction) acts as a filter for the high frequency On-Off commands given to the motor, and a smooth continuous motor speed profile can be obtained. Arduino library provides a PWM output generator block to generate the PWM with a fixed period and varying duty cycle on a selected PWM output pin on the Arduino Mega board.

The built-in quadrature encoder of the DC motor will be used in this experiment to measure the speed of the motor. The Arduino Support Package library does not provide a block to read the encoder output from the motor, so a custom S-function block is used to read the speed sensor.

9.3.1 Running DC motor in open-loop steady-state points using Simulink®

In order to observe and analyze the behavior of the DC motor, we will first run the DC motor in open loop to collect input/output data by sweeping the PWM signal across its full operating range and recording the resultant motor speed. This will allow us to plot the response curve of input versus output (steady-state response). The DC motor is a single-input single-output system as shown in Fig. 9.14. The PWM duty cycle command is the input and the motor speed is the output. The input to Arduino PWM block is a value between 0 and 255. The block generates a PWM duty cycle between 0% and 100% on the configured hardware pin. The frequency of the waveform is approximately constant at 490 Hz.

We will sweeps the PWM command from 0 25 to 255 with increments of 25 (except for the last step which ends in 255). The command will be changed every 50 s and corresponding motor speed output is logged. Fig. 9.15 shows the PWM input waveform generated for response curve generation.

Follow the below steps to create and configure a Simulink® model to run on Arduino Mega to sweep the PWM input and record corresponding motor speed output:

1. Open a new Simulink model and save it with name that we want to use for the model.
2. Goto ***Model*** >> ***Simulation*** >> ***Model Configuration Parameters*** >> ***Solver*** setting page and make the highlighted changes as shown in Fig. 9.16. Simulation ***Stop Time*** of 500 s is selected because as per Fig. 9.15 we need 500 s to sweep the input changes. ***Solver Type*** is changed to ***Fixed-step*** because all the blocks and subsystems in this model is expected to run at a certain discrete rate 0.02 s specified in the **Fixed-step size** field.
3. Next go to the Hardware Implementation configuration setting and select ***Arduino Mega 2560*** as the ***Hardware board*** as shown in Fig. 9.17. After making these changes, click ***Apply*** and ***Ok*** on the configuration setting GUI.

Figure 9.14 Input/output of a DC Motor speed control system.

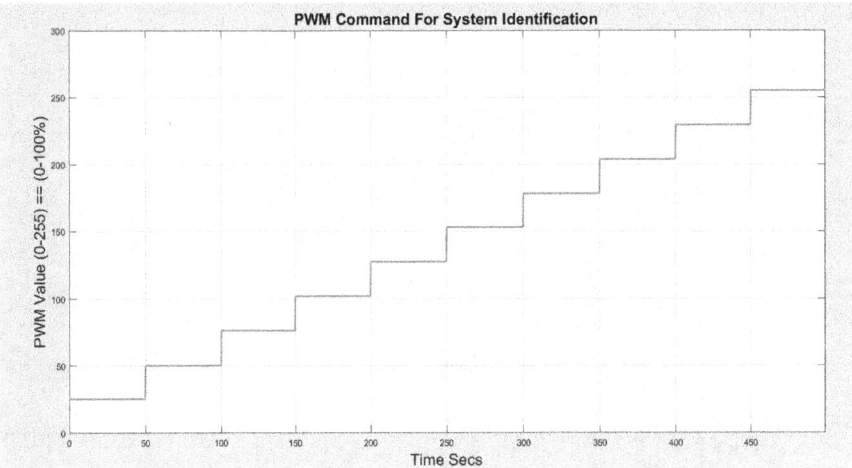

Figure 9.15 System ID input PWM Value signal.

4. Add a ***Repeating Sequence*** block from the Simulink library to the newly created model, double click on the ***Repeating Sequence*** block and enter the ***Time values*** and ***Output values*** to match the input sweep waveform showed in Fig. 9.18.

   Time Values: [0 49.9 50 99.9 100 149.9 150 199.9 200 249.9 250 299.9 300 349.9 350 399.9 400 449.9 450 500] ***Output Values***: [25.5 25.5 50.1 50.1 76.5 76.5 102 102 127.5 127.5 153 153 178.5 178.5 204 204 229.5 229.5 255 255].

5. Type ***arduinolib*** on the MATLAB command window to open up the Arduino hardware support library. From the library, add the ***PWM*** block in to the model as shown in Fig. 9.19 and connect the output of the ***Repeating Sequence*** block to the input of the ***PWM*** block. Also double click on the ***PWM*** block and enter the ***Pin number*** as 5 (if not already). This is to configure the Digital pin 5 of the Arduino to drive the PWM output.
6. Just to make sure this Simulink® model can communicate with the Hardware setup we can try to build and download this simple model into the hardware. Make sure all the connections in Section 9.2.1 are proper. Now connect the Arduino Mega to computer running MATLAB® using the USB cable which came with the Arduino board. Change the ***Simulation mode*** of the model to ***External*** as shown in Fig. 9.20. The external mode lets us tune the parameters in and monitor data from the Simulink model while it is running on the Arduino hardware. Also connect a scope and a display block to the output of the ***Repeating Sequence*** block through a Rate transition block to monitor the value while it is running on the Arduino board.
7. Click on the Simulation button on the model. Now MATLAB® and Simulink® will generate code from this model; compile it and download the executable into the Arduino Mega hardware. If all the connections are proper after the download is completed, you can see the Motor starts to run an initial PWM command of 25.5 and every 50 s the motor speed will increase. If you get any errors during the simulation build process, most probably it will be because the

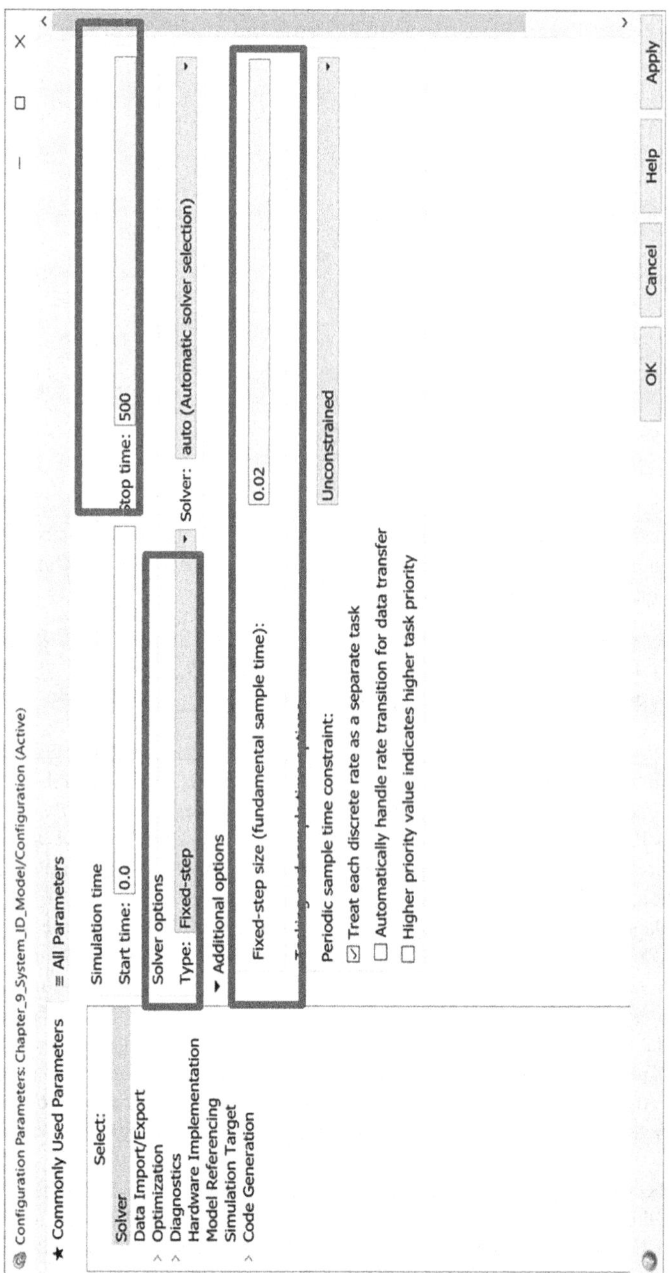

Figure 9.16 System ID Model solver settings.

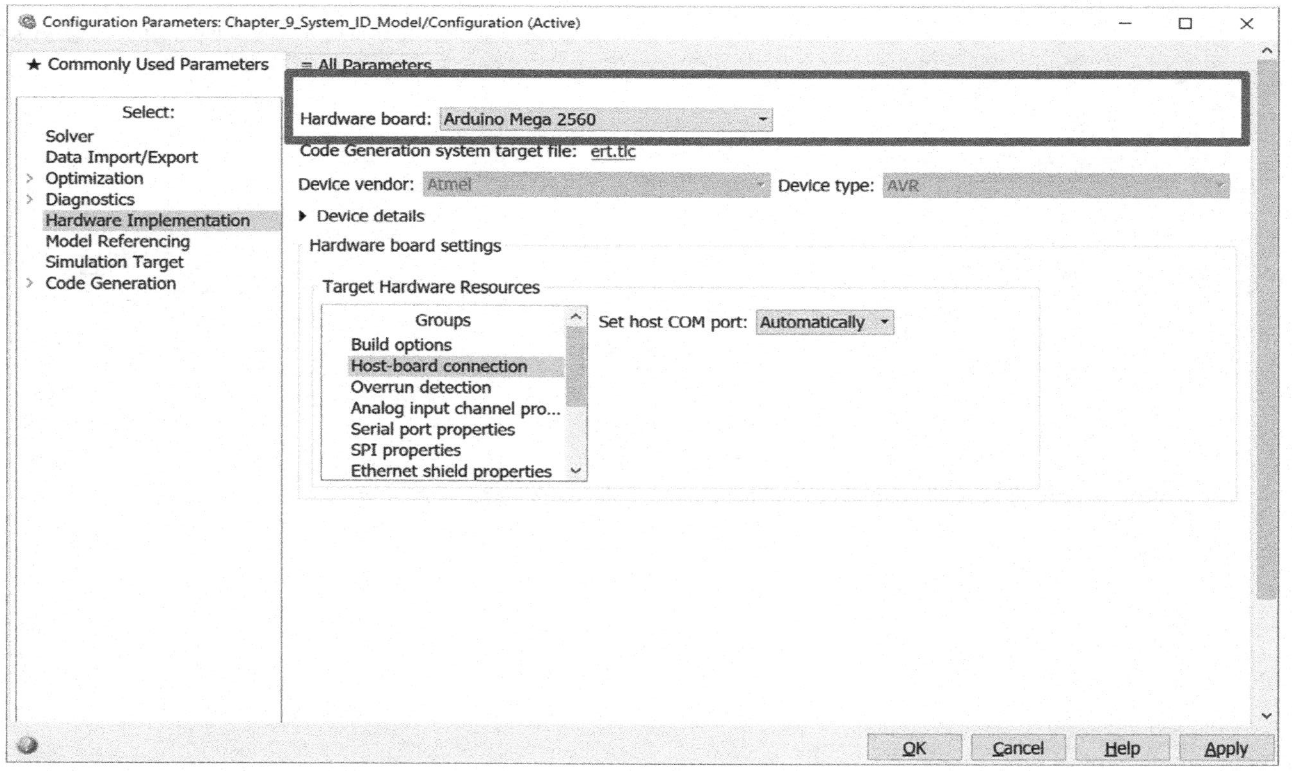

Figure 9.17 Hardware board selection.

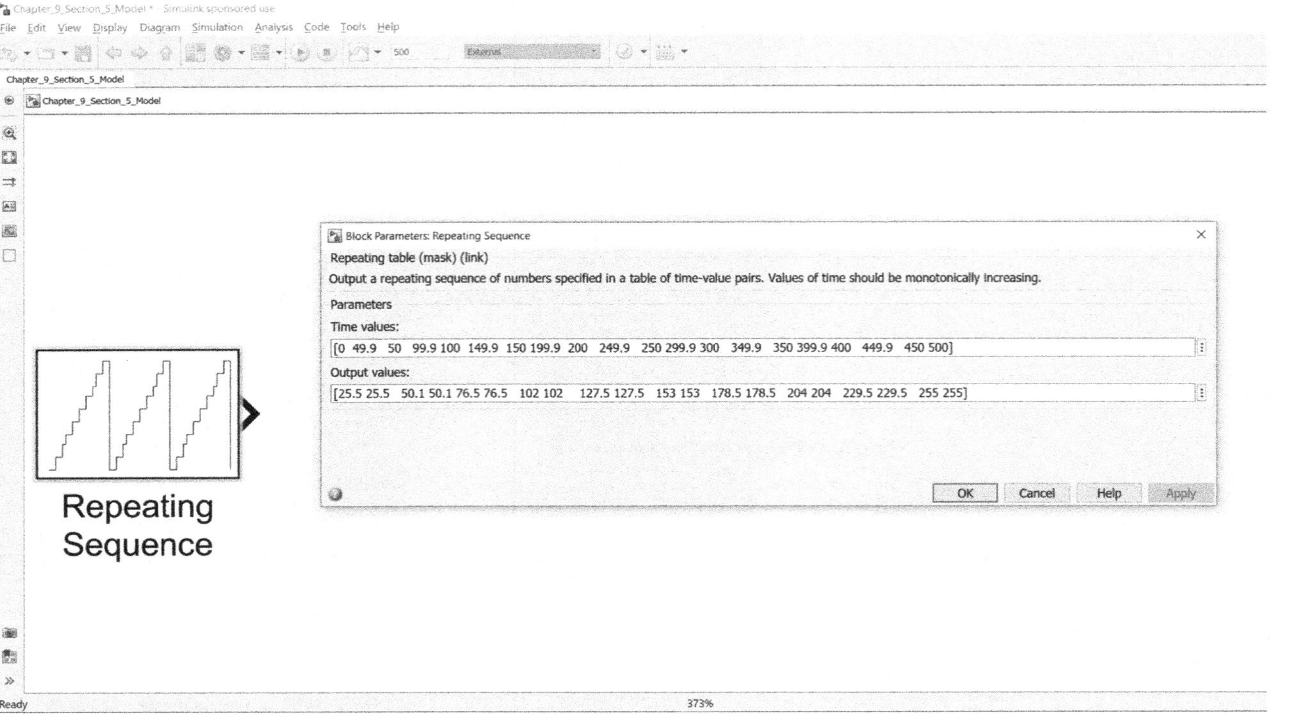

Figure 9.18 System ID input values in Repeating Sequence block.

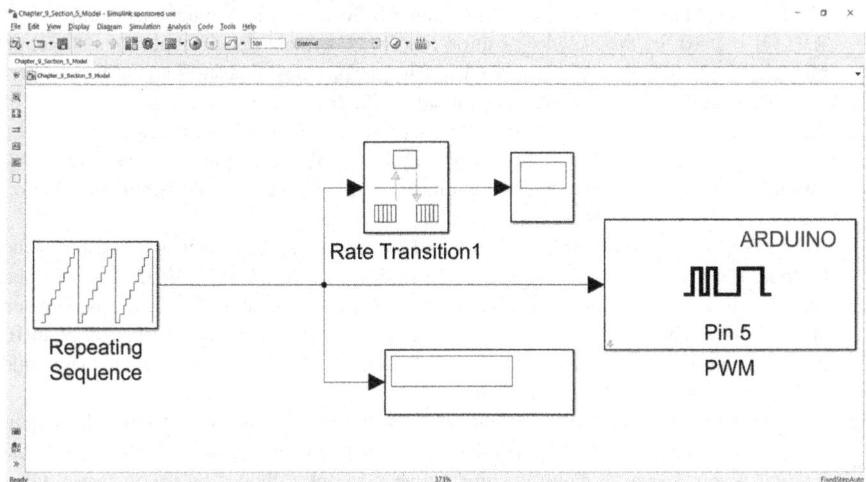

Figure 9.19 Configuring the PWM block on Arduino digital pin 5.

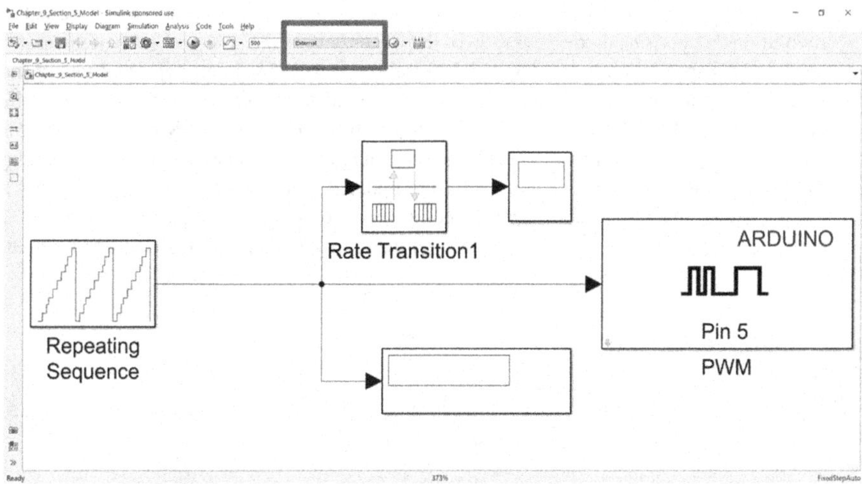

Figure 9.20 Selecting external mode simulation on the model.

Arduino board is not properly connected to the computer. If the simulation successfully starts but the motor is not spinning, it probably is because the connections are not proper as recommended in Section 9.2.1.

8. Proceed further if the Step 7 is successful and the motor is running with an increasing speed after every 50 s until 500 s.
9. Now the Simulink® logic required to sense the speed of the motor needs to be added. As mentioned earlier in this chapter, the Arduino Support Package does not provide an in-built

block in the library to read the encoder output. So a custom S-function obtained from Ref. [2] is used for the encoder position reading. For more details about the S-function please refer [2]. Add the encoder block from the package downloaded from the above reference link (or copy it from the **arduino_encoder_lib.*slx*** available in the attachment folder of this chapter, which is essentially the same S-function, but copied into a new library for the use of this chapter). Add a scope and display blocks to the output of the encoder block through a rate transition block as shown in Fig. 9.21 and click on the ***Simulation*** button again.

10. Now when the simulation runs, the Scope and Display blocks connected to the output of the Encoder block will show the encoder count readings as shown in Fig. 9.22. But it can be seen that the encoder counts starts going up from zero, but after about 13 s, it reaches a value and rolls over. This is because the S-function uses int16 as the data type for the output, int16 can only hold values between -2^{15} (−32,768) and 2^{15} (32,768) (15 bits for the magnitude and 1 bit is used to hold the sign of the value).

11. The speed of the motor will be calculated from the Encoder counts. The encoder output indicates the motor's position. The motor speed can be approximated over a specific time interval as the change in motor position between samples divided by the change in time. This will give the average speed of the motor over the time interval. The highlighted logic in Fig. 9.23 first takes a difference between the encoder positions between two samples using the ***Unit Delay*** block and divides the output by the sample time of the model which is 0.02 s to give the encoder counts/seconds. These counts/seconds need to be converted to revolutions/seconds using the gear ratio of the motor being used. The gear ratio of the motor used for this application is 1/1856, which corresponds to 1856 counts per revolution of the gearbox's output shaft. So in order to convert the counts/seconds to revolutions/seconds, the count is divided by the 1856 to give the motor speed in revolutions/seconds. Then it is multiplied by 60 to convert the revolutions/seconds to revolutions/minutes. After adding the logic, hit the Simulation again to run the updated logic on the Hardware. Also add a Scope block to monitor and log the motor speed output. Fig. 9.24 shows the PWM commanded by the Arduino Mega board and the corresponding motor speed RPM output.

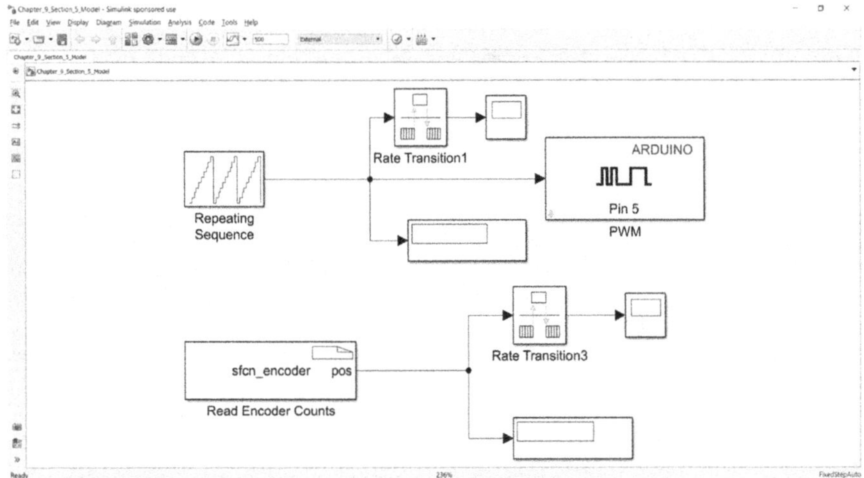

Figure 9.21 Adding the Encoder Read S-function.

Figure 9.22 Output of Encoder block.

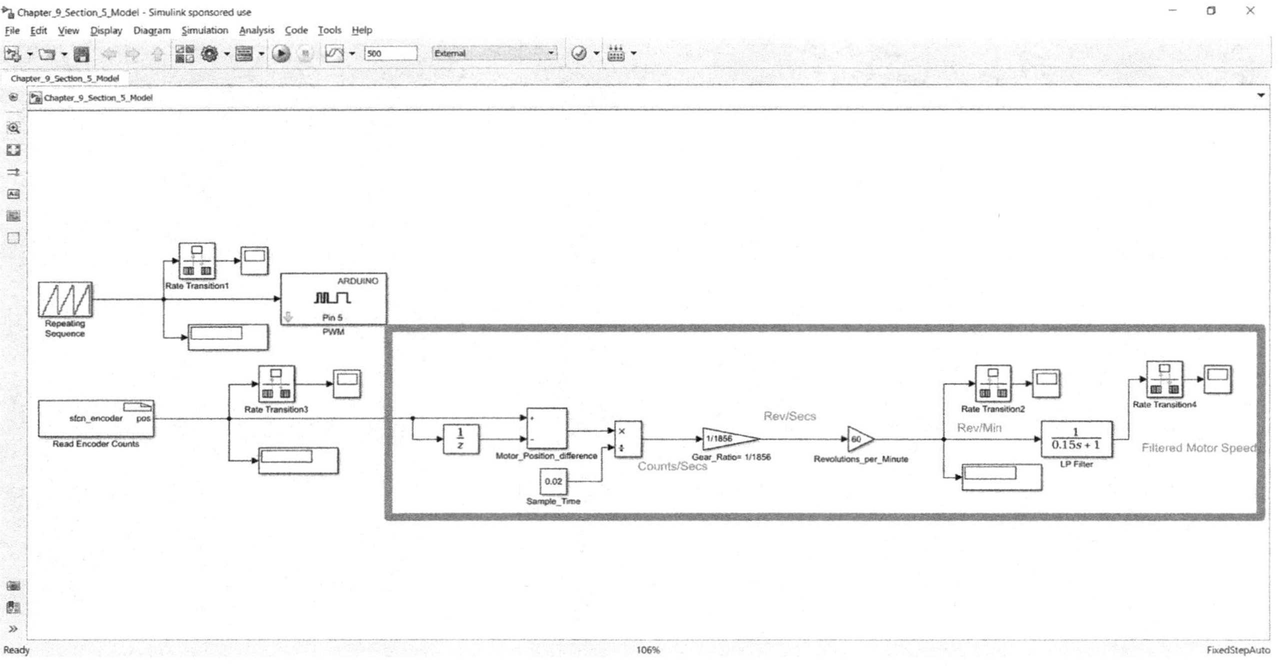

Figure 9.23 Motor speed calculation from Encoder Counts.

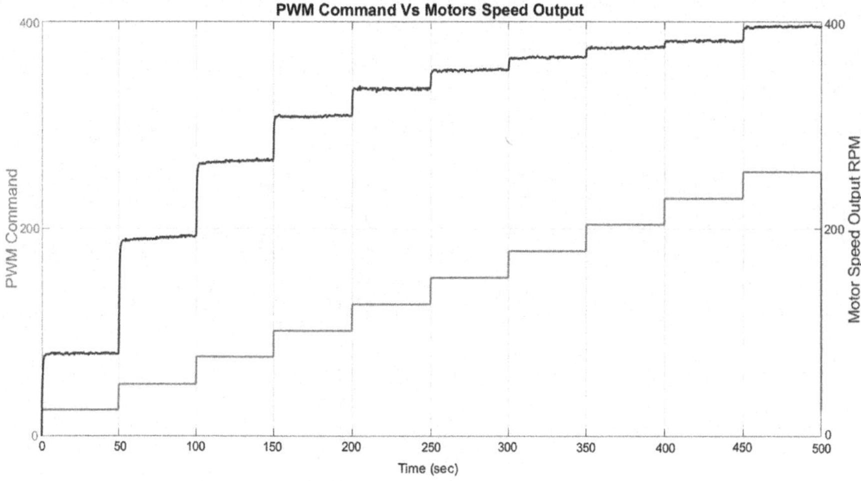

Figure 9.24 PWM Input Command value versus Motor Speed RPM.

12. On a closer zoomed in look of the sensed motor speed signal, even for the fixed steady PWM input command, there are some variations in the speed. This is attributed for noises in the sensing mechanism. It is recommended to use a filter to smooth out any noises in the measurements before the sensed value is used for controls purpose. A filtering was not really required in the previous chapters while the data for system identification are collected because only simulation models were used in those chapters and the measurement noise was not really modeled there. A simple first order filter logic using the Simulink ***Transfer Function*** block is added to filter the motor speed as shown in Fig. 9.25. A filter will cause a lag in the filtered value. The filter constant needs to be carefully chosen such that there is a balance between the output to input signal lag and reduction of noise. A filter constant of 0.15 s is used in this example. Fig. 9.26 shows the response of the filter.

9.3.2 DC motor nonlinearity analysis using collected data

In this section, the data collected from Section 9.3.1 are analyzed to assess nonlinearity of the DC motor. This analysis will enable us later in this chapter to do parameter estimation for the Simscape® model for different linear operating points of the DC Motor. The steady-state points of Fig. 9.24 have been captured in Table 9.2. These points have been plotted in Fig. 9.27. Based on Fig. 9.27, we observe three regions of slope changes (marked as red (gray in print version), green (light gray in print version), and blue (dark gray in print version) vertical lines). Parameter estimation of the DC Motor Simscape® model will be done later in this chapter based these linear regions which will then be used to deploy into the Cloud as Digital Twin. Run ***Chapter_9_Section_3_2_Script.m*** to replot Fig. 9.27.

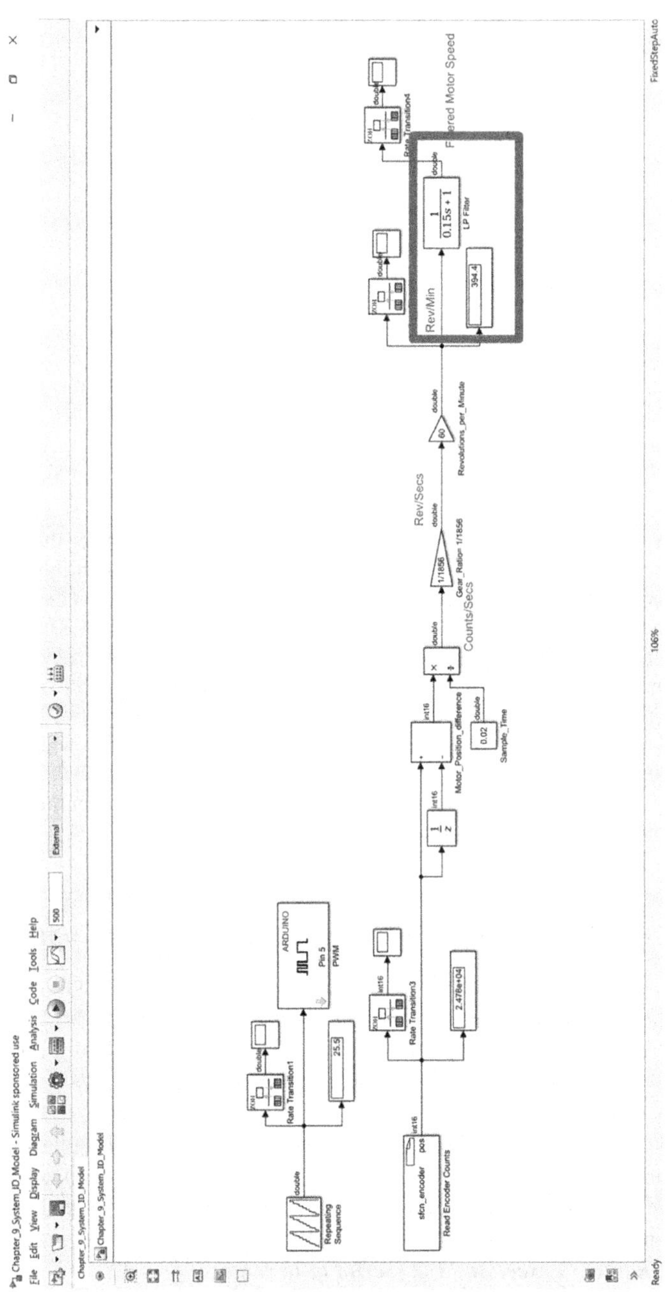

Figure 9.25 Filtering Motor Speed.

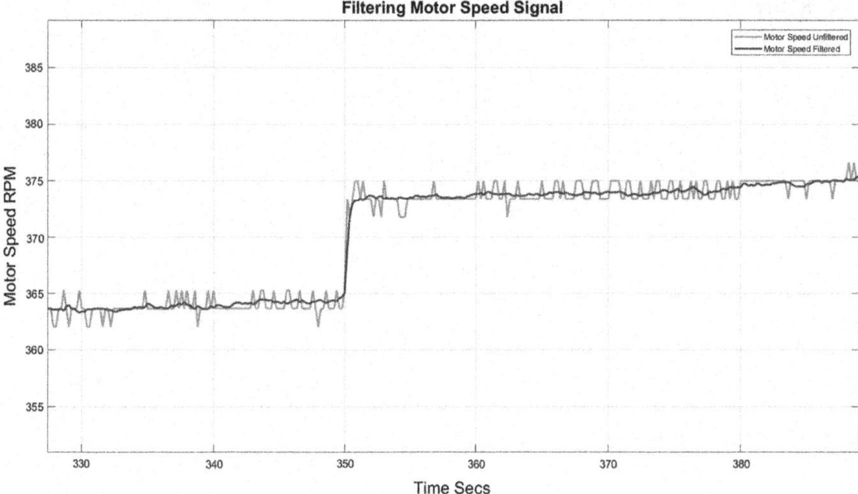

Figure 9.26 Comparing Unfiltered and Filtered Motor Speed.

Table 9.2 Steady-state operating points.

PWM command SS value	25.5	50.1	76.5	102	127.5	153	178.5	204	229.5	255
Motor speed SS value	78.1	185.9	260.2	305.1	331.5	351	363.3	374.1	382.7	394.4

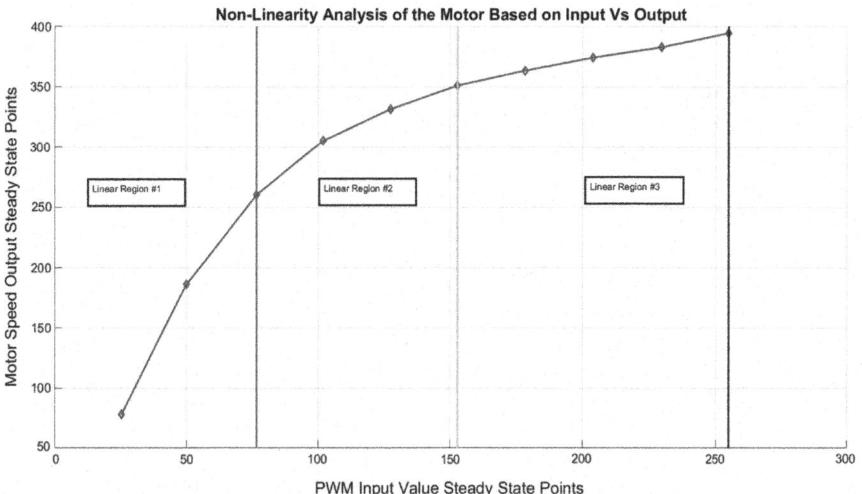

Figure 9.27 Non-Linearity Assessment of the Motor System.

Chapter_9_Section_3_2_Script.m

```
%Book Title: Digital Twin Development and Deployment On Cloud Using Matlab
%SimScape
%Chapter: 9
%Section: 3.2
%Authors: Nassim Khaled and Bibin Pattel
%Last Modified: 09/10/2019
close all
fig=figure;
hax=axes;
hold on
input_pwm_steady_state_points = [25.5 50.  176.5 102 127.5 153 178.5
204 229.5 255];
output_motor_speed_points     = [78.1 185.9 260.2 305.1 331.5 351 363.3
374.1 382.7 394.4];
plot(input_pwm_steady_state_points,output_motor_speed_points,
'linewidth',2,'Marker','diamond');
line([76.5 76.5],get(hax,'YLim'),'Color',[1 0 0],'linewidth',2)
line([153 153],get(hax,'YLim'),'Color',[0 1 0],'linewidth',2)
line([255 255],get(hax,'YLim'),'Color',[0 0 1],'linewidth',2)
grid on
xlabel('PWM Input Value Steady State Points','FontSize',18);
ylabel('Motor Speed Output Steady State Points','FontSize',18)
set(gcf,'color',[1 1 1]);
title('Non-Linearity  Analysis  of  the  Motor  Based  on  Input  Vs
Output','FontSize',18)
annotation(fig,'textbox',
    [0.1635 0.577991452991452 0.0948333333333331 0.0512820512820514],
    'String',{'Linear Region #1'},
    'LineWidth',2,
    'FitBoxToText','off');
annotation(fig,'textbox',
    [0.389802083333333      0.579594017094016      0.0948333333333332
0.0512820512820514],
    'String',{'Linear Region #2'},
    'LineWidth',2,
    'FitBoxToText','off');
annotation(fig,'textbox',
    [0.649697916666666      0.582799145299144      0.0948333333333332
0.0512820512820514],
    'String',{'Linear Region #3'},
    'LineWidth',2,
    'FitBoxToText','off');
```

=====================================

9.3.3 Open-Loop Feedforward and Closed-Loop PID controller design and deployment for DC Motor Speed Control

In this section, we will design and deploy an Open-Loop Feedforward and a Closed-Loop Feedback controllers in Simulink® to control a time-varying reference speed target of the DC Motor. First, we will develop an Open-Loop Feed Forward controller that does not have any feedback to control the output which is the speed of the motor. This controller commands the control signal in a predefined way without knowing how the output signal behaves. This predefined behavior can be derived from a physical model of the system or steady-state data collected from the real system. In our case, we have the data collected from Section 9.3.1 and the steady-state points are identified in Section 9.3.2. A top-level view of the simulation model is shown in Fig. 9.28.

The main components of the model are

- Motor speed reference system
- Motor speed feedback system
- Controller system
- PWM actuation system

The reference system generates a reference signal for the motor speed. The reference is constantly varying and is designed to push the controller across multiple regions of the operating space. Fig. 9.29 shows a time series plot for the reference signal. Figs. 9.30 and 9.31 show the Simulink® subsystem and the logic for the reference generation.

Figs. 9.32 and 9.33 show the Simulink® logic for the motor speed sensing system. The same logic developed in Section 9.3.1 for the motor speed sensing for system identification data collection is used here as well.

The Feedforward controller top-level subsystem is shown in Fig. 9.34. This block takes the motor speed reference and motor speed feedback as input and outputs the PWM command value. A look inside the subsystem as shown in Fig. 9.35 has the speed reference going to a lookup table derived in Table 9.2 to output steady-state PWM command. The PWM command is saturated between 0 and 255 before applying to Arduino PWM pin. Rate transition blocks are inserted at the input and output.

Figs. 9.36 and 9.37 show the Simulink® logic for command the controller PWM signal to the Arduino hardware.

The model configuration settings can be kept same as that of the System ID model shown in Fig. 9.16. Now the model is ready to deploy to the Arduino hardware. Click on the Simulation button on the model. Now MATLAB® and Simulink® will generate code from this model, compile it, and download the executable into the Arduino Mega hardware. Using the scopes added to the model, we can observe the motor speed command, actual motor speed, and also the PWM command coming out of the Feedforward controller. The Feedforward controller speed tracking and PWM command are shown in Figs. 9.38 and 9.39. It can be observed that there is some speed tracking error which the Feedforward controller is not able to compensate for because it doesn't look at the feedback motor speed. This is the reason why we need a feedback controller. Further in this section, we will be designing a PID feedback controller for the same control objective.

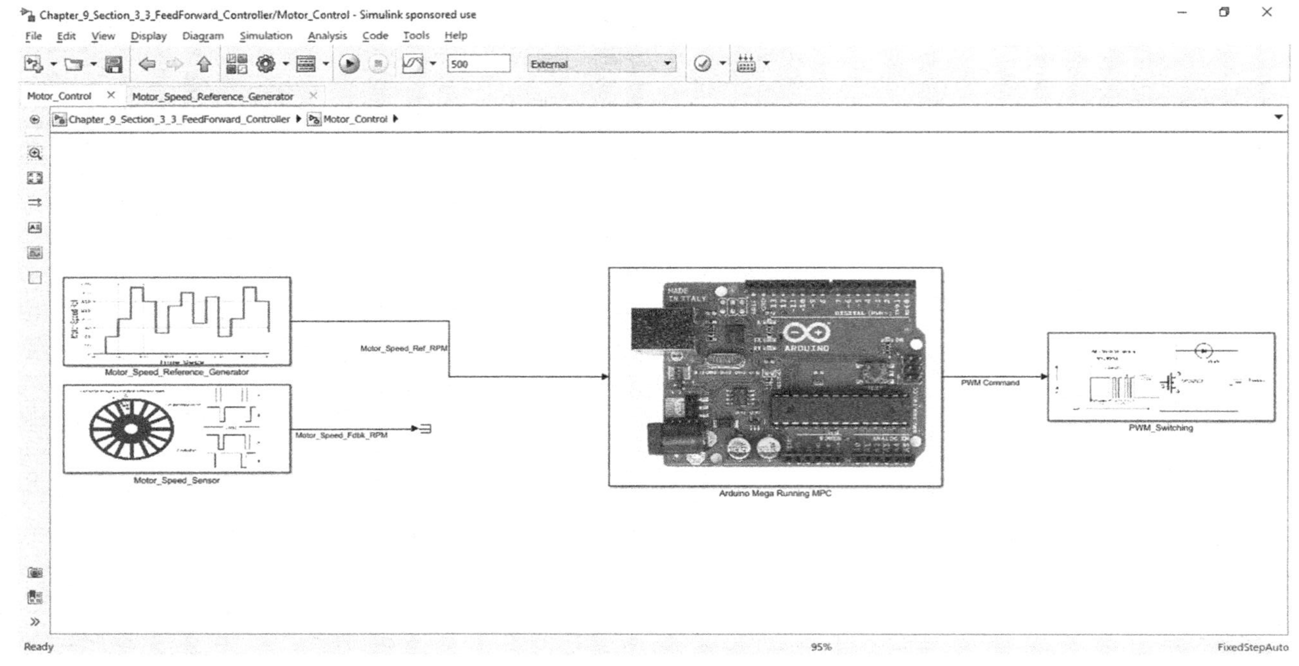

Figure 9.28 Top-level Simulink Model of Real Time Motor Speed Controller using Open Loop Feedforward Controller.

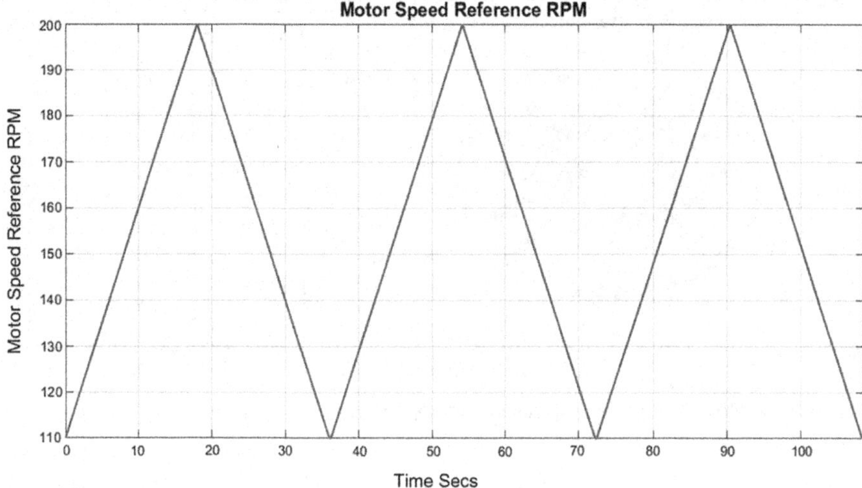

Figure 9.29 Motor Speed Reference for MPC controller.

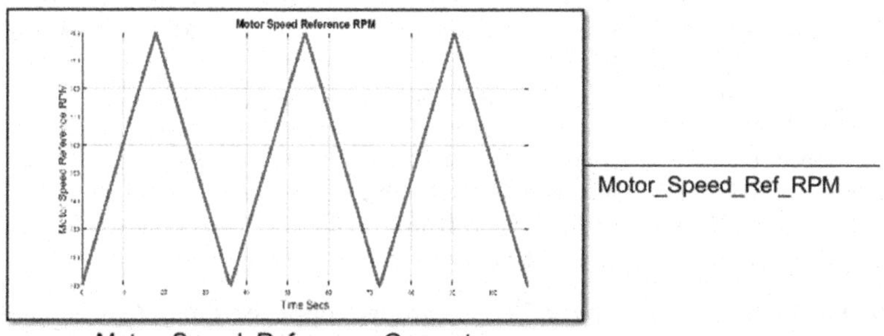

Figure 9.30 Reference Generator Subsystem top view.

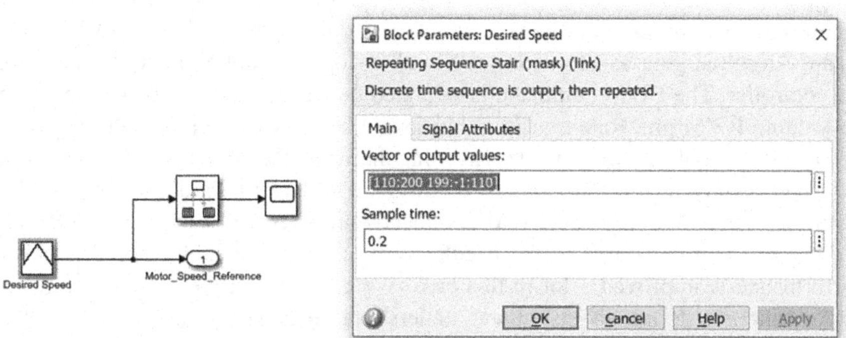

Figure 9.31 Reference generation using Simulink Repeating sequence block.

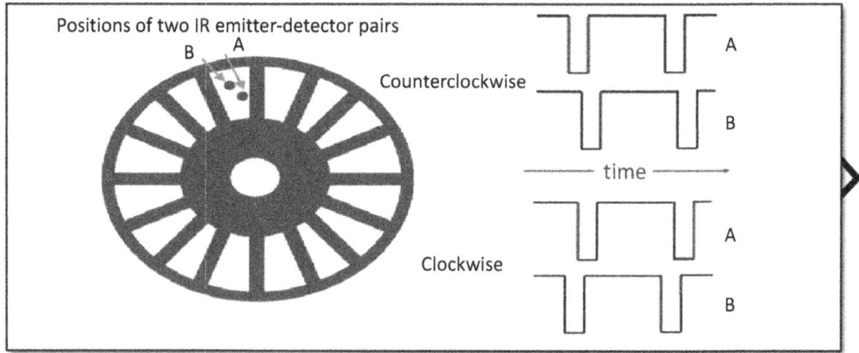

Figure 9.32 Motor Speed Sensing Subsystem top view.

A top-level view of the simulation model with Feedforward + Feedback controller is shown in Fig. 9.40. The main difference in the overall controller structure compared to the Feedforward setup is that the sensed Motor speed is taken into consideration by the controller for determining the PWM command. In this setup, the feedforward controller guarantees a faster response to the reference change, and feedback loop provides better tracking and robustness.

Similar to Feedforward controller, the main components of the model are

- Motor speed reference system (same as feedforward)
- Motor speed feedback system (same as feedforward)
- Feedforward + PID feedback controller system
- PWM actuation system (same as feedforward)

The Feedforward + Feedback PID controller top-level subsystem is shown in Fig. 9.41. This block takes the motor speed reference and actual-sensed motor speed and outputs the PWM command value. A look inside the subsystem as shown in Fig. 9.42 has the Feedforward controller developed and tested earlier in this section, an error calculation logic for the motor speed tracking, and applying proportional, integral, and derivative gains for the error, integral of error and derivative of error accordingly and summing all these terms together to apply to the final PWM command. An initial gain as shown in the Simulink logic is applied to start evaluating the controller. The PWM command is saturated between 0 and 255 before applying to Arduino PWM pin. Rate transition blocks are inserted at the input and output.

With the rest of the components remaining the same, the model is now ready to be built and deployed into the Arduino hardware. The Feedforward + Feedback PID controller speed tracking and PWM commands are shown in Figs. 9.43 and 9.44. We can see that compared to the Feedforward-only controller, the speed tracking performance is improved a lot in this case. We can achieve the desired controller performance in terms of rise time, undershoot, overshoot, settling time, etc., by systematically tuning the PID controller. But since the focus of this book and chapter is not the controller design, we will not be spending more time to return the controller.

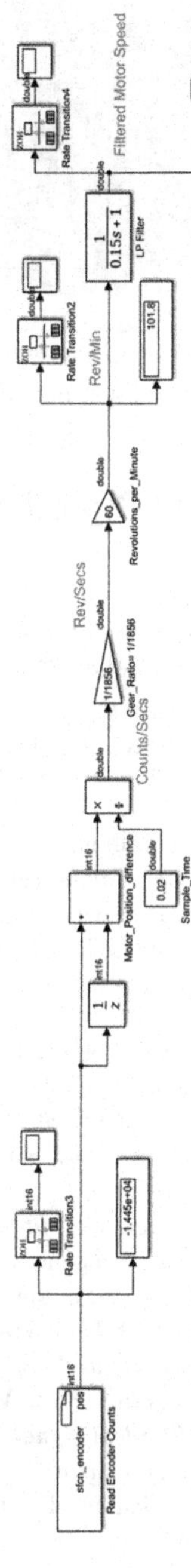

Figure 9.33 Motor Speed Sensing logic.

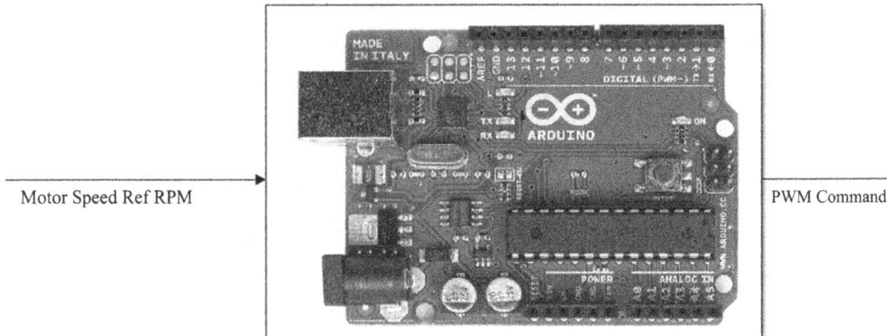

Arduino Mega running MPC

Figure 9.34 Motor Feedforward Controller Subsystem top view.

9.4 Developing Simscape™ Digital Twin model for the DC motor

In this section, we will develop a physical plant model of the DC motor system with MATLAB®, Simulink, and Simscape™ toolbox. The input out structure of the model will be same as that of the Motor hardware, where it takes the PWM command as the input and Motor speed is the output. Follow the below steps to replicate the Simulink plant model using Simscape™:

1. Open a new Simulink model and save it with name that we want to use for the model.
2. Go to ***Model*** >> ***Simulation*** >> ***Model Configuration Parameters*** >> ***Solver*** setting page and make the highlighted changes as shown in Fig. 9.45 DC Motor Simscape™ Plant Model Configuration Settings. ***Solver Type*** is changed to ***Fixed-step*** because all the blocks and subsystems in this model are expected to run at a certain discrete rate 0.1 s specified in the **Fixed-step size** field.
3. On the Simulink model add an input port for the PWM input for the DC Motor. The PWM input value that we applied to the Arduino Pin earlier in this chapter ranges from 0 to 255, but for the Simscape™ model blocks we want to scale that PWM signal to the range of 0–1, so a gain block with a scaling of 1/255 is added as shown in Fig. 9.46, Scaling PWM Input to the range 0–1.
4. Add a ***Controlled PWM Voltage*** block from ***elec_lib/Actuators & Drivers/Drivers*** model provided by the Simscape™ toolbox to our Simulink model as shown in shown in Fig. 9.47 Controlled PWM Voltage and H-Bridge Block in SimScape™.
5. In order to feed the PWM input to the controlled PWM Voltage block, we need to change the block setting from ***Electrical Input Ports*** to ***Physical Input Port.*** This can be done by right clicking on the newly added ***Controlled PWM Voltage*** block and select the option ***Simscape™*** >> ***Block Choices*** >> ***Physical signal input*** as shown in Fig. 9.48. This will now change the block layout from two inputs to one input.
6. Connect the output of the PWM Scaling Gain block to the ***Controlled PWM Voltage*** block using a ***Simulink to PS Converter*** block.

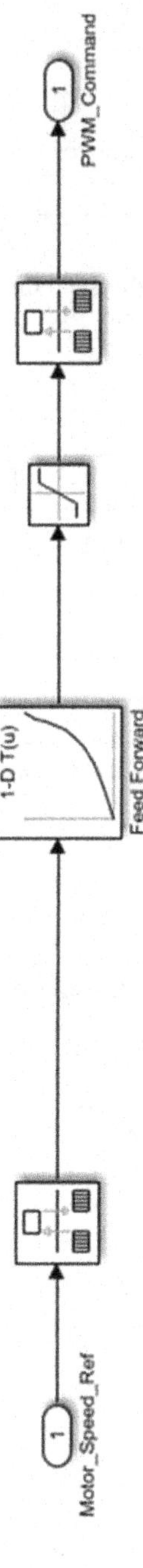

Figure 9.35 Motor Feedforward controller logic.

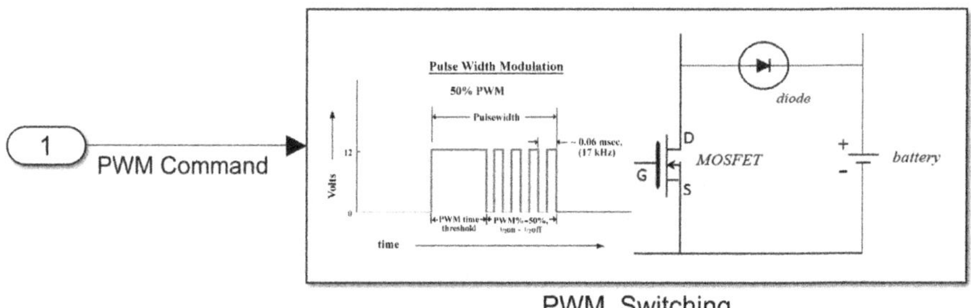

Figure 9.36 Arduino PWM Command Subsystem top view.

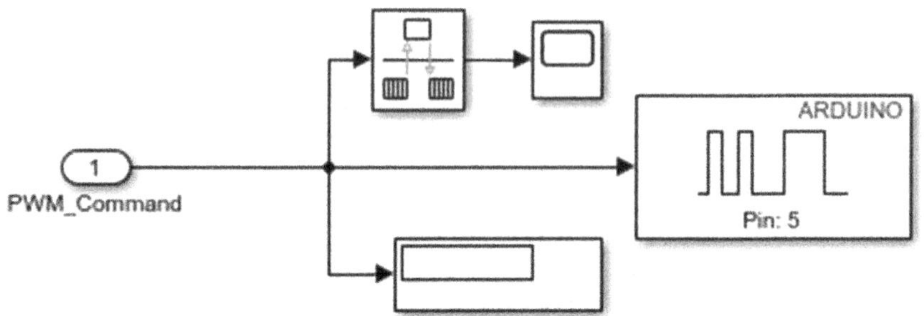

Figure 9.37 Commanding PWM to PIN 5.

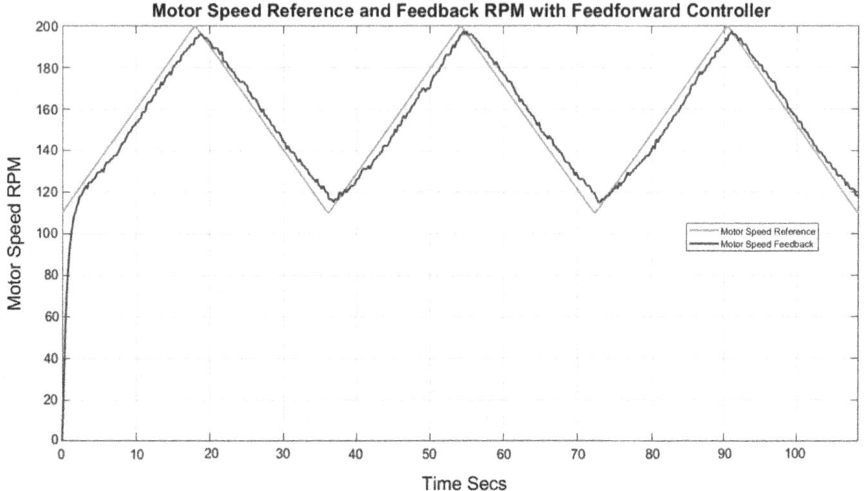

Figure 9.38 Feedforward Controller Motor Speed tracking.

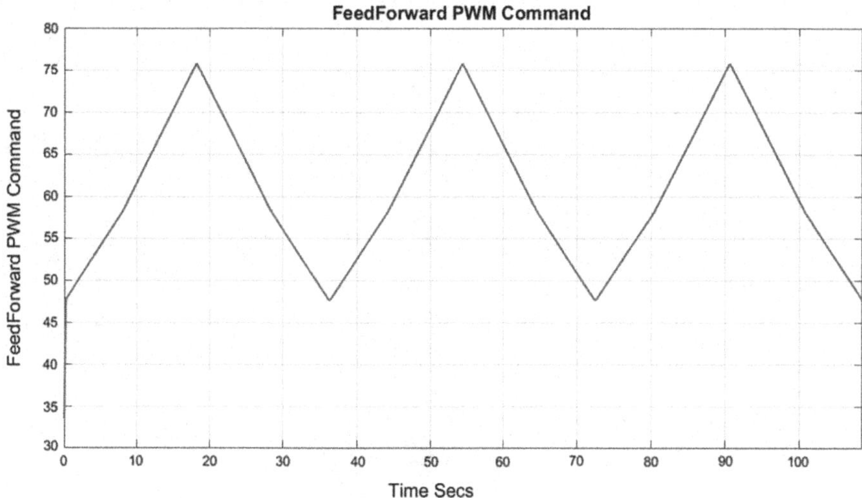

Figure 9.39 Feedforward Controller PWM Command.

7. Double click on the Controlled PWM Voltage block, on the PWM tab, select the ***Simulation mode*** as ***Averaged***, and set the ***PWM frequency*** to 4000 Hz, which is the frequency of the signal generated from Arduino hardware. On the ***Input Scaling*** tab, enter the input value for the 0% and 100% duty cycle to be 0 and 1, respectively. And on the ***Output Voltage*** tab, enter the ***Output voltage amplitude*** to be 5 V. Check Figs. 9.49–9.51 for more details.
8. Add an ***H-Bridge*** block as shown in Fig. 9.47 Controlled PWM Voltage and H-Bridge Block in SimScape™ from ***elec_lib/Actuators & Drivers/Drivers*** model provided by the Simscape™ toolbox to our Simulink model and make the connections as shown in Fig. 9.52.
9. Double click on the newly added H-Bridge block and make the setting changes as shown in Figs. 9.53–9.55.
10. Add an ***DC Motor*** block as shown in Fig. 9.56 from ***elec_lib/Actuators & Drivers/Rotational Actuators*** model provided by the Simscape™ toolbox to our Simulink model and make the connections as shown in Fig. 9.57. We will tune the parameters of this DC Motor block based on the data collected from real hardware later.
11. Double click on the newly added ***DC Motor*** block and make the setting changes as shown in Figs. 9.58 and 9.59. Basically these are tunable electrical and mechanical parameter properties of the DC Motor.
12. Add an ***Ideal Rotational Motion Sensor*** block as shown in Fig. 9.60 from ***fl_lib/Mechanical/Mechanical Sensors*** model provided by the Simscape™ toolbox to our Simulink model and make the connections as shown in Fig. 9.61. This motion sensor block is used to measure the rotational speed of the DC motor in the model. Connect an output port to the ***W*** angular speed output of the ***Ideal Rotational Motion Sensor*** block.
13. Add ***Solver Configuration, Electrical Reference, and Mechanical Rotational Reference*** blocks as shown in Figs. 9.62–9.64 provided, respectively, by ***nesl_utility, fl_lib/Electrical/Electrical Elements, and fl_lib/Mechanical/Rotational Elements*** models in the Simscape™ toolbox to our Simulink model and make the connections as shown in

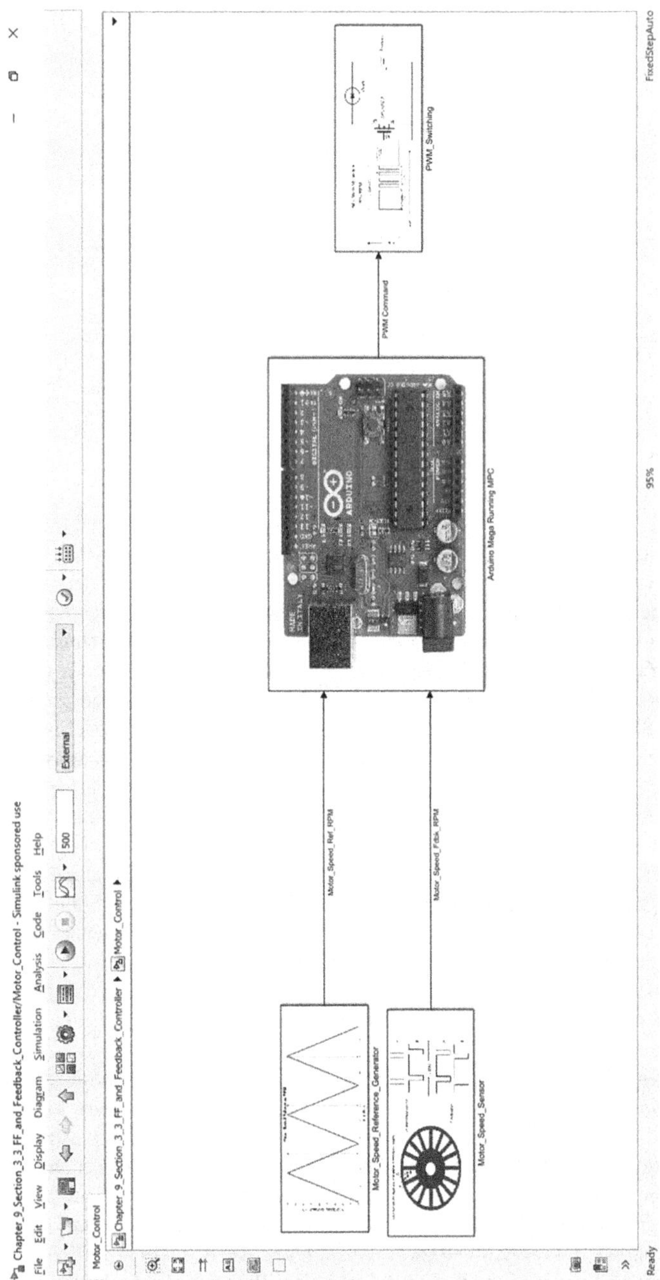

Figure 9.40 Top Level Simulink Model of Real Time Motor Speed Controller using Feedforward + Feedback PID Controller.

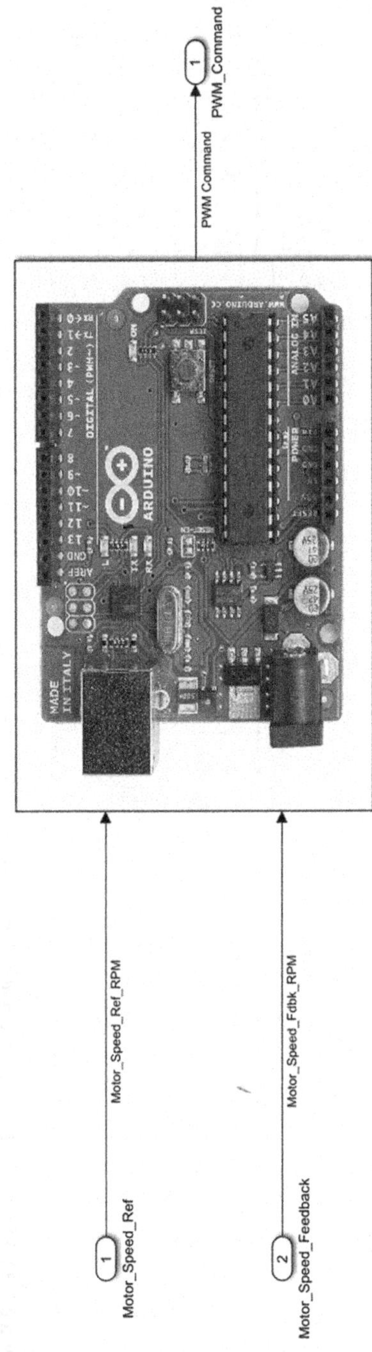

Figure 9.41 Motor Feedforward + Feedback PID Controller Subsystem top view.

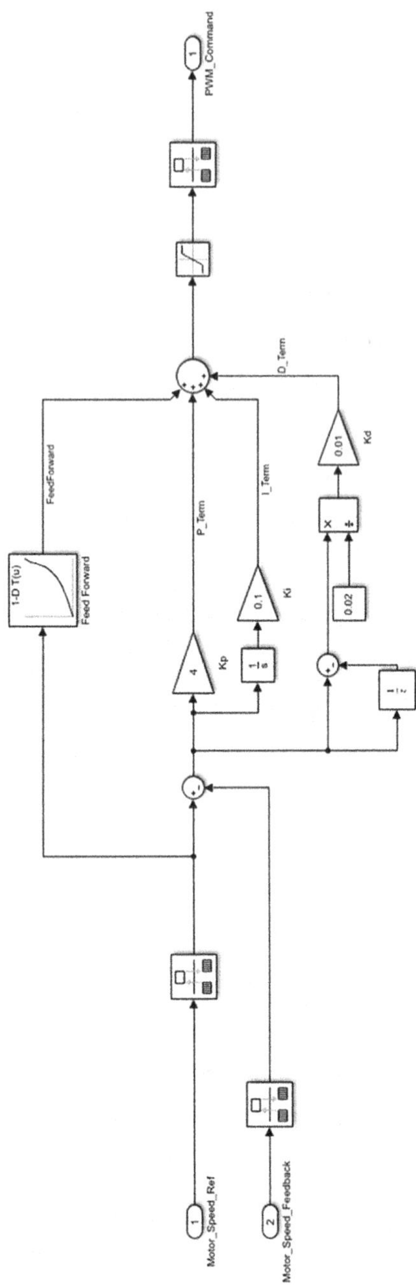

Figure 9.42 Motor Feedforward + Feedback PID Controller Subsystem logic.

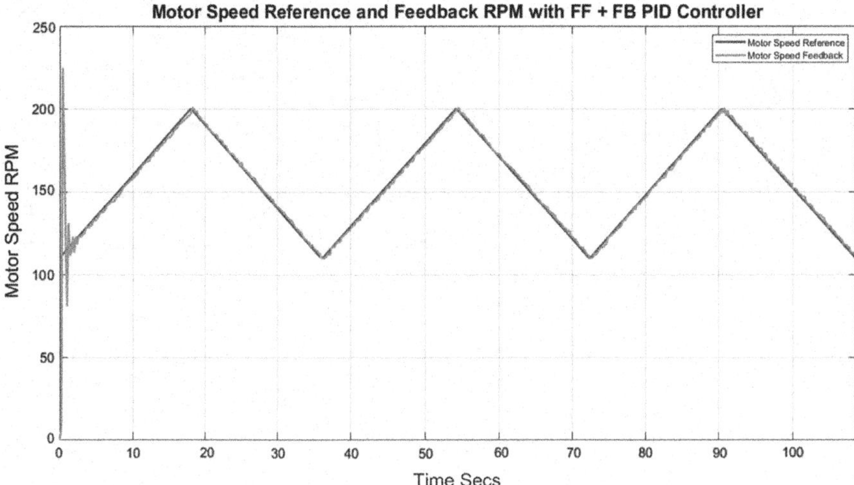

Figure 9.43 Feedforward + Feedback PID Controller Motor Speed tracking.

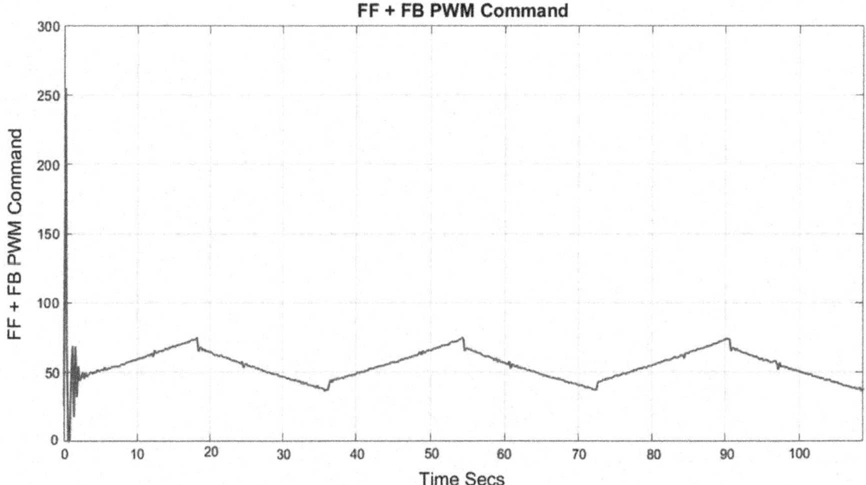

Figure 9.44 Feedforward + Feedback PID Controller PWM Command.

Fig. 9.65. Add input and output Scopes to the model to complete the model creations steps as shown on Fig. 9.66. Though the model development is completed, this model will not simulate as is because the tunable parameters for the DC Motor's electrical and mechanical properties need to be initialized in the MATLAB workspace. At this point we don't know what the actual or close to actual values of those tunable parameters are. In the next section of this chapter, we will use the model developed in this section and the DC Motor hardware data collected in the earlier section to tune these DC Motor block parameters, with which we will be able to run the Simscape™ DC Motor plant model.

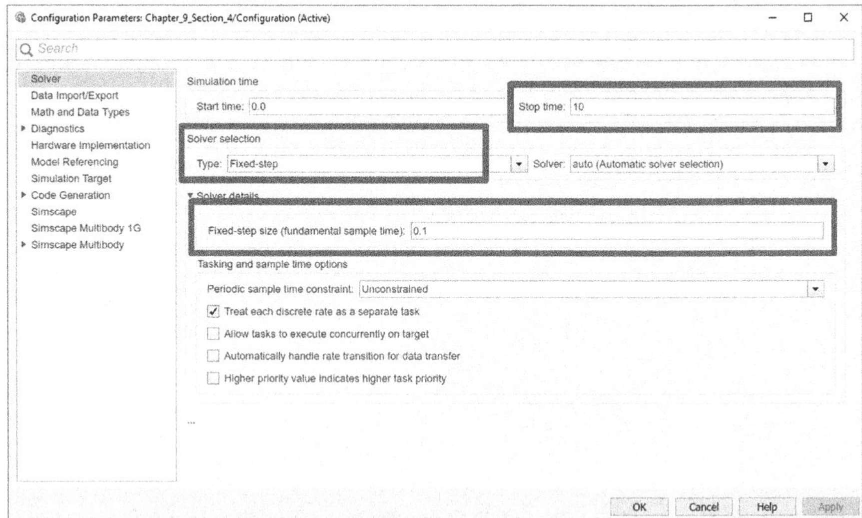

Figure 9.45 DC Motor Simscape™ Plant Model Configuration Settings.

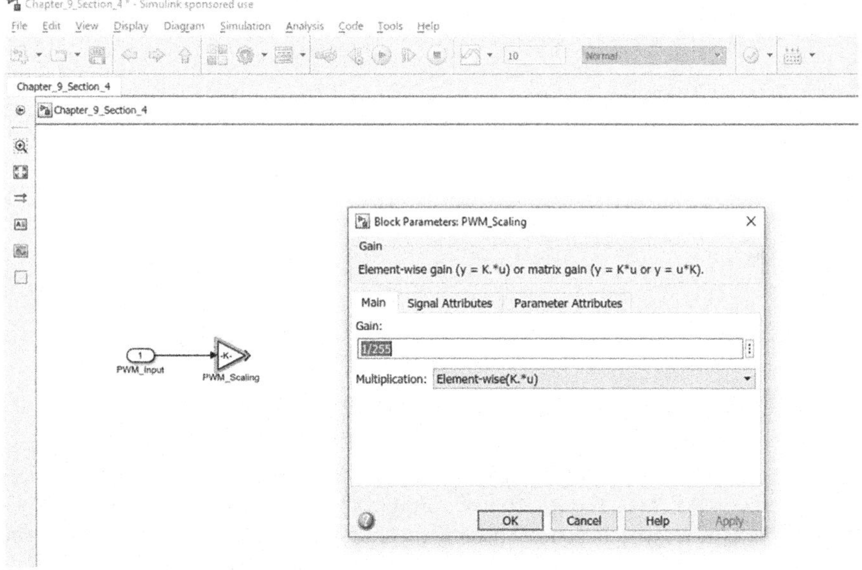

Figure 9.46 Scaling PWM input to the range 0–1.

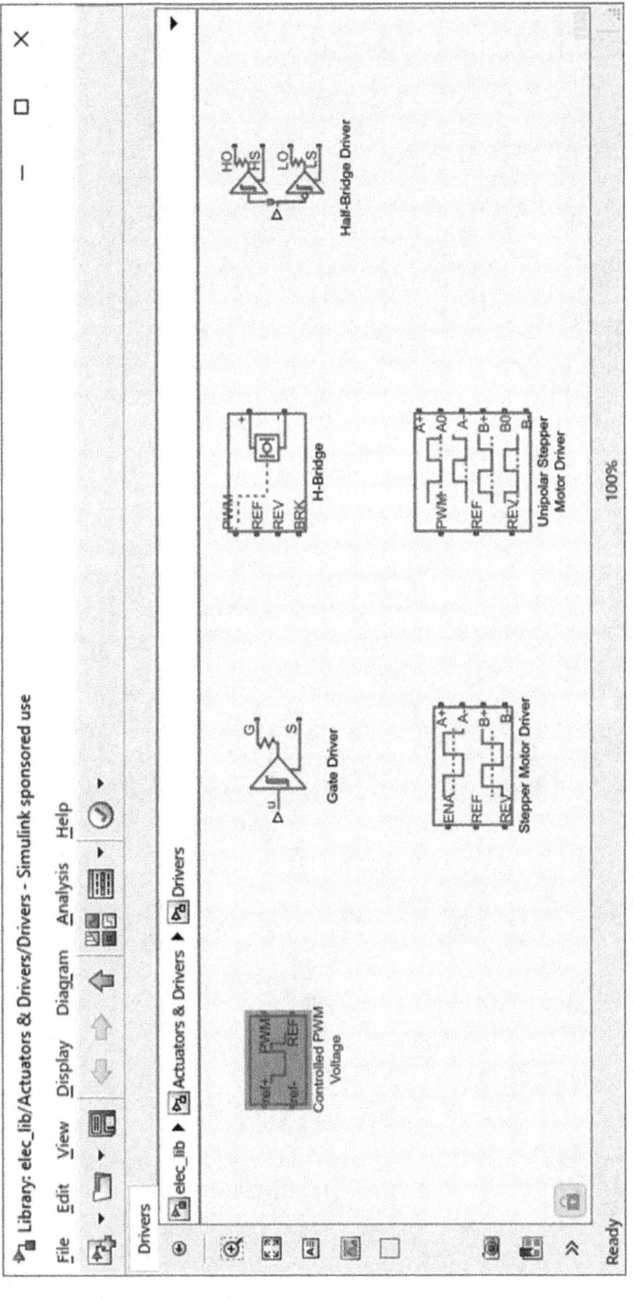

Figure 9.47 Controlled PWM Voltage and H-Bridge block in SimScape™.

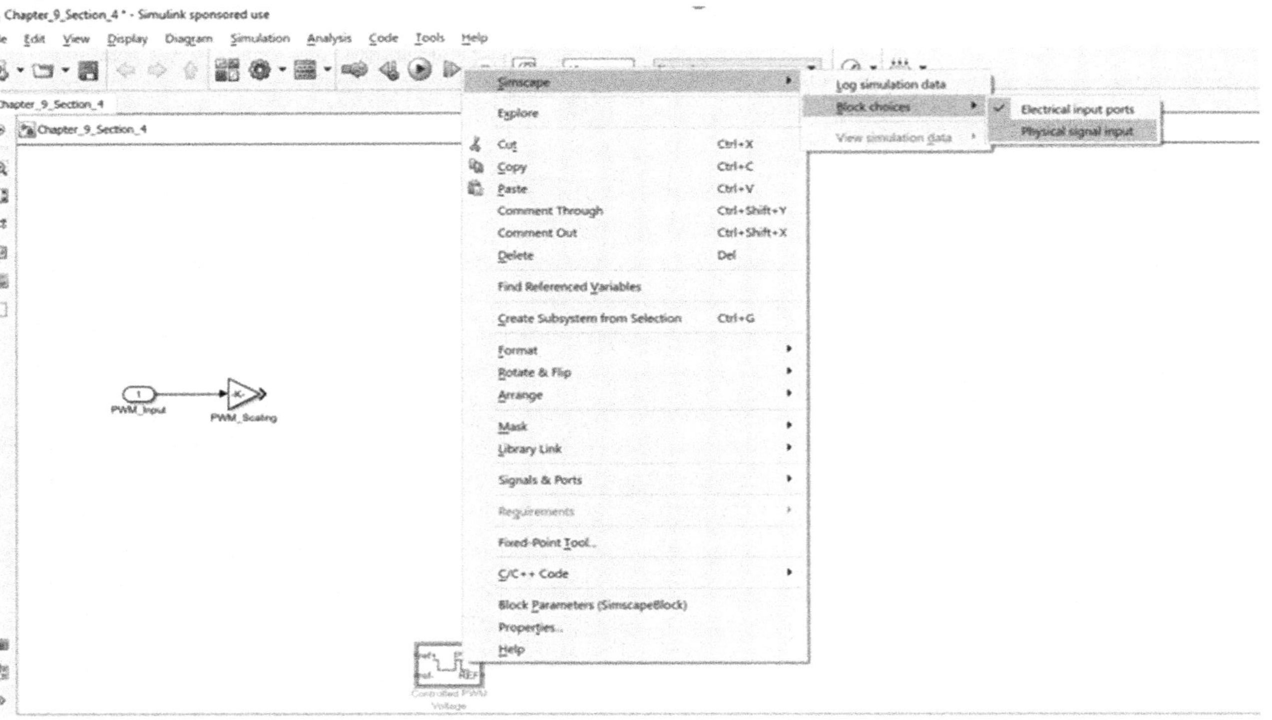

Figure 9.48 Selecting physical input property for the PWM Voltage block.

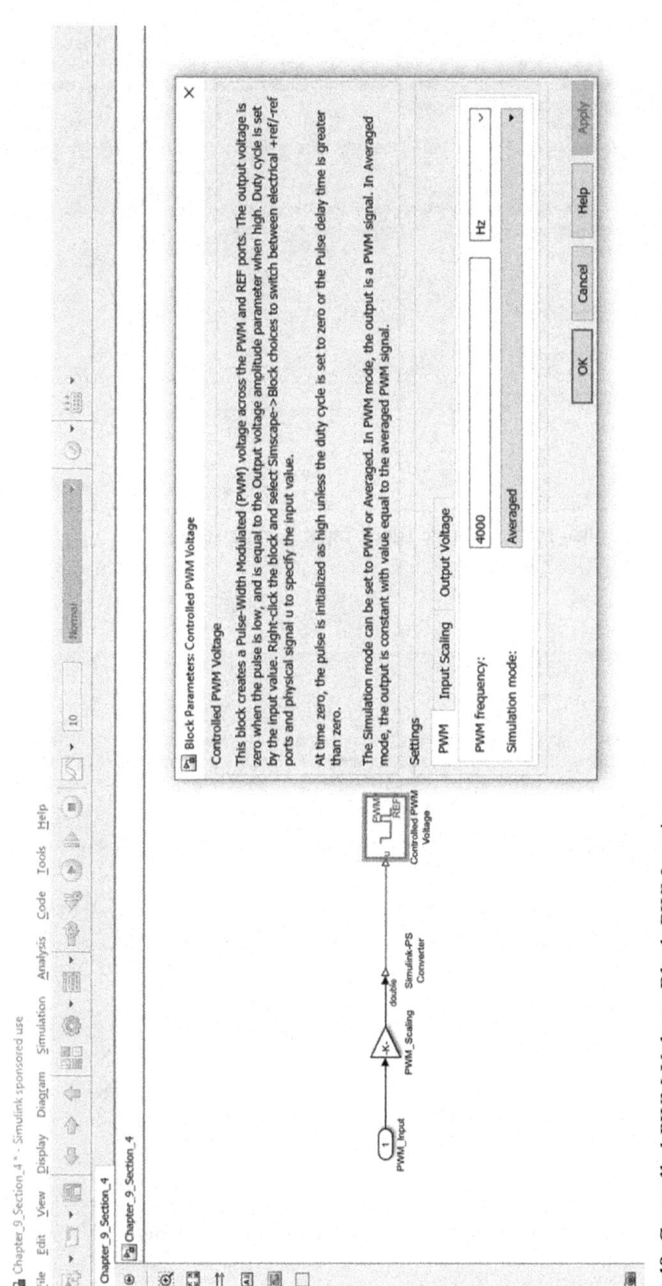

Figure 9.49 Controlled PWM Voltage Block PWM setting.

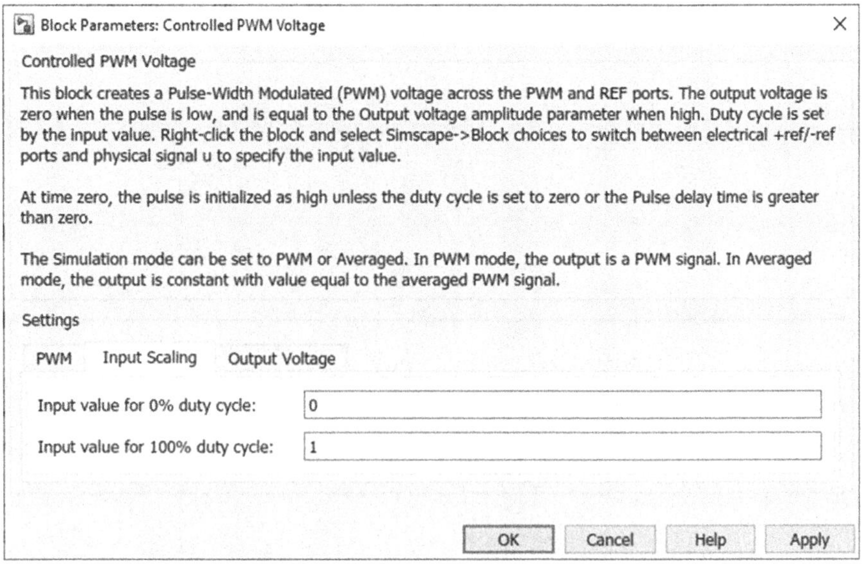

Block Parameters: Controlled PWM Voltage

Controlled PWM Voltage

This block creates a Pulse-Width Modulated (PWM) voltage across the PWM and REF ports. The output voltage is zero when the pulse is low, and is equal to the Output voltage amplitude parameter when high. Duty cycle is set by the input value. Right-click the block and select Simscape->Block choices to switch between electrical +ref/-ref ports and physical signal u to specify the input value.

At time zero, the pulse is initialized as high unless the duty cycle is set to zero or the Pulse delay time is greater than zero.

The Simulation mode can be set to PWM or Averaged. In PWM mode, the output is a PWM signal. In Averaged mode, the output is constant with value equal to the averaged PWM signal.

Settings

PWM | Input Scaling | Output Voltage

Input value for 0% duty cycle: 0

Input value for 100% duty cycle: 1

OK | Cancel | Help | Apply

Figure 9.50 Controlled PWM Voltage Block Input Scaling setting.

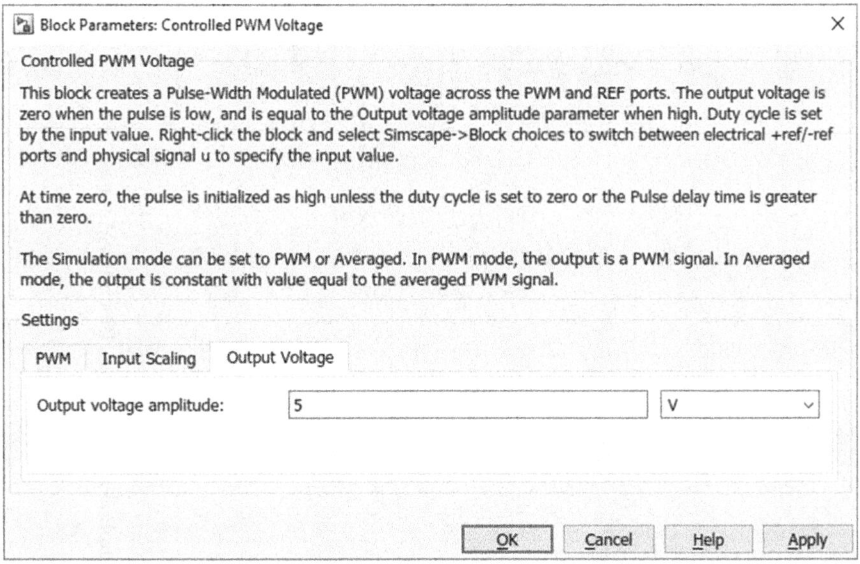

Block Parameters: Controlled PWM Voltage

Controlled PWM Voltage

This block creates a Pulse-Width Modulated (PWM) voltage across the PWM and REF ports. The output voltage is zero when the pulse is low, and is equal to the Output voltage amplitude parameter when high. Duty cycle is set by the input value. Right-click the block and select Simscape->Block choices to switch between electrical +ref/-ref ports and physical signal u to specify the input value.

At time zero, the pulse is initialized as high unless the duty cycle is set to zero or the Pulse delay time is greater than zero.

The Simulation mode can be set to PWM or Averaged. In PWM mode, the output is a PWM signal. In Averaged mode, the output is constant with value equal to the averaged PWM signal.

Settings

PWM | Input Scaling | Output Voltage

Output voltage amplitude: 5 V

OK | Cancel | Help | Apply

Figure 9.51 Controlled PWM Voltage Block Output Voltage setting.

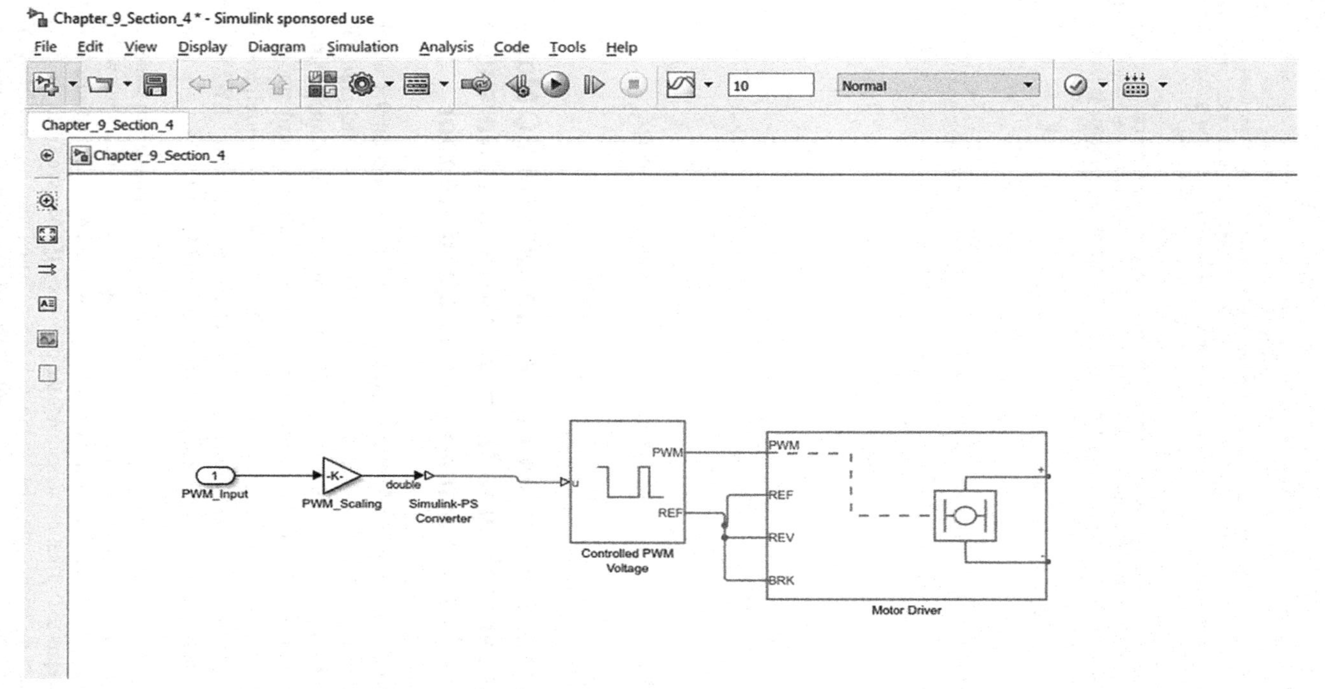

Figure 9.52 Adding H-Bridge PWM Driver block.

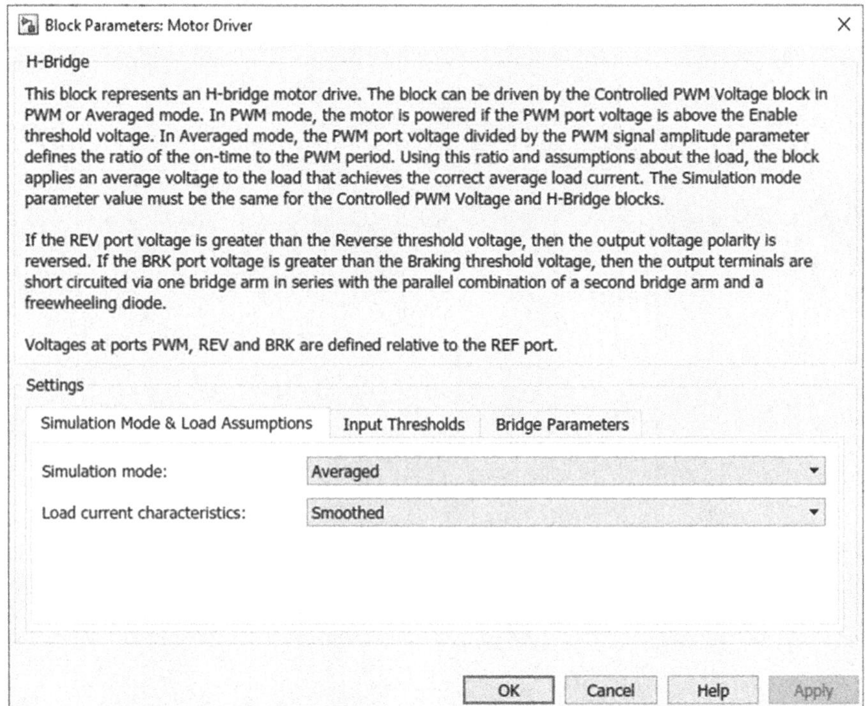

Figure 9.53 H-Bridge Driver Block Simulation Mode and Load settings.

9.5 Parameter tuning of the SimscapeTM DC Motor Model with data from DC motor hardware using Simulink® parameter estimation TM tool

In this section, the Simscape™ plant model developed in Section 9.4 will be used to tune its Electrical and Mechanical parameters with the open-loop actuator PWM sweep data we collected in Section 9.3.1. Simulink® provides a tool named Parameter Estimation which is very useful in tuning the parameters of the physical models if we have validation data. From Fig. 9.27, we saw that the DC motor is a nonlinear system and we partitioned three distinct linear regions from the open-loop PWM sweep data. Follow the below steps for parameter estimation and tuning:

1. We will use the script shown below to initialize the tunable parameters to some default value. Ideally, it can all be initialized to zeros, but because the model will not run with zero Armature resistance and Back EMF constant only those two parameters are initialized to 0.1 and rest all parameters to 0. Then the script loads the open-loop PWM sweep data, from Section 9.3.1, and extracts the linear region data and saves into time-based data array variables ***PWM_Input_-Linear_Region_1*** and ***Motor_Speed_RPM_Filtered_Linear_Region_1.*** The script will also make a plot of the extracted data for the linear region 1 as shown in Fig. 9.67.

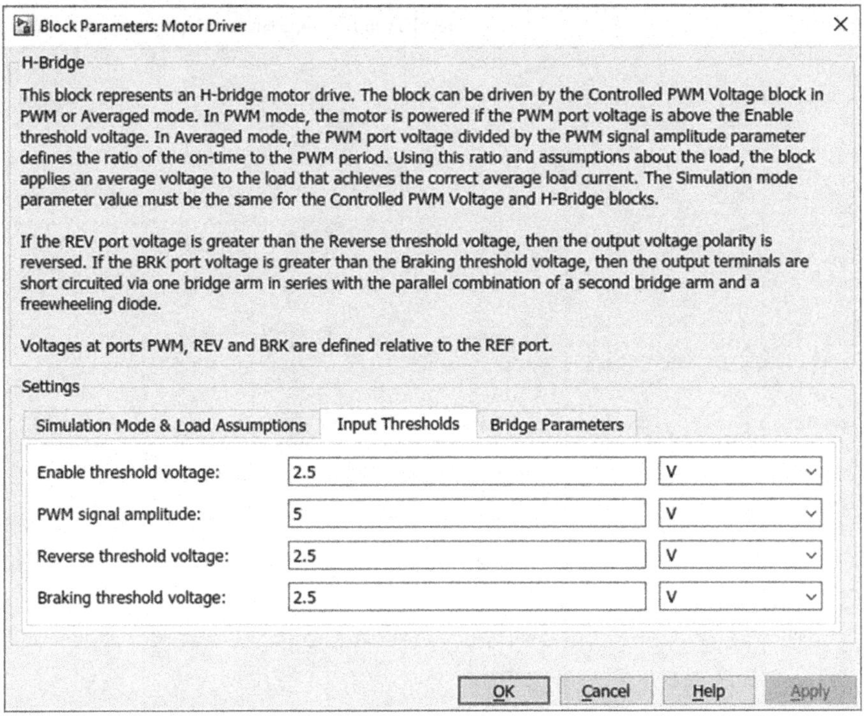

Block Parameters: Motor Driver

H-Bridge

This block represents an H-bridge motor drive. The block can be driven by the Controlled PWM Voltage block in PWM or Averaged mode. In PWM mode, the motor is powered if the PWM port voltage is above the Enable threshold voltage. In Averaged mode, the PWM port voltage divided by the PWM signal amplitude parameter defines the ratio of the on-time to the PWM period. Using this ratio and assumptions about the load, the block applies an average voltage to the load that achieves the correct average load current. The Simulation mode parameter value must be the same for the Controlled PWM Voltage and H-Bridge blocks.

If the REV port voltage is greater than the Reverse threshold voltage, then the output voltage polarity is reversed. If the BRK port voltage is greater than the Braking threshold voltage, then the output terminals are short circuited via one bridge arm in series with the parallel combination of a second bridge arm and a freewheeling diode.

Voltages at ports PWM, REV and BRK are defined relative to the REF port.

Settings

Simulation Mode & Load Assumptions | Input Thresholds | Bridge Parameters

Enable threshold voltage:	2.5	V
PWM signal amplitude:	5	V
Reverse threshold voltage:	2.5	V
Braking threshold voltage:	2.5	V

OK | Cancel | Help | Apply

Figure 9.54 H-Bridge Driver Block Input Threshold setting.

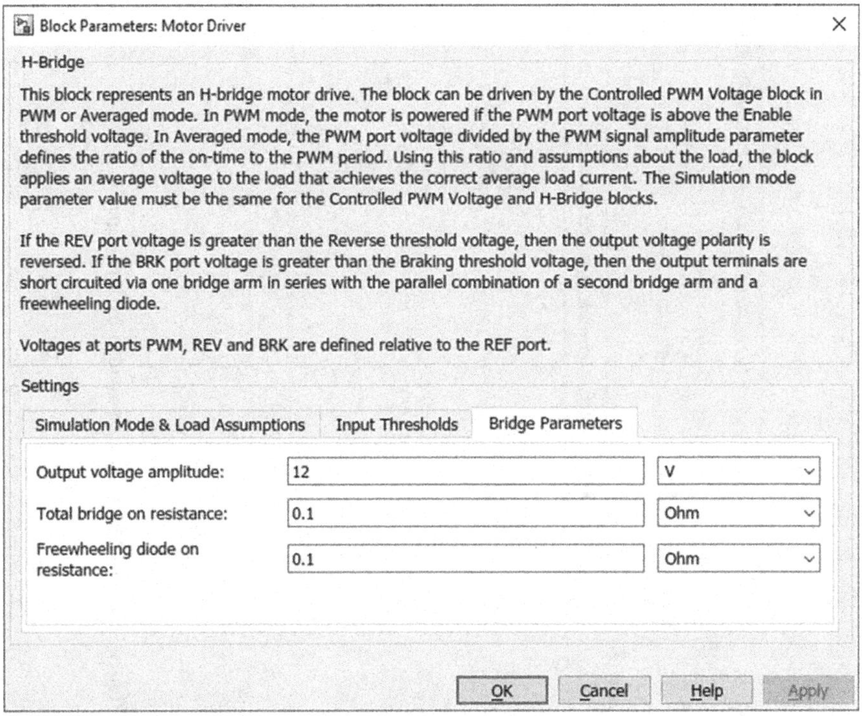

Block Parameters: Motor Driver

H-Bridge

This block represents an H-bridge motor drive. The block can be driven by the Controlled PWM Voltage block in PWM or Averaged mode. In PWM mode, the motor is powered if the PWM port voltage is above the Enable threshold voltage. In Averaged mode, the PWM port voltage divided by the PWM signal amplitude parameter defines the ratio of the on-time to the PWM period. Using this ratio and assumptions about the load, the block applies an average voltage to the load that achieves the correct average load current. The Simulation mode parameter value must be the same for the Controlled PWM Voltage and H-Bridge blocks.

If the REV port voltage is greater than the Reverse threshold voltage, then the output voltage polarity is reversed. If the BRK port voltage is greater than the Braking threshold voltage, then the output terminals are short circuited via one bridge arm in series with the parallel combination of a second bridge arm and a freewheeling diode.

Voltages at ports PWM, REV and BRK are defined relative to the REF port.

Settings

Simulation Mode & Load Assumptions | Input Thresholds | Bridge Parameters

Output voltage amplitude:	12	V
Total bridge on resistance:	0.1	Ohm
Freewheeling diode on resistance:	0.1	Ohm

OK | Cancel | Help | Apply

Figure 9.55 H-Bridge Driver Block Bridge Parameters settings.

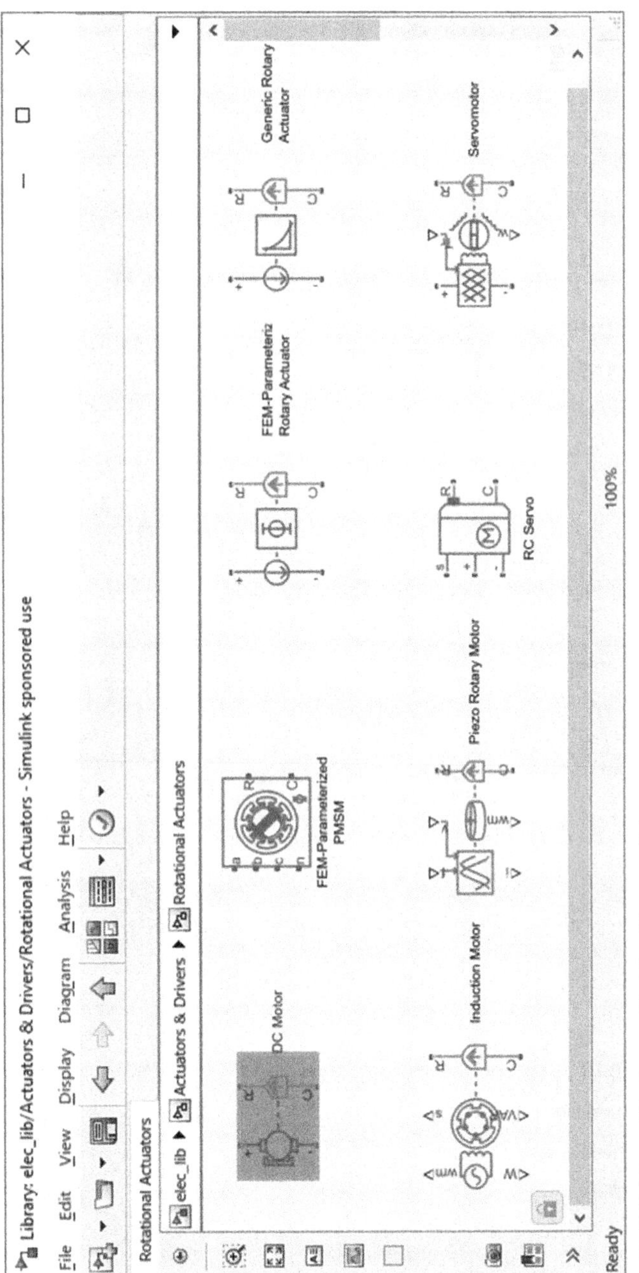

Figure 9.56 DC Motor Block provided by Simscape™.

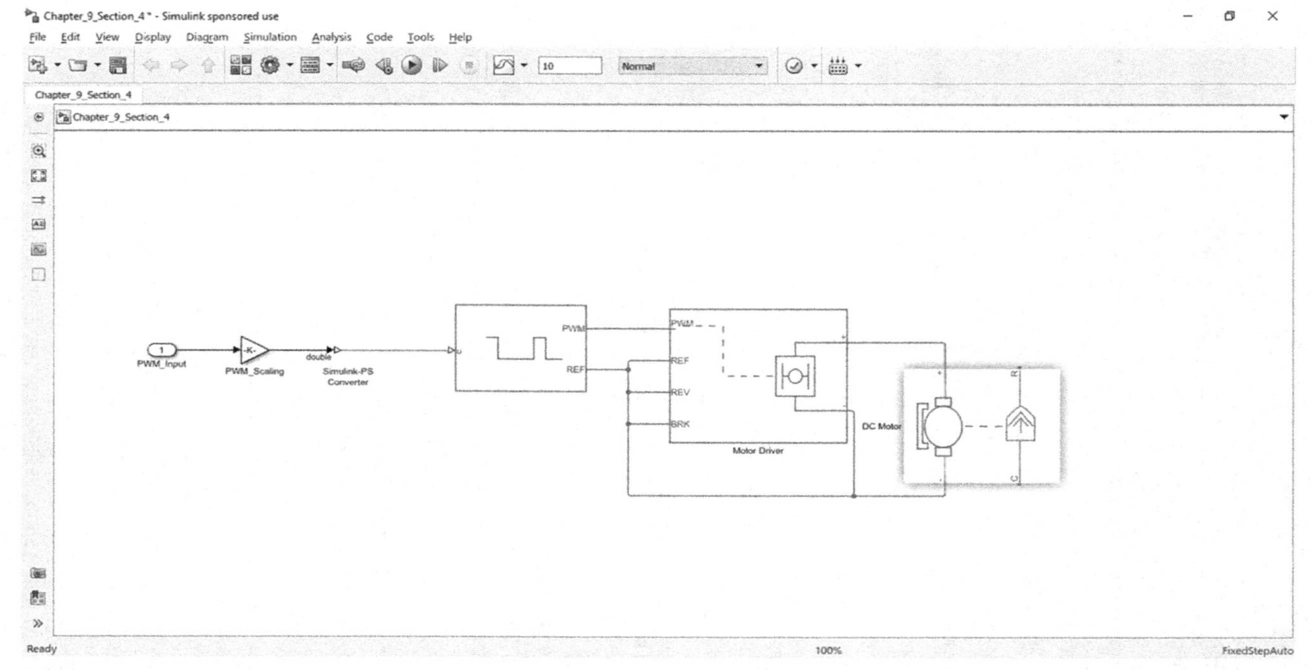

Figure 9.57 Adding DC Motor block.

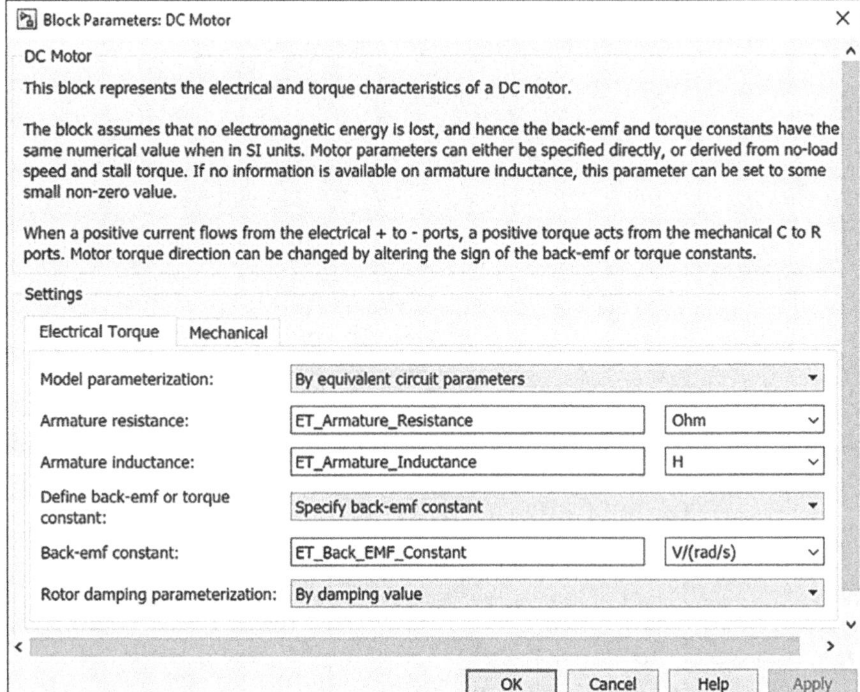

Figure 9.58 Tunable Electrical Torque properties of DC Motor block.

Chapter_9_Section_5_Script.m

```
%Book Title: Digital Twin Development and Deployment On Cloud Using Matlab
%SimScape
%Chapter: 9
%Section: 5
%Authors: Nassim Khaled and Bibin Pattel
%Last Modified: 09/10/2019
clc
clear all
%% Initialize the Simscape Motor Plant Tunable Parameters
ET_Armature_Resistance = 0.1;
ET_Armature_Inductance = 0;
ET_Back_EMF_Constant = 0.1;
M_Rotor_Inertia = 0;
M_Rotor_Damping = 0;
M_Initial_Rotor_Speed = 0;
%% Load the Open Loop PWM Sweep Data from Section 9.3
load Chapter_9_Section_3_1_System_ID_Data.mat
```

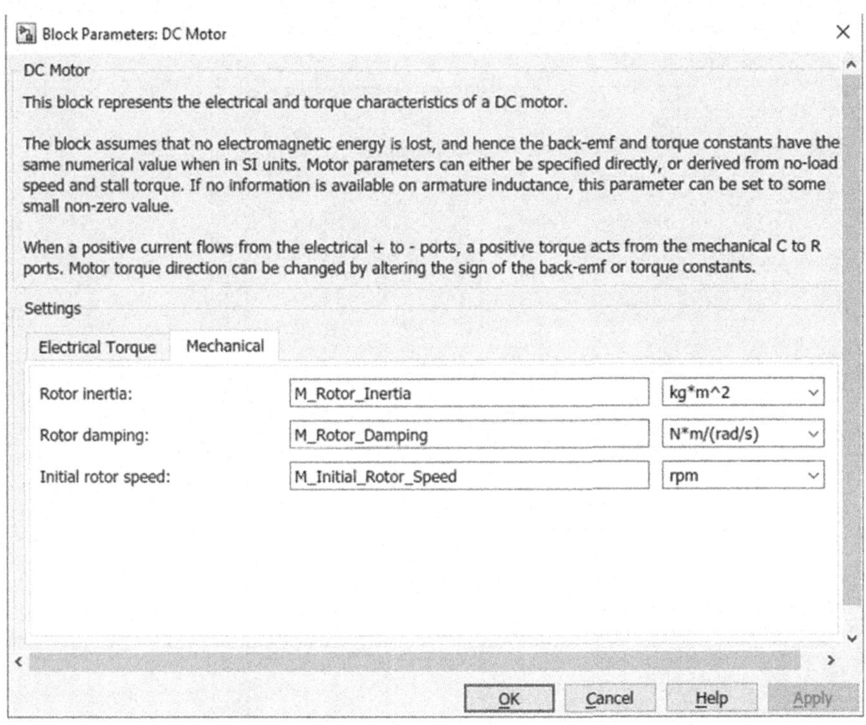

Figure 9.59 Tunable Mechanical properties of DC Motor block.

```
%% Separate the Data into Three Regions

%% Linear Region 1
% Save the PWM_Input and Filtered Motor Speed data without Time data
into
% temporary variable
time = PWM_Input(:,1);
pwm_input_data = PWM_Input(:,2);
filtered_motor_speed_data = Motor_Speed_RPM_Filtered(:,2);
% From time T =1 to 700 is the data for Linear Region 1. Extract just
that
% data and store into a time series data array
PWM_Input_Linear_Region_1 = [time(1:700)-time(1),pwm_input_data(1:
700)];
Motor_Speed_RPM_Filtered_Linear_Region_1 =[time(1:700)-
time(1),filtered_motor_speed_data(1:700)];
% Plot the Data
[hAx,hLine1,hLine2] = plotyy(PWM_Input_Linear_Region_1(:
,1),PWM_Input_Linear_Region_1(:,2),Motor_Speed_RPM_Filtered_
Linear_Region_1(:,1),Motor_Speed_RPM_Filtered_Linear_Region_1(:,2));
title('PWM Command Vs Motors Speed Output for Linear Region 1')
xlabel('Time [sec]')
```

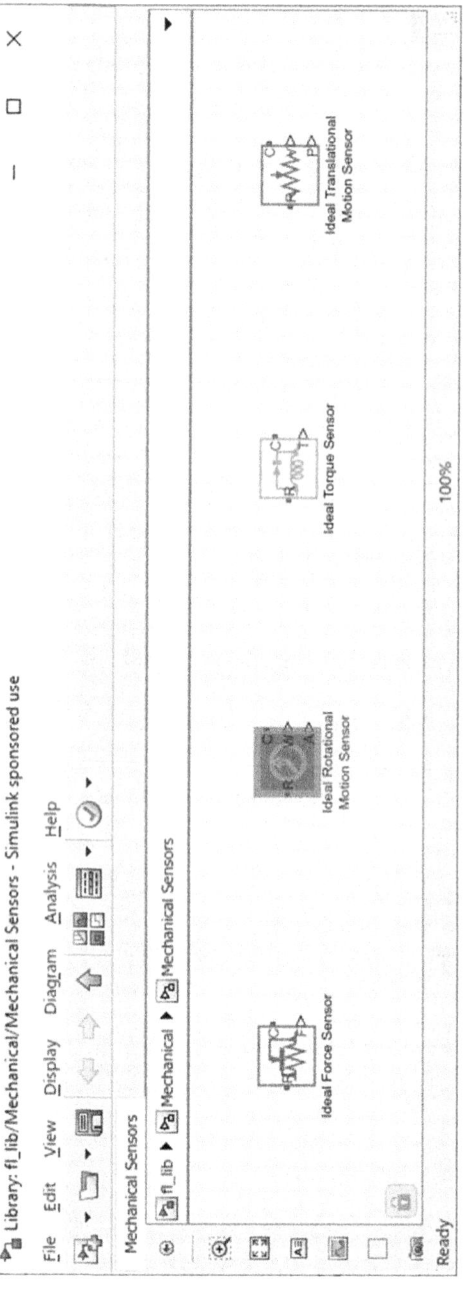

Figure 9.60 Ideal Rotational Motion Sensor block.

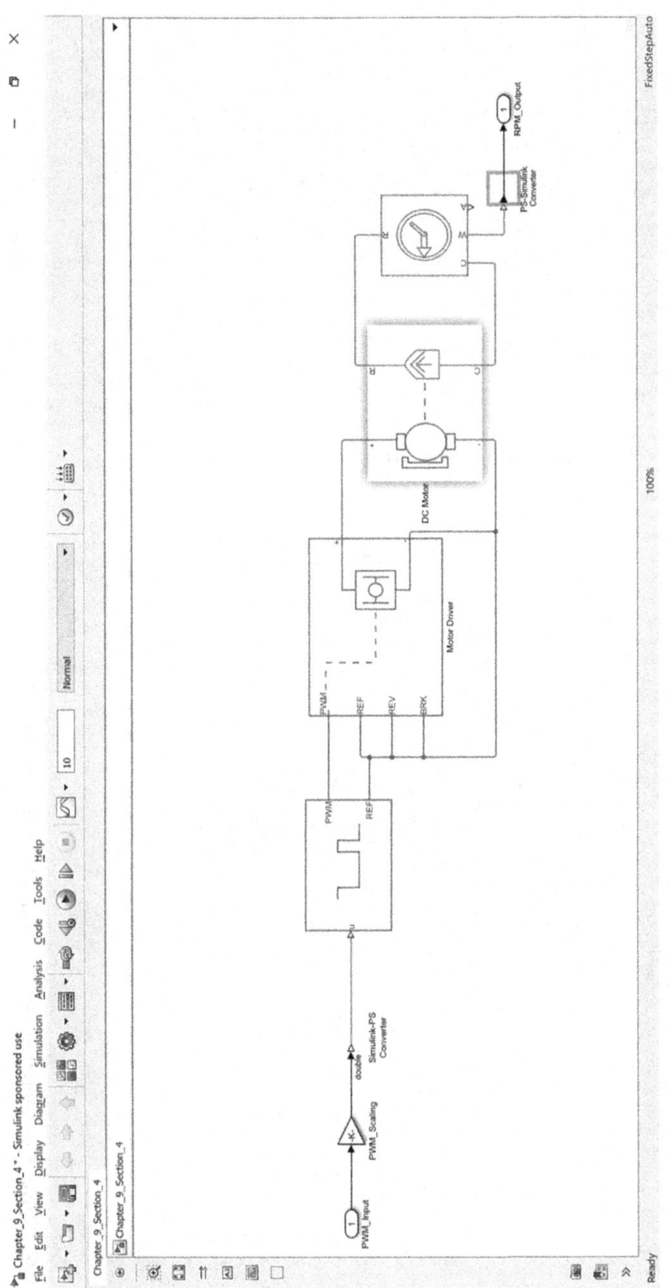

Figure 9.61 Adding Ideal Rotational Motion Sensor to DC Motor.

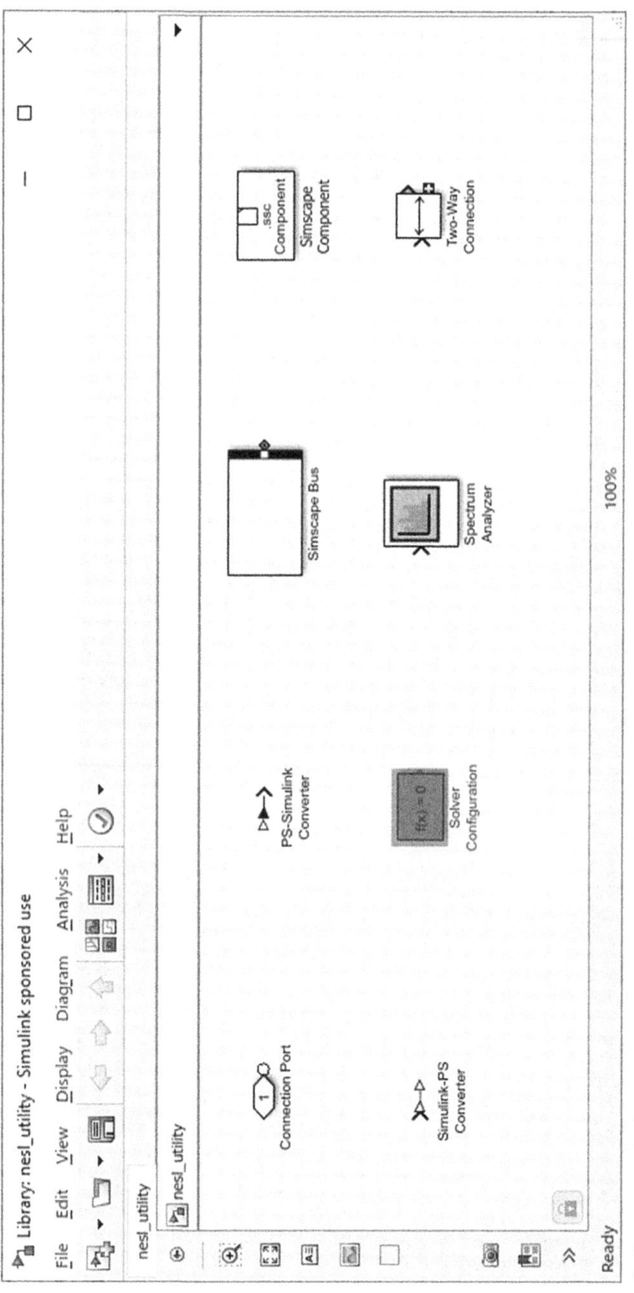

Figure 9.62 Solver Configuration block provided by Simscape™.

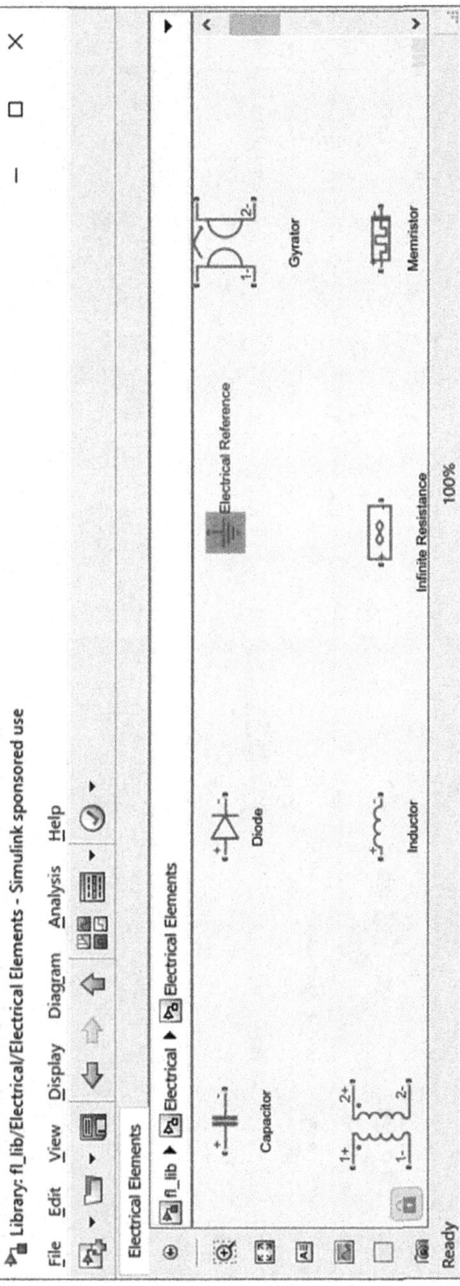

Figure 9.63 Electrical Reference block provided by Simscape™.

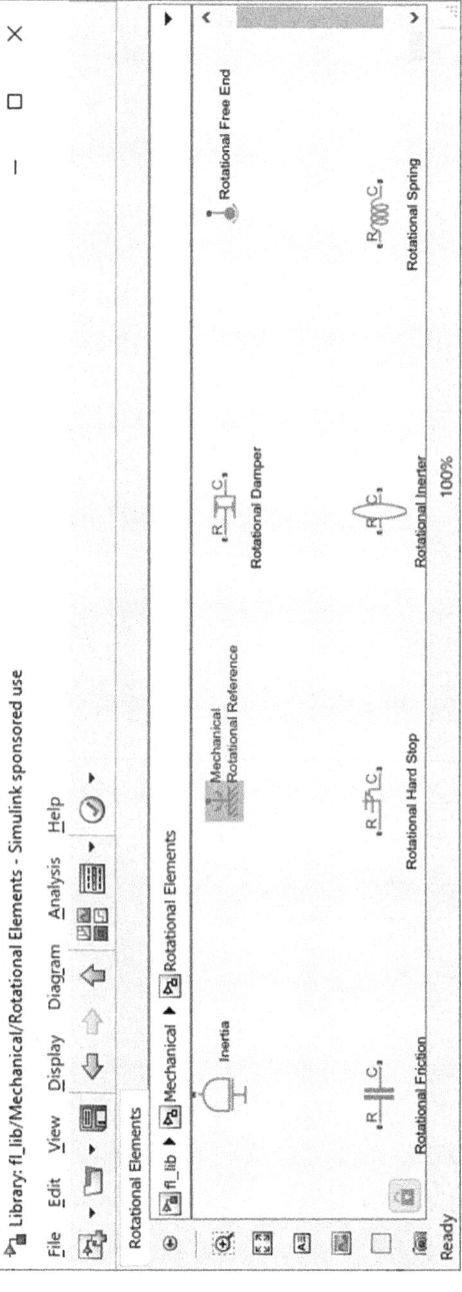

Figure 9.64 Mechanical Rotational Reference block provided by Simscape™.

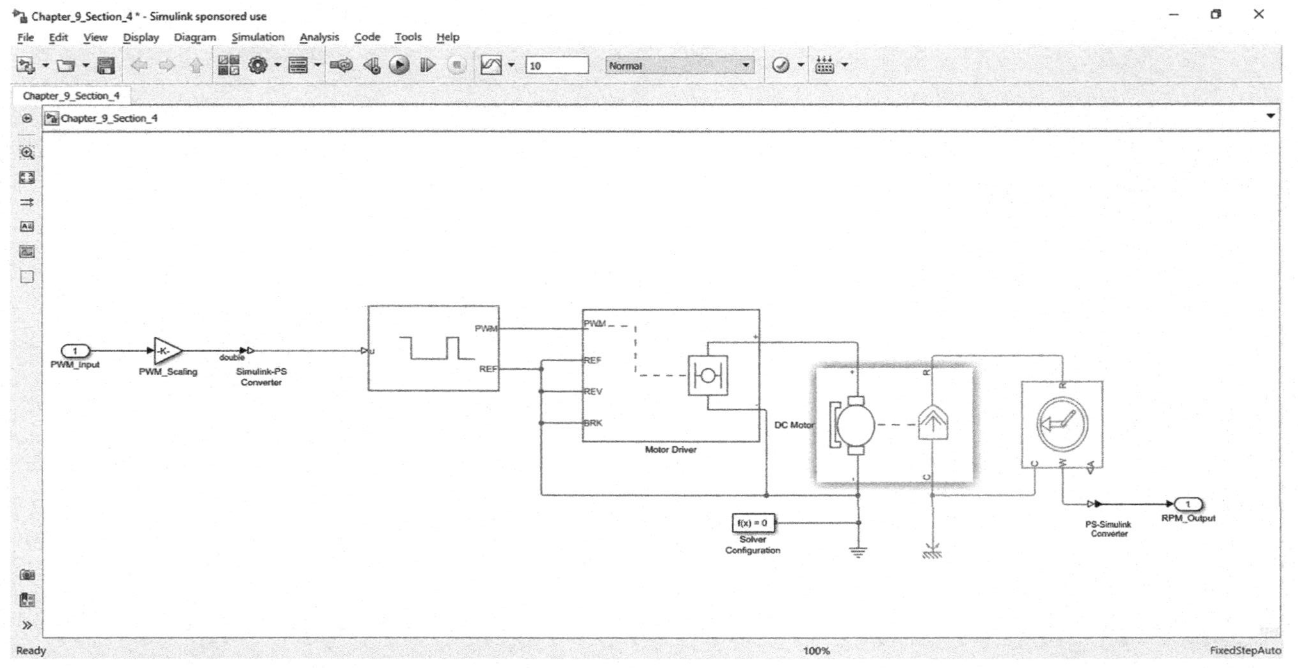

Figure 9.65 Adding Solver Configuration, Mechanical, and Electrical Reference blocks.

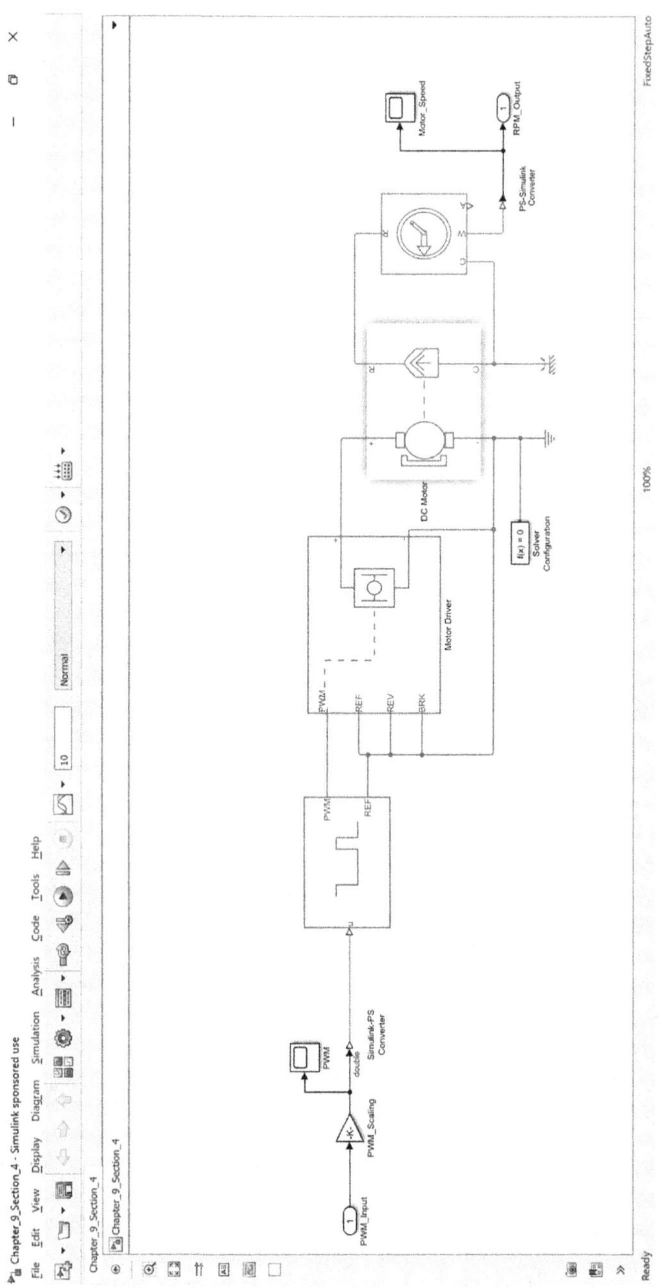

Figure 9.66 Completed Simscape™ Plant Model for DC Motor.

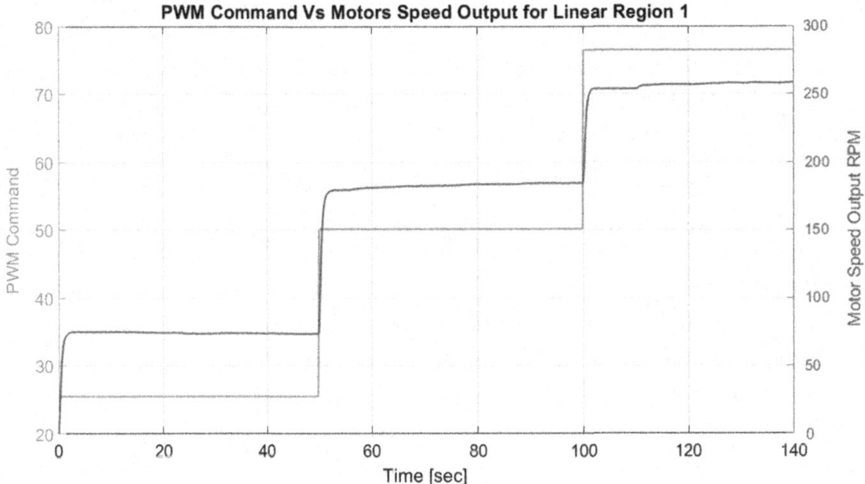

Figure 9.67 Extracted Linear Region 1 Data for Model Parameter Estimation.

```
ylabel(hAx(1),'PWM Command') % left y-axis
ylabel(hAx(2),'Motor Speed Output RPM') % right y-axis
grid on
set(gcf,'color',[1 1 1]);
hLine1.LineWidth = 2;
hLine2.LineWidth = 2;
% Save the Data
save  Chapter_9_Section_5_Linear_Region_1_Data.mat  PWM_Input_Linear_
Region_1 Motor_Speed_RPM_Filtered_Linear_Region_1
```

2. Open the Simscape™ DC motor plant model that we have developed in Section 9.4. The same model is renamed to Chapter_9_Section_5.slx for this section. Open the ***Parameter Estimation*** tool by going to the model's ***Analysis*** >> ***Parameter Estimation*** option as shown in Fig. 9.68. The Parameter Estimation GUI will be launched as shown in Fig. 9.69.
3. In the ***Parameter Estimation*** GUI, click on ***New Experiment*** to open up the experiment setup window. On this window, enter/copy paste the names ***PWM_Input_Linear_Region_1*** and ***Motor_Speed_RPM_Filtered_Linear_Region_1***, respectively, to the input and output text box options and hit enter. It can be seen that the size of the data will get populated in the text box field for input and output. Press ***Ok*** on the ***New Experiment*** GUI. Check Fig. 9.70 for details.
4. On the main GUI on the parameters section, right click and choose the ***Edit*** option, see Fig. 9.71. Click on the ***Select Parameters*** option as highlighted in Fig. 9.72. From the ***Select Model Variables*** window select all listed parameters for tuning using the checkboxes and click ***Ok*** as shown in Fig. 9.73. It will show the default values of all the selected parameters on the window shown in Fig. 9.74; we have the option to change the initial value on this window as well, but for now we will go with the same default values.

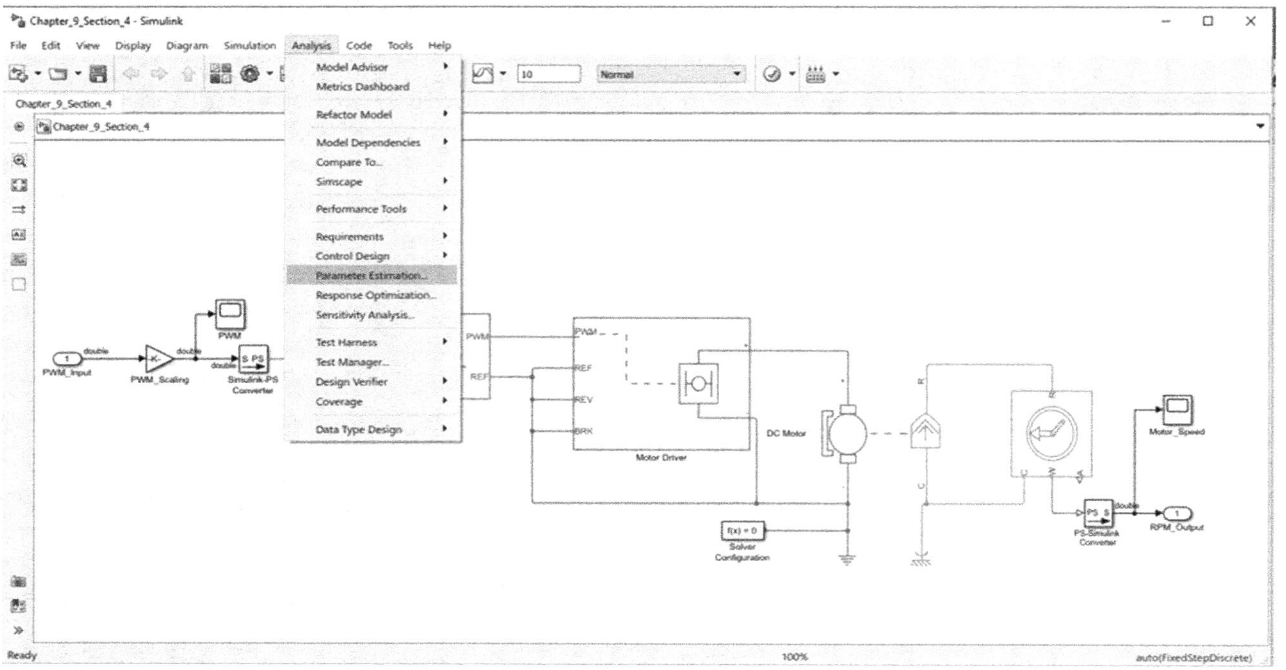

Figure 9.68 Launching Parameter Estimation Tool.

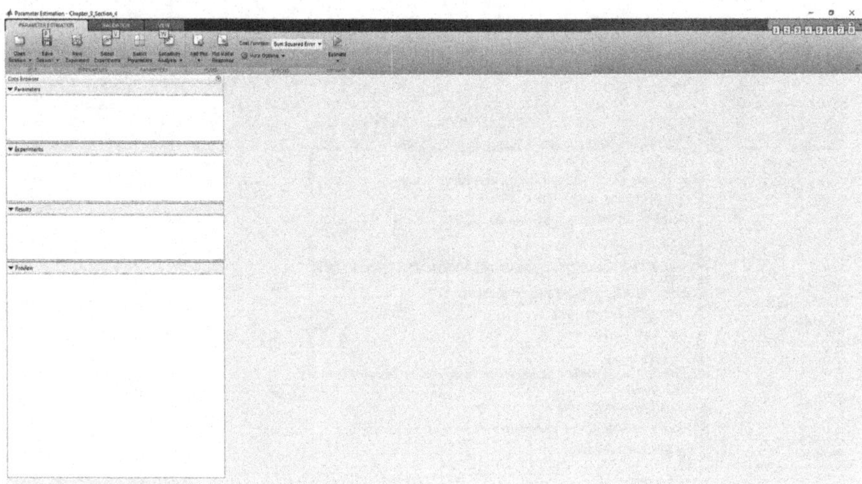

Figure 9.69 Parameter Estimation tool GUI.

5. On the main GUI from the ***Add Plot*** submenu, select the options ***Parameter Trajectory*** to show the trajectory of change in value of the parameters as the optimization progresses. Also select the ***Experiment Plots*** to show how well the simulation data with retuned parameters are matching the test data. Fig. 9.75 shows the newly added plots window.
6. On the main GUI, click on the ***Estimate*** button. This will start the iterative optimization process to estimate the parameters, which minimizes a cost function which in this case is the Sum Squared Error between the model simulated and actual test data. We can also see as the optimization finishes how well the simulation results with optimized parameters are matching the actual test data and also trajectory of the parameter changing after each optimization iteration as shown in Fig. 9.76 and also optimization cost function value is reported as shown in Fig. 9.77. The final optimized values of the parameters can be seen in the ***Preview*** window of the main GUI as shown in Fig. 9.78. The values shown in Fig. 9.78 will be used with the Digital Twin when we deploy the model into the cloud.
7. User can save this Parameter Estimation session will all its settings and updated parameters as a MATLAB data file using the save option provided in the GUI, to use it later.

9.6 Adding AWS cloud connectivity to real-time embedded hardware for DC Motor Speed Control

In the previous section, we have developed the plant model of the DC Motor and tuned its parameters to match closely with the DC Motor hardware data. In this section, we will prepare the DC Motor hardware board to communicate with the AWS Cloud. We will use a very popular Embedded WI-FI controller module named ESP32, to communicate between Arduino controller board, which controls the DC motor and the AWS cloud. The Arduino controller will communicate with the ESP32 Wi-Fi module using the wired serial data Transmit and Receive ports of

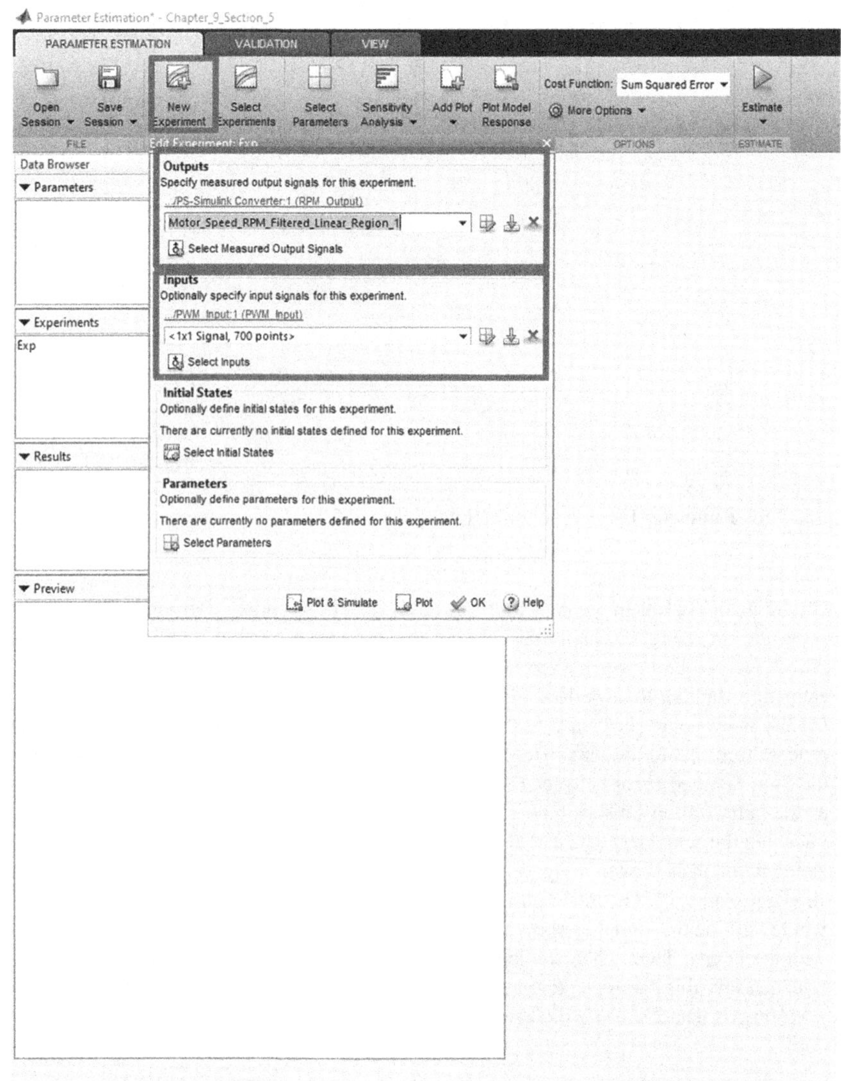

Figure 9.70 New parameter tuning experiment input/output selection window.

Arduino and ESP32. It should be noted that we cannot connect the serial Transmit pin of the Arduino to serial Receive pin of the ESP32 because Arduino works with DC 5 V and ESP32 works with DC 3.3 V. So a proper DC voltage level shifting has to be placed between the two modules for proper communication and safe operation of the modules. ESP32 can connect to the Wi-Fi Router or any Wi-Fi Hotspot, which is available near range of the module. And if the Wi-Fi Router or the Wi-Fi Hotspot is connected to Internet, ESP32 can establish connection to the AWS cloud. Once the data from the DC motor hardware is received in ESP32,

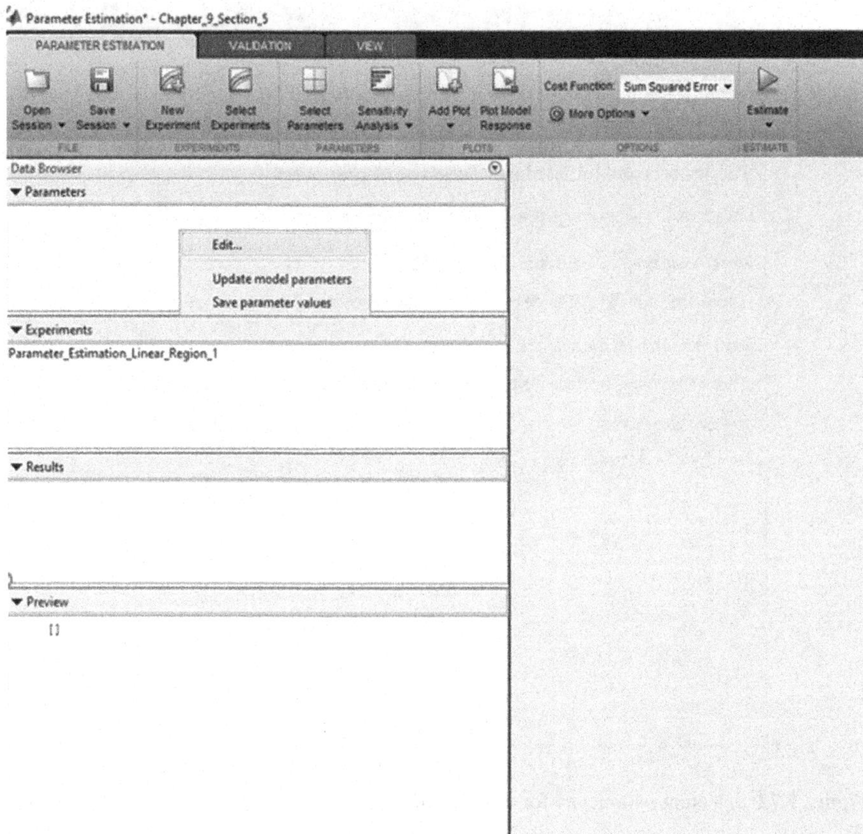

Figure 9.71 Launching Parameter Selection GUI.

ESP32 will convert the data into specific JSON structure format and send it to AWS cloud using the WI-FI and Internet connectivity. Fig. 9.79 shows the overall bock diagram of the entire setup for digital twin deployment. It can be seen that the Arduino module sends the PWM command to control the DC motor speed and it senses the Motor speed. This PWM command and actual Motor speed feedback will be transmitted to the ESP32 module through the 5−3.3 V Level shifter. The ESP32 module is connected to the Wi-Fi router using the Wi-Fi protocol using a program which is written and deployed into the ESP32 module using the Arduino IDE. We have to install and configure the Arduino IDE in a specific way to support programming the ESP32 module, which is described later in this section.

9.6.1 Hardware setup to communicate between Arduino and ESP32

On top of the existing connections and hardware setup shown in Fig. 9.6, now we need to interface Arduino microcontroller with the ESP32 module and transmit data from

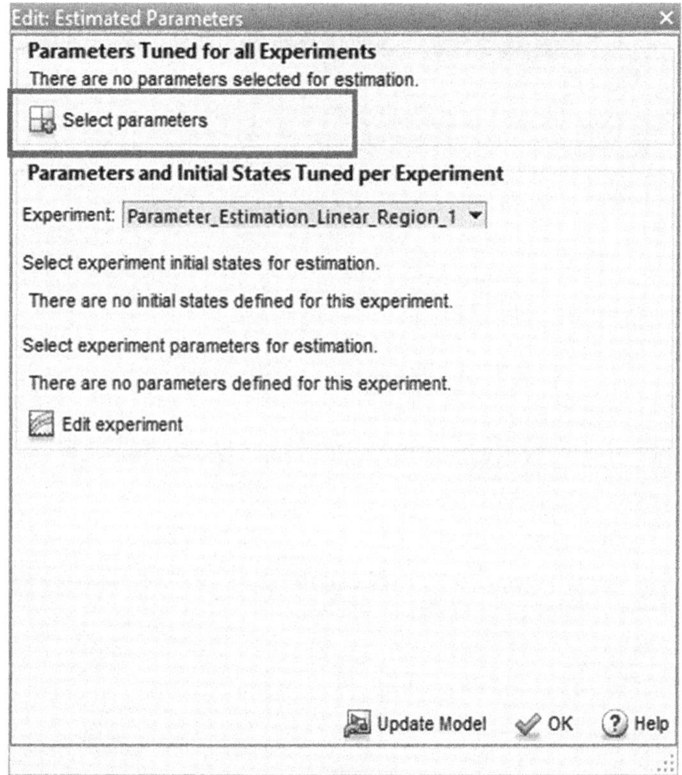

Figure 9.72 Selecting parameters for estimation.

Select model variables

Filter by variable name

▼	Variable	Curr...	Used By
☑	ET_Armature_Inductance	0	... Section 5/DC Motor
☑	ET_Armature_Resistance	0.1	... Section 5/DC Motor
☑	ET_Back_EMF_Constant	0.1	... Section 5/DC Motor
☑	M_Initial_Rotor_Speed	0	... Section 5/DC Motor
☑	M_Rotor_Damping	0	... Section 5/DC Motor
☑	M_Rotor_Inertia	0	... Section 5/DC Motor

▶ Specify expression indexing if necessary (e.g., a(3) or s.x)

OK Cancel Help

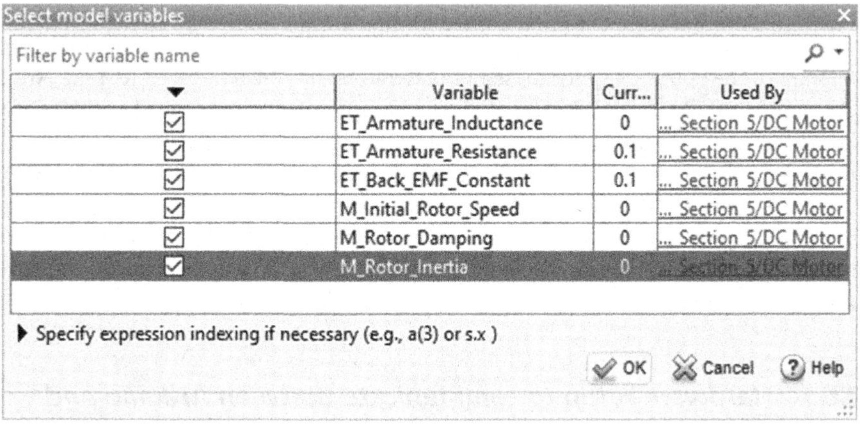

Figure 9.73 Select all model parameters for estimation.

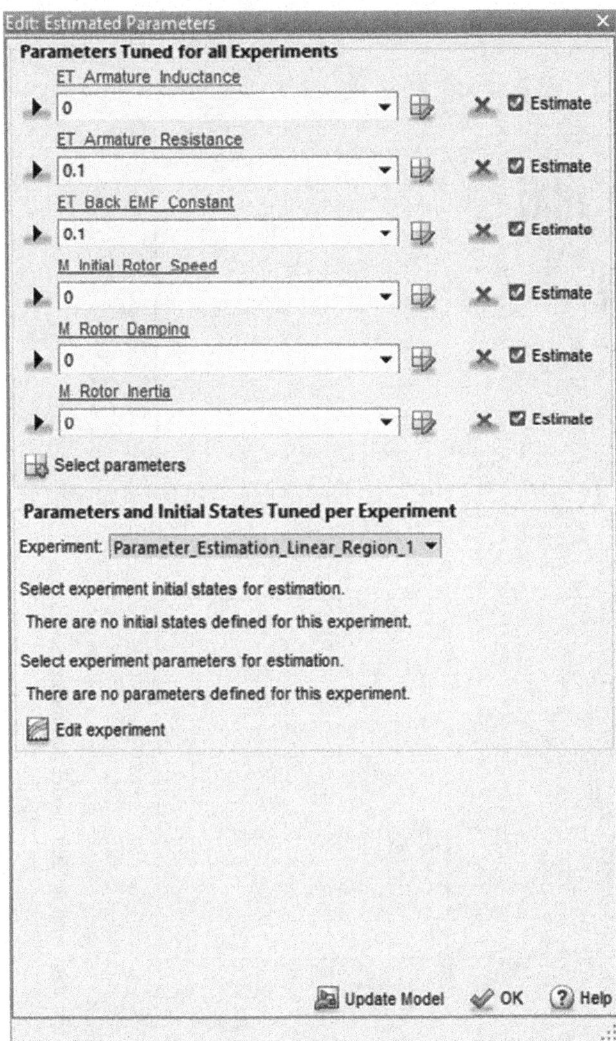

Figure 9.74 Tunable parameters showing the default values specified.

Arduino to ESP32 using wired serial communication. Make the connections like shown in Fig. 9.80 and Table 9.3 between Arduino and ESP32. Note that the 5 to 3.3 V DC level shifter is needed in order to compensate for the operating voltage differences between both the controller modules.

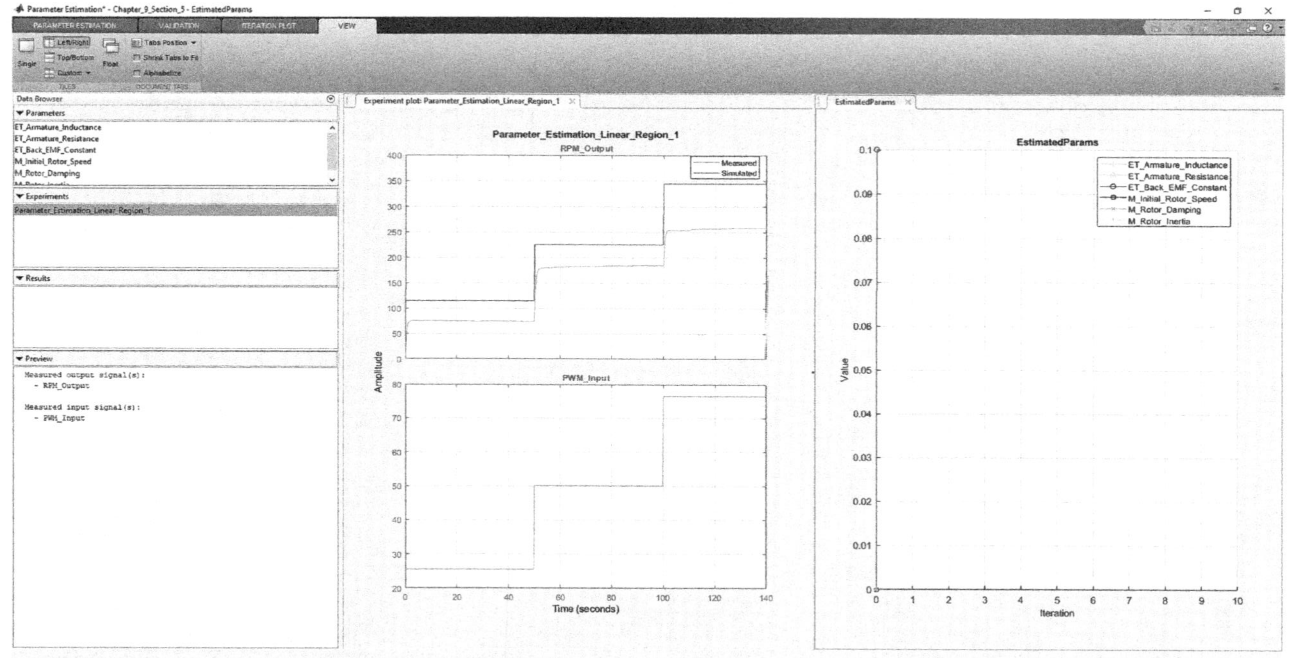

Figure 9.75 Comparing simulation and test data with default model parameter values.

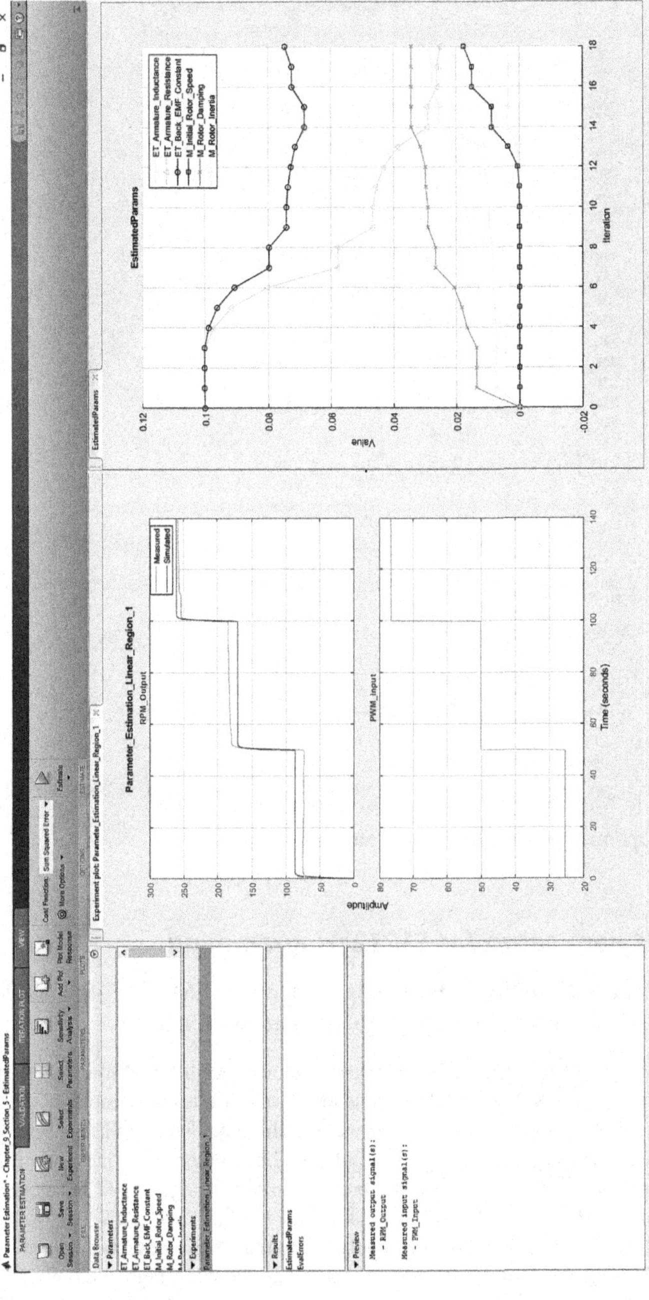

Figure 9.76 Showing model versus actual data and parameter trajectory after optimization finished.

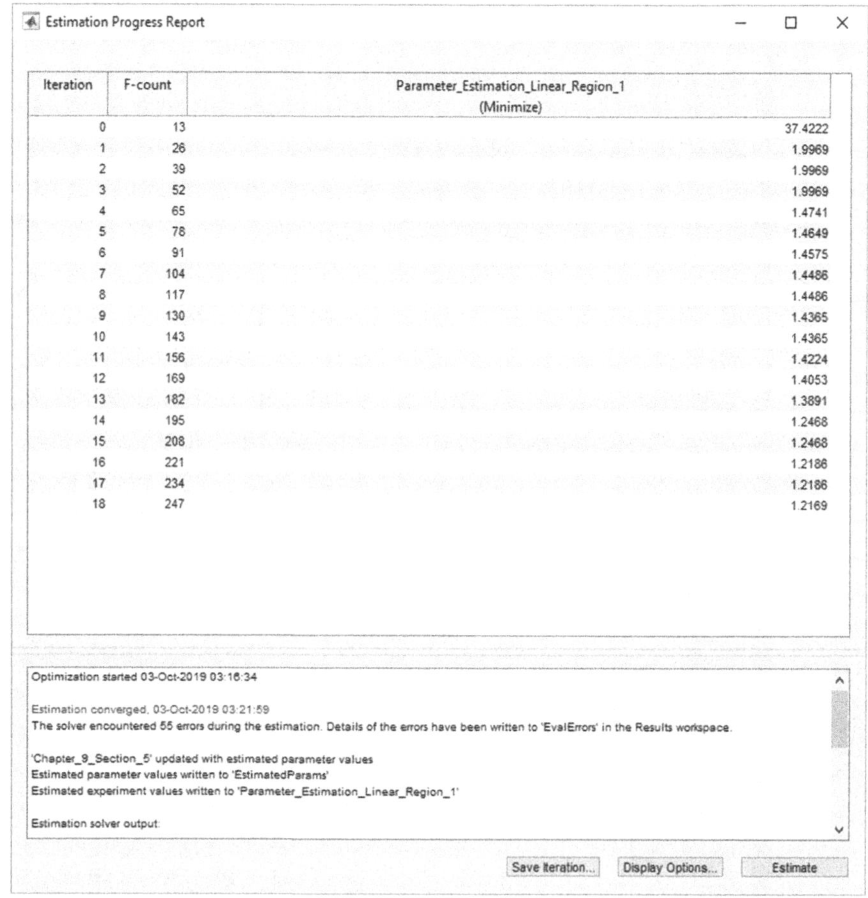

Iteration	F-count	Parameter_Estimation_Linear_Region_1 (Minimize)
0	13	37.4222
1	26	1.9969
2	39	1.9969
3	52	1.9969
4	65	1.4741
5	78	1.4649
6	91	1.4575
7	104	1.4486
8	117	1.4486
9	130	1.4365
10	143	1.4365
11	156	1.4224
12	169	1.4053
13	182	1.3891
14	195	1.2468
15	208	1.2468
16	221	1.2186
17	234	1.2186
18	247	1.2169

Figure 9.77 Optimization Progress Report and Summary.

9.6.2 *Software setup for ESP32 programming*

We will be using the Arduino IDE for programming the ESP32 Wi-Fi module. Follow the below steps to set up Arduino IDE to program the ESP32:

1. Install the Arduino IDE first. The installer can be downloaded from https://www.arduino.cc/en/main/software. Download the executable and install it into the computer.
2. Once the base Arduino IDE is installed, we will upgrade that installation with an add-on board manager package to program ESP32. Open the Arduino IDE, and go to **File >> Preferences** and enter the URL https://dl.espressif.com/dl/package_esp32_index.json into the "Additional Board Manager URLs" field as shown in Figs. 9.81 and 9.82.
3. Open the Boards Manager by going to **Tools >> Board >> Boards Manager** and search for ESP32 and install the Add-on package for ESP32 as shown in Figs. 9.83–9.85.
4. Just to make sure the ESP32 add-on package is installed properly, we will do a quick test by running one of the installed Wi-Fi scan programs. Now connect the ESP32 board with a

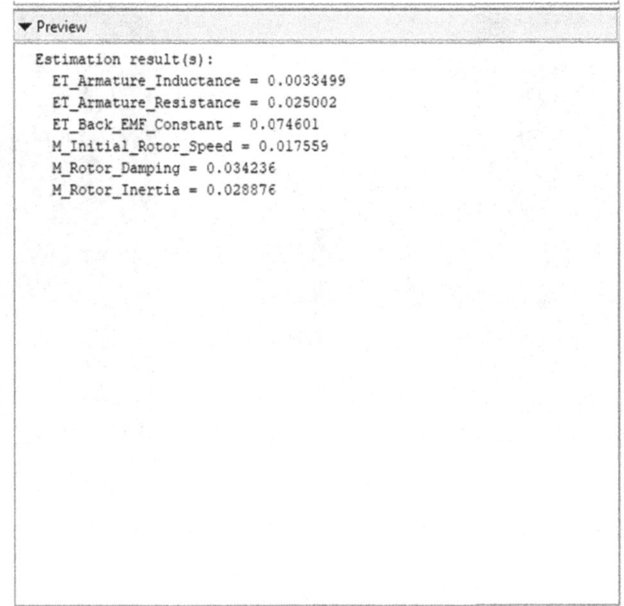

Figure 9.78 Parameter values after optimization.

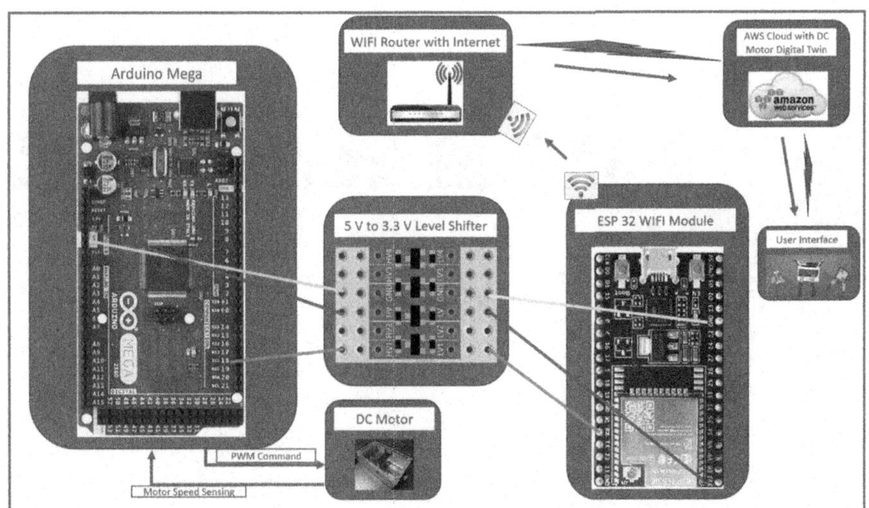

Figure 9.79 DC Motor Control Digital Twin Deployment Overall Block Diagram.

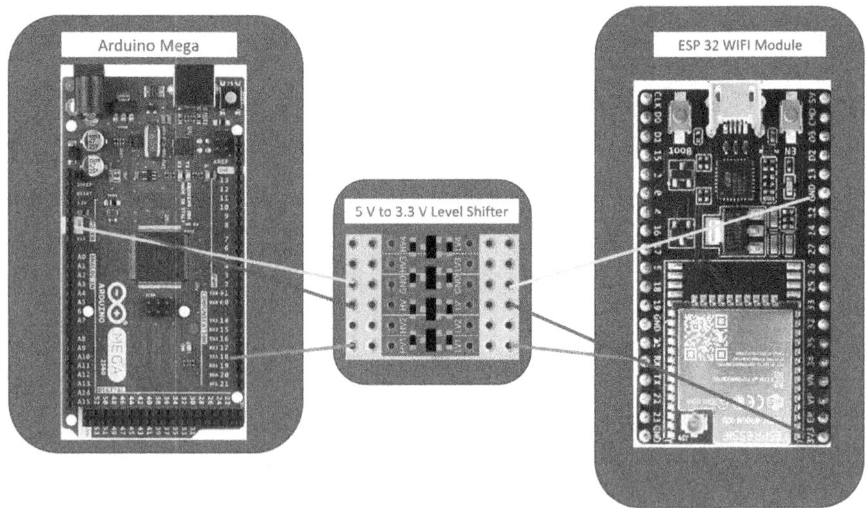

Figure 9.80 Interface Arduino with ESP32 for wired serial communication.

Table 9.3 Wiring and connection table.

Signal transmit line connection	Ground connection	VCC connection
Arduino PIN 18 (Tx1)	Arduino Ground (GND)	Arduino 5 V
Level shifter HV1	Level shifter GND	Level shifter HV
ESP32 RX	ESP32 GND	ESP32 3.3 V

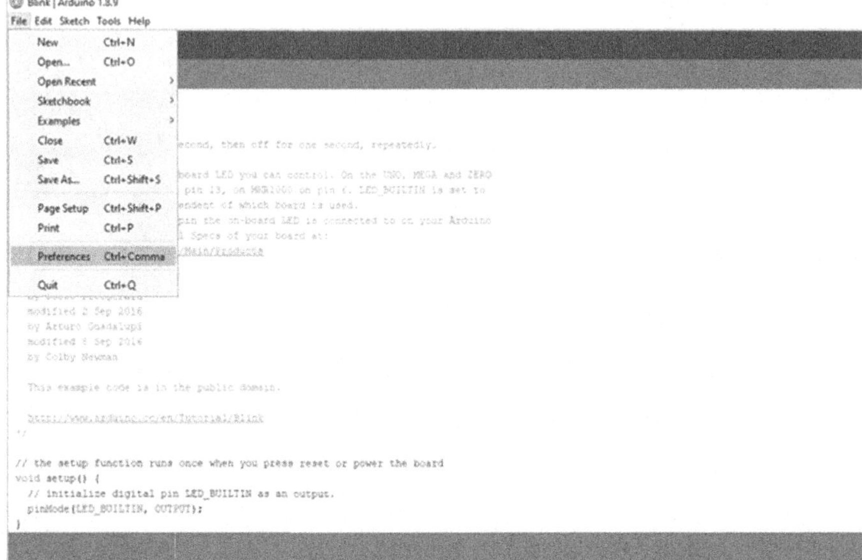

Figure 9.81 Selecting and updating Arduino IDE preference for ESP32 Wi-Fi module.

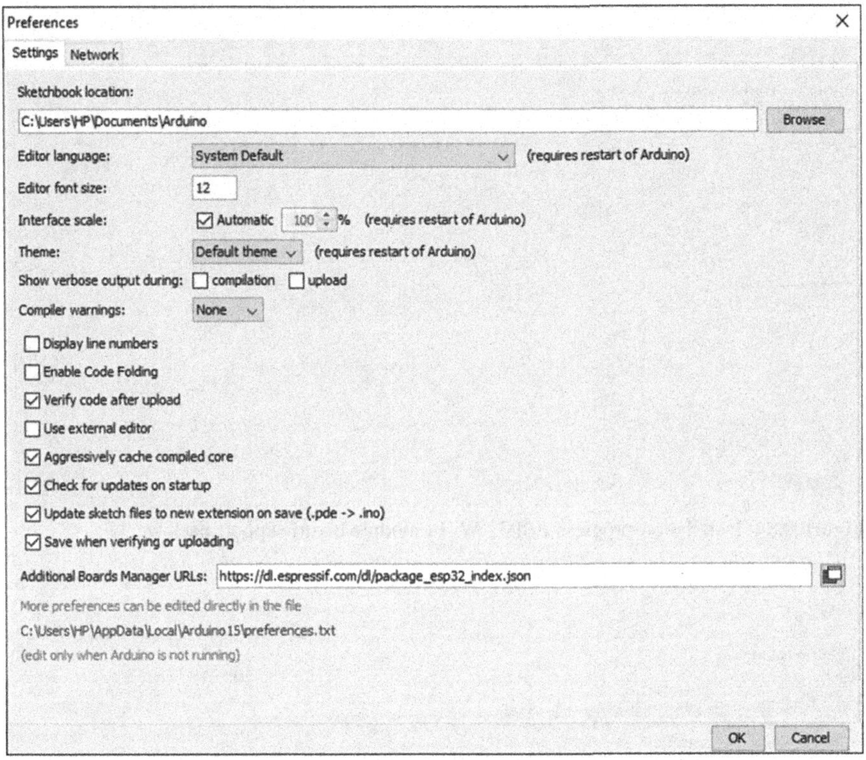

Figure 9.82 Updating additional board manager in preferences.

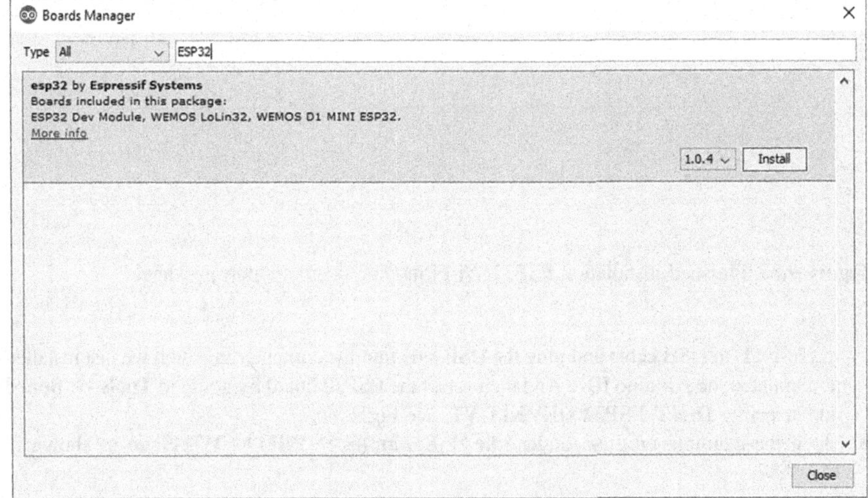

Figure 9.83 Installing ESP32 Wi-Fi module board support package.

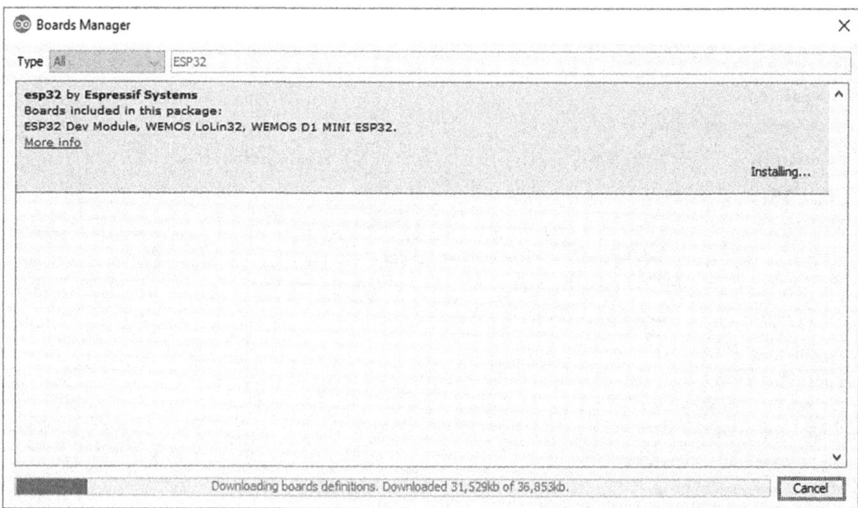

Figure 9.84 Installation progress ESP32 Wi-Fi module board support package.

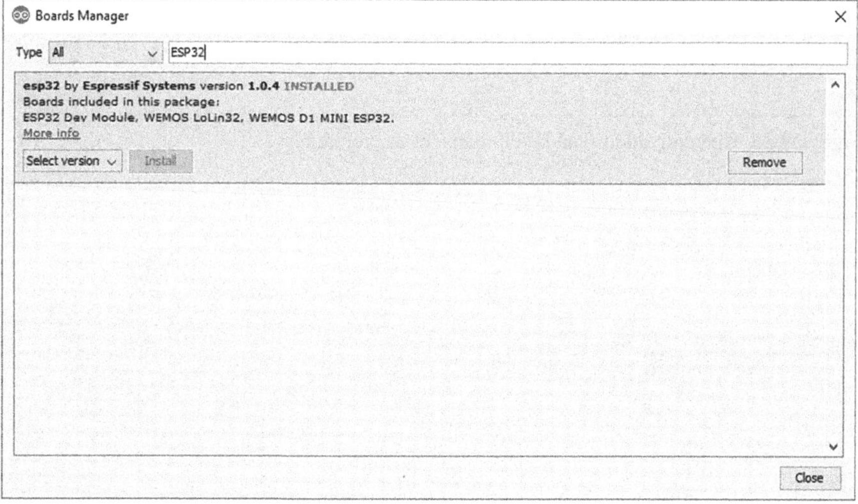

Figure 9.85 Finished installation ESP32 Wi-Fi module board support package.

micro-USB to USB cable and plug the USB side into the computer in which we just installed and updated the Arduino IDE. And then select the ESP32 board by going to **Tools >> Board** and selecting **DOIT ESP32 DEVKIT V1.** See Fig. 9.86.

5. Open the example program under **File >> Examples >> WiFi >> WiFiScan** as shown in Fig. 9.87.

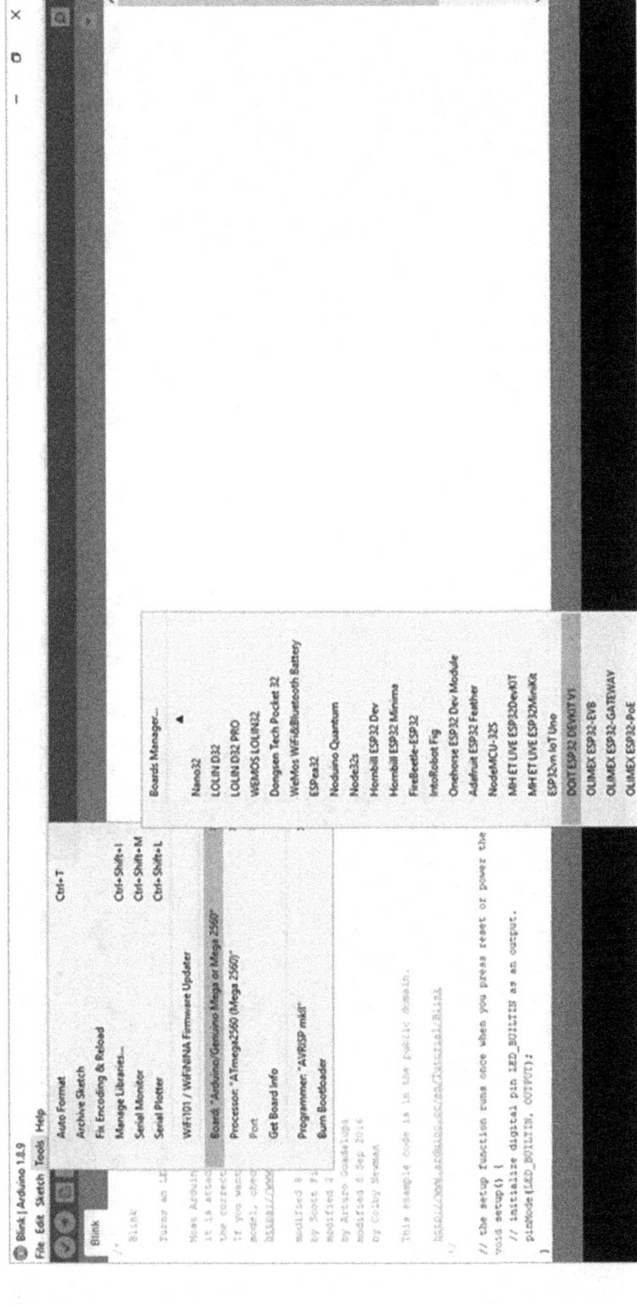

Figure 9.86 Selecting the ESP32 development board.

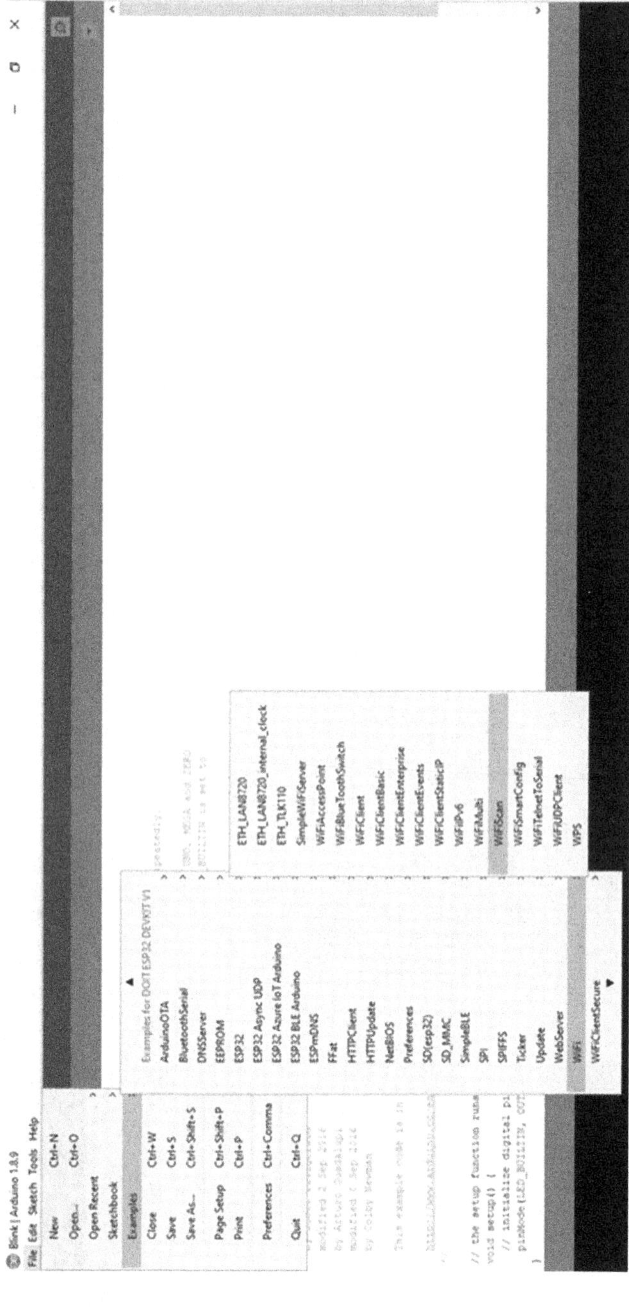

Figure 9.87 Opening sample Wifi scan program code.

6. Select the COM port in which the ESP32 board is connected by going to **Tools >> Port.** In this case, it is COM5, please note that it may not be the same COM5 when this is tried on a different machine. Check Fig. 9.88.
7. Select the **Sketch >> Upload** option from the WiFiScan program sketch file. This will take a few seconds to compile and upload the code into ESP32 hardware as shown in Figs. 9.89 and 9.90.
8. In order to see the program output, open the Arduino Serial Monitor from the Tools menu. Make sure the Baud Rate is set to 115,200, as this is the baud rate used in the WiFiScan program. If the program is successfully downloaded, the ESP32 should scan the Wi-Fi access points in the range of it and display in the Arduino serial monitor as shown in Fig. 9.91.

9.6.3 Simulink model updates for communication between Arduino and ESP32

We will extend the Simulink model capabilities developed in Section 9.3.3 to control the DC motor using the Feedforward and Feedback PID controller, to transmit the PWM_Command and Actual_Motor_Speed over the wired serial transmit pin of the Arduino to the serial receive pin of the ESP32. Follow the below steps:

1. Make a copy of the Simulink model developed in Section 9.3.3, open Simulink library browser, and add a **Function-Call** subsystem block to the model as shown in Figs. 9.92 and 9.93.
2. Add a **Function Call Generator** block from Simulink library and configure its Sample Time to be 0.1 s as shown in Fig. 9.94. We will be sending the PWM_Command and the Actual_Motor_Speed from the DC motor hardware every 0.1 s (100 ms) to the ESP32. The reason we choose 0.1 s is because the Simscape™ Digital Twin model will be running at a discrete step size of 0.1 s.
3. In the **Called Function** subsystem added, delete the one output port which is added by default and add one more input port and name the ports PWM_Command and Actual_Motor_Speed as shown in Fig. 9.95.
4. At the top level of the model, make the connections from the signals PWM command and Sensed Motor Speed as shown in Fig. 9.96. Insert rate transition blocks in the connection lines to make sure the data are propagated properly between the subsystems running at different rates.
5. Now we will add a Stateflow chart block from Simulink® to the model to control when to send the data to the Arduino serial port. See Fig. 9.97.
6. Double click on the newly added chart, and right click on the blank chart and select the **Explore** option as shown in Fig. 9.98. This will open up the model explorer where we can enter the input/output names and properties of the data and trigger used in this chart. The model explorer window is shown in Fig. 9.99.
7. In the model explorer window go to **Add** menu. We can see the options for **Data** and **Event.** Add three **Data** and one **Event** by clicking one by one. Name the first two Data as PWM_Command and Actual_Motor_Speed, and select their **Scope** to be **Input.** This will add two input ports to the Stateflow chart. Name the third Data as Serial_Write_Value, and select its **Scope** to be **Output.** Name the Event added as Serial_Send_Trigger and select its **Scope** also to be **Output.** See Figs. 9.100 and 9.101 for details. Now the Stateflow chart will have two input two output ports as shown in Fig. 9.102.

Figure 9.88 Selecting the COM port in to communicate to the ESP32 Wi-Fi module connected.

Figure 9.89 Compiling and deploying the sample Wi-Fi scan program into ESP32 Wi-Fi module.

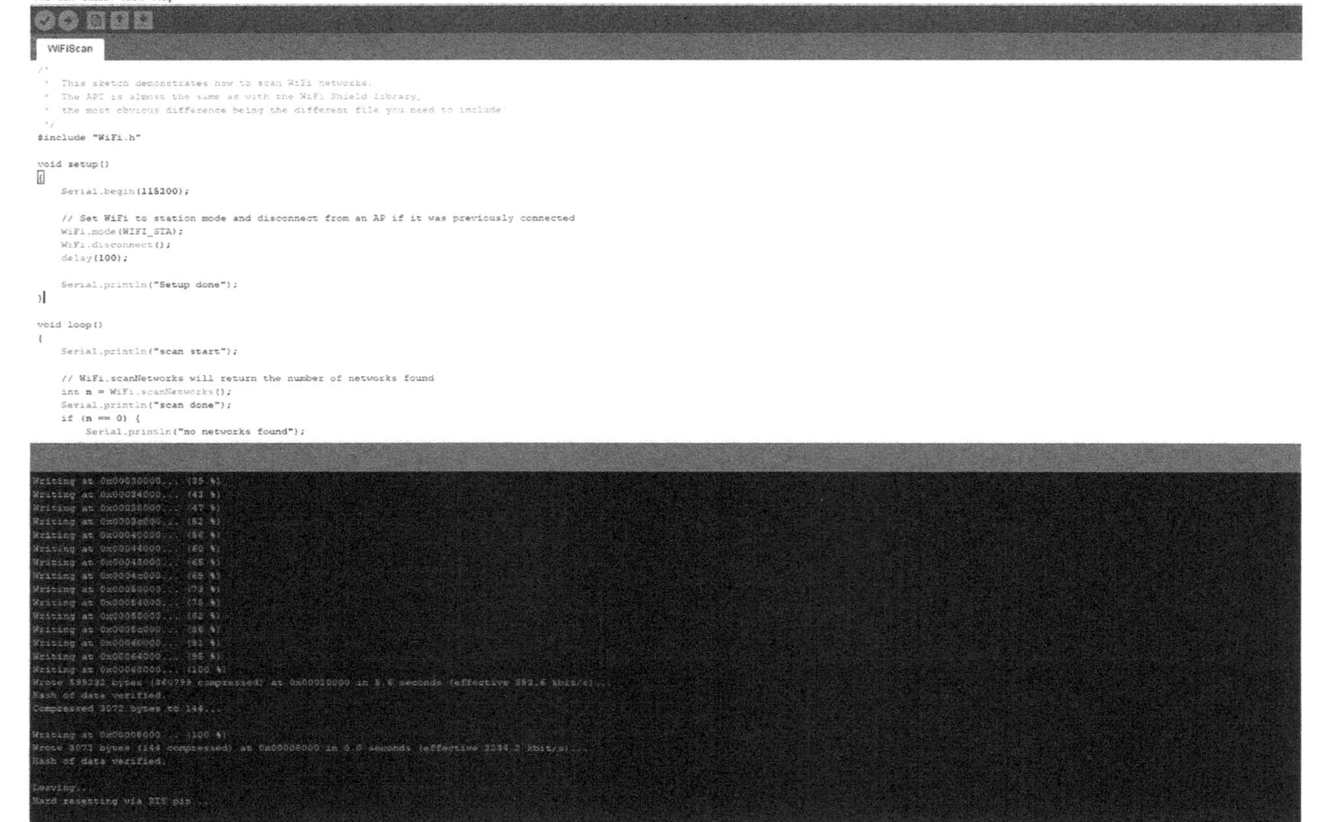

Figure 9.90 Code upload status to the ESP32 Wifi module.

Figure 9.91 Wi-Fi scan sample program output shown in Arduino serial monitor.

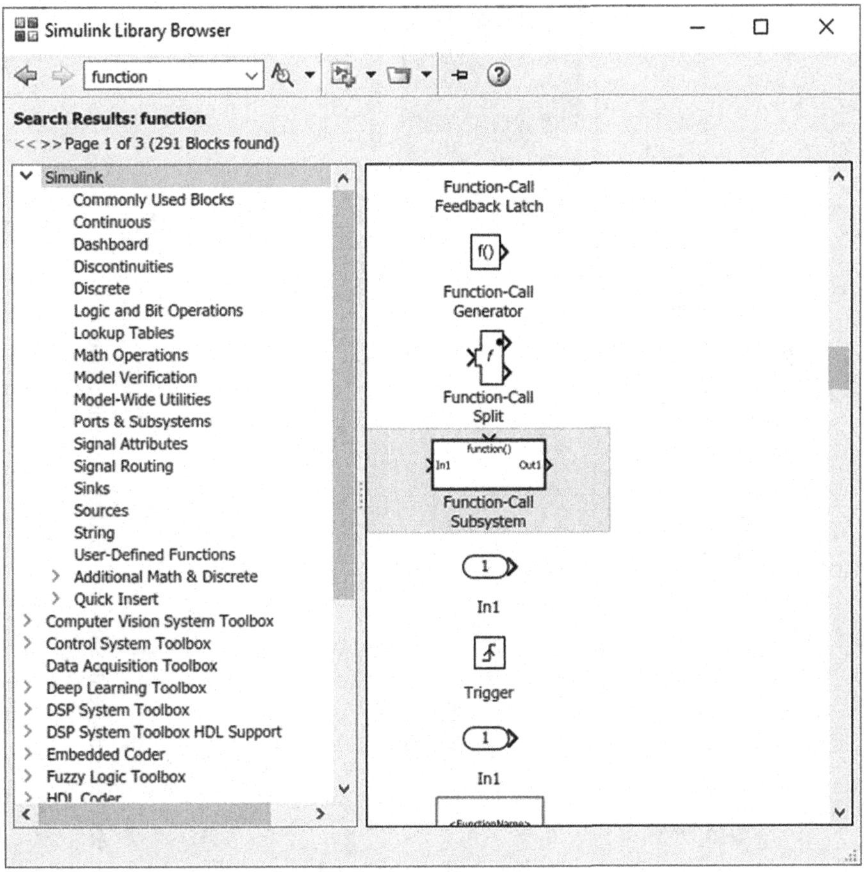

Figure 9.92 Adding Function Call Subsystem from Simulink Library.

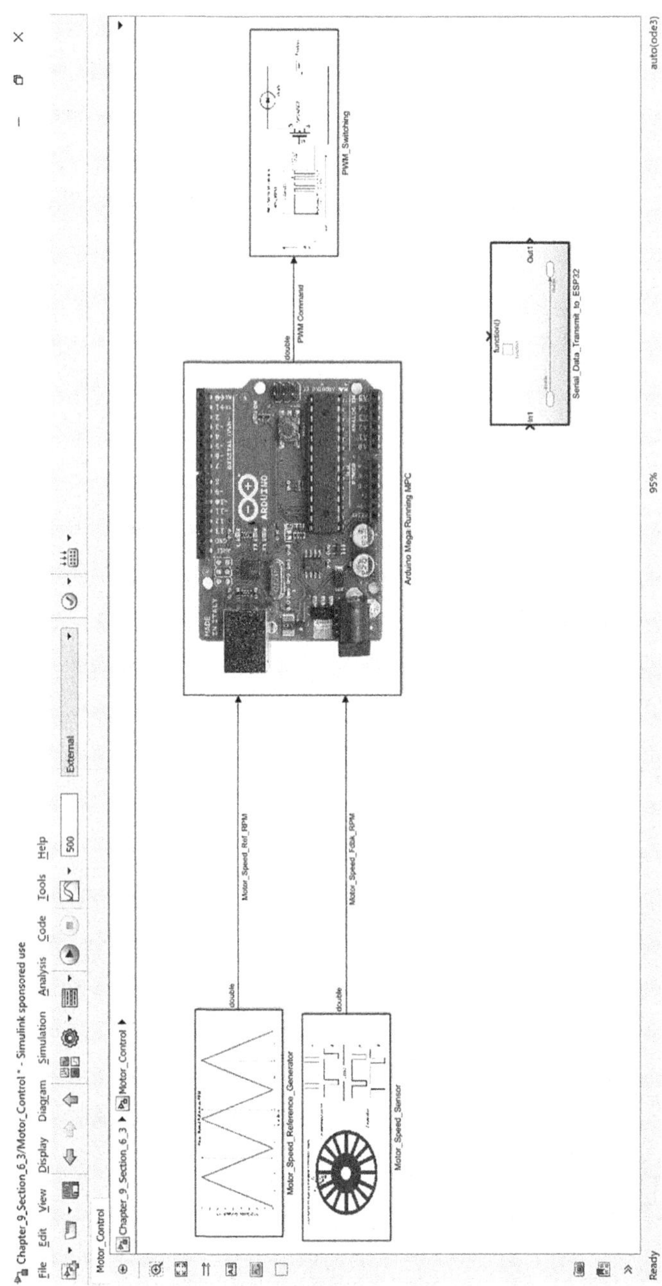

Figure 9.93 Function Call Subsystem added to Simulink model.

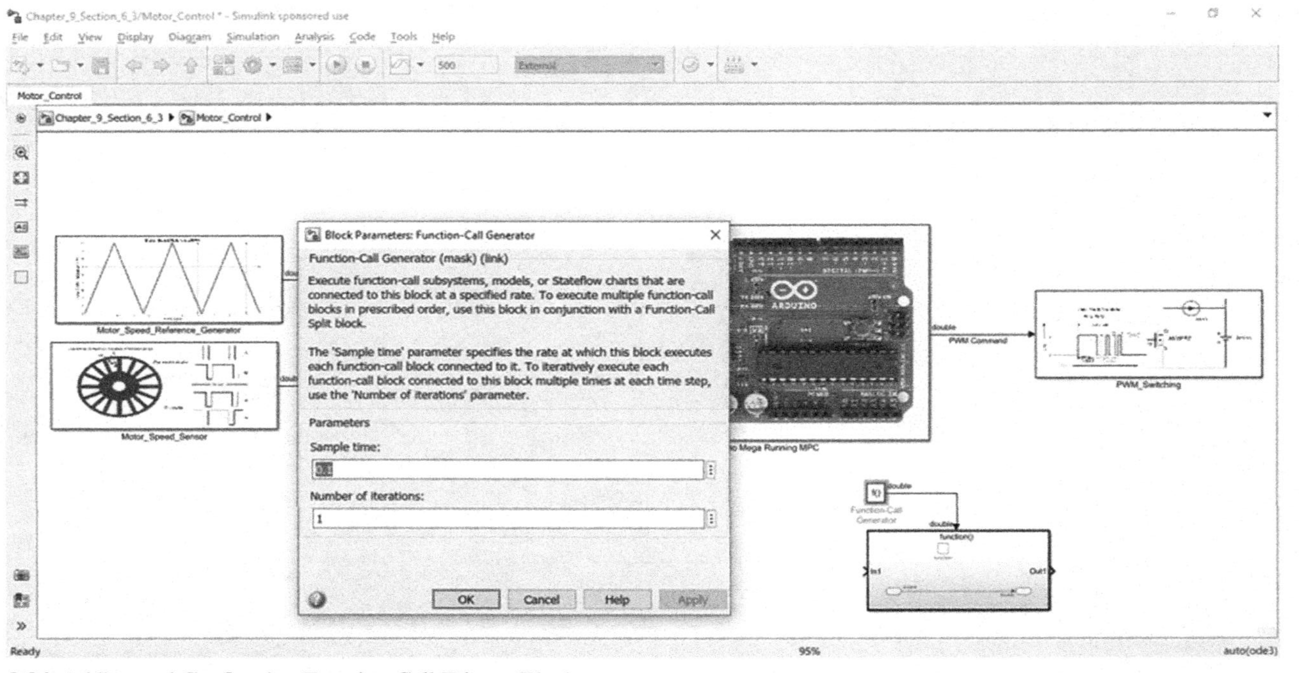

Figure 9.94 Adding and Configuring Function Call Trigger Block.

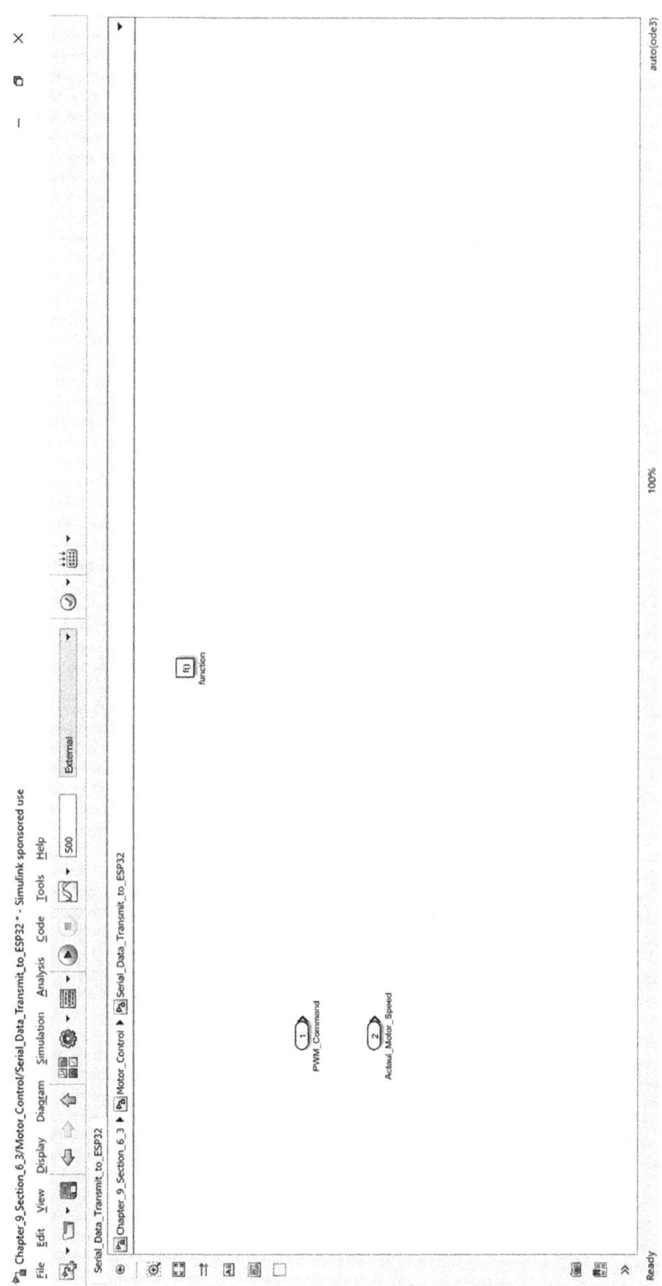

Figure 9.95 Adding input ports for PWM_Command and Actual_Motor_Speed.

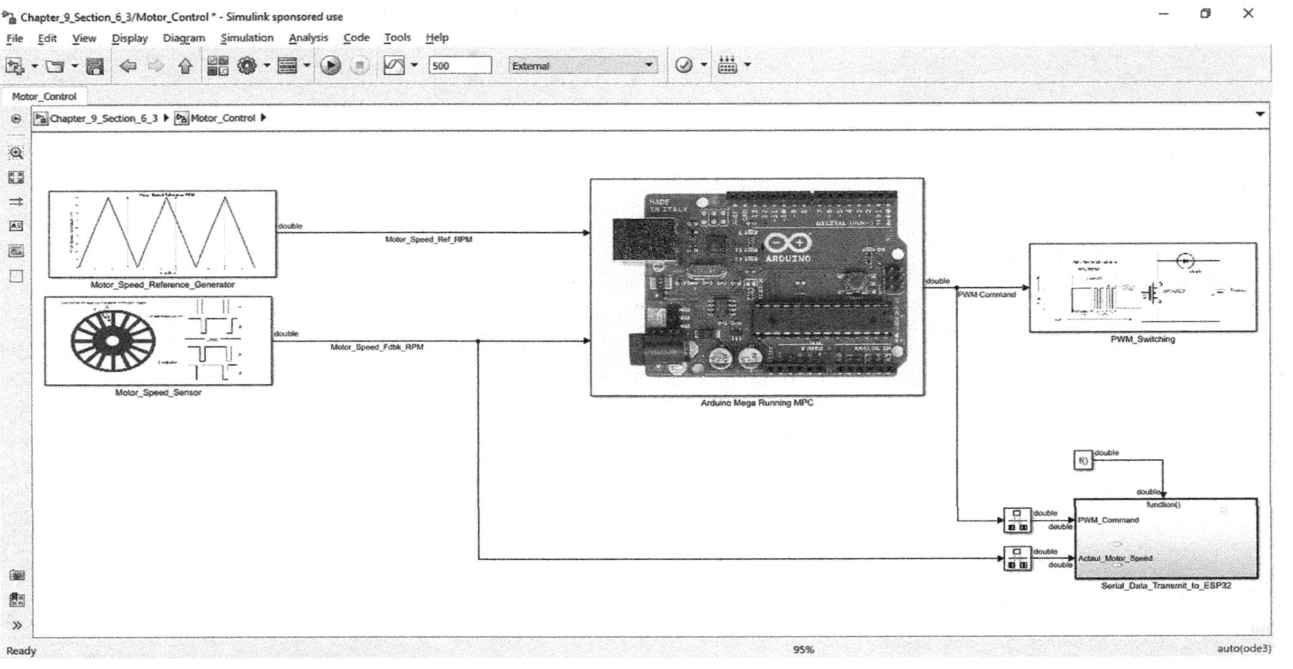

Figure 9.96 Making connections at the top level.

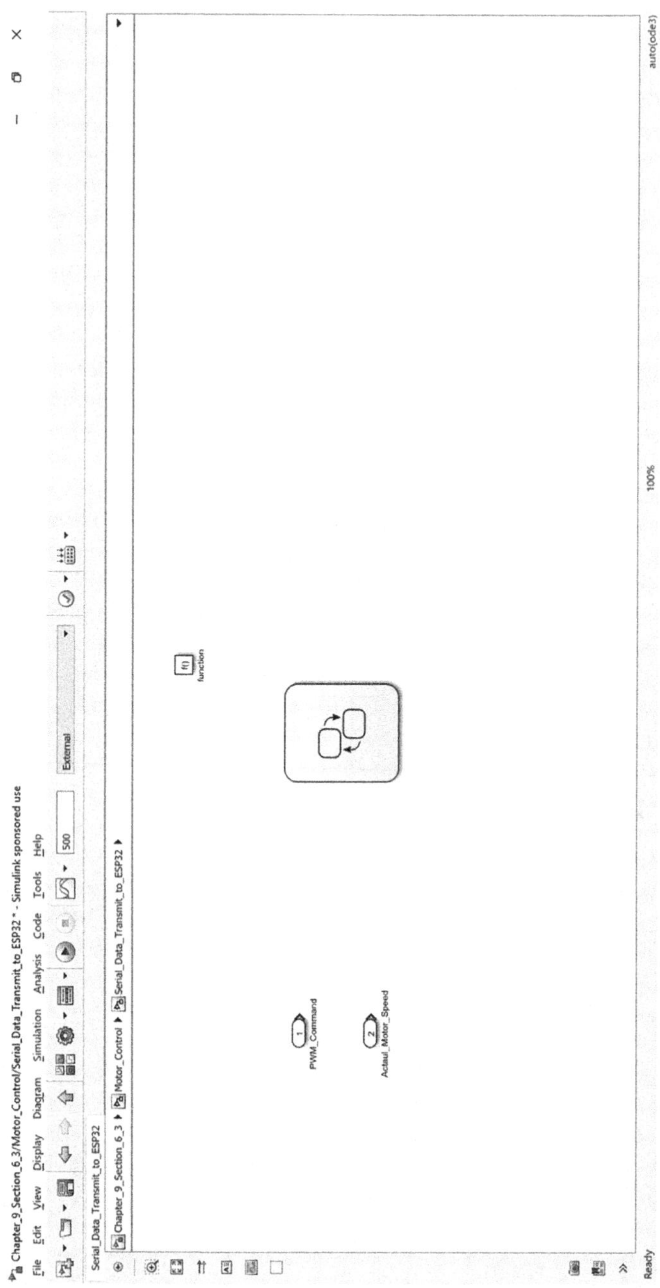

Figure 9.97 Adding Stateflow chart for the Serial Data Sending Logic.

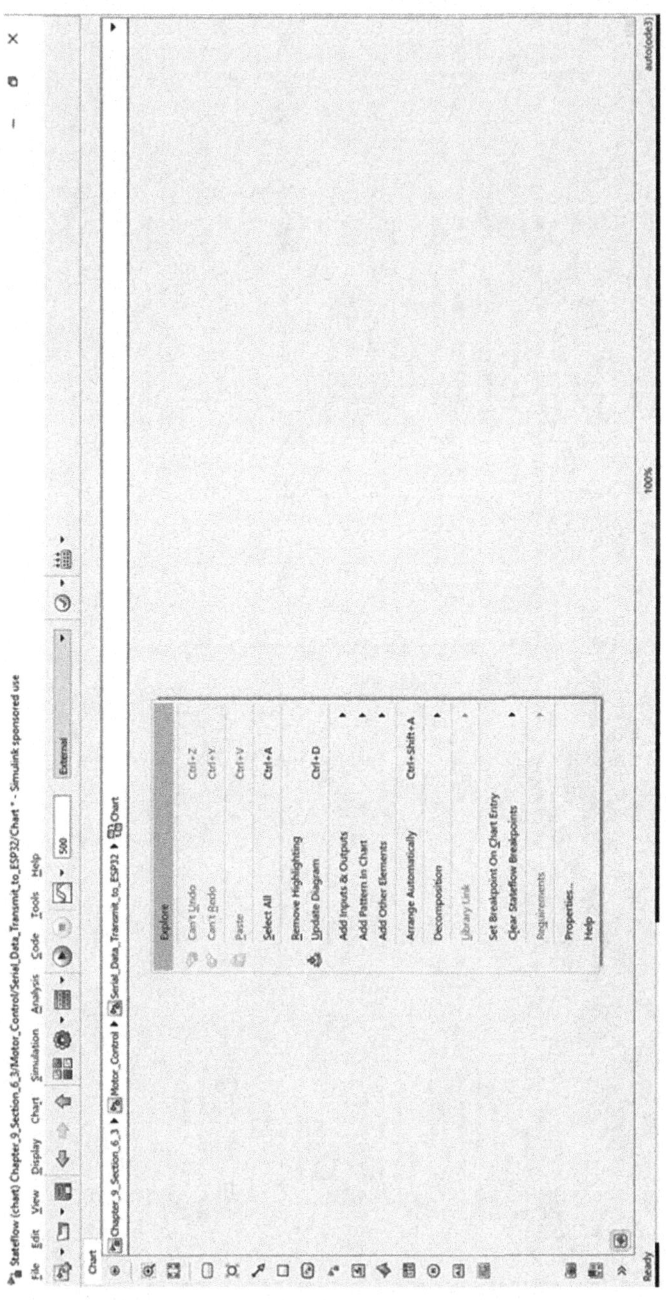

Figure 9.98 Opening model explorer for defining stateflow interface.

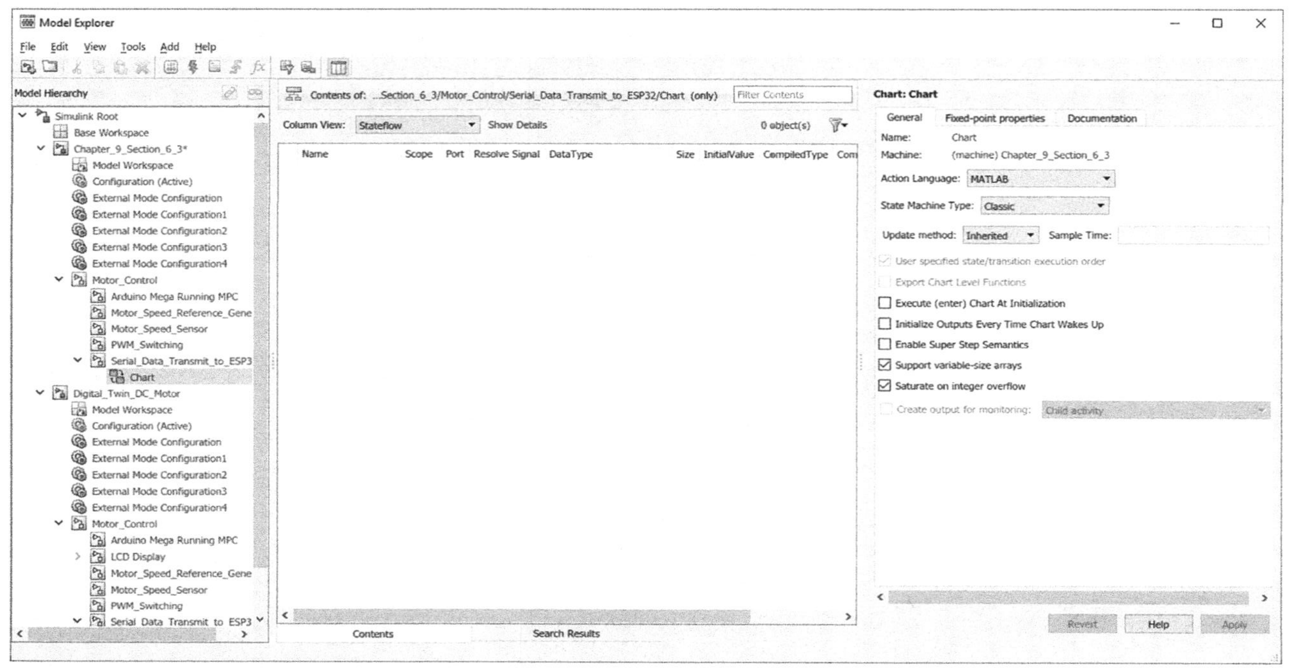

Figure 9.99 Model Explorer Window.

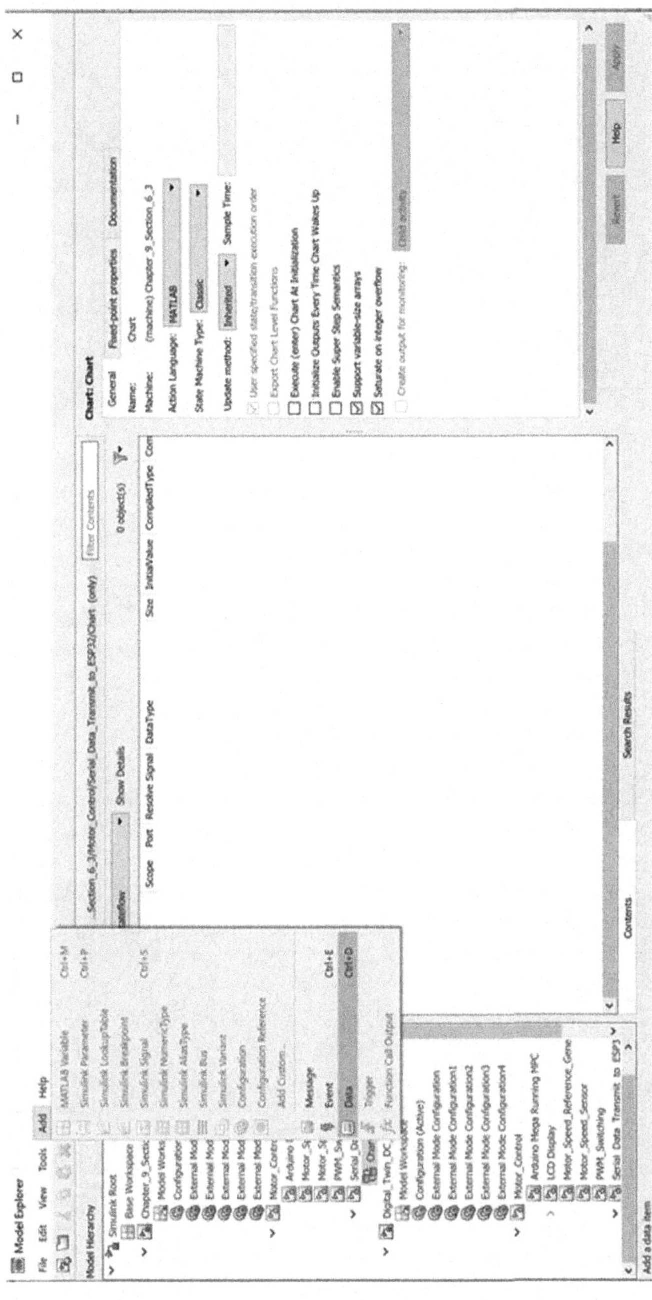

Figure 9.100 Option to input/output data and trigger to the stateflow chart.

Figure 9.101 Model Data Explorer Window after adding the input/output data and trigger.

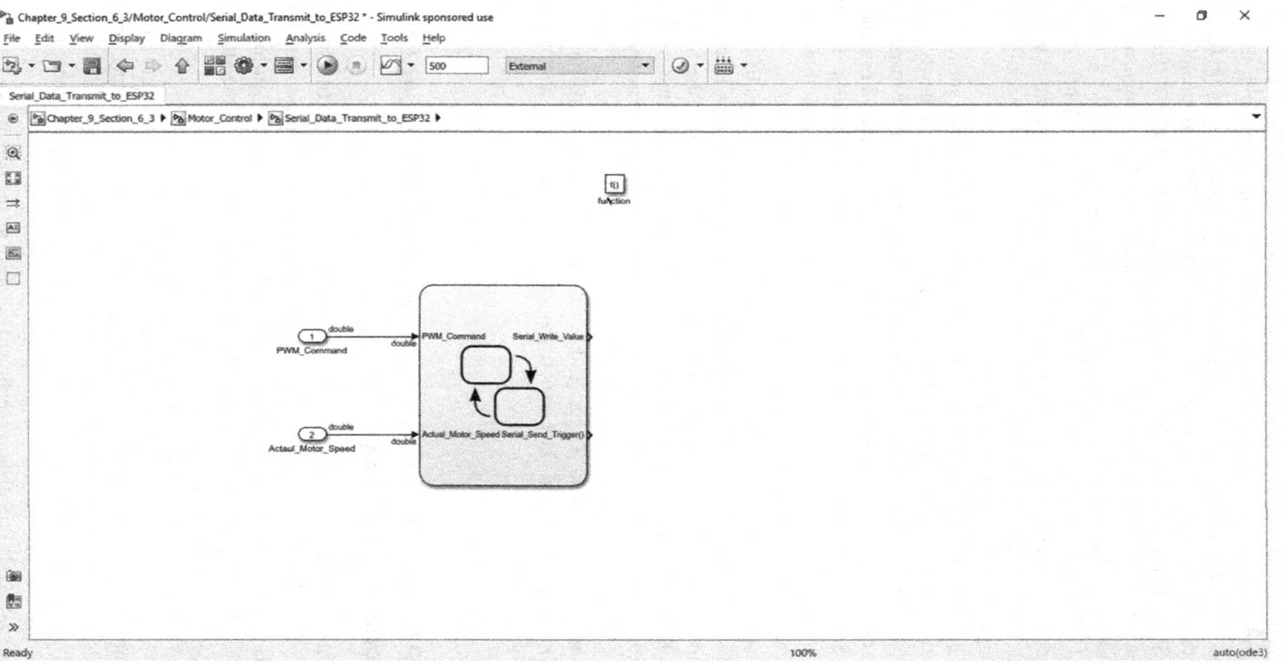

Figure 9.102 Stateflow chart showing the input/outputs added.

8. Now we will add logic to the Stateflow chart to direct the output data and control the event trigger. Add a **Default Transition** to the chart as shown in Fig. 9.103. And on the transition line first assign the ***Serial_Write_Value*** output with the input ***PWM_Command*** and call the ***Serial_Send_Trigger.*** Then assign the ***Serial_Write_Value*** output with the ***Actual_Motor_Speed*** input and call the ***Serial_Send_Trigger*** again. Ideally, we could have send these two data directly using the Arduino Serial Transmit block. But unfortunately we can place only one Serial Transmit block with a specific port in the Simulink model. That is the reason why we have to make this Stateflow chart logic and send the PWM_Command and Actual_Motor_Speed separately instead of sending them together. So every 0.1 s it will first send the PWM_Command and immediately followed by the Actual_Motor_Speed. We will have to keep this in mind when we receive this message at the ESP32 end as well to separate and handle these data. We will discuss about that in the later section. Check Fig. 9.104 to see the new logic added in the Stateflow chart.
9. Now add one more **Called Function** block inside the **Called Function** subsystem we added earlier, next to the Stateflow chart as shown in Fig. 9.105. Inside the newly added subsystem, delete the default output port and connect the **Arduino Serial Transmit** block to the input port as shown in Fig. 9.106 and 9.107. Double click on the **Serial Transmit** block and enter the **Serial Port** number as 1, and select the **Serial Mode** as **println.** The **println** option will send a newline character at the end of every data transmission, we will use this newline character to read data from the serial buffer at the ESP32 end.
10. We will have to configure the serial transmit data baud rate in the Simulink model configuration. Go to **Simulation >> Model Configuration Parameters >> Hardware Implementation >> Serial Port Properties** and change the baud rate of the Serial port 1 to 115,200. Keep the other baud rates unchanged. We will use the same baud rate at the ESP32 programming side as well to make sure the transmit and receive is working properly. See Fig. 9.108.
11. Now connect the Serial_Send_Trigger and Serial_Write_Value outputs from the Stateflow chart to the Trigger input and Data input of the Called Function subsystem, respectively, as shown in Fig. 9.109.

9.6.4 ESP32 Arduino software for communication between Arduino and ESP32

In the previous section, we setup the Simulink model for Arduino to transmit the Motor control data to the ESP32. In this section, we will create an Arduino sketch file in the Arduino IDE to program the ESP32 to receive the data sent from Arduino motor control hardware. The program described in this section is available in the attachment section of this chapter. Follow the steps below:

1. We can start from a blank sketch by going to **File >> New** from the Arduino IDE and save it with a name. By default when we create a new sketch, it will add a **setup()** function which gets executed one time during the powerup of the ESP32 and a **loop()** function which runs periodically during ESP32 is powered. See Fig. 9.110. We will update these functions later with some functionality.
2. We will add a variable initialization section to initialize all the variables we are going to use in the sketch. The goal of this program is to listen to the serial receive port of the ESP32 for incoming data sent by the Arduino motor control hardware and create JSON data structures

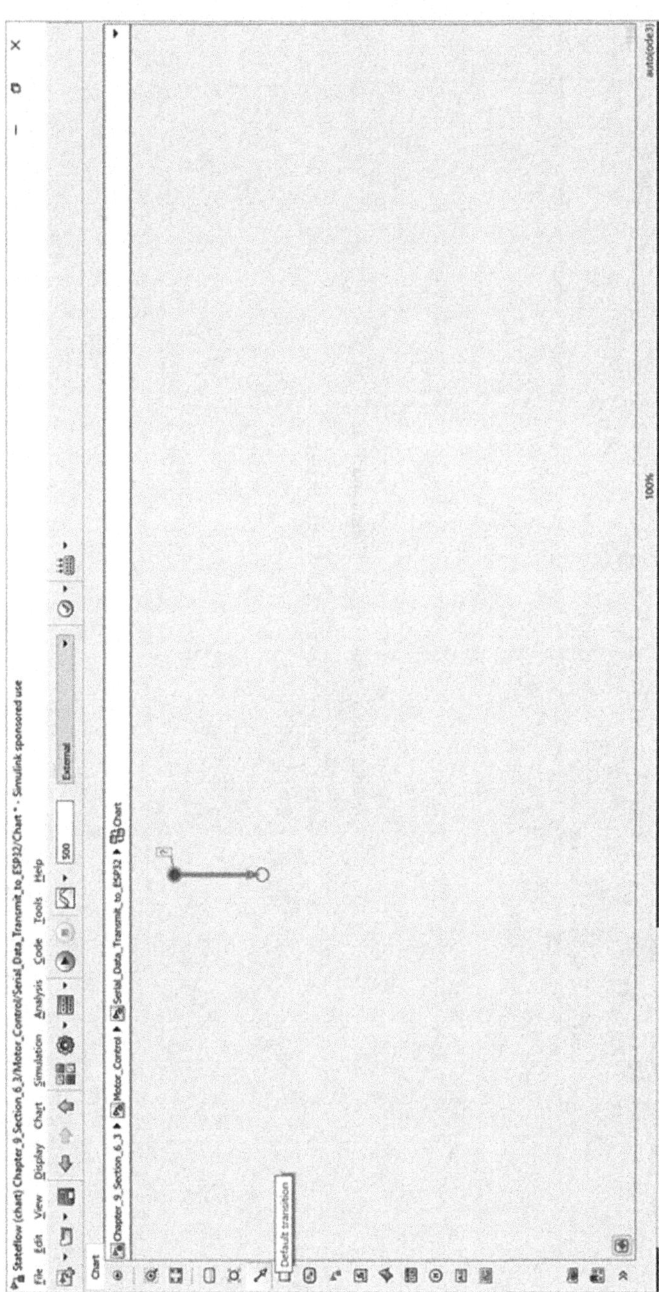

Figure 9.103 Adding Transition flow to Stateflow chart.

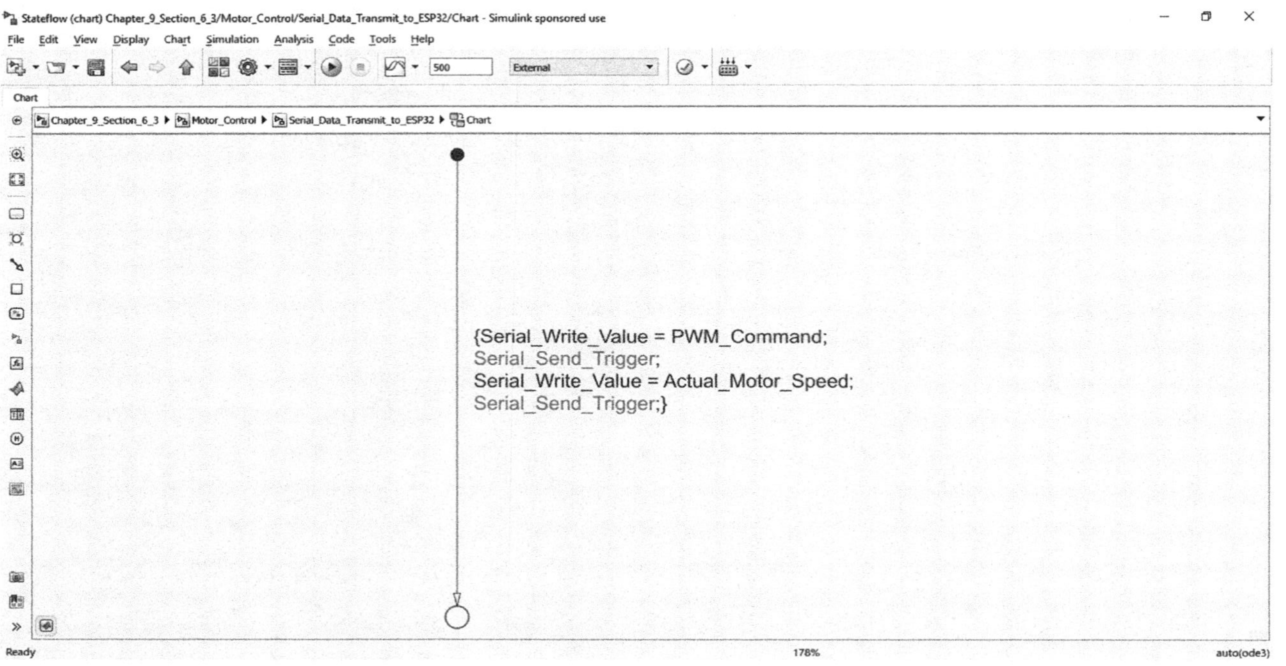

Figure 9.104 Adding logic to output the data and trigger from Stateflow chart.

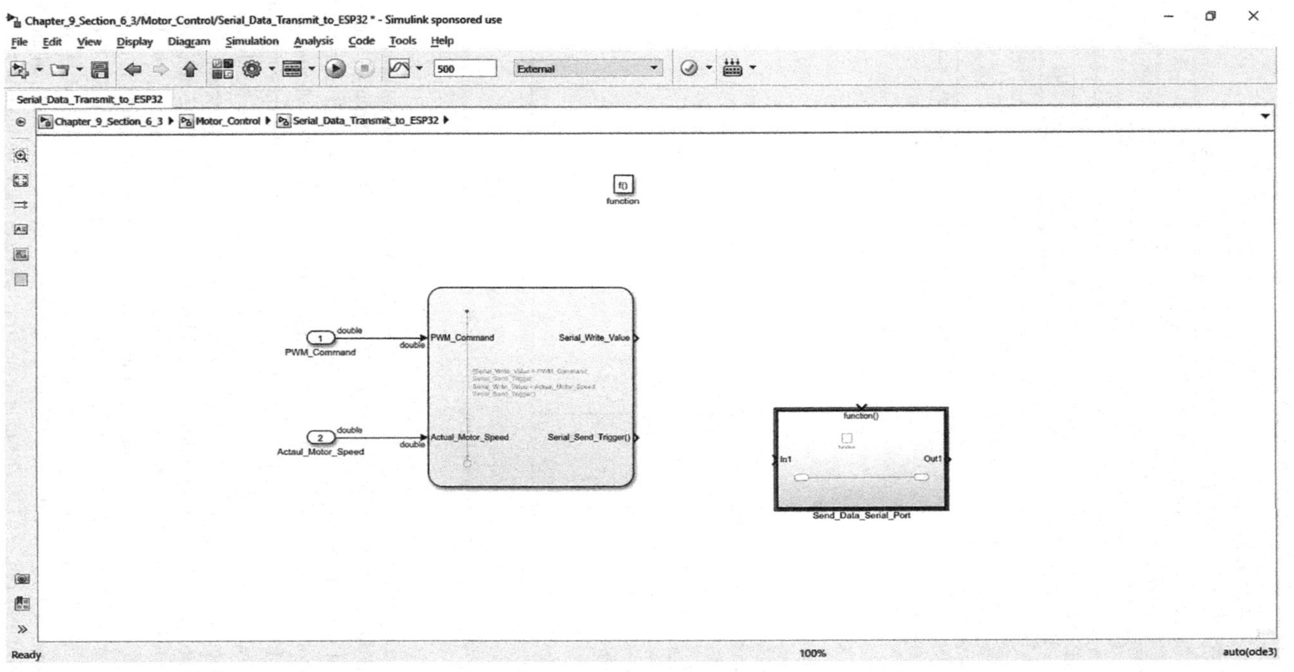

Figure 9.105 Adding Called Function to Send Serial Data using Arduino Serial Transmit.

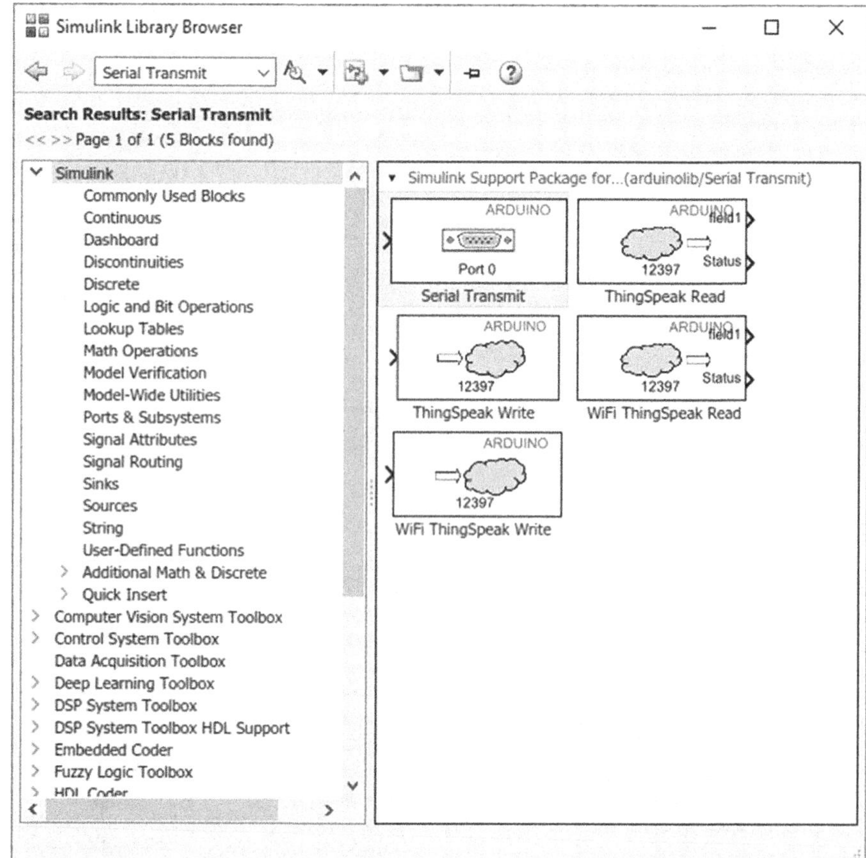

Figure 9.106 Adding Arduino serial transmit block to the model.

with data available at every 3 s. It should also be noted that we are sending PWM_Command and Actual_Motor_Speed as two different messages from Arduino motor control, so the first message expected is for PWM_Command and the second message would be Actual_Motor_Speed. Each message will end with a newline "\n" character which we will look for in the program to segregate the data bytes from the message buffer. The data are transmitted at a rate of 0.1 s, so for gathering 3 s data, we need arrays of size 30. After every 3 s, the last known data for PWM_Command and Actual_Motor_Speed will be stored into the corresponding prev_value variables, we will use this for initializing the Digital Twin model to bring it to a steady state before running it with the newly collected 3 s data. We will use the variable **message_counter** to count the incoming messages and when it becomes 30 we will conclude that 3 s data have arrived. The flag **first_message** is used and updated internally to decide whether the data received are PWM_Command or Actual_Motor_Speed. All these variables are declared and initialized first as shown in Fig. 9.111. The character array **payload** will be used to form the JSON structure array with the incoming data.

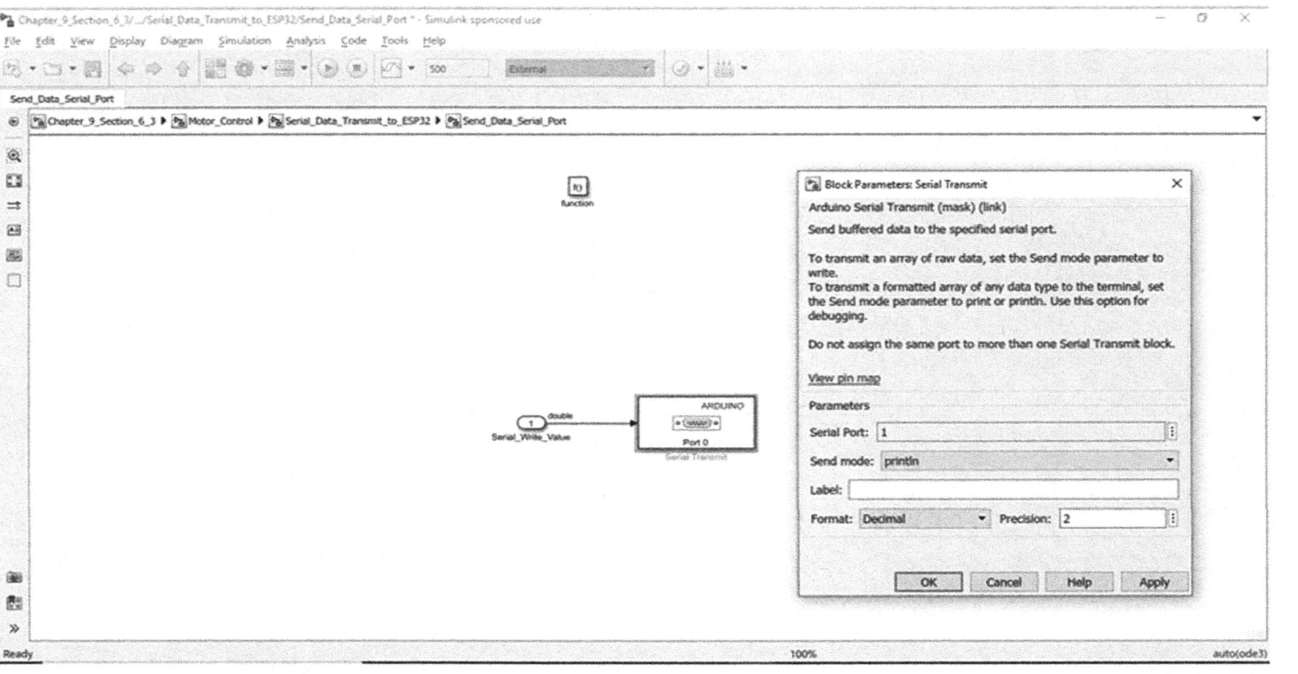

Figure 9.107 Connecting and configuring Arduino Serial Transmit block.

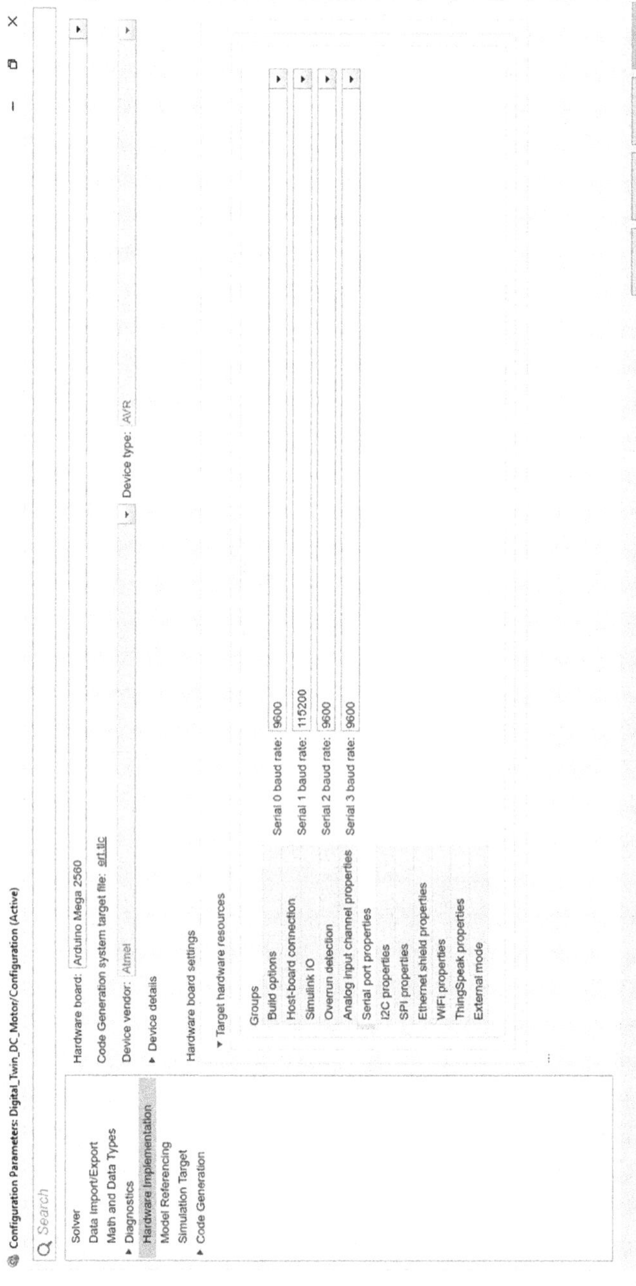

Figure 9.108 Setting serial port baud rate properties.

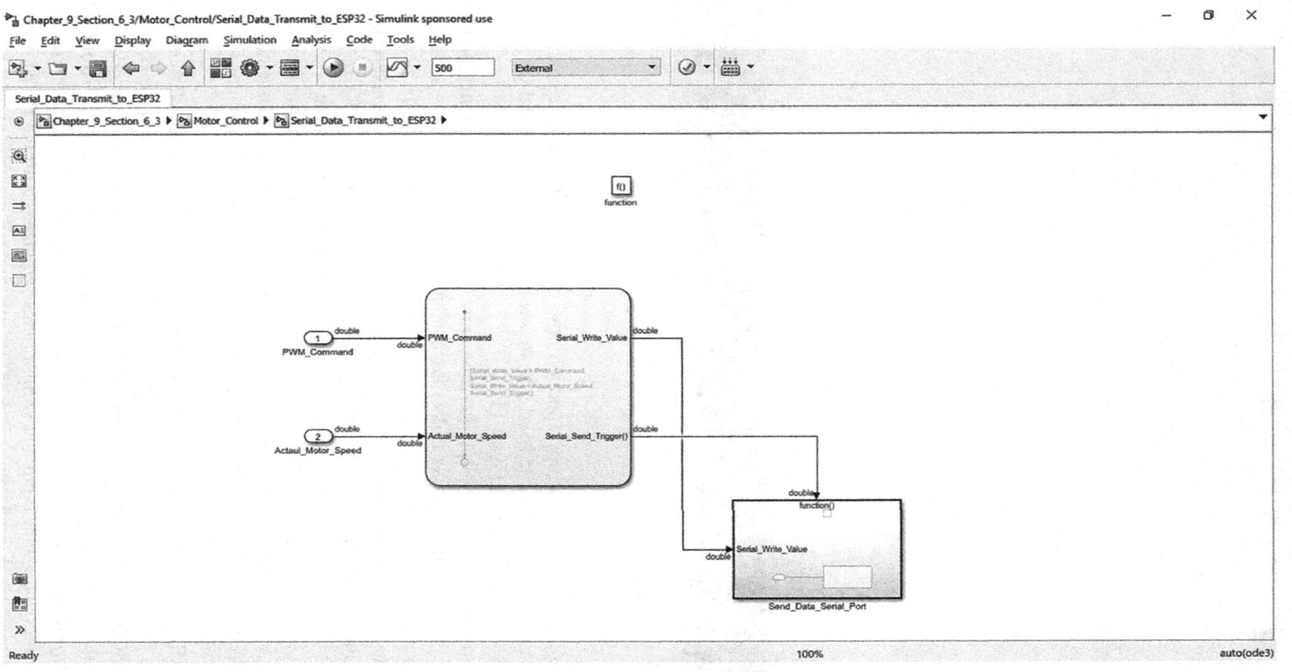

Figure 9.109 Connecting the data and trigger line for the subsystem with serial data transmit.

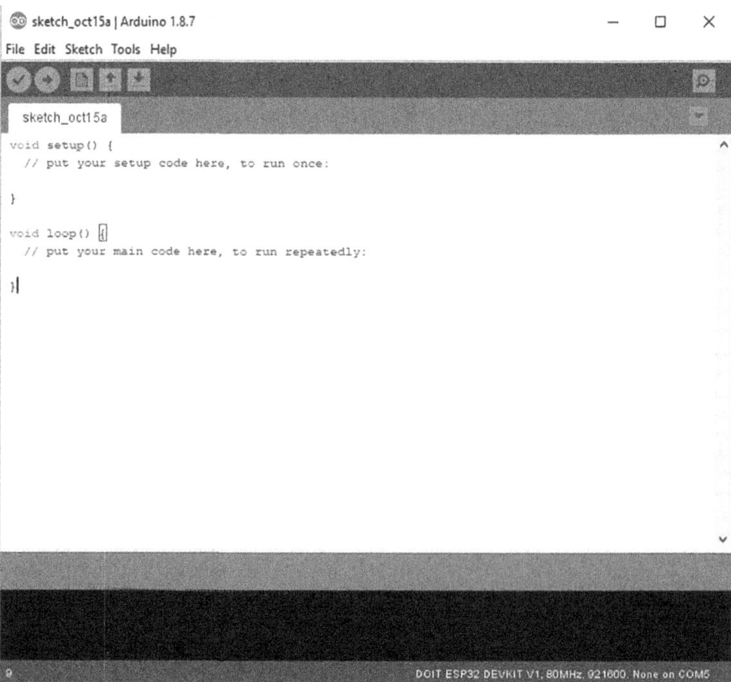

Figure 9.110 Creating a new sketch for ESP32 programming.

3. We will update the default setup() function now with a serial baud rate setting of 115,200. Please note that this is the same baud rate we set in the Simulink model for Arduino motor control hardware Port 1. See Fig. 9.112.
4. We will update the default loop() function now. This function executes periodically and every time it runs it first checks if data are available or received on the serial port. If there are data available, we will copy that data into a temporary buffer until the newline character "\n" is detected. And we check if this is the first_message, if so we convert the data to an integer value and store into the PWM_Command array. Once PWM_Command array is updated, it will make the first_message variable 1 so that the next time we will update the Actual_Motor_Speed array in the else condition which will then set this first_message variable back to 0 and this cycle repeats. Also everytime both PWM_Command and Actual_Motor_Speed are received for one sample we update the message_counter variable. Check Fig. 9.113 for more details.
5. Next we will check for the message_conter variable to see if it reached 30 to indicate that we have collected data for 3 s and the data arrays are full for usage. When the counter is 30, we create a JSON structure using the **sprint** built in function into the variable payload. The format of the JSON structure will look like **"{\"state\":{\"desired\":{\"Input\":[data],\"Output\":[data]}}}.** So in summary, the JSON structure is reporting the current state of the DC Motor hardware with Input and Output fields populated with the PWM_Command and Actual_Motor_Speed array values. The JSON structure will be printed on the Serial Monitor window. The first data for Input and Output in the JSON structure will be the prev_values

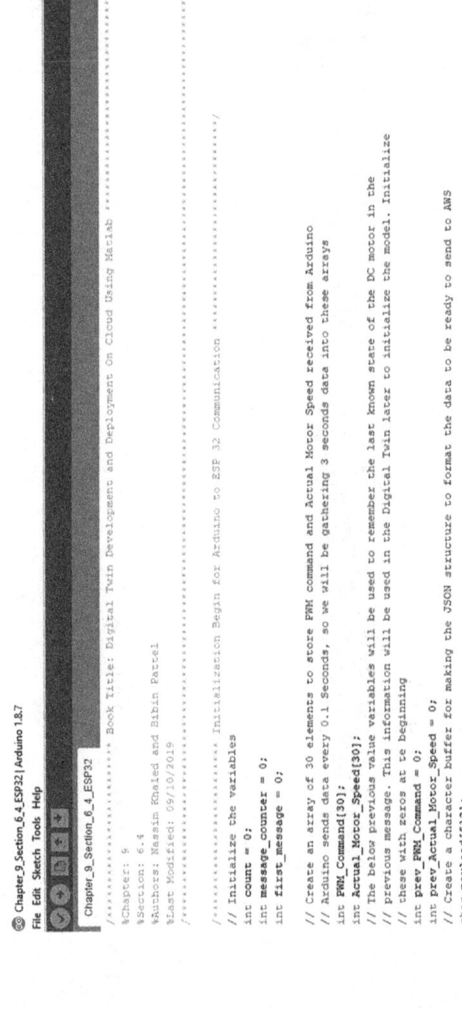


```
/*********************** Book Title: Digital Twin Development and Deployment On Cloud Using Matlab ***********************
%Chapter: 9
%Section: 6.4
%Authors: Nassim Khaled and Bibin Pattel
%Last Modified: 09/10/2019
*************************************************************************************************************************/

/*********************** Initialization Begin for Arduino to ESP 32 Communication ***********************/
// Initialize the variables
int count = 0;
int message_counter = 0;
int first_message = 0;

// Create an array of 30 elements to store PWM command and Actual Motor Speed received from Arduino
// Arduino sends data every 0.1 Seconds, so we will be gathering 3 seconds data into these arrays
int PWM_Command[30];
int Actual_Motor_Speed[30];
// The below previous value variables will be used to remember the last known state of the DC motor in the
// previous message. This information will be used in the Digital Twin later to initialize the model. Initialize
// these with zeros at te beginning
int prev_PWM_Command = 0;
int prev_Actual_Motor_Speed = 0;
// Create a character buffer for making the JSON structure to format the data to be ready to send to AWS
char payload[512];

/*********************** Initialization End for Arduino to ESP 32 Communication ***********************/
```

Figure 9.111 Initializing the variables used in the sketch.

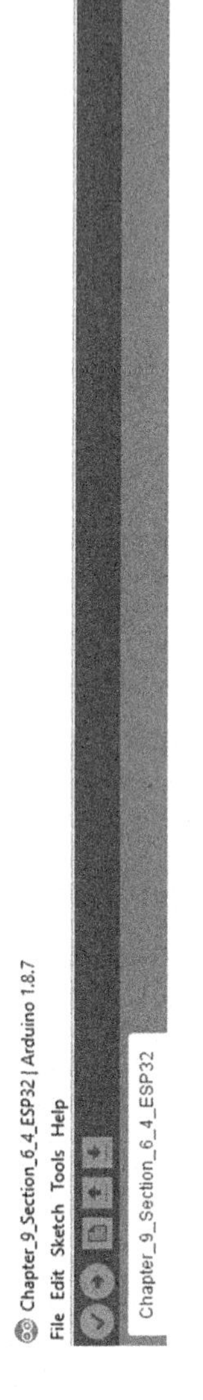

Figure 9.112 Powerup setup() function.

Chapter_9_Section_6_4_ESP32 | Arduino 1.8.7

File Edit Sketch Tools Help

Chapter_9_Section_6_4_ESP32

```
// The loop function runs periodically
void loop()
{
  // If there is any serial data available this if condition gets triggered
  if (Serial.available() > 0)
    {
      // Increament the counter to indicate the total message count
      count = count + 1;
      // Initialize a temporary character buffer to copy the incoming serial data
      char bfr[501];

      memset(bfr, 0, 501);
      // From the buffer read the bytes until there is a new line character '\n'. Note that from the Simulink model we are sending a newline
      // character for each data
      Serial.readBytesUntil('\n', bfr, 500);
      // COpy the message data until newline character into a string
      String string_data ;
      string_data = bfr;
      // We are sending the PWM_Command and Actual_Motor_Speed data as two separate packets from Simulink. The first data sent is PWM_Command and the
      // second data sent is Actual_Motor_Speed. If first_message flag is 0, then it indicates it is the PWM_Command data, and first_message flag is 1,
      // then it indicates it is the Actual_Motor_Speed data. We need to a string to integer conversion using the toInt() function as shown below. The
      // variable message_counter is incremented every time we receive both the PWM_Command and Actual_Motor_Speed data. Once message_counter counted up
      // to 30 means the data array is full and we recieved 3 seconds data from the Arduino hardware
      if (first_message == 0)
        {
          first_message = 1;
          PWM_Command[message_counter] = string_data.toInt();
        }
       else
        {
          first_message = 0;
          Actual_Motor_Speed[message_counter] = string_data.toInt();
          message_counter = message_counter + 1;
        }
```

Figure 9.113 Runtime periodic loop() function Part 1.

memorized from the last 3 s data. So the prev_values will be updated and message_counter will be set back to 0 to collect next 3 s data and the cycle continues. See Fig. 9.114.

6. Now the ESP32 program to receive data from Arduino DC Motor Controller is ready. From the hardware setup we finished in Section 9.6.1 connect the ESP32 controller board to the computer using a USB to Micro USB cable and build the ESP32 Arduino program by selecting the option **Sketch >> Upload** as shown in Fig. 9.115 and also open a Serial monitor window and set its baud rate to 115,200. After the ESP32 sketch is successfully downloaded, now connect the Arduino Mega microcontroller board which controls the DC motor to a different USB port of the computer and from MATLAB build and download the Simulink model we completed in Section 9.6.3.

   Please note whenever we program ESP32, disconnect the Arduino Mega USB connection to the PC to avoid interactions through the wired serial port between Arduino and ESP32 boards.
   Also it was observed that when we added the stateflow block in the Simulink model, we are no longer able to run the simulation in external mode with the Arduino Mega hardware. It gives an "nmake" compiler error. This seems like a MATLAB bug at this point. So do not try hit the run button instead hit the build button, which will still compile and download the Simulink program to Arduino Mega. Just that we will not be able to do interactive simulation.

7. After the ESP32 and Arduino Mega boards are successfully programmed we can observe that the DC Motor will now start running and in the ESP32 Arduino Serial monitor we can see every 3 s it updates a JSON string with the input PWM_Command and Actual_Motor_Speed for the past 3 s as shown in Fig. 9.116. A close in view of one of the JSON data structure is shown in Fig. 9.117. We can see that because of the time varying Motor speed reference, the PID controller is steadily increasing the PWM_Command and the Actual_Motor_Speed is getting increased gradually.

9.6.5 DC motor hardware connectivity to AWS cloud and real-time data transfer

So far we have the data for the DC Motor sensor and actuator ready in the ESP32 Wi-Fi module. In this section, we will send that data to AWS Cloud using the Wi-Fi capability of ESP32. We will first establish a Wi-Fi connection to a Router which is connected to Internet. And then using an AWS-IoT library for ESP32 in Arduino IDE, we will establish a connection to AWS cloud and exchange data collected from DC Motor hardware. First, we will start AWS side steps, and we have to get some specific files when we set up AWS which will then be used to program the ESP32 for AWS-IoT connection. We will use the AWS service called AWS IoT Core in this section, which lets the connected devices securely interact with cloud or other connected devices. For more information about the AWS IoT Core services follow the link https://aws.amazon.com/iot-core/. Fig. 9.118 shows the setup we are trying to establish in this section, which establishes a connection between ESP32 and AWS IoT Core and transmit data from DC Motor hardware that is connected to the ESP32.

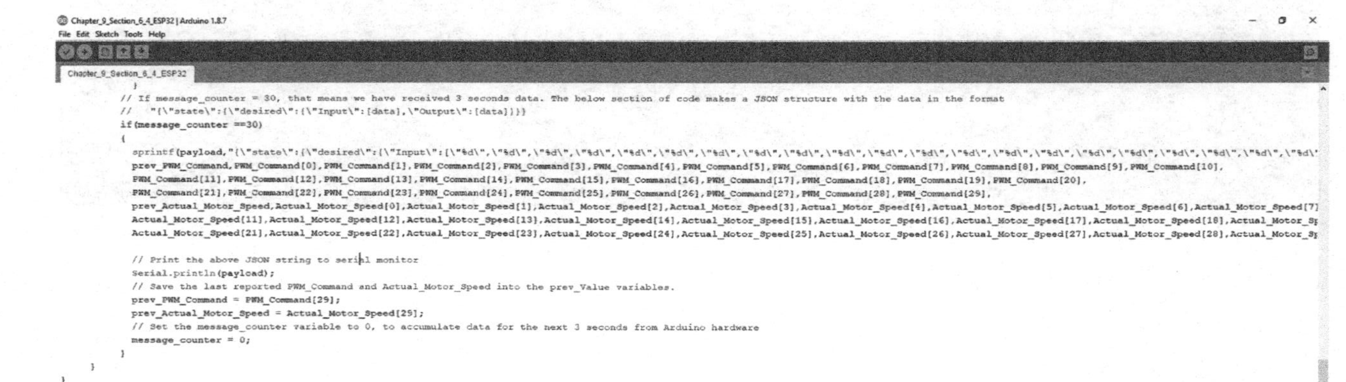

Figure 9.114 Runtime Periodic loop() function Part 2.

Figure 9.115 Downloading ESP32 program.

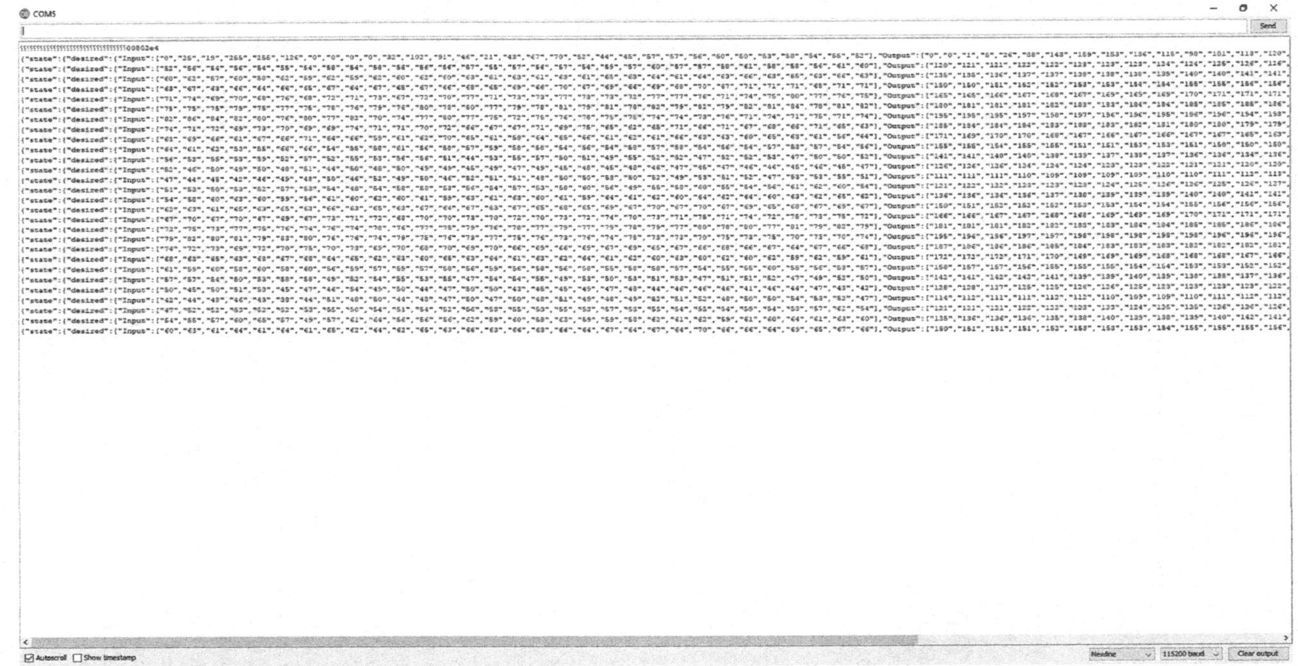

Figure 9.116 ESP32 serial monitor showing formatted JSON data received from Arduino Mega.

{"state":{"desired":{"In-
put":["52","56","54","56","54","55","54","58","54","58","56","56","56","57","55"
,"57","56","57","54","59","57","60","57","57","58","61","58","58","56","61","60"
],"Out-
put":["120","121","121","122","122","123","123","123","124","124","125","126",
"126","127","127","128","128","129","129","130","130","130","131","132","132"
,"132","133","134","135","135","135"]}}}

Figure 9.117 One of the 3 s JSON data structure showing Input PWM_Command and Output Actual_Motor_Speed.

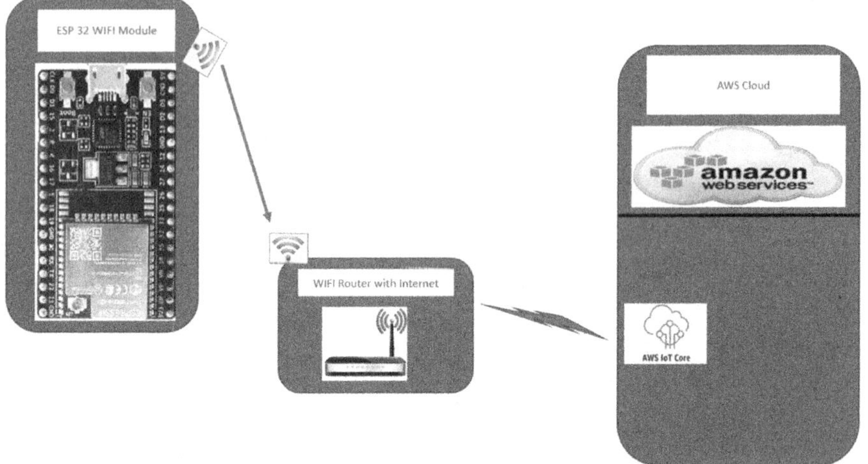

Figure 9.118 Establishing ESP32 and AWS IoT connection.

1. As a first step, User needs to create an AWS Management Console account if you don't already have one. If you already have an Amazon account, you can use the same account information to login. If you don't have an account please create an account. Follow this link https://aws.amazon.com/console/ on a web browser for logging in, creating a new account, etc. Use the ***Sign in to the Console*** option in the main page as shown in Fig. 9.119 and login using the credentials as shown in Fig. 9.120.
2. On the logged in Console window search for **IoT** and select the **IoT Core** as shown in Fig. 9.121. The AWS IoT window will be loaded and it will display all the **AWS IoT Things** available in the User account. An **AWS IoT Thing** represents an **IoT** connected device, in this case it will represent our ESP32 Wi-Fi module which is connected to DC Motor and AWS cloud. The Author already has few **IoT Things** created previously on the account, which is the reason why it displays some **IoT Things** already as shown in Fig. 9.122. For security reasons, the already existing **IoT Things** are not shown in the images. We will be following similar graying out for some parts of the images for protecting Author's information, but for the new additions that we are doing for this book, we will display it in the images.
3. Before creating a new **IoT Thing**, we need to create a **Policy.** A **Policy** is a JSON formatted document which allows access to the AWS services for the connected devices. We need to

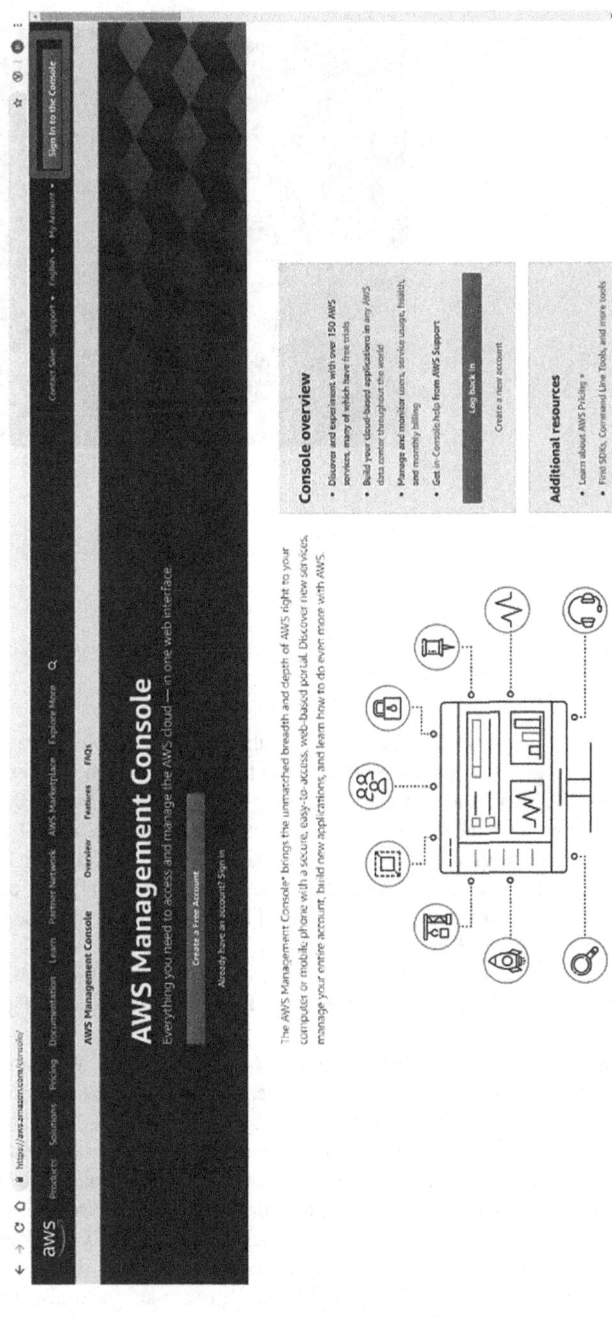

Figure 9.119 AWS Console login page.

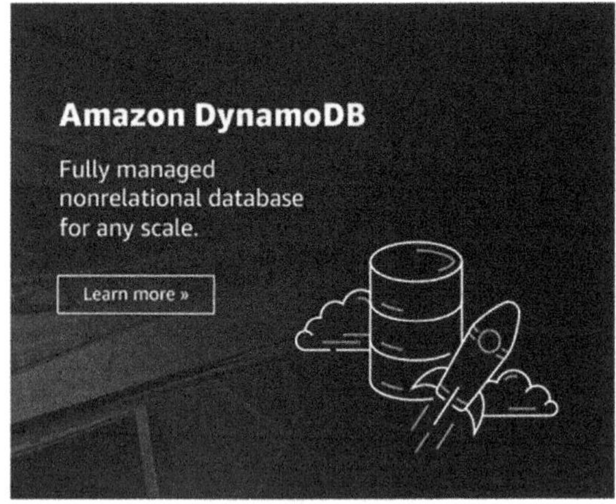

Figure 9.120 AWS Console Login.

create a **Policy** and link it with the **IoT Thing** for the ESP32 to establish a connection with the **AWS IoT Thing.** Go to **Secure >> Policies** on the left side tabbed window of the AWS Console under **AWS IoT** and click on the **Create** button as shown in Fig. 9.123.

4. On the new window, enter the Policy name. This case we will name it as **digital_twin_policy.** But User can choose to give a different name if needed. For the **Action** field enter **iot***, **Resource ARN** enter *, and for **Effect** check the **Allow** option and click on the Create button as shown in Fig. 9.124. The newly created Policy will show up under the Policies section as shown in Fig. 9.125.
5. The next step is to create an **AWS IoT Thing.** Go to **Manage >> Things** menu and click on **Create.** From the next window, click on **Create a Single Thing** as shown in Figs. 9.126 and 9.127.
6. Enter the name of the new **AWS IoT Thing** as shown in Fig. 9.128. We name it **digital_-twin_thing** but User can choose a different name. We can leave the rest of the fields as default. Click **Next** at the bottom of the page.
7. We will use a certificate-based authentication for the ESP32 device to connect to the **AWS IoT Thing** for secure connection. So we have to generate and download the certificates and use it to program the ESP32 while establishing connection with **AWS IoT.** From the IoT Thing creation window, shown in Fig. 9.129, click on the **Create Certificate** option. It will generate a private key, a public key, and a certificate files as shown in Fig. 9.130. Download the certificate for the Thing, public key, private key, and also the **Root CA** file to somewhere safely into the computer. We will be using these files later for

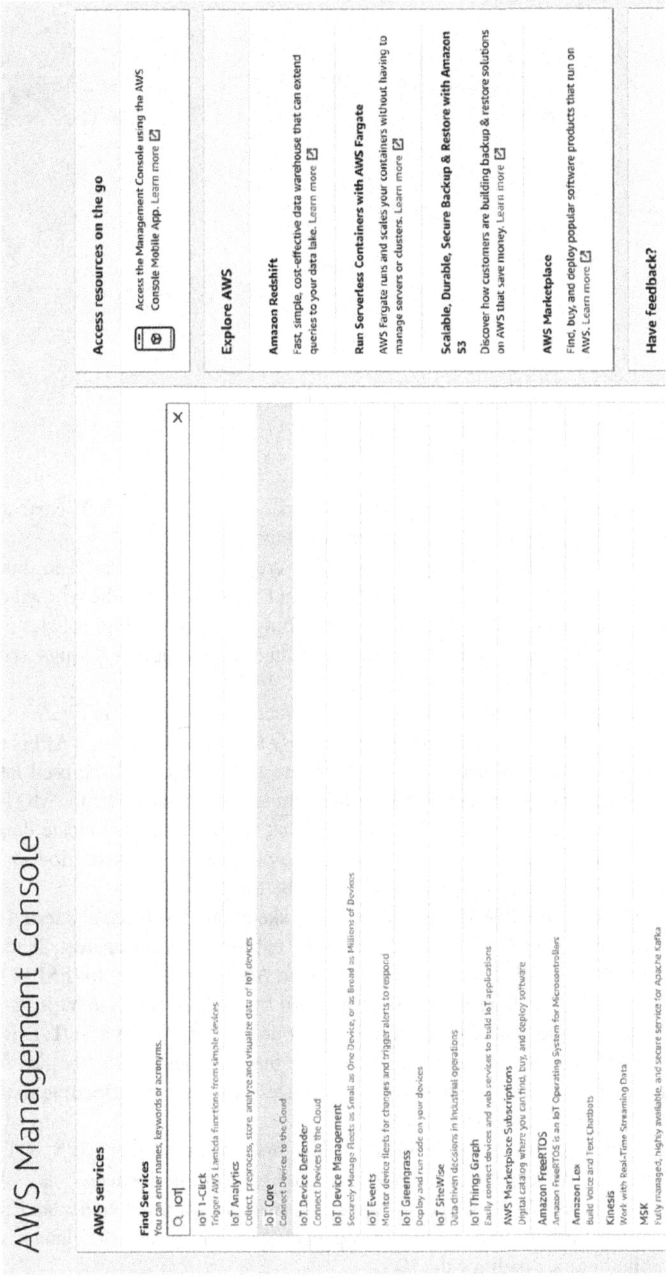

Figure 9.121 Selecting IoT core services from AWS Console.

Figure 9.122 AWS IoT things main page.

programming the ESP32 to establish secure connection with **AWS IoT.** Next we need to **Activate** the Thing by clicking on the **Activate** button.

8. Next we need to attach the **Policy** that we have created in Step 3 and 4 to this new **IoT Thing.** Click on the Attach Policy option shown in Fig. 9.130 and in the new window select the **digital_twin_policy** and click on Register Thing as shown in Fig. 9.131.
9. The newly created **IoT Thing** will show up in the **Manage >> Things** section. See Fig. 9.132.
10. Click on the digital_twin_thing and go to the Interact menu as shown in Fig. 9.133 and note down the **Rest API** for the **IoT Thing**. For security reasons, the Author's API is not shown in the screenshot, but you need to copy the entire string which will be used later for the ESP32 programming. Also note down the **IoT Thing** topic name under the MQTT section. MQTT is a fast, secure, and efficient protocol that we use to communicate data between ESP32 and **IoT Thing.** MQTT nodes or end points publish and subscribe to the topics and exchange information between them using the topics.
11. Next we will extend the ESP32 Arduino program sketch we developed and tested in Section 9.6.4 and connect to Wi-Fi and to AWS cloud and publish the data that are gathered from the DC Motor to the **AWS IoT Thing.** We will use an **AWS_IoT** library for ESP32 to be used with Arduino IDE. This library can be downloaded from the GitHub link https://github.com/ExploreEmbedded/Hornbill-Examples/tree/master/arduino-esp32/AWS_IOT. Download the latest library repository and keep the zipped file in the Arduino IDE libraries folder. For Windows-based machines, mostly it will be **C:\Users\<user_name>\Documents\Arduino\-libraries.** See Figs. 9.134 and 9.135.
12. Extract the zipped folder and go inside the folder, and copy the folder AWS_IOT to the top level folder **C:\Users\<user_name>\Documents\Arduino\libraries** as shown in Figs. 9.136 and 9.137. Also we can delete the original extracted folder and the zipped file now. If any Arduino IDE window is opened right now, please close and restart again to update the libraries path for the IDE.
13. We need to update **aws_iot_certificates.c** file under **C:\Users\<user_name>\Document-s\Arduino\libraries\AWS_IOT\src** with the character contents from the Root CA,

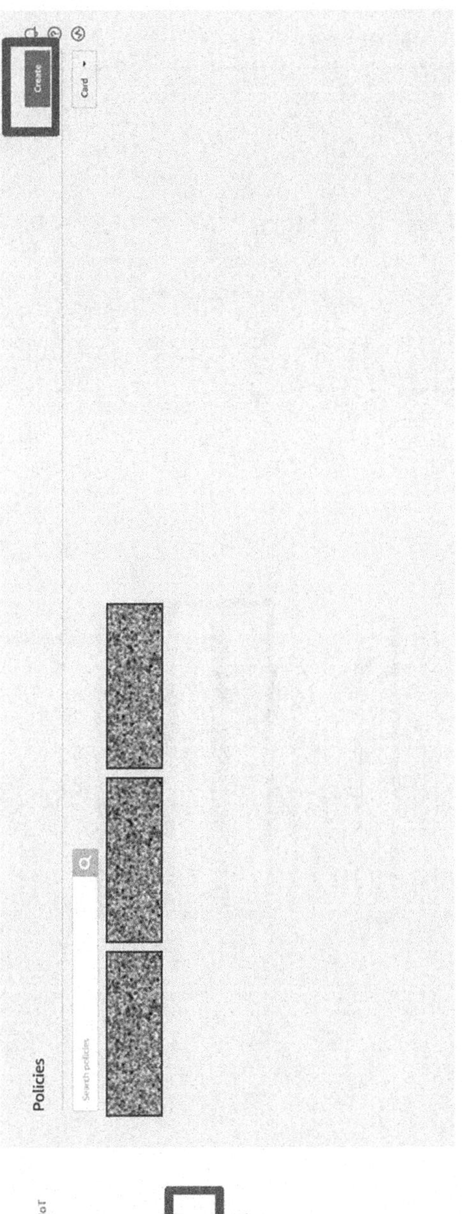

Figure 9.123 Creating a new policy.

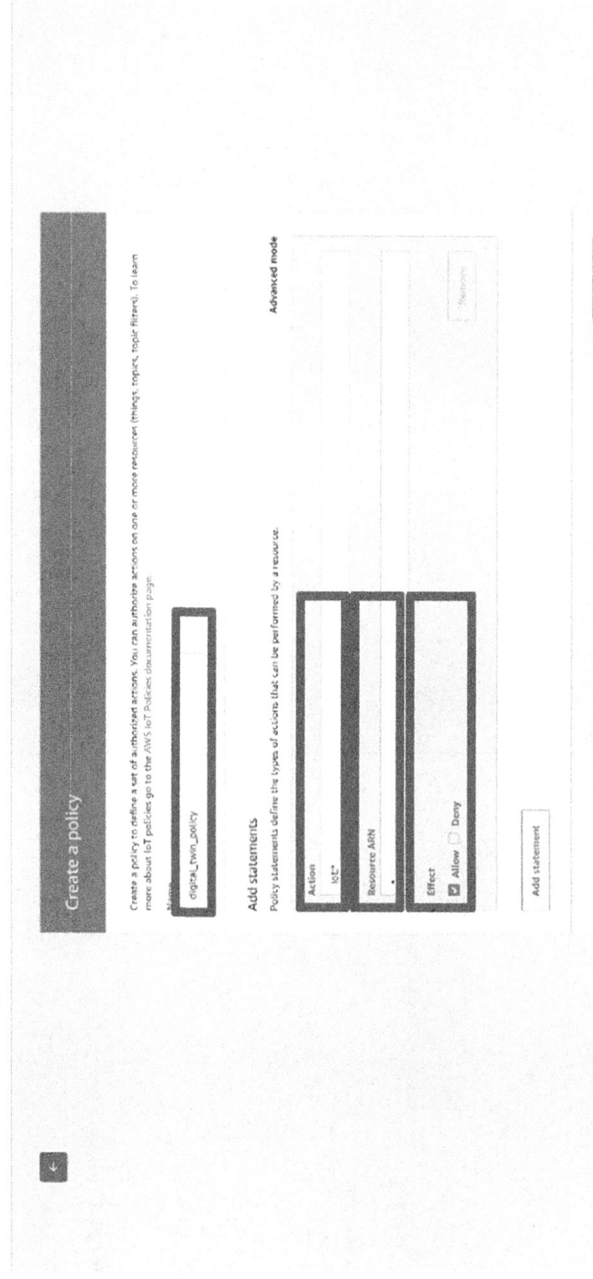

Figure 9.124 Entering new policy details.

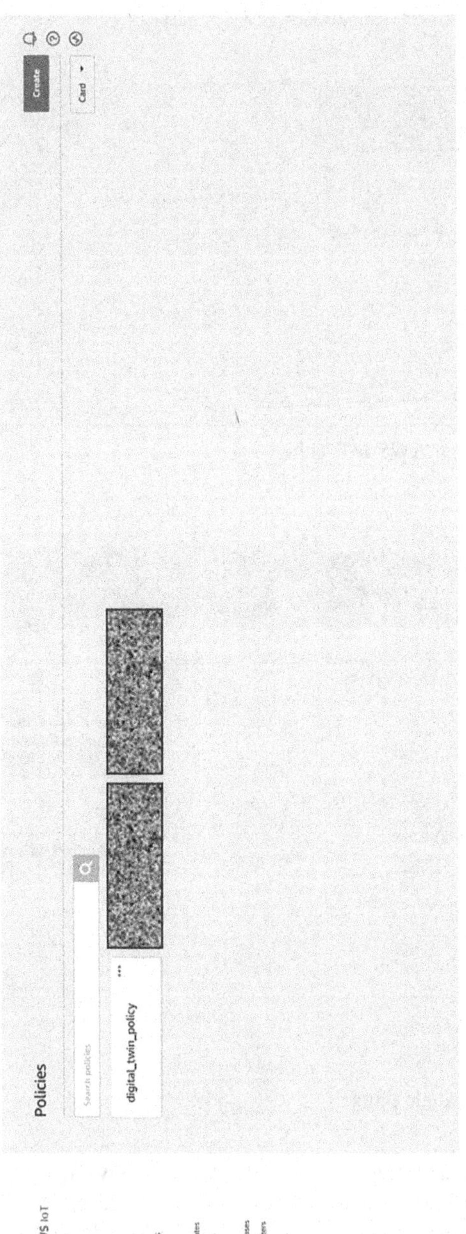

Figure 9.125 Newly created policy.

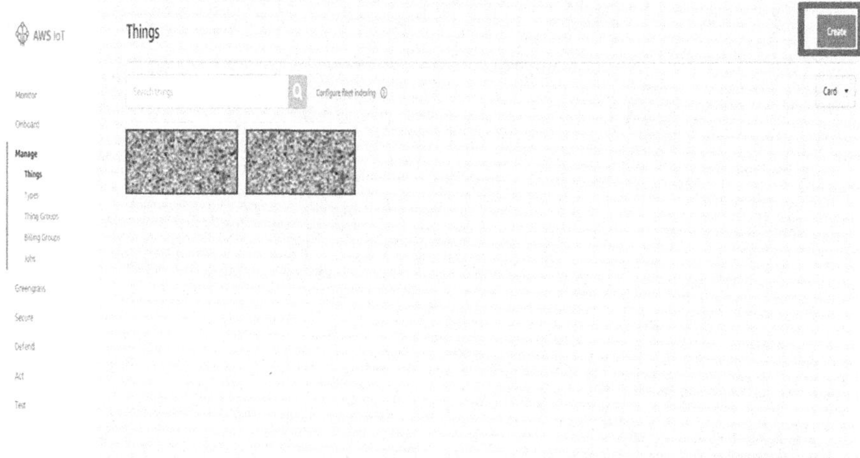

Figure 9.126 Creating an AWS IoT Thing.

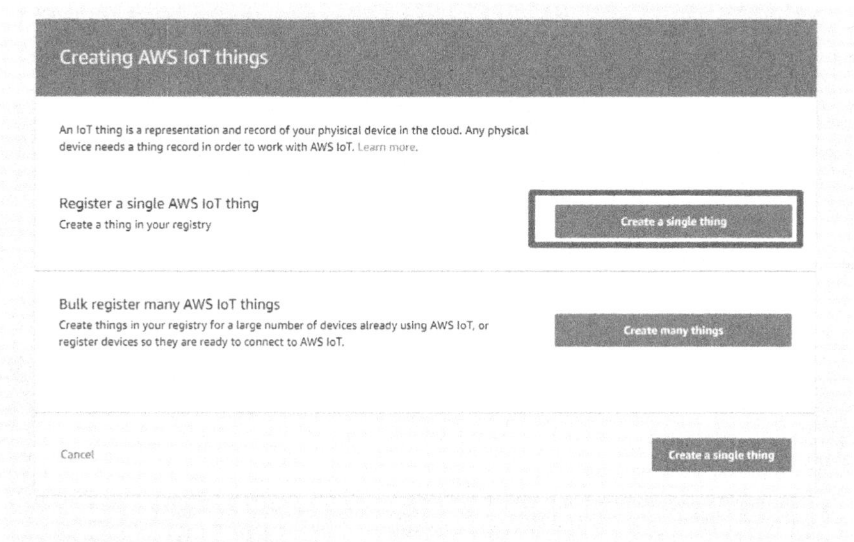

Creating AWS IoT things

An IoT thing is a representation and record of your phyisical device in the cloud. Any physical device needs a thing record in order to work with AWS IoT. Learn more.

Register a single AWS IoT thing
Create a thing in your registry

Create a single thing

Bulk register many AWS IoT things
Create things in your registry for a large number of devices already using AWS IoT, or register devices so they are ready to connect to AWS IoT.

Create many things

Cancel

Create a single thing

Figure 9.127 Create a single thing.

certificate, and private key files we downloaded from **AWS IoT Thing** Certificate generation process **Step 7** above. The c file arrays aws_root_ca_pem[ ], certificate_pem_crt[ ], and private_pem_key[ ] in the file **aws_iot_certificates.c** should be updated with the character contents from the **Root CA** file, the **xxxxxxxx.cert.pem** file, and the **xxxxxxxx.private.key** file we downloaded in **Step 7.** A screenshot of the updated file is shown in Fig. 9.138. Some portion of the screenshot is masked for security. The contents have to be carefully entered; otherwise, we may face difficulty to connect to the **AWS IoT**

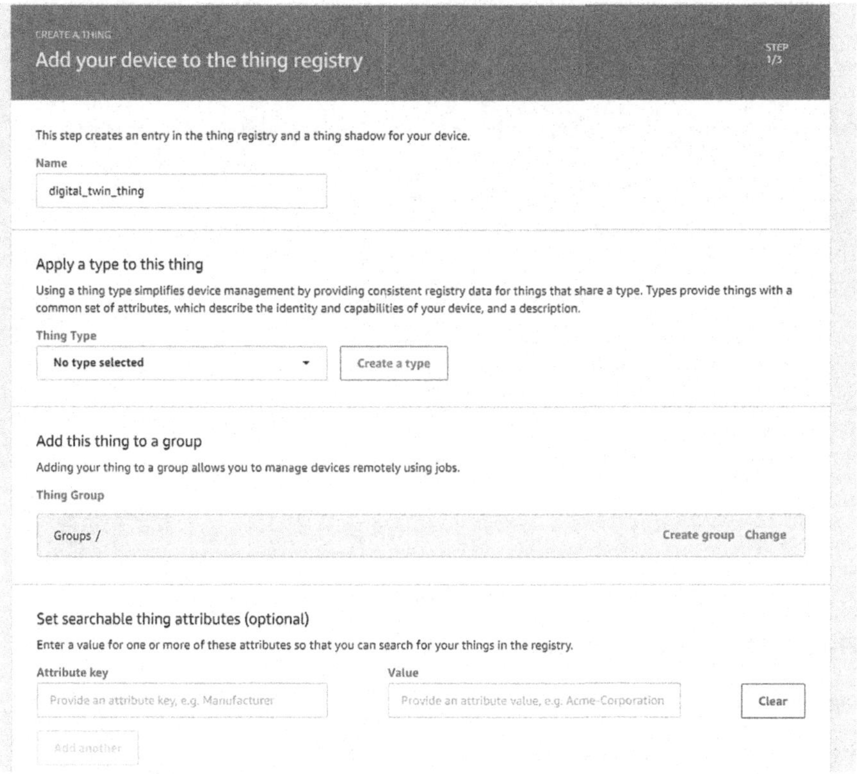

CREATE A THING

Add your device to the thing registry

STEP 1/3

This step creates an entry in the thing registry and a thing shadow for your device.

Name

digital_twin_thing

Apply a type to this thing

Using a thing type simplifies device management by providing consistent registry data for things that share a type. Types provide things with a common set of attributes, which describe the identity and capabilities of your device, and a description.

Thing Type

No type selected

Create a type

Add this thing to a group

Adding your thing to a group allows you to manage devices remotely using jobs.

Thing Group

Groups /

Create group Change

Set searchable thing attributes (optional)

Enter a value for one or more of these attributes so that you can search for your things in the registry.

Attribute key

Provide an attribute key, e.g. Manufacturer

Value

Provide an attribute value, e.g. Acme-Corporation

Clear

Add another

Figure 9.128 Naming the newly created AWS IoT thing.

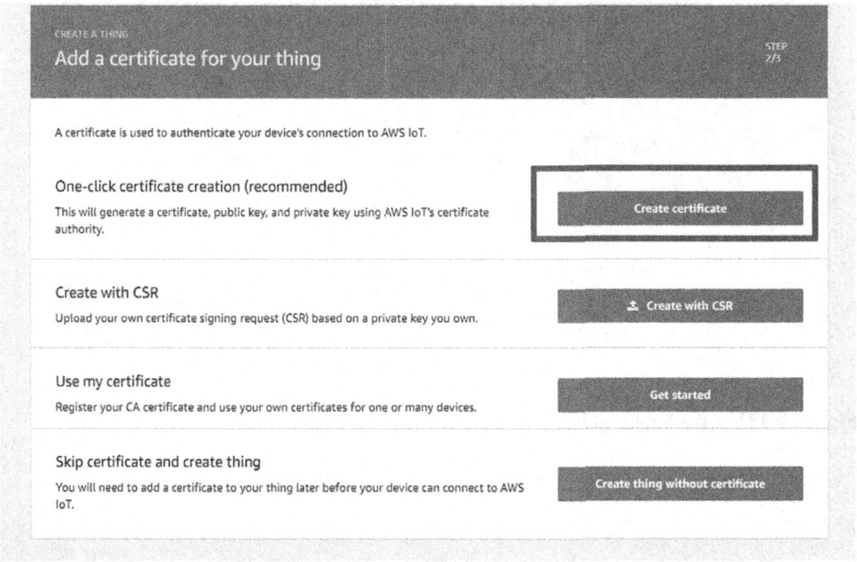

CREATE A THING

Add a certificate for your thing

STEP 2/3

A certificate is used to authenticate your device's connection to AWS IoT.

One-click certificate creation (recommended)

This will generate a certificate, public key, and private key using AWS IoT's certificate authority.

Create certificate

Create with CSR

Upload your own certificate signing request (CSR) based on a private key you own.

Create with CSR

Use my certificate

Register your CA certificate and use your own certificates for one or many devices.

Get started

Skip certificate and create thing

You will need to add a certificate to your thing later before your device can connect to AWS IoT.

Create thing without certificate

Figure 9.129 IoT thing certificate creation window.

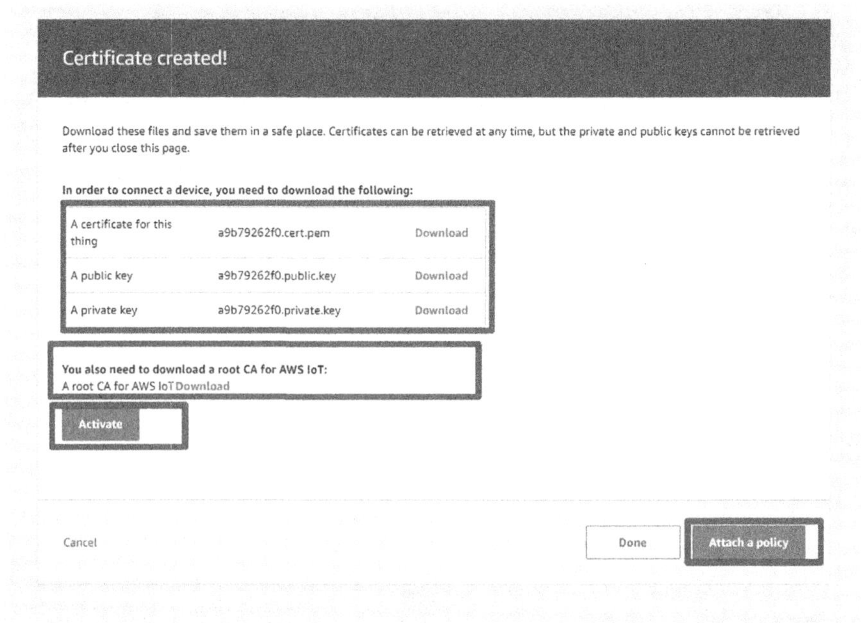

Figure 9.130 Certificates generated for IoT Thing.

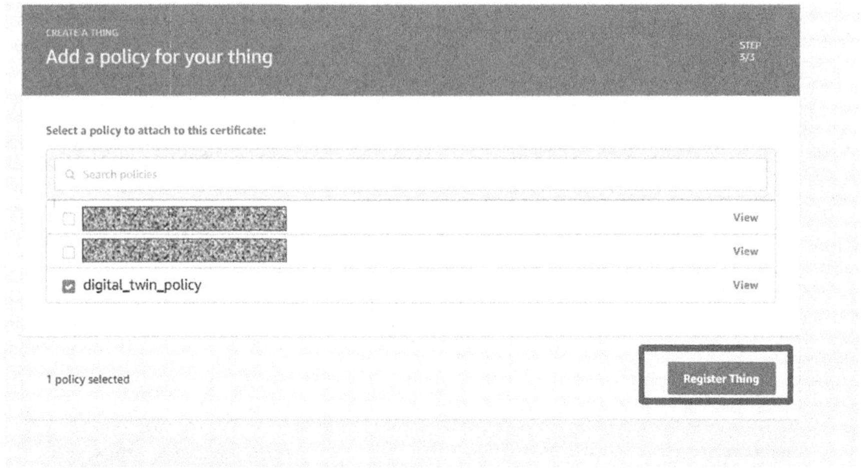

Figure 9.131 Attaching a policy and registering the IoT Thing.

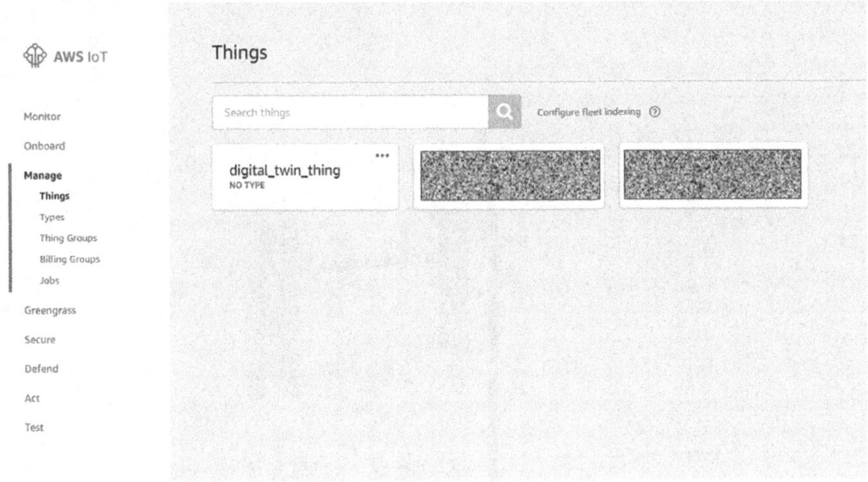

AWS IoT

Things

Monitor
Onboard
Manage
Things
Types
Thing Groups
Billing Groups
Jobs
Greengrass
Secure
Defend
Act
Test

Search things
Configure fleet indexing

digital_twin_thing
NO TYPE

Figure 9.132 Newly created thing.

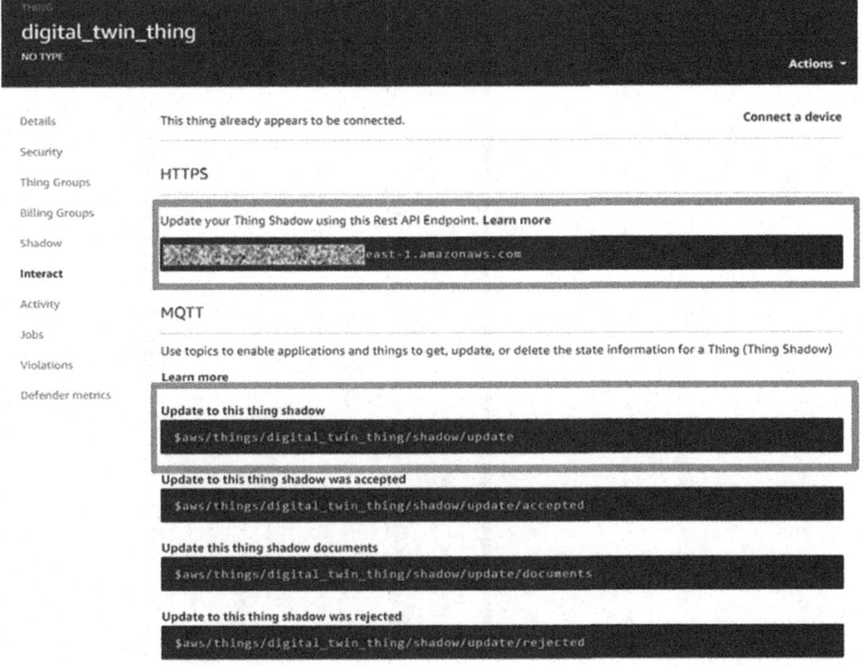

Figure 9.133 IoT Thing interact options.

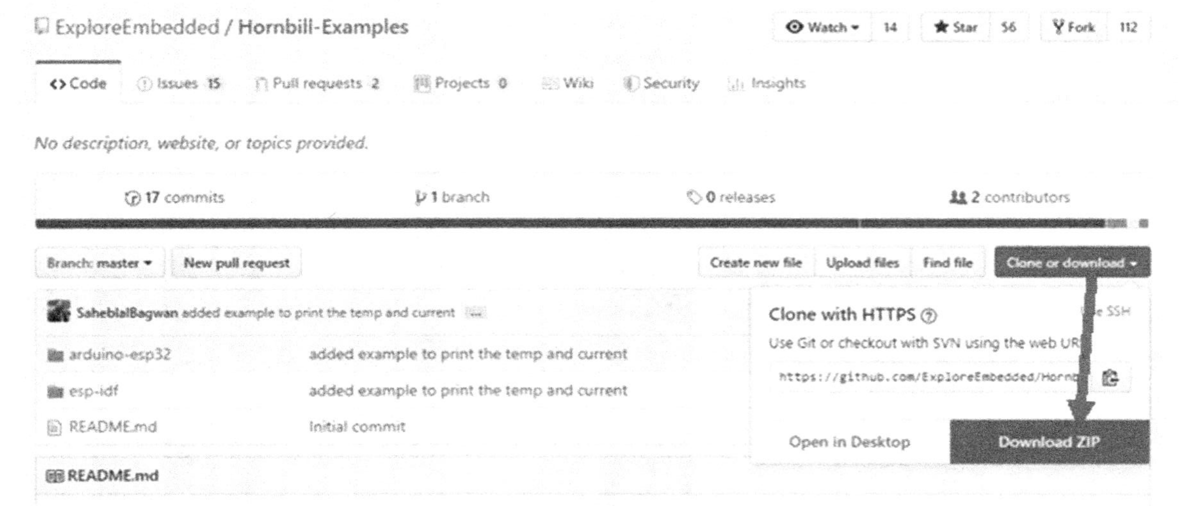

Figure 9.134 Downloading ESP32 Arduino Library from GitHub.

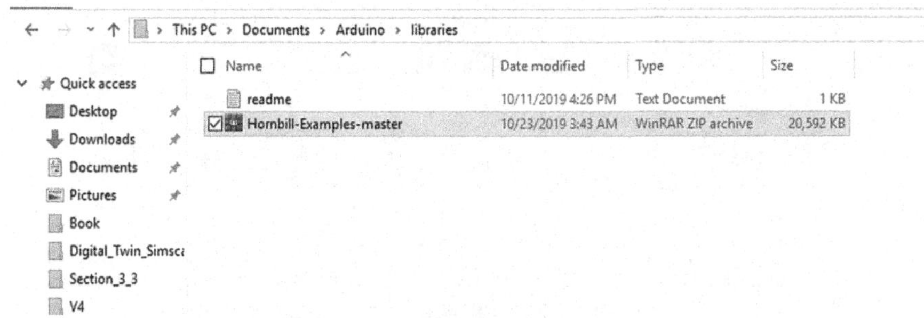

Figure 9.135 Copying the ESP32 Arduino Library to Arduino Libraries folder.

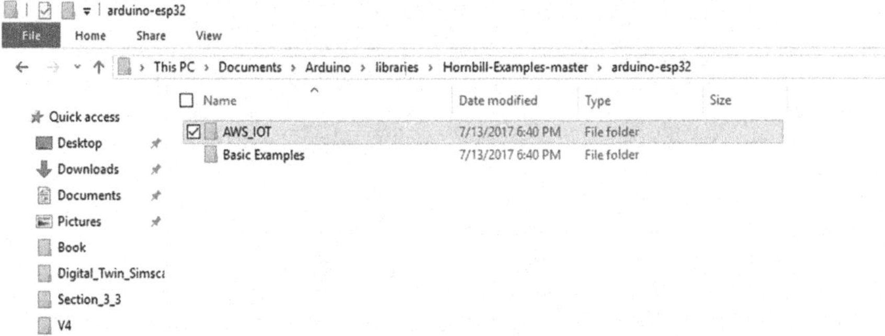

Figure 9.136 Unzip and bringing the AWS_IOT folder to Arduino\libraries folder.

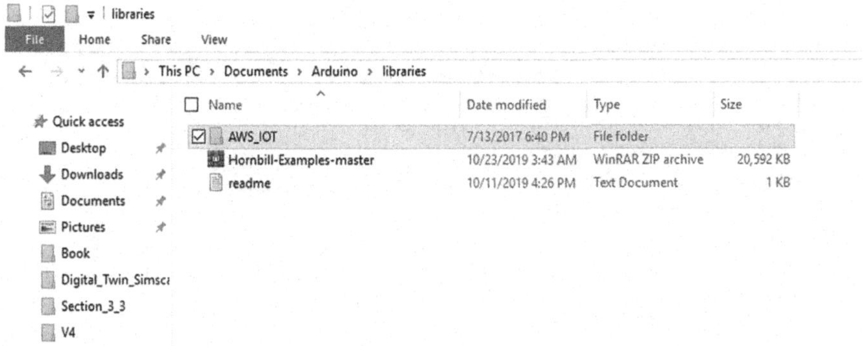

Figure 9.137 AWS_IOT copied to Arduino\libraries.

with ESP32 and it can be a painful process to debug. The rules to remember while entering the contents are

- Put all the contents in the certificate and key files within double quotes in the C array.
- Add a "\n\" at the end of each line and add a "\n" at the end of the last line.
- Do not add any white space, tab, or any other characters other than the content in the certificate and key files.

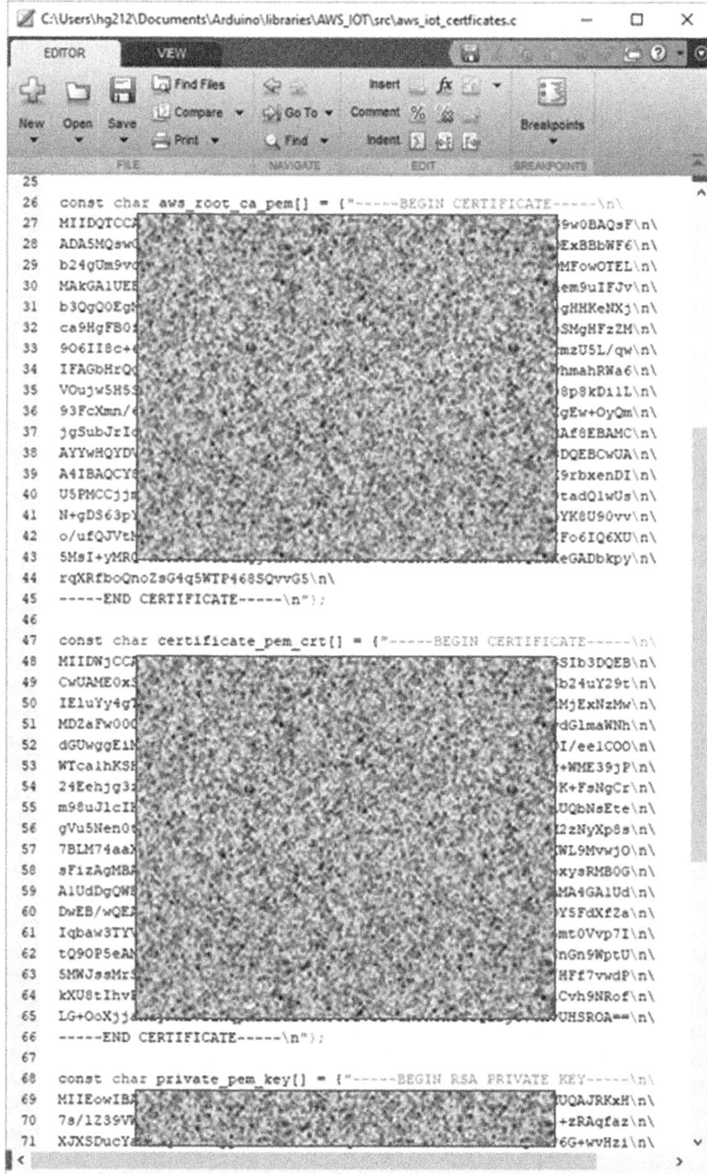

Figure 9.138 Updating the aws_iot_certificates.c file with our Certificate and Key.

14. Open and rename the ESP32 sketch we tested in Section 9.6.4. At the top of the sketch include the AWS_IOT.h and WiFi.h header file to include those libraries and functions in this program sketch. The WiFi.h is included from a default Arduino library; we can identify that from the color differences in the include statements. Then we will initialize an **AWS_IOT** object and name it as **esp32_aws_iot.** User may choose to give a different

name for this object, but wherever the object is referenced use the same name. And then on the character arrays enter the Wi-Fi router's **SSID** and **Password** to which we will be connecting our ESP32. Also create arrays for **HOST_ADDRESS** which will be the REST API we noted from our **digital_twin_thing Thing** in **Step 10**, **CLIENT_ID** will be our **Thing** name, and the **TOPIC_NAME** is the topic we noted from the **Thing** in **Step 10.** The ESP32 will be publishing and subscribing to the same topic in this example. A subscription handler function is added named **mySubCallBackHandler.** Whenever the ESP32 receives a message from the **IoT Thing**, this function will be triggered. Please see Fig. 9.139 for more details.

15. Now the **setup** () function is updated to connect to the Wi-Fi Router using the **SSID** and **Password** stored in the program. Also a code logic is written to connect to the AWS IoT using the **HOST_ADDRESS** and **CLIENT_ID** mentioned above using the **connect** function in the **AWS_IOT** ESP32 library. Once the connection is established, we will subscribe to the TOPIC using the **subscribe** function. Please see Fig. 9.140.
16. Now in the **loop()** function where we had the logic to print the JSON structure for the 3 s data received from DC Motor Arduino Mega, we will call a **publish** function from the **AWS_IOT** ESP32 library to publish the JSON data to **AWS IoT Thing** as shown in Fig. 9.141.
17. The ESP32 Arduino code to send data to the AWS cloud is now ready, and we will proceed to test the whole interface with the hardware. We will repeat the same **Step 6** in Section 9.6.4 to program both the Arduino Mega and the ESP32. From the hardware setup we finished in Section 9.6.1, connect the ESP32 controller board to the computer using a USB to Micro USB cable and build the ESP32 Arduino program we updated in

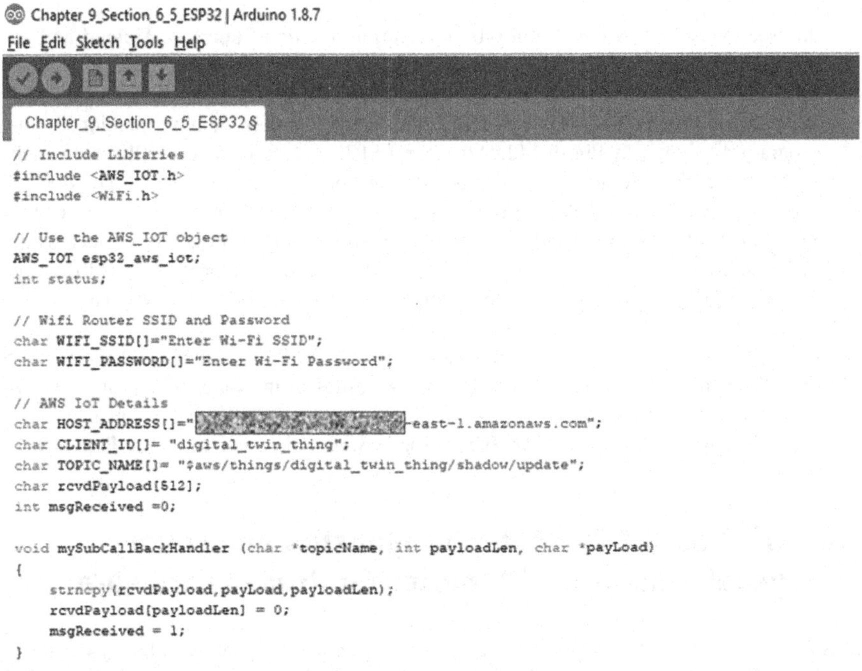

Chapter_9_Section_6_5_ESP32 | Arduino 1.8.7

File Edit Sketch Tools Help

Chapter_9_Section_6_5_ESP32 §

```
// Include Libraries
#include <AWS_IOT.h>
#include <WiFi.h>

// Use the AWS_IOT object
AWS_IOT esp32_aws_iot;
int status;

// Wifi Router SSID and Password
char WIFI_SSID[]="Enter Wi-Fi SSID";
char WIFI_PASSWORD[]="Enter Wi-Fi Password";

// AWS IoT Details
char HOST_ADDRESS[]="[illegible]-east-1.amazonaws.com";
char CLIENT_ID[]= "digital_twin_thing";
char TOPIC_NAME[]= "$aws/things/digital_twin_thing/shadow/update";
char rcvdPayload[512];
int msgReceived =0;

void mySubCallBackHandler (char *topicName, int payloadLen, char *payLoad)
{
    strncpy(rcvdPayload,payLoad,payloadLen);
    rcvdPayload[payloadLen] = 0;
    msgReceived = 1;
}
```

Figure 9.139 Storing Wi-Fi and AWS IoT details in the program.

```
Chapter_9_Section_6_5_ESP32 | Arduino 1.8.7
File Edit Sketch Tools Help
Chapter_9_Section_6_5_ESP32

// The setup() function just begins the serial communication at a baud rate of 115200.
// It connects ESP 32 to Wi-Fi Router
// This function is executed only once per every power up of the ESP 32
void setup()
{
  Serial.begin(115200);
  while (status != WL_CONNECTED)
  {
      Serial.print("Attempting to connect to SSID: ");
      Serial.println(WIFI_SSID);
      // Connect to WPA/WPA2 network. Change this line if using open or WEP network:
      status = WiFi.begin(WIFI_SSID, WIFI_PASSWORD);
      // wait 5 seconds for connection:
      delay(5000);
  }
  Serial.println("Connected to wifi");
  // Connect to AWS IoT
  if(esp32_aws_iot.connect(HOST_ADDRESS,CLIENT_ID)== 0)
  {
      Serial.println("Connected to AWS");
      delay(1000);

      if(0==esp32_aws_iot.subscribe(TOPIC_NAME,mySubCallBackHandler))
      {
          Serial.println("Subscribe Successfull");
      }
      else
      {
          Serial.println("Subscribe Failed, Check the Thing Name and Certificates")
          while(1);
      }
  }
  else
  {
      Serial.println("AWS connection failed, Check the HOST Address");
      while(1);
  }

  delay(2000);
}
Done Saving
```

Figure 9.140 Updating the loop function to connect to Wi-Fi and AWS IoT Thing.

this section for AWS connectivity by selecting the option **Sketch >> Upload** and also open a Serial monitor window and set its baud rate to 115,200. When the ESP32 powers up in the Serial monitor we can observe that the ESP32 will be first trying to connect to the Wi-Fi hotspot and then to the **AWS IoT** with the credentials we have mentioned in the program as shown in Fig. 9.142. After the ESP32 sketch is successfully downloaded now connect the Arduino Mega microcontroller board which controls the DC motor to a different USB port of the computer and from MATLAB® build and download the Simulink model we completed in Section 9.6.3. After both the programs are successfully downloaded to the hardware, we can observe that the DC motor will starts to run and the Arduino Mega will send JSON formatted data to the ESP32 as we have seen in Section 9.6.4.

18. In order to verify that the published message from the ESP32 is reaching AWS IoT, go to the AWS console goto **IOT Core >> Things >> digital_twin_thing >> Activity** and it will show the same JSON structure we are displaying on the ESP32 serial command window which is being published to **AWS IoT Thing** as well at the same time. See Fig. 9.143.

9.7 Off-Board Diagnostics/prognostics algorithm development for DC Motor Controller Hardware

In this section, we will develop an algorithm to diagnose and detect a faulty condition for the DC Motor Controller Hardware and test the diagnostic algorithm with the

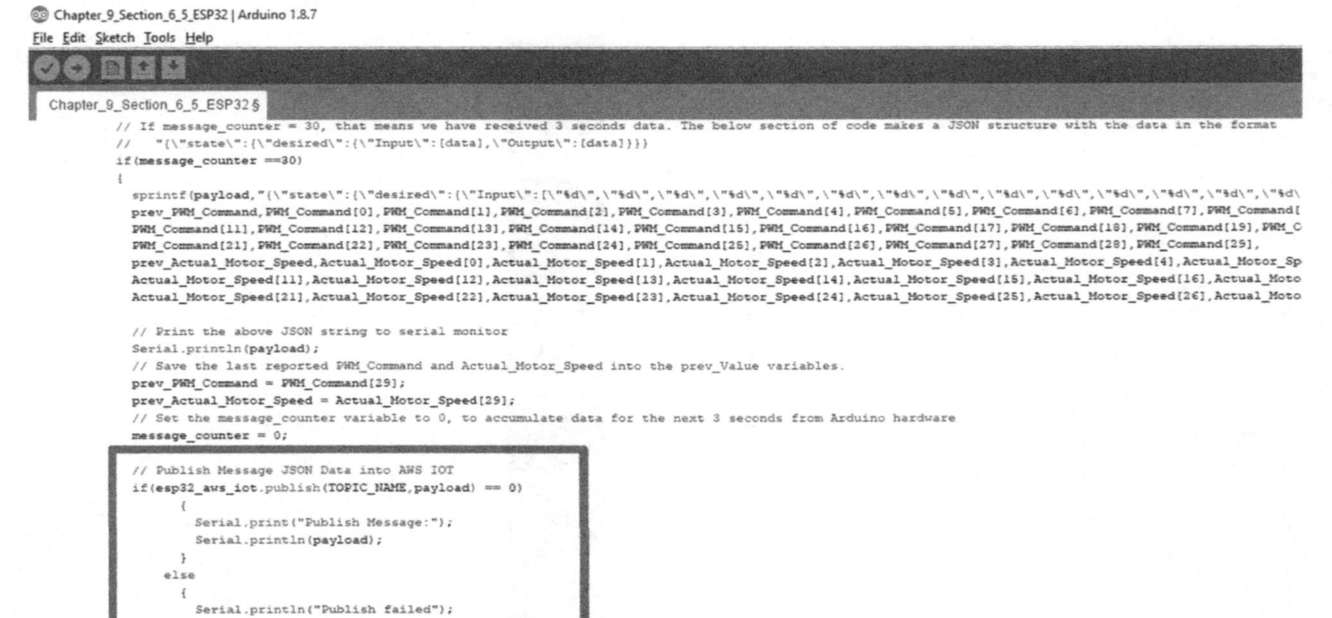

Figure 9.141 Publishing the JSON data structure to AWS IoT Thing.

Figure 9.142 ESP32 connecting to Wi-Fi HotSpot and AWS.

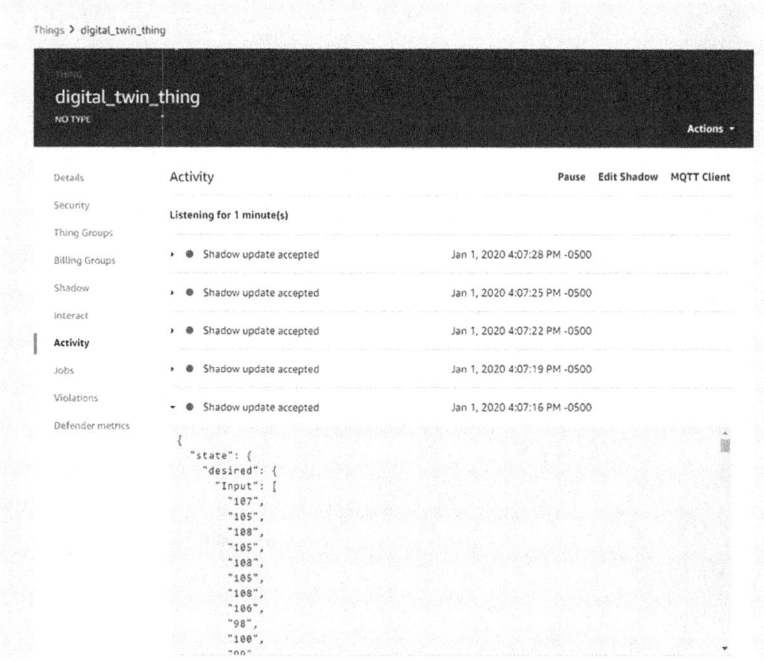

Figure 9.143 AWS IoT Thing receiving data from ESP32 Wi-Fi module.

Digital Twin model developed and parameter tuned in Section 9.5. We will first test the diagnostic algorithm locally on a computer before deploying it on the cloud. The failure condition considered is

1. Power Supply Failure of the DC Motor Hardware

The flow chart shown in Fig. 9.144 will be used to diagnose both the above failure conditions where we collect the input and output data from the hardware, run the Digital Twin model of the DC Motor with the input data to get the expected Motor Speed, calculate the RMSE between the actual speed and predicted speed, and compare the RMSE value with a threshold to detect if there is a failure. The E-mail/Text notification will not be tested in this section, but will be covered later when we deploy the Off-BD and Digital Twin to the AWS cloud.

We will be running the Off-BD algorithm using a well-known programming language Python. One of the main reason why Python is chosen is that we eventually have to deploy this Off-BD algorithm to AWS cloud. We will be using **AWS Lambda** services for running the Off-BD algorithm on the cloud and **AWS Lambda** supports Python. Note that the Digital Twin model we have is developed using MATLAB®, Simulink®, and Simscape™, so in order to call the Digital Twin model from Python, we will have to generate "C" code from the Simscape™ model and compile it into an executable and then call the executable from Python program. **One other thing to note is that in the backend the AWS is running on Linux Operating System. So the compiled executables that we have to make for the Digital Twin model should also be in Linux so that the executable can be deployed and ran from AWS Linux machines.** So we will use Linux operating system just for this section, to generate code and compile an executable for the Digital Twin.

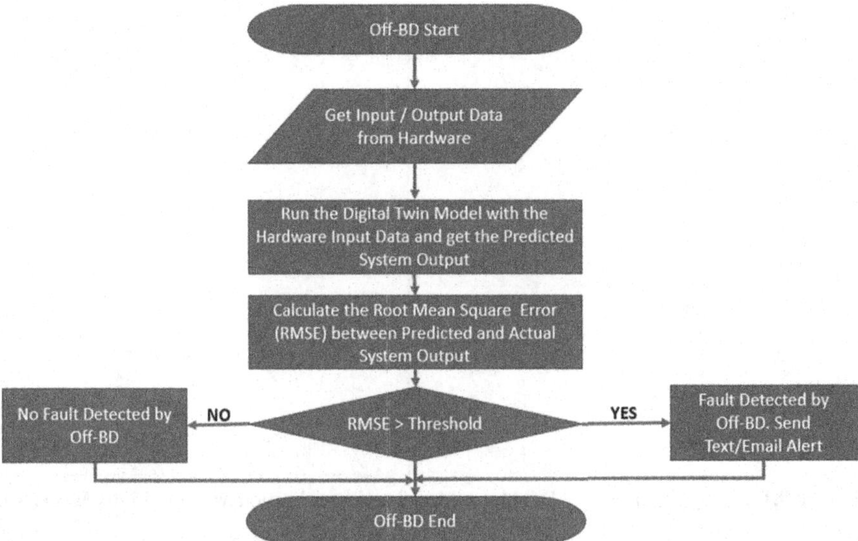

Figure 9.144 Off-BD algorithm flow chart.

Follow the below steps to develop the Off-BD algorithm and test it locally on a machine:

1. The first step is to install Linux on a Machine. The Authors have tested the process with Ubuntu Linux. The installation process for Ubuntu Linux will not be covered in this chapter; there are plenty of online resources available for Linux installation. Authors recommend to install Ubuntu as a Dual-Boot setup if the User already has Windows installed on the machine. Dual-Boot installation allows Users to switch between Windows and Linux when required by just restarting one operating system and selecting other from the boot up menu. One of the working links to install Ubuntu at the time of writing of this chapter is here: https://itsfoss.com/install-ubuntu-1404-dual-boot-mode-windows-8-81-uefi/
2. The rest of the steps in this section is using Linux. So make sure Linux is installed and setup. We will also need to install MATLAB® on Linux to generate code and compile the Simscape™ Digital Twin model. Once again, since it is outside the scope of the book, we will not go through the MATLAB® installation process in Linux. An instructions link to install MATLAB® on Linux that worked at the time of writing of this chapter is given here: https://www.cmu.edu/computing/software/all/matlab/matlabinstall-linux.html.
3. Open MATLAB® in Linux and create a new folder and copy the DC Motor Simscape™ plant model created in Section 9.5, and also create an initialization M file and put the optimized model parameters in that script as shown in Figs. 9.145 and 9.146.

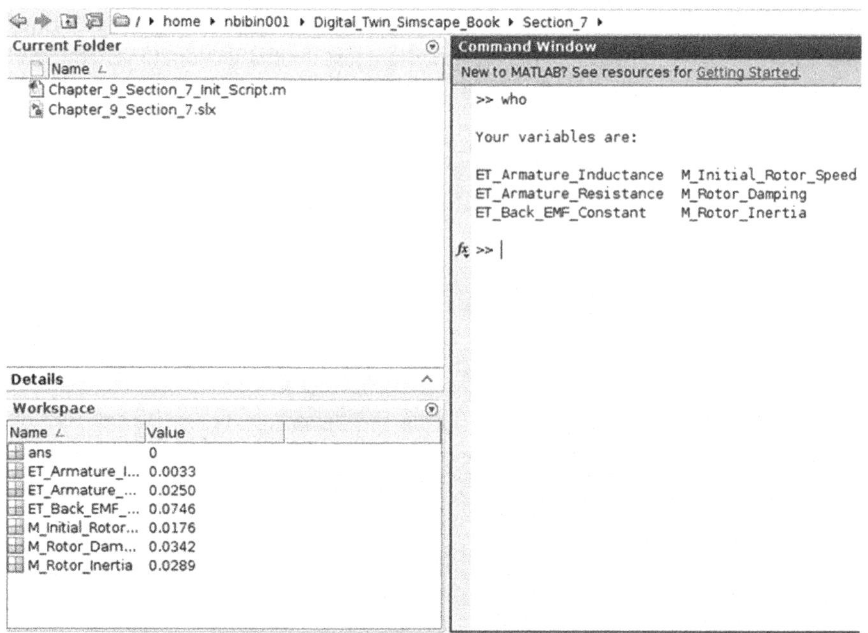

Figure 9.145 Copy Simscape™ DC Motor Model and Initialization M file to Linux MATLAB Folder.

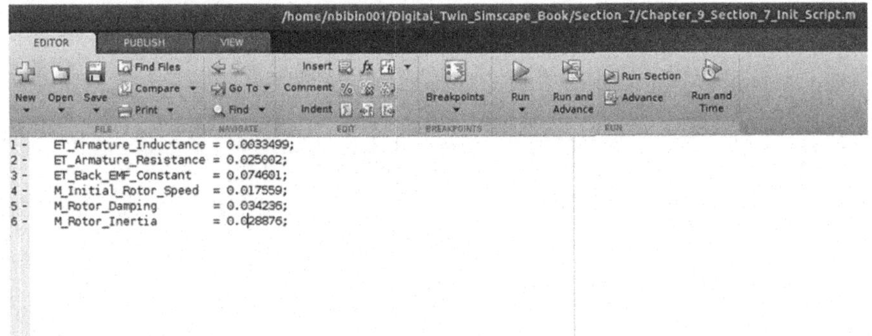

Figure 9.146 Optimized Motor Model parameters copied to an initialization script.

4. Run the above parameter init file to initialize the MATLAB workspace and open the Simscape™ DC Motor model and make sure the System Target File under the **Model >> - Configuration Parameters >> Code Generation >> System Target File** is set to "*ert.tlc*." Then click on the "***Build***" button on the Simulink Model Menu as highlighted in Fig. 9.147. This will now update the model, generate "*C*" source code from the model, create a Makefile to compile the generated code, compile the code, and create an executable application from the Model logic. MATLAB® will show the progress of this process as shown in Fig. 9.148. Since we are trying this on a Linux Machine, it will create a Linux executable application. The "*C*" code will be generated into a folder named **Model_Name_ert_rtw** under the working folder and the name of the executable will be same as that of the model name. In this case, our Model Name is **Chapter_9_Section_7.slx**, so the generated code will be placed in **Chapter_9_Section_7_ert_rtw** and the generated executable application name will be **Chapter_9_Section_7.** Fig. 9.149 shows the code generated folder and the executable application in the MATLAB® working folder.
5. Go inside the codegen folder **Chapter_9_Section_7_ert_rtw** and open the file ert_main.c as shown in Fig. 9.150 in an editor. Just double click on the file to open it in MATLAB® editor itself. As the User might know, every "*C*" code application requires an entry function named **main()** which is the entry point function when the application starts to run. The **main()** function is generated by MATLAB® Code Generator into this **ert_main.c** file.
6. The ert_main.c file has two functions mainly in it. One is the **main()** and the other is the **rt_OneStep()** function as shown in Figs. 9.151 and 9.152. As mentioned above, **main()** is the entry point function, and this function initializes the model states if any using the **Model_Name_initialize** function and just enters and stays into a **while loop.** So we can see that even when the application runs, it is not really running our model logic yet in this framework as is now. So we will edit the **main()** function to suit our purpose in later steps mentioned below. The **rt_OneStep()** function calls another function **Model_Name_step (Chapter_9_Section_7_step()** in this case; this step function contains the actual logic that we have implemented in the Simulink® model. We can see the **Model_step()** function takes the arguments to receive input and return the output. When MATLAB® generates code the **rt_OneStep()** function call is commented out from the **main()** function, this is mainly because the Users can integrate the **Model_step()**

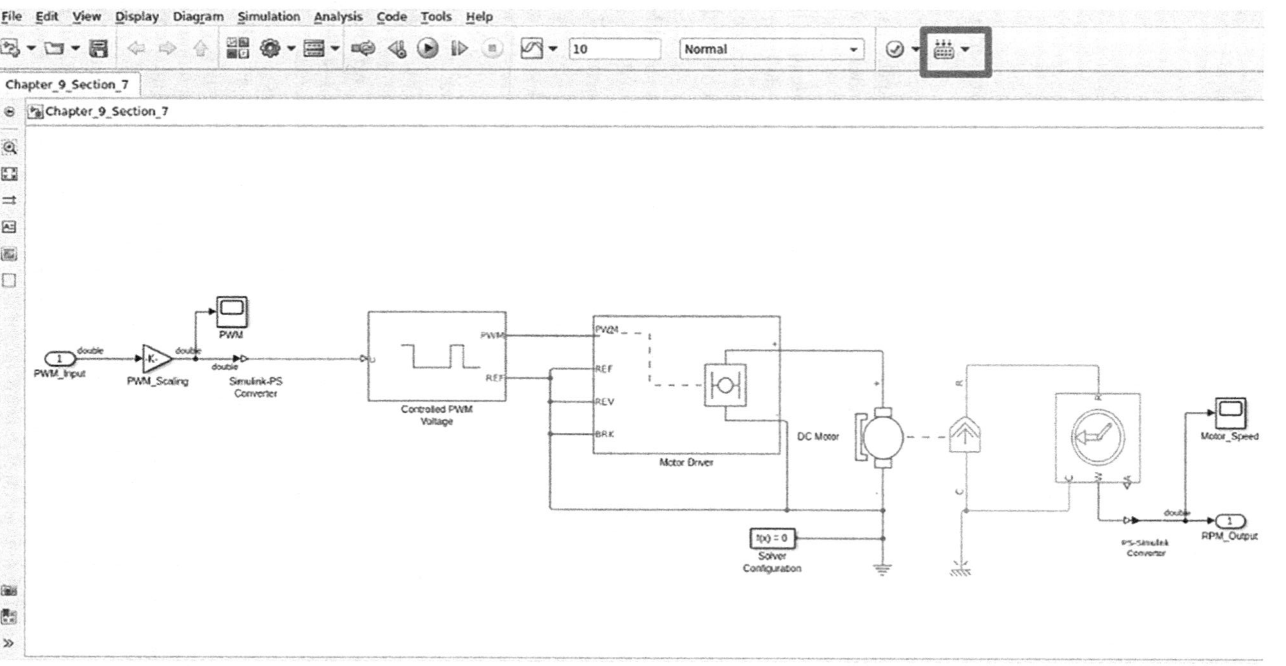

Figure 9.147 Building the Simscape™ model.

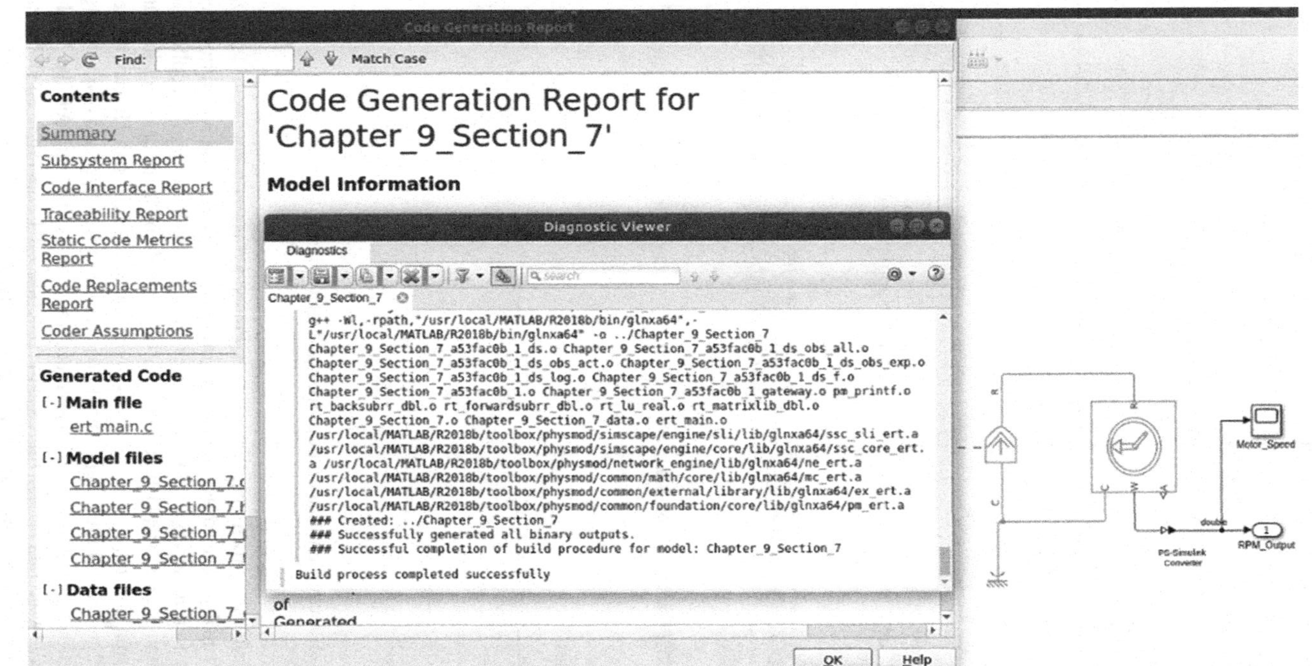

Figure 9.148 Model Codegen and build progress.

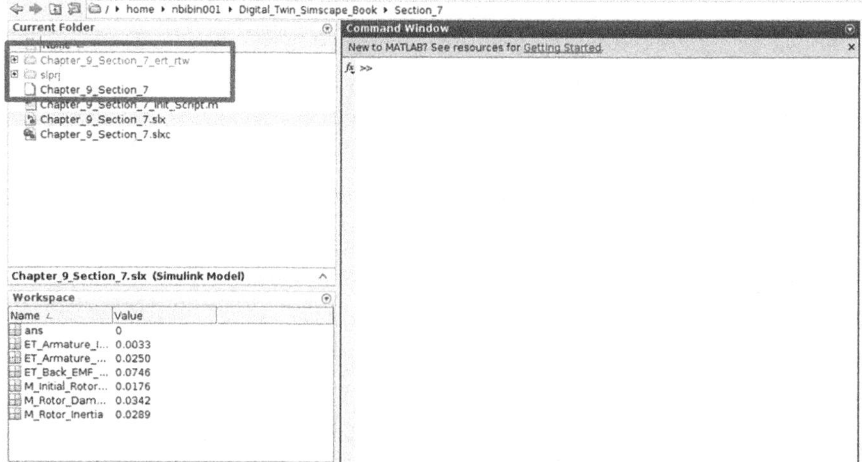

Figure 9.149 Generated code folder and executable application.

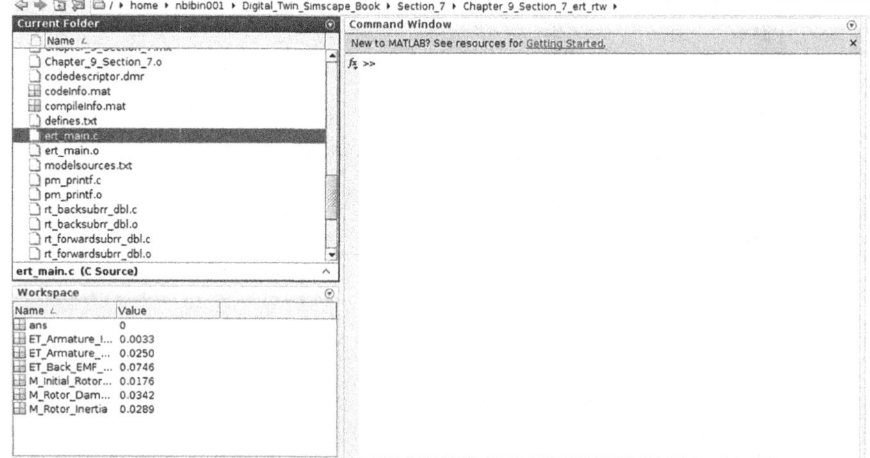

Figure 9.150 Selecting ert_main.c File from codegen folder.

function with their own operating system, embedded target software, or scheduler, etc., and don't really have to use the **ert_main.c** in their application software. The way we are going to run this application is we get the input data, we run the application, which will run the **Model_step()** function with one sample of input data, get its output, and then we run the application with the next sample data again. It can be seen from Fig. 9.152 that the **Model_step()** function takes the model input and output as arguments, but sometimes depending on the configuration settings of the model, the **Model_step()** may be void-void function, in that case the function will be operating on global data structures for accessing inputs and setting the outputs.

```
 */
int_T main(int_T argc, const char *argv[])
{
  /* Unused arguments */
  (void)(argc);
  (void)(argv);

  /* Initialize model */
  Chapter_9_Section_7_initialize();

  /* Attach rt_OneStep to a timer or interrupt service routine with
   * period 0.1 seconds (the model's base sample time) here.  The
   * call syntax for rt_OneStep is
   *
   *  rt_OneStep();
   */
  printf("Warning: The simulation will run forever. "
         "Generated ERT main won't simulate model step behavior. "
         "To change this behavior select the 'MAT-file logging' option.\n");
  fflush((NULL));
  while (rtmGetErrorStatus(Chapter_9_Section_7_M) == (NULL)) {
    /*  Perform other application tasks here */
  }

  /* Disable rt_OneStep() here */

  /* Terminate model */
  Chapter_9_Section_7_terminate();
  return 0;
}

/*
 * File trailer for generated code.
 *
```

Figure 9.151 main() function in ert_main.c.

7. Now we will start editing the ert_main.c to suit our application purpose. We will change the main() function to read an input file "**input.csv**" from a Linux user temporary folder **/tmp.** The **/tmp** folder is selected for the input and output files because **/tmp** is the only folder which is writable from AWS. The **input.csv** file will be automatically created into the **/tmp** folder when we receive data from Hardware system, we will discuss that later when we deploy the Digital Twin model and Off-BD algorithm into AWS cloud. This input file will contain the actual DC Motor data, PWM_Command, and Motor_Speed, one line for each sample of data at every 0.1 s, for 3 s. There will be total 31 entries in this file. The first entry in the file will be the last known state of the hardware, which will be used to first initialize the application to a steady state, before running the new 3 s data. A sample of **input.csv** file is shown in Fig. 9.153, which has 31 lines, each line has Actual PWM Command and Motor Speed separated by commas. The first line entry is the last known previous state to initialize the model application with, and the next 30 samples are the 3 s data to run the model application to get the predicted output. The sample time of 0.1 s and prediction horizon of 3 s are chosen specifically for this application because 3 s seems to be a good enough time to capture the dynamics and time constant of the DC Motor hardware we are working with. For a different application and hardware, we may want to choose these sampling time and number of samples for prediction differently.
8. First in the **ert_main.c**, a new hand written C function **Parse_CSV_Line()** is added as shown in Fig. 9.154 to read the lines in the **input.csv** file and splits the values between the delimiter comma in each lines. Add this function above the **main()** function. So essentially this splits and returns the actual PWM_Command and Motor_Speed.

```
void rt_OneStep(void)
{
  static boolean_T OverrunFlag = false;

  /* '<Root>/PWM_Input' */
  static real_T arg_PWM_Input = 0.0;

  /* '<Root>/RPM_Output' */
  static real_T arg_RPM_Output;

  /* Disable interrupts here */

  /* Check for overrun */
  if (OverrunFlag) {
    rtmSetErrorStatus(Chapter_9_Section_7_M, "Overrun");
    return;
  }

  OverrunFlag = true;

  /* Save FPU context here (if necessary) */
  /* Re-enable timer or interrupt here */
  /* Set model inputs here */

  /* Step the model */
  Chapter_9_Section_7_step(&arg_PWM_Input, &arg_RPM_Output);

  /* Get model outputs here */

  /* Indicate task complete */
  OverrunFlag = false;

  /* Disable interrupts here */
```

Figure 9.152 rt_OneStep() function in ert_main.c.

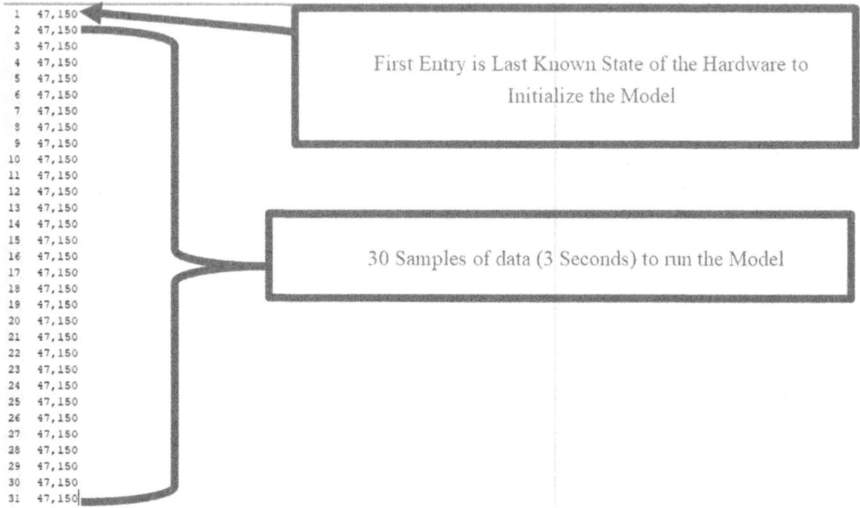

Figure 9.153 Sample input.csv file with Actual PWM Command and Motor Speed.

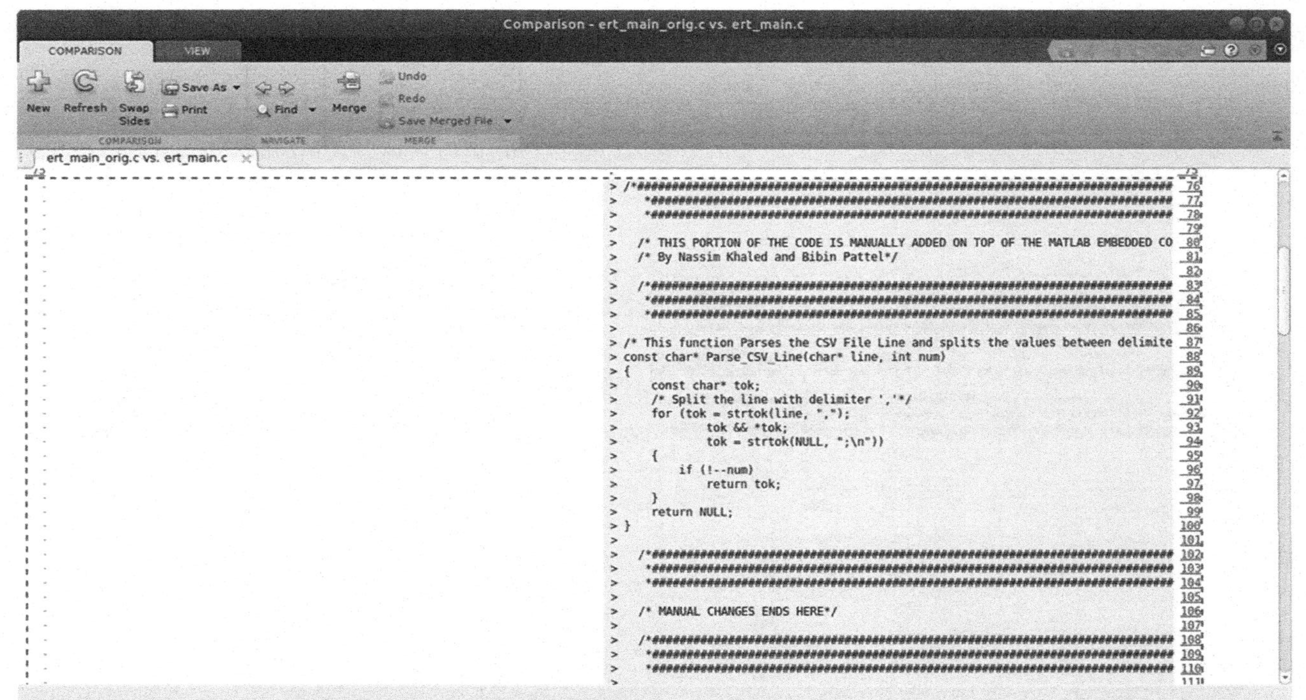

Figure 9.154 Function to Parse CSV File Lines and get the PWM Command and Motor Speed.

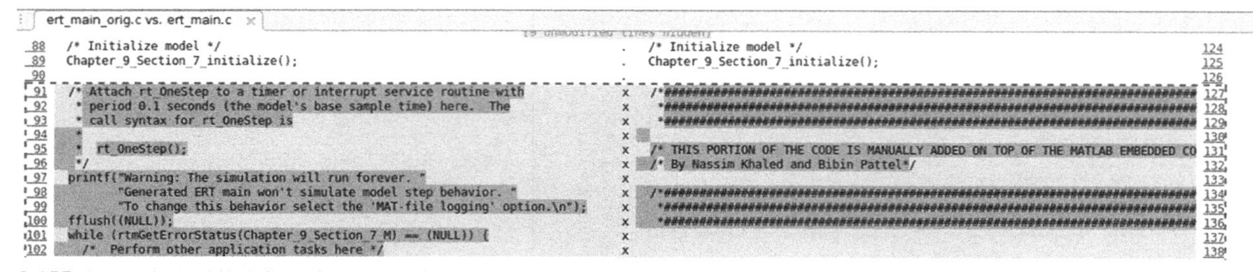

Figure 9.155 Removing while() loop from ert_mian.c.

9. Next is to take out the infinite **while()** loop function from **main()** function of the **ert_main.c.** The **Model_step()** function will be called explicitly, so we don't need this **while()** loop. See Fig. 9.155.
10. Next is to add a portion of code to open the input.csv file from the path **/tmp/input.csv** and read the actual PWM_Input and Motor_Speed into two arrays. The first line values in the **input.csv** file which is the last known state will be stored separately into previous state variable, and the rest of the line entries will be stored into arrays actual_hw_input_pwm[] and actual_hw_output_rpm[].
11. After all the lines are read and stored into the arrays from the input file, a for loop is ran for 130 iterations to fill two new arrays padded_hw_input_pwm[] and padded_hw_output_rpm [] of size 130 elements each. The first 100 elements of these arrays are filled with the previous state variable values, so when we run the compiled executable model, we first run thees 100 samples of data (with the model sample time of 0.1 s, 100 samples are equivalent to initializing the model for 10 s) and rest 30 array locations are filled with the actual 3 s data from arrays actual_hw_input_pwm[] and actual_hw_output_rpm[] populated in the **Step 10.** See Figs. 9.156 and 9.157 for more details.
12. The next step is to add a code run a for loop for 130 times to call the Model_step function **Chapter_9_Section_7_step()** in each iteration with the values from padded_hw_input_pwm[] array sequentially and store the input PWM Command and Predicted Motor Speed, which is output by the step function in to a file **output.csv** under the **/tmp** folder itself. See the code to do this in Fig. 9.158. **Note that if the Model_step() function which is generated is a void-void function with no input/output arguments, we have to find the global data structure for input and output which the step function is working on and assign values to the input structure before calling the step function and retrieve values from the output structure after calling the step function.**
13. The changes to the ert_main.c file is completed now. Save and close the file. From the Chapter_9_Section_7_ert_rtw folder, look for the Makefile with the extension "Chapter_9_Section_7.mk." Copy this file to a new file named Makefile using the Linux command ***!cp Chapter_9_Section_7.mk Makefile*** from MATLAB command prompt. We can try the same from a Linux command window as well without the "!" at the beginning. "!" tells the MATLAB command window that what follows is an operating system command. Since we have changed the ert_main.c file, let us just delete the previously created object file ert_mian.o using the rm command ***!rm ert_main.o.*** Now rebuild the executable application using the command **!make -f Makefile.** Since only the ert_main.c file is changed and all the other files are same, the **make** command only recompiled **ert_mian.c.** All of the commands tried in this step and their outputs are shown in Fig. 9.159. The recompiled executable application **Chapter_9_Section_7** is shown in Fig. 9.160.
14. Let us just make a quick MATLAB®-based program to test the executable application. The MATLAB® program will take the DC Motor data collected from the hardware in Section 3.1 and chunk it up to 3 s data as if it is coming from the actual hardware in real time, crate the "**input.csv**" file copy it to **/tmp** folder, call the executable application which will run with the input PWM data from the input.csv, and create the "**output.csv**" file under **/tmp** folder. The MATLAB® program will then look at the "**output.csv**" file and get the predicted motor speed and calculate RMSE between the actual and predicted motor speed

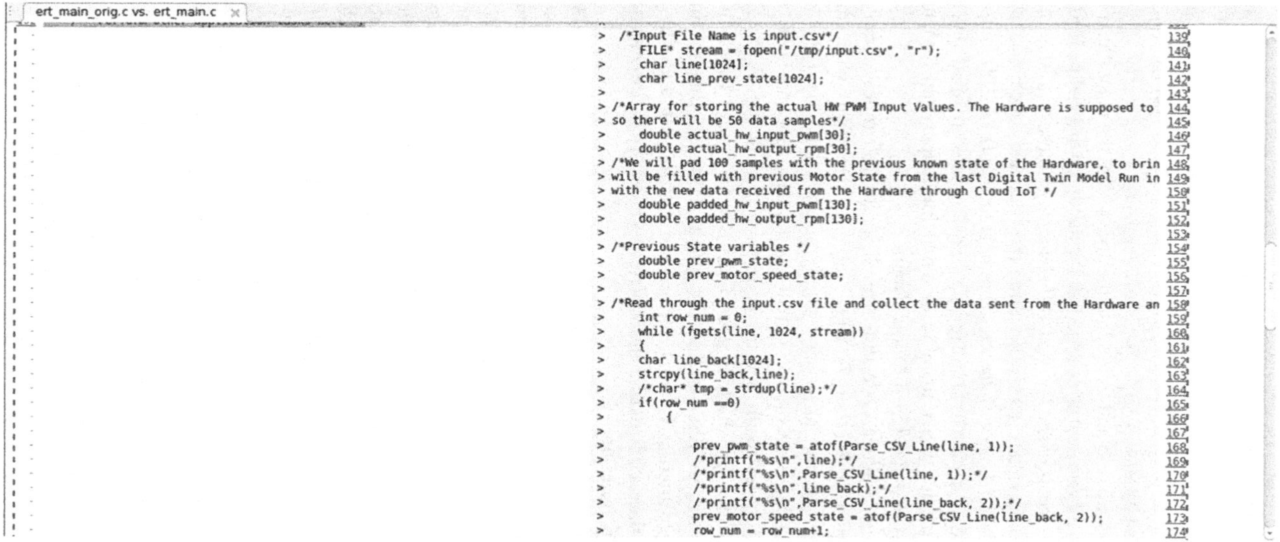

ert_main_orig.c vs. ert_main.c

```
>  /*Input File Name is input.csv*/
>     FILE* stream = fopen("/tmp/input.csv", "r");
>     char line[1024];
>     char line_prev_state[1024];
>
> /*Array for storing the actual HW PWM Input Values. The Hardware is supposed to
> so there will be 50 data samples*/
>     double actual_hw_input_pwm[30];
>     double actual_hw_output_rpm[30];
> /*We will pad 100 samples with the previous known state of the Hardware, to brin
> will be filled with previous Motor State from the last Digital Twin Model Run in
> with the new data received from the Hardware through Cloud IoT */
>     double padded_hw_input_pwm[130];
>     double padded_hw_output_rpm[130];
>
> /*Previous State variables */
>     double prev_pwm_state;
>     double prev_motor_speed_state;
>
> /*Read through the input.csv file and collect the data sent from the Hardware an
>     int row_num = 0;
>     while (fgets(line, 1024, stream))
>     {
>     char line_back[1024];
>     strcpy(line_back,line);
>     /*char* tmp = strdup(line);*/
>     if(row_num ==0)
>         {
>
>             prev_pwm_state = atof(Parse_CSV_Line(line, 1));
>             /*printf("%s\n",line);*/
>             /*printf("%s\n",Parse_CSV_Line(line, 1));*/
>             /*printf("%s\n",line_back);*/
>             /*printf("%s\n",Parse_CSV_Line(line_back, 2));*/
>             prev_motor_speed_state = atof(Parse_CSV_Line(line_back, 2));
>             row_num = row_num+1;
```

Figure 9.156 Reading input.csv and storing PWM Command and Motor Speed into Arrays Part 1.

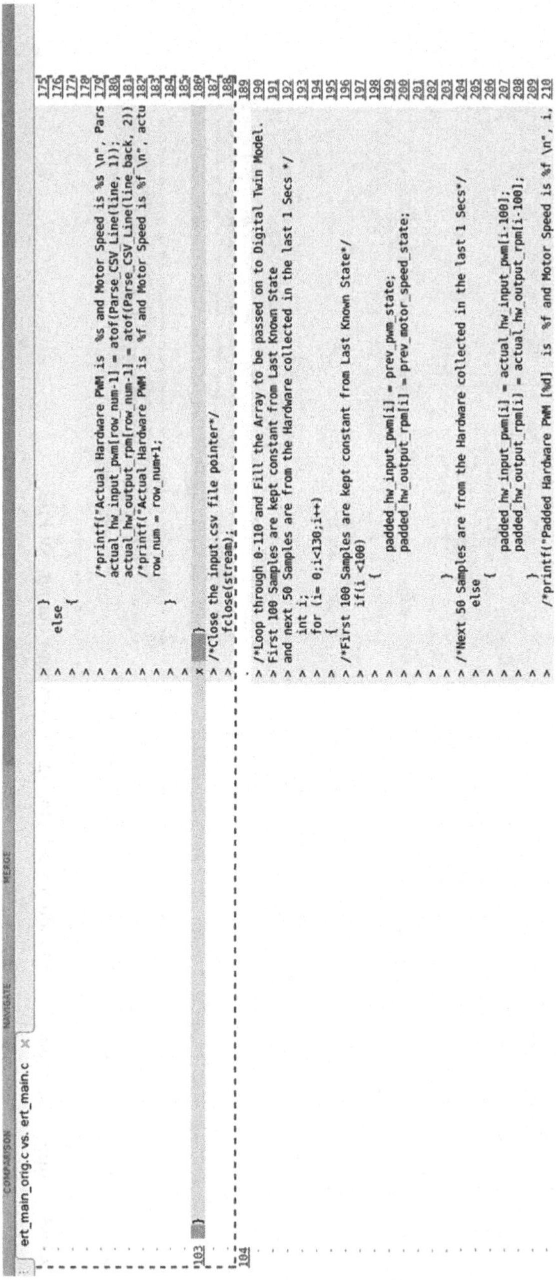

COMPARISON
NAVIGATE
MERGE
ert_main_orig.c vs. ert_main.c
}
else
{
/*printf("Actual Hardware PWM is %s and Motor Speed is %s \n", Pars
actual_hw_input_pwm[row_num-1] = atof(Parse_CSV_Line(line, 1));
actual_hw_output_rpm[row_num-1] = atof(Parse_CSV_Line(line_back, 2))
/*printf("Actual Hardware PWM is %f and Motor Speed is %f \n", actu
row_num = row_num+1;
}
}
/*Close the input.csv file pointer*/
fclose(stream);
/*Loop through 0-110 and Fill the Array to be passed on to Digital Twin Model.
First 100 Samples are kept constant from Last Known State
and next 50 Samples are from the Hardware collected in the last 1 Secs */
int i;
for (i= 0;i<130;i++)
{
/*First 100 Samples are kept constant from Last Known State*/
if(i <100)
{
padded_hw_input_pwm[i] = prev_pwm_state;
padded_hw_output_rpm[i] = prev_motor_speed_state;
}
/*Next 50 Samples are from the Hardware collected in the last 1 Secs*/
else
{
padded_hw_input_pwm[i] = actual_hw_input_pwm[i-100];
padded_hw_output_rpm[i] = actual_hw_output_rpm[i-100];
}
/*printf("Padded Hardware PWM [%d] is %f and Motor Speed is %f \n", i,

Figure 9.157 Reading input.csv and storing PWM Command and Motor Speed into Arrays Part 2.

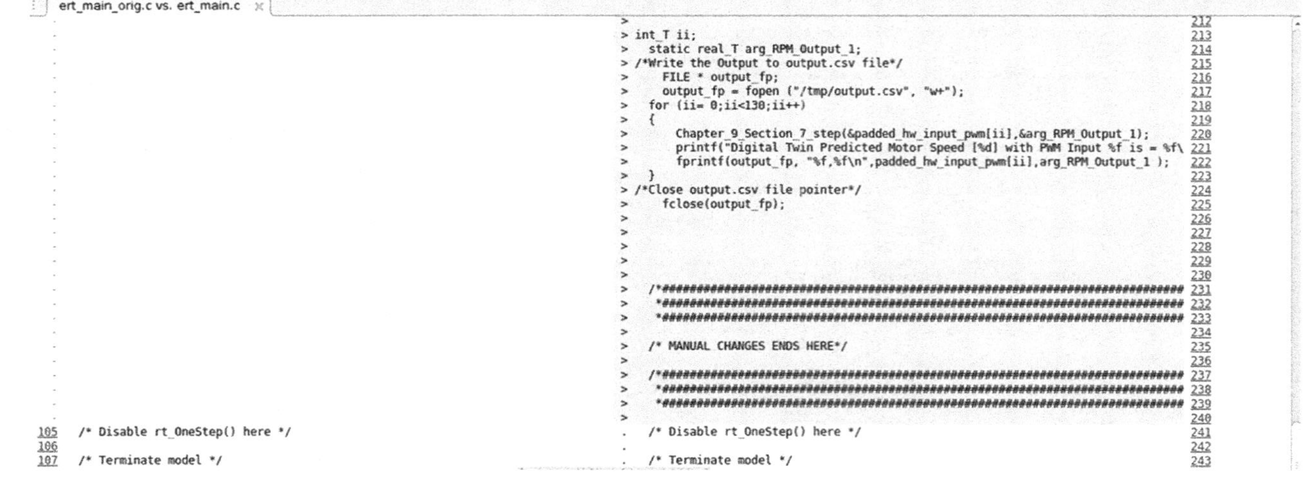

Figure 9.158 Calling the Model_step() function and Storing Results into output.csv File.

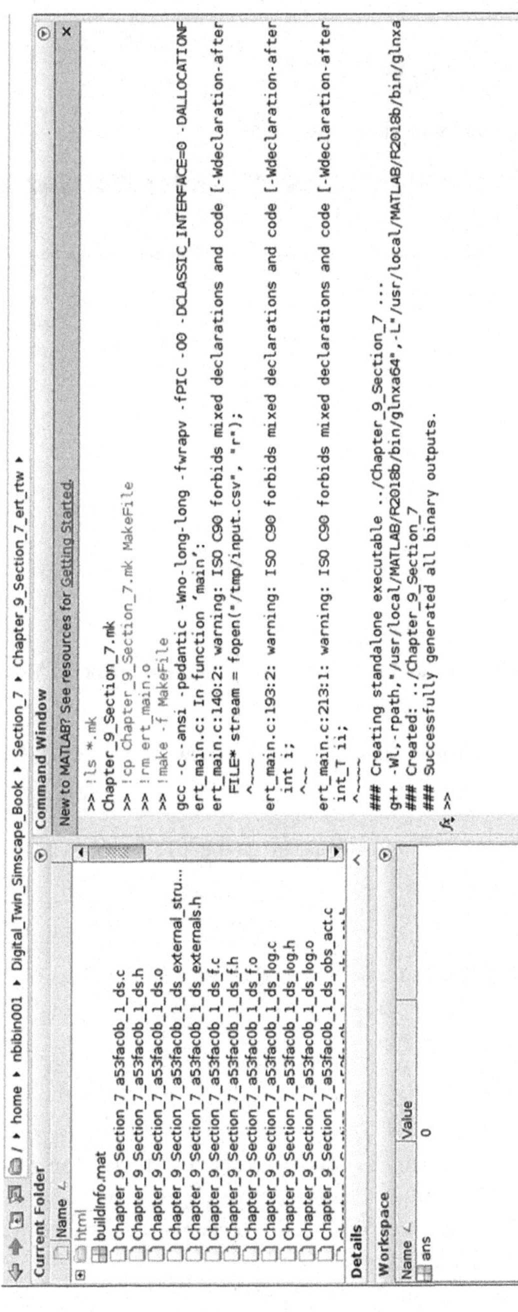

Figure 9.159 Recompiling the Application with the updated ert_mian.c.

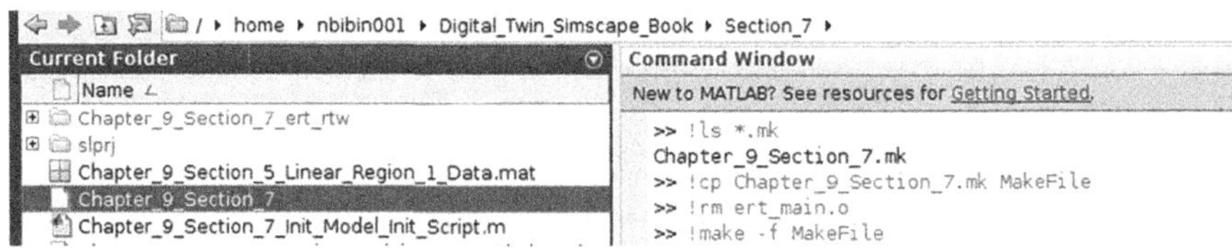

Figure 9.160 Recompiled executable application.

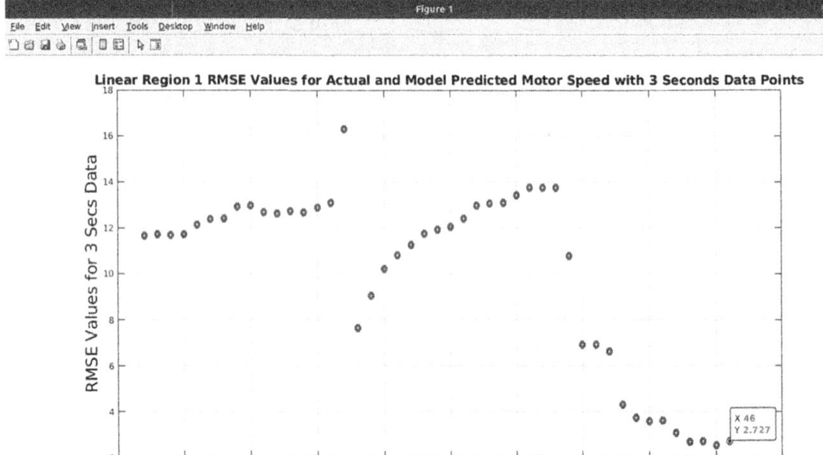

Figure 9.161 Root Mean Square Error Values for Actual and Predicted Motor Speeds with 3 s Data Points.

for each 3 s data and mark a point in the plot as shown in Fig. 9.161. See the MATLAB® program below:

```
%Book Title: Digital Twin Development and Deployment On Cloud Using Matlab
%SimScape
%Chapter: 9
%Section: 7
%Authors: Nassim Khaled and Bibin Pattel
%Last Modified: 09/10/2019
%%
clc
clear all
close all
% load the DC Motor Data collected from Hardware for Linear Region 1
load Chapter_9_Section_5_Linear_Region_1_Data.mat
% The Data collected from DC Motor Hardware is logged at 0.2 Secs.
% Resample PWM_Command to 0.1 Seconds
pwm_data_time = PWM_Input_Linear_Region_1(:,1);
pwm_resampled_time = (0:0.1:pwm_data_time(end));
pwm_resampled_data = interp1(pwm_data_time,PWM_Input_Linear_Region_1
(:,2),pwm_resampled_time);
```

```
% Resample Motor_Speed to 0.1 Seconds
motor_speed_time = Motor_Speed_RPM_Filtered_Linear_Region_1(:,1);
motor_speed_resampled_time = (0:0.1:motor_speed_time(end));
motor_speed_resampled_data = interp1(motor_speed_time,Motor_Speed_RPM_
Filtered_Linear_Region_1(:,2),motor_speed_resampled_time);
% Remove the input and output CSV files
!rm input.csv
!rm output.csv
% Initialize the Arrays and Variables
count = 1;
first_time = 1;
PWM_Command = [];
Actual_Motor_Speed = [];
prev_PWM_Command = pwm_resampled_data(1);
prev_Actual_Motor_Speed = motor_speed_resampled_data(1);
call_model_index =1;
% Run a loop for each of the PWM_Input data
for i =1:length(pwm_resampled_time)
    % The very first PWM_Command and Actual_Motor_Speed should be
    % initialized with the previous Value for initializing the Digital
Twin
    % model to a steady state
    if(first_time)
      PWM_Command(count) = prev_PWM_Command;
      Actual_Motor_Speed(count) = prev_Actual_Motor_Speed;
      count = count +1;
      first_time = 0;
    end
    % Continue adding the PWM_Command and Motor_Speed to the array to
gather
    % 3 Seconds sata
    PWM_Command(count) = pwm_resampled_data(i);
    Actual_Motor_Speed(count) = motor_speed_resampled_data(i);
    count = count+1;
    % When count becomes 32 the we have gathered 3 Seconds data
    if(count ==32)
      % Update the previous values
      prev_PWM_Command = PWM_Command(count-1);
      prev_Actual_Motor_Speed = Actual_Motor_Speed(count-1);
      count =1;
      first_time = 1;
      % Print the collected 3 Seconds data to input.csv file
      fid = fopen('input.csv','w+');
      for j =1:31
      fprintf(fid,'
%s,%s\n',num2str(PWM_Command(j)),num2str(Actual_Motor_Speed(j)));
```

```
        end
        fclose(fid);
        pause(1);
        % Move the input.csv file to /tmp folder
        !cp input.csv /tmp
        % Run the Digital Twin Model
        !./Chapter_9_Section_7
        % Digital Twin Model creates an output.csv file , copy it back
from
        % the /tmp folder to the current working directory
        !cp /tmp/output.csv .
        pause(1);
        % Read the output.csv file
        model_predicted_data = csvread('output.csv');
        temp = model_predicted_data(:,2);
        % Calculate RMSE between Actual Hardware and Model Predicted
Motor
        % Speed
        Error = Actual_Motor_Speed(2:end) - temp(101:130)';
        Squared_Error = Error.^2;
        Mean_Squared_Error = mean(Squared_Error);
        Root_Mean_Squared_Error = sqrt(Mean_Squared_Error);
        % Plot the RMSE of Speed Prediction for every 3 Seconds when the
        % Digital Twin Model Runs
        figure(101);
        plot(call_model_index,Root_
Mean_Squared_Error,'b','Marker','Diamond','linewidth',3);
        hold all
        call_model_index = call_model_index+1;
        % Delete the input.csv and output.csv for the next 3 Seconds
data
        !rm input.csv
        !rm output.csv
        pause(1);
      end
end
figure(101)
grid on
set(gcf,'color',[1 1 1]);
ylabel('RMSE Values for 3 Secs Data','Fontsize',18);
title('Linear Region 1 RMSE Values for Actual and Model Predicted Motor
Speed with 3 Seconds Data Points','Fontsize',18);
```

15. Now that we have validated the executable application in MATLAB® let us quickly develop a Python program to test one of the "**input.csv**" files. Please see the Python program shown in Fig. 9.162, which runs the Chapter_9_Section_7 and calculates the

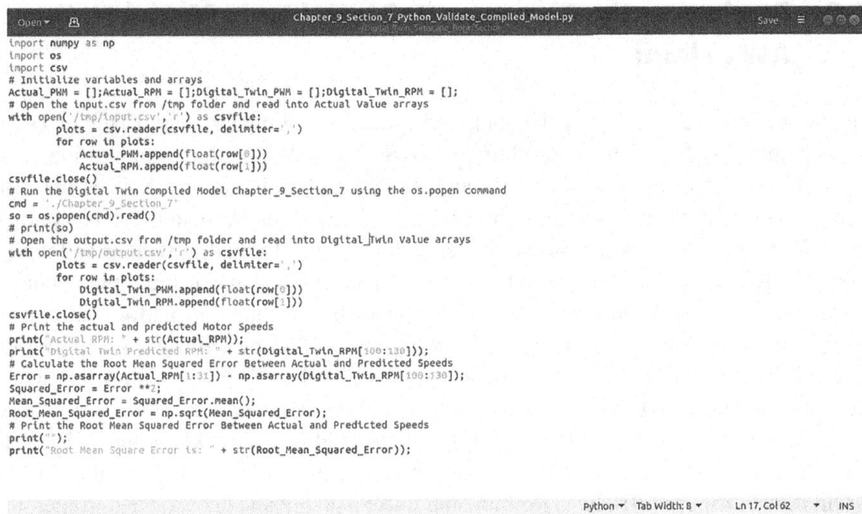

Figure 9.162 Python program to test the executable application.

RMSE value for just one instance (the last set) of "**input.csv**" and "**output.csv**" data for actual and predicted motor speed. Run the Python script using the command **python Chapter_9_Section_7_Python_Validate_Compiled_Model.py** from a Linux terminal as shown in Fig. 9.163. It can be noted that the RMSE value reported by the Python script matches the last test data RMSE value shown in Fig. 9.161 generated from the MATLAB® validation process.

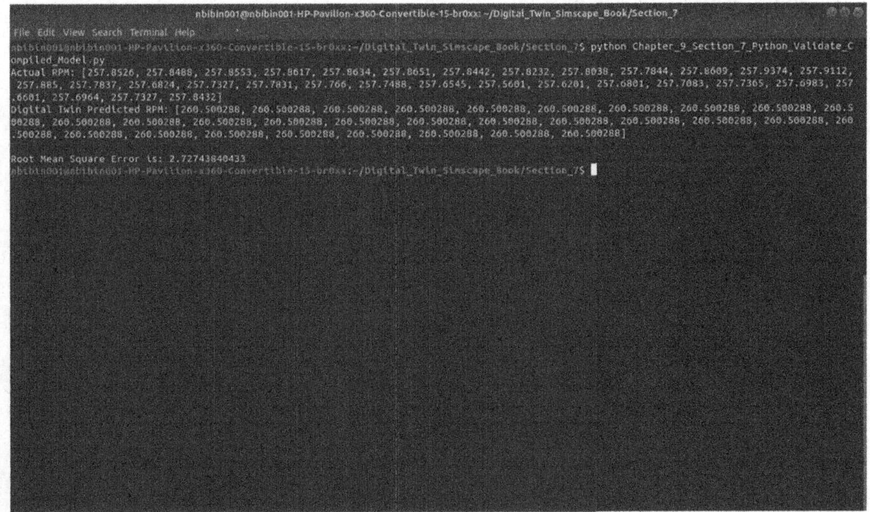

Figure 9.163 Running the python validation script.

9.8 Deploying the Simscape™ Digital Twin Model to the AWS cloud

In this section, we will deploy the compiled executable of the DC Motor Digital Twin model and the diagnostic algorithm to AWS cloud. We will be configuring and using two AWS in this section, SNS and AWS Lambda Functions. SNS is used to send Text/Email messages to the subscribers, and the subscriber details can be configured. We can create topics for SNS and add subscriber information such as cell number, Email-ID, etc., so that the subscribers will be notified when the topic event is triggered. AWS Lambda is an event-driven serverless computing platform that runs a specific code which we can develop and deploy. This code will run in response to a specific event, in our case the event is triggered by the AWS IoT Core when it receives data from the ESP32. Please check the AWS documentation for more details about these services. Fig. 9.164 shows the high-level diagram, where the AWS IoT Core will be triggering a Lambda function written in Python, which will run the Digital Twin model, run the Off-BD algorithm, and make a diagnostic decision and trigger the SNS service to notify the subscribers about the status of Off-BD algorithm decision. We will first get started with the SNS configuration, please follow the below steps:

1. On the AWS Management Console, search for SNS and select the SNS as shown in Fig. 9.165.
2. Click on the ***Create Topic*** button. See Fig. 9.166.
3. Give a Name to the topic on the Create topic form. We used **digital_twin_topic** in this example. Also give a Display name, which will be displayed in the SMS and E-mail. Click on the **Create topic** button to finish creating the new SNS topic as shown in Fig. 9.167.
4. The newly created topic is shown in Fig. 9.168. Now we need to create a Text and E-mail subscriptions for this topic. Click on the Create subscription button as shown in Fig. 9.168. Note down the topic ARN here, we will be using it later in the Lambda function to interact with the SNS.

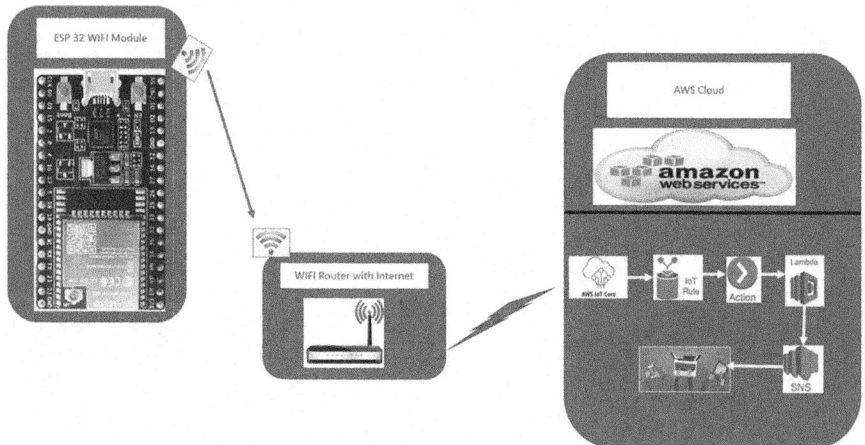

Figure 9.164 Digital Twin Deployment High Level Diagram.

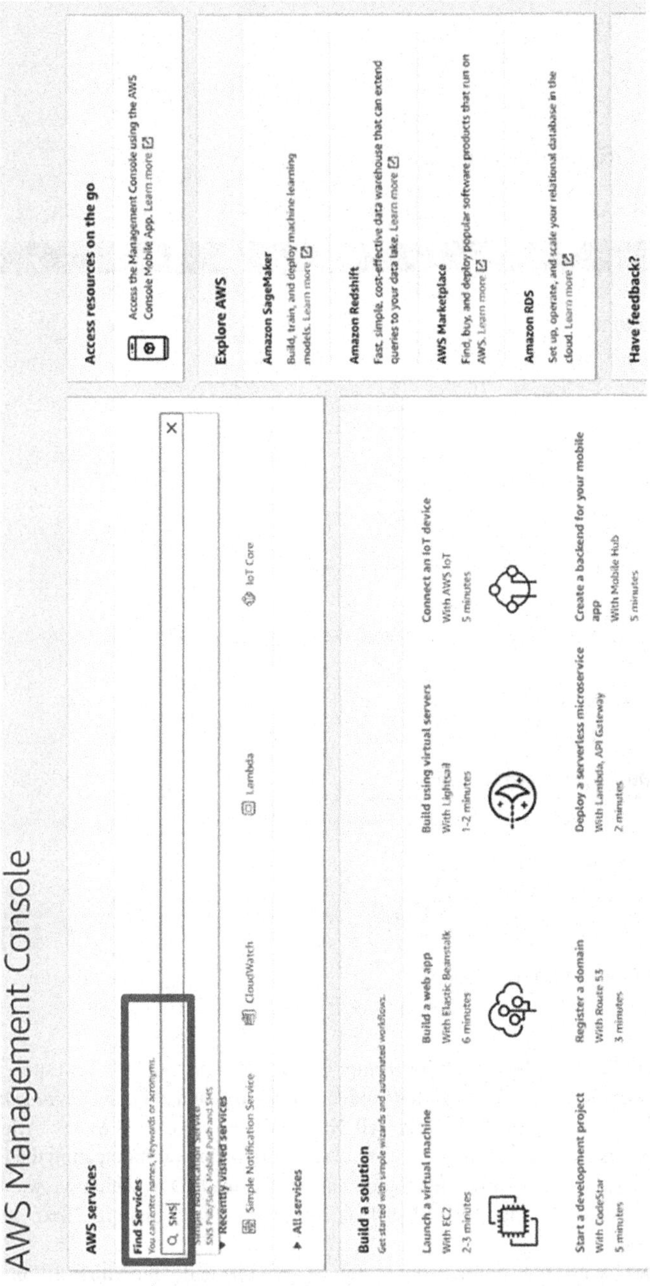

Figure 9.165 Launching Simple Notification Services (SNSs).

Amazon SNS
Dashboard
Topics
Subscriptions
Mobile
Push notifications
Text messaging (SMS)
Amazon SNS > Topics
Topics (0)
Name
ARN
No topics
To get started, create a topic
Create topic
Create topic

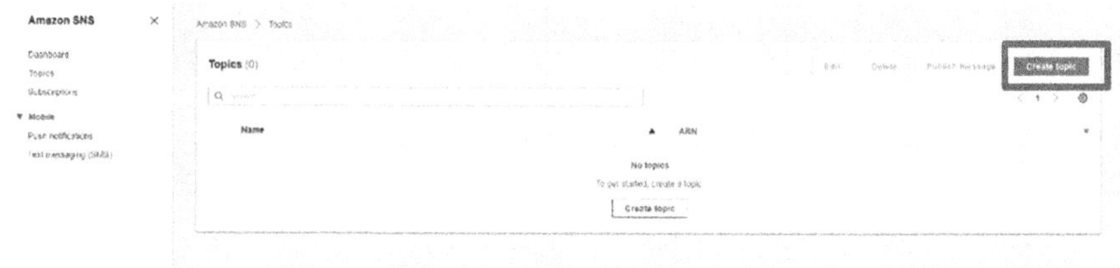

Figure 9.166 Creating an SNS topic.

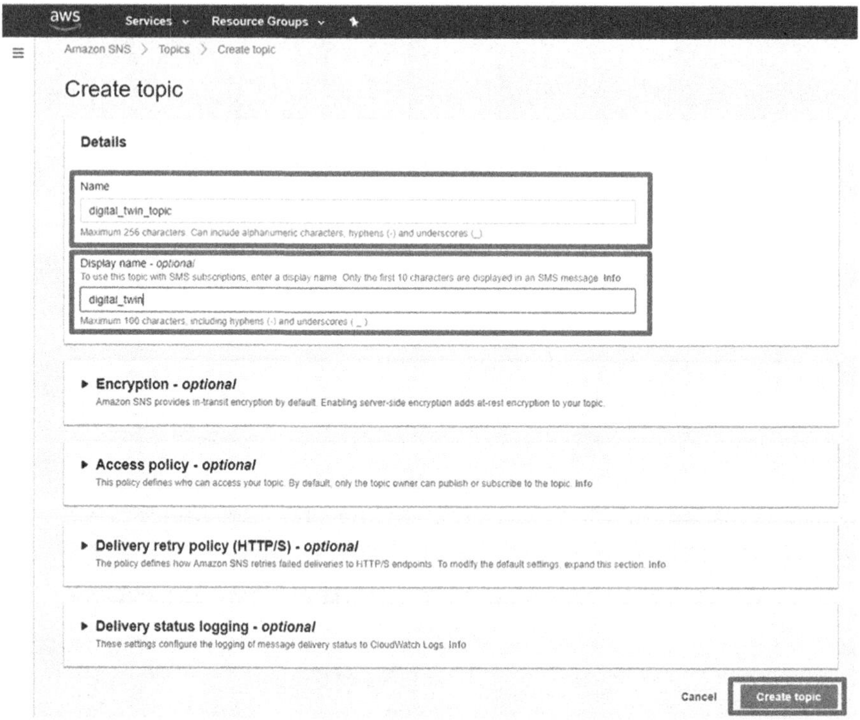

aws
Services
Resource Groups
Amazon SNS > Topics > Create topic
Create topic
Details
Name
digital_twin_topic
Maximum 256 characters. Can include alphanumeric characters, hyphens (-) and underscores (_)
Display name - optional
To use this topic with SMS subscriptions, enter a display name. Only the first 10 characters are displayed in an SMS message. Info
digital_twin
Maximum 100 characters, including hyphens (-) and underscores (_)
Encryption - optional
Amazon SNS provides in-transit encryption by default. Enabling server-side encryption adds at-rest encryption to your topic.
Access policy - optional
This policy defines who can access your topic. By default, only the topic owner can publish or subscribe to the topic. Info
Delivery retry policy (HTTP/S) - optional
The policy defines how Amazon SNS retries failed deliveries to HTTP/S endpoints. To modify the default settings, expand this section. Info
Delivery status logging - optional
These settings configure the logging of message delivery status to CloudWatch Logs. Info
Cancel
Create topic

Figure 9.167 New SNS topic details.

5. On the new ***Create subscription*** form enter first the SMS protocol for Text subscription and enter the Cell number starting with + and Country Code and click the ***Create subscription*** button. For example, **+1<10 Digit Cell Number>** for the United States. After the SMS subscription is created, use the same Create subscription button for creating the E-mail subscription from the Drop down menu and enter the E-mail ID to which we want to get the notification. Check Figs. 9.169–9.171 for details for creating Text and E-Mail subscriptions.
6. Though we created the subscriptions, it can be seen that under the **Subscriptions** menu the status shows Pending confirmation as shown in Fig. 9.172. When we created the subscriptions, AWS will automatically send an Email to the Email address given in the subscription

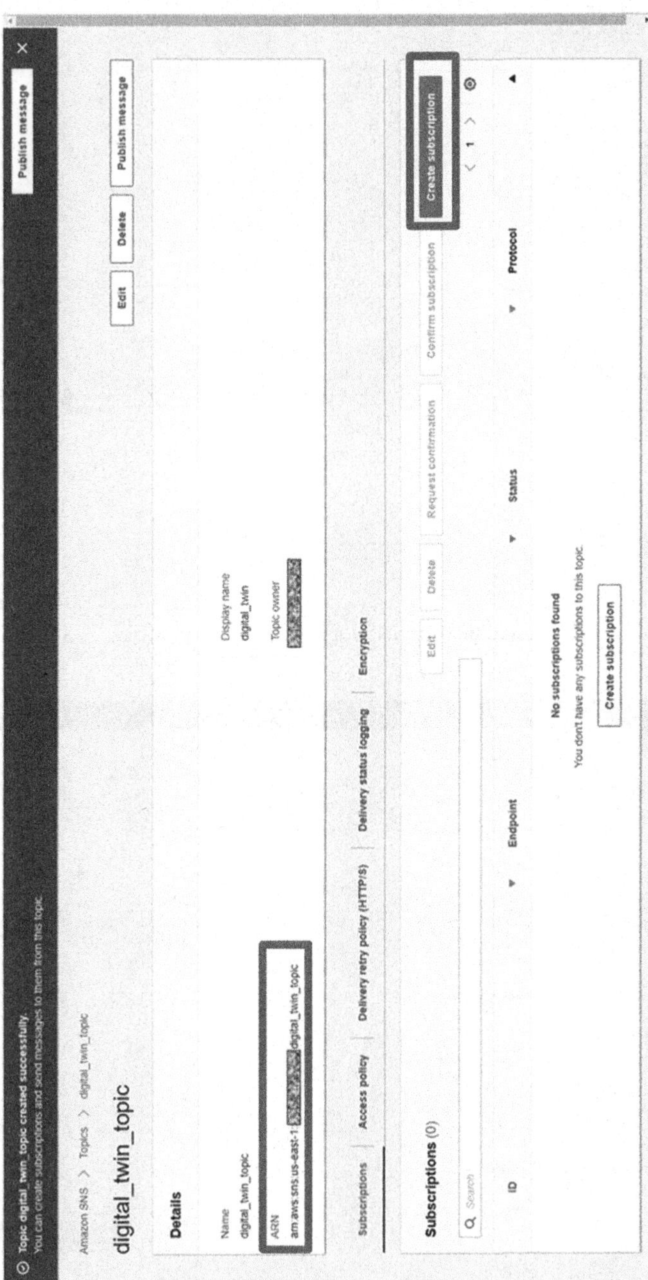

Figure 9.168 Newly created SNS topic.

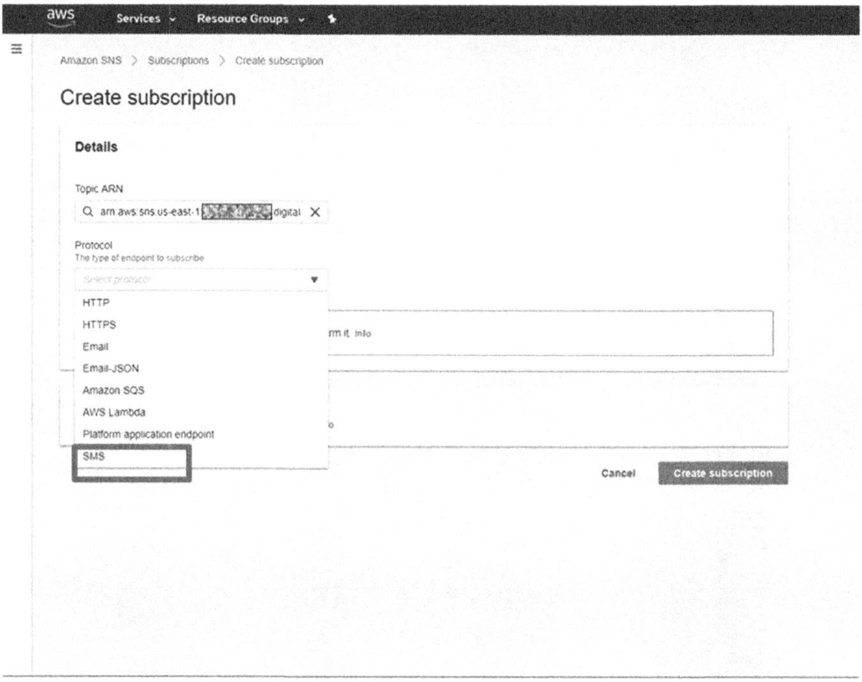

Figure 9.169 Creating subscription for SNS topic.

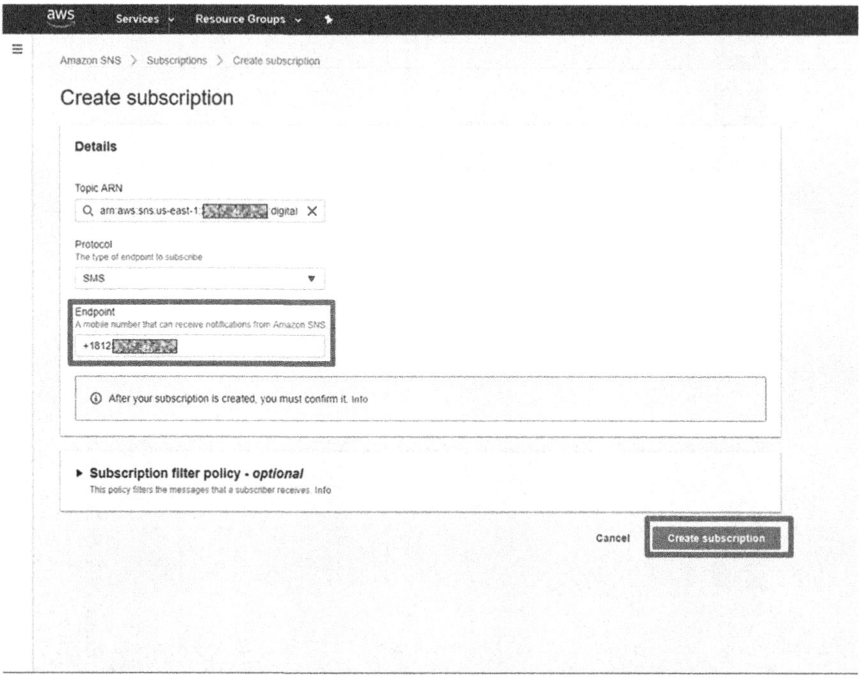

Figure 9.170 Creating Text/SMS subscription.

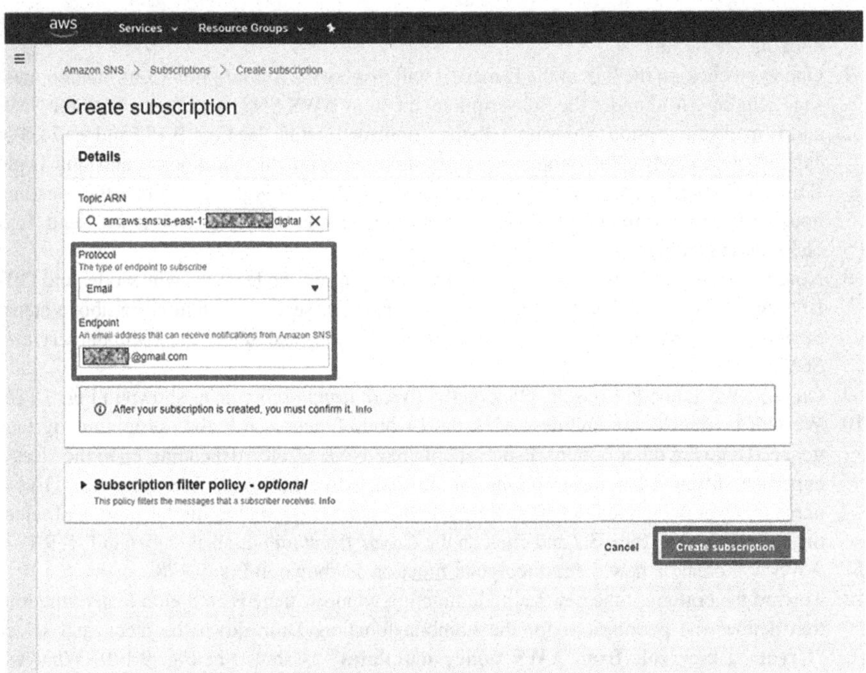

Figure 9.171 Creating Email subscription.

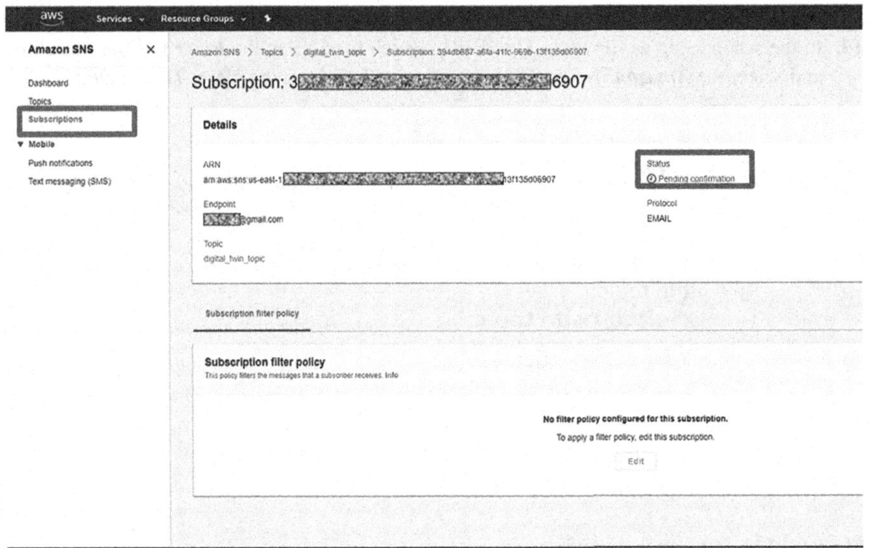

Figure 9.172 Pending subscription.

with a link to confirm the subscription. Check the Email and confirm the subscription by clicking on the link.

7. Once you click on the link in the E-mail, it will display the Subscription Confirmation message window. And under the Subscriptions menu in AWS SNS, we can see both the SMS and E-mail subscriptions show up with the Confirmed status. See Figs. 9.173 and 9.174. We can test these newly created topics are working by selecting the checkbox against the Topic ID and clicking the ***Publish message*** button highlighted in Fig. 9.174. Enter the Message body in the new form and publish it, based on the Topic ID we selected it will send Text or Email accordingly.
8. Now let us configure the AWS Lambda Functions to run the Digital Twin Model and Off-BD algorithm and also notify the user using the SNS service configured in above steps. Search for lambda in the AWS Management Console and select the Lambda services. See Fig. 9.175.
9. On the AWS Lambda Console, click on the ***Create function*** button as shown in Fig. 9.176.
10. We will be developing and deploying the Lambda function in Python programming language. There are other options also available like Node js, etc.; depending upon the User's expertise, different language options can be selected in the ***Create function*** form. Give a name to the new Lambda function, we chose the name as dc_motor_digital_twin, and select the ***RunTime*** as **Python 3.7** and click on the ***Create function*** button as shown in Fig. 9.177.
11. AWS will create a new default template function as shown in Fig. 9.178.
12. Toward the bottom of the new Lambda function window, there is an option to give the role that defines the permissions for the Lambda function. Drop down the menu and select "**Create a new role from AWS policy templates**" as shown in Fig. 9.179. What we want to do is allow the Lambda function to publish messages to the SNS topics that we created and configured in the above steps. The permission has to be explicitly given to the Lambda function; otherwise, it will not be able to interact with the SNS service.
13. The new **Create role** form will show up, which involves a four-step process. On the first step, select the ***AWS service*** option and click on the ***Next: permissions*** button as shown in Fig. 9.180.
14. In the second step as shown in Fig. 9.181, search for SNS in the ***Filter policies*** search bar and select the ***AmazonSNSFullAccess*** policy and click on the ***Next: Tags*** button.

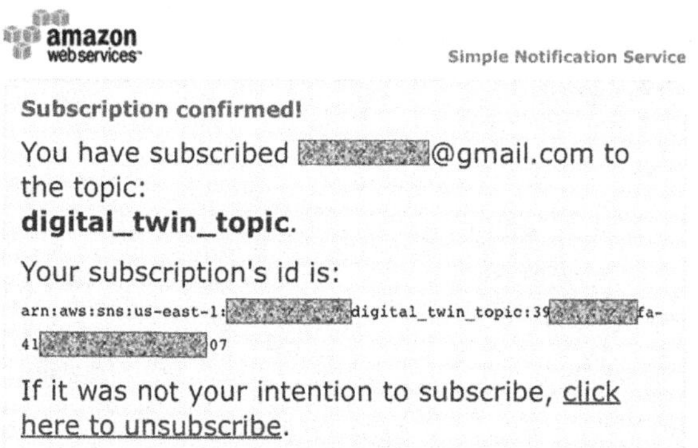

Figure 9.173 Subscription confirmation.

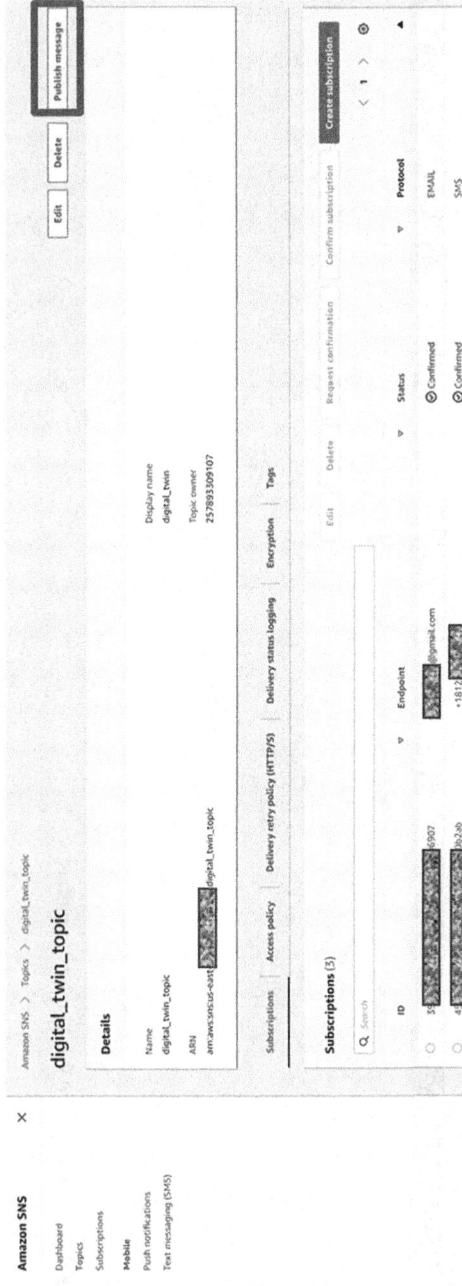

Figure 9.174 Confirmed subscriptions.

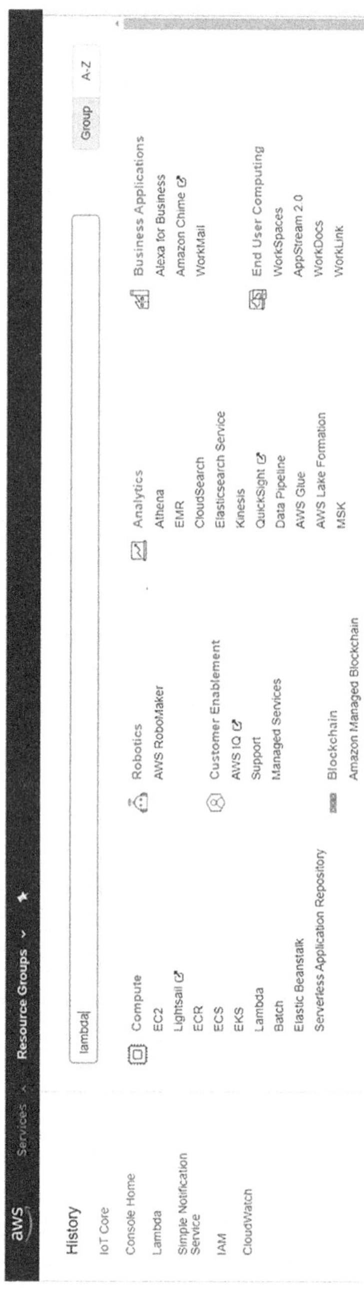

Figure 9.175 Launching AWS Lambda functions.

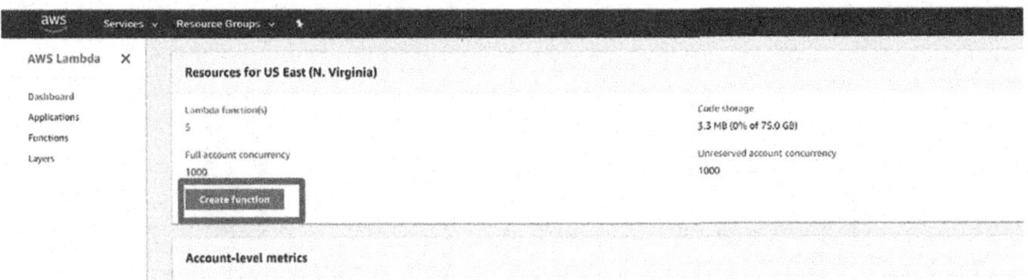

Figure 9.176 Creating Lambda function.

15. Next Adding the Tags is an optional step. In this case, we just gave a tag name ***digital_twin_role***; we can even skip this step. See Fig. 9.182. Click on the ***Next: Review*** button at the bottom.
16. The final step is the review, see Fig. 9.183. On the Review window, give a Role name, we chose ***digital_twin_role*** in this case. We can give any name to the Role. It can be seen that the policy ***AmazonSNSFullAccess*** we selected in the previous step is attached to this role. The idea is that we will attach this role to the Lambda function which will give the Lambda function access to the SNS service and publish Text/E-mails to the topics. Finish the role creation by clicking on the ***Create role*** button.
17. The new Role will be created and shown under the AWS service Identity and Access Management or called as IAM. The new Role that we created is shown under the IAM Console as shown in Figs. 9.184 and 9.185.
18. After the new Execution Role is created, go back to the AWS Lambda function console, under the lambda function we created earlier, select the "***Use an existing role***" option and browse and select the new role we created named ***digital_twin_role.*** If the User chose to give a different name for the role in the above step, chose the role name accordingly. See Fig. 9.186 for details.
19. Save the Lambda function by clicking on the ***Save*** button at the top as shown in Fig. 9.187.
20. It can be seen in Fig. 9.188 that the newly created Lambda function has full access to the AWS SNS because we linked the Execution Role which has the SNS policy attached to it. Next step is to add a trigger for the Lambda function, to decide when the Lambda function will run. Click on the Add trigger button shown in Fig. 9.188. We will be triggering the Lambda function whenever the IoT Thing receives data from the DC Motor Hardware.
21. The Add trigger window will be opened as shown in Fig. 9.189. Select the AWS IoT from the dropdown as the trigger source. We need to create a Custom IoT Rule and use it for the trigger. An IoT routes the IoT Thing updates to a specific AWS service. In this case, we are going to create a rule to trigger the AWS Lambda service. Check the box "***Custome IoT rule***" and select the option "***Create a new rule.***" Give a name to the new rule, in this case we used ***digital_twin_rule.*** The ***Rule description*** is optional. In the ***Rule query statement***, enter the string **SELECT * FROM '$aws/things/digital_twin_thing/shadow/update/accepted'**. This query statement will pull all values from the IoT Thing update and pass it to the AWS Lambda. Also check the "***Enable trigger***" option and click on the ***Add*** button.
22. The Lambda function will now show the AWS IoT as the Trigger source as shown in Fig. 9.190.
23. Next step is developing the Lambda function. At this point, we only have a skeleton template Lambda function, so we will have to expand on that. For convenience of

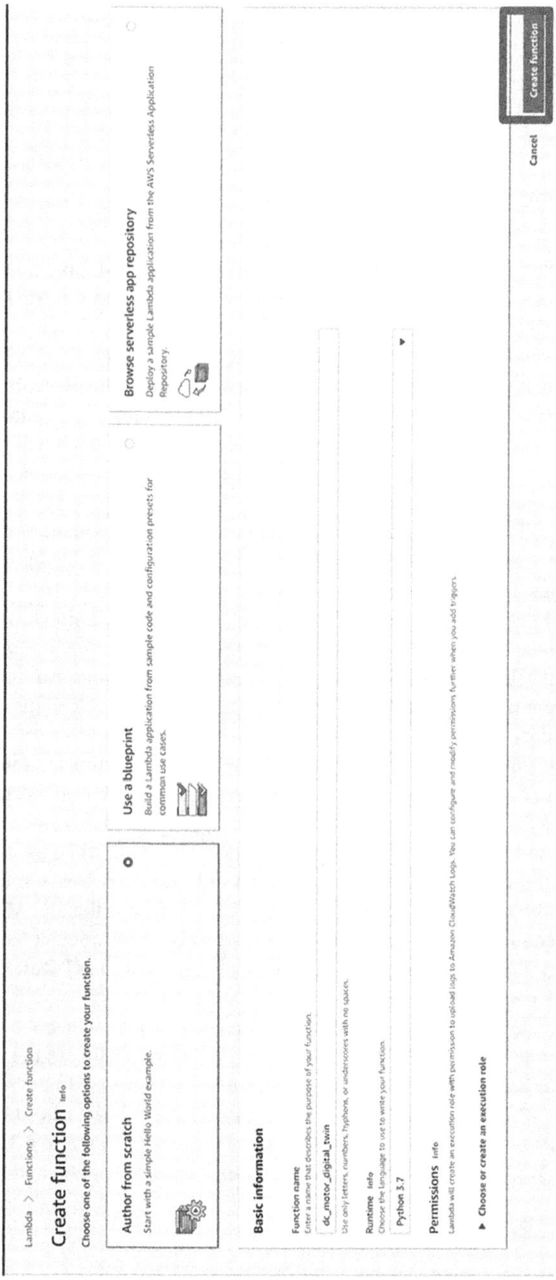

Figure 9.177 New Lambda function details.

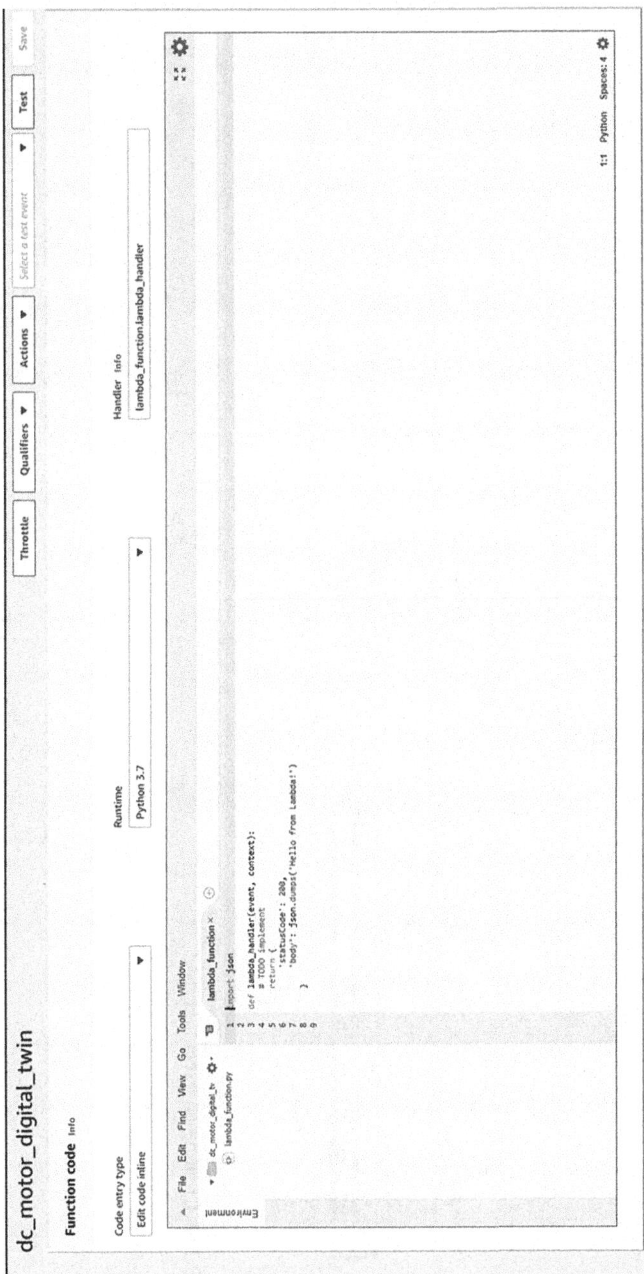

Figure 9.178 Default Lambda function body.

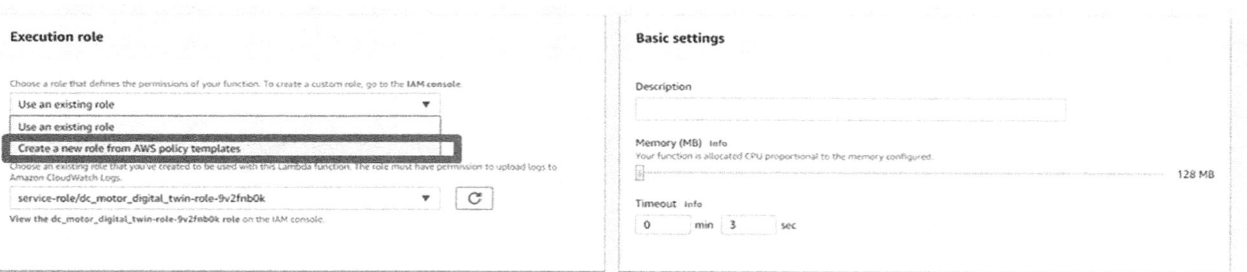

Figure 9.179 Creating an execution role.

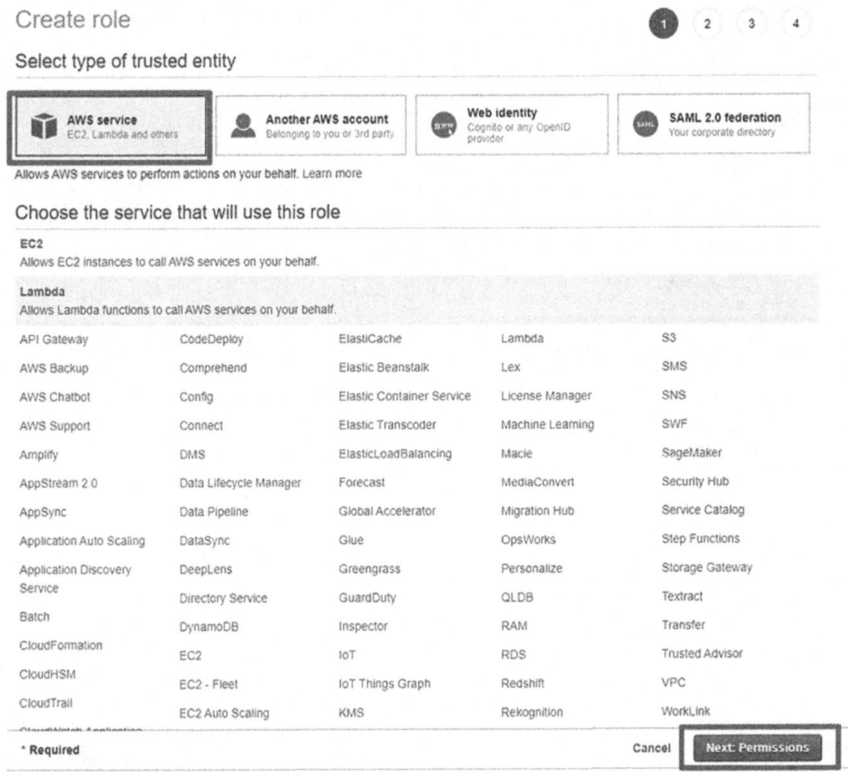

Figure 9.180 Create a new role.

testing the Lambda function along with the Digital Twin compiled executable and to package the Lambda function and Digital Twin executable to deploy to AWS, we have developed the Lambda function in Linux. From the Ubuntu Linux (which we used for compiling the Digital Twin model earlier), open an editor and name the file as **lambda_function.py.** At the top of the python script, add all the packages we will be using in this program. So here the package **json** is used to deal with the incoming json string from AWS IoT, **os** is used to call the compiled Digital Twin executable, **boto3** is used to interact with the AWS SNS, **csv** for creating csv file with incoming data, and **math** is for some math calculations for **Off-BD** algorithm. See Fig. 9.191.

24. Next we will expand the Lambda handler function. We can see from Fig. 9.178 that when we create the AWS Lambda function, it creates a default handler function named **lambda_handler.** The input argument **event** contains the JSON string that we sent from the ESP32 Wi-Fi module with the DC Motor Hardware data. As shown in Fig. 9.192, we will parse the JSON string and store it into variables for input and output. So here input_list will store the PWM_Input and output_list will store the Actual_Motor_Speed reported from the hardware.

25. Next step is to write the input and output data in a CSV format to a file under the **/tmp** folder. As we have seen earlier, the compiled executable for the Digital Twin model will be looking for the **/tmp** folder for the input file with data reported from the Hardware. See Fig. 9.193.

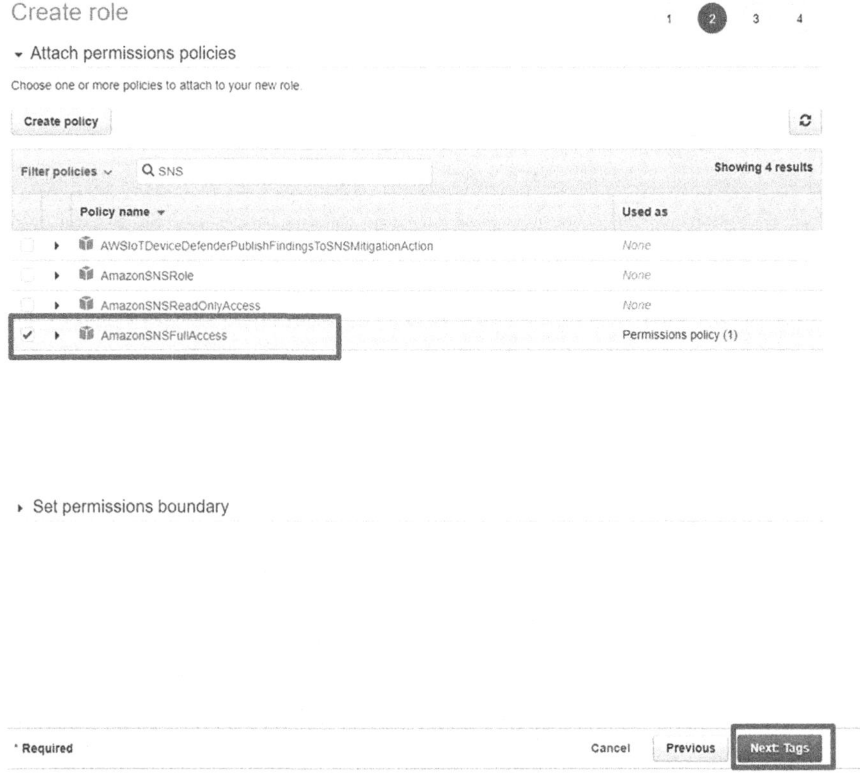

Figure 9.181 Attach SNS permission policy to the new role.

26. Once the input.csv file is created we will trigger the compiled executable Digital Twin model. Note that we are calling the compiled executable **DC_Motor_Digital_Twin**, so we will have to rename the executable we made in the earlier section, where the executable is created with the Simulink model name. See Fig. 9.194. When the executable is run, it will create the **output.csv** file in the **/tmp** folder as we have seen and tested before.
27. Now we have the actual and Digital Twin model predicted data for the DC Motor and we are ready to perform the Off-BD algorithm. First read the **input.csv** and **output.csv** files from the **/tmp** folder and store it into arrays as shown in Fig. 9.195.
28. The section shown in Fig. 9.196 calculates the RMSE between the actual and predicted DC Motor speed. Remember from previous section that we have padded the input data for the first 100 locations inside the compiled executable for initial steady state of the model. So from the output data, we have to avoid the first 100 points to get the exact predicted output speed, which is why there is some indexing done in the logic before the RMSE calculation.
29. The Off-BD algorithm has calculated the RMSE value and now we need to compare the RMSE value against a threshold and take a decision. As shown in Fig. 9.197, we picked a threshold of 105. So if the RMSE value is less than 105, Off-BD algorithm decides there is no failure detected, and if the RMSE exceeds 105, there is a failure. Finding out the right threshold value may involve a little bit of a tuning exercise generally. For publishing the

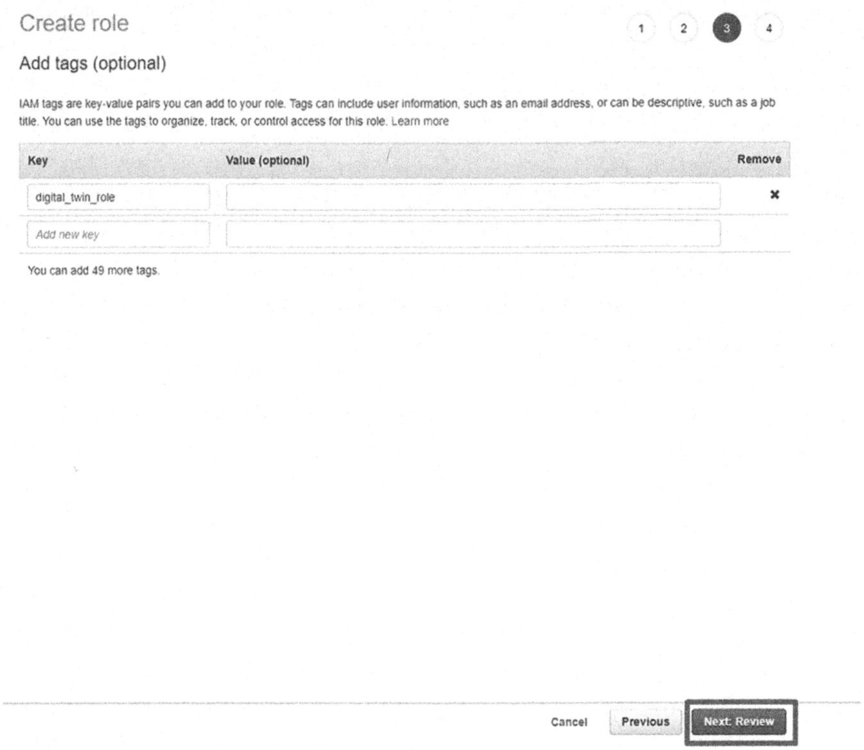

Figure 9.182 Optional Tag for the new role.

Off-BD decision to the SNS, we create an SNS object using the boto3 package. We will use the topic ARN noted from Step 4 to publish message. Call the function **sns.publish** with the topic ARN and custom message based on the Off-BD diagnostic decision. User can edit the Message string however they want, this is the message we will see in the TEXT and E-mail messages when the Lambda function runs. The Lambda function development is now finished.

30. Save the above-developed Lambda function into a new folder. In this case, we named the folder to be **Section_8** under the folder **Digtal_Twin_Simscape_book.** Copy the Digital Twin compiled executable from the previous section also to this folder. Fig. 9.198 shows a set of Linux commands to give the necessary executable permissions and packaging of the Lambda function and Digital Twin executable. The commands are explained below. Open a Linux terminal and run the below commands in steps.

   a. The ***ls*** command shows the folder has the compiled executable **Chapter_9_Section_7** and the Lambda function **lambda_function.py.**

   b. Using the ***cp*** command, copy the executable to a new file name **DC_Motor_Digital_Twin**, which we used in the Lambda function. Now when we run the ***ls*** command, there is also the newly copied file in the folder.

   c. We need to give full read/write/executable permissions to all files in this folder for the AWS to be able to run it. Use the command ***sudo chmod -R 777 <folder_name>***.

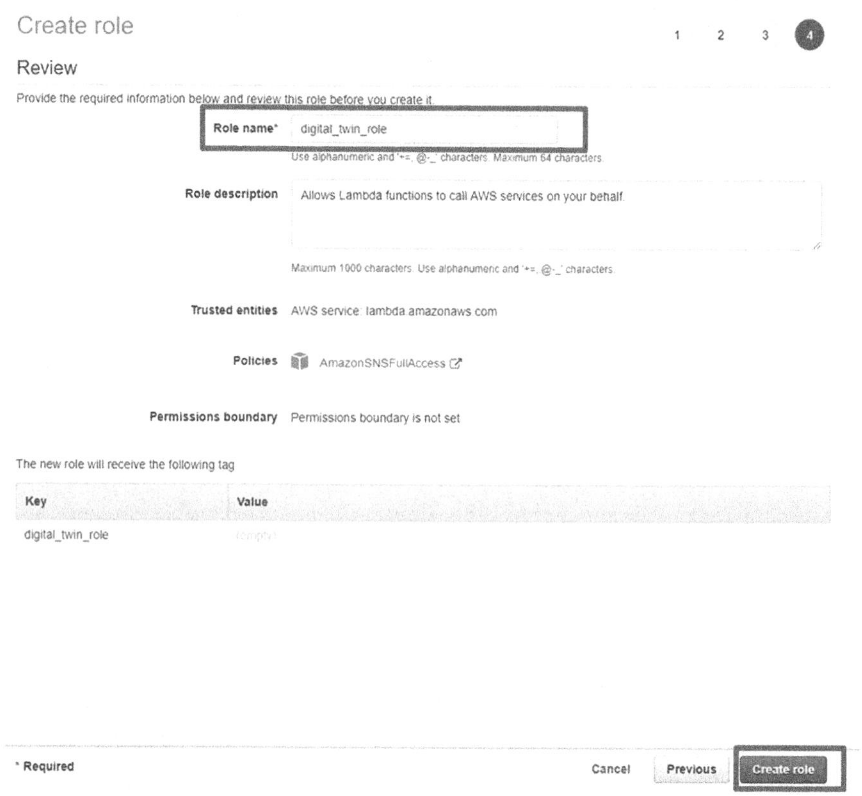

Figure 9.183 Naming the new role.

d. Run the command ***ls -l***. This will list all the files and folders in the current folder and their permissions. Everything should be showing ***rwxrwxrwx***.
e. Now package the Lambda function and compiled executable using the command ***zip bundle.zip lambda_function.py DC_Motor_Digital_Twin***. This will zip the files ***lambda_function.py*** and ***DC_Motor_Digital_Twin*** into a file named ***bundle.zip***.
f. After the bundle.zip is created, once again repeat the Step c to allow read/write/executable permissions to the ***bundle.zip*** file as well.
g. The ***ls*** command will show the newly created ***bundle.zip*** file as well. Now we are ready to deploy the Lambda function to AWS and do the final testing.

31. From the Linux operating system itself, open the web browser, the AWS Management Console, and the Lambda function **dc_motor_digtal_twin** we created. Select the option from the drop down "**Upload a .zip file**" as shown in Fig. 9.199. Browse and select the ***bundle.zip*** file we packaged earlier as shown in Fig. 9.200.
32. After the bundle.zip file is successfully uploaded click on the ***Upload*** and ***Save*** button as shown in Fig. 9.201.
33. We can see in Fig. 9.202 the AWS Lambda function editor now shows our updated Lambda function and also the Digital Twin compiled executable. Congratulations!!! We are ready to test the whole Off-BD algorithm for the DC Motor Speed Diagnostics in real time now.

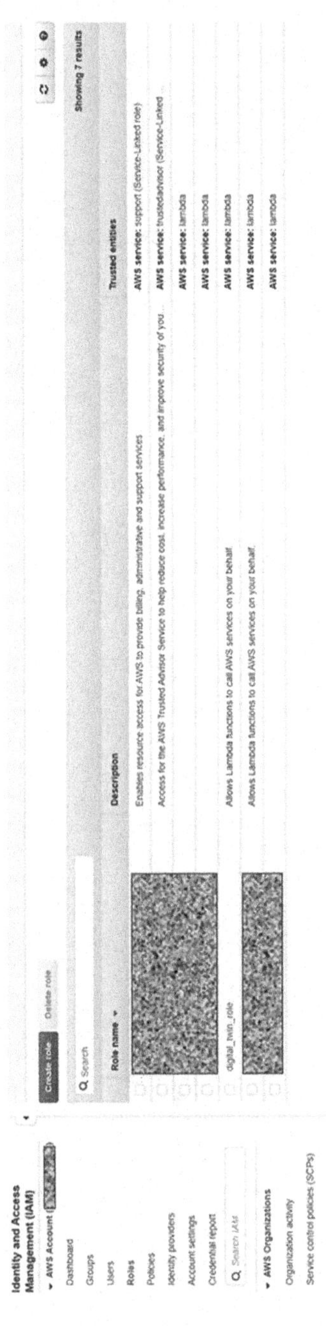

Figure 9.184 New role created.

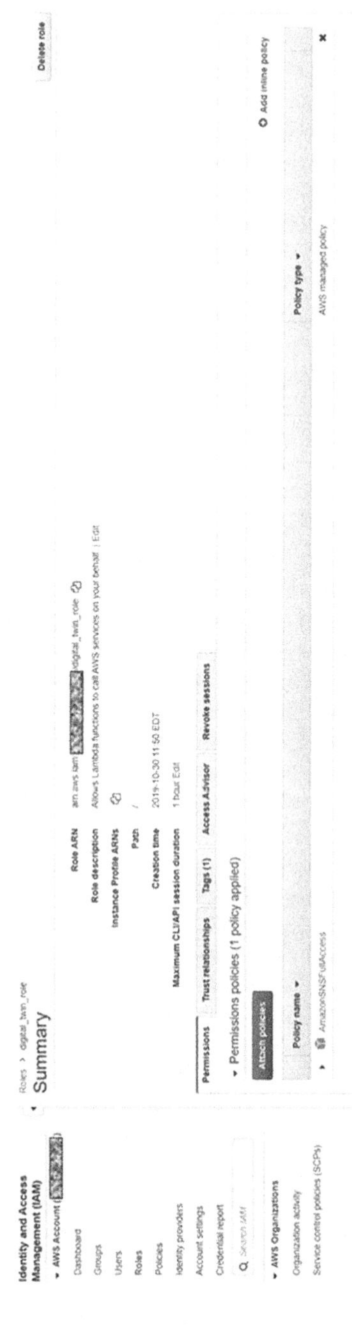

Figure 9.185 New role created showing the SNS policy attached.

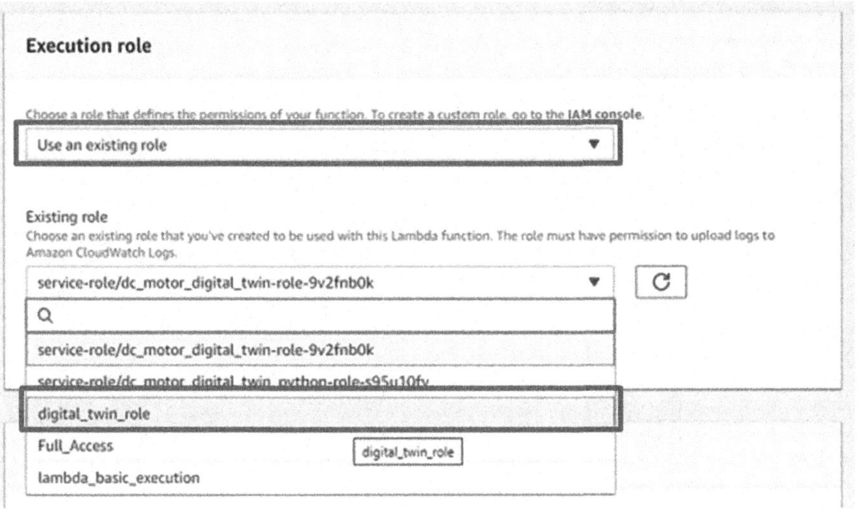

Figure 9.186 Selecting the new execution role in the Lambda function.

34. First, we will test the No Fault condition. Repeat the Step 8 from Section 9.6.5 to run the DC Motor Feedback Speed Controller on the Arduino Mega and the ESP32 Wi-Fi sketch on the ESP32 module to communicate between Arduino Mega and AWS. As the motor starts to run, we can see from the SNS we configured in the Lambda function will send Text and Email Notifications on the Off-BD status based on the Digital Twin prediction, decision logic in the Lambda function. Figs. 9.203 and 9.204 shows the Email messages received from the Digital Twin notification service, note in this case no failure condition is detected.
35. Next we will test a Fault condition where the DC Motor is not powered. So though the Arduino Mega is trying to control the speed of the motor, since the motor is not getting supply voltage, it will not spin and the speed sensor in Arduino software will always read 0, but the feedback controller will try to compensate by increasing the PWM command until it saturates to its maximum value. So clearly for the same PWM value, the motor speed from the hardware and predicted by Digital Twin (which doesn't know about the power supply failure) will be different and the Off-BD algorithm will detect it as a faulty condition. Repeat the Step 8 from Section 9.6.5 to run the DC Motor Feedback Speed Controller on the Arduino Mega and the ESP32 Wi-Fi sketch on the ESP32 module to communicate between Arduino Mega and AWS, but this time with the power supply to the DC Motor disconnected/unplugged. As the program starts to run on the Arduino Mega, we can see from the SNS we configured in the Lambda function will send Text and Email Notifications on the Off-BD status based on the Digital Twin prediction, decision logic in the Lambda function. Figs. 9.205 and 9.206 show the Email messages received from the Digital Twin notification service, note in this case no failure condition is detected.

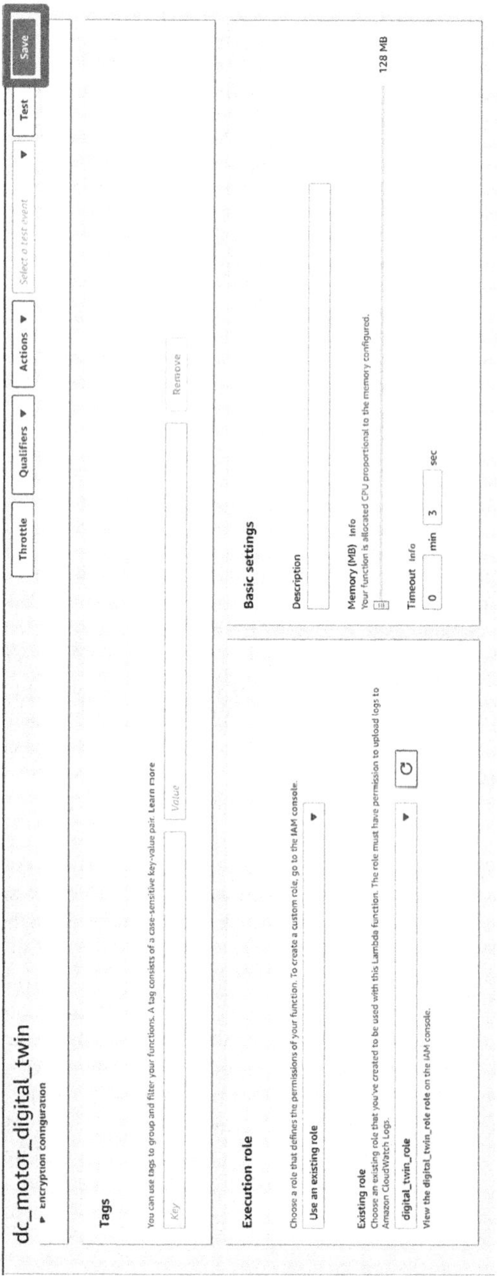

Figure 9.187 Save Lambda function with updated execution role.

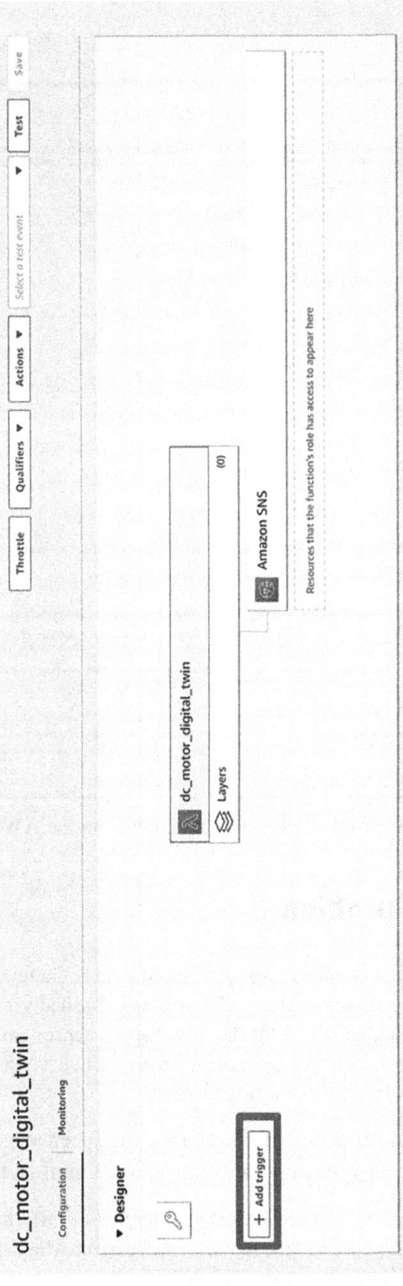

Figure 9.188 Lambda function showing its access to AWS SNS.

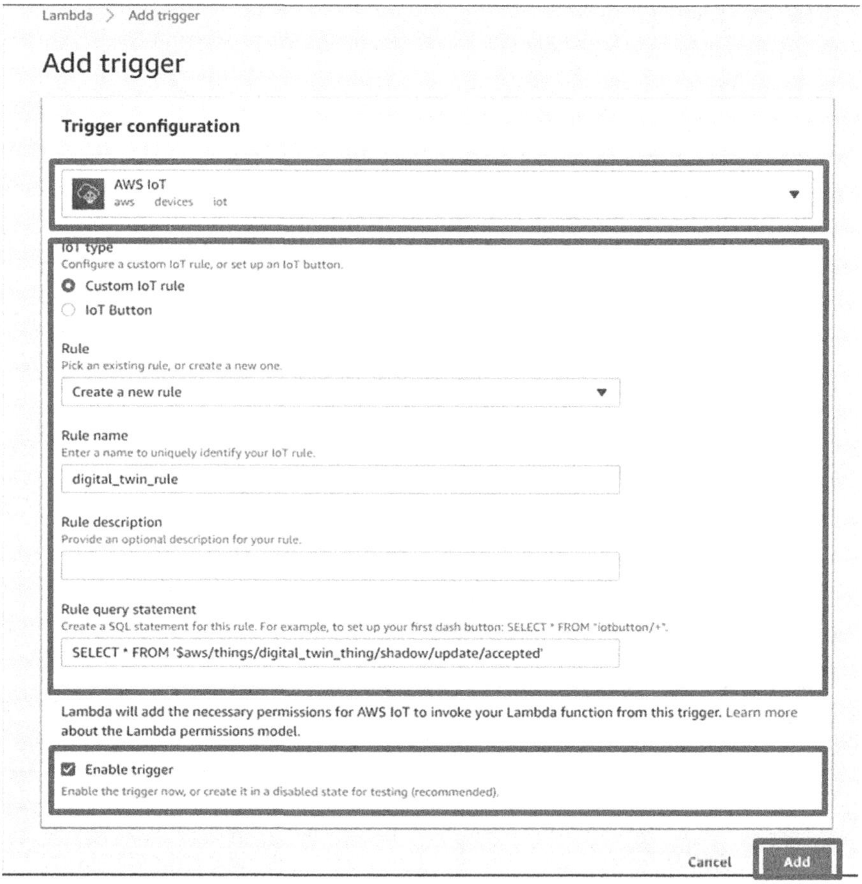

Figure 9.189 Adding a trigger for the Lambda function from the AWS IoT.

9.9 Application problem

1. Develop an Off-BD process to detect a Motor Speed Sensor Gain issue (factor of 1.5) while sensing the speed of the motor in the Arduino Mega. Basically in the Arduino Simulink controller model, insert a gain of 1.5 for the motor speed sensor, so for a given PWM command the sensed motor speed will be more than the actual value. Use the Digital Twin running on the cloud to detect the sensor gain issue.

Hint: Repeat the whole process, but just change the speed sensing logic in the Arduino Simulink feedback controller model to introduce a gain to the speed sensing.

2. Develop an Off-BD process to detect a Motor Speed Sensor Offset issue (constant offset of 50 RPM) while sensing the speed of the motor in the Arduino Mega. Basically in the Arduino Simulink controller model, insert a sum of 50 RPM for the motor speed sensor, so for a given PWM command, the sensed motor speed will be more than the actual value. Use the Digital Twin running on the cloud to detect the sensor gain issue.

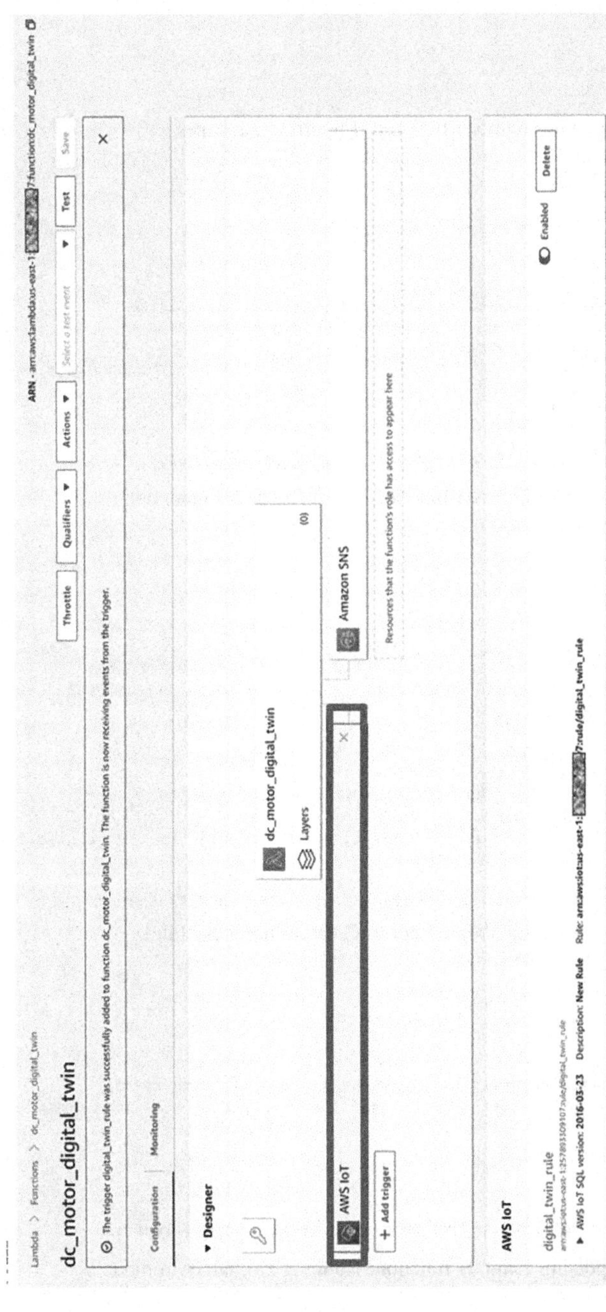

Figure 9.190 Lambda function showing the AWS IoT trigger added.

```
lambda_function.py
~/Digital_Twin_Simscape_Book/Section_8
##############################################################
# Import the packages
import json
import os
import boto3
import csv
import math
```

Figure 9.191 Updating Lambda function: Importing packages.

```
##############################################################
# Lambda Handler function which runs when IoT Trigger Happens
def lambda_handler(event, context):
# The input argument event has the JSON structure , separate the Input and Output from the
# input argument "event"
    input_list = event['state']['desired']['Input']
    output_list = event['state']['desired']['Output']

##############################################################
```

Figure 9.192 Updating Lambda function: Reading trigger input data.

```
##############################################################
# Create a file "input.csv" under /tmp folder and write the input and output values
# separated by commas in each row
    data_count = 0;
    with open('/tmp/input.csv', 'w') as writeFile:
        writer = csv.writer(writeFile)
# This for loop runs for the total number of input samples
        for item in input_list:
            writer.writerows([[(input_list[data_count]),(output_list[data_count])]])
            data_count = data_count +1
    writeFile.close()
##############################################################
```

Figure 9.193 Updating Lambda function: Creating input.csv File.

```
##############################################################
# Create an OS command string to trigger the DC Motor Digital Twin Compiled Executable
# Application
    cmd = './DC_Motor_Digital_Twin'
    so = os.popen(cmd).read()
# Print the output
    print(so)
##############################################################
```

Figure 9.194 Updating Lambda function: Running Digital Twin model.

```
###################################################################
# Initialize variables and arrays
    Actual_PWM = [];Actual_RPM = [];Digital_Twin_PWM = [];Digital_Twin_RPM = [];
# Open the input.csv from /tmp folder and read into Actual Value arrays
    with open('/tmp/input.csv','r') as csvfile:
        plots = csv.reader(csvfile, delimiter=',')
        for row in plots:
            Actual_PWM.append(float(row[0]))
            Actual_RPM.append(float(row[1]))
    csvfile.close()
# Open the output.csv from /tmp folder and read into Digital_Twin Value arrays
    with open('/tmp/output.csv','r') as csvfile:
        plots = csv.reader(csvfile, delimiter=',')
        for row in plots:
            Digital_Twin_PWM.append(float(row[0]))
            Digital_Twin_RPM.append(float(row[1]))
    csvfile.close()
###################################################################
```

Figure 9.195 Updating Lambda function: Reading Input and Output files for Actual and Predicted Data.

```
###################################################################
# Print the actual and predicted Motor Speeds
    print("Actual RPM: " + str(Actual_RPM));
    print("Digital Twin Predicted RPM: " + str(Digital_Twin_RPM[100:130]));
# Calculate the Root Mean Squared Error Between Actual and Predicted Speeds
    Error = [None]*30
    Squared_Error = [None]*30
    Mean_Squared_Error = 0
    Root_Mean_Squared_Error = 0
    for index in range(30):
        Error[index] = Actual_RPM[index+1] - Digital_Twin_RPM[index+99];
        Squared_Error[index] = Error[index]*Error[index];
        Mean_Squared_Error = Mean_Squared_Error + Squared_Error[index]
    Mean_Squared_Error = Mean_Squared_Error/30;
    Root_Mean_Squared_Error = math.sqrt(Mean_Squared_Error)

# Print the Root Mean Squared Error Between Actual and Predicted Speeds
    print("");
    print("Root Mean Square Error is: " + str(Root_Mean_Squared_Error));
###################################################################
```

Figure 9.196 Updating Lambda function: Calculating Root Mean Squared Error for Actual and Predicted Motor Speed.

```
######################################################################
# Set the AWS Simple Notification Service client and Topic for sending Text/Email from
# this Python program
    sns = boto3.client(service_name="sns")
    topicArn = 'arn:aws:sns:us-east-1:257893309107:digital_twin_topic'
    if Root_Mean_Squared_Error > 20 :
        sns.publish(
        TopicArn = topicArn,
        Message = 'Digital Twin Off-BD for DC Motor Detected a Failure !!! The Root Mean Square Error Actual and Predicted Motor Speed is  ' +
str(Root_Mean_Squared_Error) + ' which is Greater than the Set Threshold of 20.'
        )
        cmd_line_Message = 'Digital Twin Off-BD for DC Motor Detected a Failure !!! The Root Mean Square Error Actual and Predicted Motor
Speed is  ' + str(Root_Mean_Squared_Error) + ' which is Greater than the Set Threshold of 20.'
        print(cmd_line_Message)
    else :
        sns.publish(
        TopicArn = topicArn,
        Message = 'All is Well. No Failure Conditions Detected Between Digital Twin Predicted DC Motor Speed and Actual Motor Speed Reported'
        )
        cmd_line_Message = 'All is Well. No Failure Conditions Detected Between Digital Twin Predicted DC Motor Speed and Actual Motor Speed
Reported'
        print(cmd_line_Message)

    return 0
######################################################################
```

Figure 9.197 Updating Lambda function: Off-BD algorithm with SMS/Text SNS alert.

```
nbibin001@nbibin001-HP-Pavilion-x360-Convertible-15-br0xx: ~/Digital_Twin_Simscape_Book/Section_8
File  Edit  View  Search  Terminal  Help
nbibin001@nbibin001-HP-Pavilion-x360-Convertible-15-br0xx:~/Digital_Twin_Simscape_Book/Section_8$ ls
Chapter_9_Section_7  lambda_function.py
nbibin001@nbibin001-HP-Pavilion-x360-Convertible-15-br0xx:~/Digital_Twin_Simscape_Book/Section_8$ cp Chapter_9_
Section_7 DC_Motor_Digital_Twin
nbibin001@nbibin001-HP-Pavilion-x360-Convertible-15-br0xx:~/Digital_Twin_Simscape_Book/Section_8$ ls
Chapter_9_Section_7  DC_Motor_Digital_Twin  lambda_function.py
nbibin001@nbibin001-HP-Pavilion-x360-Convertible-15-br0xx:~/Digital_Twin_Simscape_Book/Section_8$ sudo chmod -R
 777 /home/nbibin001/Digital_Twin_Simscape_Book/Section_8
nbibin001@nbibin001-HP-Pavilion-x360-Convertible-15-br0xx:~/Digital_Twin_Simscape_Book/Section_8$ ls -l
total 6324
-rwxrwxrwx 1 nbibin001 nbibin001 3234648 Oct 26 05:13 Chapter_9_Section_7
-rwxrwxrwx 1 nbibin001 nbibin001 3234648 Oct 31 03:44 DC_Motor_Digital_Twin
-rwxrwxrwx 1 root      root         4087 Oct 31 03:40 lambda_function.py
nbibin001@nbibin001-HP-Pavilion-x360-Convertible-15-br0xx:~/Digital_Twin_Simscape_Book/Section_8$ zip bundle.zi
p lambda_function.py DC_Motor_Digital_Twin
  adding: lambda_function.py (deflated 69%)
  adding: DC_Motor_Digital_Twin (deflated 65%)
nbibin001@nbibin001-HP-Pavilion-x360-Convertible-15-br0xx:~/Digital_Twin_Simscape_Book/Section_8$ sudo chmod -R
 777 /home/nbibin001/Digital_Twin_Simscape_Book/Section_8
nbibin001@nbibin001-HP-Pavilion-x360-Convertible-15-br0xx:~/Digital_Twin_Simscape_Book/Section_8$ ls -l
total 7424
-rwxrwxrwx 1 nbibin001 nbibin001 1124434 Oct 31 03:53 bundle.zip
-rwxrwxrwx 1 nbibin001 nbibin001 3234648 Oct 26 05:13 Chapter_9_Section_7
-rwxrwxrwx 1 nbibin001 nbibin001 3234648 Oct 31 03:44 DC_Motor_Digital_Twin
-rwxrwxrwx 1 root      root         4087 Oct 31 03:40 lambda_function.py
nbibin001@nbibin001-HP-Pavilion-x360-Convertible-15-br0xx:~/Digital_Twin_Simscape_Book/Section_8$
```

Figure 9.198 Linux command line showing packaging the Lambda function and Digital Twin executable.

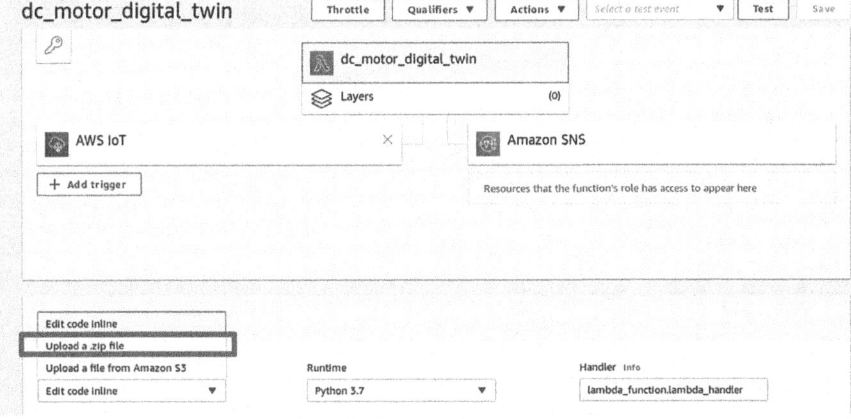

Figure 9.199 Uploading the packaged Lambda function to AWS.

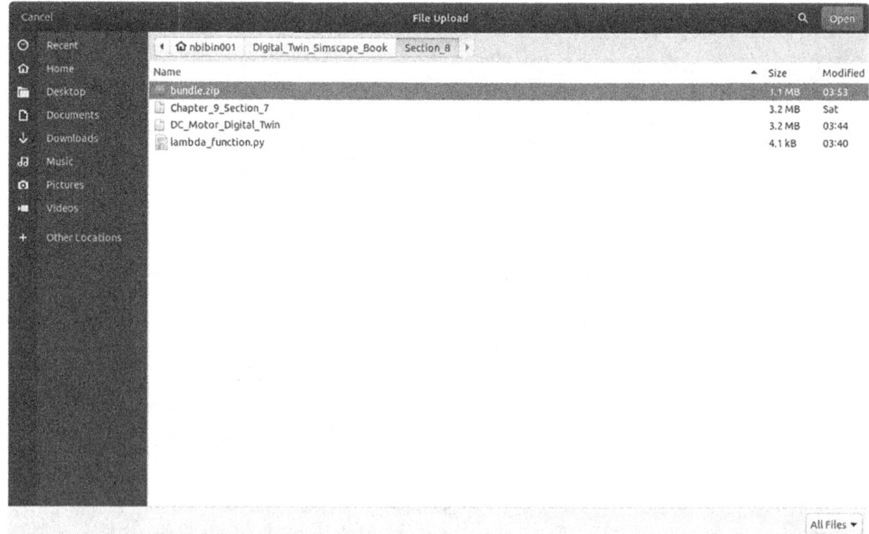

Figure 9.200 Browse and select the Lambda function Zip File package.

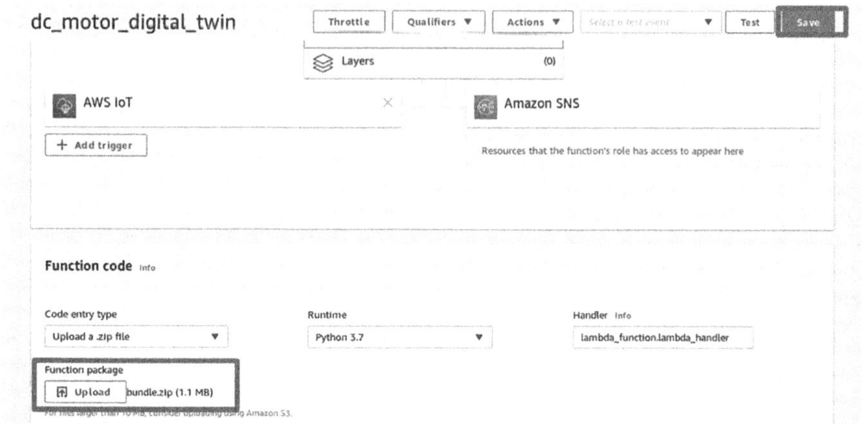

dc_motor_digital_twin

Throttle Qualifiers Actions Select a test event Test Save

Layers (0)

AWS IoT

Add trigger

Amazon SNS

Resources that the function's role has access to appear here

Function code Info

Code entry type: Upload a .zip file

Runtime: Python 3.7

Handler Info: lambda_function.lambda_handler

Function package

Upload bundle.zip (1.1 MB)

Figure 9.201 Saving the uploaded Lambda function Package.

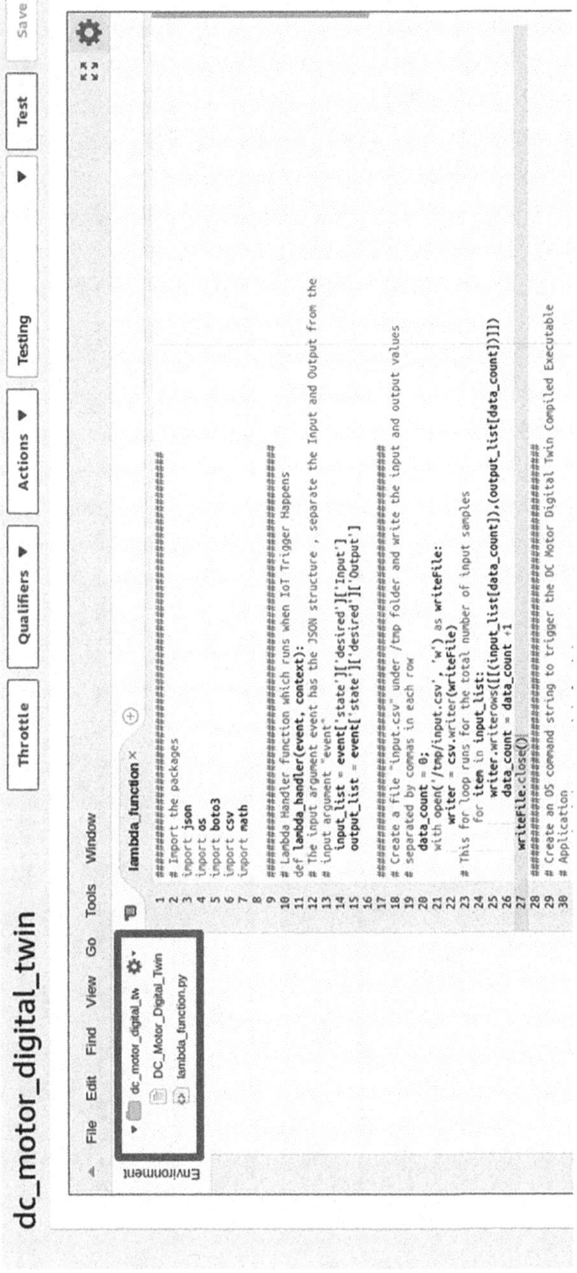

Figure 9.202 Lambda function inline editor showing the uploaded function and executable.

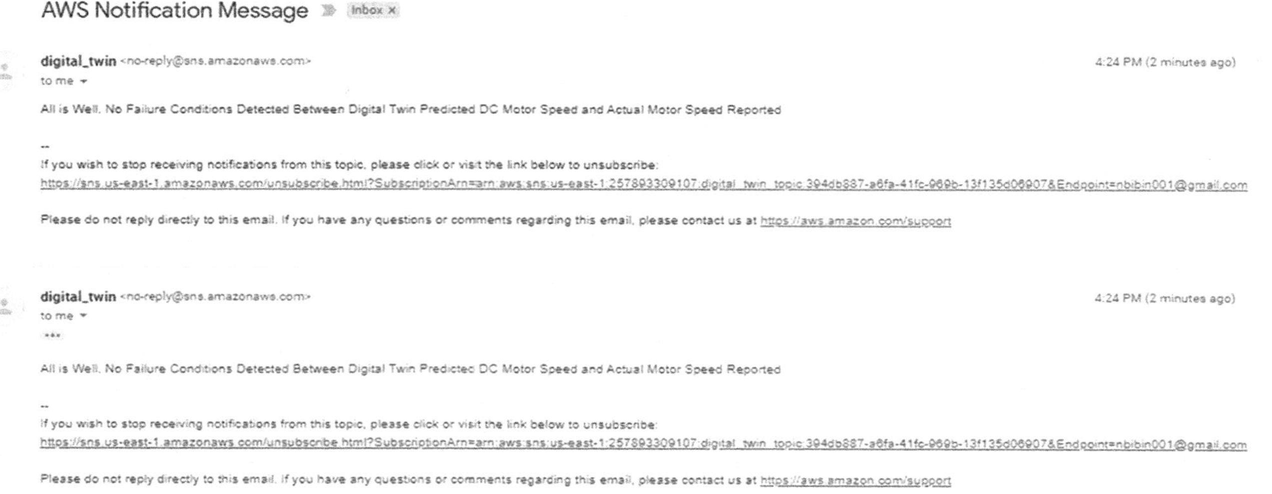

Figure 9.203 No fault condition Email notifications from the Digital Twin about the Off-BD status.

Figure 9.204 No fault condition detailed Email message information from the Digital Twin about the Off-BD status.

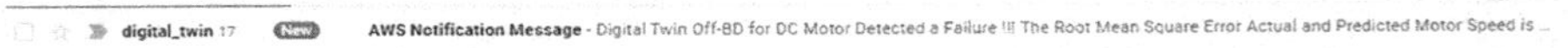

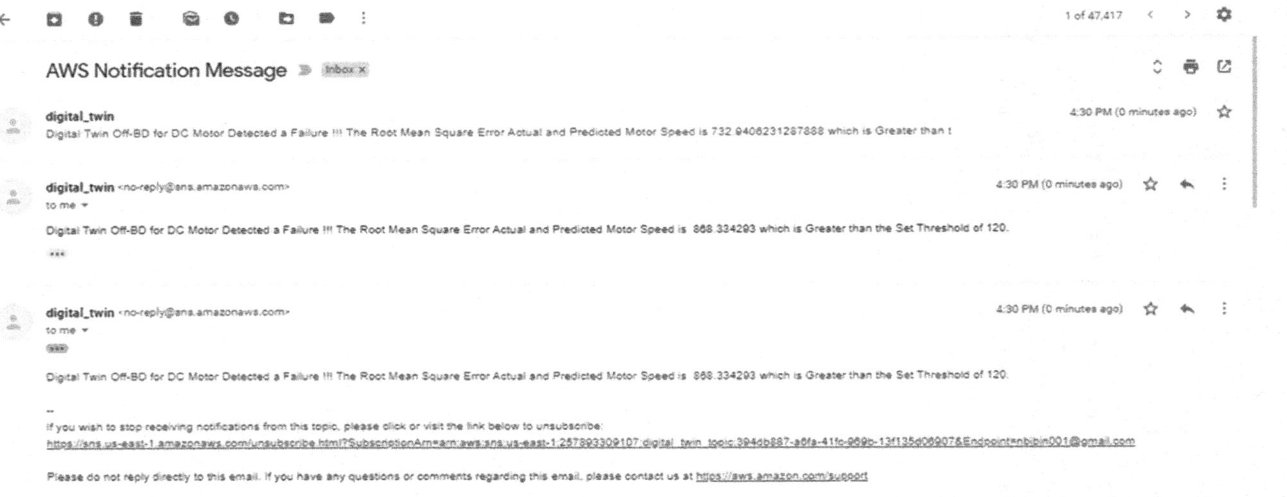

Figure 9.205 DC Motor power supply fault condition, Email notifications from the Digital Twin about the Off-BD status.

Figure 9.206 DC Motor power supply fault condition, detailed Email message information from the digital twin about the Off-BD status.

Hint: Repeat the whole process, but just change the speed sensing logic in the Arduino Simulink feedback controller model to introduce a sum block to add a constant block with 50 RPM to the speed sensing.

References

[1] PI Control of a DC Motor. http://ctms.engin.umich.edu/CTMS/index.php?aux=Activities_DCmotorB.

[2] Developing Simulink Device Driver Blocks: Step-By-Step Guide and Examples. http://www.mathworks.com/matlabcentral/fileexchange/39354-device-drivers.

Digital twin development and deployment for a wind turbine

10

10.1 Introduction

This chapter guides the user in developing a digital twin of a wind turbine using MATLAB Simscape™ and then deploying it into Amazon Web Services (AWS) cloud. There will be an equivalent hardware prototype representing the wind turbine. In this case, the hardware is the DC motor hardware along with the ESP32 used in Chapter 9. The data from the DC motor speed sensor for a given wind speed input shall be compared with the digital twin on the cloud for off-board diagnostics (Off-BD). Fig. 10.1 shows the Off-BD process for the wind turbine, whereas Fig. 10.2 shows the boundary diagram of the wind turbine system along with the interaction with its twin model. The outline of this chapter is shown below:

1. Physical asset setup and considerations; wind turbine hardware
2. Understanding the input−output behavior of the wind turbine Simscape™ model
3. Developing the driver Simscape™ model for the hardware and communicating to AWS
4. Deploying the Simscape™ digital twin model to the AWS cloud and performing Off-BD

With the exception of the digital twin model, the rest of the files and settings can be carried forward from Chapter 9. This chapter will focus on utilizing an existing MATLAB Simscape™ Wind Turbine example and converting that model to a digital twin for deployment on AWS.

All the codes used in the chapter can be downloaded for free from MATLAB File Exchange. Follow the link below and search for the ISBN or title of this book:

https://www.mathworks.com/matlabcentral/fileexchange/

Alternatively, the reader can also download the material and other resources from the dedicated website or contact the authors for further help:

https://www.practicalmpc.com/

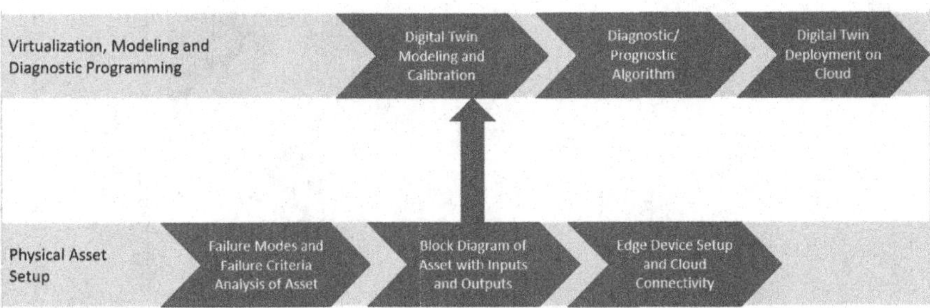

Figure 10.1 Off-board diagnostics process for wind turbine system.

Digital Twin Development and Deployment on the Cloud. https://doi.org/10.1016/B978-0-12-821631-6.00010-4

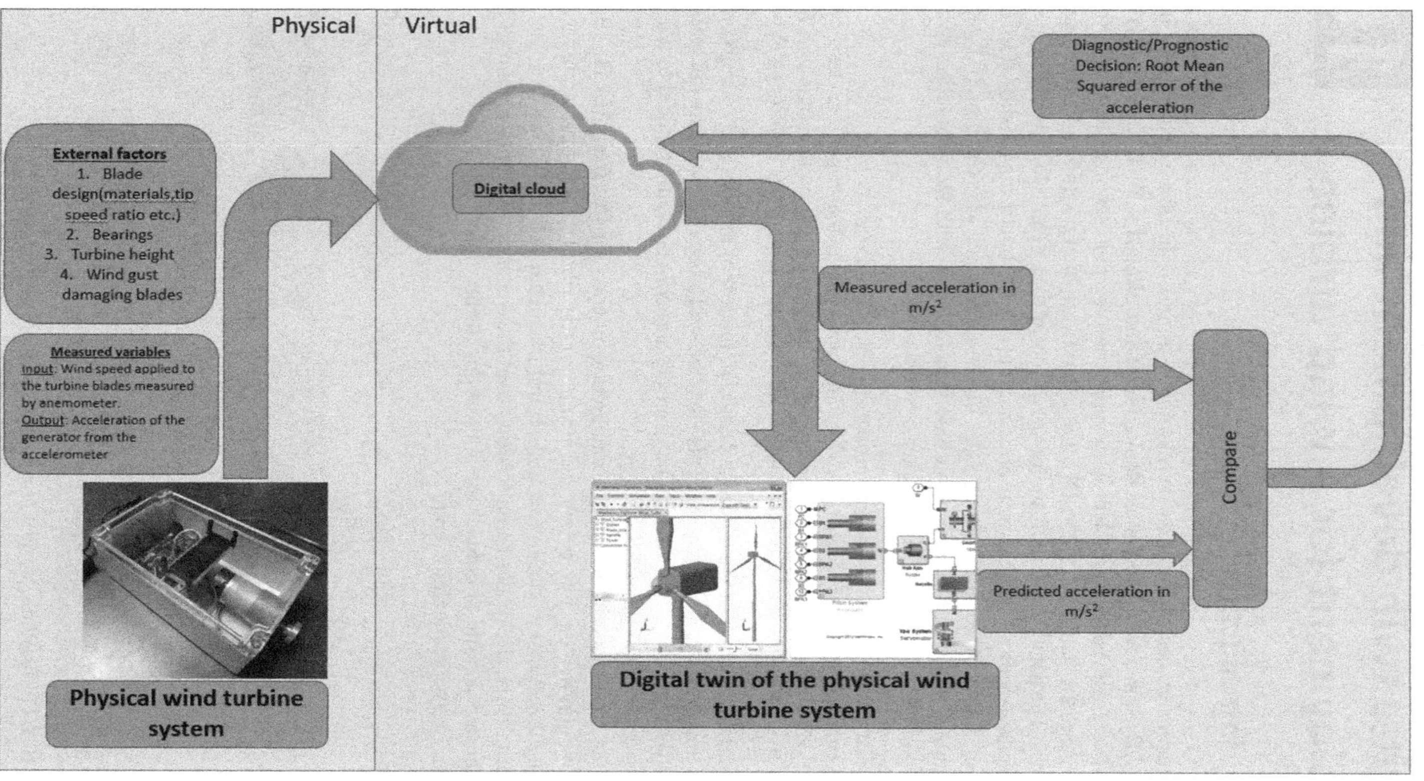

Figure 10.2 Block diagram of a wind turbine system.

10.2 Physical asset setup and considerations: wind turbine hardware

As mentioned in Section 10.1, the DC motor hardware along with the ESP32 module in Chapter 9 (Fig. 10.3) can be used for wind turbine purposes as well. The DC motor hardware is driven by the DC motor driver model described in Chapter 9. The goal of this chapter is to drive the DC motor using wind speed instead of predefined desired RPM as an input in the driver model. The DC motor is analogous to a generator in the wind turbine. There would be a relationship between wind speed and the RPM of the DC motor. Based on the RPM calculated in the model, a respective PWM command can be sent to the Arduino board. Fig. 10.4 shows the illustration of the Chapter 10 concept. Note that conversion from wind speed to motor speed would not be as straightforward as a Gain block as explained later in Section 10.3. The next section shall elaborate on this conversion process.

10.3 Understanding the input–output behavior of the wind turbine Simscape™ model

The digital twin for the wind turbine would be upon an existing MATLAB Simscape™ of a wind turbine model on MATLAB File Exchange [1]. The file may be

Figure 10.3 DC motor hardware with ESP32 module.

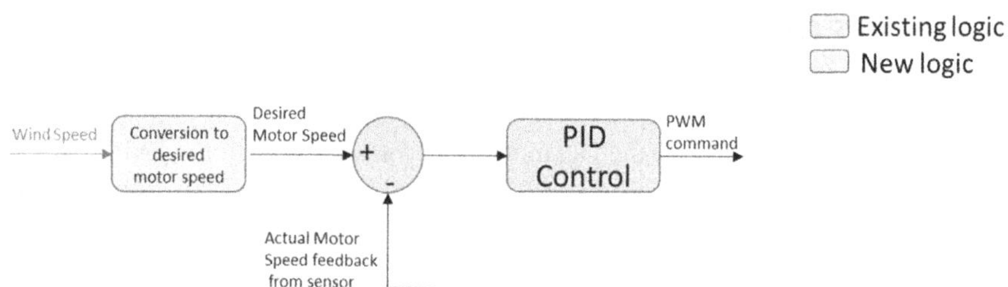

Figure 10.4 Control structure for wind speed to PWM command.

downloaded from https://www.mathworks.com/matlabcentral/fileexchange/25752-wind-turbine-model. It is recommended to read the instructions on how to download and setup this model given in the link. There is also a four-part webinar series based on the design of this model also shared in the link. The following paragraph below shall give a brief overview of the model as well as the key inputs and outputs that will correlate with the DC motor driver model.

The Wind Turbine model is shown in Fig. 10.5 with key areas highlighted. As seen in the figure, the inputs to the model are the wind speed along with the states of the wind turbine (explained below). The output is the RPM of the generator located in the Nacelle subsystem. The solver options are set to phasor 60 Hz as default.

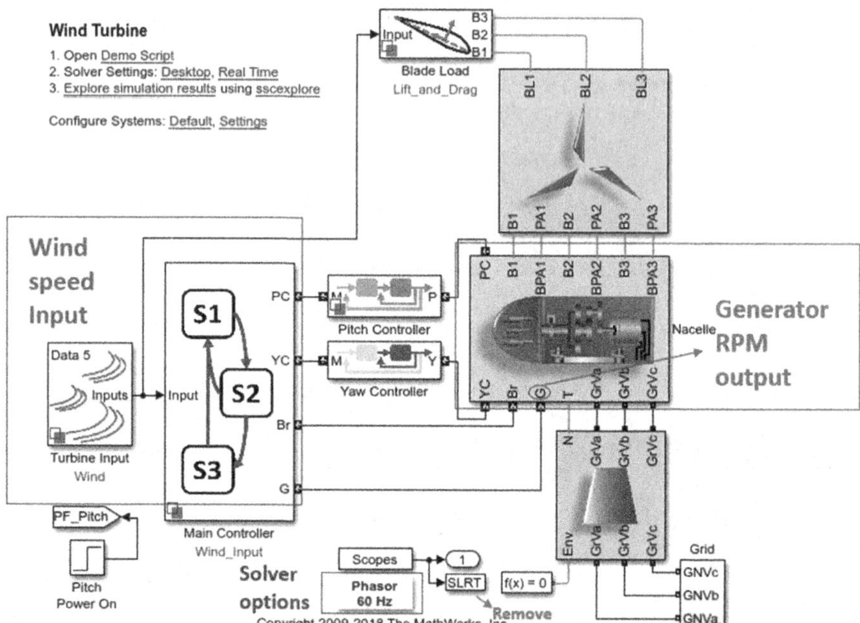

Figure 10.5 Wind turbine Simscape model.

10.3.1 Wind turbine model components

1. Firstly, the wind speed and wind direction is input to a stateflow logic. The different states for the wind turbine are PARK, STARTUP, GENERATING, and BRAKE. The stateflow always initializes in PARK state. The states can be seen in the model path Wind_Turbine/Main Controller/Wind Input/Turbine Stateflow also shown in Fig. 10.6.
   a. If the wind speed is less than a "cut in lower" threshold, the wind turbine will remain in PARK state where the rotor for the generator is stopped.
   b. Once the wind speed goes above the threshold in point 1a, and the wind speed is below a "cut out threshold", then the wind turbine will enter STARTUP state. The wind turbine blades will start rotating, hence energizing the generator until the rotor speed has achieved a steady-state value.
   c. Once the rotor has reached a steady-state speed, the wind turbine will enter GENERATING state where the generator is able to deliver power to the power grid.
   d. If the wind speed goes below the "cut in lower" threshold when the state is in STARTUP or GENERATING, then the state will transition to BRAKE where a pitch of the blades are adjusted to stall the motion. Note that there are multiple conditions that can cause the state to transition from GENERATING to BRAKE, but it is beyond the scope of this chapter.
   e. If the rotor speed is less than the park speed threshold in BRAKE state, then the state will transition from BRAKE to PARK.
2. Depending on the state of the turbine, a pitch is applied to the turbine blade to move the blades when the state is in STARTUP or GENERATING and reduce the movement when the state is in PARK or BRAKE state.
3. The Blade Load subsystem in Fig. 10.5 is to calculate lift and drag loads on the turbine blade. In order to determine the lift and drag loads, the pitch and yaw commands are needed from the pitch and yaw controllers. This information is sent to the Simscape portion to simulate the movement in the Mechanics Explorer and then to the Nacelle subsystem.
4. The Nacelle subsystem includes the gear train that converts the low turbine blades RPM to high-speed RPM for the generator as per the gear ratio. The output RPM of the generator is used in the stateflow logic, and it is this output that will be compared with the actual DC motor RPM.
5. The yaw controller subsystem is used to turn the turbine around the Z axis (yaw) to make sure the wind is perpendicular to the blades for movement.
6. Lastly, the tower subsystem is used to deliver the power generated by the generator to the power grid subsystem.

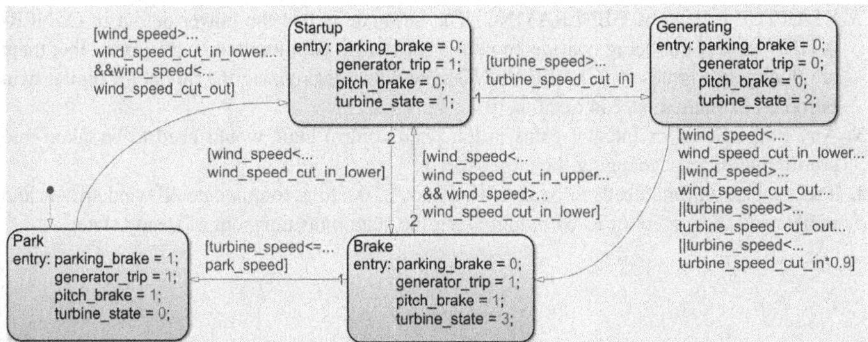

Figure 10.6 States of the wind turbine.

10.3.2 Input–output response of the model

The model is ready to be simulated with variable time step. Before running the simulation, the SLRT block adjacent to the Phasor 60 Hz block in the top level of the model needs to be removed as shown in Fig. 10.5. SLRT is for Simulink Real-Time Workshop for code generation purposes; however, in this chapter, embedded coder ert.tlc is used instead for code generation. Once the SLRT block is deleted, the model can be simulated by pressing the play button. As the simulation is running, the system performance of the wind turbine can be observed graphically in the Mechanics Explorer window. The key outputs to observe can be found in Wind_Turbine/Scopes/Generator Scopes path of the model. Double click on the Rotor Turbine speed scope to observe the RPM response of the generator as shown in Fig. 10.7. As seen from the second subplot in Fig. 10.7, the generator ramps up to a steady state speed of 1200 RPM after approximately 10 seconds. This behavior can be understood as a response to the wind speed input located in Wind_Turbine/Turbine Input/Wind shown in Fig. 10.8.

The input wind speed has a direct relationship to the generator RPM as per the stateflow logic shown in Fig. 10.6. It can be observed that when the wind speed is less than wind_speed_cut_in_lower (tuned to 4 m/s), the turbine is PARK mode; hence, the generator RPM is 0. After a few seconds when the wind speed ramps above 4 m/s, the generator RPM starts to ramp as the state is now in STARTUP. The PI control in the Pitch Control subsystem will now generate a pitch command as per the desired target rotor speed, which is set to 1200 RPM. Once the steady state has been achieved, the generator can output power. The generator RPM output plot can be divided into three sections as shown in Fig. 10.9.

10.3.3 Customizing the model as per the digital twin requirements

The input–output relationship can now be customized as per the digital twin and physical hardware needs of this chapter. Four factors need to be kept in mind:

1. The target speed of 1200 RPM needs to be changed to a smaller value to accommodate for the physical asset DC motor hardware limits.
2. The inputs to the stateflow would need to be modified to now allow steady-state speed in STARTUP instead of GENERATING. The rationale is that the power output in GENERATING state is not being considered in this chapter. In addition to this, it is found that there are divergence issues for GENERATING state when discretizing the model for digital twin purposes explained later in Section 10.4.
3. Any Proportional or Integral gains in the pitch control logic would need to be taken into consideration after changing the target speed.
4. The simulation time needs to be increased above 70 s to accommodate all wind speeds and understand the generator RPM output when the state transitions out of steady state.

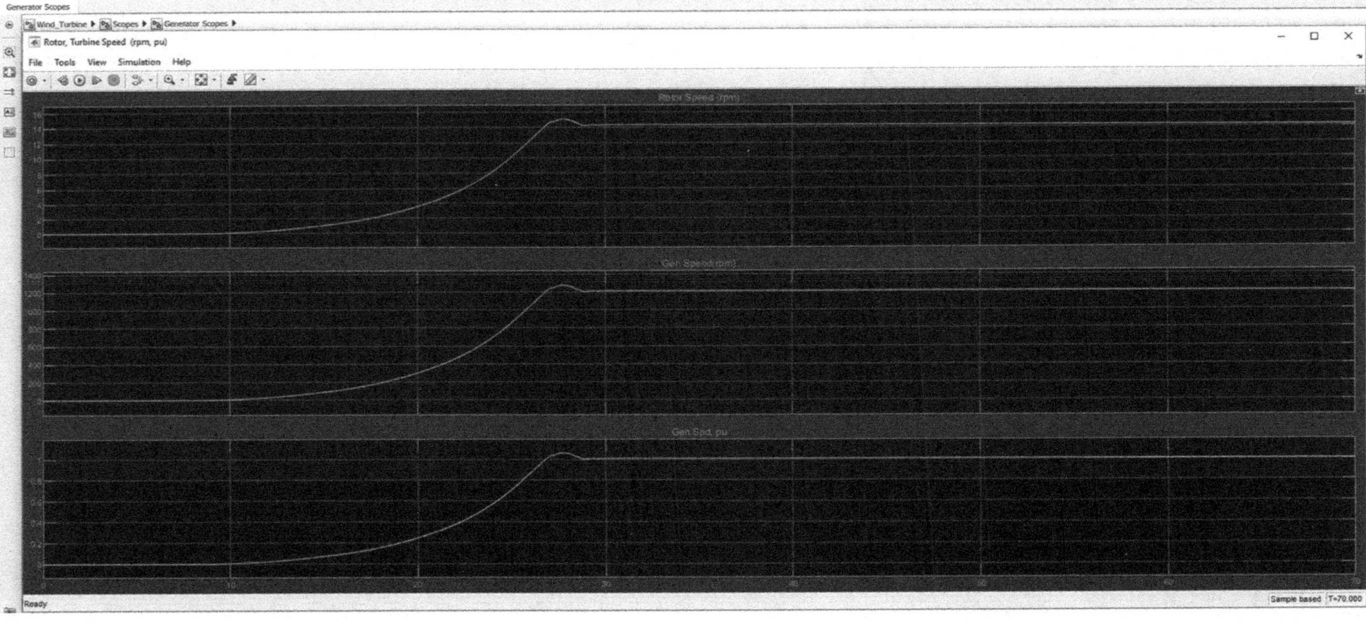

Figure 10.7 Output response of the generator.

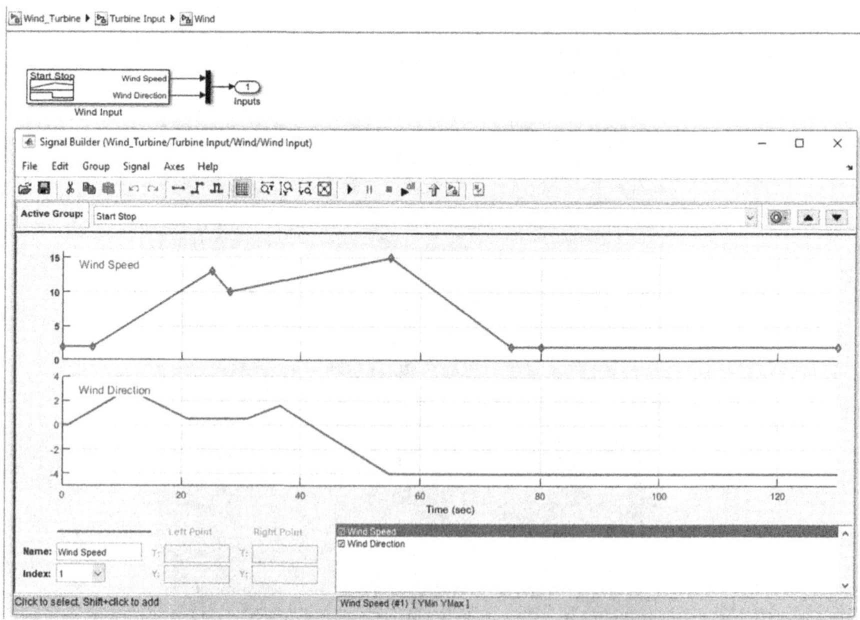

Figure 10.8 Wind speed and wind direction input.

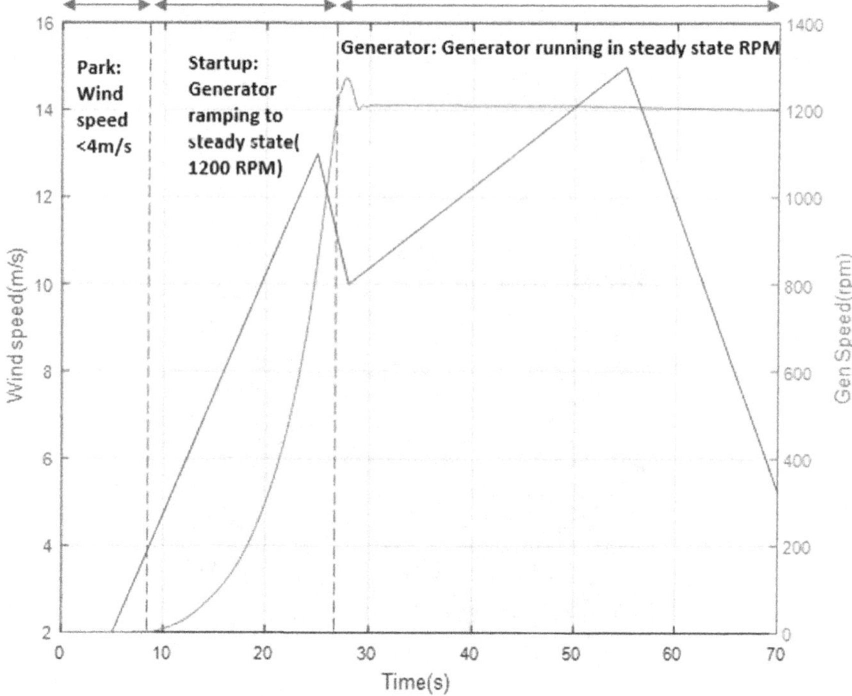

Figure 10.9 Relationship between wind speed and generator RPM.

10.3.3.1 Target speed

To change the target speed, navigate to Wind_Turbine/Pitch Controller/PI on AoA/ Pitch Controller/Determine Pitch Command/Determine Desired Angle of Attack and change the Nominal RPM value to the value shown in Fig. 10.10. WT_Params.Rotor.-nominal_rpm is set as 1200 as per the Wind_Turbine_Parameters.m script. This script is run automatically when the model is run as per its design. Dividing this number by 5 will set the desired RPM as 240, which is within the acceptable range of speeds of the DC motor.

10.3.3.2 Steady-state speed in STARTUP state

With the target speed changed, navigate to the stateflow logic Wind_Turbine/Main Controller/Wind Input/Turbine Stateflow. Observe that the speed changed in 10.3.3.1 is not part of the stateflow. To make sure the state remains in STARTUP when the wind speed has reached the target speed, verify the turbine_speed_cut_in is set to 1200 RPM. This should be the default value in the Wind_Turbine_Parameters.m. Notice the condition to transition from STARTUP to GENERATING is turbine_speed>turbine_speed_cut_in. Since the turbine_speed stead state is 240 RPM, it will not transition to GENERATING state.

10.3.3.3 Proportional (P) and I (Integral) gains change in the pitch actuation control logic

Changing the target speed may affect the transient behavior of the pitch actuation, which is needed to derive drag and lift forces for the turbine blades. The P and I gains can be tuned by using the MATLAB Control System Tuner and Parameter Estimator. The P and I gains for the pitch control are located in Wind_Turbine/Pitch Controller/PI on AoA/Pitch Controller/Actuator Controller/PI Controller. Change only the P gain as shown in Fig. 10.11.

10.3.3.4 Increasing the simulation time

As seen in Fig. 10.8, the wind speed input is set for 130 s. Increase the simulation time from 70 to 130 s next to the play button in Simulink. This way, the transition from STARTUP to BRAKE is captured when wind speed falls below wind_speed_cut_in_lower.

Run the simulation again and navigate to Wind_Turbine/Scopes/Generator Scopes. Double click on the Rotor Turbine Speed scope. It can be observed that the generator speed now stabilizes its speed around 240 RPM with an overshoot shown in Fig. 10.12. The reader may tune the P and I gains to reduce overshoot; however, fine tuning these gains is outside the scope of this chapter. Also, look into the turbine states scope in Wind_Turbine/Scopes/Main Controller Scopes. As seen in Fig. 10.13, the state now transitions from 0 (PARK) to 1 (STARTUP) to 3 (BRAKE).

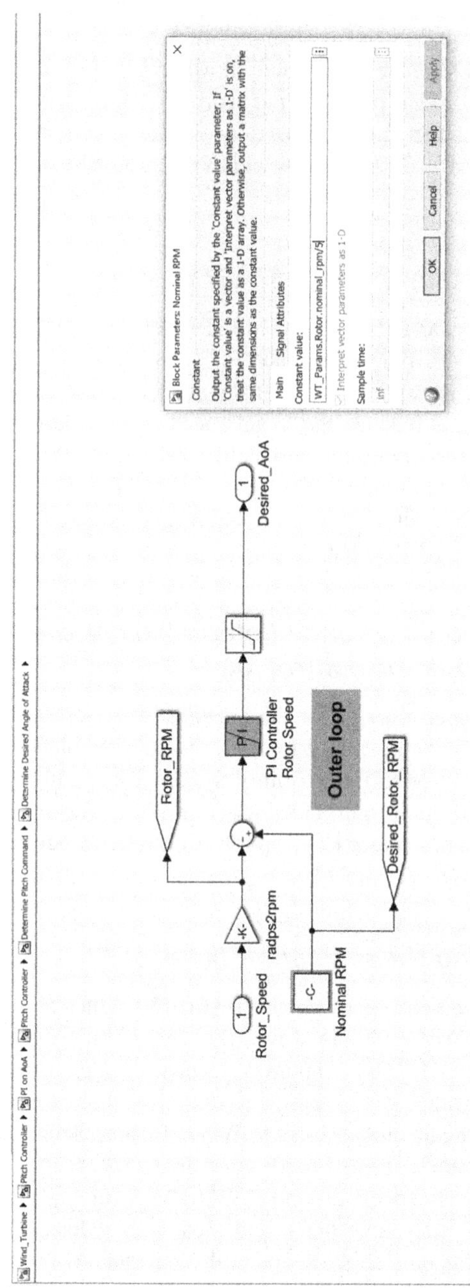

Figure 10.10 Changing the target speed.

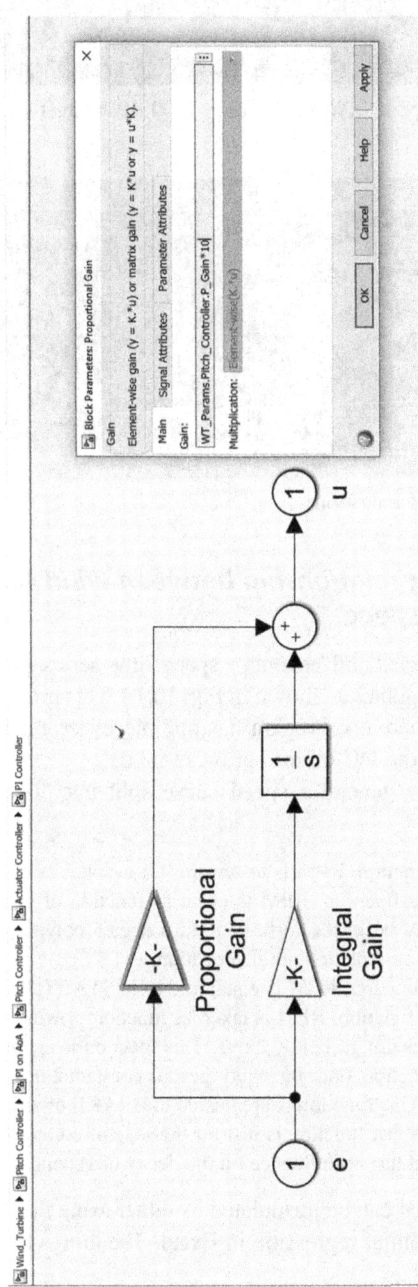

Figure 10.11 Changing the P gain.

Figure 10.12 Generator speed RPM profile for a 1200 RPM target setpoint.

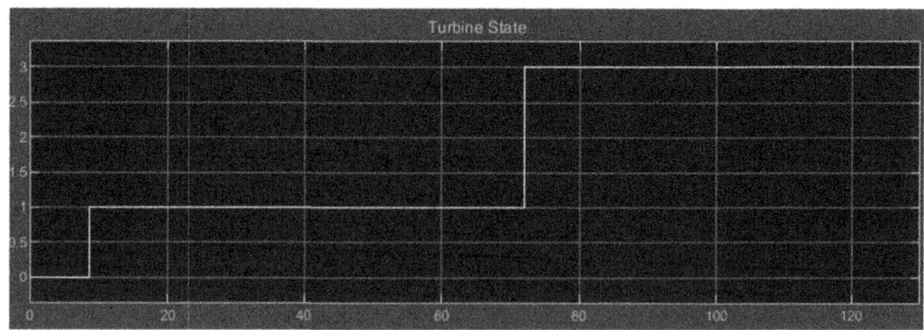

Figure 10.13 Turbine state transitions.

10.3.4 Developing relationship between wind speed and generator speed

The wind speed time series and generator speed time series according to the digital twin needs are plotted together as shown in Fig. 10.14. This profile can now be divided into separate regions where each region has one piecewise function. This is done to replicate this profile on the DC motor hardware model.

Fig. 10.15 shows the generator speed curve split into five piecewise functions described below:

1. In the first piecewise function, RPM is to remain 0 if turbine state is in PARK state.
2. In the second piecewise function, RPM is taken as function of wind speed in STARTUP state. This function may be in the form of a third-degree polynomial equation. Note that the overshoot is not being modeled in this equation.
3. In the third piecewise function, RPM is equal to 240 in STARTUP state.
4. In the fourth piecewise function, RPM is taken as function of wind speed provided the fact that the wind speed does not go below 2 m/s. This function is applicable to BRAKE state.
5. In the fifth piecewise function, once the wind speed is constant 2 m/s, then RPM is a function of the simulation time. This function is applicable to BRAKE or when state transitions from BRAKE to PARK. This last function is just for the sake of completing the generator speed profile since wind speed has no influence on the decay of generator speed.

The piecewise function can be formulated by either using the "polyfit" command in MATLAB or the polynomial regression in Excel. The things to keep in mind are

1. For the second piecewise function, use only the points when wind speed starting point is 4 m/s and end point would be when the generator speed reaches 240 RPM. This way the overshoot is not modeled for the DC motor driver model. Use wind speed as the x input and the generator speed as the y output.

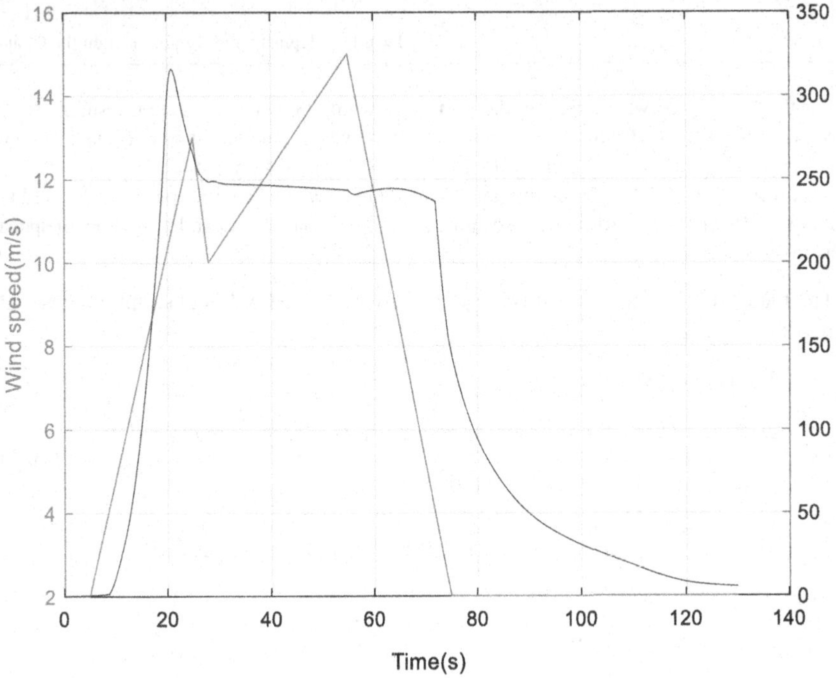

Figure 10.14 Wind speed input and generator speed output.

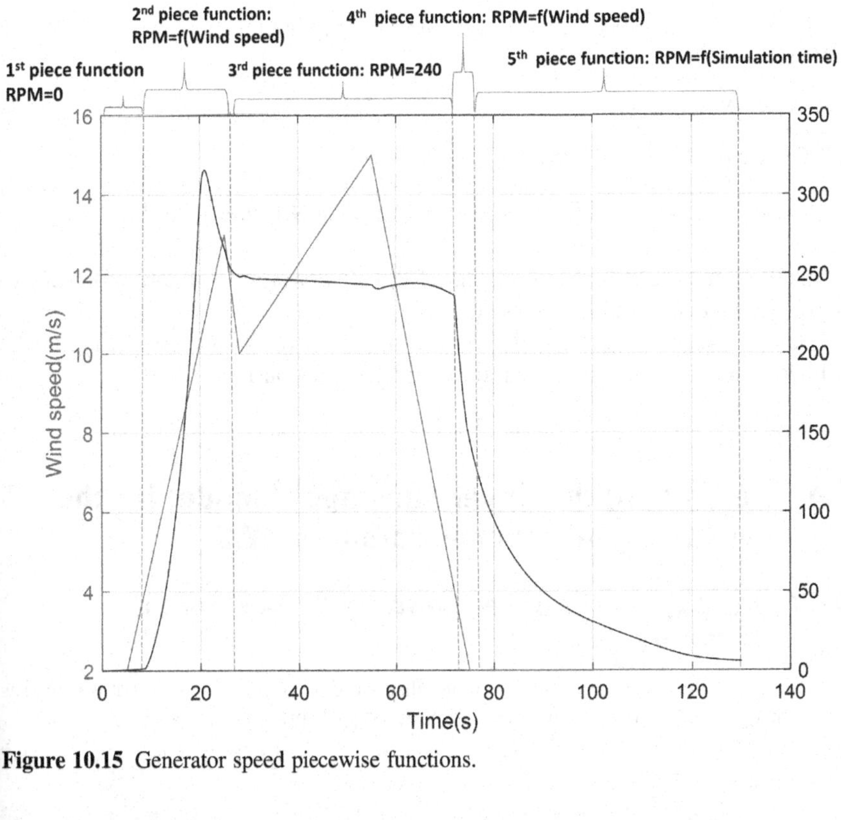

Figure 10.15 Generator speed piecewise functions.

2. For the fourth piecewise function, the start point would be when the state transitions from STARTUP to BRAKE and the end point would be when wind speed has decayed to 2 m/s. Use wind speed as the x input and the generator speed as the y output.
3. For the fifth piecewise function, the start point would be when the wind speed has decayed to 2 m/s in BRAKE or PARK state. Use simulation time as the x input and the generator speed as the y output.

The piecewise functions for this chapter were developed in Excel using polynomial regression.

1. Second piecewise function:

$$\begin{aligned}\text{Generator RPM} = 0.4362 * \left(\text{windspeed}^3\right) + \left(-1.7756 * \text{windspeed}^2\right) \\ + 0.3608 * \text{windspeed}\end{aligned} \tag{10.1}$$

Eq. (10.1): Wind speed to generator RPM when state is STARTUP

2. Fourth piecewise function:

$$\begin{aligned}\text{Generator RPM} = 0.219 * \left(\text{windspeed}^3\right) + \left(-9.5573 * \left(\text{windspeed}^2\right)\right) \\ + 95.988 * \text{windspeed}\end{aligned} \tag{10.2}$$

Eq. (10.2): Wind speed to generator RPM when state has transition from STARTUP to BRAKE

3. Fifth piecewise function:

$$\begin{aligned}\text{Generator RPM} = -0.0021 * \left(\text{time}^3\right) \\ + \left(0.6993 * \left(\text{time}^2\right)\right) - 79.429 * \text{time} + 3041.3\end{aligned} \tag{10.3}$$

Eq. (10.3): Simulation time to generator RPM relationship when state is BRAKE or PARK and wind speed has decayed to 2 m/s.

With the piecewise functions developed, the DC motor driver model can now be modified to include the wind speed to motor RPM relationship.

10.4 Developing the driver Simscape™ model for the hardware and communicating to AWS

10.4.1 Developing the driver Simscape™ model for the hardware

1. Firstly, open the driver model for DC motor that was developed in Chapter 9, it is named as Chapter_9_Section_6_3.slx in Chapter 9. Save it as Chapter_10.slx.

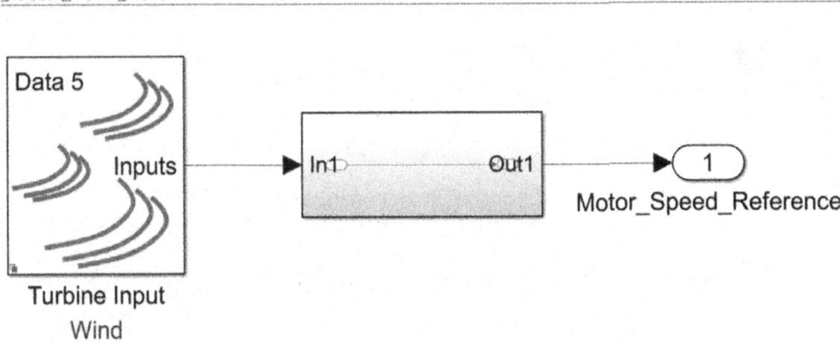

Figure 10.16 Changes inside Motor_Speed_Reference subsystem.

2. Double click the Motor_Speed_Reference_Generator subsystem and delete everything except the Motor_Speed_Reference outport. Then copy the Turbine Input subsystem block from the Wind_Turbine.slx Simulink model and paste into the Chapter_10.slx model in the Motor_Speed_Reference_Generator subsystem. Then add an empty subsystem and connect the Turbine Input to the inport of the subsystem. The outport of the empty subsystem should be connected to the Motor_Speed_Reference outport. Fig. 10.16 shows the changes.
3. Rename the empty subsystem as Main Controller, the In1 inport as Input, and Out1 outport as RPM. Delete the connection between Input and RPM. Then click anywhere on the model and type **Demux** as shown in Fig. 10.17. Note for the rest of this section, any Simulink blocks that needed to be added can be added in this way. A **Demux** block separates the wind speed and wind direction from the Turbine Input subsystem.
4. Connect Input to the Demux. Connect the second output of the Demux to a **terminator.** A **terminator** can be found in the same manner as a **Demux.** Then copy the Turbine Stateflow logic in Wind_Turbine/Main Controller/Wind Input in Wind_Turbine.slx model and paste it to Chapter_10/Motor_Control/Motor_Speed_Reference_Generator/Main_Controller in Chapter 10.slx. Connect the first output of the Bus selector to the wind_speed input of the stateflow logic. The changes are shown in Fig. 10.18.
5. Before proceeding with the stateflow logic, the parameters inside the stateflow need to be defined. These parameters are defined in Wind_Turbine_Parameters.m file of the Wind_Turbine.slx. Enter Wind_Turbine_Parameters in the MATLAB Command window to define the variables. In addition to this, go to File->Model Properties->Model Properties and enter Wind_Turbine_Parameters in the PreLoadFcn as shown in Fig. 10.19. This shall run the m file before the Chapter_10.slx model is opened.

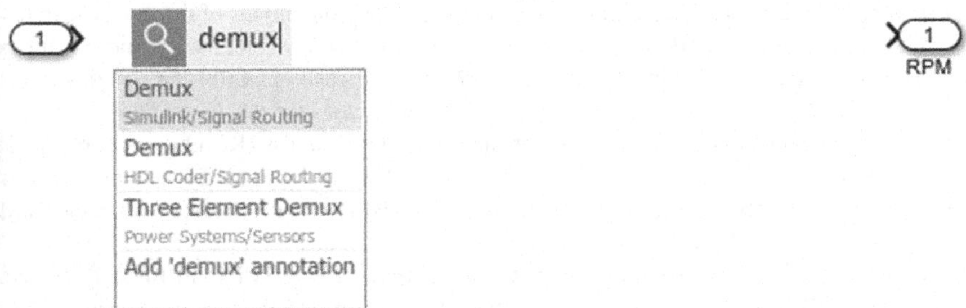

Figure 10.17 Demux to separate wind speed and wind direction.

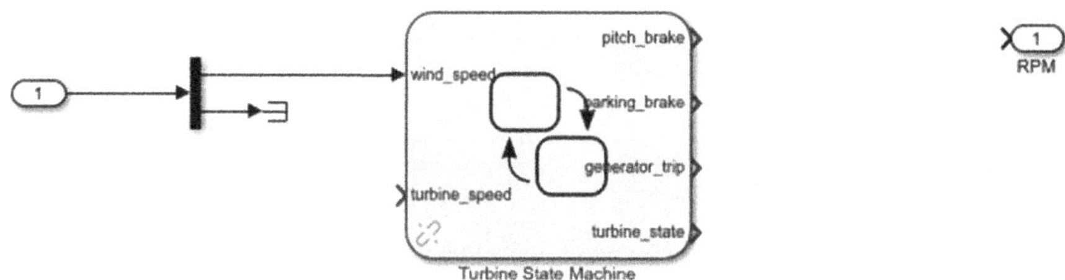

Figure 10.18 Demux to separate wind speed and wind direction.

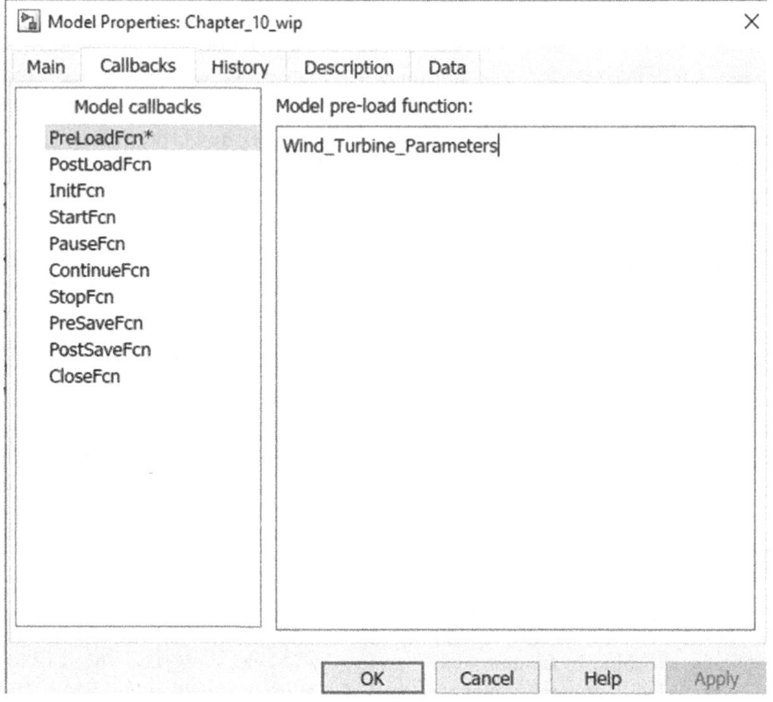

Figure 10.19 PreLoad function for the model.

6. Connect the pitch_brake, parking_brake, and generator_trip outputs of the stateflow to terminators. Then add an **If** block. Double click on the **If** block and set the Parameter settings as shown in Fig. 10.20. In this case, u1 is the turbine_state output of the stateflow logic whereas u2 is the simulation time. There will be four cases for this if condition:
   a. u1==0 and u2<60 is for the case when turbine state is in PARK and simulation time is less than 60 s. This is for the case when motor RPM needs to be 0. The reason time is used here is because the turbine state can go from BRAKE to PARK state when the RPM is decaying and it is a nonzero value.
   b. u1==1|u1==2 is for the case when turbine state is in STARTUP or PREPARING state. This will have the second piecewise function developed in the previous section.

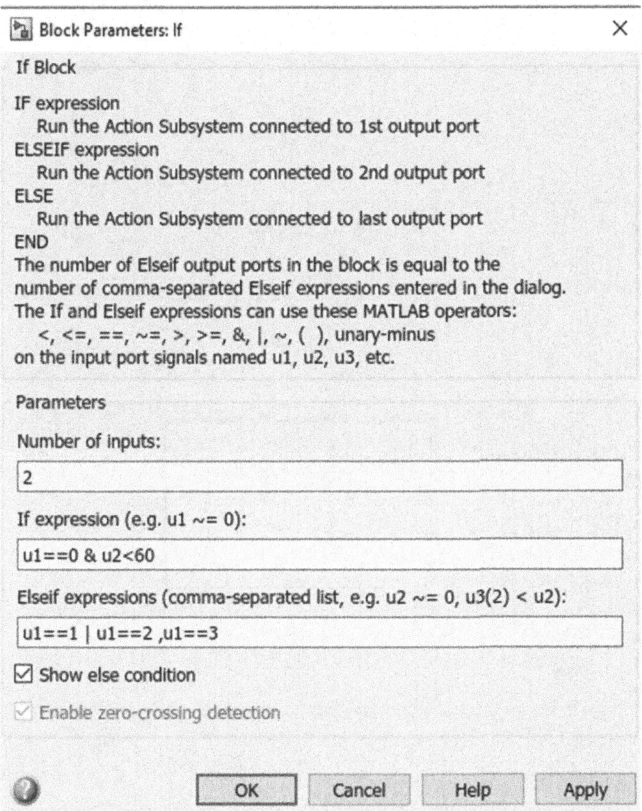

Figure 10.20 Parameter settings for If block.

c. u1==3 is for the case when turbine state is in BRAKE state. This will have the fourth and fifth piecewise function.

d. The else condition is for the case when u1==0 only. This case activates when state changes from BRAKE to PARK only for the generator RPM profile shown in the previous section.

7. Connect the turbine_state output of the stateflow logic to u1 of the If block. Then insert a **clock** and connect it to u2 of the **If** block. The changes can be seen in Fig. 10.21.
8. Insert four **If Action Subsystem** blocks and align them vertically to the right of the **If** block. Connect the case of the **If** block to the respective **If Action Subsystem** block. Next, delete the inport of the **If Action Subsystem** connected to the u1==0 and u2<60 case and also the else case. Name outports of **all the If Action Subsystem** as RPM and the remaining inports as Wind_speed. The updated model is shown in Fig. 10.22.
9. Now, the logic for each subsystem can be put in place. Go inside the **If action subsystem** block for the u1==0 and u2>60 and connect a **Constant** block of 0 to the RPM outport as shown in Fig. 10.23
10. Go inside the **If action subsystem** block for the u1==1|u1==2 case and follow these steps:

    a. Insert a **Fcn** block, a **Switch** block, and two **Unit delay** blocks.

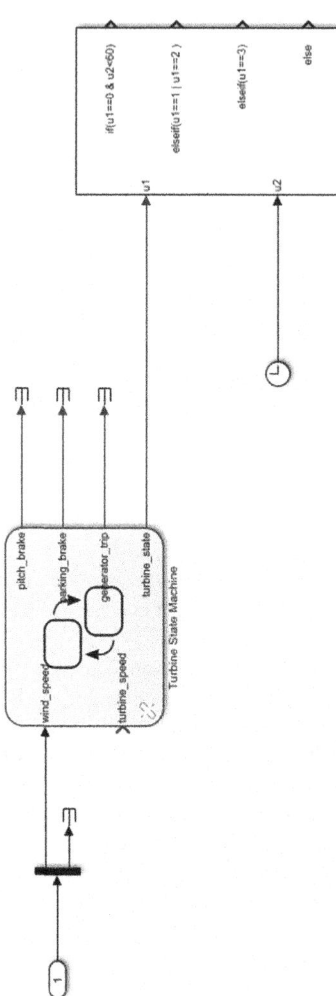

Figure 10.21 Inputs for the If block.

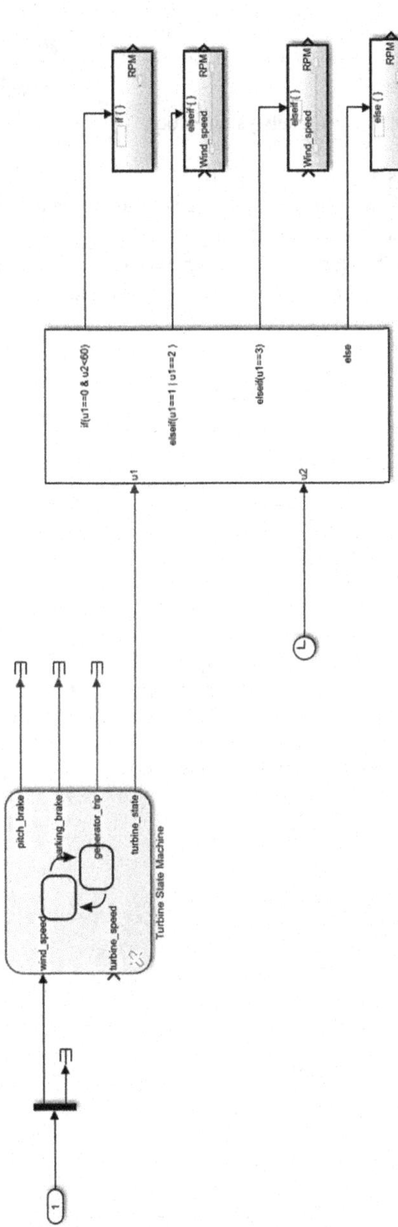

Figure 10.22 If action subsystems for the If cases.

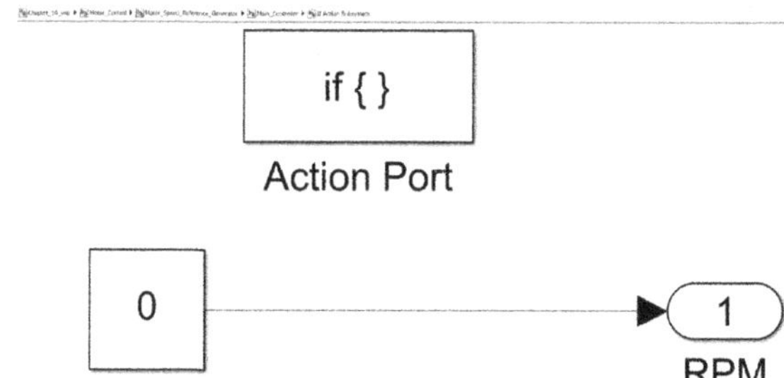

Figure 10.23 Inside the first If action subsystem block.

b. Connect these blocks as shown in Fig. 10.24.
c. Double click on the **Fcn** block and insert the second piecewise function developed in the previous section as shown in Fig. 10.25.
d. Double click on the **Switch** block and set the u2>=Threshold as 240 as shown in Fig. 10.26. This will ensure that the piecewise function is followed until motor speed reaches 240 RPM.

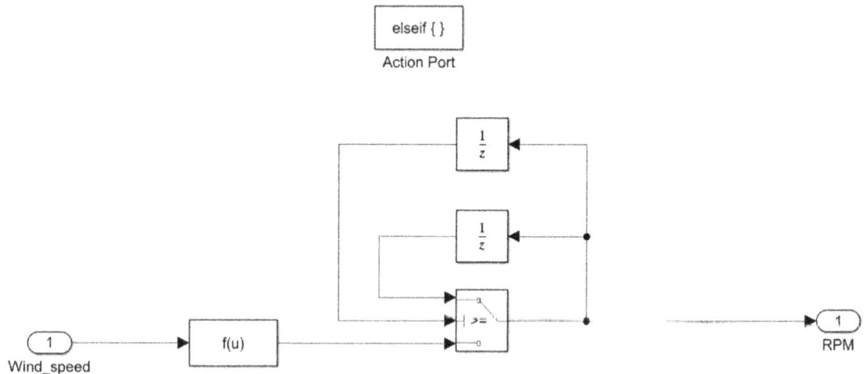

Figure 10.24 Inside the second If action subsystem block.

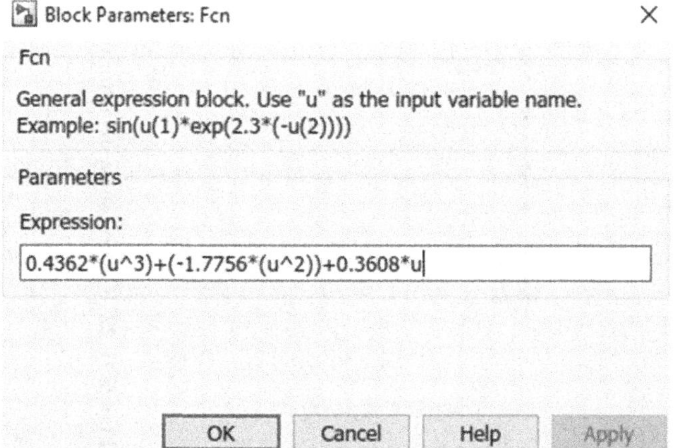

Figure 10.25 Second piecewise function.

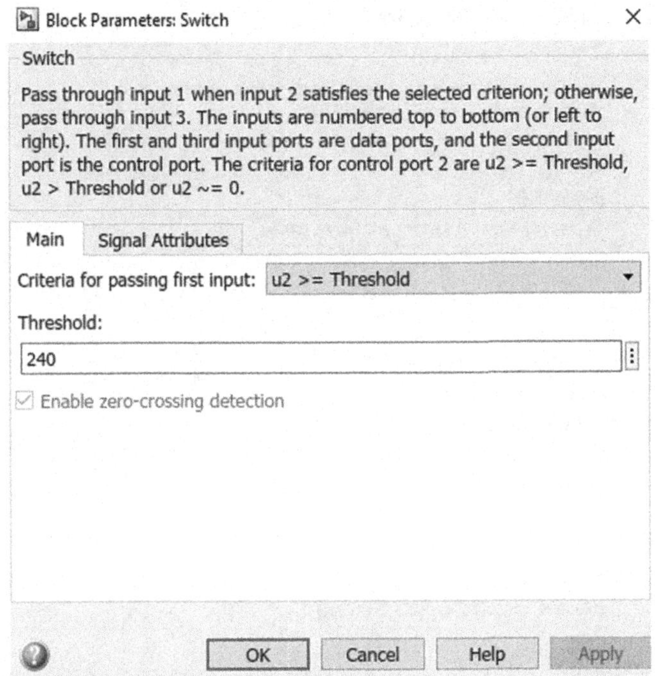

Figure 10.26 Switch threshold for second If action subsystem.

11. Go inside the **If action subsystem** block for the u1==3 case and follow these steps:
   a. Insert a Clock block, two Fcn blocks, a Switch block, and a Saturation block.
   b. Connect them as shown in Fig. 10.27. Rename the Fcn blocks as WindFcn and TimeFcn.
   c. Double click on the Wind Fcn block as insert the fourth piecewise function as shown in Fig. 10.28.

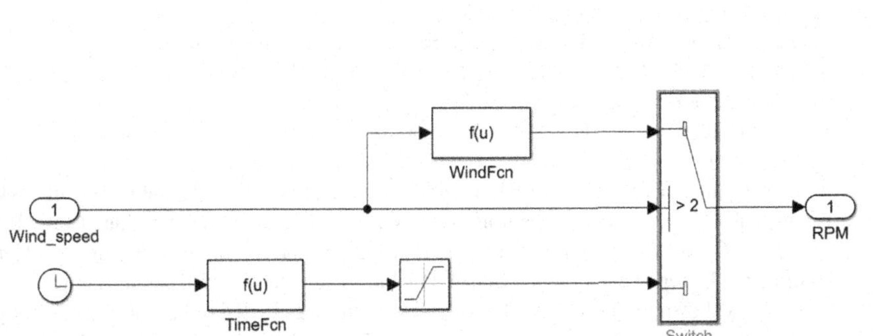

Figure 10.27 Inside the third If action subsystem block.

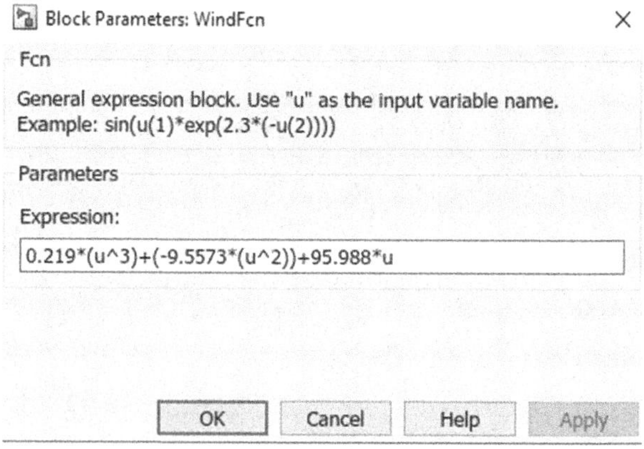

Block Parameters: WindFcn

Fcn

General expression block. Use "u" as the input variable name.
Example: sin(u(1)*exp(2.3*(-u(2))))

Parameters

Expression:

0.219*(u^3)+(-9.5573*(u^2))+95.988*u

OK Cancel Help Apply

Figure 10.28 Piecewise function for wind to RPM.

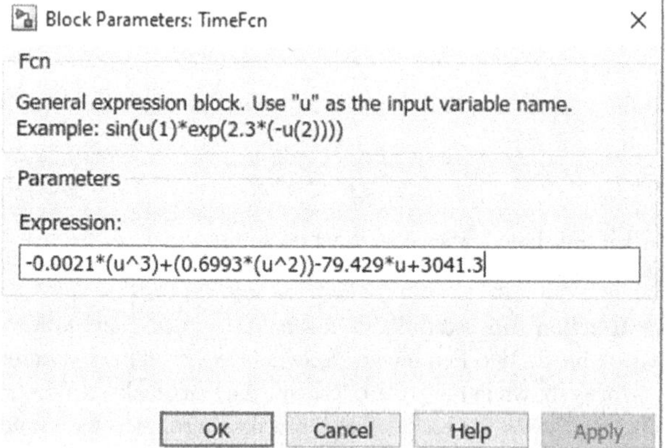

Figure 10.29 Piecewise function for time to RPM.

   d. Insert the fifth piecewise function for the TimeFcn block as shown in Fig. 10.29.
   e. Double click on the Saturation block and set the upper limit as 240 and lower limit as 0.
   f. Double click on the **Switch** block and for the "Criteria for passing first input" field, set it as u2>threshold where the threshold is 2.

12. Go inside the **If action subsystem** block for the else case and follow these steps:
   a. Insert a **Clock** block, a **Fcn** block, and a **Saturation** block.
   b. Connect them as shown in Fig. 10.30.
   c. Double click on the Fcn block as insert the fifth piecewise function as shown in Fig. 10.31.
   d. Double click on the **Saturation** block and set the upper limit as 240 and lower limit as 0.

13. Navigate to Chapter10/Motor_Control/Motor_Speed_Reference_Generator/Main_Controller.
Place a **Goto** flag and connect it to the top signal of the two signals coming out of the **Demux** block. Rename it as Wind_Speed. Double click on the flag and change tag visibility to global as shown in Fig. 10.32. The following **Goto** flags shall also have global visibility. Then, place two **From** flags and connect it to the second and third **If action subsystems**.

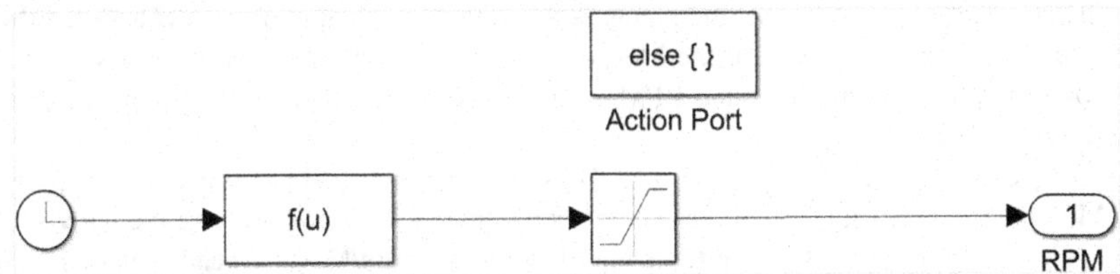

Figure 10.30 Inside the fourth If action subsystem block.

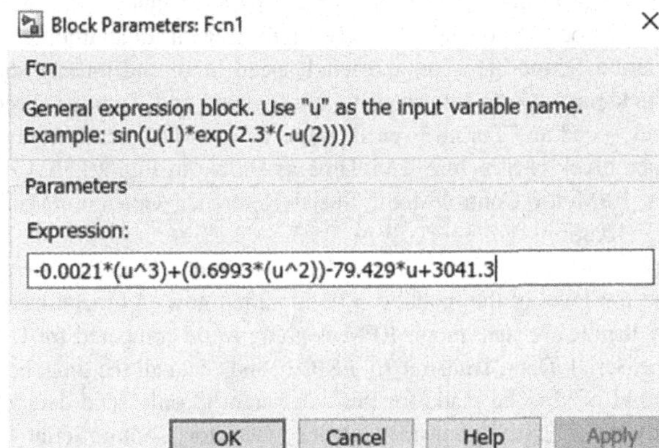

Figure 10.31 Piecewise function for time to RPM.

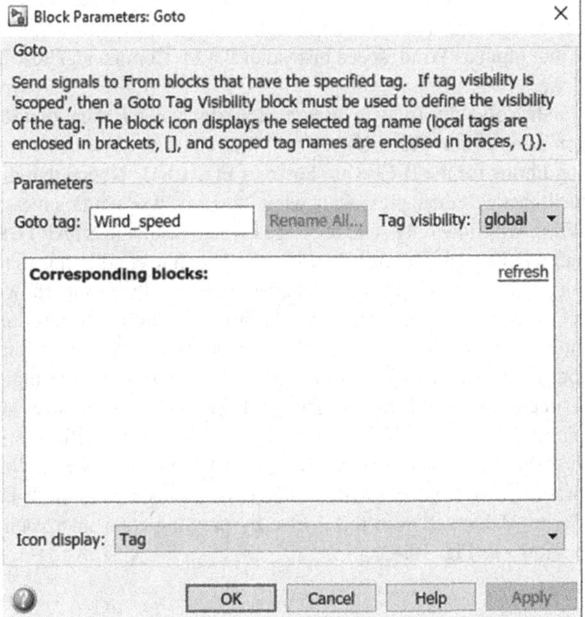

Figure 10.32 Wind_speed GoTo flag global Tag visibility.

14. Insert a **Merge** block and set the inputs as 4. Connect the output of each **If action subsystem** to the respective input of the **Merge** block. The output of the Merge block should be connected to the outport named RPM. The changes in this step and the previous one are shown in Fig. 10.33.
15. For the turbine_speed input, place a **Goto** flag named Turbine_speed in Chapter_10/Motor_Control/Motor_Speed_Sensor as shown in Fig. 10.34. Make sure the tag visibility is global. Place a **From** flag in Chapter_10/Motor_Control/Motor_Speed_Reference_Generator/Main_Controller with the name Turbine_speed.
16. Place a **Goto** flag named state for the turbine_state output of the stateflow logic in Chapter_10/Motor_Control/Motor_Speed_Sensor and name it State. Make sure tag visibility is global as this will be used outside this subsystem later on. In addition to this, add a **Rate Transition** and a **Scope** block for the wind_speed input and turbine_state output. Also add a **ToWorkspace** block connected to wind_speed and Turbine_speed inputs. Name them as Wind_speed and Turbine_speed, respectively. Make sure the save format for the **ToWorkspace** block is Structure with Time as shown in Fig. 10.35. Completed section for Chapter_10/Motor_Control/Motor_Speed_Reference_Generator/Main_Controller is shown in Fig. 10.36.
17. With the Main_Controller subsystem completed, the Serial_Data_Transmit_to_ESP32 subsystem in the top level of the model can be modified now. As mentioned in the previous section, only the steady-state motor RPM is going to be compared for Off-BD detection. Currently, the Serial_Data_Transmit_to_ESP32 sends data all the time; hence, a few modifications would need to be made for this subsystem to only send data when system has entered steady state. Firstly, go inside Chapter_10/Motor_Control/Serial_Data_Transmit_-to_ESP32/Chart, right click anywhere, and click Explore. It should open the Model Explorer. In the top bar menu, click on "Add Data" two times to add two parameters as shown in Fig. 10.37.
18. Name the two new parameters as time and State, respectively. Click on each of the two new parameters and enter the settings on the right side as shown in Figs. 10.38 and 10.39. Rename the first input as Wind_speed instead of PWM_Command. Close the explorer window once done.
19. Right click on the solid blue (gray in print version) dot which is the initial state and select the If-Else option as shown in Fig. 10.40.
20. Insert the conditions for the If-Else as shown in Fig. 10.41. Notice that the If action is the same statement that was there previously when stateflow was sending messages all the time. The If condition will now only send messages when state is in STARTUP, time is greater than 30 s, and wind speed is greater than 10 m/s. The time condition is to make sure we are not capturing any overshoot. This is initial guess based on the results shown in the previous section. The wind speed greater than 10 m/s in conjunction with the time condition will make sure only the wind increasing speed option is chosen. Since this wind speed is going to be an input for the digital twin and it will take some time for the digital twin to reach steady state, decreasing wind speeds may yield lower RPM value than steady-state value as per the piecewise function. The If-Else loop is shown in Fig. 10.42.
21. Delete the first (blue) and second node (transparent) that has the Serial_Write_Value action as this is not needed. The Serial_Write_Value condition is now in the If-Else loop. Format the Serial_Write_Value action in the If-Else by pressing enter after each semicolon. The changes are shown in Fig. 10.43.

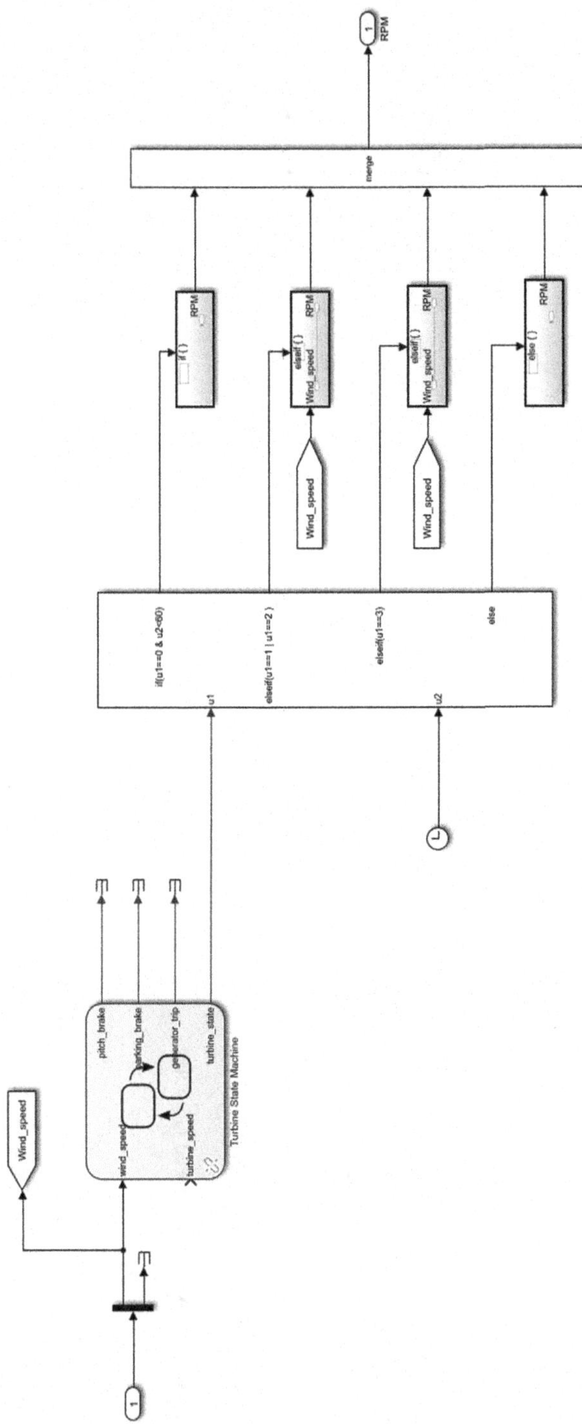

Figure 10.33 Merge block inputs and output.

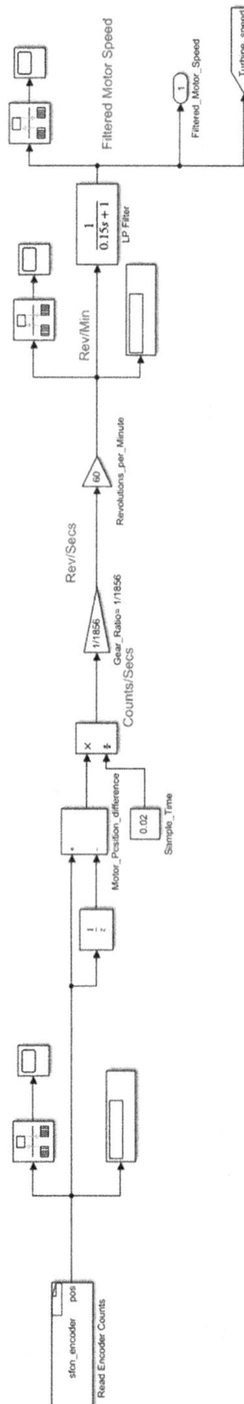

Figure 10.34 Turbine_speed Goto flag.

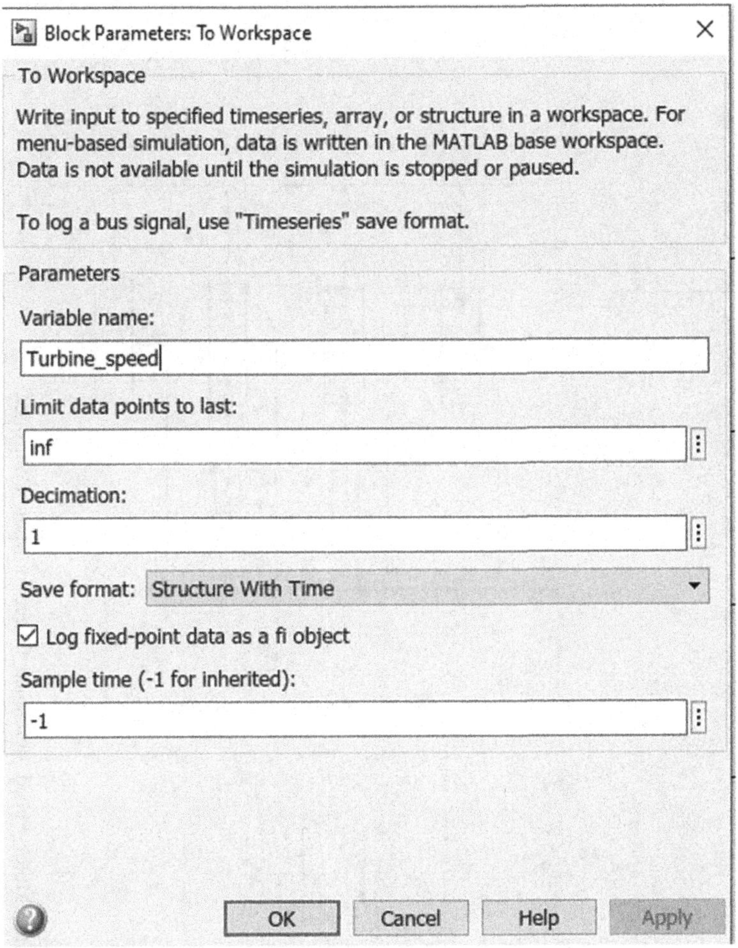

Figure 10.35 Save format for ToWorkspace variables.

22. Go outside the stateflow chart and add two inports. Rename the first inport as Wind_speed, third inport as time, and the fourth one as State as shown in Fig. 10.44.
23. Navigate to the top level of the model and add a **Clock** block, a **From** block, and two **Rate Transition** blocks. Rename the **From** block as State and connect them to the time and State inputs of the Serial_Data_Transmit_to_ESP32 subsystem. In addition to this, delete the arrow coming from the Arduino Mega Running MPC subsystem to the Wind_speed input of the Serial_Data_Transmit_to_ESP32 subsystem. Add another **From** block and rename it as Wind_speed. The changes are shown in Fig. 10.45.

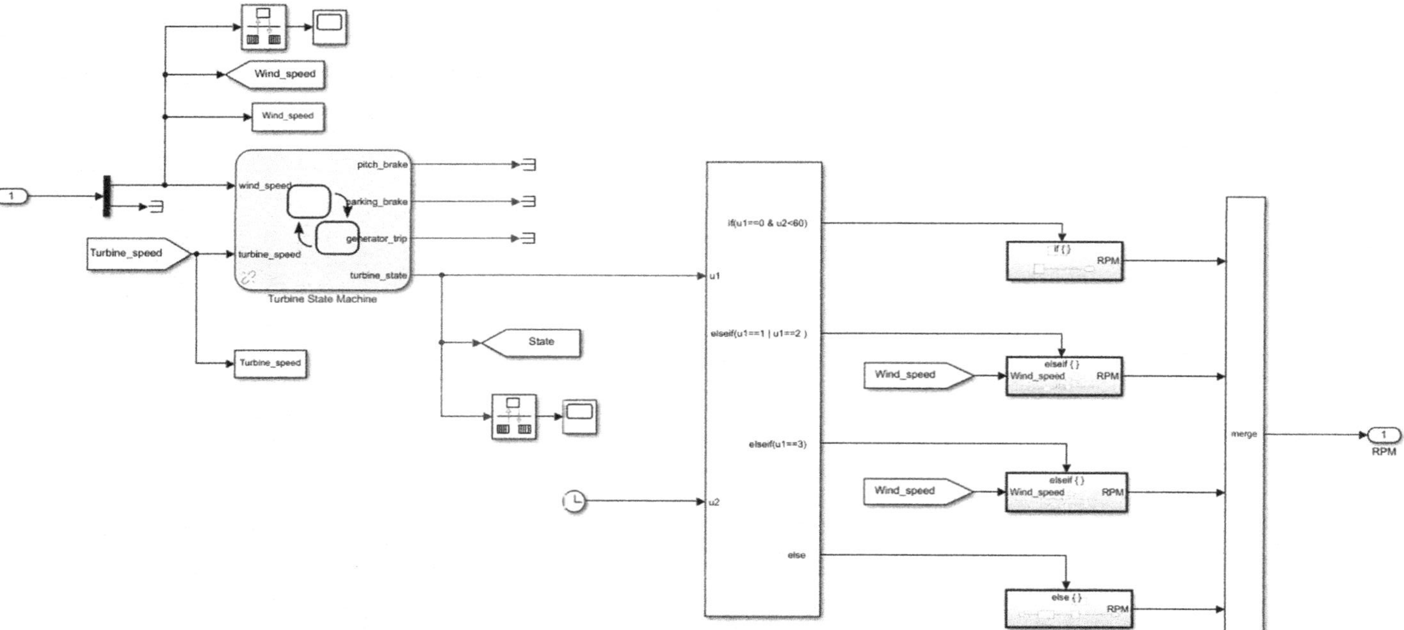

Figure 10.36 Completed Chapter_10/Motor_Control/Motor_Speed_Reference_Generator/Main_Controller.

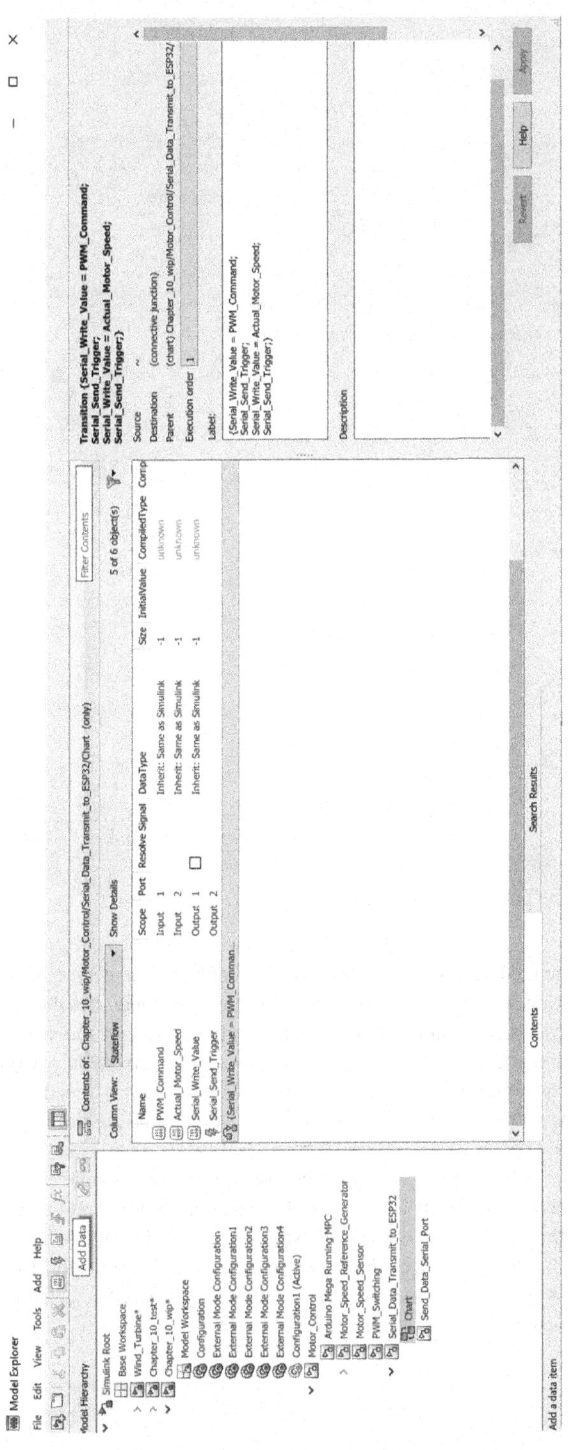

Figure 10.37 Add Data option in Model Explorer.

Figure 10.38 "Time" parameter settings.

24. The changes for the DC motor driver model are complete. Connect the Arduino board to the USB port making sure the right port is selected for this model. Set the simulation time to 130 s, and press play to simulate the results. Double click on the wind speed scope in Chapter_10/Motor_Control/Motor_Speed_Reference_Generator/Main_Controller and the filtered motor speed scope in Chapter_10/Motor_Control/Motor_Speed_Sensor to visualize the results. The scope outputs for wind speed and motor speed are shown in Figs. 10.46 and 10.47, respectively.

25. Notice that with the hardware prototype, the overshoot impact is smaller due to how the PID logic was tuned for the DC motor driver. Also, the Wind_speed and the Turbine_speed are now saved to the MATLAB workspace as structure due to the **ToWorkspace** variables that were created in step 16. Save these two structures into a separate .mat file named as chapt10.mat as it would be needed for the digital twin C code generation verification. This concludes the DC motor driver model changes section.

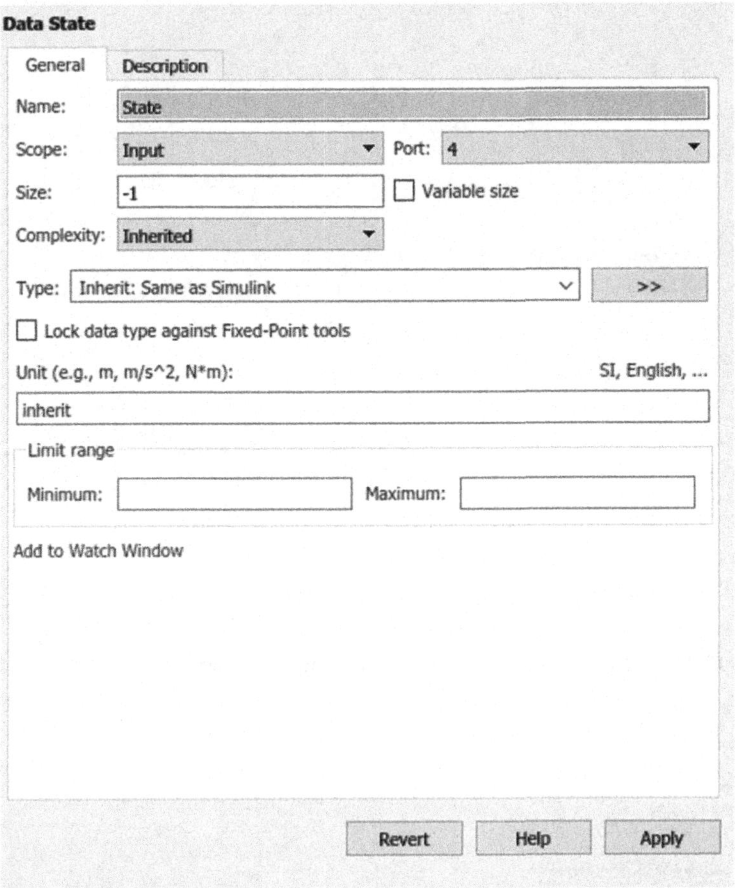

Figure 10.39 "State" parameter settings.

10.4.2 DC motor hardware communication to AWS

The communication between the DC motor hardware model (now renamed as Chapter_10.slx) to ESP32 and then to AWS cloud has already been handled in the DC motor chapter. However, there are some modifications that need to be made in the Arduino script for ESP32 to send messages to AWS.

1. Open Arduino IDE script that was used to send messages to AWS in the previous chapter. Rename it as Chapter_10_ESP32_AWS.
2. Press Ctrl+F and replace PWM_Command with Wind_speed.
3. Press Ctrl+F and find "if(message_counter==30)". Then under that statement, add the following statements as shown in Fig. 10.48. This makes sure that the previous wind speed and motor speed values are not 0 since the messages sent to ESP32 are only sent when system has entered steady state; hence, previous messages cannot be set as 0 as before in the DC motor chapter.

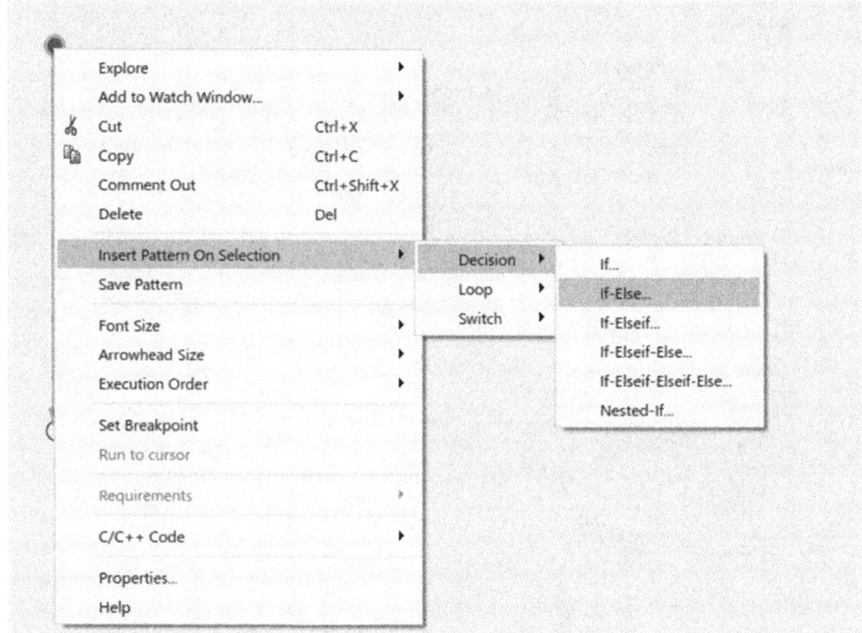

Figure 10.40 Selecting the If-Else decision pattern for stateflow.

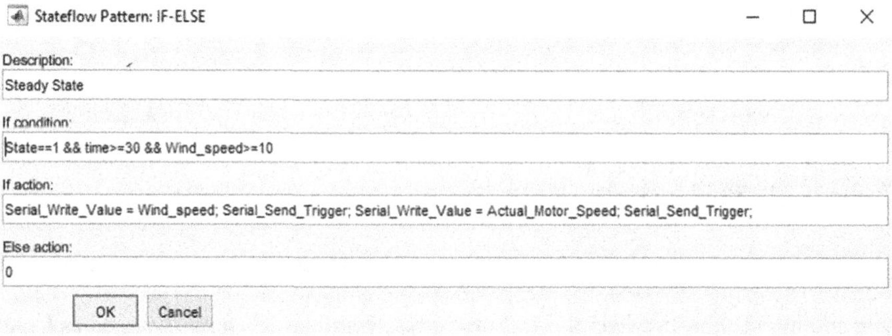

Stateflow Pattern: IF-ELSE

Description:

Steady State

If condition:

State==1 && time>=30 && Wind_speed>=10

If action:

Serial_Write_Value = Wind_speed; Serial_Send_Trigger; Serial_Write_Value = Actual_Motor_Speed; Serial_Send_Trigger;

Else action:

0

OK Cancel

Figure 10.41 Conditions for the If-Else.

Note: The only time the "count=count+1" is referenced is in Fig. 10.48. If the variable "count" is not defined, set "int count=0" before the void setup function.

4. Connect the ESP32 to the USB port on the computer and run the Arduino script. Once, the messages "Connected to AWS" and "Subscribe Successful" are seen in the COM serial monitor of the Arduino IDE, connect the Arduino Mega 2560 to a USB port and run the "Chapter_10.slx" on the microcontroller by pressing play. The messages in the COM Serial Monitor will come up as shown in Fig. 10.49.

Figure 10.42 Added If-Else loop.

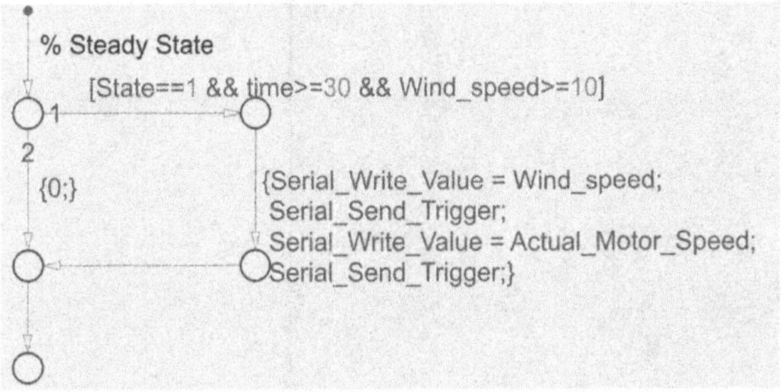

Figure 10.43 Formatted If-Else loop.

Note: Ignore any email notification you may receive by AWS Lambda function. The AWS Lambda is still using the C code for the DC motor digital twin; hence, that C Code needs to be changed to the Wind_Turbine.slx generated code.

10.5 Deploying the Simscape™ digital twin model to the AWS cloud and performing Off-BD

This section requires a Linux OS to generate the C code for the Wind_Turbine digital twin model. The Linux OS used in this chapter is Ubuntu. Once booted into Ubuntu and after launching MATLAB, follow these steps below:

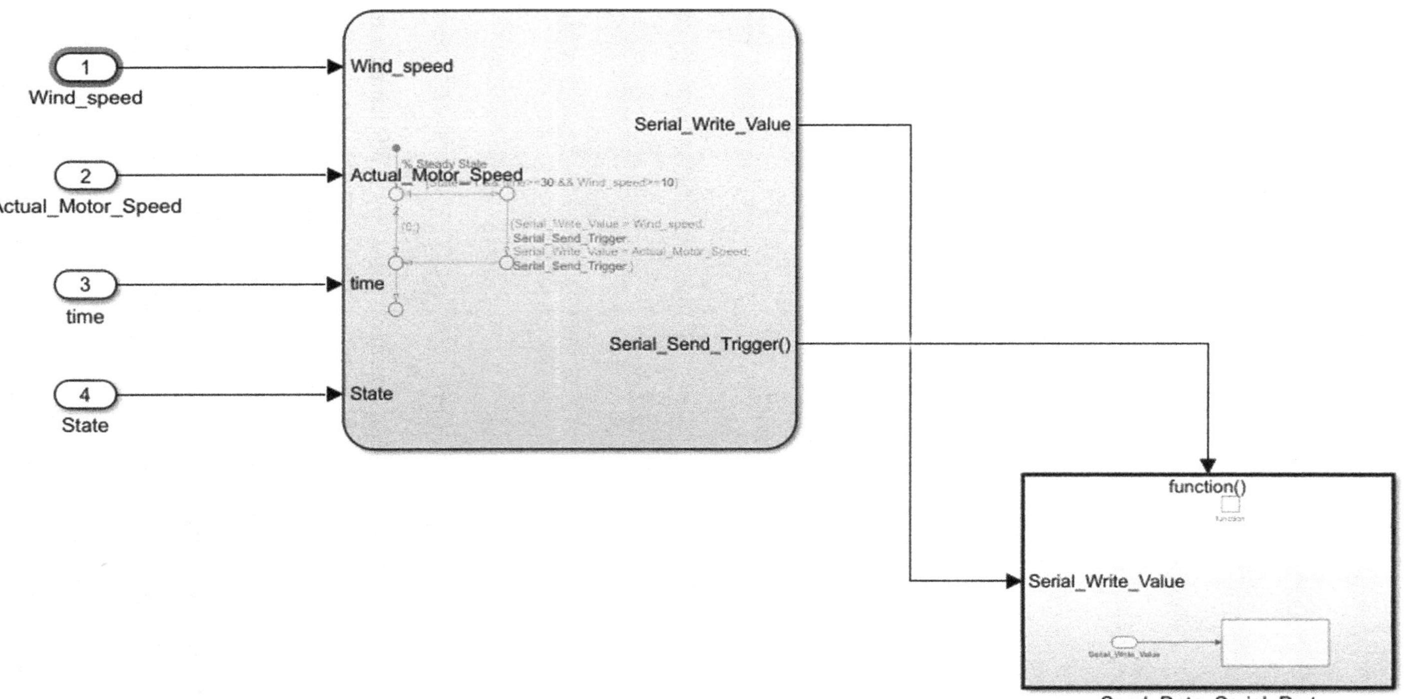

Figure 10.44 First, third, and fourth input for stateflow.

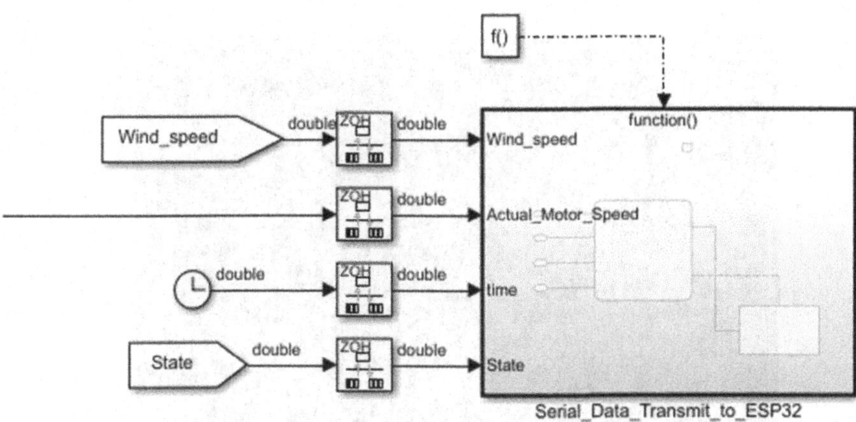

Figure 10.45 Connections to the time and state ports of the Serial_Data_Transmit_to_ESP32 subsystem.

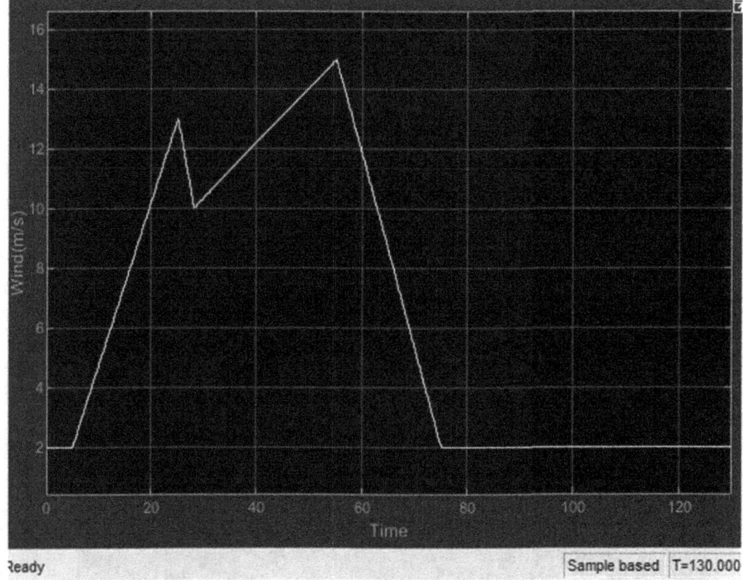

Figure 10.46 Wind speed input on DC motor driver model.

1. Open the Wind_Turbine.slx model and perform the changes done in Section 10.3.3 if not already done so for this model.
2. In the top level of the model, delete the Scopes block and outport 1. This is not needed for code generation. Also delete the SLRT block as mentioned in Section 10.3.1.
3. Go to Wind_Turbine/Turbine Input/Wind and double click on the Wind Input Signal Generator block. Right click on the Wind speed signal and select delete to delete the signal. The modified Signal builder block should only have the Wind Direction signal as shown in Fig. 10.50.

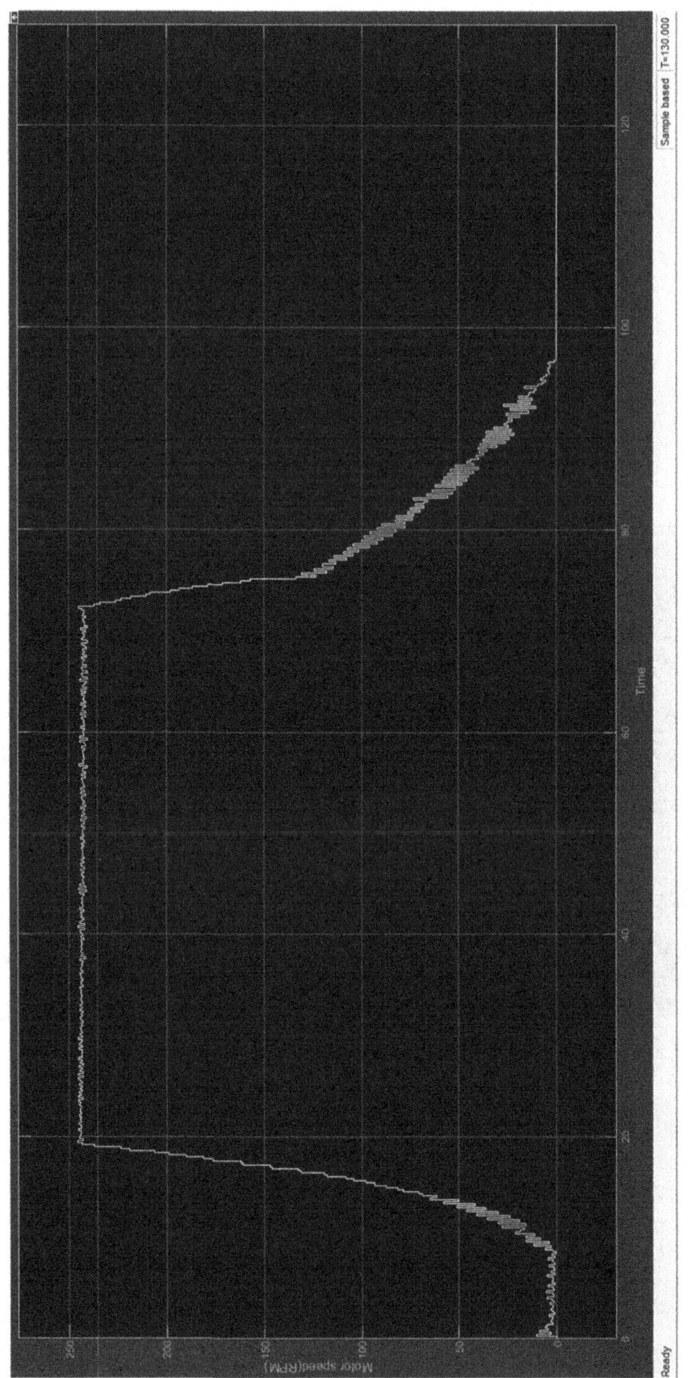

Figure 10.47 DC motor speed RPM on DC motor driver model.

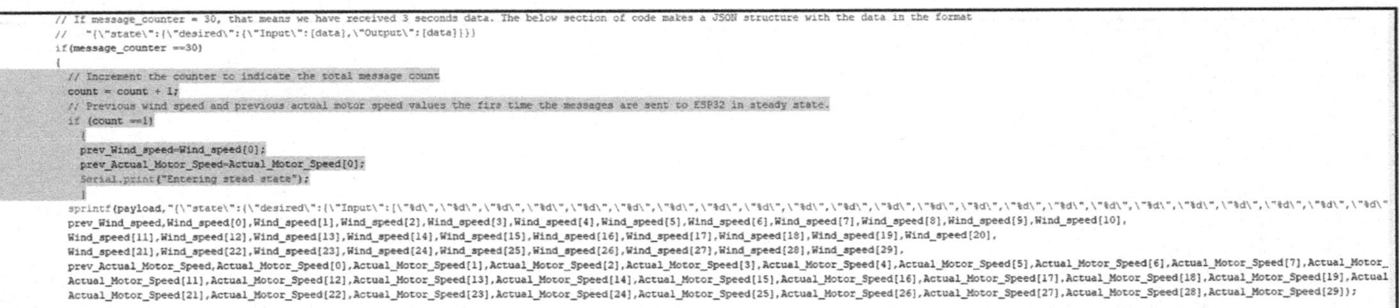

Figure 10.48 Modification to ESP32 script to set previous input and output messages.

```
Connected to wifi
Connected to AWS
Subscribe Successfull
Entering stead state{"state":{"desired":{"Input":["10","10","10","10","10","10","10","10","10","10","10","10","10","10","10","10","10","10","10","10","10","10","10","10","10","10","10","10","10","10"],"Output":["244","244",
Publish failed
{"state":{"desired":{"Input":["10","10","10","10","10","11","11","11","11","11","11","11","11","11","11","11","11","11","11","11","11","11","11","11","11","11","11","11","11","11"],"Output":["243","242","244","246","242","2
Publish Message:{"state":{"desired":{"Input":["10","10","10","10","10","11","11","11","11","11","11","11","11","11","11","11","11","11","11","11","11","11","11","11","11","11","11","11","11","11"],"Output":["243","242","244
{"state":{"desired":{"Input":["11","11","11","11","11","11","11","11","11","11","11","11","11","11","11","11","11","11","11","11","11","11","11","11","11","11","11","11","12","12"],"Output":["244","242","241","243","244","2
Publish Message:{"state":{"desired":{"Input":["11","11","11","11","11","11","11","11","11","11","11","11","11","11","11","11","11","11","11","11","11","11","11","11","11","11","11","11","12","12"],"Output":["244","242","241
{"state":{"desired":{"Input":["12","12","12","12","12","12","12","12","12","12","12","12","12","12","12","12","12","12","12","12","12","12","12","12","12","12","12","12","12","12"],"Output":["242","243","243","243","242","2
Publish Message:{"state":{"desired":{"Input":["12","12","12","12","12","12","12","12","12","12","12","12","12","12","12","12","12","12","12","12","12","12","12","12","12","12","12","12","12","12"],"Output":["242","243","243
{"state":{"desired":{"Input":["12","12","12","12","12","12","12","12","12","12","12","12","12","12","12","12","12","12","12","12","12","12","13","13","13","13","13","13","13","13"],"Output":["241","242","244","243","242","2
Publish Message:{"state":{"desired":{"Input":["12","12","12","12","12","12","12","12","12","12","12","12","12","12","12","12","12","12","12","12","12","12","13","13","13","13","13","13","13","13"],"Output":["241","242","244
{"state":{"desired":{"Input":["13","13","13","13","13","13","13","13","13","13","13","13","13","13","13","13","13","13","13","13","13","13","13","13","13","13","13","13","13","13"],"Output":["243","242","241","242","243","2
Publish Message:{"state":{"desired":{"Input":["13","13","13","13","13","13","13","13","13","13","13","13","13","13","13","13","13","13","13","13","13","13","13","13","13","13","13","13","13","13"],"Output":["243","242","241
{"state":{"desired":{"Input":["13","13","13","13","13","13","13","13","13","13","13","13","13","13","13","13","13","14","14","14","14","14","14","14","14","14","14","14","14","14"],"Output":["241","242","244","242","241","2
Publish Message:{"state":{"desired":{"Input":["13","13","13","13","13","13","13","13","13","13","13","13","13","13","13","13","13","14","14","14","14","14","14","14","14","14","14","14","14","14"],"Output":["241","242","244
{"state":{"desired":{"Input":["14","14","14","14","14","14","14","14","14","14","14","14","14","14","14","14","14","14","14","14","14","14","14","14","14","14","14","14","14","14"],"Output":["243","243","242","241","242","2
Publish Message:{"state":{"desired":{"Input":["14","14","14","14","14","14","14","14","14","14","14","14","14","14","14","14","14","14","14","14","14","14","14","14","14","14","14","14","14","14"],"Output":["243","243","242
{"state":{"desired":{"Input":["14","14","14","14","14","14","14","14","14","14","14","15","14","14","14","14","14","14","14","14","14","14","14","14","14","14","13","13","13","13"],"Output":["241","241","243","243","241","2
Publish Message:{"state":{"desired":{"Input":["14","14","14","14","14","14","14","14","14","14","14","15","14","14","14","14","14","14","14","14","14","14","14","14","14","14","13","13","13","13"],"Output":["241","241","243
{"state":{"desired":{"Input":["13","13","13","13","13","13","13","13","13","13","13","13","12","12","12","12","12","12","12","12","12","12","12","12","12","12","12","11","11","11"],"Output":["242","244","242","241","241","2
Publish Message:{"state":{"desired":{"Input":["13","13","13","13","13","13","13","13","13","13","13","13","12","12","12","12","12","12","12","12","12","12","12","12","12","12","12","11","11","11"],"Output":["242","244","242
```


Figure 10.49 Messages sent to AWS.

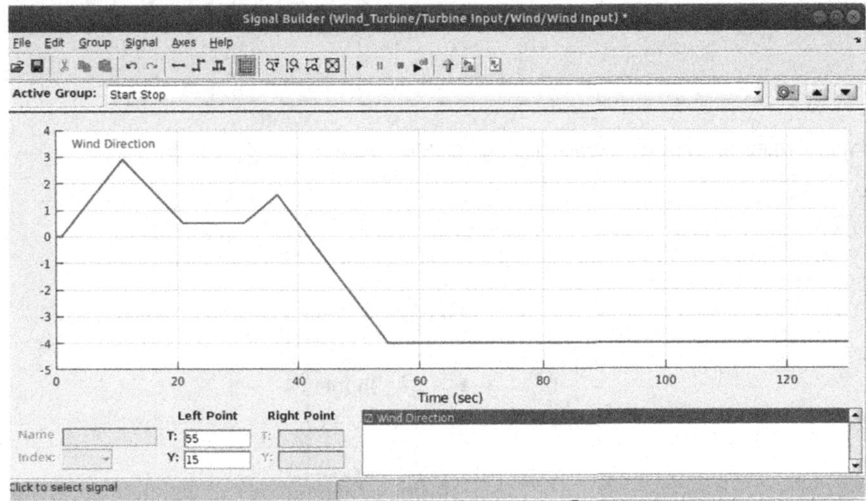

Figure 10.50 Modified signal builder for wind direction.

4. For the first input of the **Mux** block in Wind_Turbine/Turbine Input/Wind, add an **inport**, rename it as WS, and connect it to the first input of the **Mux** block as shown in Fig. 10.51. Copy the WS **inport** and paste it in Wind_Turbine/Turbine Input as shown in Fig. 10.52. Repeat the same for Wind_Turbine top level. Connect the WS **inport** to the Turbine Input subsystem as shown in Fig. 10.53.
5. In the top level of the model, place a **From** and an **outport** block. Rename the **From** flag as Gen_Spd_rpm and the **outport** block as Turb. The **Goto** flag for this is already in Wind_Turbine/Nacelle/Generator/Full. Connect the **From** block to the **outport** and place them above the Phasor 60 Hz block as shown in Fig. 10.54. This shall serve as the turbine speed outport for the given wind speed input from WS inport.
6. In the top level of the model, double click on the phasor 60 Hz block and input the settings as shown in Fig. 10.55. Notice that the discrete sampling time is set as 0.0015 s, which is different than the 0.1 s execution rate of the function block sending messages to ESP32 in Chapter 10.slx. The reason 0.1 s cannot be set as a sampling rate for this Wind_Turbine digital twin model is because the Asynchronous Machine block in Wind_Turbine/Nacelle/Generator will not be able to converge on a solution if the sample time is too large. There would be some further modifications needed in the generated C code to address the difference between the sampling rates of the motor hardware and the digital twin shown later in this chapter.

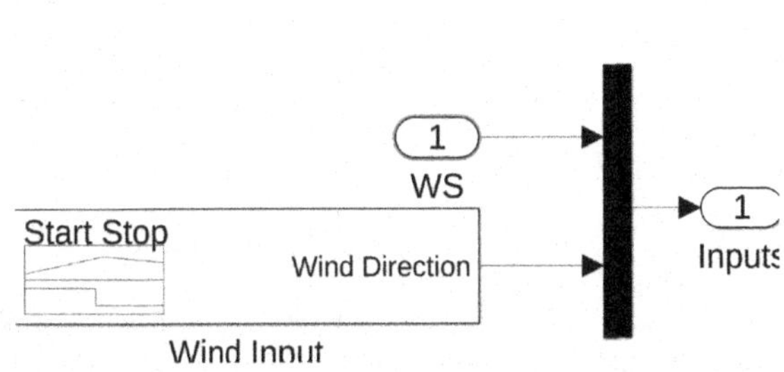

Figure 10.51 WS inport to Mux block.

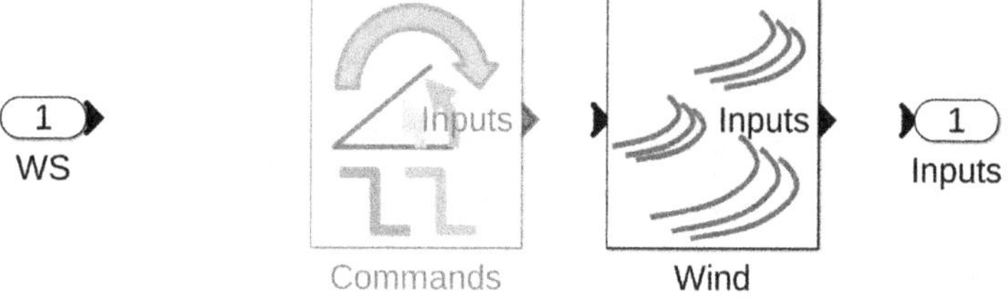

Figure 10.52 WS inport in Wind_Turbine/Turbine_Input.

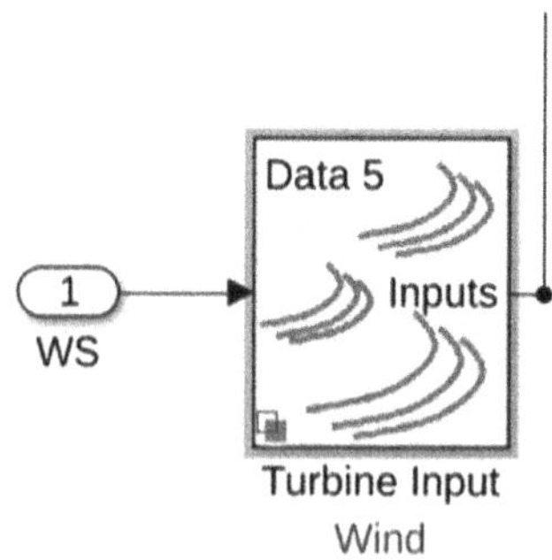

Figure 10.53 WS inport in top level.

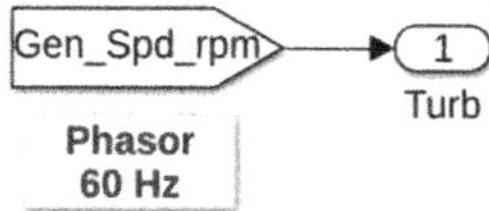

Figure 10.54 Turb outport in top level.

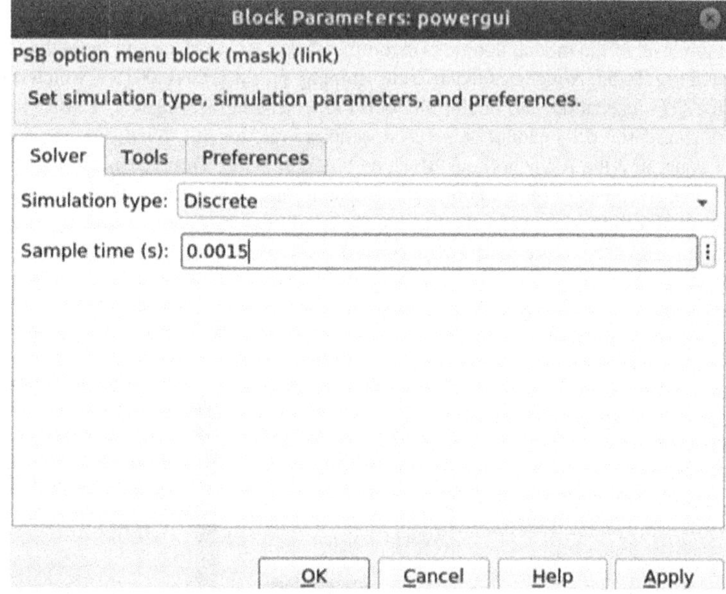

Figure 10.55 Discrete sampling time settings.

7. In the top level of the model, in the bottom right corner, double click on the Pitch Power On block and set the sample time as 0.0015 s as shown in Fig. 10.56.
8. Save and close the model. Open the model again, and in the top bar menu of Simulink, go to Code->C/C++ Code->Embedded Coder as shown in Fig. 10.57. The quick start shall guide you to the process of generating C code for the model while selecting the correct settings for the C Code. Once the C code is generated, open the ert_main.c file in the Wind_Turbine_ ert_rtw folder that is created after code generation. The two main functions of the generated ert_main.c file are shown in Figs. 10.58 and 10.59.
9. For this step, replace the contents of the Wind_Turbine ert_main.c after the header file definitions with the contents of the ert_main.c that was used in the DC motor chapter. Some modifications need to be made in the Wind_Turbine ert_main.c after the replacement as shown on Figs. 10.60–10.63. The new changes in the Wind_Turbine ert_main.c are shown in the left half of each figure.

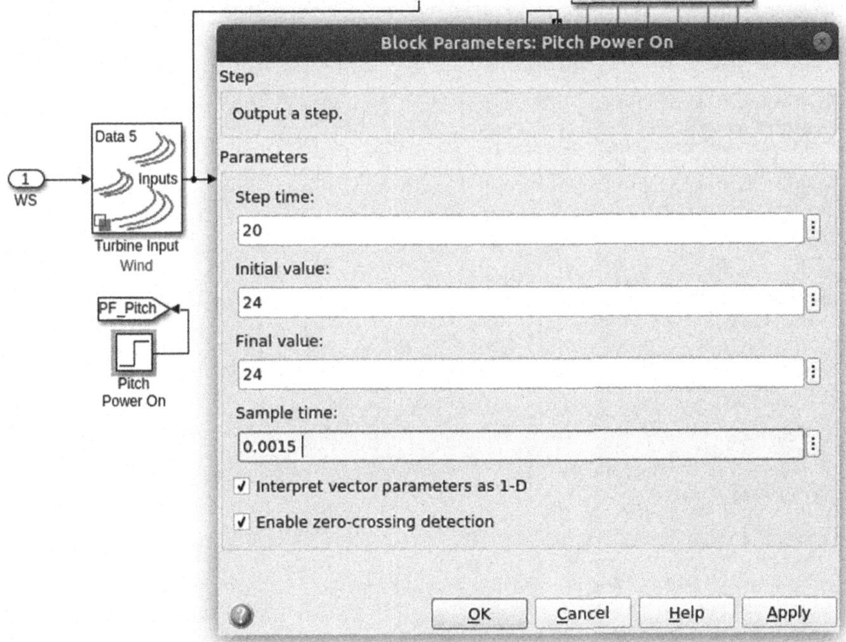

Figure 10.56 Pitch Power On settings.

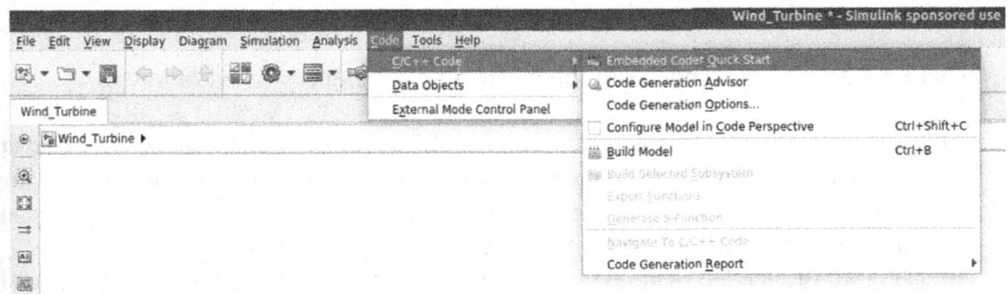

Figure 10.57 Embedded coder quick start.

```
#include <stddef.h>
#include <stdio.h>                    /* This ert_main.c example uses printf/fflush */
#include "Wind_Turbine.h"             /* Model's header file */
#include "rtwtypes.h"

/*
 * Associating rt_OneStep with a real-time clock or interrupt service routine
 * is what makes the generated code "real-time".  The function rt_OneStep is
 * always associated with the base rate of the model.  Subrates are managed
 * by the base rate from inside the generated code.  Enabling/disabling
 * interrupts and floating point context switches are target specific.  This
 * example code indicates where these should take place relative to executing
 * the generated code step function.  Overrun behavior should be tailored to
 * your application needs.  This example simply sets an error status in the
 * real-time model and returns from rt_OneStep.
 */
void rt_OneStep(void);
void rt_OneStep(void)
{
  static boolean_T OverrunFlag = false;

  /* Disable interrupts here */

  /* Check for overrun */
  if (OverrunFlag) {
    rtmSetErrorStatus(rtM, "Overrun");
    return;
  }

  OverrunFlag = true;

  /* Save FPU context here (if necessary) */
  /* Re-enable timer or interrupt here */
  /* Set model inputs here */

  /* Step the model for base rate */
  Wind_Turbine_step();

  /* Get model outputs here */

  /* Indicate task complete */
  OverrunFlag = false;

  /* Disable interrupts here */
  /* Restore FPU context here (if necessary) */
  /* Enable interrupts here */
}

/*
 * The example "main" function illustrates what is required by your
```

Figure 10.58 void rt_OnStep function in generated ert_main.c file.

There are a few observations in the ert_main.c differences listed below:

(a) In Fig. 10.60, it can be seen for the DC motor hardware ert_main.c that there were input and output arguments to pass into the step function in parenthesis. This may not be done for the Wind_Turbine ert_main.c due to how the code was generated. The input and output argument instead will be explicitly mentioned in Fig. 10.63.

(b) In Fig. 10.61, line 148 of the Wind_Turbine ert_main.c file, 7100 samples are being used instead of 100 because Wind_Turbine needs about 7100 samples at a sampling rate of

```
/*
 * The example "main" function illustrates what is required by your
 * application code to initialize, execute, and terminate the generated code.
 * Attaching rt_OneStep to a real-time clock is target specific.  This example
 * illustrates how you do this relative to initializing the model.
 */
int_T main(int_T argc, const char *argv[])
{
  /* Unused arguments */
  (void)(argc);
  (void)(argv);

  /* Initialize model */
  Wind_Turbine_initialize();

  /* Simulating the model step behavior (in non real-time) to
   *  simulate model behavior at stop time.
   */
  while ((rtmGetErrorStatus(rtM) == (NULL)) && !rtmGetStopRequested(rtM)) {
    rt_OneStep();
  }

  /* Disable rt_OneStep() here */
  return 0;
}

/*
 * File trailer for generated code.
 *
 * [EOF]
 */
```

Figure 10.59 int main function in generated ert_main.c file.

0.0015 s to reach steady state. This would be equivalent to 10 s data as it was in the DC motor chapter. The next 2000 samples are the new data received from the hardware equivalent to 3 s of data as before.

Note: Even though we are asking for more samples, we will still only send 30 samples sent from ESP32 to AWS. An interpolating function would need to be used in the lambda function to interpolate data to 9100 samples.

(c) In Fig. 10.62, line 199 of the Wind_Turbine ert_main.c file, if the sample count is less than 7100, then the padd_hw_input_wind and padded_hw_output_rpm arrays are fed with previous data.

(d) In Fig. 10.63, line 222 of the Wind_Turbine ert_main.c file, the input wind speed is explicitly defined as rTU.WS=padded_hw_input_wind[ii] where rTU.WS is the WS inport in the model. The same is done for equating arg_Turb_speed to rtY.Turb where rtY.Turb is the Turb outport in the model. Note that line 225 is “printf(“Digital Twin Predicted Motor speed [%d] with Wind Input %f is = %f\n”,ii,padded_hw_input_wind[ii],arg_Turb_speed_1”); In Fig. 10.63, there is no terminate function for the Wind_Turbine ert_main.c file.

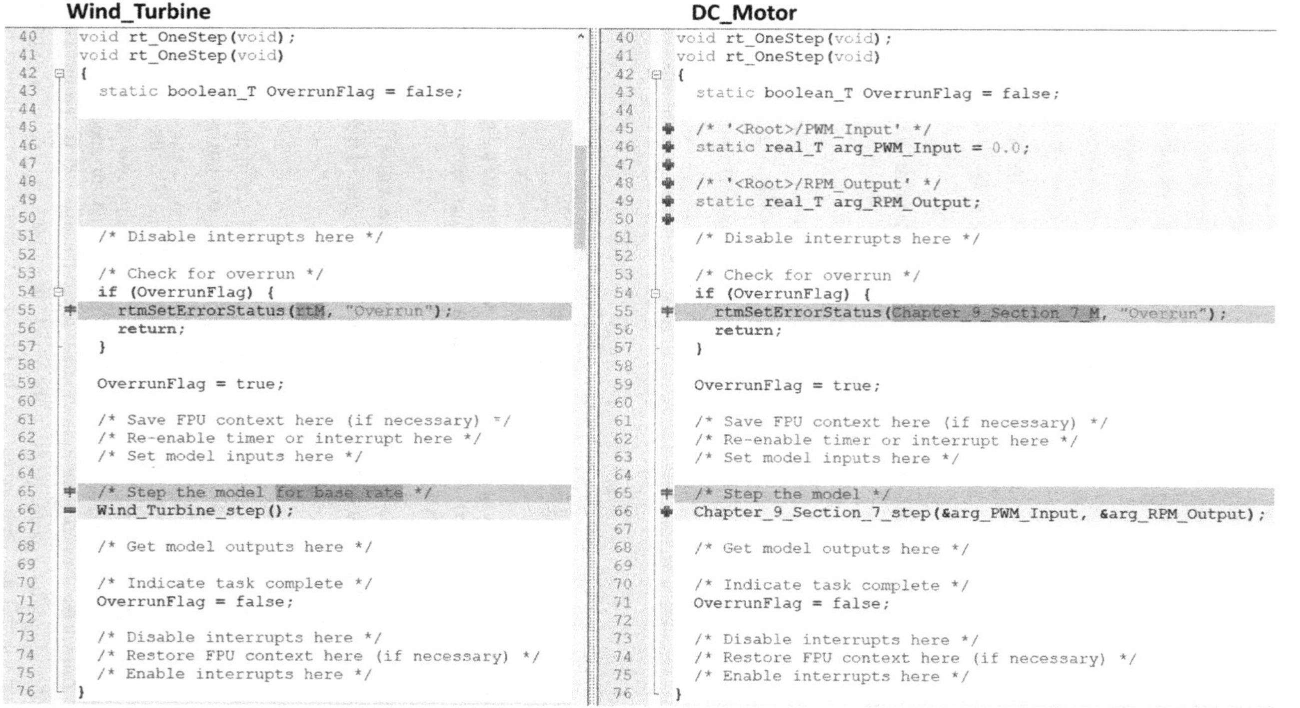

Figure 10.60 ert_main.c differences between DC motor chapter and wind turbine chapter with regard to input and output arguments.

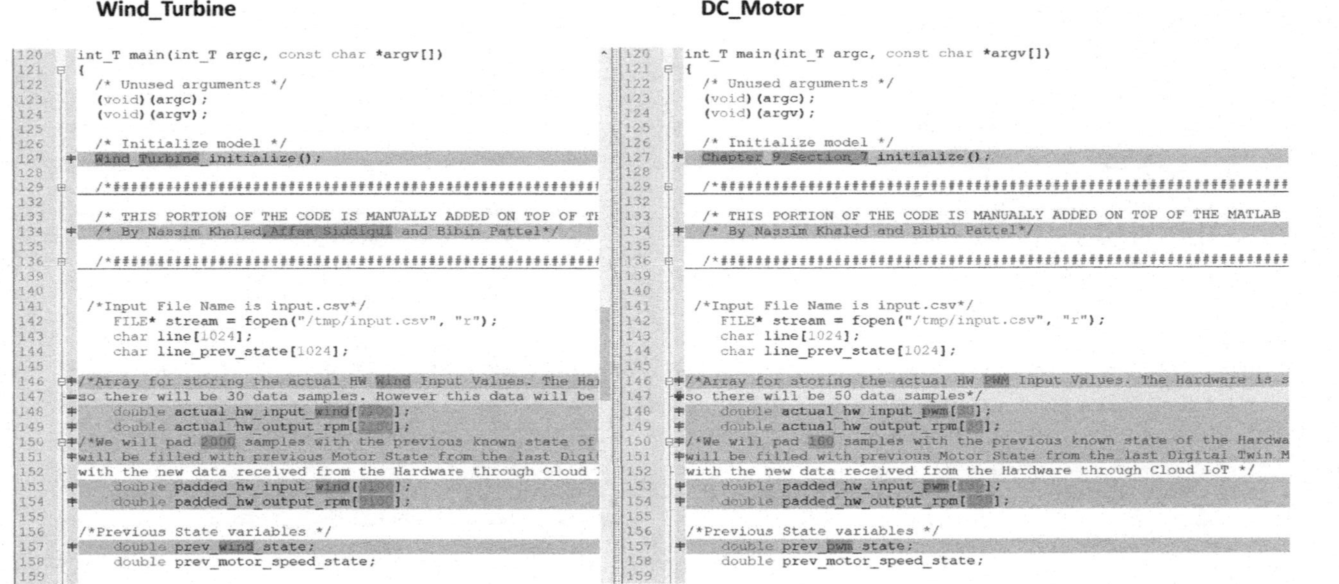

Figure 10.61 ert_main.c differences between DC motor chapter and wind turbine chapter with regard to storing information.

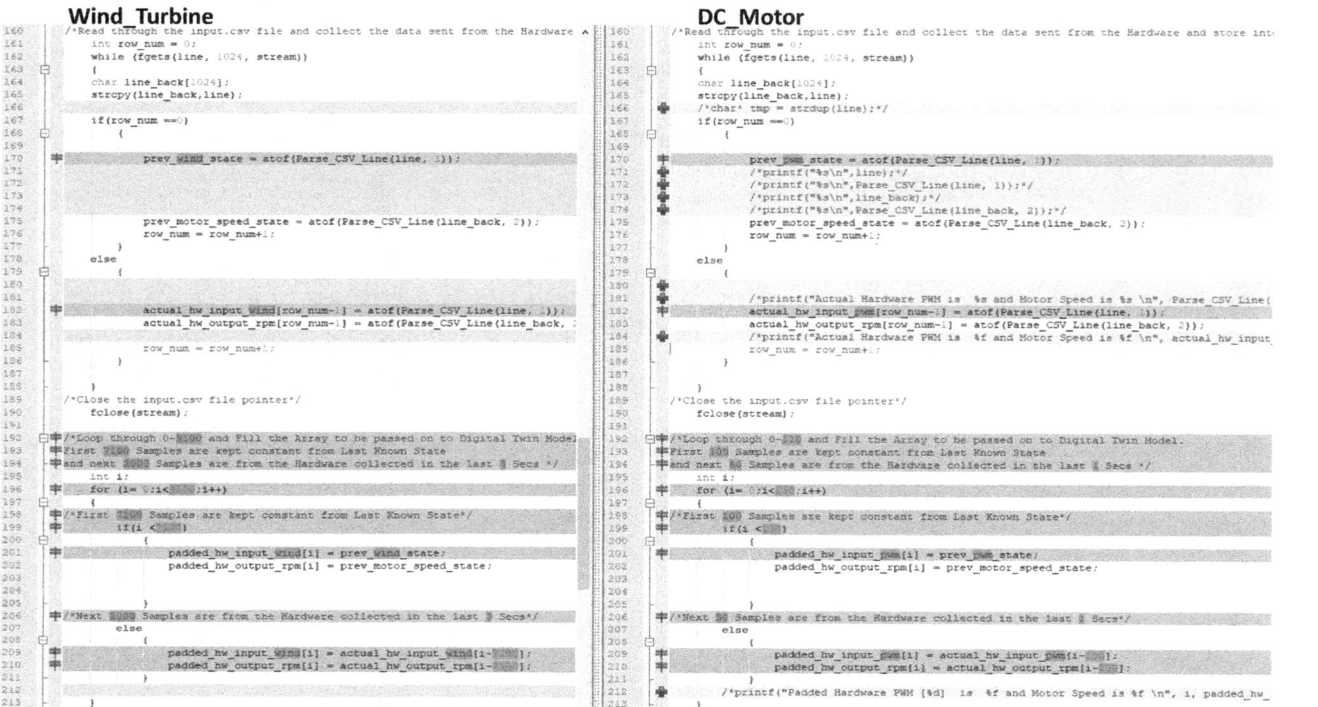

Figure 10.62 ert_main.c differences between DC motor chapter and wind turbine chapter with regard to processing input and output.

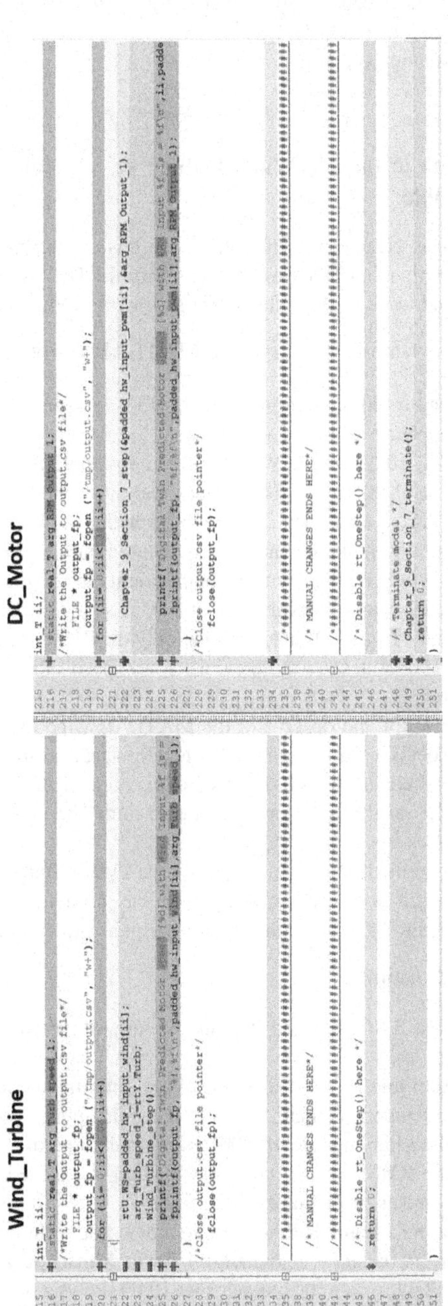

Figure 10.63 ert_main.c differences between DC motor chapter and wind turbine chapter with regard to terminate function.

10. Next, navigate to the instrumented folder in the Wind_Turbine_ert_rtw folder and type the following commands in sequence in the MATLAB workspace to generate a new executable based on the modifications made in step 9:
    i. !cp Wind_Turbine.mk Makefile
    ii. !rm ert_main.o
    iii. !make -f Makefile

    Once done, navigate to outside Wind_Turbine_ert_rtw folder and notice a new Wind_Turbine executable.

11. After the executable is created, it is time to verify this executable through a MATLAB script. Use the same .m script that was used to verify the DC motor chapter. There are a few changes that need to be made to this script as shown in Figs. 10.64 and 10.65.

    There are a few observations to see in the MATLAB script:

    (a) In Fig. 10.64, data is only read from 1501 to 3135 elements as this is the range where the turbine is supposed to enter steady state.
    (b) Notice that in Fig. 10.64 for the Wind_Turbine, resample function is used instead of itnerp1 function as there will be NaN values that may come up with interp1 function for the Wind_Turbine code. The resampling function will set the time divisions to 0.0015 s and appropriately compute a y value based on the resampled time.
    (c) In line 55 of the Wind_Turbine, notice that the count threshold is set as 2002 instead of 32 since 3 s of data would be equivalent to 0.0015 s*2002.
    d) In Fig. 10.65, line 81 of the Wind_Turbine, only the 7100 to 9100 elements are used since the first 7100 values are the previous wind speed input to make the system to go to steady state.

12. Run the verification script and observe in the MATLAB window, for 7100 samples, the same wind speed is being used bringing the generator speed to steady-state value. Notice that there is an overshoot in the RPM values as mentioned in Section 10.3. The script will finish plotting the results after approximately 5 min. The plotted RMSE is shown in Fig. 10.66.

13. With the executable verified, it is time to develop the Python function that will be used in AWS Lambda. Use the same python script that was used for the DC motor chapter. Modifications are need to be made to this python script as shown in Figs. 10.67 and 10.68.

    Here are a few observations:

    (a) In Fig. 10.67, line 8 of the Wind_Turbine, numpy library would be needed for the interpolation.
    (b) In Fig. 10.67, from line 17 onwards of the Wind_Turbine, the data from the hardware are first sent to a temporary input_1.csv file. Then the input_1.csv is read and interpolated to match the sampling rate of 0.0015 s sampling rate of the digital twin.
    (c) In Fig. 10.68, the RMSE error processing function is adjusted to read data from the 7100 to 9100 elements of the interpolated data.

14. Save the above developed Lambda function into a new folder. In this case, we named the folder to be **Chapter_10** under the folder **Digtal_Twin_Simscape_book.** Copy the Digital Twin compiled executable also to this folder. Fig. 10.69 shows a set of Linux commands to give the necessary executable permissions and packaging of the Lambda function and

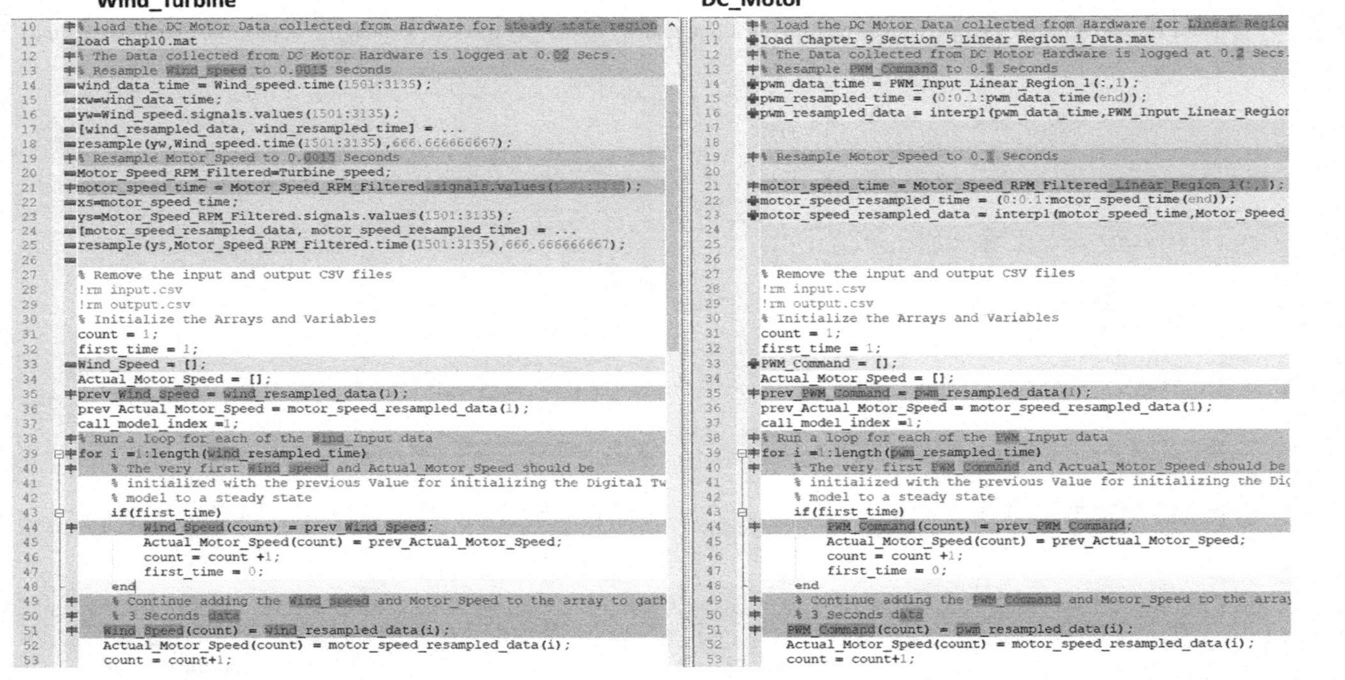

Figure 10.64 Executable verification script differences between DC motor chapter and wind turbine chapter with regard to processing input and output.

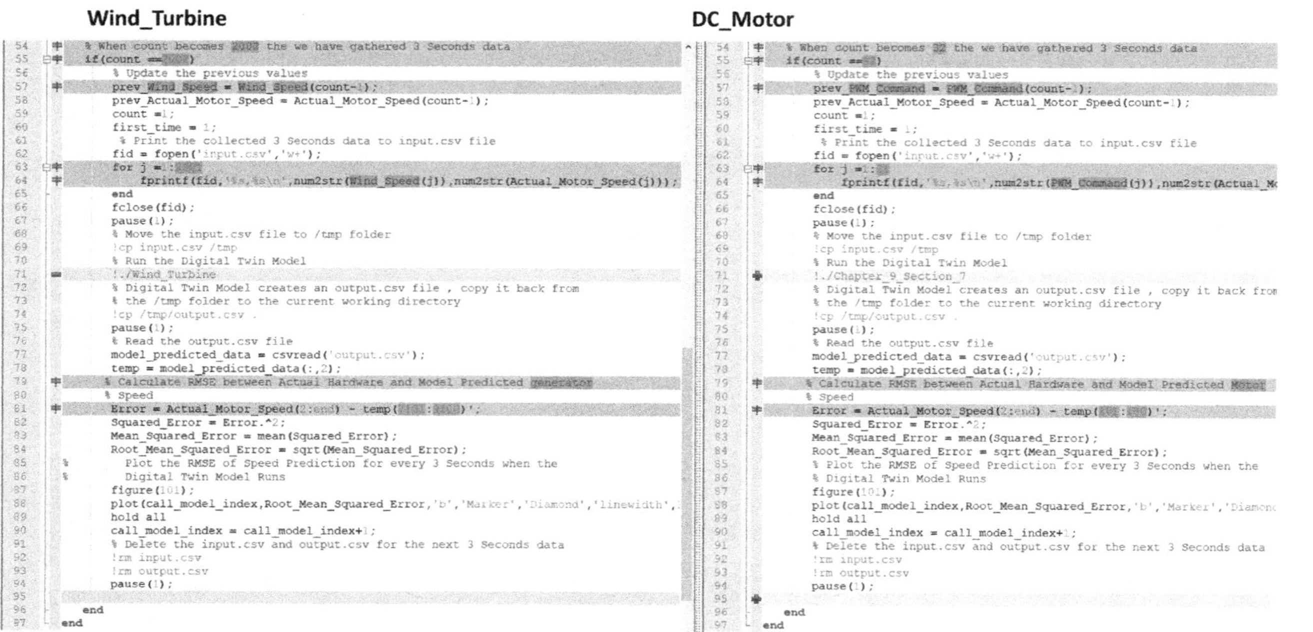

Figure 10.65 Executable verification differences between DC motor chapter and wind turbine chapter with regard to terminate function.

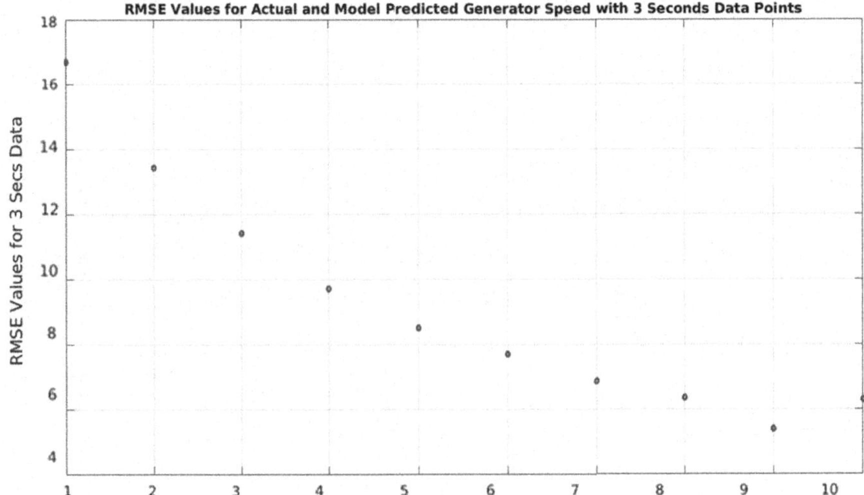

Figure 10.66 RMSE plot for verification script.

Digital Twin executable. The commands are explained below. Open a Linux terminal and run the below commands in steps:

a. The ***ls*** command shows the folder has the compiled executable **Chapter_9_Section_7** and the lambda function **lambda_function.py**.

b. Using the ***cp*** command, copy the executable to a new file name **Wind_Turbine_Digital_Twin,** which we used in the Lambda function.

c. We need to give full read/write/executable permissions to all files in this folder for the AWS to be able to run it. Use the command ***sudo chmod -R 777 <folder_name>***.

d. Run the command ***ls -l.*** This will list all the files and folders in the current folder and their permissions. Everything should be showing ***rwxrwxrwx***.

e. Now package the lambda function and compiled executable using the command ***zip bundle.zip lambda_function.py Wind_Turbine***. This will zip the files ***lambda_function.py*** and ***Wind_Turbine*** into a file named ***bundle.zip***.

f. After the bundle.zip is created once again repeat the step c to allow read/write/executable permissions to the ***bundle.zip*** file as well.

g. The ***ls*** command will show the newly created ***bundle.zip*** file as well. Now we are ready to deploy the Lambda function to AWS and do the final testing.

15. Now open a web browser and then the AWS Management Console and then open the dc_motor_digital_twin lambda function created in the last chapter. Select the option from the drop down "Upload a .zip file." Browse and select the bundle.zip file that was packaged in step 14. Click on "Save" when done.

16. There are a couple of more steps to execute before running the lambda function. The lambda python script includes numpy as a library; however, AWS on its own does not recognize that library unless that library was zipped as well. However, there is now a way in AWS such that his library can be used without the need of zipping it. In the Designer area, click on Layers and then click on Add a layer as shown in Fig. 10.70.

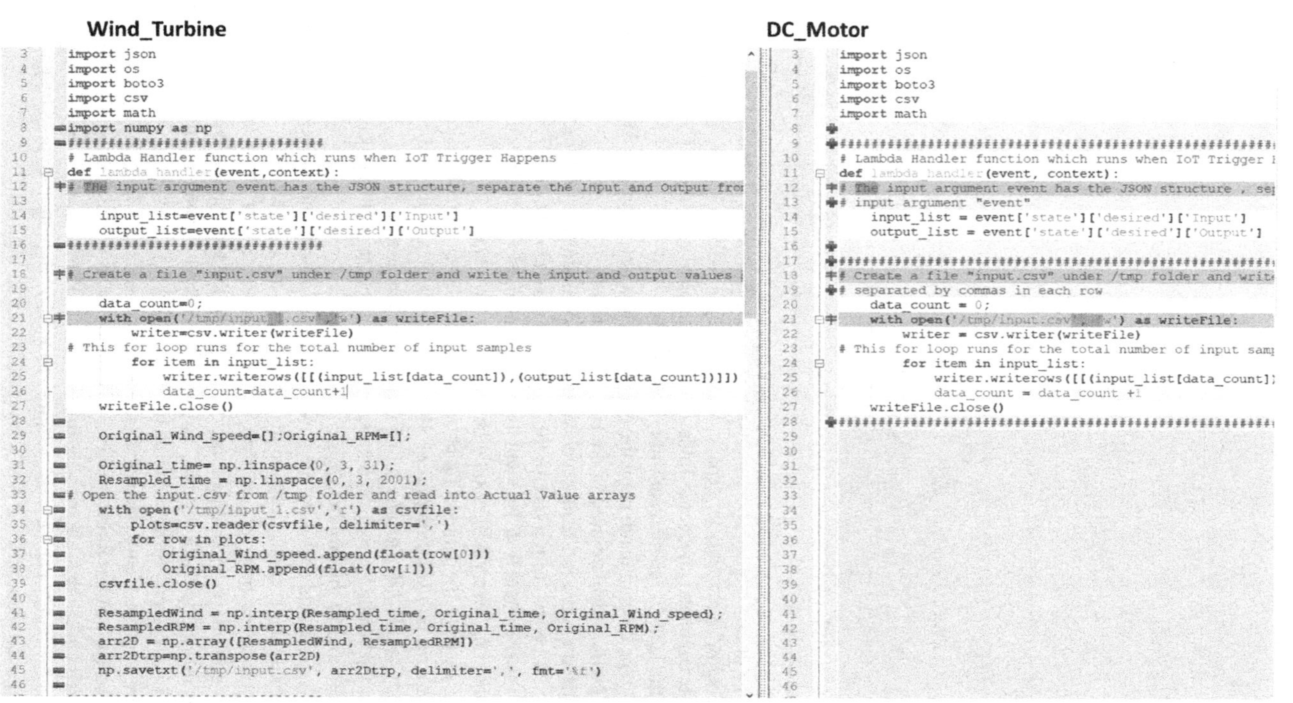

Figure 10.67 Lambda function differences between DC motor chapter and wind turbine chapter with regard to interpolating the input data.

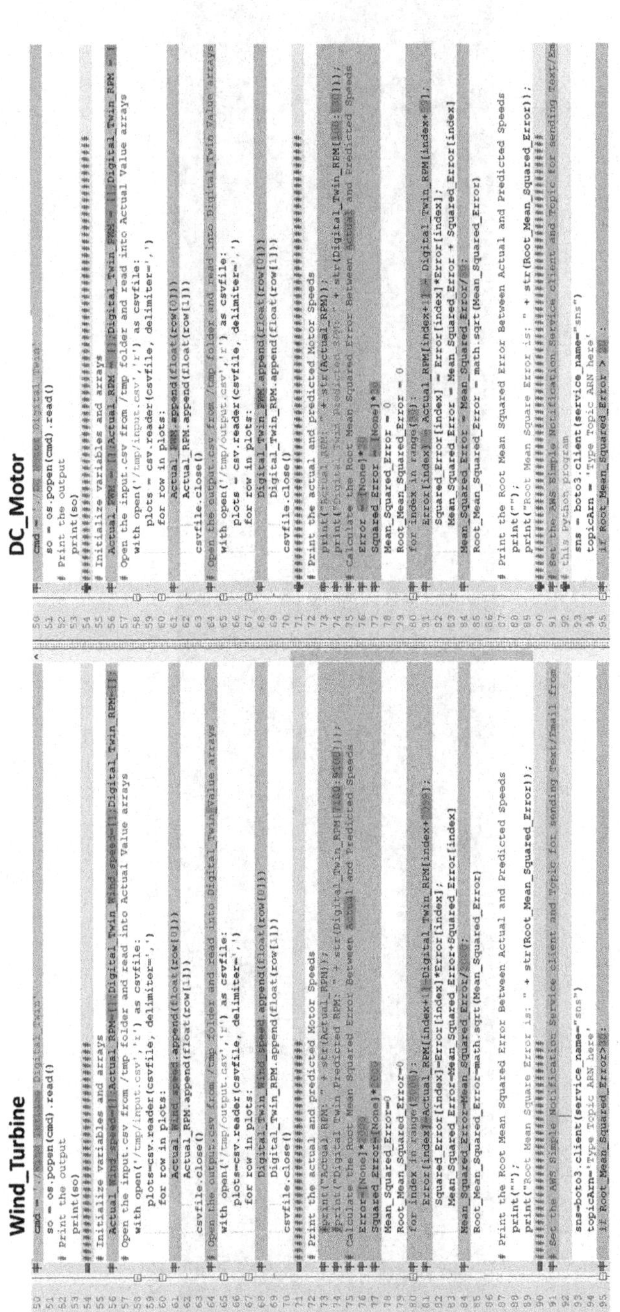

Figure 10.68 Lambda function differences between DC motor chapter and wind turbine chapter with regard to processing RMSE.

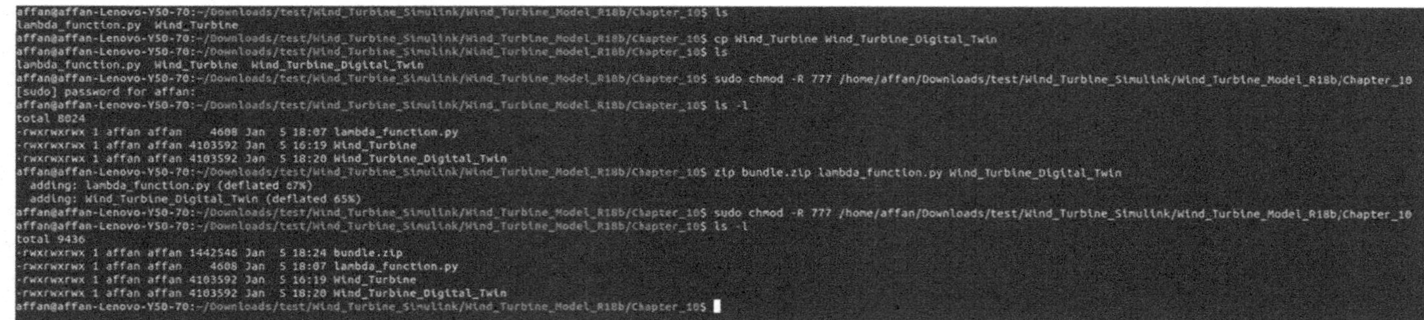

Figure 10.69 Linux terminal commands for packaging of lambda function and digital twin executable.

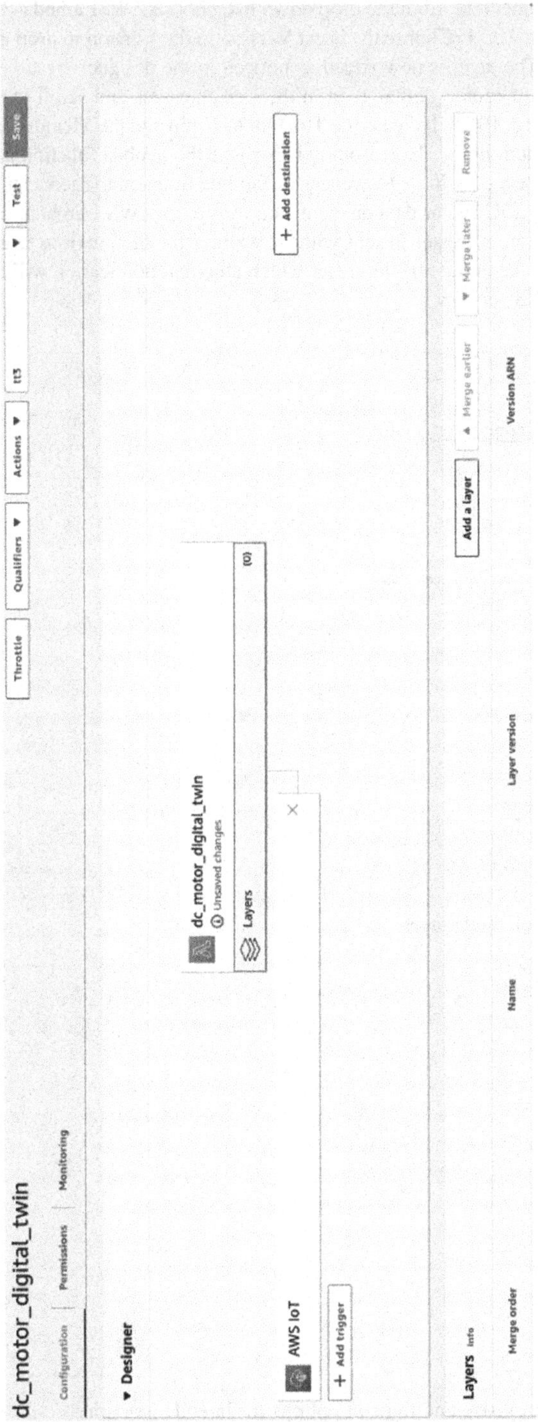

Figure 10.70 Layer options in AWS Lambda.

17. Then, in the Name field, from the drop down list, choose AWSLambda-Python37-SciPy1x as shown in Fig. 10.71. Choose the latest Version in the Version in drop down menu. Then click on Add. The layer is now added as noticed in the designer area.
18. Next, click on dc_motor_digital_twin in the Designer area and scroll to the Basic settings area shown in Fig. 10.72. Increase the Timeout to 1 min and the Memory (MB) to 3008 MB or to the maximum limit. The reasoning is because the lambda function takes some time to run (approximately 30–40 s). However, the lambda function triggers can run concurrently. As in, when the DC motor data are sent as a trigger to AWS Lambda, it will execute the function each time a trigger is sent without waiting for the previous trigger to end. Each execution will take about 30–40 s after which an email notification will be sent according to the AWS SNS service.

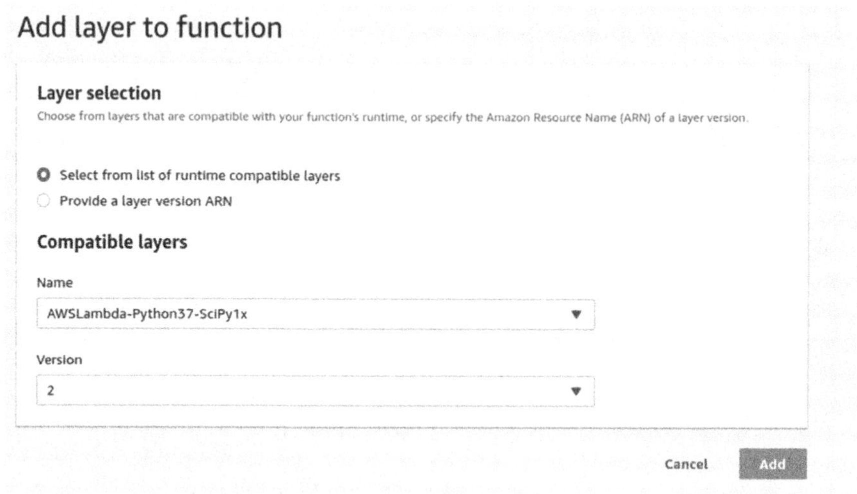

Figure 10.71 SciPy lambda layer.

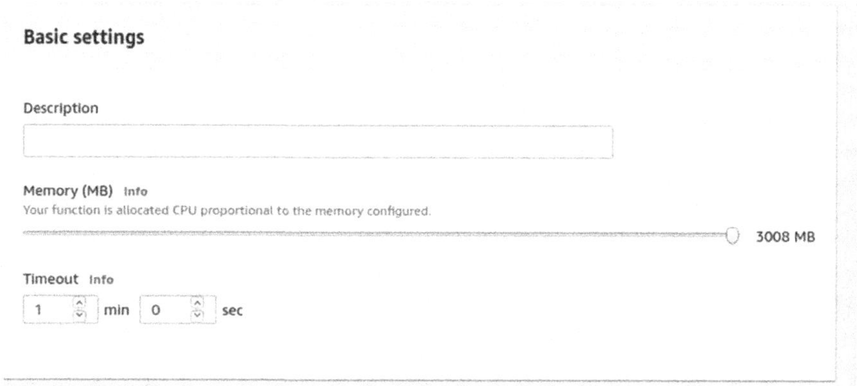

Figure 10.72 Memory size and timeout options for lambda function.

Figure 10.73 Shadow update in AWS.

19. This brings a conclusion to the deployment of the digital twin on AWS. We are now ready to test the hardware and digital twin. You can now switch to Windows OS where you saved the DC motor hardware model and the Arduino script. Connect the ESP32 and 2560 to the respective COM port and run the Arduino IDE script and the DC Motor hardware model on MATLAB. You should be seeing the messages now being sent to AWS when hardware has achieved steady state in AWS IoT Core digital_twin_thing shadow update as shown in Fig. 10.73. Once the lambda function finishes executing one trigger, you should receive an email according to SNS stating there is no failure since we have not introduced any failure into the system. Fig. 10.74 shows the messages for a healthy system.

20. Now introduce a failure into the system by disconnecting the power supply to the DC motor. When the DC motor hardware model is run, it will still send as per the conditions in Fig. 10.41 since there is no circular reference to motor speed. However, motor speed will show 0 RPM. Once the lambda function finishes executing, you will see the failure email messages as shown in Fig. 10.75.

10.6 Application problem

Modify the output of this chapter to use acceleration instead of RPM and compute the RMSE between the acceleration of the DC motor and digital twin generator when the system has entered steady state. You may observe that this comparison is easier than RPM as you will get an acceleration close to 0 when system as entered steady state.

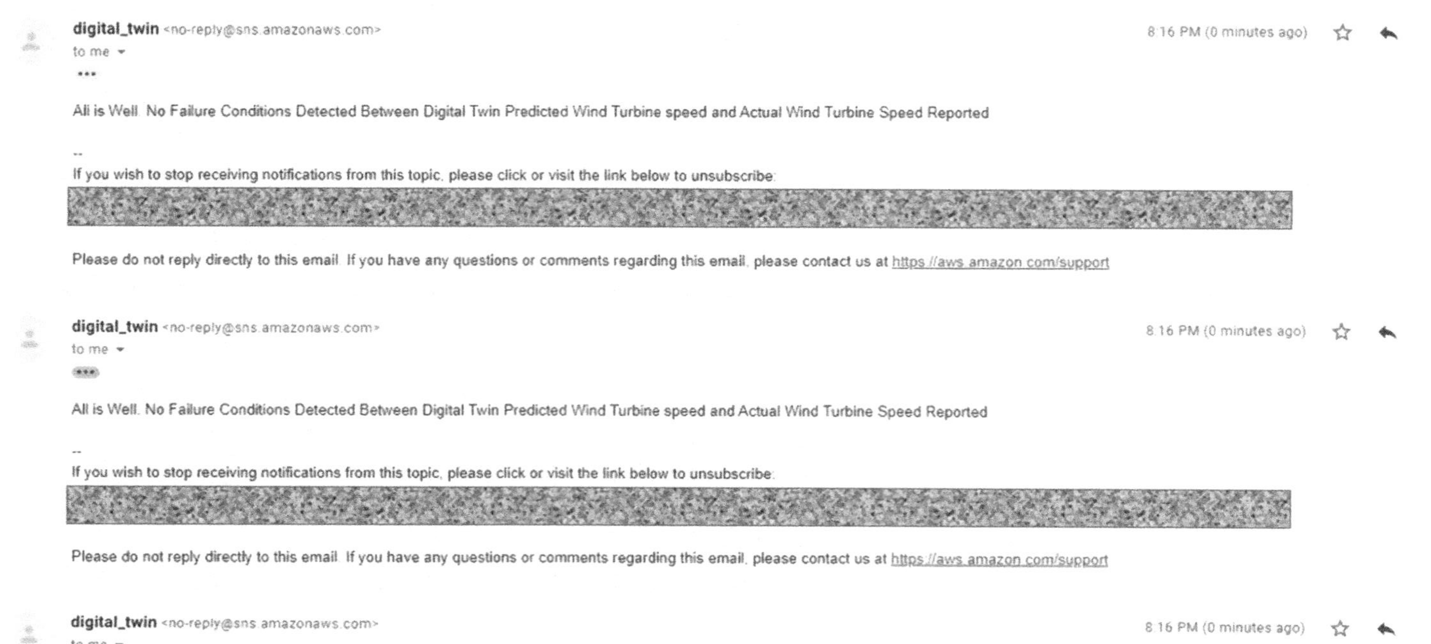

Figure 10.74 Email messages for a healthy system.

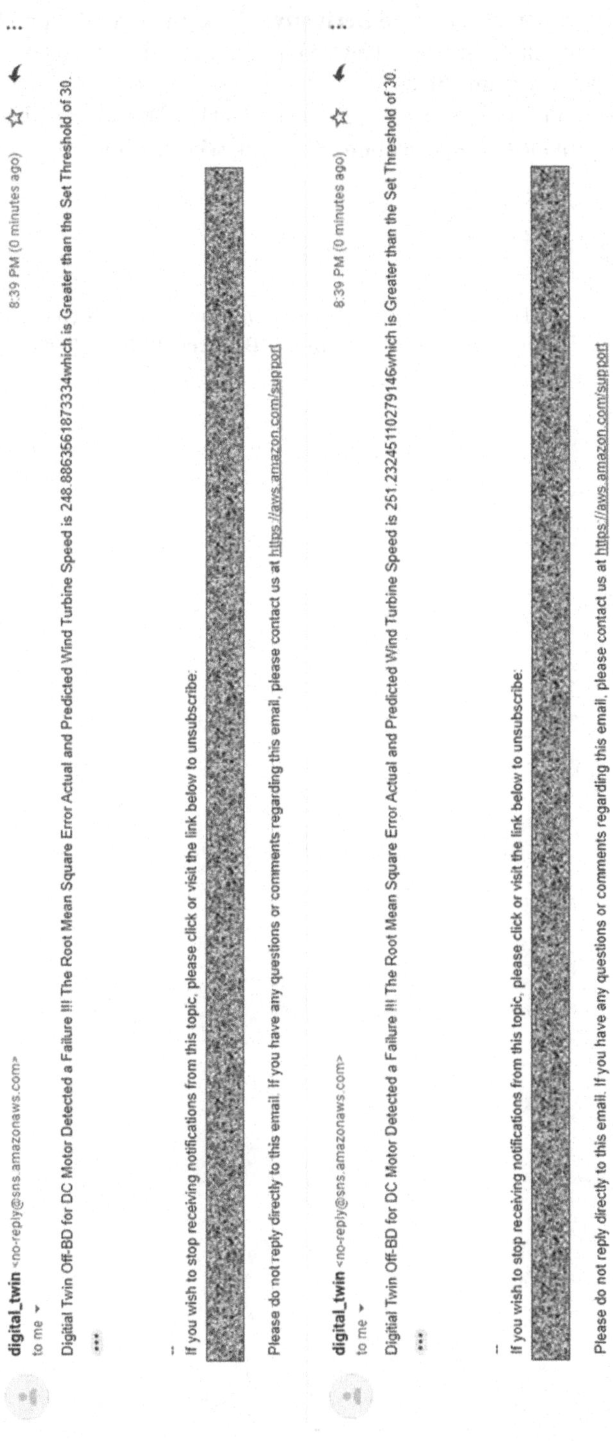

digital_twin <no-reply@sns.amazonaws.com> 8:39 PM (0 minutes ago)
to me

Digitial Twin Off-BD for DC Motor Detected a Failure !!! The Root Mean Square Error Actual and Predicted Wind Turbine Speed is 248.8863561873334which is Greater than the Set Threshold of 30.

...

--
If you wish to stop receiving notifications from this topic, please click or visit the link below to unsubscribe:

Please do not reply directly to this email. If you have any questions or comments regarding this email, please contact us at https://aws.amazon.com/support

digital_twin <no-reply@sns.amazonaws.com> 8:39 PM (0 minutes ago)
to me

Digitial Twin Off-BD for DC Motor Detected a Failure !!! The Root Mean Square Error Actual and Predicted Wind Turbine Speed is 251.23245110279146which is Greater than the Set Threshold of 30.

...

--
If you wish to stop receiving notifications from this topic, please click or visit the link below to unsubscribe:

Please do not reply directly to this email. If you have any questions or comments regarding this email, please contact us at https://aws.amazon.com/support

Figure 10.75 Email messages when a fault is introduced.

Hint: You may find the **Discrete Derivative** block useful for this problem. Keep in mind the difference in sample rates between the hardware driver model and the digital twin model when using this block.

Download the material for the chapter from MATLAB® Central. The final models and codes are provided in **Application_Problem_windturbine** folder.

Reference

[1] S. Miller, Wind Turbine Model, 2020. https://www.mathworks.com/matlabcentral/fileexchange/25752-wind-turbine-model. MATLAB Central File Exchange. Retrieved November 11, 2019.

Index

Note: 'Page numbers followed by "*f*" indicate figures and "*t*" indicate tables'.

E

Printed in the United States
By Bookmasters

Printed in the United States
By Bookmasters